biology
today and tomorrow
with physiology

biology

Today and Tomorrow With Physiology

Cecie Starr

Christine A. Evers

Lisa Starr

BROOKS/COLE
CENGAGE Learning

Australia · Brazil · Japan · Korea · Mexico · Singapore · Spain · United Kingdom · United States

BROOKS/COLE
CENGAGE Learning

Biology: Today and Tomorrow With Physiology, Third Edition
Cecie Starr, Christine A. Evers, Lisa Starr

Publisher: Yolanda Cossio

Senior Acquisitions Editor: Peggy Williams

Assistant Editor: Elizabeth Momb

Editorial Assistant: Alexis Glubka

Technology Project Manager: Kristina Razmara

Executive Marketing Manager: Stacy Best

Marketing Assistant: Katherine Malatesta

Marketing Communications Manager: Linda Yip

Project Manager, Editorial Production: Andy Marinkovich

Creative Director: Rob Hugel

Art Director: John Walker

Print Buyer: Karen Hunt

Permissions Editor: Sarah D'Stair

Production Service: Grace Davidson & Associates

Text Design: John Walker, tani hasegawa, Jeanne Calabrese

Cover Design: John Walker

Photo Researcher: Paul Forkner, Myrna Engler Photo Research Inc.

Copy Editor: Anita Wagner

Illustrators: Gary Head, ScEYEnce Studios, Lisa Starr, Precision Graphics

Compositor: Lachina Publishing Services

Cover Image: © Buddy Mays/ Corbis. Green tree frog catching a hawk moth.

For product information and technology assistance, contact us at
Cengage Learning Customer & Sales Support, 1-800-354-9706.

For permission to use material from this text or product, submit all requests online at **cengage.com/permissions**. Further permissions questions can be emailed to **permissionrequest@cengage.com**.

Library of Congress Control Number: 2008940277

Paper Edition:
ISBN-13: 978-0-495-56157-6
ISBN-10: 0-495-56157-6

Hardcover Edition:
ISBN-13: 978-0-495-82753-5
ISBN-10: 0-495-82753-3

Brooks/Cole
10 Davis Drive
Belmont, CA 94002-3098
USA

Cengage Learning is a leading provider of customized learning solutions with office locations around the globe, including Singapore, the United Kingdom, Australia, Mexico, Brazil, and Japan. Locate your local office at: **www.cengage.com/international**.

Cengage Learning products are represented in Canada by Nelson Education, Ltd.

To learn more about Brooks/Cole visit **www.cengage.com/brookscole**.

Purchase any of our products at your local college store or at our preferred online store **www.ichapters.com**.

Printed in the United States of America
1 2 3 4 5 6 7 12 11 10 09 08

Contents in Brief

Contents

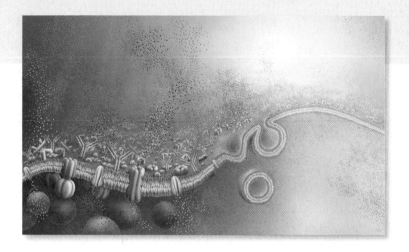

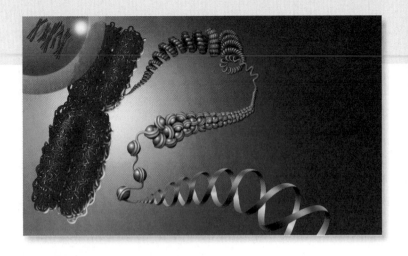
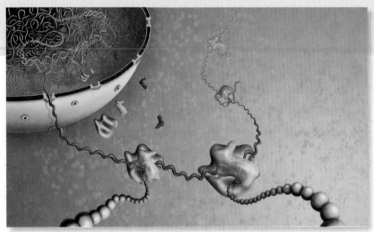

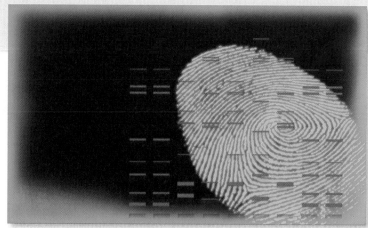

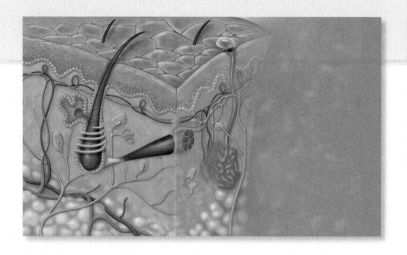

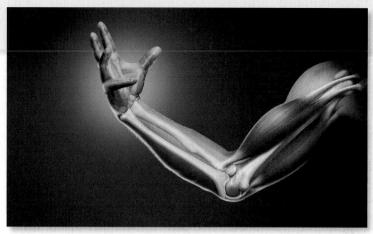

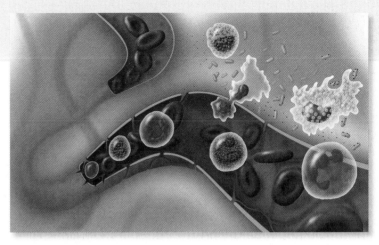

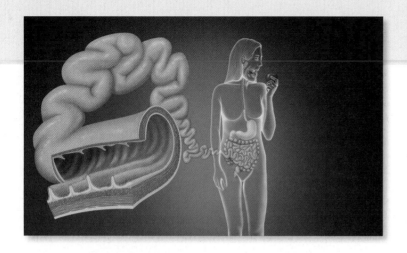

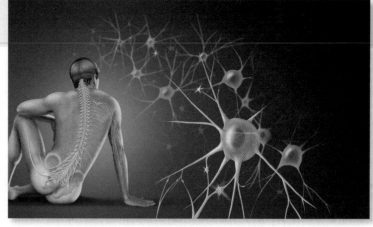

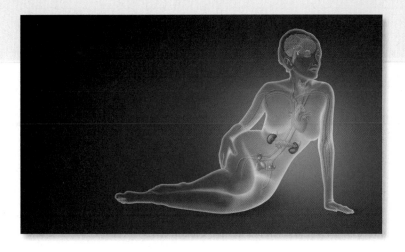
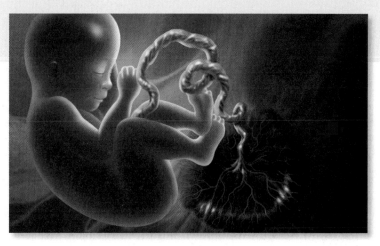

Preface

Throughout our lives, we are called upon to make important decisions based on our understanding of science, whatever that understanding may be. This book was designed and written specifically for students who do not intend to become biologists and may never again take another science course. It is an accessible and engaging introduction to biology that provides future decision-makers with a basic understanding of biology and the process of science.

Accessible Text Understanding stems mainly from making connections between concepts and details, so a text with too little detail reads as a series of facts that beg to be memorized. However, too much detail overwhelms the introductory student. Thus, we authors are constantly trying to strike the perfect balance between level of detail and accessibility. We revised every page of this edition to pare detail to the minimum level we felt was necessary for the student to be able to make connections. We also know that English is a second language for many students, so we wrote in a clear and straightforward style.

This edition includes many analogies to familiar objects and phenomena that will help students understand abstract concepts. For example, in the discussion of covalent bonding in Chapter 2 (Molecules of Life), we explain that sharing electrons links two atoms, just as sharing earphones links two friends. A graphic of linked listeners (*right*) reinforces the explanation in the text.

A Wealth of Applications This book is packed with examples of applications that are relevant to a student's daily life (as marked by red squares in the index). At every opportunity, discussions of biological processes are enlivened by references to their effects on human health and the environment. Each chapter begins with an IMPACTS/ISSUES section

that explores a current event or controversy directly related to the chapter's content. For example, in Chapter 2, a discussion of *trans* fats and their effects on health introduces the structure–function relationship of molecules. The introductory topic is integrated into the chapter material and the new IMPACTS/ ISSUES REVISITED section that ends each

chapter. This ending section uses chapter content to reinforce and expand the topic presented in the opening essay. In chapter 2, for example, we link the function of hydrolysis enzymes to the unhealthy effects of *trans* fats. We also pose an associated HOW-WOULD-YOU-VOTE QUESTION that gives students the opportunity to practice making an informed choice about a

science-related controversy. The student web site provides access to articles and other material that the student can use when considering the question.

Emphasis on the Process of Science Throughout the book, descriptions of current research and photos of the scientists who do it drive home the message that science is an ongoing endeavor carried out by a diverse community of people. For example, Chapter 5 (Capturing and Releasing Energy) discusses the biofuels research of two young scientists at North Carolina State University (*right*). Research topics include not only what researchers have already discovered, but also how discoveries were made, how our understanding has changed over time, and what remains under investigation.

To strengthen a student's analytical skills and offer further insight into contemporary research, each chapter now includes an exercise on DIGGING INTO DATA. The exercise consists of a short text passage—usually about a published scientific experiment—and a table, chart, or other graphic that presents experimental data. The student can use information in the text and graphic to answer a series of questions.

In-Text Learning Tools To facilitate learning, each chapter consists of several numbered sections that contain a manageable chunk of information. Every section ends with a boxed TAKE-HOME MESSAGE in which we pose a question that reflects the critical content of the section, and then answer the question in bulleted list format. Every chapter has at least one FIGURE IT OUT QUESTION with an answer immediately following. These questions allow students to quickly check their understanding as they read. Mastering scientific vocabulary challenges many students, so we have included an ON-PAGE GLOSSARY of key terms introduced in each two-page spread, in addition to a complete glossary at the book's end. The end-of-chapter material features a VISUAL SUMMARY that reinforces each chapter's key concepts. A SELF-QUIZ poses multiple choice and other short answer questions for self-assessment (answers are in Appendix I). A set of more challenging CRITICAL THINKING QUESTIONS provides thought-provoking exercises for the motivated student.

Design and Content Revisions Previous adopters of *Biology: Today and Tomorrow* will notice that this edition features a dramatically different design. Each chapter now opens with a stunning piece of art designed to encourage student interest in the chapter's topic. The art is also used to illustrate one of the concepts within the chapter. A one-column design gives this book a simple look, and it allows placement of the on-page glossary. The following list highlights some of the revisions to each chapter. A page-by-page guide to revised content and figures is available upon request from your sales representative.

Introduction - Chapter 1 has a new Impacts/Issues section about the discovery of new species, and a new section on critical thinking. Expanded discussions of philosophy and process of science.

Unit I How Cells Work
• Chapter 2 has a new Impacts/Issues section about *trans* fats.
• Chapter 3 has a new Impacts/Issues section about foodborne *E. coli* O157:H7.
• Chapter 4 sections on energy and metabolism have been reorganized and rewritten. The new Impacts/Issues Revisited section covers alcohol metabolism and alcoholic liver disease.
• Chapter 5 now has integrated coverage of photosynthesis and energy-releasing pathways. A new Impacts/Issues about biofuels opens the chapter, and the Impacts/Issues Revisited section introduces the problem of rising atmospheric carbon dioxide.

Unit II Genetics - Material within this unit has been rearranged.
• Chapters 6 describes the structure of chromosomes, and discusses the structure and function of DNA.
• Chapter 7 covers gene expression. Gene control material reworked to focus on human examples such as breast cancer.
• Chapter 8 covers mitosis and meiosis. Revised mitosis spread juxtaposes photos of mitosis in animal and plant cells.
• Chapter 9 covers inheritance. Revised Impacts/Issues section shows a panel of young victims of cystic fibrosis. Tracking traits section has a new essay about human skin color.
• Chapter 10 covers genetic engineering. Content has been simplified and updated with current information and examples.

Unit III Evolution and Diversity - Material within this unit has been rearranged.
• Chapter 11 covers Darwin's theory and evidence of evolution. Transitional fossils and comparative embryology photo series added. New spread visually correlates sedimentary rock layers in the Grand Canyon with periods in the geologic time scale.
• Chapter 12 covers processes of evolution (microevolution, speciation, macroevolution) and classification systems. Revised and expanded coverage of reproductive isolation, cladistics.
• Chapter 13 covers prokaryotes, protists, and viruses. Impacts/Issues section now focuses on the causes of AIDS and malaria.
• Chapter 14 now covers plants and fungi. It opens with an Issues/Impacts section about Nobel Prize winner Wangari Maathai's efforts to replant trees and slow deforestation.
• Chapter 15 has a revised Impacts/Issues section that includes information about additional early bird fossils. New material added about placazoans. Birds are now discussed under the Reptile subhead. Human evolution has been updated to include a discussion of *H. floresiensis*.

• The chapter comparing plants and animals has been deleted.

Unit IV Ecology - The order of material within this unit has been altered and the chapter about animal behavior has been deleted.
• Chapter 16 covers population ecology and has a new Impacts/Issues about population explosions of Canada geese.
• Chapter 17 covers communities and ecosystems, previously covered in two separate chapters.
• Chapter 18 covers the biosphere and human effects. It opens with a new Impacts/Issues about threats to the Arctic and includes increased coverage of global climate change.

Unit V How Animals Work
• Chapter 19 has an updated Impacts/Issues section about stem cells. Homeostasis and thermoregulation now covered here.
• Chapter 20 has a new Impacts/Issues section about myostatin and muscle bulk. Improved coverage of joint, muscle disorders.
• Chapter 21 has more concise coverage of circulation and respiration.
• Chapter 22 has simplified coverage of immunology and a new Impacts/Issues section about the HPV vaccine.
• Chapter 23 has simplified coverage of urine formation.
• Chapter 24 discussion of the spinal cord now includes reflex pathways. Discussion of memory deleted.
• Chapter 25 includes information about role of thymus in AIDS, melatonin's cancer-suppressing effects, effects of phthalates.
• Chapter 26 has expanded and updated coverage of pregnancy, fertility, and sexually transmitted diseases.

Unit VI How Plants Work
• Chapter 27 has a new Impacts/Issues section about phytoremediation.
• Chapter 28 has a new Impacts/Issues section about colony collapse disorder. Expanded, revised asexual reproduction section.

Acknowledgments

No list can convey our thanks to the team of dedicated people who made this book happen, including our academic advisors listed on the following page. We could not have done this book without Grace Davidson, whose heroic contribution coordinated our efforts and kept us on schedule. Paul Forkner's tenacious photo research helped us achieve our creative vision. At Cengage Learning, Yolanda Cossio and Peggy Williams unwaveringly supported us and our ideals. Andy Marinkovich guided the production effort, Stacy Best helped us learn what works best for instructors, Kristina Razmara continues to refine our amazing technology package, Samantha Arvin helped us stay organized, Alexis Glubka organized and tracked our review feedback, and Elizabeth Momb managed all of the print ancillaries.

LISA STARR, CHRISTINE EVERS, AND CECIE STARR *November, 2008*

Academic Advisors

We owe a special debt to the members of our advisory board, listed below. They helped us shape the book's design and to choose appropriate content. We appreciate their guidance.

Andrew Baldwin, Mesa Community College
Charlotte Borgeson, University of Nevada, Reno
Gregory A. Dahlem, Northern Kentucky University
Gregory Forbes, Grand Rapids Community College
Hinrich Kaiser, Victor Valley Community College
Lyn Koller, Pierce College
Terry Richardson, University of North Alabama

We also wish to thank the reviewers listed below. Special thanks to Barbara Boss for taking the time to discuss with us how best to improve upon the prior edition.

Meghan Andrikanich, Lorain County Community College
Lena Ballard, Rock Valley College
Barbara D. Boss, Keiser University, Sarasota
Susan L. Bower, Pasadena City College
James R. Bray Jr., Blackburn College
Mimi Bres, Prince George's Community College
Randy Brewton, University of Tennessee
Evelyn K. Bruce, University of North Alabama
Steven G. Brumbaugh, Green River Community College
Chantae M. Calhoun, Lawson State Community College
Thomas F. Chubb, Villanova University
Julie A. Clements, Keiser University, Melbourne
Francisco Delgado, Pima Community College
Elizabeth A. Desy, Southwest Minnesota State University
Brian Dingmann, University of Minnesota, Crookston
Josh Dobkins, Keiser University, online
Pamela K. Elf, University of Minnesota, Crookston
Johnny El-Rady, University of South Florida
Patrick James Enderle, East Carolina University
Jean Engohang-Ndong, BYU Hawaii
Ted W. Fleming, Bradley University
Edison R. Fowlks, Hampton University
Martin Jose Garcia Ramos, Los Angeles City College
J. Phil Gibson, University of Oklahoma
Judith A. Guinan, Radford University
Carla Guthridge, Cameron University
Robert H. Inan, Inver Hills Community College
Dianne Jennings, Virginia Commonwealth University
Ross S. Johnson, Chicago State University
Paul Kaseloo, Virginia State University
Ronald R. Keiper, Valencia Community College West
Dawn G. Keller, Hawkeye Community College
Ruhul H. Kuddus, Utah Valley State College
Vic Landrum, Washburn University
Lisa Maranto, Prince George's Community College
Kevin C. McGarry, Keiser University, Melbourne
Timothy Metz, Campbell University
Alexander E. Olvido, John Tyler Community College
Joshua M. Parke, Community College of Southern Nevada
Elena Pravosudova, Sierra College
Nathan S. Reyna, Howard Payne University
Carol Rhodes, Cañada College
Todd A. Rimkus, Marymount University
Laura H. Ritt, Burlington County College
Lynette Rushton, South Puget Sound Community College
Erik P. Scully, Towson University
Marilyn Shopper, Johnson County Community College
Jennifer J. Skillen, Community College of Southern Nevada
Jo Ann Wilson, Florida Gulf Coast University

We were also fortunate to have conversations with the following workshop attendees. The insights they shared proved invaluable.

Robert Bailey, Central Michigan University
Brian J. Baumgartner, Trinity Valley Community College
Michael Bell, Richland College
Lois Borek, Georgia State University
Heidi Borgeas, University of Tampa
Charlotte Borgenson, University of Nevada
Denise Chung, Long Island University
Sehoya Cotner, University of Minnesota
Heather Collins, Greenville Techincal College
Joe Conner, Pasadena Community College
Gregory A. Dahlem, Northern Kentucky University
Juville Dario-Becker, Central Virginia Community College
Jean DeSaix, University of North Carolina
Carolyn Dodson, Chattanooga State Technical Community College
Kathleen Duncan, Foothill College, California
Dave Eakin, Eastern Kentucky University
Lee Edwards, Greenville Technical College
Linda Fergusson-Kolmes, Portland Community College
Kathy Ferrell, Greenville Technical College
April Ann Fong, Portland Community College
Kendra Hill, South Dakota State University
Adam W. Hrincevich, Louisiana State University
David Huffman, Texas State University, San Marcos
Peter Ingmire, San Francisco State
Ross S. Johnson, Chicago State University
Rose Jones, NW-Shoals Community College
Thomas Justice, McLennan Community College
Jerome Krueger, South Dakota State University
Dean Kruse, Portland Community College
Dale Lambert, Tarrant County College
Debabrata Majumdar, Norfolk State University
Vicki Martin, Appalachian State University
Mary Mayhew, Gainesville State College
Roy Mason, Mt. San Jacinto College
Alexie McNerthney, Portland Community College
Brenda Moore, Truman State University
Alex Olvido, John Tyler Community College
Molly Perry, Keiser University
Michael Plotkin, Mt. San Jacinto College
Amanda Poffinbarger, Eastern Illinois University
Johanna Porter-Kelley, Winston-Salem State University
Sarah Pugh, Shelton State Community College
Larry A. Reichard, Metropolitan Community College
Darryl Ritter, Okaloosa-Walton College
Sharon Rogers, University of Las Vegas
Lori Rose, Sam Houston State University
Matthew Rowe, Sam Houston State University
Cara Shillington, Eastern Michigan University
Denise Signorelli, Community College of Southern Nevada
Jennifer Skillen, Community College of Southern Nevada
Jim Stegge, Rochester Community and Technical College
Andrew Swanson, Manatee Community College
Megan Thomas, University of Las Vegas
Kip Thompson, Ozarks Technical Community College
Steve White, Ozarks Technical Community College
Virginia White, Riverside Community College
Lawrence Williams, University of Houston
Michael L. Womack, Macon State College

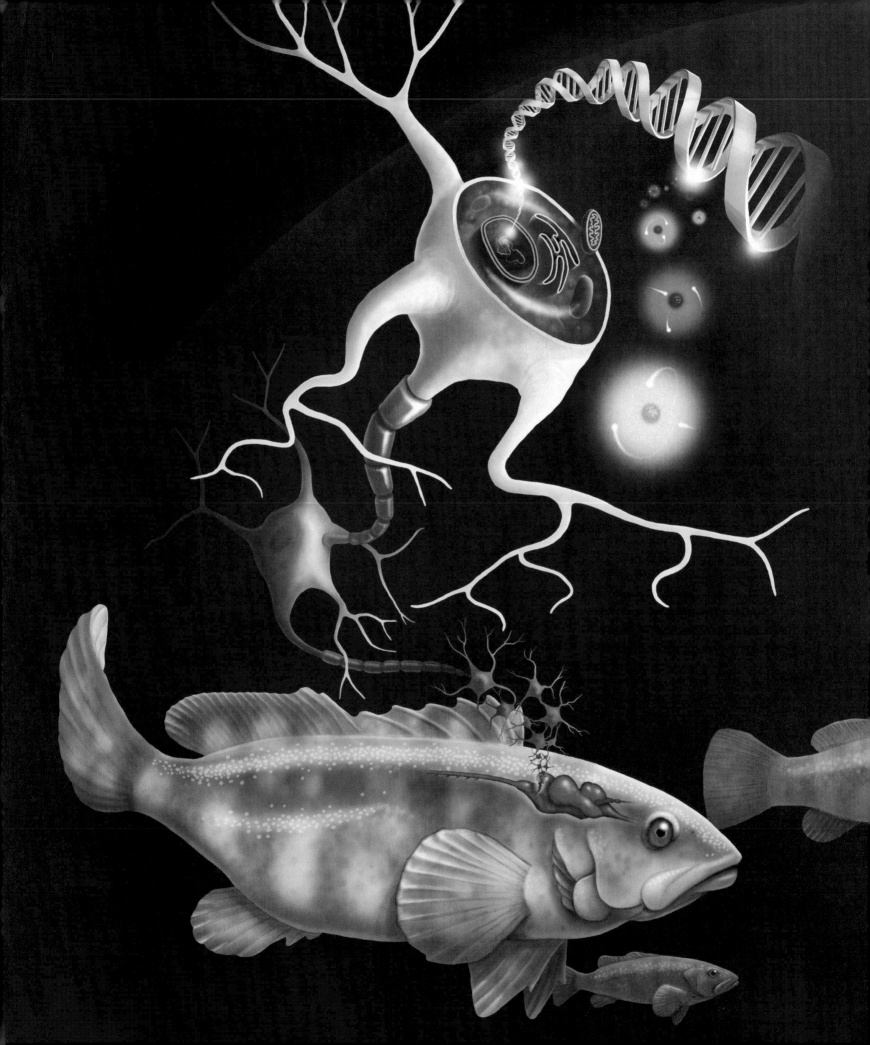

1 Invitation to Biology

> Ironically, the more we learn about nature,
> the more we realize we have yet to learn.

1.1 Impacts/Issues: The Secret Life of Earth

In this era of satellites and global positioning systems, submarines and sonar, could there possibly be any more places on Earth that we have not explored? Well, yes, actually. In 2005, for instance, helicopters dropped a team of explorers into the middle of a vast and otherwise inaccessible Indonesian cloud forest. Within minutes, the explorers realized that their landing site, a dripping, moss-covered swamp, was home to plants and animals that had been unknown to science. Over the next month, they discovered dozens of new species there, including a giant treetop rhododendron with flowers the size of plates and a frog the size of a pea. They also came across hundreds of species that are on the brink of extinction in other parts of the world, some that supposedly were extinct, and one that had not been seen in so long it had been forgotten.

The animals in the forest had never learned to be afraid of humans, so they could be approached and even picked up (Figure 1.1). A few new species were discovered as they casually wandered through the campsite. Team member Bruce Beehler remarked, "Everywhere we looked, we saw amazing things we had never seen before. I was shouting. This trip was a once-in-a-lifetime series of shouting experiences."

How do we know whether a particular organism belongs to a new species? What is a species, anyway, and why should discovering a new one matter to anyone other than a scientist? You will find the answers to such questions in this book. They are part of the scientific study of life, **biology**, which is one of many ways we humans try to make sense of the world around us.

Trying to understand the immense scope of life on Earth gives us some perspective on where we fit into it. For example, we routinely discover hundreds of species every year, but about 20 species become extinct *every minute* in rain forests alone. The current rate of extinctions is about 1,000 times faster than normal, and human activities are responsible for the acceleration. At this rate, we will never know about most of the species that are alive on Earth today. Does that matter? Biologists think so.

Whether or not we are aware of it, we humans are intimately connected with the world around us. We are profoundly changing the entire fabric of life on Earth. The changes are, in turn, affecting us in ways we are just beginning to fathom. Ironically, the more we learn about nature, the more we realize we have yet to learn.

But don't take our word for it. Find out what biologists know, and what they do not, and you will have a solid foundation upon which to base your own opinions about our place in this world. By reading this book, you are choosing to learn about the human connection—your connection—with all life on Earth.

FIGURE 1.1 Biologist Kris Helgen holds a rare golden-mantled tree kangaroo he and his colleagues found in a cloud forest in the Foja Mountains of New Guinea.

Life's Levels of Organization

If you are reading this book, you are starting to explore how a subset of scientists—biologists—think about nature. **Nature** is every substance and energy in the universe except what humans have manufactured. It includes flowers, water, animals, rocks, thunder, and so on. Biologists study the parts of nature that have to do with life, past and present. Through their work, we glimpse a great pattern of organization (Figure 1.2). The pattern starts with **atoms**, which are basic building blocks of all matter ❶. At the next level of organization, atoms join as **molecules** ❷. Only living cells make the molecules of life—complex carbohydrates and lipids, proteins, and nucleic acids—in nature. The pattern crosses the threshold to life when many molecules organize as a cell ❸. A **cell** is the smallest unit of life that can survive and reproduce on its own, given information in its DNA, energy, and raw materials. An **organism** is an individual that consists of one or more cells. In larger multicelled organisms, trillions of cells may be organized as tissues, organs, and organ systems that interact to keep the individual's body working properly ❹. At the next level of organization, a **population** is a group of individuals of the same type, or species, living in a given area ❺. An example would be all of the lake trout living in Lake Tahoe, California. At the next level, a **community** consists of all populations of all species in a given area ❻. An underwater ocean community, for example, may include many kinds of organisms that make their home in or on a particular reef. The next level of organization is the **ecosystem**: a community interacting with its environment ❼. The most inclusive level, the **biosphere**, encompasses all regions of Earth's crust, waters, and atmosphere in which organisms live ❽.

Remember that life is more than the sum of its individual parts. In other words, some emergent property occurs at each successive level of life's organization. An **emergent property** is a characteristic of a system that does not appear in any of a system's component parts. For example, the molecules of life are themselves not alive. Considering them separately, no one would be able to predict that a particular arrangement of molecules would form a living cell. Life—an emergent property—appears first at the level of the cell.

FIGURE 1.2 Animated!
Levels of organization in nature, from simpler to more complex.

❶ Atoms
❷ Molecules
❸ Cells
❹ Organisms
❺ Populations
❻ Communities
❼ Ecosystems
❽ The biosphere

atom Fundamental building block of all matter.
biology Systematic study of life.
biosphere All regions of Earth where organisms live.
cell Smallest unit of life.
community All populations of all species in a given area.
ecosystem A community interacting with its environment.
emergent property A characteristic of a system that does not appear in any of a system's component parts.
molecule An association of two or more atoms.
nature Everything in the universe except what humans have manufactured.
organism Individual that consists of one or more cells.
population Group of individuals of the same species that live in a given area.

Take-Home Message How does "life" differ from "nonlife"?

■ The building blocks—atoms—that make up all living things are the same ones that make up all nonliving things.

■ Atoms join as molecules. The unique properties of life emerge as certain kinds of molecules become organized into cells.

■ Higher levels of organization include multicelled organisms, populations, communities, ecosystems, and the biosphere.

Overview of Life's Unity

"Life" is not easy to define: It is just too big, and it has been changing for billions of years. Even so, we know that all living things have similar characteristics. All living things require energy and raw materials; they sense and respond to change; and they reproduce with the help of DNA.

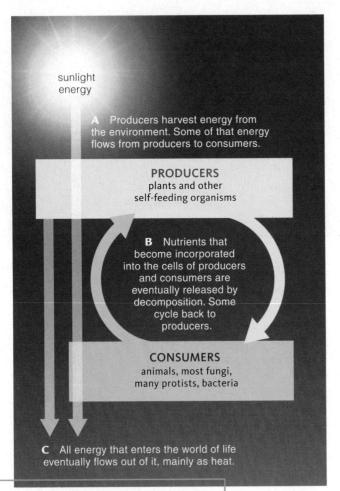

sunlight energy

A Producers harvest energy from the environment. Some of that energy flows from producers to consumers.

PRODUCERS
plants and other
self-feeding organisms

B Nutrients that become incorporated into the cells of producers and consumers are eventually released by decomposition. Some cycle back to producers.

CONSUMERS
animals, most fungi,
many protists, bacteria

C All energy that enters the world of life eventually flows out of it, mainly as heat.

FIGURE 1.3 Animated! The one-way flow of energy and the cycling of materials in the world of life. *Below*, a consumer eating a producer.

Energy Sustains Life's Organization Eating supplies your body with energy and nutrients that keep it organized and functioning. **Energy** is the capacity to do work. A **nutrient** is a substance that an organism needs for growth and survival, but cannot make for itself.

All organisms spend a lot of time acquiring energy and nutrients, although different species get them from different sources. The differences allow us to classify organisms into one of two broad categories: producers and consumers. **Producers** make their own food using energy and simple raw materials they get directly from their environment. Plants are producers that use the energy of sunlight to make sugars from water and carbon dioxide (a gas in air). **Consumers** cannot make their own food; they get energy and nutrients indirectly—by feeding on other organisms. Animals are consumers. So are decomposers, which feed on the wastes or remains of other organisms. The leftovers of their meals end up in the environment, where they serve as nutrients for producers. Said another way, nutrients cycle between producers and consumers.

Energy, however, is not cycled. It flows through the world of life in one direction: from the environment to organisms. This flow maintains the organization of individual organisms. It is also the basis of how organisms interact with one another and their environment. The flow is one-way, because with each transfer, some energy escapes as heat. Cells do not use heat to do work. Thus, the energy that enters the world of life eventually leaves it, permanently (Figure 1.3).

Organisms Sense and Respond to Change An organism senses and responds to change both inside and outside of its body by way of receptors. A **receptor** is a molecule or cellular structure that responds to a specific form of stimulation (Figure 1.4).

A receptor that has been stimulated can trigger changes in an organism's activities. For example, after you eat, the sugars from your meal enter your bloodstream. The added sugars in your blood bind to molecular receptors on cells of the pancreas (an organ). Binding sets in motion a series of events that causes cells throughout the body to take up sugar faster, so the sugar level in your blood quickly falls. The interplay of processes keeps your blood sugar level within a certain range, which in turn helps keep your cells alive and your body functioning.

The fluid in your blood is part of your body's internal environment, which consists of all body fluids outside of cells. Unless the internal environment is kept within certain ranges of composition, temperature, and other conditions, your body cells will die. By sensing and adjusting to change, you and all other organisms keep conditions in the internal environment within a range that

FIGURE 1.4 Organisms have receptors that allow them to sense and respond to stimuli such as the mechanical energy of a bite.

favors cell survival. **Homeostasis** is the name for this process, and it is a defining feature of life.

Organisms Grow and Reproduce Individuals of every natural population are alike in certain aspects of their body form, function, and behavior, but the details of such traits often differ from one individual to the next. For example, humans characteristically have two eyes, but those eyes occur in a range of color among individuals. Eye color and most other traits are the outcome of information encoded in **DNA**, or deoxyribonucleic acid. DNA is the signature molecule of life. It carries information that guides **growth**—increases in cell number, size, and volume—and **development**, the process by which the first cell of a new individual becomes a multicelled adult.

Only multicelled species undergo development, but all organisms inherit DNA from parents. Lions look like lions and not like peas because they inherited lion DNA, which differs from pea DNA in the information it carries. **Inheritance** refers to the transmission of DNA from parents to offspring. Such transmission occurs by processes of **reproduction**, which produce new individuals.

Take-Home Message How are all living things alike?

- A one-way flow of energy and a cycling of nutrients sustain life's organization.
- Organisms sense and respond to change. They make adjustments that keep conditions in their internal environment within a range that favors cell survival, a process called homeostasis.
- Organisms grow, develop, and reproduce based on information encoded in their DNA, which they inherit from their parents.

consumer Organism that gets energy and carbon by feeding on tissues, wastes, or remains of other organisms.

development Multistep process by which the first cell of a new individual becomes a multicelled adult.

DNA Deoxyribonucleic acid; molecule that carries hereditary information about traits.

energy The capacity to do work.

growth Increases in the number, size, and volume of cells in multicelled species.

homeostasis Set of processes by which an organism keeps its internal conditions within tolerable ranges.

inheritance Transmission of DNA from parents to offspring.

nutrient Substance that an organism needs for growth and survival, but cannot make for itself.

producer Organism that makes its own food using energy and simple raw materials from the environment.

receptor Molecule or structure that responds to a specific form of stimulation.

reproduction Process by which parents produce offspring.

1.4 Introduction to Life's Diversity

Each time we discover a new **species**, or kind of organism, we give it a two-part name. The first part of the name specifies the **genus** (plural, genera), which is a group of species that share a unique set of features. It designates one species when combined with the second part, the species name. Individuals of a species share one or more heritable traits, and they can interbreed successfully if the species is a sexually reproducing one. Genus and species names are always italicized. For example, *Panthera* is a genus of big cats. The lions in Figure 1.4 are of the species *Panthera leo*. Tigers, or *P. tigris*, are a different species in the same genus. Note that the genus name may be abbreviated after it has been spelled out once.

We use various classification systems to organize and retrieve information about species. Most systems sort species into groups on the basis of their traits. Figure 1.5 shows a common system in which all species are grouped into three domains: Bacteria, Archaea, and Eukarya. Protists, plants, fungi, and animals make up domain Eukarya. We return to systems of naming and grouping organisms in more detail in Chapter 12.

All **bacteria** (singular, bacterium) and **archaeans** are single-celled organisms. All of them are **prokaryotes**, which means that their DNA is not contained within a nucleus. A nucleus is a membrane-enclosed sac that protects a cell's DNA. As a group, prokaryotes are the most diverse organisms. Different kinds are producers or consumers that inhabit nearly all of the biosphere, including extreme environments such as frozen desert rocks, boiling sulfur-clogged lakes, and nuclear reactor waste. The first cells on Earth may have faced similarly hostile challenges to survival.

Cells of **eukaryotes** have a nucleus. Structurally, **protists** are the simplest kind of eukaryote. Different protist species are producers or consumers. Many are single cells that are larger and more complex than prokaryotes. Some of them are tree-sized, multicelled seaweeds. Cells of fungi, plants, and animals are also eukaryotic. Most **fungi** (singular, fungus), such as the types that form mushrooms, are multicelled. Many are decomposers, and all secrete enzymes that digest food outside the body. Their cells then absorb the released nutrients.

Plants are multicelled species that live on land or in freshwater environments. Most are producers. By a process called photosynthesis, they harness the energy of sunlight to drive the production of sugars from carbon dioxide and water. Besides feeding themselves, plants and other photosynthesizers serve as food for most of the other organisms in the biosphere.

Animals are multicelled consumers that ingest tissues or juices of other organisms. Herbivores graze, carnivores eat meat, scavengers eat remains of other organisms, and parasites withdraw nutrients from the tissues of a host. Animals grow and develop through a series of stages that lead to the adult form, and most kinds actively move about during at least part of their lives.

Bacteria

Archaea

Eukarya

FIGURE 1.5 Animated! Three-domain classification system with a few representatives of life's diversity.

Take-Home Message

How do living things differ from one another?

- Organisms differ in their details; they show tremendous variation in traits.
- Classification systems by which species are grouped according to traits help us organize information about species. One common system sorts all species into three domains: Bacteria, Archaea, and Eukarya.

The Nature of Scientific Inquiry

Most of us assume that we do our own thinking—but do we, really? You might be surprised to find out just how often we let others think for us. For instance, a school's job, which is to impart as much information as possible to students, meshes with a student's job, which is to acquire as much knowledge as possible. In this rapid-fire exchange of information, it is easy to forget about the quality of what is being exchanged. If you accept information without question, you allow someone else to think for you.

Critical thinking means judging information before accepting it. "Critical" comes from the Greek *kriticos* (discerning judgment). When you think this way, you move beyond the content of information. You look for underlying assumptions, evaluate the supporting statements, and think of possible alternatives.

How does the busy student manage this? Be aware of what you intend to learn from new information. Be conscious of bias or underlying agendas in books, lectures, or online. Decide whether ideas are based on opinion or evidence. Question authority figures (respectfully). Consider what you want to believe, and realize that those biases influence your learning. Such practices will help you decide whether to accept or reject information.

The Scope—and the Limits—of Science Because each of us is unique, there are as many ways to think about the natural world as there are people. **Science**, the systematic study of nature, is one way. It helps us be objective about our observations of nature, in part because of its limitations. We limit science to a subset of the world: only that which is observable.

Science does not address some questions, such as "Why do I exist?" Most answers to such questions are subjective—they come from within as an integration of the personal experiences and mental connections that shape our consciousness. This is not to say subjective answers have no value, because no human society can function for long unless its individuals share standards for making judgments, even if they are subjective. Moral, aesthetic, and philosophical standards vary from one society to the next, but all help people decide what is important and good. All give meaning to what we do.

Also, science does not address the supernatural, or anything that is "beyond nature." Science does not assume or deny that supernatural phenomena occur, but scientists may still cause controversy when they discover a natural explanation for something that was thought to be supernatural. Such controversy often arises when a society's moral standards have become interwoven with its understanding of nature.

For example, Nicolaus Copernicus studied the planets centuries ago in Europe, and concluded that Earth orbits the sun. Today his conclusion seems obvious, but at the time it was heresy. The prevailing belief was that the Creator made Earth—and, by extension, humans—as the center of the universe. Galileo Galilei, another scholar, found evidence for the Copernican model of the solar system and published his findings. He was forced to publicly recant his publication, and to put Earth back at the center of the universe.

Exploring a traditional view of the natural world from a scientific perspective might be misinterpreted as questioning morality even though the two are not the same. As a group, scientists are no less moral, less lawful, or less spiritual than anyone else. As you will see in the next section, however, their work follows a particular standard: Their explanations must be testable in ways that others can repeat.

animal Multicelled consumer that develops through a series of embryonic stages and moves about during part or all of the life cycle.

archaean A member of the prokaryotic domain Archaea.

bacterium A member of the prokaryotic domain Bacteria.

critical thinking Mental process of judging information before accepting it.

eukaryote Organism whose cells characteristically have a nucleus.

fungus Type of eukaryotic consumer that obtains nutrients by digestion and absorption outside the body.

genus A group of species that share a unique set of traits.

plant A multicelled, typically photosynthetic producer.

prokaryote Single-celled organism in which the DNA is not contained in a nucleus.

protist Diverse group of simple eukaryotes.

science Systematic study of nature.

species A type of organism.

Science helps us communicate our experiences without bias, so it may be as close as we can get to a universal language. We are fairly sure, for example, that laws of gravity apply everywhere in the universe. Intelligent beings on a distant planet would likely understand the concept of gravity. We might well use gravity or another scientific concept to communicate with them, or anyone, anywhere. The point of science, however, is not to communicate with aliens. It is to find common ground here on Earth.

Take-Home Message **What is science?**

- Science is the study of the observable—those objects or events for which objective evidence can be gathered. It does not address the supernatural.

1.6 How Science Works

Science helps us communicate our experiences without bias.

Observations, Hypotheses, and Tests To get a sense of how science works, consider this list of common research practices:

1. Observe some aspect of nature.

2. Frame a question about your observation.

3. Read what others have discovered concerning the subject, then propose a **hypothesis**, a testable answer to your question.

4. Using the hypothesis as a guide, make a **prediction**—a statement of some condition that should exist if the hypothesis is not wrong. Making predictions is called the if–then process: "if" is the hypothesis, and "then" is the prediction.

5. Devise ways to test the accuracy of the prediction by conducting experiments or gathering information. Tests may be performed on a **model**, or similar system, if testing an object or event directly is not possible.

6. Assess the results of the tests or observations—the data. Data that confirm the prediction are evidence in support of the hypothesis. Data that disprove the prediction are evidence that the hypothesis may be flawed.

7. Report all the steps of your work, along with any conclusions you drew, to the scientific community.

You might hear someone refer to these practices as "the scientific method," as if all scientists march to the drumbeat of a fixed procedure. They do not. There are different ways to do research, particularly in biology (Figure 1.6). Some biologists survey and observe without making hypotheses. Others make hypotheses and leave the testing to others. A few stumble onto valuable information they are not even looking for. Regardless of the variation, one thing is constant: Scientists do not accept information simply because someone says it is true. They evaluate the supporting evidence and find alternative explanations. Does this practice sound familiar? It should—it is critical thinking.

Theories and Laws Suppose a hypothesis has not been disproven even after years of tests. It is consistent with all evidence gathered to date, and it has helped us to make successful predictions about other phenomena. When a hypothesis meets these criteria, it is considered a **scientific theory** (Table 1.1). For

FIGURE 1.6 Making observations for a field study is one of many ways of doing science.

example, scientists no longer spend time testing the hypothesis that all matter is composed of atoms, for the compelling reason that no one has ever detected matter composed of anything else. This hypothesis is now called atomic theory.

A **law of nature** describes a phenomenon that has been observed to occur in every circumstance without fail, but for which we currently do not have a complete scientific explanation. The laws of thermodynamics, which describe energy, are examples. We know how energy behaves, but not why it behaves the way it does.

Most scientists carefully avoid the word "truth" when discussing science, preferring instead to use "accurate" in reference to data. In science, there is only evidence that supports a hypothesis because an infinite number of tests would be necessary to confirm that a theory holds under every possible circumstance. A single observation or result that is not consistent with a theory would open it to revision. For example, if someone discovered a type of matter not composed of atoms, the atomic theory would be revised. The theory of evolution, which states that change occurs in a line of descent over time, still holds after a century of testing and scrutiny. As with all other scientific theories, we cannot be sure that it will hold under all possible conditions, but we can say it has a very high probability of not being wrong. If evidence turns up that is inconsistent with evolution, biologists will revise the theory.

You may hear people apply the word "theory" to a speculative idea, as in the phrase "It's just a theory." Speculation is opinion or belief, a personal conviction that is not necessarily supported by evidence. A scientific theory is not an opinion: By definition, it must be supported by a large body of evidence.

Unlike theories, many beliefs and opinions cannot be tested. Ideas that cannot be tested cannot be disproven. Personal conviction has tremendous value in our lives, but it is not the same as scientific theory.

Table 1.1	Examples of Scientific Theories
Theory	**Main Premises**
Cell theory	All organisms consist of one or more cells, the cell is the basic unit of life, and all cells arise from existing cells.
Atomic theory	All substances consist of atoms.
Evolution	Change occurs in inherited traits of a population over generations.
Germ theory	Microorganisms cause many diseases.
Global warming	Human activities are causing Earth's average temperature to increase.
Plate tectonics	Earth's crust is cracked into pieces that move in relation to one another.
Big Bang	The universe originated with an explosion and continues to expand.

Take-Home Message **How does science work?**

- Scientific inquiry involves making observations and asking questions about some aspect of nature; formulating hypotheses; making predictions and testing them using real or model systems; and reporting the results and conclusions.

hypothesis Testable explanation of a natural phenomenon.

law of nature Generalization that describes a consistent natural phenomenon for which there is incomplete scientific explanation.

model System similar to an object or event that cannot itself be tested directly.

prediction A statement, based on a hypothesis, about a condition that should exist if the hypothesis is not wrong.

scientific theory Hypothesis that has not been disproven after many years of rigorous testing, and is useful for making predictions about other phenomena.

The Power of Experiments

Careful observations are one way to test predictions that flow from a hypothesis. So are experiments. **Experiments** are tests that can support or falsify a prediction. They are usually designed to determine the effects of a single **variable**, which is a characteristic or event that differs among individuals. Biological systems are complex, with many interacting variables. It can be difficult to study one variable separately from the rest. Thus, biology researchers often test two groups of individuals, side by side. An **experimental group** is a set of individuals that have a certain characteristic or receive a certain treatment. This group is tested side by side with a **control group**, which is identical to the experimental group except for one variable—the characteristic or the treatment being tested. Thus, any differences in experimental results between the two groups should be an effect of changing the variable. We will look at two different experiments to see how these groups are used.

Potato Chips and Stomachaches In 1996 the FDA approved Olestra®, a fat replacement manufactured from sugar and vegetable oil, as a food additive. Potato chips were the first Olestra-containing food product on the market in the United States. Controversy soon raged. Some people complained of intestinal problems after eating the chips and concluded that the Olestra was at fault. Two years later, researchers at Johns Hopkins University School of Medicine designed an experiment to test the hypothesis that this food additive causes cramps. They predicted that *if* Olestra causes cramps, *then* people who eat Olestra will be more likely to get cramps than people who do not.

To test the prediction, they used a Chicago theater as a "laboratory." They asked 1,100 people between the ages of thirteen and thirty-eight to watch a movie and eat their fill of potato chips. Each person got an unmarked bag that contained 13 ounces of chips. The individuals who got Olestra-containing potato chips were the experimental group, and individuals who got regular chips were the control group. The variable consisted of the presence of Olestra in the chips. Afterward, the researchers contacted all of the people and tabulated any reports of gastrointestinal problems. Of 563 people making up the experimental group, 89 (15.8 percent) complained about cramps. However, so did 93 of the 529 people (17.6 percent) making up the control group—who had munched on the regular chips! This experiment disproved the prediction that people who eat Olestra are more likely to get cramps than people who do not (Figure 1.7).

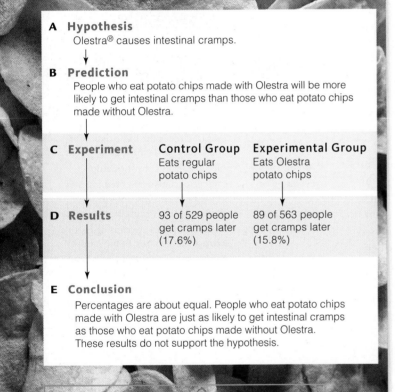

A Hypothesis
Olestra® causes intestinal cramps.

B Prediction
People who eat potato chips made with Olestra will be more likely to get intestinal cramps than those who eat potato chips made without Olestra.

C Experiment

	Control Group	Experimental Group
	Eats regular potato chips	Eats Olestra potato chips

D Results

	93 of 529 people get cramps later (17.6%)	89 of 563 people get cramps later (15.8%)

E Conclusion
Percentages are about equal. People who eat potato chips made with Olestra are just as likely to get intestinal cramps as those who eat potato chips made without Olestra. These results do not support the hypothesis.

FIGURE 1.7 Animated! The steps in a scientific experiment to determine if Olestra causes cramps. A report of this study was published in the *Journal of the American Medical Association* in January of 1998.

Butterflies and Birds Consider the peacock butterfly, a winged insect that was named for the large, colorful spots on its wings. In 2005, researchers published a report on their tests to identify factors that help peacock butterflies defend themselves against insect-eating birds. The researchers made two observations. First, when a peacock butterfly rests, it folds its wings, so only the dark underside shows. Second, when a butterfly sees a predator approaching, it repeatedly flicks its paired forewings and hindwings open and closed. At the same time, each forewing slides over the hindwing, which produces a hissing sound and a series of clicks.

Table 1.2	Results of Peacock Butterfly Experiment*				
Wing Spots	Wing Sound	Total Number of Butterflies	Number Eaten	Number Survived
Spots	Sound	9	0	9 (100%)
No spots	Sound	10	5	5 (50%)
Spots	No sound	8	0	8 (100%)
No spots	No sound	10	8	2 (20%)

Proceedings of the Royal Society of London, Series B (2005) 272: 1203–1207.

FIGURE 1.8 Peacock butterfly defenses.

A When a bird approaches, this butterfly repeatedly flicks its wings open and closed, which exposes brilliant spots and produces hissing and clicking sounds.

The researchers tested whether the behavior deters blue tits **B**. They painted over the spots of some butterflies, cut the sound-making part of the wings on others, and did both to a third group; then the biologists exposed each butterfly to a hungry bird. The results are listed *above*, in Table 1.2.

Figure It Out: What percentage of butterflies with no spots and no sound survived the test?

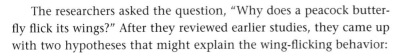 *Answer: 20 percent*

The researchers asked the question, "Why does a peacock butterfly flick its wings?" After they reviewed earlier studies, they came up with two hypotheses that might explain the wing-flicking behavior:

1. Although the wing-flicking probably attracts predatory birds, it also exposes brilliant spots that resemble owl eyes (Figure 1.8**A**). Anything that looks like owl eyes is known to startle small, butterfly-eating birds, so exposing the wing spots might scare off predators.

2. The hissing and clicking sounds produced when the peacock butterfly rubs the sections of its wings together may deter predatory birds.

The researchers used their hypotheses to make the following predictions:

1. *If* brilliant wing spots of peacock butterflies deter predatory birds, *then* individuals with no wing spots will be more likely to get eaten by predatory birds than individuals with wing spots.

2. *If* the sounds that peacock butterflies produce deter predatory birds, *then* individuals that do not make the sounds will be more likely to be eaten by predatory birds than individuals that make the sounds.

The next step was the experiment. The researchers painted the wing spots of some butterflies black, cut off the sound-making part of the hindwings of others, and did both to a third group. They put each butterfly into a large cage with a hungry blue tit (Figure 1.8**B**) and then watched the pair for thirty minutes.

Table 1.2 lists the results of the experiment. All of the butterflies with unmodified wing spots survived, regardless of whether they made sounds. By contrast, only half of the butterflies that had spots painted out but could make sounds survived. Most of the butterflies with neither spots nor sound structures were eaten quickly. The test results confirmed both predictions, so they support the hypotheses. Birds are deterred by peacock butterfly sounds, and even more so by wing spots.

control group A group of objects or individuals that is identical to an experimental group except for one variable.
experiment A test designed to support or falsify a prediction.
experimental group A group of objects or individuals that display or are exposed to a variable under investigation.
variable A characteristic or event that differs among individuals.

FIGURE 1.9 Example of error bars in a graph. This particular graph was adapted from the peacock butterfly research described on the previous page.

The researchers recorded the number of times each butterfly flicked its wings in response to an attack by a bird.

The *orange* squares represent the average frequency of wing flicking for each sample set of butterflies. *Black* error bars that extend above and below the squares indicate the range of values—the sampling error—for each set of butterflies.

Sampling Error

Sampling Error Researchers can rarely observe all individuals of a group. For example, the explorers you read about in Section 1.1 did not—and could not—survey every uninhabited part of New Guinea. The cloud forest itself cloaks more than 2 million acres of the Foja Mountains. Surveying all of it would take unrealistic amounts of time and effort. Besides, tromping about even in a small area can damage delicate forest ecosystems.

Given such limitations, researchers often look at subsets of an area, a population, event, or some other aspect of nature. They test or survey the subset, then use the results to make generalizations. However, generalizing from a subset is risky because a subset may not be representative of the whole.

For example, the golden-mantled tree kangaroo pictured in Figure 1.1 was first discovered in 1993 on a single forested mountaintop in New Guinea. For more than a decade the species was never seen outside of that habitat, which is getting smaller every year because of human activities. Thus, the golden-mantled tree kangaroo was considered to be one of the most endangered animals on the planet. Then, in 2005, the explorers featured in Section 1.1 discovered that this kangaroo species is fairly common in the Foja Mountain cloud forest. As a result, biologists now believe its future is secure, at least for the moment.

Sampling error is a difference between results from a subset and results from the whole. It happens most often when a sample size is small. Starting with a large sample or repeating an experiment many times helps minimize sampling error. Sampling error is an important consideration in the design of most experiments. However, it is sometimes unavoidable, as illustrated by the example of the golden-mantled tree kangaroo. Error bars on a graph may indicate sampling error or other variation in experimental results. For example, the error bars in Figure 1.9 show the range of results for each sample set of butterflies.

Probability

Probability Probability is a measure of the chance that a particular outcome will occur. That chance depends on the total number of possible outcomes. We usually express probability as a percentage. For instance, if 10 million people enter a drawing, each has the same probability of winning: 1 in 10 million, or a very improbable 0.00001 percent.

Sample size is important in probability. To understand why, imagine flipping a coin. There are two possible outcomes: The coin lands heads up, or it lands tails up. Every time you flip the coin, the chance that it will land, say, heads up, is one in two (1/2)—a probability of 50 percent. However, when you actually flip the coin, it often lands heads up, or tails up, several times in a row. With just 10 flips, the proportion of times that heads actually land up may be very far from 50 percent. With 1,000 flips, that proportion is more likely to be near 50 percent.

An experimental result is said to be **statistically significant** if it is unlikely to have occurred by chance. In this context, the word "significant" does not refer to the result's importance. It means that the result has been subjected to a rigorous statistical analysis that shows it has a very low probability (usually 5 percent or less) of being skewed by sampling error.

Variation in data is often shown as error bars on a graph. Depending on the graph, error bars may indicate variation around an average for one sample set (Figure 1.9), or the difference between two sample sets.

Asking Useful Questions Researchers try to design experiments that have a single variable and that will yield counts or some other data that can be measured or otherwise gathered objectively. Even so, they risk designing experiments and interpreting results in terms of what they want to find out. Particularly when studying humans, isolating a single variable is not often possible. For example, by thinking critically, we may realize that the people who participated in the Olestra experiment were chosen randomly. That means the study was not controlled for gender, age, weight, medications taken, and so on. Such variables may well have influenced the results.

Scientists expect one another to put aside bias. If one individual does not, others will, because science is both cooperative and competitive.

Take-Home Message **What is the purpose of experiments?**

- Natural processes are often influenced by many interacting variables.
- Experiments help researchers unravel causes of complex natural processes by focusing on the effects of changing a single variable.
- Researchers try to design experiments carefully in order to minimize sampling error and the potential for bias.

FIGURE 1.10 Discovered in Madagascar in 2005, this tiny mouse lemur was named *Microcebus lehilahytsara* in honor of primatologist Steve Goodman (lehilahytsara is a combination of the Malagasy words for "good" and "man"). Lemurs are evolutionary cousins of monkeys and apes.

1.8 Impacts/Issues Revisited: The Secret Life of Earth

Just since 2002, biologists have discovered many hundreds of species that had been previously unknown to science. A return expedition to New Guinea's Foja Mountains turned up a mouse-sized opossum and a cat-sized rat. Expeditions elsewhere revealed lemurs (Figure 1.10) and sucker-footed bats in Madagascar; birds in the Philippines; monkeys in Tanzania, Brazil, and India; cave-dwelling spiders and insects in two of California's national parks; carnivorous sponges near Antarctica; whales, sharks, giant jellylike animals, fishes, and other aquatic wildlife; and scores of plants and single-celled organisms. Most were discovered by biologists who were simply trying to find out what lives where.

Biologists make discoveries every day, though we may never hear of them. Each new species they discover is another reminder that we do not yet know all of the organisms on our own planet. We don't even know how many to look for. The vast information about the 1.8 million species we do know about changes so quickly that collating it has been impossible—until now. A new web site, titled the Encyclopedia of Life, is intended to be an online reference source and database of species information that is maintained by collaborative effort. See it at www.eol.org.

How Would YOU ☑ Vote? There is a possibility that substantial populations of some species currently listed as endangered may exist in unexplored areas. Should we wait to protect endangered species until all of Earth has been surveyed? See CengageNow for details, then vote online (cengagenow.com).

sampling error Difference between results derived from testing an entire group of events or individuals, and results derived from testing a subset of the group.

statistically significant Refers to a result that is statistically unlikely to have occurred by chance.

Summary

Section 1.1 Biology is the systematic study of life. New species are discovered on a continuing basis. We have encountered only a fraction of the organisms that live on Earth, in part because we have explored only a fraction of its habitable regions.

Section 1.2 Biologists think about nature at different levels of organization. All matter consists of atoms, which combine as molecules. Organisms consist of one or more cells, the smallest units of life. A population is a group of individuals of a species in a given area; a community is all populations of all species in a given area. An ecosystem is a community interacting with its environment. The biosphere includes all regions of Earth that hold life.

CENGAGENOW™ Explore life's levels of organization with the interaction on CengageNow.

Section 1.3 Life has underlying unity in that all living things have similar features. All organisms require energy and nutrients to sustain themselves. Producers such as plants make their own food; animals and other consumers eat other organisms, or their wastes and remains. Receptors help organisms keep the conditions in their internal environment within ranges that their cells tolerate—a process called homeostasis. DNA, which is inherited from parents, contains information that is the source of an individual's traits.

A one-way flow of energy through the biosphere and the cycling of nutrients among organisms sustains life, and life's organization.

CENGAGENOW™ See energy flow and materials cycling with the animation on CengageNow.

Section 1.4 The millions of species (types of organisms) that currently exist on Earth show tremendous diversity in that they differ greatly in details of body form and function. Each species is given a unique two-part name that includes genus and species name. Various classification systems sort species into groups on the basis of subsets of shared traits. One system categorizes all organisms into three domains: Bacteria, Archaea, and Eukarya. Plants, protists, fungi, and animals are the eukaryotes.

CENGAGENOW™ Explore characteristics of the three domains of life with the interaction on CengageNow.

Section 1.5 Critical thinking means judging the quality of information before deciding whether or not to accept it. Science is one of many ways of looking at the natural world. It helps us to communicate our experiences without bias by focusing on only testable ideas about observable phenomena. Science does not address the supernatural.

Section 1.6 Researchers make and test potentially falsifiable predictions about how the natural world works. Generally, scientific inquiry involves forming a hypothesis (a testable assumption) about an observation, then making and testing predictions based on the hypothesis. A hypothesis that is not consistent with the results of scientific tests (evidence) is modified or discarded.

A scientific theory is a long-standing hypothesis that is useful for making predictions about other phenomena. A law of nature describes a consistent and universal phenomenon (such as gravity) for which we do not yet have a complete scientific explanation.

Section 1.7 Natural processes are often influenced by many interacting variables. Experiments simplify interpretations of complex biological systems by focusing on the effect of one variable at a time. Biology researchers necessarily experiment on subsets of a group, which may result in sampling error.

CENGAGENOW™ See the steps in an experiment and how sample size affects accuracy with the interactions on CengageNow.

Self-Quiz   Answers in Appendix I

1. _____ are fundamental building blocks of all matter.
   a. Cells
   b. Atoms
   c. Nutrients
   d. Molecules

2. The smallest unit of life is the _____ .
   a. atom
   b. molecule
   c. cell
   d. organism

3. _____ move around for at least part of their life.
   a. Organisms
   b. Plants
   c. Animals
   d. Prokaryotes

4. Organisms require _____ and _____ to maintain themselves, grow, and reproduce.

5. _____ is a process that maintains conditions in the internal environment within ranges that cells can tolerate.
   a. Sampling error
   b. Development
   c. Homeostasis
   d. Critical thinking

6. Bacteria, Archaea, and Eukarya are three _____ .
   a. organisms
   b. domains
   c. consumers
   d. producers

7. DNA _____ .
   a. guides growth and development
   b. is a nucleic acid
   c. is transmitted from parents to offspring
   d. all of the above

8. An animal is a(n) _____ (choose all that apply).
   a. organism
   b. domain
   c. species
   d. eukaryote
   e. consumer
   f. producer
   g. prokaryote
   h. trait

9. Plants are _____ (choose all that apply).
   a. organisms
   b. a domain
   c. a species
   d. eukaryotes
   e. consumers
   f. producers
   g. prokaryotes
   h. traits

10. _____ is the transmission of DNA to offspring.
   a. Reproduction
   b. Development
   c. Homeostasis
   d. Inheritance

11. A process by which an organism produces offspring is called _____ .
   a. reproduction
   b. inheritance
   c. development
   d. homeostasis

12. Science only addresses that which is _____ .
   a. alive
   b. observable
   c. variable
   d. indisputable

13. A control group is _____ .
   a. a set of individuals that have a certain characteristic or receive a certain treatment
   b. the standard against which an experimental group is compared
   c. the experiment that gives conclusive results

14. Match the terms with the most suitable description.
   _____ emergent property
   _____ species
   _____ scientific theory
   _____ hypothesis
   _____ prediction

   a. statement of what a hypothesis leads you to expect to see
   b. type of organism
   c. occurs at a higher organizational level in nature, not at levels below it
   d. time-tested hypothesis
   e. testable explanation

Additional questions are available on CENGAGENOW™

Critical Thinking

1. Why would you think twice about ordering from a restaurant menu that lists only the second part of the species name (not the genus) of its offerings? *Hint:* Look up *Ursus americanus*, *Ceanothus americanus*, *Bufo americanus*, *Homarus americanus*, *Lepus americanus*, and *Nicrophorus americanus*.

2. Witnesses in a court of law are often asked to "swear to tell the truth, the whole truth, and nothing but the truth." Can you think of a less subjective alternative for this oath?

3. Procter & Gamble makes Olestra and financed the study described in Section 1.7. The main researcher was a consultant to Procter & Gamble during the study. What do you think about scientific information that comes from tests financed by companies with a vested interest in the outcome?

4. Suppose an outcome of some event has been observed to happen with great regularity. Can we predict that the same outcome will always occur? Not really, if there is no way to test all of the variables that affect the outcome. To illustrate the point, Garvin McCain and Erwin Segal offer the following parable.

Digging Into Data

Peacock Butterfly Predator Defenses

The photographs *below* represent experimental and control groups used in the peacock butterfly experiment that was discussed in Section 1.7.

See if you can identify the experimental groups, and match them up with the relevant control group(s). *Hint*: Identify which variable is being tested in each group (each variable has a control).

 A Wing spots painted out

D Wings painted but spots visible

 B Wing spots visible; wings silenced

E Wings cut but not silenced

 **C** Wing spots painted out; wings silenced

F Wings painted but spots visible; wings cut but not silenced

Once there was a highly intelligent turkey that lived in a pen and was attended by a kind, thoughtful master. The turkey had nothing to do but reflect on the world's wonders and regularities. It observed some major regularities. Morning always started out with the sky turning light, followed by the master's footsteps, which was always followed by the appearance of food. Other things varied—sometimes the morning was warm and sometimes it was cold, for example—but food always followed footsteps. The sequence of events was so predictable that it eventually became the basis of the turkey's theory about the goodness of the world. One morning, after more than 100 confirmations of the goodness theory, the turkey listened for the master's footsteps, heard them, and had its head chopped off.

Any scientific theory is modified or discarded when contradictory evidence becomes available. The absence of absolute certainty has led some people to conclude that "facts are irrelevant—facts change." If that is so, should we stop doing scientific research? Why or why not?

5. In 2005 a South Korean scientist, Woo-suk Hwang, reported that he had made immortal stem cells from eleven human patients. His research was hailed as a breakthrough for people affected by currently incurable degenerative diseases, because such stem cells might be used to repair a person's own damaged tissues. Hwang published his results in a respected scientific journal. In 2006, the journal retracted his paper after other scientists discovered that Hwang and his colleagues had faked their results. Does the incident show that the results of scientific studies cannot be trusted? Or does it confirm the usefulness of a scientific approach, because other scientists quickly discovered and exposed the fraud?

2 Molecules of Life

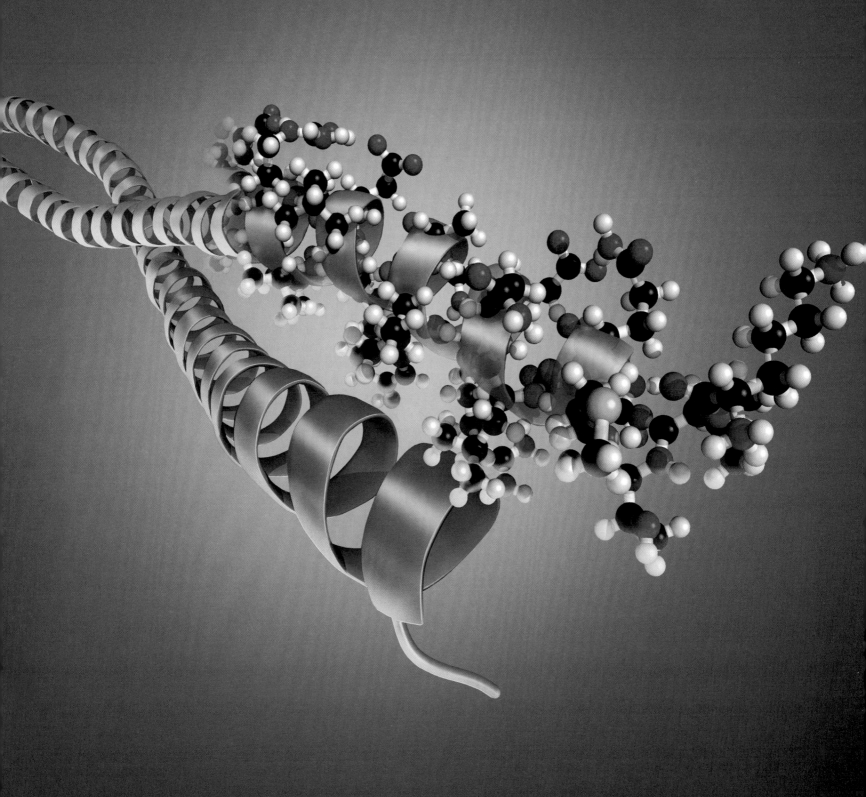

Small differences in the way molecules are put together can have big effects in a living organism.

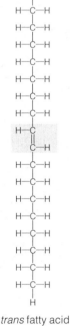

trans fatty acid

2.1

Impacts/Issues: Fear of Frying

The human body requires only about a tablespoon of fat each day to stay healthy, but most people in developed countries eat far more than that. The average American consumes the equivalent of one stick of butter per day—about 100 pounds of fat per year—which may be part of the reason why the average American is overweight. Being overweight increases one's risk for many health conditions. However, the total quantity of fat we eat may be of less importance to health than the *kinds* of fats we eat. Fats are more than just inert molecules that accumulate in strategic areas of our bodies. They are major constituents of cell membranes, and as such they have powerful effects on cell function.

The typical fat molecule has three fatty acids, which have long chains of carbon atoms. Different fats consist of different fatty acids. Fats with a certain arrangement of hydrogen atoms around that carbon chain are called *trans* fats. Small amounts of *trans* fats occur naturally in red meat and dairy products, but most of the *trans* fats that humans eat come from partially hydrogenated vegetable oil, an artificial food product.

Hydrogenation, a manufacturing process that adds hydrogen atoms to a substance, changes liquid vegetable oils into solid fats. Procter & Gamble Co. developed partially hydrogenated soybean oil in 1908 as a substitute for the more expensive solid animal fats they had been using to make candles. However, the demand for candles began to wane as more households in the United States became wired for electricity, and P & G began to look for another way to sell its proprietary fat. Partially hydrogenated vegetable oil looks a lot like lard, so in 1911 the company began aggressively marketing it as a revolutionary new food: a solid cooking fat with a long shelf life, mild flavor, and lower cost than lard or butter. By the mid-1950s, hydrogenated vegetable oil had become a major part of the American diet. It was (and still is) found in a tremendous number of manufactured and fast foods (Figure 2.1).

For decades, hydrogenated vegetable oil was considered more healthy than animal fats because it was made from plants. We now know otherwise. The *trans* fats in hydrogenated vegetable oils raise the level of cholesterol in our blood more than any other fat, and they directly alter the function of our arteries and veins. The effects of such changes are quite serious. Eating as little as 2 grams a day of hydrogenated vegetable oils increases a person's risk of atherosclerosis (hardening of the arteries), heart attack, and diabetes. A small serving of french fries made with hydrogenated vegetable oil contains about 5 grams of *trans* fats.

All organisms consist of the same kinds of molecules, but small differences in the way those molecules are put together can have big effects in a living organism. With this concept, we introduce you to the chemistry of life. This is your chemistry. It makes you far more than the sum of your body's molecules.

FIGURE 2.1 *Trans* fats. *Left*, the particular arrangement of hydrogen atoms around two carbon atoms (*yellow* box) makes a *trans* fat very unhealthy as a food. *Right*, french fries and other fast foods are often cooked with *trans* fats.

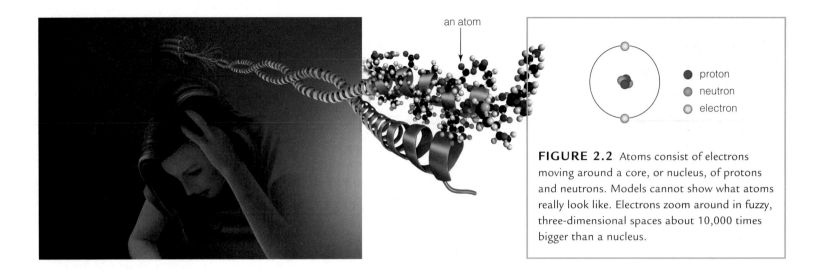

FIGURE 2.2 Atoms consist of electrons moving around a core, or nucleus, of protons and neutrons. Models cannot show what atoms really look like. Electrons zoom around in fuzzy, three-dimensional spaces about 10,000 times bigger than a nucleus.

2.2 Start With Atoms

Life's unique characteristics start with the properties of different **atoms**, tiny particles that are building blocks of all substances. Even though they are about 20 million times smaller than a grain of sand, atoms consist of even smaller subatomic particles: positively charged **protons** (p^+), uncharged **neutrons**, and negatively charged **electrons** (e^-). **Charge** is an electrical property: Opposite charges attract, and like charges repel. Protons and neutrons cluster in an atom's central core, or **nucleus**, and electrons move around the nucleus (Figure 2.2).

Atoms differ in the number of subatomic particles. The number of protons in an atom's nucleus is called the **atomic number**, and it determines the type of atom, or element. **Elements** are pure substances, each consisting only of atoms that have the same number of protons in their nucleus. For example, the atomic number of carbon is 6, so all atoms with six protons in their nucleus are carbon atoms, no matter how many electrons or neutrons they have. A chunk of carbon consists only of carbon atoms, and all of those atoms have six protons. Each of the 118 known elements has a symbol that is an abbreviation of its Latin name (Appendix VI shows a periodic table of the elements). Carbon's symbol, C, is from *carbo*, the Latin word for coal. Coal is mostly carbon.

The same elements that make up a living body also occur in, say, dirt or seawater, but the proportions of different elements differ between living and nonliving things. For example, a human body contains a lot of carbon, but seawater and Earth's crust contain very little. Why? A body contains a greater proportion of carbon atoms because only living things make the molecules of life, which have a greater proportion of carbon atoms than other molecules do.

Carbon and all other elements occur in different forms, or **isotopes**, that differ in their number of neutrons. We refer to an isotope by its total number of protons and neutrons, which is the isotope's **mass number**. Mass number is shown as a superscript to the left of an element's symbol. For example, atoms of the most common carbon isotope, ^{12}C, have six protons and six neutrons; those of ^{14}C have six protons and eight neutrons ($6 + 8 = 14$).

^{14}C (carbon 14) is a radioactive isotope, or **radioisotope**, which means that the nucleus of a ^{14}C atom is unstable. Many isotopes are like this. Atoms of radioisotopes emit subatomic particles and energy when their nucleus spontaneously disintegrates. This process, **radioactive decay**, can transform one element into another. For example, ^{14}C decays when one of its neutrons splits into a

atom Particle that is a fundamental building block of matter.

atomic number Number of protons in the atomic nucleus; determines the element.

charge Electrical property of some subatomic particles. Opposite charges attract; like charges repel.

electron Negatively charged subatomic particle that occupies orbitals around the atomic nucleus.

element A pure substance that consists only of atoms with the same number of protons.

isotopes Forms of an element that differ in the number of neutrons their atoms carry.

mass number Total number of protons and neutrons in the nucleus of an element's atoms.

neutron Uncharged subatomic particle in the atomic nucleus.

nucleus Core of an atom; occupied by protons and neutrons.

proton Positively charged subatomic particle that occurs in the nucleus of all atoms.

radioactive decay Process by which atoms of a radioisotope spontaneously emit energy and subatomic particles when their nucleus disintegrates.

radioisotope Isotope with an unstable nucleus.

FIGURE 2.3 **Animated!** Shell models. Each circle (shell) represents all orbitals at one energy level. Atoms with vacancies in their outermost shell can interact with other atoms.

proton and an electron. The proton stays in the nucleus, and the electron is emitted. Thus, an atom of ^{14}C (with six protons and eight neutrons) becomes an atom of ^{14}N, which is nitrogen (with seven protons and seven neutrons).

An atomic nucleus cannot be altered by heat or any other ordinary means. Thus, a radioisotope decays at a constant rate into predictable products, independently of external factors such as temperature, pressure, or whether the atoms are part of molecules. For example, we know that half of the ^{14}C atoms in any sample will be ^{14}N atoms after 5,730 years. This predictability can be used to estimate the age of rocks and fossils (we return to this topic in Section 11.4).

Researchers and clinicians also introduce radioisotopes into living organisms. All isotopes of an element generally have the same chemical properties. This consistency means that organisms use atoms of one isotope (such as ^{14}C) the same way that they use atoms of another (such as ^{12}C). Thus, radioisotopes are often used in **tracers**, which are molecules with a detectable substance attached. Typically, a radioactive tracer is a molecule in which radioisotopes have been swapped for one or more atoms. Researchers deliver radioactive tracers into a biological system such as a cell or a body, then use an instrument that can detect radioactivity to follow the tracer as it moves through the system.

Why Electrons Matter Electrons are really, really small. How small are they? If they were as big as apples, you would be about 3.5 times taller than our solar system is wide. Simple physics explains the motion of, say, an apple falling from a tree, but electrons are so tiny that such everyday physics cannot explain their behavior. However, that behavior underlies atomic interactions.

A typical atom has about as many electrons as protons, so a lot of electrons may be zipping around one nucleus. Those electrons never collide, despite moving at nearly the speed of light (300,000 kilometers per second, or 670 million miles per hour). Why not? They avoid one another because they travel in different orbitals, which are defined volumes of space around the nucleus.

A The first shell corresponds to the first energy level, and it can hold up to 2 electrons. Hydrogen has one proton, so it has one vacancy. A helium atom has 2 protons, and no vacancies. The number of protons in each shell model is shown.

first shell

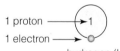
1 proton
1 electron
hydrogen (H)

helium (He)

B The second shell corresponds to the second energy level, and it can hold up to 8 electrons. Carbon has 6 protons, so its first shell is full. Its second shell has 4 electrons, and four vacancies. Oxygen has 8 protons and two vacancies. Neon has 10 protons and no vacancies.

second shell

carbon (C)

oxygen (O)

neon (Ne)

C The third shell, which corresponds to the third energy level, can hold up to 8 electrons, for a total of 18. A sodium atom has 11 protons, so its first two shells are full; the third shell has one electron. Thus, sodium has seven vacancies. Chlorine has 17 protons and one vacancy. Argon has 18 protons and no vacancies.

third shell

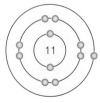
sodium (Na)

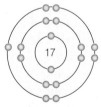
chlorine (Cl)

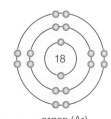
argon (Ar)

Imagine that an atom is a multilevel apartment building, with rooms available for rent by electrons. The nucleus is in the basement, and each "room" is an orbital. No more than two electrons can share a room at the same time. Each "floor" of the building corresponds to one energy level, and each has a certain number of rooms available to rent to electrons. Electrons fill orbitals from lower to higher energy levels. In other words, they populate rooms from the ground floor up. The farther an electron is from the basement, the greater its energy. An electron can move to a room on a higher floor if an energy input gives it a boost, but it immediately emits its extra energy and moves back down.

We use a **shell model** to visualize how electrons populate an atom. In this model, nested "shells" correspond to successively higher energy levels. Thus, each shell includes all rooms on one floor of our atomic apartment building. We draw a shell model of an atom by filling its shells with electrons (represented as balls or dots), from the innermost shell out, until there are as many electrons as the atom has protons. For example, there is only one room on the first floor, one orbital at the lowest energy level. It fills up first. In hydrogen, the simplest atom, a single electron occupies that room (Figure 2.3**A**). Helium has two electrons that fill its first shell. In larger atoms, more electrons rent the second-floor rooms (Figure 2.3**B**). When the second floor fills, more electrons rent third-floor rooms (Figure 2.3**C**), and so on.

If an atom's outermost shell is full of electrons, we say that it has no vacancies. Helium is an example. Atoms of such elements are chemically inactive, which means they are most stable as single atoms. By contrast, if an atom's outermost shell has room for another electron, it has a vacancy. Hydrogen has one vacancy. Atoms with vacancies tend to interact with other atoms: They give up, acquire, or share electrons until they have no vacancies in their outermost shell. Any atom is in its most stable state when it has no vacancies.

Ions The negative charge of an electron is the same magnitude as the positive charge of a proton, so the two charges cancel one another. Thus, an atom that has as many electrons as protons has no charge. An atom with different numbers of electrons and protons is called an **ion**. An ion carries a charge; either it acquired a positive charge by losing an electron, or it acquired a negative charge by pulling an electron away from another atom (Figure 2.4). For example, an uncharged chlorine atom has 17 protons and 17 electrons. Seven electrons are in its outermost (third) shell, which can hold eight, so it has one vacancy. Chlorine tends to pull an electron away from another atom to fill that vacancy. When that happens, the atom becomes a chloride ion (Cl⁻) with 17 protons, 18 electrons, and a net negative charge (Figure 2.4**A**). As another example, an uncharged sodium atom has 11 protons and 11 electrons. This atom has one electron in its outer (third) shell, which can hold eight, so it has seven vacancies. A sodium atom tends to lose the single electron in its third shell. When that happens, the atom has two full shells and no vacancies. It is a sodium ion (Na⁺), with 11 protons, 10 electrons, and a net positive charge (Figure 2.4**B**).

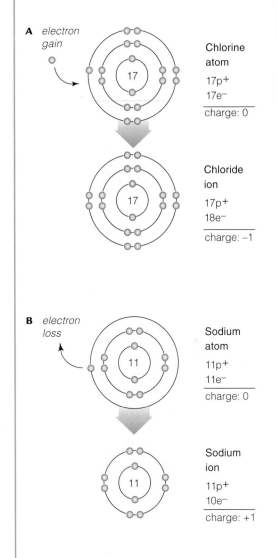

FIGURE 2.4 Animated! Ion formation.

A A chlorine atom becomes a negatively charged chloride ion (Cl⁻) when it gains an electron and fills the vacancy in its third, outermost shell.

B A sodium atom becomes a positively charged sodium ion (Na⁺) when it loses the electron in its third shell. The atom's full second shell is now its outermost, so it has no vacancies.

Take-Home Message

What are the basic building blocks of all matter?

- Tiny particles called atoms are the building blocks of all substances. Atoms consist of electrons moving around a nucleus of protons and neutrons.
- An element is a pure substance that consists only of atoms with the same number of protons.
- Atoms tend to get rid of vacancies by gaining or losing electrons (thereby becoming ions), or by sharing electrons with other atoms.

ion Atom that carries a charge because it has an unequal number of protons and electrons.
shell model Model of electron distribution in an atom.
tracer A molecule with a detectable label.

Table 2.1	Different Ways of Referring to a Molecule	
Common name	Water	Familiar term.
Chemical name	Dihydrogen monoxide	Systematically describes elemental composition.
Chemical formula	H_2O	Indicates unvarying proportions of elements. Subscripts show number of atoms of an element per molecule. The absence of a subscript means one atom.
Structural formula	H—O—H	Represents each covalent bond as a single line between atoms.
Structural model	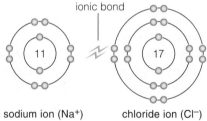	Shows the positions and relative sizes of atoms.
Shell model		Shows how pairs of electrons are shared in covalent bonds.

Two atoms with vacancies can join in a **chemical bond**, which is an attractive force that arises when their electrons interact. The result is a **molecule**, a union of two or more atoms joined by a chemical bond. Molecules that consist of two or more different elements are called **compounds**. Table 2.1 lists some of the ways we represent molecules.

The same atoms bonded together in different ways make up different molecules. For example, carbon atoms bonded one way form a soft, slippery mineral called graphite. Carbon atoms bonded a different way make the hardest mineral, diamond. Carbon atoms bonded to oxygen and hydrogen atoms make sugar. The idea that different structures can be assembled from the same basic building blocks is a recurring theme in our world, and also in biology (Figure 2.5).

Although bonding applies to a range of interactions among atoms, we can categorize most bonds into distinct types based on their different properties. Which type forms—an ionic, covalent, or hydrogen bond—depends on the atoms that take part in it.

Ionic Bonds An **ionic bond** is a strong mutual attraction of two oppositely charged ions. Such bonds do not usually form by the direct transfer of an electron from one atom to another. Instead, atoms that have already become ions stay close together because of their opposite charges. Table salt (sodium chloride, or NaCl) is an example of an ionic solid. Ionic bonds hold sodium and chloride ions together in salt crystals:

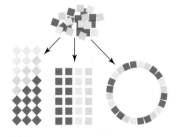

FIGURE 2.5 Animated! Example of how different objects can be assembled using the same set of starting materials.

ionic bond

sodium ion (Na^+) chloride ion (Cl^-)

Covalent Bonds In a **covalent bond**, two atoms share a pair of electrons, so each atom's vacancy becomes partially filled. Sharing electrons links the two atoms, just as sharing a pair of earphones links two friends (Figure 2.6). Covalent bonds can be stronger than ionic bonds, but they are not always so. Structural formulas show how covalent bonds connect atoms. A line between two atoms represents a single covalent bond, in which two atoms share one pair of electrons. An example is molecular hydrogen (H_2), with a single covalent bond between two hydrogen atoms (H—H, *right*).

molecular hydrogen (H_2)

Two lines between atoms represent a double covalent bond, in which two atoms share two pairs of electrons. A double covalent bond links two oxygen atoms in molecular oxygen (O=O, *right*). Three lines indicate a triple covalent bond, in which two

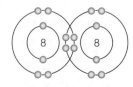

molecular oxygen (O_2)

atoms share three pairs of electrons. A triple covalent bond links two nitrogen atoms in molecular nitrogen ($N\equiv N$, or N_2).

Some covalent bonds are **nonpolar**, meaning that the atoms participating in the bond are sharing electrons equally. There is no difference in charge between the two ends of such bonds. The bonds in molecular hydrogen (H_2), oxygen (O_2), and nitrogen (N_2) mentioned earlier are examples. These molecules are some of the gases that make up air. The bonds in methane (CH_3), another gas, are also nonpolar.

Atoms participating in **polar** covalent bonds do not share electrons equally. One atom pulls the electrons a little more toward its "end" of the bond, so that atom bears a slightly negative charge. The atom at the other end of the bond bears a slightly positive charge. For example, a water molecule (H—O—H, *right*) has two covalent bonds; both are polar. The oxygen atom in a water molecule carries a slight negative charge, and each of the hydrogen atoms carries a slight positive charge. Any separation of charge into distinct positive and negative regions is called **polarity**. As you will see in the next section, the polarity of the water molecule is important for the world of life.

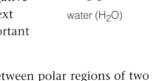

water (H_2O)

Hydrogen Bonds Hydrogen bonds form between polar regions of two molecules, or between two regions of the same molecule. A **hydrogen bond** is a weak attraction between a hydrogen atom and another atom taking part in a separate polar covalent bond. Hydrogen bonds form between water molecules:

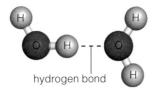

hydrogen bond

Like ionic bonds, hydrogen bonds form by the mutual attraction of opposite charges: The hydrogen atom has a slight positive charge and the other atom has a slight negative charge. However, unlike ionic bonds, hydrogen bonds do not make molecules out of atoms, so they are not chemical bonds.

Hydrogen bonds form and break much more easily than covalent or ionic bonds do. Even so, many of them form. Their collective strength imparts unique properties to many substances such as water. Hydrogen bonds that form among the atoms of biological molecules such as DNA hold these molecules in their characteristic shapes.

FIGURE 2.6 Sharing electrons links two atoms just as sharing earphones links two friends.

chemical bond An attractive force that arises between two atoms when their electrons interact.

compound Type of molecule that has atoms of more than one element.

covalent bond Chemical bond in which two atoms share a pair of electrons.

hydrogen bond Attraction that forms between a covalently bonded hydrogen atom and another atom taking part in a separate covalent bond.

ionic bond Type of chemical bond in which a strong mutual attraction forms between ions of opposite charge.

molecule Group of two or more atoms joined by chemical bonds.

nonpolar Having an even distribution of charge.

polar Having an uneven distribution of charge.

polarity Any separation of charge into distinct positive and negative regions.

Take-Home Message How do atoms interact?

- A chemical bond forms when the electrons of two atoms interact. Depending on the atoms, the bond may be ionic or covalent.

- An ionic bond is a strong mutual attraction between ions of opposite charge.

- Atoms share a pair of electrons in a covalent bond. When the atoms share electrons equally, the bond is nonpolar; when they share unequally, it is polar.

- A hydrogen bond is an attraction between a hydrogen atom and another atom taking part in a different polar covalent bond.

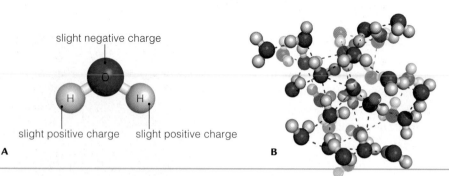

slight negative charge

slight positive charge   slight positive charge

A

B

C

FIGURE 2.7 Animated! Water. **A** Polarity of a water molecule: Each of the hydrogen atoms has a slight positive charge, and the oxygen atom has a slight negative charge. **B** The many hydrogen bonds (dashed lines) that keep water molecules clustered together impart special properties to liquid water. **C** Visible effect of cohesion: a wasp drinking (not sinking). Cohesion imparts surface tension to liquid water, which means that the surface of liquid water behaves a bit like a sheet of elastic.

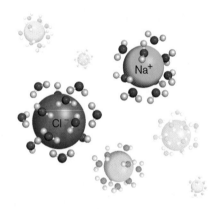

FIGURE 2.8 Animated!
Water molecules that surround an ionic solid pull its atoms apart, thereby dissolving them.

cohesion Tendency of molecules to stick together.
evaporation Transition of a liquid to a gas.
hydrophilic Describes a substance that dissolves easily in water.
hydrophobic Describes a substance that resists dissolving in water.
pH Measure of the number of hydrogen ions in a fluid.
salt Compound that dissolves easily in water and releases ions other than H^+ and OH^-.
solute A dissolved substance.
solvent Liquid that can dissolve other substances.
temperature Measure of molecular motion.

Water

Life evolved in water. All living organisms are mostly water, many of them still live in it, and all of the chemical reactions of life are carried out in water. What makes water so fundamentally important for life?

Water's special properties as a liquid begin with the two polar covalent bonds in each water molecule. Overall, the molecule has no charge, but the oxygen pulls the shared electrons a bit more than the hydrogen atoms do. Thus, each of the atoms in a water molecule carries a slight charge: The oxygen atom is slightly negative, and the hydrogen atoms are slightly positive (Figure 2.7**A**). The separation of charge means that the water molecule itself is polar. The polarity is very attractive to other water molecules, and hydrogen bonds form between them in tremendous numbers (Figure 2.7**B**). Extensive hydrogen bonding between water molecules imparts unique properties to liquid water, and those properties make life possible.

First, water is an excellent solvent. A **solvent** is a liquid that can dissolve other substances. When a substance dissolves, its individual molecules or ions become **solutes** as they disperse. Salts, sugars, and many other compounds that dissolve easily in water are polar, so many hydrogen bonds form between them and water molecules. A **salt** is a compound that dissolves easily in water and releases ions other than H^+ and OH^- when it does. Hydrogen bonding with water dissolves such **hydrophilic** (water-loving) substances by pulling their individual molecules away from one another and keeping them apart (Figure 2.8).

You can see how water interacts with **hydrophobic** (water-dreading) substances if you shake a bottle filled with water and salad oil, then set it on a table and watch what happens. Salad oil consists of nonpolar molecules, and water molecules do not form many hydrogen bonds with nonpolar molecules. Shaking breaks some of the hydrogen bonds that keep water molecules together. However, the water quickly begins to cluster into drops as new hydrogen bonds form among its molecules. The bonding excludes molecules of oil and pushes them together into droplets that rise to the surface of the water. The same interaction occurs at the thin, oily membrane that separates the water inside of cells from the water outside of them. The organization of membranes—and of life—starts with such interactions (you will read more about membranes in Chapter 3).

A second property of water is temperature stability. **Temperature** is a way to measure the energy of molecular motion: All molecules jiggle nonstop, and they jiggle faster as they absorb heat. However, extensive hydrogen bonding restricts the movement of water molecules—it keeps them from jiggling as much as they would otherwise. Thus, compared with other liquids, water absorbs much more heat before its temperature rises. Temperature stability is an important component of homeostasis, because most of the molecules of life function properly only within a certain range of temperature.

Below 0°C (32°F), water molecules do not jiggle enough to break hydrogen bonds, and they become locked in the rigid, lattice-like bonding pattern of ice. Individual water molecules pack less densely in ice than they do in water, so ice floats on water. During cold winters, ice sheets may form near the surface of ponds, lakes, and streams. Such ice "blankets" insulate liquid water under them, so they help keep fish and other aquatic organisms from freezing.

A third life-sustaining property of liquid water is **cohesion**, which means that water molecules resist separating from one another (Figure 2.7**C**). This property is important in many processes that sustain multicelled bodies. As one example, water molecules constantly escape from the surface of liquid water as vapor, a process called **evaporation**. Evaporation is resisted by the hydrogen bonding that keeps water molecules together. In other words, overcoming water's cohesion takes energy. Thus, evaporation sucks energy in the form of heat from liquid water, which decreases its surface temperature. Evaporative water loss can help you and some other mammals cool off when you sweat in hot, dry weather. Sweat, which is about 99 percent water, cools the skin as it evaporates.

Take-Home Message   Why is water essential to life?

- Being polar, water molecules hydrogen-bond to one another and to other polar (hydrophilic) substances, and repel nonpolar (hydrophobic) substances.

- Extensive hydrogen bonding between water molecules gives water unique properties that make life possible: cohesion, temperature stability, and a capacity to dissolve many substances.

2.5 Acids and Bases

In liquid water, some water molecules spontaneously separate into hydrogen ions (H⁺) and hydroxide ions (OH⁻). These ions combine again to form water:

$$H_2O \longrightarrow H^+ + OH^- \longrightarrow H_2O$$

water   hydrogen hydroxide   water
ions   ions

pH is a measure of the number of hydrogen ions in water or any other liquid. When the number of H⁺ ions is the same as the number of OH⁻ ions, the pH of the solution is 7, or neutral. The pH of pure water (not rainwater or seawater) is like this. The more hydrogen ions, the lower the pH. A one-unit decrease in pH corresponds to a tenfold increase in the amount of H⁺ ions, and a one-unit increase corresponds to a tenfold decrease in the amount of H⁺ ions (Figure 2.9). One way to get a sense of the scale is to taste dissolved baking soda (pH 9),

FIGURE 2.9 Animated! A pH scale. Here, *red* dots signify hydrogen ions (H⁺) and *blue* dots signify hydroxyl ions (OH⁻). Also shown are approximate pH values for some common solutions. This pH scale ranges from 0 (most acidic) to 14 (most basic). A change of one unit on the scale corresponds to a tenfold change in the amount of H⁺ ions.

Figure It Out: What is the approximate pH of cola?

Answer: 2.5

distilled water (pH 7), and lemon juice (pH 2). Nearly all of life's chemistry occurs near pH 7. Most of your body's internal environment (tissue fluids and blood) is between pH 7.3 and 7.5.

Substances called **acids** give up hydrogen ions when they dissolve in water, so they lower the pH of fluids and make them acidic (below pH 7). **Bases** accept hydrogen ions, so they can raise the pH of fluids and make them basic, or alkaline (above pH 7). Acids or bases that accumulate in the environment can be harmful to organisms because they change the pH of fluids in an ecosystem. For instance, fossil fuel emissions and nitrogen-containing fertilizers release strong acids into the atmosphere (Figure 2.10). The acids make rain acidic, which in turn can drastically change the pH of water and soil (we return to this topic in Section 18.5). Such changes are harmful because most enzymes and other biological molecules function properly only within a narrow range of pH. Even a slight deviation from that range can halt cellular processes.

Under normal circumstances, fluids in cells and bodies stay within a consistent range of pH because they are buffered. A **buffer** is a set of chemicals, often a weak acid or base and a salt, that can keep the pH of a solution stable. It works because the two chemicals can alternately donate and accept ions that contribute to pH changes. For example, when a base is added to an unbuffered fluid, the number of OH^- ions increases, so the pH rises. However, if the fluid is buffered, the addition of base causes the buffer to release H^+ ions. These combine with OH^- ions to form water, which does not affect pH. So, the buffered fluid's pH stays the same even when base is added.

For example, carbon dioxide gas takes part in an important buffer system. It becomes carbonic acid when it dissolves in human blood:

$$H_2O \ + \ CO_2 \ \longrightarrow \ H_2CO_3$$

carbon dioxide   carbonic acid

The carbonic acid can separate into hydrogen ions and bicarbonate ions, which can turn around and recombine to form carbonic acid:

$$H_2CO_3 \ \longrightarrow \ H^+ \ + \ HCO_3^- \ \longrightarrow \ H_2CO_3$$

carbonic acid   bicarbonate   carbonic acid

Together, carbonic acid and bicarbonate constitute a buffer. Any excess OH^- in blood combines with the H^+ to form water, which does not contribute to pH. Any excess H^+ in blood combines with the bicarbonate. Thus bonded, the hydrogen does not affect pH. The exchange of ions keeps the blood pH between 7.3 and 7.5, but only up to a point. A buffer can neutralize only so many ions. Even slightly more than that limit and the pH of the fluid changes dramatically. Buffer failure can be catastrophic in a biological system. For example, too much carbonic acid forms in blood when breathing is impaired suddenly. The resulting decline in blood pH may cause an individual to enter a coma, which is a dangerous level of unconsciousness. Hyperventilation (sustained rapid breathing) causes the body to lose too much CO_2. The loss can result in a rise in blood pH. Prolonged muscle spasm (tetany) or coma may result.

FIGURE 2.10 Sulfur dioxide emissions from a coal-burning power plant. Some airborne pollutants, including sulfur dioxide, dissolve in water vapor to form acids that end up in rain.

acid Substance that releases hydrogen ions in water.

base Substance that accepts hydrogen ions in water.

buffer Set of chemicals that can keep the pH of a solution stable by alternately donating and accepting ions that contribute to pH.

organic Type of molecule that consists primarily of carbon and hydrogen atoms.

Take-Home Message **Why are hydrogen ions important to life?**

- Hydrogen ions contribute to pH. Acids release hydrogen ions in water; bases accept them. Salts release ions other than H^+ and OH^-.

- Buffers keep the pH of body fluids stable. They are part of homeostasis.

Organic Molecules

Living things consist mainly of hydrogen, oxygen, and carbon atoms (Table 2.2). Most of the oxygen and hydrogen are in the form of water. Put water aside, and carbon makes up more than half of what is left. The carbon in living organisms is part of the molecules of life: complex carbohydrates, lipids, proteins, and nucleic acids. These molecules consist primarily of hydrogen and carbon atoms, so they are **organic**. The term is a holdover from a time when such molecules were thought to be made only by living things, as opposed to the "inorganic" molecules that formed by nonliving processes. The term persists, even though we now know that organic compounds were present on Earth long before organisms were, and we can also make them in laboratories.

Carbon's importance to life starts with its versatile bonding behavior. Each carbon atom can form covalent bonds with up to four other atoms. Depending on the other elements in the resulting molecule, such bonds may be polar or nonpolar. Many organic compounds have a backbone—a chain of carbon atoms—to which other atoms attach. The ends of a backbone may join so that the carbon chain forms one or more ring structures. The versatility means that carbon atoms can be assembled into a variety of organic molecules.

Using different kinds of models to represent the structure of organic compounds allows us to focus on different characteristics of a particular molecule. For example, structural formulas show how all the atoms in a molecule connect to one another (Figure 2.11**A**). In such models, each line indicates one covalent bond. A double line (═) indicates a double bond; a triple line (≡) indicates a triple bond. Some of the atoms or bonds in a molecule may be implied but not shown (Figure 2.11**B**). Hydrogen atoms bonded to a carbon backbone may also be omitted, and other atoms as well.

For simplicity, carbon ring structures such as the ones that occur in glucose and other sugars are often represented as polygons that imply carbon atoms at their corners (Figure 2.11**C**). Ball-and-stick models show the positions of the atoms in three dimensions (Figure 2.11**D**). Single, double, and triple covalent bonds are all shown as one stick connecting two balls, which represent atoms. Ball size reflects relative size of the atoms. Space-filling models show the overall shape of a molecule (Figure 2.11**E**). Note that the elements in ball-and-stick and space-filling models are typically coded using the same color scheme.

Table 2.2	Some Elemental Abundances*		
Element	Human	Earth	Seawater
Hydrogen	62.0%	3.1%	66.0%
Oxygen	24.0	60.0	33.0
Carbon	12.0	0.3	<0.1
Nitrogen	1.2	<0.1	<0.1
Phosphorus	0.2	<0.1	<0.1
Calcium	<0.1	<0.1	<0.1
Sodium	<0.1	<0.1	0.3
Potassium	<0.1	0.8	<0.1
Chlorine	<0.1	<0.1	0.3

* As a percentage of the total number of atoms in each source

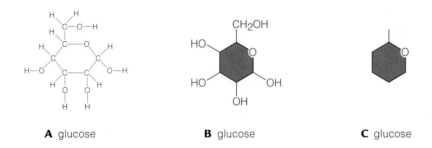

A glucose   **B** glucose   **C** glucose

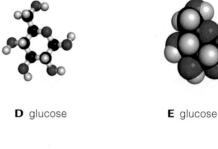

D glucose   **E** glucose

FIGURE 2.11 Modeling an organic molecule. **A** A structural formula shows atoms and bonds. **B,C** Structural formulas are often abbreviated to omit labels for some atoms such as the carbons at the corners of ring structures. **D** A ball-and-stick model shows the arrangement of atoms in three dimensions. **E** A space-filling model shows a molecule's overall shape. *Right*, typical color code for elements in molecular models.

carbon   hydrogen   oxygen   nitrogen   phosphorus

A Condensation. Cells build large molecules from smaller ones by this reaction. An enzyme removes a hydroxyl group from one molecule and a hydrogen atom from another. Water forms as a covalent bond forms between the two molecules.

B Hydrolysis. Cells split molecules into smaller ones by this water-requiring reaction. An enzyme splits a molecule by attaching a hydroxyl group and a hydrogen atom (both from water) to each half.

FIGURE 2.12 Animated!
Two common metabolic processes by which cells build and break down organic molecules.

From Structure to Function All biological systems are based on the same organic molecules, a similarity that is a legacy of life's common origin. However, the details of those molecules differ among organisms. Just as atoms bonded in different numbers and arrangements form different molecules, so do simple organic building blocks bonded in different numbers and arrangements form different versions of the molecules of life.

Cells maintain pools of small organic molecules that they assemble into complex carbohydrates, lipids, proteins, and nucleic acids. When used as subunits of larger molecules, the small organic molecules (simple sugars, fatty acids, amino acids, and nucleotides) are called **monomers**. Molecules that consist of multiple monomers are called **polymers**.

Cells build polymers from monomers, and break down polymers to release monomers. **Metabolism** refers to activities by which cells acquire and use energy as they make and break apart organic compounds. These activities help cells stay alive, grow, and reproduce. Metabolism also requires enzymes, which are molecules that make reactions proceed faster than they would on their own.

Cells commonly construct large molecules from smaller ones by **condensation**, a process in which an enzyme covalently bonds two molecules together. Water usually forms as a product of condensation when a hydroxyl group (—OH) from one of the molecules combines with a hydrogen atom from the other molecule (Figure 2.12**A**). Large molecules are typically broken down into smaller ones by **hydrolysis**, which is the reverse of condensation (Figure 2.12**B**). Enzymes break a bond by attaching a hydroxyl group to one atom and a hydrogen atom to the other. The —OH and —H are derived from a water molecule.

Take-Home Message **How are all of the molecules of life alike?**

■ Carbohydrates, lipids, proteins, and nucleic acids are all organic molecules, which consist mainly of carbon and hydrogen atoms.

■ Cells assemble large polymers from smaller monomers of simple sugars, fatty acids, amino acids, and nucleotides. They also break apart polymers into their component monomers.

2.7 Carbohydrates

Carbohydrates consist of carbon, hydrogen, and oxygen atoms in a 1:2:1 ratio. Cells use different kinds of carbohydrates as structural materials and for energy. With one sugar unit, monosaccharides are the simplest carbohydrates. "Saccharide" is from a Greek word that means sugar. Common monosaccharides have a backbone of five or six carbon atoms. Most are water soluble, so they are easily transported through the internal fluids of all organisms. The sugars that are part of nucleic acids are five-carbon monosaccharides. Cells use six-carbon glucose as an energy source or as a monomer of larger molecules.

An oligosaccharide is a short chain of covalently bonded monosaccharides (*oligo*– means a few). Disaccharides consist of two sugar monomers. Sucrose, the most plentiful sugar in nature, has one unit of glucose and one of fructose, another monosaccharide. Sucrose extracted from sugarcane or sugar beets is our table sugar (*left*). Oligosaccharides with three or more sugar units are often attached to lipids or proteins that have important functions in immunity.

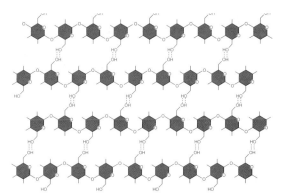

A Cellulose, a structural component of plants. Chains of glucose units stretch side by side and hydrogen-bond at many —OH groups. The hydrogen bonds stabilize the chains in tight bundles that form long fibers. Very few types of organisms can digest this tough, insoluble material.

B In starch, a series of glucose units form a chain that coils. Starch is the main energy reserve in plants, which store it in their roots, stems, leaves, fruits, and seeds (such as coconuts).

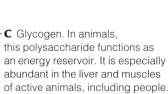

C Glycogen. In animals, this polysaccharide functions as an energy reservoir. It is especially abundant in the liver and muscles of active animals, including people.

FIGURE 2.13 Three of the most common complex carbohydrates and their locations in a few organisms. Each polysaccharide consists only of glucose units, but different bonding patterns that link the subunits result in substances with very different properties.

"Complex" carbohydrates, or polysaccharides, are straight or branched chains of hundreds or thousands of sugar monomers. The most common polysaccharides (cellulose, starch, and glycogen) consist only of glucose monomers, but as substances their properties are very different. Why? The answer begins with differences in patterns of covalent bonding that link their glucose units. Cellulose, the major structural material of plants, may be the most abundant organic molecule in the biosphere. Glucose chains that stretch side by side are linked by hydrogen bonds that stabilize them in tight, sturdy bundles (Figure 2.13**A**). By contrast, the covalent bonding pattern of starch makes the molecule coil like a spiral staircase (Figure 2.13**B**). Starch does not dissolve easily in water, so it resists hydrolysis. This stability is a reason why starch is used to store chemical energy in the watery, enzyme-filled interior of plant cells. In animals, the long, branched chains of glycogen are the energy-storage equivalent of starch in plants (Figure 2.13**C**). Muscle and liver cells make and store glycogen. When the glucose level in blood falls, liver cells break glycogen into its component glucose monomers by hydrolysis. The released glucose enters the bloodstream, and the blood glucose level rises again.

Take-Home Message **What are carbohydrates?**

- Subunits of simple carbohydrates (sugars), arranged in different ways, form various types of complex carbohydrates.

- Cells use carbohydrates for energy or as structural materials.

carbohydrate Molecule that consists primarily of carbon, hydrogen, and oxygen atoms in a 1:2:1 ratio.

condensation Process by which enzymes build large molecules from smaller subunits; water also forms.

hydrolysis Process by which an enzyme breaks a molecule into smaller subunits by attaching a hydroxyl group to one part and a hydrogen atom to the other.

metabolism All the enzyme-mediated chemical reactions by which cells acquire and use energy as they build and break down organic molecules.

monomer Molecule that is a subunit of polymers.

polymer Molecule that consists of multiple monomers.

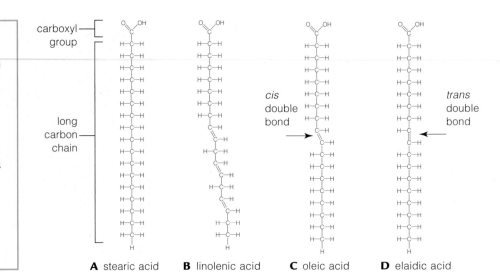

FIGURE 2.14 Fatty acids.

A The backbone of stearic acid is fully saturated with hydrogen atoms.

B The backbone of linolenic acid, with three double bonds, is unsaturated. The first double bond occurs at the third carbon from the end, so linoleic acid is called an omega-3 fatty acid. Omega-3 and omega-6 fatty acids are "essential fatty acids." Your body does not make them, so they must come from food.

The only difference between oleic acid **C**, a *cis* fatty acid, and elaidic acid **D**, a *trans* fatty acid, is the arrangement of hydrogens around the one double bond in the backbone.

carboxyl group
long carbon chain
cis double bond
trans double bond

A stearic acid   **B** linolenic acid   **C** oleic acid   **D** elaidic acid

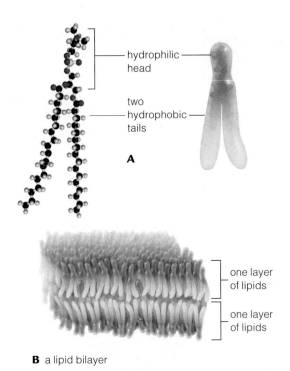

hydrophilic head
two hydrophobic tails

A

one layer of lipids
one layer of lipids

B a lipid bilayer

FIGURE 2.15 Phospholipids. **A** Each phospholipid has a hydrophilic head and two hydrophobic tails. **B** A double layer of phospholipids is the structural foundation of all cell membranes.

2.8 Lipids

Lipids are fatty, oily, or waxy organic compounds. Many lipids incorporate **fatty acids**, small organic compounds that consist of a long carbon chain with an acidic carboxyl group at one end (Figure 2.14). **Fats** are lipids with one, two, or three fatty acids that dangle like tails from a small alcohol called glycerol. Most neutral fats, including butter and vegetable oils, have three fatty acid tails, so they are called **triglycerides**. Triglycerides are the most abundant and richest energy source in vertebrate bodies. They are concentrated in adipose tissue that insulates and cushions body parts.

The fatty acid tails of **saturated fats** have only single bonds (Figure 2.14**A**). Animal fats tend to remain solid at room temperature because their saturated fatty acid tails can pack tightly. By contrast, the fatty acid tails of **unsaturated fats** have one or more double bonds that kink their carbon chains (Figure 2.14**B**). Kinky tails keep unsaturated fats from packing tightly. Most vegetable oils are liquid at room temperature because they are unsaturated (Figure 2.14**C**). The partially hydrogenated vegetable oils that you learned about in Section 2.1 are an exception. The double bond in these *trans* fatty acids keeps them straight. *Trans* fats pack tightly, so they are solid at room temperature (Figure 2.14**D**).

A **phospholipid** has two fatty acid tails attached to a phosphate-containing head (Figure 2.15**A**). The tails are hydrophobic, but the phosphate makes the head hydrophilic. Phospholipids are the most abundant lipids in cell membranes, which have two layers of lipids (Figure 2.15**B**). The heads of one layer are dissolved in the cell's watery interior, and the heads of the other layer are dissolved in the cell's fluid surroundings. All of the hydrophobic tails are sandwiched between the heads.

Waxes are complex, varying mixtures of lipids with long fatty acid tails bonded to long-chain alcohols or carbon rings. The molecules pack tightly, so the resulting substance is firm and water-repellent. A layer of secreted waxes that covers the exposed surfaces of plants helps restrict water loss and keep out parasites and other pests. Other types of waxes protect, lubricate, and soften skin and hair. Waxes, together with fats and fatty acids, make feathers waterproof. Bees store honey and raise new generations of bees inside honeycomb, which they make from beeswax.

FIGURE 2.16 Estrogen and testosterone, steroid hormones that cause different traits to arise in males and females of many species such as wood ducks (*Aix sponsa*).

Steroids are lipids with a rigid backbone of four carbon rings and no fatty acid tails. All eukaryotic cell membranes contain them. Cholesterol, the most common steroid in animal tissue, is also a starting material that cells remodel into many molecules, such as bile salts (which help digest fats) and vitamin D (required to keep teeth and bones strong). Steroid hormones are also derived from cholesterol. Estrogens and testosterone, hormones that govern reproduction and secondary sexual traits, are steroid hormones (Figure 2.16).

Take-Home Message What are lipids?

- Lipids are fatty, waxy, or oily organic compounds.
- Triglycerides are lipids that serve as energy reservoirs in vertebrate animals.
- Phospholipids are the main lipid component of cell membranes.
- Waxes are lipid components of water-repelling and lubricating secretions.
- Steroids are lipids that occur in cell membranes. Some are remodeled into other molecules.

2.9 Proteins

A **protein** is an organic compound composed of one or more chains of amino acids. An **amino acid** is a small organic compound with an amine group, a carboxyl group (the acid), and one or more atoms called an "R group." Typically, these groups are all attached to the same carbon atom:

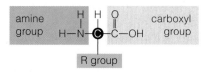

Of all biological molecules, proteins are the most diverse. Structural proteins support cell parts and multicelled bodies. Spiderwebs, feathers, hooves, and hair, as well as bones and other body parts consist mainly of structural proteins. Foods such as seeds and eggs are nutritious in part because they are high in protein. Most enzymes are proteins. Proteins move substances, help cells communicate, and defend the body. Amazingly, cells can make thousands of different kinds of proteins from only twenty kinds of amino acids.

amino acid Small organic compound with a carboxyl group, an amine group, and a characteristic side group (R).

fat Lipid with one, two, or three fatty acid tails.

fatty acid Organic compound that consists of a long chain of carbon atoms with an acidic carboxyl group at one end.

lipid Fatty, oily, or waxy organic compound.

phospholipid A lipid with a phosphate group in its hydrophilic head, and two nonpolar fatty acid tails.

protein Organic compound that consists of one or more chains of amino acids.

saturated fat Fatty acid with no double bonds in its carbon tail.

steroid A type of lipid with four carbon rings and no fatty acid tails.

triglyceride A lipid with three fatty acid tails attached to a glycerol backbone.

unsaturated fat Lipid with at least one double bond in a fatty acid tail.

wax Water-repellent lipid with long fatty acid tails bonded to long-chain alcohols or carbon rings.

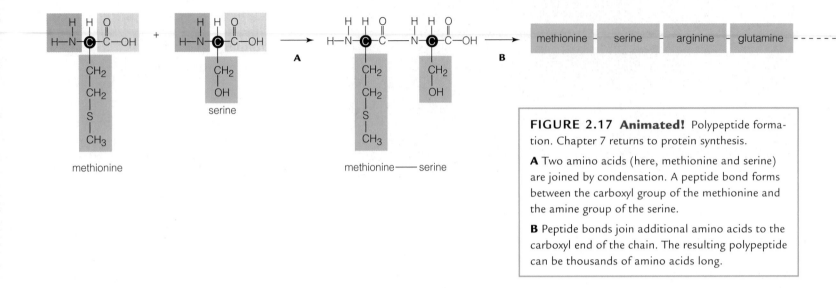

FIGURE 2.17 Animated! Polypeptide formation. Chapter 7 returns to protein synthesis.

A Two amino acids (here, methionine and serine) are joined by condensation. A peptide bond forms between the carboxyl group of the methionine and the amine group of the serine.

B Peptide bonds join additional amino acids to the carboxyl end of the chain. The resulting polypeptide can be thousands of amino acids long.

The shape of each protein defines its biological activity.

Protein synthesis involves covalently bonding amino acids into a chain. For each type of protein, instructions coded in DNA specify the order in which any of the twenty kinds of amino acids will occur at every place in the chain. By condensation, an enzyme joins the amine group of an amino acid with the carboxyl group of the next in a bond called a **peptide bond** (Figure 2.17). The process is repeated hundreds or thousands of times, so a long chain of amino acids called a **polypeptide** forms (Figure 2.18). The linear sequence of amino acids in each type of protein is unique. The amino acid sequence is the protein's primary structure ❶. Enzymes may attach linear or branched oligosaccharides to polypeptide chains, thus forming sugar-proteins such as those that allow a tissue or a body to recognize its own cells.

Even before a polypeptide is finished, it begins to acquire secondary structure by twisting and folding. Hydrogen bonds that form between amino acids may hold parts of the polypeptide chain in flat sheets, or in coils (helices) that are a bit like spiral staircases ❷. Similar patterns of sheets and coils occur in most types of proteins.

Much as an overly twisted rubber band coils back on itself, the sheets and coils of a protein can fold up even more into compact domains. A "domain" is a part of a protein that is organized as a structurally stable unit. Such units constitute a protein's third level of organization, its tertiary structure ❸. Tertiary structure is what makes a protein a working molecule. As an example, the barrel-shaped domains of some proteins function as tunnels through which ions cross cell membranes.

Many proteins also have a fourth level of organization, or quaternary structure: They consist of two or more polypeptide chains bonded together or in close association ❹. Most enzymes and many other proteins are globular, with several polypeptide chains folded into shapes that are roughly spherical.

Some proteins aggregate by many thousands into much larger structures, with their polypeptide chains organized into strands or sheets. The keratin in your hair is an example ❺. Some fibrous proteins contribute to the structure and organization of cells and tissues. Others, such as the actin and myosin that form filaments in muscle cells, are part of the mechanisms that help cells and cell parts move.

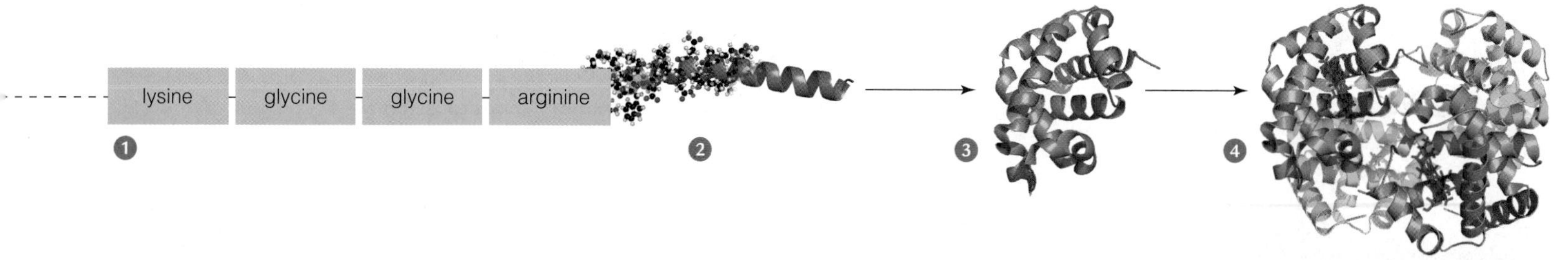

lysine — glycine — glycine — arginine

1 **2** **3** **4**

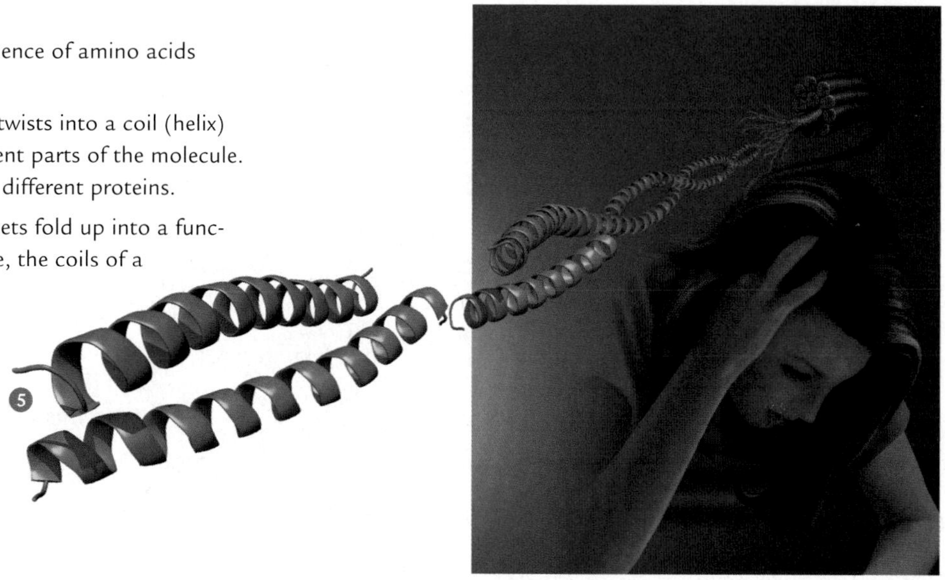

FIGURE 2.18 Animated! Protein structure.

1 A protein's primary structure consists of a linear sequence of amino acids (a polypeptide chain).

2 Secondary structure arises when a polypeptide chain twists into a coil (helix) or sheet held in place by hydrogen bonds between different parts of the molecule. The same patterns of secondary structure occur in many different proteins.

3 Tertiary structure occurs when a chain's coils and sheets fold up into a functional domain such as a barrel or pocket. In this example, the coils of a globin chain form a pocket.

4 Some proteins have quaternary structure, in which two or more polypeptide chains associate as one molecule. Hemoglobin, shown here, consists of four globin chains (*green* and *blue*). Each globin pocket now holds a heme group (*red*).

5 Many proteins aggregate by the thousands into larger structures, such as the keratin filaments that make up hair.

The Importance of Protein Structure

An enzyme speeds up a process, a receptor protein receives an energy signal, hemoglobin molecules in your blood carry oxygen—you and all other organisms consist of and depend on protein function. However, a protein only functions properly if it stays coiled, folded, and packed in a particular way, because the shape of each protein defines its biological activity. That shape depends on many hydrogen bonds and other interactions that heat, shifts in pH, or detergents can disrupt. At such times, proteins **denature**, which means they unwind and otherwise lose their three-dimensional shape. Once a protein's shape unravels, so does its function.

You can see denaturation in action when you cook an egg. A protein called albumin is a major component of egg white. Heat does not disrupt the covalent bonds of albumin's primary structure, but it does destroy the weaker hydrogen bonds that maintain the protein's shape. When a translucent egg white turns opaque, the albumin has been denatured. For very few proteins, denaturation is reversible if normal conditions return, but albumin is not one of them. There is no way to uncook an egg.

Prion diseases, including mad cow disease (bovine spongiform encephalitis, or BSE) in cattle, Creutzfeldt–Jakob disease in humans, and scrapie in sheep, are the dire aftermath of a misfolded protein. These infectious diseases may be inherited, but more often they arise spontaneously. All are characterized by relentless deterioration of mental and physical abilities that eventually causes death of the individual.

All prion diseases begin with a protein that occurs normally in mammals. One, PrP^C, is found in cell membranes throughout the body. This copper-binding protein is especially abundant in brain cells, but we still know very little about what it does. Very rarely, a PrP^C protein spontaneously misfolds so that it loses some of its coiled character. In itself, a single misfolded protein molecule

denature To unravel the shape of a protein or other large biological molecule.

peptide bond A bond between the amine group of one amino acid and the carboxyl group of another.

polypeptide Chain of amino acids linked by peptide bonds.

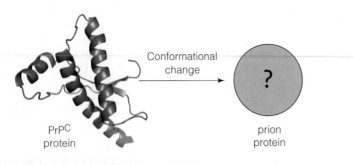

PrP^C
protein

Conformational
change

?

prion
protein

A

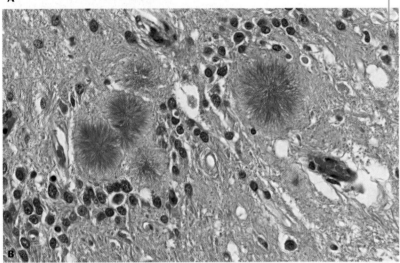

B

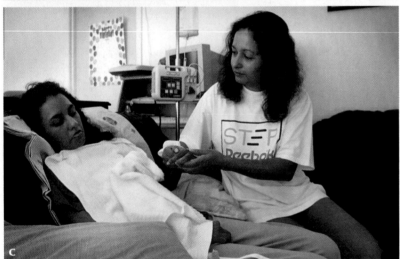

C

FIGURE 2.19 Variant Creutzfeldt–Jakob disease (vCJD).

A The PrP^C protein becomes a prion when it misfolds into an as-yet unknown conformation. Prions cause other PrP^C proteins to misfold, and the misfolded proteins aggregate into long fibers.

B Slice of brain tissue from a person with vCJD. Characteristic holes and prion protein fibers radiating from several deposits are visible.

C Charlene Singh, here being cared for by her mother, was one of three people who developed symptoms of the disease while living in the United States. Like the others, Singh most likely contracted the disease elsewhere; she spent her childhood in Britain. She was diagnosed in 2001, and she died in 2004.

would not pose much of a threat. However, when this particular protein misfolds, it becomes a **prion**, or infectious protein (Figure 2.19**A**). The altered shape of a misfolded PrP^C protein somehow causes normally folded PrP^C proteins to misfold too. Because each misfolded protein becomes infectious, the number of prions increases exponentially.

The shape of misfolded PrP^C proteins allows them to align tightly into long fibers. Fibers composed of prion protein begin to accumulate in the brain as large, water-repellent patches. The patches grow as more prions form, and they begin to disrupt brain cell function, an effect that is the source of prion disease symptoms. Tiny holes form in the brain as its cells die (Figure 2.19**B**). Eventually, the brain becomes so riddled with holes that it looks like a sponge.

In the mid-1980s, an epidemic of mad cow disease in Britain was followed by an outbreak of a new variant of Creutzfeldt–Jakob disease (vCJD) in humans. Researchers isolated a prion similar to the one in scrapie-infected sheep from cows with BSE, and also from humans affected by the new type of Creutzfeldt–Jakob disease (Figure 2.19**C**). How did the prion get from sheep to cattle to people? Prions are not denatured by cooking or typical treatments that inactivate other types of infectious agents. The cattle became infected by the prion after eating feed prepared from the remains of scrapie-infected sheep, and people became infected by eating beef from the infected cattle.

Two hundred people have died from vCJD since 1990. The use of animal parts in livestock feed is now banned, and the number of cases of BSE and vCJD has since declined. Cattle with BSE still turn up, but so rarely that they pose little threat to the general human population.

·Take-Home
Message   Why is protein structure important?

- A protein's function depends on its structure, which consists of chains of amino acids that twist and fold into functional domains.

- Changes in a protein's structure may also alter its function.

Nucleic Acids

Nucleotides are small organic molecules, various kinds of which function as energy carriers, enzyme helpers, chemical messengers, and subunits of DNA and RNA. Each nucleotide consists of a sugar with a five-carbon ring, bonded to a nitrogen-containing base and one or more phosphate groups. The nucleotide **ATP** (adenosine triphosphate) has a row of three phosphate groups attached to a ribose sugar (Figure 2.20**A**). When ATP transfers the outermost phosphate group to other molecules, it also transfers energy. You will read about such phosphate-group transfers and their important metabolic role in Chapter 5.

Nucleic acids are chains of nucleotides in which the sugar of one nucleotide is joined to the phosphate group of the next. An example is **RNA**, or ribonucleic acid, named after the ribose sugar of its component nucleotides. RNA consists of four kinds of nucleotide monomers, one of which is ATP. RNA molecules carry out protein synthesis, which we discuss in Chapter 7.

DNA, or deoxyribonucleic acid, is a nucleic acid named after the deoxyribose sugar of its component nucleotides. A DNA molecule consists of two nucleotide chains twisted into a double helix (Figure 2.20**B**). Hydrogen bonds between the nucleotides hold the two strands of DNA together. Each cell starts out life with DNA inherited from a parent cell. That DNA contains all of the information necessary to build a new cell and, in the case of multicelled organisms, an entire individual. The cell uses DNA's information to make RNA and proteins. Parts of a DNA molecule are identical or nearly so in all organisms, but most is unique to a species or an individual (Chapter 6 returns to DNA structure and function).

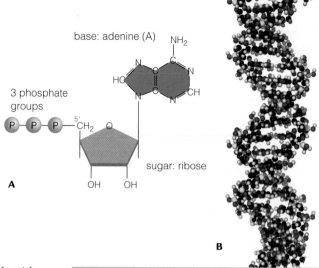

base: adenine (A)

3 phosphate groups

sugar: ribose

A

B

FIGURE 2.20 Animated! A nucleotide and a nucleic acid. **A** The nucleotide ATP. **B** DNA consists of two chains of nucleotides, twisted into a double helix held together by hydrogen bonds.

Take-Home Message
What are nucleotides and nucleic acids?

- Nucleotides are monomers of the nucleic acids DNA and RNA, coenzymes, energy carriers, and messengers.
- DNA's nucleotide sequence encodes heritable information.
- Different types of RNA have roles in the processes by which a cell uses the information in its DNA.

How Would

YOU
☑ **Vote?**

All packaged foods in the United States now list *trans* fat content, but may be marked "zero grams of *trans* fats" even when a single serving contains up to half a gram. Should hydrogenated vegetable oils be banned from all food? **See** CengageNow for details, then vote online (cengagenow.com).

2.11
Impacts/Issues Revisited:
Fear of Frying

Trans fatty acids are relatively rare in unprocessed foods, so it makes sense from an evolutionary standpoint that our bodies may not have enzymes to deal with them efficiently. The enzymes that hydrolyze *cis* fatty acids have difficulty breaking down *trans* fatty acids, a problem that may be a factor in the ill effects of *trans* fats. Several countries have already restricted the use of *trans* fats in food. In 2004, Denmark passed a law prohibiting importation of foods that contain partially hydrogenated vegetable oils. French fries and chicken nuggets that the Danish import from the United States contain almost no *trans* fats; the same foods sold to consumers in the United States contain 5 to 10 grams of *trans* fats per serving.

ATP Nucleotide that consists of an adenine base, five-carbon ribose sugar, and three phosphate groups. Also functions as an energy carrier.

DNA Nucleic acid that carries hereditary material; consists of two nucleotide chains twisted in a double helix.

nucleic acid Chain of nucleotides joined by sugar–phosphate bonds.

nucleotide Monomer of nucleic acids; has five-carbon sugar, nitrogen-containing base, and phosphate groups.

prion Infectious protein.

RNA Typically single-stranded nucleic acid; roles in protein synthesis.

Summary

Section 2.1 All living things consist of the same kinds of molecules, put together in slightly different ways.

Section 2.2 All substances consist of atoms. Atoms are composed of electrons traveling around a nucleus of protons and neutrons. Elements are pure substances that consist only of atoms with the same number of protons. Researchers make tracers with detectable substances such as radioisotopes, which emit subatomic particles and energy as they decay spontaneously.

carbon (C)

CENGAGENOW™ Study electron distribution and the shell model, and learn how radioisotopes are used in a clinical setting, with the animations and interaction on CengageNow.

Section 2.3 Atoms tend to fill their vacancies by gaining or losing electrons, or by sharing electrons with other atoms. Ions are atoms with unequal numbers of protons and electrons.

Ionic and covalent bonds are chemical bonds, which join atoms in molecules. An ionic bond is a strong mutual attraction of oppositely charged ions. Atoms share a pair of electrons in a covalent bond, which is polar if the electrons are not shared equally, and nonpolar if the sharing is equal. Hydrogen bonds form between a hydrogen atom and another atom taking part in a separate polar covalent bond.

CENGAGENOW™ See how different objects are assembled from the same materials, and compare the types of bonds in biological molecules with the animations on CengageNow.

Section 2.4 Polar covalent bonds join two hydrogen atoms to one oxygen atom in each water molecule. The molecule's polarity invites extensive hydrogen bonding between water molecules, and this bonding is the basis of unique properties that make life possible: cohesion, temperature stability, and a capacity to dissolve many substances.

CENGAGENOW™ Learn how water's structure gives rise to its unique properties with animation on CengageNow.

Section 2.5 pH reflects the number of hydrogen ions (H^+) in a solution. At neutral pH (7), the amounts of H^+ and OH^- ions are the same. Salts release ions other than H^+ and OH^- in water. Acids release H^+; bases accept H^+. A buffer keeps a solution within a consistent range of pH. Most cell and body fluids are buffered because the molecules of life work only within a narrow range of pH.

CENGAGENOW™ Discover the pH of common substances with the interaction on CengageNow.

Section 2.6 Carbohydrates, proteins, lipids, and nucleic acids are organic molecules, which consist primarily of carbon and hydrogen atoms. Carbon atoms bond covalently with up to four other atoms, often forming long chains or rings. Enzyme-driven reactions construct large molecules from smaller subunits, and break large molecules into smaller ones.

glucose

CENGAGENOW™ Use the animation on CengageNow to see how sucrose forms by a condensation reaction.

Section 2.7 Enzymes assemble complex carbohydrates such as cellulose, glycogen, and starch from simple carbohydrate (sugar) subunits. Cells use carbohydrates for energy, and as structural materials.

Section 2.8 Lipids are greasy or oily nonpolar molecules, often with one or more fatty acid tails. Phospholipids are the main structural component of cell membranes. Waxes are lipids that are part of water-repellent and lubricating secretions. Steroids are lipids that occur in cell membranes, and some are remodeled into other molecules.

CENGAGENOW™ See a triglyceride form with animation on CengageNow.

Sections 2.9 Protein structure begins as a linear sequence of amino acids called a polypeptide chain (primary structure). The chains form sheets and coils (secondary structure), which may pack into functional domains (tertiary structure). Many proteins, including most enzymes, consist of two or more polypeptides (quaternary structure). Fibrous proteins aggregate into large chains or sheets. A protein's structure dictates its function, so changes in a protein's structure may alter its function.

CENGAGENOW™ Explore amino acid structure and peptide bond formation with the animations on CengageNow.

Section 2.10 Nucleotides consist of a sugar, a nitrogen-containing base, and phosphate groups. Nucleic acids DNA and RNA are polymers of nucleotide monomers. DNA encodes heritable information about a cell's proteins and RNAs. Different RNAs interact with DNA and with one another to carry out protein synthesis.

CENGAGENOW™ Explore DNA structure with the animation on CengageNow.

Self-Quiz   Answers in Appendix I

1. A(n) _____ is a molecule into which a radioisotope has been incorporated.

2. A(n) _____ forms when atoms of two or more elements bond covalently.

3. Atoms share electrons unequally in a(n) _____ bond.

4. A(n) _____ substance repels water.

5. Hydrogen ions (H^+) are _____ .
   a. indicated by pH   c. in blood
   b. protons   d. all of the above

6. When dissolved in water, a(n) _____ donates H^+.

7. A(n) _____ is a chemical partnership between a weak acid or base and its salt.

8. Each carbon atom can share pairs of electrons with up to _____ other atom(s).

Digging Into Data

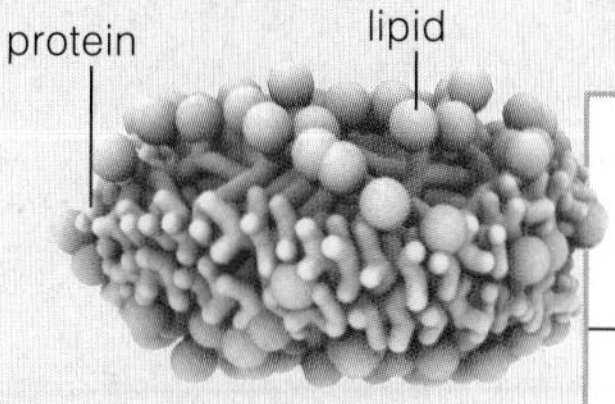

Effects of Dietary Fats on Lipoprotein Levels

Cholesterol does not dissolve in blood, so it is carried through the bloodstream by lipid–protein aggregates called lipoproteins. Lipoproteins vary in structure. Low-density lipoprotein (LDL) carries cholesterol to body tissues such as artery walls, where it can form deposits associated with cardiovascular disease. LDL is often called "bad" cholesterol. High-density lipoprotein (HDL) carries cholesterol away from tissues to the liver for disposal, so HDL is often called "good" cholesterol.

In 1990, Ronald Mensink and Martijn Katan published a study that tested the effects of different dietary fats on blood lipoprotein levels. Their results are shown in Figure 2.21.

1. In which group was the level of LDL ("bad" cholesterol) highest?

2. In which group was the level of HDL ("good" cholesterol) lowest?

3. An elevated risk of heart disease has been correlated with increasing LDL-to-HDL ratios. In which group was the LDL:HDL ratio highest? Rank the three diets according to their potential effect on cardiovascular health.

protein   lipid

an HDL particle

| | Main Dietary Fats | | | |
	cis-fatty acids	trans-fatty acids	saturated fats	optimal level
LDL	103	117	121	<100
HDL	55	48	55	>40
ratio	1.87	2.44	2.2	<2

FIGURE 2.21 Effect of diet on lipoprotein levels. Researchers placed 59 men and women on a diet in which 10 percent of their daily energy intake consisted of *cis* fatty acids, *trans* fatty acids, or saturated fats. Blood LDL and HDL levels were measured after three weeks on the diet; averaged results are shown in mg/dL (milligrams per deciliter of blood). All subjects were tested on each of the diets. The ratio of LDL to HDL is also shown.

9. _____ is a simple sugar (a monosaccharide).
   a. Glucose   c. Ribose   e. a and c
   b. Sucrose   d. Starch   f. a, b, and c

10. Unlike saturated fats, the fatty acid tails of unsaturated fats incorporate one or more _____ .

11. Is this statement true or false? Unlike saturated fats, all of the unsaturated fats are beneficial to health because their fatty acid tails do not pack tightly together.

12. Which of the following is a class of molecules that encompasses all of the other molecules listed?
   a. triglycerides   c. waxes   e. lipids
   b. fatty acids   d. steroids   f. phospholipids

13. _____ are to proteins as _____ are to nucleic acids.
   a. Sugars; lipids   c. Amino acids; hydrogen bonds
   b. Sugars; proteins   d. Amino acids; nucleotides

14. A denatured protein has lost its _____ .
   a. hydrogen bonds   c. function
   b. shape   d. all of the above

15. _____ consist(s) of nucleotides.
   a. sugars   b. DNA   c. RNA   d. b and c

16. Match each molecule with its most suitable description.
   _____ chain of amino acids   a. polar
   _____ energy carrier in cells   b. phospholipid
   _____ glycerol, fatty acids, phosphate c. temperature
   _____ two strands of nucleotides   d. DNA
   _____ one or more sugar monomers   e. ATP
   _____ richest source of energy   f. triglycerides
   _____ hydrophilic   g. atomic number
   _____ number of protons in nucleus   h. carbohydrate
   _____ molecular jiggling   i. polypeptide

Additional questions are available on CENGAGENOW™

Critical Thinking

1. Alchemists were medieval scholars and philosophers who were the forerunners of modern-day chemists. Many spent their lives trying to transform lead (atomic number 82) into gold (atomic number 79). Explain why they never did succeed in that endeavor.

2. Draw a shell model of a nitrogen atom, which has 7 protons.

3. In 1976, a team of chemists in the UK was developing new insecticides by modifying sugars with chlorine (Cl_2), phosgene (Cl_2CO), and other toxic gases. One young member of the team misunderstood his verbal instructions to "test" a new molecule. He thought he had been told to "taste" it. Luckily, the molecule was not toxic, but it was very sweet. It became the food additive sucralose.

Sucralose has three chlorine atoms substituted for three hydroxyl groups of sucrose. It binds so strongly to the sweet-taste receptors on the tongue that the human brain perceives it as 600 times sweeter than sucrose (table sugar). Sucralose was originally marketed as an artificial sweetener called Splenda, but it is now available under several other brand names.

Researchers proved that the body does not recognize sucralose as a carbohydrate by feeding sucralose labeled with ^{14}C to volunteers. Analysis of the radioactive molecules in the volunteers' urine and feces showed that 92.8 percent of the sucralose passed through the body without being altered. Many people are worried that the chlorine atoms impart toxicity to sucralose. How would you respond to that concern?

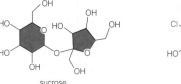

sucrose

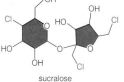

sucralose

3 Cell Structure

3.1 Impacts/Issues: Food for Thought

We find bacteria at the bottom of the ocean, high up in the atmosphere, miles underground—essentially anywhere we look. Mammal intestines typically harbor fantastic numbers of them, but bacteria are not just stowaways there. Intestinal bacteria make vitamins that mammals cannot, and they crowd out more dangerous germs. Cell for cell, bacteria that live in and on a human body outnumber the person's own cells by about ten to one.

Escherichia coli is one of the most common intestinal bacteria of warm-blooded animals. Only a few of the hundreds of types, or strains, of *E. coli*, are harmful. One, O157:H7, makes a potent toxin that can severely damage the lining of the human intestine. After ingesting as few as ten O157:H7 cells, a person may become ill with severe cramps and bloody diarrhea that lasts up to ten days. In some people, complications of O157:H7 infection result in kidney failure, blindness, paralysis, and death. About 73,000 people in the United States become infected with *E. coli* O157:H7 each year, and more than 60 of them die.

E. coli O157:H7 lives in the intestines of other animals—mainly cattle, deer, goats, and sheep—apparently without sickening them. Humans are exposed to the bacteria when they come into contact with feces of animals that harbor it, for example, by eating contaminated ground beef. During slaughter, meat occasionally comes into contact with feces. Bacteria in the feces stick to the meat, then get thoroughly mixed into it during the grinding process. Unless contaminated meat is cooked to at least 71°C (160°F), live bacteria will enter the digestive tract of whoever eats it.

People also become infected by eating fresh fruits and vegetables that have come into contact with animal feces. For example, in 2006, more than 200 people became ill and 3 died after eating fresh spinach. The spinach was grown in a field close to a cattle pasture, and water contaminated with manure may have been used to irrigate the field. Washing contaminated produce with water does not remove *E. coli* O157:H7, because the bacteria are sticky.

The economic impact of such outbreaks, which occur with some regularity, extends beyond the casualties. Growers lost $50–100 million recalling fresh spinach after the 2006 outbreak. In 2007, about 5.7 million pounds of ground beef were recalled after 14 people were sickened. Food growers and processors are beginning to implement new procedures that they hope will reduce *E. coli* O157:H7 outbreaks. Some meats and produce are now tested for pathogens before sale, and improved documentation should allow a source of contamination to be pinpointed more quickly.

What makes bacteria sticky? Why do people but not cows get sick with *E. coli* O157:H7? You will begin to find answers to these and many more questions that affect your health in this chapter, as you learn about cells and how they work.

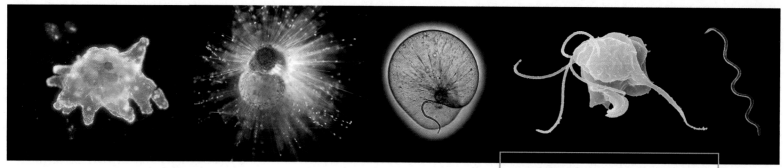

What, Exactly, Is a Cell?

There are many different kinds of cells, a few of which are shown in Figure 3.1. Despite their differences, however, all cells have certain organizational and functional features. For example, every cell has an outer membrane, or **plasma membrane**, that separates its metabolic activities from events outside of the cell. At its most basic structural level, a cell membrane consists of a **lipid bilayer**, a double layer of lipids (*right*). In addition to a plasma membrane, many cells also have membranes that divide their interior into compartments with various functions. A cell would die very quickly without continuously exchanging substances such as raw materials and wastes with its environment, so a plasma membrane is necessarily permeable to certain substances. Gases such as oxygen and carbon dioxide freely cross lipid bilayers, as does water. Ions and other substances can only cross with the assistance of proteins embedded in the membrane. Other proteins carry out different functions, as you will see in Section 3.4.

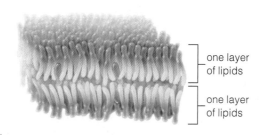

one layer of lipids

one layer of lipids

a lipid bilayer

A plasma membrane encloses a fluid or jellylike mixture of water, sugars, ions, and proteins called **cytoplasm**. An important part of homeostasis consists of maintaining the composition of cytoplasm, which differs—often dramatically—from the composition of fluid outside the cell. Some or all of a cell's metabolism occurs in the cytoplasm, and the cell's internal components, including organelles, are suspended in it. **Organelles** are structures that carry out special metabolic functions inside a cell.

All cells start out life with DNA, although a few types of cells lose it as they mature. We categorize cells into two major categories, prokaryotic and eukaryotic, based on whether their DNA is housed in a nucleus or not. The **nucleus** (plural, nuclei) is an organelle with a double membrane that contains the cell's DNA. Only eukaryotic cells have a nucleus. In most prokaryotic cells, the DNA is suspended directly in the cytoplasm. We call the region of cytoplasm in which prokaryotic DNA is most concentrated a **nucleoid**.

Almost all cells are too small to see with the naked eye. Why? The answer begins with the processes that keep a cell alive. A living cell must exchange substances with its environment at a rate that keeps pace with its metabolism. Nutrients have to enter the cell fast enough to supply its molecule-building activities, and wastes have to exit at a rate that prevents the cell from being poisoned. Both processes occur across the plasma membrane, which can handle

cytoplasm Semifluid substance enclosed by a cell's plasma membrane.

lipid bilayer Structural foundation of cell membranes; mainly phospholipids arranged tail-to-tail in two layers.

nucleoid Region of cytoplasm where the DNA is concentrated inside a prokaryotic cell.

nucleus Organelle with two membranes that holds a eukaryotic cell's DNA.

organelle Structure that carries out a specialized metabolic function inside a cell.

plasma membrane A cell's outermost membrane.

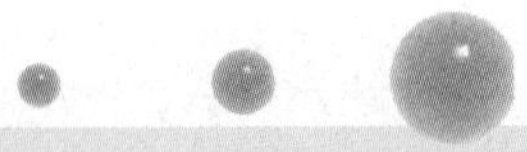

FIGURE 3.2 Animated! Examples of surface-to-volume ratio. This physical relationship between increases in volume and surface area limits the size and influences the shape of cells.

Diameter (cm)	2	3	6
Surface area (cm²)	12.6	28.2	113
Volume (cm³)	4.2	14.1	113
Surface-to-volume ratio	3:1	2:1	1:1

only so many exchanges at a time between the cytoplasm and the external environment. Thus, cell size is limited by a physical relationship called the **surface-to-volume ratio**. By this ratio, an object's volume increases with the cube of its diameter, but its surface area increases only with the square.

Apply the surface-to-volume ratio to a round cell. As Figure 3.2 shows, when a cell expands in diameter, its volume increases faster than its surface area does. Imagine that a round cell expands until it is four times its original diameter. The volume of the cell has increased 64 times (4^3), but its surface area has increased only 16 times (4^2). Each unit of plasma membrane must now handle exchanges with four times as much cytoplasm ($64 \div 16 = 4$). If the cell gets too big, the inward flow of nutrients and the outward flow of wastes across that membrane will not be fast enough to keep the cell alive.

Why not? A cell is filled with cytoplasm, and metabolic activities occur all through it. Molecules disperse themselves through cytoplasm by their own random motions, but this movement occurs only so quickly. Nutrients must cross the plasma membrane and get distributed through the cytoplasm fast enough to satisfy a cell's metabolic needs, and wastes must be removed fast enough to keep the cell from poisoning itself. Nutrients and wastes would not be able to move through the middle of a big, round cell fast enough to keep up with metabolism.

Surface-to-volume limits also affect the body plans of multicelled species. For example, small cells attach end to end to form strandlike algae, so that each can interact directly with its surroundings. Muscle cells in your thighs are as long as the muscle in which they occur, but each is thin, so it exchanges substances efficiently with fluids in the tissue surrounding it.

The Cell Theory Hundreds of years of observations of cell structure and function led to the way we now answer the question, What is a cell? A **cell** is the smallest unit that shows the properties of life: It carries out metabolism and homeostasis, and either reproduces on its own or it is part of a larger organism. By this definition, each cell is alive even if it is part of a multicelled body, and all living organisms consist of one or more cells. We also know that cells reproduce themselves by dividing, so it follows that all existing cells must have arisen by division of other cells. Later chapters discuss the processes by which cells divide, but for now all you need to know is that a cell passes its hereditary material—its DNA—to offspring during those processes. Taken together, these four generalizations constitute the **cell theory**, a foundation of modern biology (Table 3.1).

Table 3.1	The Cell Theory

1. Every living organism consists of one or more cells.

2. A cell is the smallest unit of life, individually alive even as part of a multicelled organism.

3. Every living cell came into existence by division of a pre-existing cell.

4. Cells contain hereditary material (DNA) that they pass along to their offspring during processes of cell division.

cell Smallest unit of life.
cell theory Fundamental theory of biology: All organisms consist of one or more cells; the cell is the smallest unit of life; each new cell arises from another cell; and a cell passes hereditary material to its offspring.
surface-to-volume ratio A relationship in which the volume of an object increases with the cube of the diameter, but the surface area increases with the square.

Take-Home Message **How are all cells alike?**

■ The cell is the fundamental unit of all life.

■ All cells start life with a plasma membrane, cytoplasm, and a region of DNA, which, in eukaryotic cells only, is enclosed by a nucleus.

■ The surface-to-volume ratio limits cell size and influences cell shape.

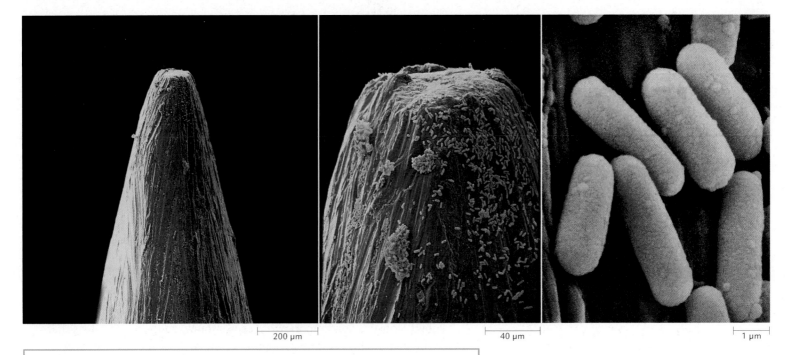

200 μm 40 μm 1 μm

FIGURE 3.3 Rod-shaped bacterial cells on the tip of a household pin, shown at increasingly higher magnifications (enlargements). The "μm" is an abbreviation for micrometers, or millionths of a meter. **Figure It Out:** About how big are these bacteria?

Answer: About 1 μm wide and 5 μm long

Measuring Cells

Do you ever think of yourself as being about 3/2000 of a kilometer (1/1000 of a mile) tall? Probably not, yet that is how we measure cells. Use the scale bars in Figure 3.3 like a ruler and you can see that the cells shown are a few micrometers "tall." One micrometer (μm) is one-thousandth of a millimeter, which is one-thousandth of a meter, which is one-thousandth of a kilometer (0.62 miles). The cells in the photos are bacteria. Bacteria are among the smallest and structurally simplest cells on Earth. The cells that make up your body are generally larger and more complex than bacteria.

Animalcules and Beasties No one even knew cells existed until well after the first microscopes were invented. Those microscopes were not very sophisticated. Given the simplicity of their instruments, it is amazing that the pioneers in microscopy observed as much as they did. By the mid-1600s, Antoni van Leeuwenhoek, a Dutch draper, was spying on the microscopic world of rainwater, insects, fabric, sperm, feces—essentially any sample he could fit into his homemade microscope (shown at *right*). He was fascinated by the tiny organisms he saw moving in many of his samples. For example, in scrapings of tartar from his teeth, Leeuwenhoek saw "many very small animalcules, the motions of which were very pleasing to behold." He (incorrectly) assumed that movement defined life, and (correctly) concluded that the moving "beasties" he saw were alive. Perhaps Leeuwenhoek was so pleased to behold his animalcules because he did not understand the implications of what he was seeing: Our world, and our bodies, teem with bacteria and other microbial life.

sample holder   focusing knob

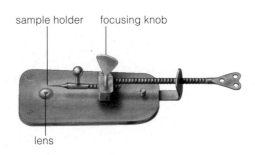

lens

Leeuwenhoek's microscope

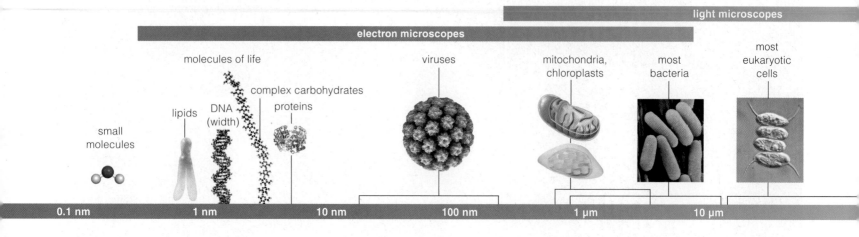

molecules of life

small molecules · lipids · DNA (width) · complex carbohydrates · proteins · viruses · mitochondria, chloroplasts · most bacteria · most eukaryotic cells

electron microscopes

light microscopes

0.1 nm · 1 nm · 10 nm · 100 nm · 1 μm · 10 μm

Robert Hooke, a contemporary of Leeuwenhoek, added another lens that made the instrument easier to use. Many of the microscopes we use today are still based on his design. Hooke magnified a piece of thinly sliced cork from a mature tree and saw tiny compartments (his drawing of them is shown at *right*). Hooke named the compartments cellulae—a Latin word for the small chambers that monks lived in—and thus coined the term "cell." Actually, they were dead plant cell walls, which is what cork consists of, but Hooke did not think of them as being dead because neither he nor anyone else knew cells could be alive. He observed cells "fill'd with juices" in green plant tissues but did not realize they were alive, either.

For nearly 200 years after Hooke discovered them, cells were assumed to be part of a continuous membrane system in multicelled organisms, not separate entities. In the mid-1800s, botanist Matthias Schleiden realized that a plant cell is an independent living unit even when it is part of a plant. Schleiden compared notes with zoologist Theodor Schwann, and together they concluded that the tissues of animals as well as plants are composed of cells and their products. The cell theory, first articulated in 1839 by Schwann and Schleiden and later revised, was a radical new interpretation of nature that underscored life's unity.

FIGURE 3.4 Different microscopes reveal different characteristics of the same organism, a green alga (*Scenedesmus*).

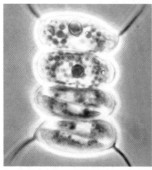

A Light micrograph. A phase-contrast microscope yields high-contrast images of transparent specimens, such as cells.

B Light micrograph. A reflected light microscope captures light reflected from opaque specimens.

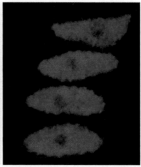

C Fluorescence micrograph. The chlorophyll molecules in these cells emitted red light (they fluoresced) naturally.

D A transmission electron micrograph reveals fantastically detailed images of internal structures.

E A scanning electron micrograph shows surface details of cells and structures. SEMs may be artificially colored to highlight certain details.

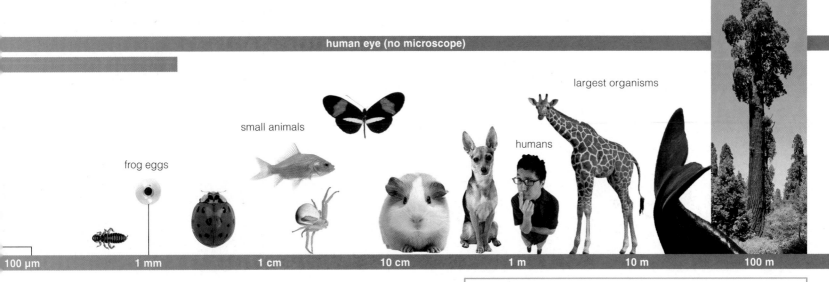

human eye (no microscope)

largest organisms

small animals

humans

frog eggs

| 100 μm | 1 mm | 1 cm | 10 cm | 1 m | 10 m | 100 m |

Modern Microscopes Like their early predecessors, many modern microscopes rely on visible light to illuminate objects. All light travels in waves, a property that makes it bend when it passes through curved glass lenses. Inside microscopes, such lenses focus light into a magnified image of a specimen. Photographs of images enlarged with any microscope are called micrographs (Figure 3.4). Figure 3.5 compares the resolving power of light and electron microscopes with that of the unaided human eye.

Phase-contrast microscopes shine light through specimens, but most cells are nearly transparent. Their internal details may not be visible unless they are first stained, or exposed to dyes that only some cell parts soak up. Parts that absorb the most dye appear darkest. Staining results in an increase in contrast (the difference between light and dark) that allows us to see a greater range of detail (Figure 3.4**A**). Surface details can be revealed by reflected light (Figure 3.4**B**).

With a fluorescence microscope, a cell or a molecule is the light source; it fluoresces, or emits energy in the form of light, when a laser beam is focused on it. Some molecules fluoresce naturally (Figure 3.4**C**). More typically, researchers attach a light-emitting tracer to the cell or molecule of interest.

Other microscopes can reveal finer details. For example, electron microscopes use electrons instead of visible light to illuminate samples. Transmission electron microscopes beam electrons through a thin specimen. The specimen's internal details appear on the resulting image as shadows (Figure 3.4**D**). Scanning electron microscopes direct a beam of electrons back and forth across a surface of a specimen, which has been coated with a thin layer of gold or another metal. The metal emits both electrons and x-rays, which are converted into an image of the surface (Figure 3.4**E**). Both types of electron microscopes can resolve structures as small as 0.2 nanometers.

FIGURE 3.5 Relative sizes. *Above*, the diameter of most cells is in the range of 1 to 100 micrometers. *Below*, converting among units of length; see Units of Measure, Appendix V. **Figure It Out: Which is smallest: a protein, a lipid, or a water molecule?**

Answer: A water molecule

1 centimeter (cm)	=	1/100 meter, or 0.4 inch
1 millimeter (mm)	=	1/1000 meter, or 0.04 inch
1 micrometer (μm)	=	1/1,000,000 meter, or 0.00004 inch
1 nanometer (nm)	=	1/1,000,000,000 meter, or 0.00000004 inch

1 meter $= 10^2$ cm $= 10^3$ mm $= 10^6$ μm $= 10^9$ nm

Take-Home Message How do we see cells?

■ Most cells are visible only with the help of microscopes.

■ Different types of microscopes reveal different aspects of cell structure.

A Phospholipids are the most abundant component of eukaryotic cell membranes. Each phospholipid molecule has a hydrophilic head and two hydrophobic tails.

B In a watery fluid, phospholipids spontaneously line up into two layers: hydrophobic tails cluster together, and hydrophilic heads face outward, toward the fluid. This lipid bilayer forms the framework of all cell membranes.

C A lipid bilayer spontaneously shapes itself into a sheet or bubble. A plasma membrane is basically a lipid bilayer balloon filled with fluid. Many types of proteins intermingle among the lipids in a cell membrane—a few that are typical of plasma membranes are shown *opposite*.

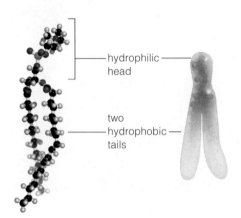

hydrophilic head

two hydrophobic tails

one layer of lipids

one layer of lipids

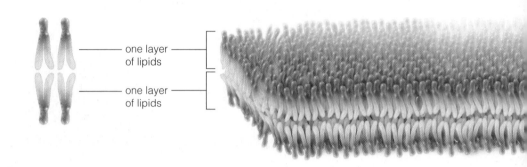

FIGURE 3.6 **Animated!** Cell membrane structure. **A–C** Organization of lipids in cell membranes. **D–G** Examples of membrane proteins.

A plasma membrane is basically a lipid bilayer balloon filled with fluid.

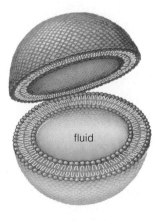

fluid

3.4 The Structure of Cell Membranes

A cell membrane is a barrier that selectively controls exchanges between the cell and its surroundings. This function emerges when certain lipids—mainly phospholipids—interact. A phospholipid molecule consists of a phosphate-containing head and two fatty acid tails. The polar head is hydrophilic, which means it interacts with water molecules. The nonpolar tails are hydrophobic, so they do not interact with water molecules. The tails do, however, interact with the tails of other phospholipids. When swirled into water, phospholipids spontaneously assemble into two layers, with all of their nonpolar tails sandwiched between all of their polar heads. Such lipid bilayers are the basic framework of all cell membranes (Figure 3.6**A–C**).

Other molecules—including steroids and proteins—are embedded in or associated with the lipid bilayer of a cell membrane. Most of these molecules flow around more or less freely. The fluidity arises from the behavior of the phospholipids, which drift sideways and spin around their long axis in a bilayer. Their tails wiggle too. The **fluid mosaic model** describes a cell membrane as a two-dimensional liquid of mixed composition.

Membrane Proteins A cell membrane physically separates an external environment from an internal one, but that is not its only task. Many types of proteins are associated with a cell membrane, and each type adds a specific function to it. Thus, even though every cell membrane consists mainly of a phospholipid bilayer, different cell membranes can have different characteristics depending on which proteins are associated with them. For example, a plasma membrane has proteins that no internal cell membrane has. Many plasma membrane proteins are **enzymes**, which accelerate chemical processes without being changed by them. **Adhesion proteins** fasten cells together in animal tissues. **Recognition proteins** function as unique identity tags for each individual or species (Figure 3.6**D**). Being able to recognize "self" means that foreign cells (harmful ones, in particular) can also be recognized.

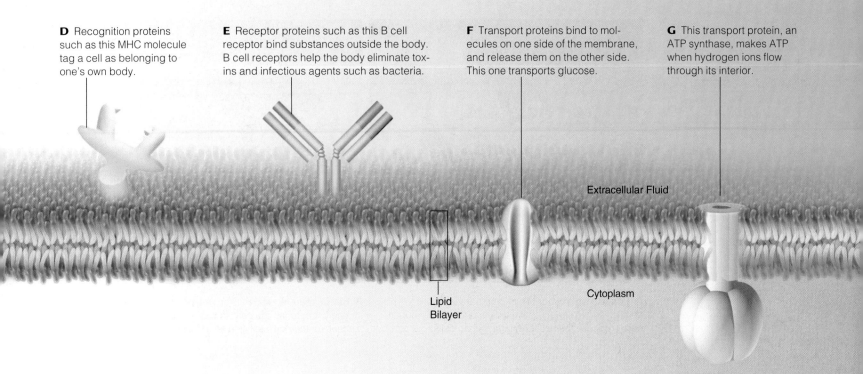

D Recognition proteins such as this MHC molecule tag a cell as belonging to one's own body.

E Receptor proteins such as this B cell receptor bind substances outside the body. B cell receptors help the body eliminate toxins and infectious agents such as bacteria.

F Transport proteins bind to molecules on one side of the membrane, and release them on the other side. This one transports glucose.

G This transport protein, an ATP synthase, makes ATP when hydrogen ions flow through its interior.

Extracellular Fluid

Cytoplasm

Lipid
Bilayer

Receptor proteins bind to a particular substance outside of the cell, such as a hormone or toxin (Figure 3.6**E**). Binding triggers a change in the cell's activities that may involve metabolism, movement, division, or even cell death. Different receptors occur on different cells, but all are critical for homeostasis.

Additional proteins occur on all cell membranes. **Transport proteins** move specific substances across a membrane, typically by forming a channel through it. These proteins are important because lipid bilayers are impermeable to most substances, including ions and polar molecules. Some transport proteins are open channels through which a substance moves on its own across a membrane (Figure 3.6**F,G**). Others use energy to actively pump a substance across. We return to the topic of transport across membranes in the next chapter.

Take-Home Message What is a cell membrane?

- A cell membrane is a mosaic of different kinds of lipids and proteins.
- The foundation of cell membranes is the lipid bilayer: two layers of phospholipids, tails sandwiched between heads.
- Many types of proteins add various functions to lipid bilayers in membranes.

adhesion protein Membrane protein that helps cells stick together in tissues.

enzyme Molecule that speeds a chemical process without being changed by it.

fluid mosaic model A cell membrane can be considered a two-dimensional fluid of mixed composition.

receptor protein Plasma membrane protein that binds to a particular substance outside of the cell.

recognition protein Plasma membrane protein that tags a cell as belonging to self (one's own body).

transport protein Protein that passively or actively assists specific ions or molecules across a membrane.

3.5 Introducing Prokaryotic Cells

The word prokaryote means "before the nucleus," a reminder that the first prokaryotes evolved before the first eukaryotes. All prokaryotes are single-celled. As a group, they are the smallest and most metabolically diverse forms of life

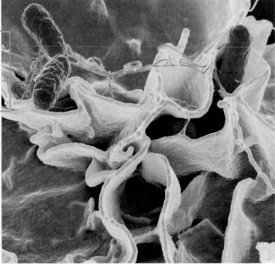

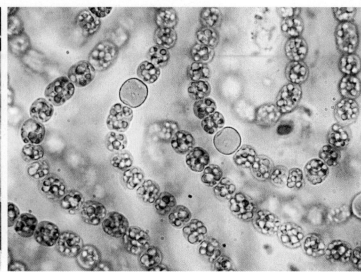

A Protein filaments, or pili, anchor bacterial cells to one another and to surfaces. Here, *Salmonella* Typhimurium cells (*red*) use their pili to invade human cells.

B Ball-shaped *Nostoc* cells are a type of freshwater photosynthetic bacteria. The cells in each strand stick together in a sheath of their own jellylike secretions.

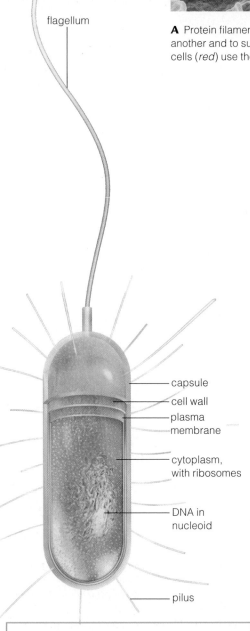

flagellum

capsule

cell wall

plasma membrane

cytoplasm, with ribosomes

DNA in nucleoid

pilus

FIGURE 3.8 Animated! Generalized body plan of a prokaryote.

that we know about. Prokaryotes inhabit nearly all of Earth's environments, including some very hostile places.

Domains Bacteria and Archaea make up the prokaryotes (Section 1.4 and Figure 3.7). The two kinds of cells may be alike in appearance and size, but they differ in structure and metabolic details. Some characteristics of archaeans indicate they are more closely related to eukaryotic cells than they are to bacteria. Chapter 13 revisits prokaryotes in more detail. Here we present a simple overview.

Most prokaryotic cells are not much bigger than a few micrometers. None has a complex internal framework, but protein filaments under the plasma membrane reinforce the cell's shape. Such filaments also act as scaffolding for internal structures.

Figure 3.8 shows a general body plan of a prokaryotic cell. The cytoplasm of these cells contains many **ribosomes** (organelles upon which polypeptides are assembled), and in some species, additional organelles. The cell's single chromosome, a circular DNA molecule, is located in the cytoplasm, in an irregularly shaped region called the nucleoid. Most nucleoids are not enclosed by a membrane. The cytoplasm of many prokaryotes also contains plasmids. These small circles of DNA carry a few genes (units of inheritance) that can provide advantages, such as resistance to antibiotics.

Many prokaryotic cells have one or more flagella projecting from their surface. **Flagella** (singular, flagellum) are long, slender cellular structures used for motion. A bacterial flagellum rotates like a propeller that drives the cell through fluid habitats, such as an animal's body fluids. Some bacteria also have protein filaments called **pili** (singular, pilus) projecting from their surface (Figure 3.7**A**). Pili help cells cling to or move across surfaces. One kind, a "sex" pilus, attaches to another bacterium and then shortens. The attached cell is reeled in, and a plasmid is transferred from one cell to the other through the pilus.

A durable **cell wall** surrounds the plasma membrane of nearly all prokaryotes. Dissolved substances easily cross this permeable layer on the way to and from the plasma membrane. The cell wall of most bacteria consists of a polymer of peptides and polysaccharides. The wall of most archaeans consists of proteins. Sticky polysaccharides form a slime layer, or capsule, around the wall

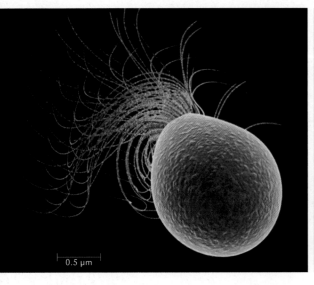

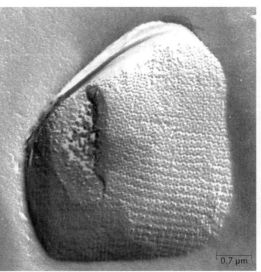

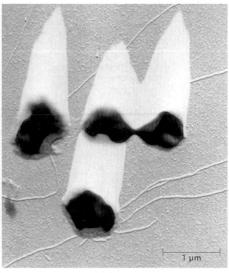

C The archaean *Pyrococcus furiosus* was discovered in ocean sediments near an active volcano. It lives best at 100°C (212°F), and it makes a rare kind of enzyme that contains tungsten atoms.

D *Ferroglobus placidus* prefers superheated water spewing from the ocean floor. The durable composition of archaean lipid bilayers (note the gridlike texture) keeps their membranes intact at extreme heat and pH.

E *Metallosphaera prunae*, an archaean discovered in a smoking pile of ore at a uranium mine, prefers high temperatures and low pH. (*White* shadows are an artifact of electron microscopy.)

of many types of bacteria. The sticky capsule helps these cells adhere to many types of surfaces (such as spinach leaves and meat), and it also offers some protection against predators and toxins.

The plasma membrane of all bacteria and archaeans selectively controls which substances move into and out of the cell, as it does for eukaryotic cells. The plasma membrane bristles with transporters and receptors, and it also incorporates proteins that carry out important metabolic processes. For example, part of the plasma membrane of cyanobacteria (Figure 3.7**B**) folds into the cytoplasm. Molecules that carry out photosynthesis are embedded in this membrane, just as they are in the inner membrane of chloroplasts, which are organelles specialized for photosynthesis in eukaryotic cells (we return to chloroplasts in Section 3.6).

Biofilms Bacterial cells often live so close together that an entire community shares a layer of secreted polysaccharides and proteins. A communal living arrangement in which single-celled organisms live in a shared mass of slime is called a **biofilm**. In nature, a biofilm typically consists of multiple species, all entangled in their own mingled secretions. It may include bacteria, algae, fungi, protists, and archaeans. Participating in a biofilm allows the cells to linger in a favorable spot rather than be swept away by fluid currents, and to reap the benefits of living communally. For example, rigid or netlike secretions of some species serve as permanent scaffolding for others; species that break down toxic chemicals allow more sensitive ones to thrive in polluted habitats that they could not withstand on their own; and waste products of some serve as raw materials for others.

Take-Home Message What do all prokaryotic cells have in common?

- All prokaryotes are single-celled organisms with no nucleus. Bacteria and archaeans are the only prokaryotes.
- Prokaryotes have a relatively simple structure, but as a group they are the most diverse forms of life. They inhabit nearly all regions of the biosphere.

biofilm Community of different types of microorganisms living within a shared mass of slime.

cell wall Semirigid but permeable structure that surrounds the plasma membrane of some cells.

flagellum Long, slender cellular structure used for motility.

pilus A protein filament that projects from the surface of some bacterial cells.

ribosome Organelle of protein synthesis.

FIGURE 3.9 Animated!

FIGURE 3.9 Animated!
Common components of eukaryotic
cells. This is an animal cell.

1 A plasma membrane controls the
kinds and amounts of substances
that move into and out of a cell.

2 The nucleus contains, protects,
and controls access to DNA.

3 Endoplasmic reticulum (ER)
modifies new polypeptides and syn-
thesizes lipids; has other tasks.

4 Different types of vesicles transport, store, or digest substances,
among other functions.

5 Golgi bodies finish, sort, and
ship lipids and proteins.

6 Mitochondria make ATP.

7 Ribosomes, either attached to
ER or free in cytoplasm, assemble
polypeptides from amino acids.

8 Centrioles produce and organize
microtubules.

9 Cytoskeletal elements provide
structural support; move cell parts
or the whole cell.

3.6 A Peek Inside a Eukaryotic Cell

All protists, fungi, plants, and animals are eukaryotes. Some of these organisms
are independent, free-living cells (Figure 3.1); others consist of many cells working together as a body.

By definition, a eukaryotic cell starts out life with a nucleus (*eu*– means
true; *karyon* means nut, or kernel). Like many other organelles, a nucleus has a
membrane. An organelle's outer membrane controls the types and amounts of
substances that cross it. Such control maintains a special internal environment
that allows the organelle to carry out its particular function. That function may
be isolating toxic or sensitive substances from the rest of the cell, transporting
substances through cytoplasm, maintaining fluid balance, or providing a favorable environment for a special process.

A typical eukaryotic cell contains a nucleus, an endomembrane system (ER,
vesicles, and Golgi bodies), mitochondria, and cytoskeletal elements. Certain
cells also have other special structures (Figures 3.9 and 3.10). Much as interactions among organs keep an animal body alive and well, interactions among
these components keep a cell alive and well.

The Nucleus A nucleus serves two important functions. First, it keeps the
cell's genetic material—its one and only copy of DNA—safe and sound. Isolated
in its own compartment, DNA stays separated from the bustling activity of the
cytoplasm, and from metabolic processes that might damage it **2**.

The second function of a nucleus is to control the passage of certain molecules between the nucleus and the cytoplasm. The nuclear membrane, which is
called the **nuclear envelope**, carries out this function. A nuclear envelope consists of two lipid bilayers folded together as a single membrane. Receptors and
transporters stud both sides of the bilayer; other proteins cluster to form tiny
pores that span it. These molecules and structures work as a system to selectively

transport various molecules across the nuclear membrane. Cells access their DNA when they make RNA and proteins, so the molecules involved in this process must pass into the nucleus and out of it. Control over their transport through the nuclear membrane is one way the cell regulates the amount of RNA and proteins it makes.

The Endomembrane System

The **endomembrane system** is a series of interacting organelles between the nucleus and the plasma membrane. Its main function is to make lipids, enzymes, and proteins for secretion or insertion into cell membranes. It also destroys toxins, recycles wastes, and has other specialized functions. The system's components vary among different types of cells, but here we present the most common ones.

Part of the endomembrane system is an extension of the nuclear envelope called **endoplasmic reticulum**, or **ER** ❸. ER forms a continuous compartment that folds into flattened sacs and tubes. Two kinds of ER, rough and smooth, are named for their appearance in electron micrographs. Thousands of ribosomes attached to the outer surface of rough ER make polypeptides that thread into the interior of the ER as they are assembled. Inside the ER, the polypeptides fold and take on their tertiary structure. Some of them become part of the ER membrane itself.

Smooth ER has no ribosomes, so it does not make protein. Some of the polypeptides made in the rough ER end up as enzymes in the smooth ER. These enzymes make most of the lipids that form the cell's membranes. They also break down carbohydrates, fatty acids, and some drugs and poisons.

Small, membrane-enclosed, saclike **vesicles** form in great numbers, in a variety of types, either on their own or by budding from other organelles or from the plasma membrane ❹. Many vesicles transport substances from one organelle to another, or to and from the plasma membrane. Those called **peroxisomes** contain enzymes that can inactivate toxins. Drink alcohol, and the peroxisomes in your liver and kidney cells break down nearly half of it. Eukaryotic cells also contain **vacuoles**. These vesicles appear empty under a microscope, but they serve an important function. Most are like trash cans: They collect waste, debris, or toxins, and dispose of these materials by fusing with other vesicles called lysosomes. **Lysosomes** contain powerful digestive enzymes that break down the contents of vacuoles.

Some vesicles fuse with and empty their contents into a **Golgi body**. This organelle has a folded membrane that typically looks a bit like a stack of pancakes ❺. Enzymes in a Golgi body put finishing touches on proteins and lipids that have been delivered from the ER. They attach phosphate groups or sugars, and cut certain polypeptides. The finished products (membrane proteins, proteins for secretion, and enzymes) are sorted and packaged into new vesicles that carry them to the plasma membrane or to lysosomes.

Mitochondria

The **mitochondrion** (plural, mitochondria) is an organelle that specializes in making ATP ❻. A mitochondrion has two membranes, one highly folded inside the other, that form its ATP-making machinery. Nearly all eukaryotic cells have mitochondria, which resemble bacteria in size, form, and

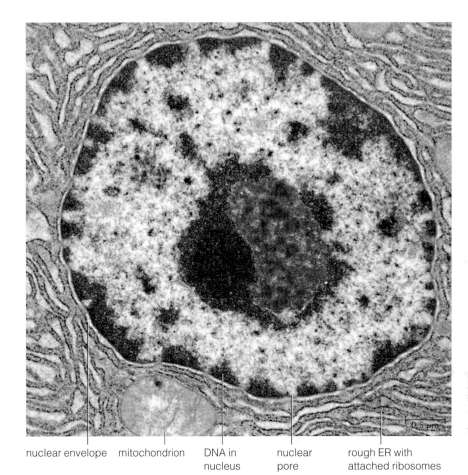

nuclear envelope　　mitochondrion　　DNA in nucleus　　nuclear pore　　rough ER with attached ribosomes

0.5 μm

FIGURE 3.10 Inside a cell taken from a mouse's pancreas.

endomembrane system Series of interacting organelles (endoplasmic reticulum, Golgi bodies, vesicles) between nucleus and plasma membrane; produces lipids, proteins.

endoplasmic reticulum (ER) Organelle that is a continuous system of sacs and tubes; extension of the nuclear envelope. Rough ER is studded with ribosomes; smooth ER is not.

Golgi body Organelle that modifies polypeptides and lipids; also sorts and packages finished products into vesicles.

lysosome Enzyme-filled vesicle that functions in intracellular digestion.

mitochondrion Double-membraned organelle that produces ATP.

nuclear envelope A double membrane that constitutes the outer boundary of the nucleus.

peroxisome Enzyme-filled vesicle that breaks down amino acids, fatty acids, and toxic substances.

vacuole A fluid-filled organelle that isolates or disposes of waste, debris, or toxic materials.

vesicle Small, membrane-enclosed, saclike organelle; different kinds store, transport, or degrade their contents.

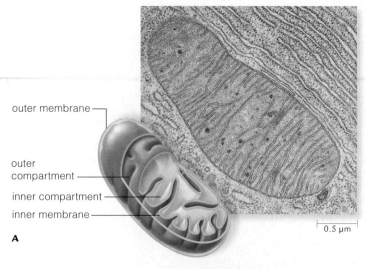

outer membrane

outer
compartment

inner compartment

inner membrane

0.5 µm

A

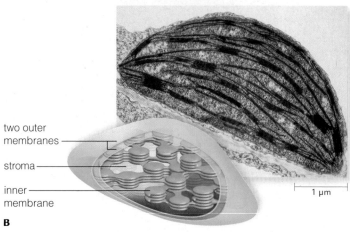

two outer
membranes

stroma

inner
membrane

1 µm

B

FIGURE 3.11 Animated! Bacteria-like organelles. **A** Mito-chondrion, specialized for producing large quantities of ATP for eukaryotic cells. **B** Chloroplast, specialized for photosynthesis.

Figure It Out: What organelle is visible to the upper right in the micrograph of the mitochondrion?

Answer: Rough ER

chloroplast Organelle of photosynthesis.

cilium Short, movable structure that projects from the plasma membrane of some eukaryotic cells.

cytoskeleton Dynamic framework of protein filaments that support, organize, and move eukaryotic cells and their internal structures.

intermediate filament Cytoskeletal element that locks cells and tissues together.

microfilament Reinforcing cytoskeletal element; fiber of actin subunits.

microtubule Cytoskeletal element involved in movement; hollow filament of tubulin subunits.

motor protein Type of energy-using protein that interacts with cytoskeletal elements to move the cell's parts or the whole cell.

pseudopod Extendable lobe of membrane-enclosed cytoplasm.

biochemistry (Figure 3.11**A**). They have their own DNA and ribo-somes, and they divide independently of the cell. Such clues led to a theory that mitochondria evolved from aerobic bacteria that took up permanent residence inside a host cell. By the theory of endosymbiosis, one cell was engulfed by another cell, or entered it as a parasite, but escaped digestion. That cell kept its plasma membrane and reproduced inside its host. In time, the cell's descendants became permanent residents that offered their hosts the benefit of extra ATP. Structures and functions once required for independent life were no longer needed and were lost over time. Later descendants evolved into mitochondria. We explore evidence for the theory of endosymbiosis in Section 13.3.

Chloroplasts Photosynthetic cells of plants and many protists contain **chloroplasts**, which are organelles specialized for photosynthesis. Most chloroplasts have an oval or disk shape formed by two outer membranes enclosing a semifluid interior called the stroma (Figure 3.11**B**). The stroma contains enzymes and the chloroplast's own DNA. Photosynthesis takes place at a third, highly folded membrane inside the stroma (we describe the process of photosynthesis in more detail in Chapter 5). In many ways, chloroplasts resemble photosynthetic bacteria, and, like mitochondria, they may have evolved by endosymbiosis.

The Cytoskeleton Between the nucleus and plasma membrane of all eukaryotic cells is a system of interconnected protein filaments collectively called the **cytoskeleton**. Elements of the cytoskeleton reinforce, organize, and move cell structures, and often the whole cell. Some are permanent; others form only at certain times.

Microtubules are long, hollow cylinders that consist of sub-units of the protein tubulin. They form a dynamic scaffolding for many cellular processes, rapidly assembling when they are needed and then disassembling when they are not. For example, some of the microtubules that assemble before a eukaryotic cell divides separate the cell's duplicated chromosomes, then dis-assemble. As another example, microtubules that form in the growing end of a young nerve cell support and guide its lengthening in a par-ticular direction (Figure 3.12).

Microfilaments are fibers that consist primarily of subunits of the globular protein actin. They strengthen or change the shape of eukaryotic cells. Cross-linked, bundled, or gel-like arrays of them make up the cell cortex, which is a reinforcing mesh under the plasma membrane. Actin microfilaments that form at the edge of a cell drag or extend it in a certain direction. In muscle cells, microfilaments of myosin and actin interact to bring about contraction.

Intermediate filaments are the most stable parts of a cell's cytoskeleton. They lock cells and tissues together. For example, some intermediate filaments called lamins form a layer that structurally supports the inner surface of the nuclear envelope.

All eukaryotic cells have similar microtubules and microfilaments. Despite the uniformity, both kinds of elements play diverse roles. How? They interact with accessory proteins, such as **motor proteins** that move cell parts when they

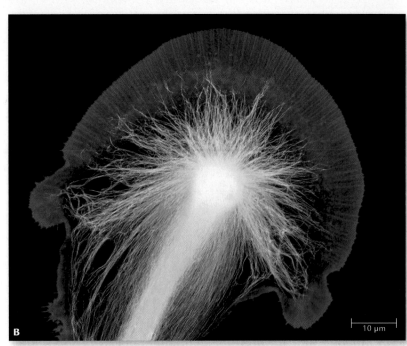

FIGURE 3.12 Cytoskeletal elements. **A** Tubulin subunits assembling into a microtubule, and actin subunits assembling into a microfilament.

B Microtubules (*yellow*) and actin microfilaments (*blue*) in the growing end of a nerve cell support and guide the cell's lengthening in certain directions.

tubulin subunit   actin subunit

25 nm

6–7 nm

A

B

10 μm

are repeatedly energized by ATP. A cell is like a train station during a busy holiday, with molecules being transported through its interior. Microtubules and microfilaments are like dynamically assembled train tracks, and motor proteins are freight engines that move along them (Figure 3.13).

Cilia, Flagella, and False Feet Organized arrays of microtubules are the basis of movement in eukaryotic flagella and cilia. Eukaryotic flagella are whiplike structures that propel cells such as sperm (*left*) through fluid. They have a different internal structure and motion than prokaryotic flagella.

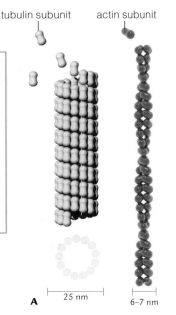

Cilia (singular, cilium) are short, hairlike structures that project from the surface of some cells. Cilia are usually shorter and more profuse than flagella. Their coordinated beating propels motile cells through fluid, and stirs fluid around stationary cells. For example, the cilia on thousands of cells lining your airways sweep inhaled particles away from your lungs.

Amoebas and other types of eukaryotic cells form **pseudopods**, or "false feet." As these temporary, irregular lobes bulge outward, they move the cell and engulf a target such as prey. Elongating microfilaments force the lobe to advance in a steady direction. Motor proteins that are attached to the microfilaments drag the plasma membrane along with them. An amoeba with multiple pseudopods is shown at the far left in Figure 3.1.

Take-Home Message **What do all eukaryotic cells have in common?**

■ All eukaryotic cells start life with a nucleus, ribosomes, and other organelles. The nucleus protects and controls access to a cell's DNA.

■ The endomembrane system, which includes ER, vesicles, and Golgi bodies, makes and modifies proteins and lipids.

■ Mitochondria are organelles that produce ATP. Some cells also contain chloroplasts, which specialize in photosynthesis.

FIGURE 3.13 Animated!
A motor protein (*tan*) drags cellular freight (here, a *pink* vesicle) as it inches along a microtubule.

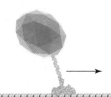

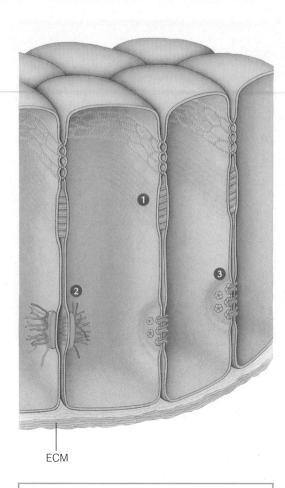

ECM

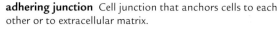

FIGURE 3.14 **Animated!** Cell junctions in animal tissues.

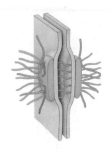

1 **Tight junctions**

Rows of proteins that run parallel with the free surface of a tissue; stop leaks between adjoining cells.

2 **Adhering junction**

A mass of interconnected proteins that welds one cell to another or to ECM; anchored under the plasma membrane by intermediate filaments.

3 **Gap junction**

Cylindrical clusters of proteins that span the plasma membrane of adjoining cells; clusters are often paired as channels that open and close.

Cell Surface Specializations

Most cells of multicelled organisms are surrounded and organized by a nonliving, complex mixture of fibrous proteins and polysaccharides called **extracellular matrix**, or **ECM**. Secreted by the cells it surrounds, ECM supports and anchors cells, separates tissues, and functions in cell signaling. Different types of cells secrete different kinds of ECM. For example, a waxy ECM secreted by plant cells forms a cuticle, or covering, that protects the plant's exposed surfaces and limits water loss. The cuticle of crabs, spiders, and other arthropods is mainly chitin, a polysaccharide.

ECM in animals typically consists of various kinds of carbohydrates and proteins; it is the basis of tissue organization, and it provides structural support. Bone is mostly an extracellular matrix composed of collagen, a fibrous protein, hardened by mineral deposits. The cell wall around the plasma membrane of plant cells and many protists and fungi is a type of ECM that is structurally different from a prokaryotic cell wall, but both types protect, support, and impart shape to a cell. Both are also porous: Water and solutes easily cross it on the way to and from the plasma membrane. Cells could not live without exchanging these substances with their environment.

A cell wall or other ECM does not prevent a cell from interacting with other cells or the surroundings. In multicelled species, such interaction occurs by way of **cell junctions**, which are structures that connect a cell to other cells and to the environment. Cells send and receive ions, molecules, or signals through some junctions. Other kinds help cells recognize and stick to each other and to extracellular matrix.

Cells in most animal tissues connect to their neighbors and to ECM by way of one or more types of cell junctions (Figure 3.14). In epithelial tissues that line body surfaces and internal cavities, rows of proteins that form **tight junctions** between plasma membranes prevent body fluids from seeping between adjacent cells **1**. To cross these tissues, fluid must pass directly through the cells. Thus, transport proteins embedded in the cell membranes control which ions and molecules cross the tissue. For example, an abundance of tight junctions in the

adhering junction Cell junction that anchors cells to each other or to extracellular matrix.

cell junction Structure that connects a cell to another cell or to extracellular matrix.

extracellular matrix (ECM) Complex mixture of substances secreted by cells; supports cells and tissues; roles in cell signaling.

gap junction Cell junction that forms a channel across the plasma membranes of adjoining animal cells.

tight junctions Arrays of fibrous proteins; join epithelial cells and collectively prevent fluids from leaking between them.

lining of the stomach normally keeps acidic fluid from leaking out. If a bacterial infection damages this lining, acid and enzymes can erode the underlying layers. The result is a painful peptic ulcer. Figure 3.15 shows how tight junctions seal the linings of ducts and tubes in the kidney.

Adhering junctions composed of adhesion proteins snap cells to each other and anchor them to extracellular matrix ❷. Skin and other tissues that are subject to abrasion or stretching have a lot of adhering junctions. These cell junctions also strengthen contractile tissues such as heart muscle.

Gap junctions form channels that connect the cytoplasm of adjoining cells, thus permitting ions and small molecules to pass directly from the cytoplasm of one cell to another ❸. By opening or closing, they allow entire regions of cells to respond to a single stimulus. For example, heart muscle and other tissues in which the cells perform some coordinated action have many of these communication channels. A signal passes instantly from cell to cell through gap junctions, so all of the connected cells can respond as a unit.

In plants, open channels called plasmodesmata (singular, plasmodesma) extend across the primary wall of adjoining cells, connecting the cytoplasm of the cells. Substances such as water, nutrients, and signaling molecules can flow quickly from cell to cell through plasmodesmata.

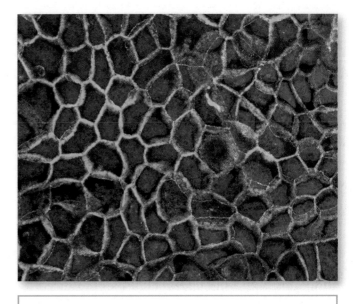

FIGURE 3.15 In this fluorescence micrograph, a continuous array of tight junctions (*green*) seals the abutting surfaces of kidney cell membranes. DNA is *red*. **Figure It Out:** Why does the DNA appear clumped in each cell?

Answer: In eukaryotic cells, DNA occurs in the nucleus.

Take-Home Message  What structures form on the outside of eukaryotic cells?

- Many cells secrete extracellular matrix that support and anchor them.

- Cells of many protists, nearly all fungi, and all plants, have a porous wall around the plasma membrane. Animal cells do not have walls.

- Via cell junctions, cells make structural and functional connections with one another and with extracellular matrix in tissues.

3.8 Impacts/Issues Revisited:
Food for Thought

The photo on the *left* shows *E. coli* O157:H7 bacteria (*red*) clustering on intestinal cells of a small child. This type of bacteria can cause a serious intestinal illness in people who eat foods contaminated with it. Meat, poultry, milk, and fruits that have been sterilized by exposure to radiation are now available in supermarkets. By law, irradiated foods must be marked with the symbol on the *right*. Items that bear this symbol have been exposed to radiation, but are not themselves radioactive. Irradiating fresh foods kills bacteria and prolongs shelf life. However, some worry that the irradiation process may alter the food and produce harmful chemicals. Whether health risks are associated with consuming irradiated foods is still unknown.

How Would YOU ☑ Vote? Some think the safest way to protect consumers from food poisoning is by exposing food to x-rays or other high-energy radiation, which kills bacteria. Others think we should tighten food safety standards instead. Would you choose irradiated food? **See CengageNow for details, then vote online (CengageNow.com).**

Summary

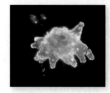

Sections 3.1–3.2 All organisms consist of one or more cells. By the cell theory, the cell is the smallest unit of life, and it is the basis of life's continuity. The surface-to-volume ratio limits cell size. All cells start out life with a plasma membrane, cytoplasm in which structures such as ribosomes are suspended, and DNA. The DNA of eukaryotic cells is contained in a nucleus; that of prokaryotes is not. The lipid bilayer is the foundation of all cell membranes.

CENGAGENOW Investigate the physical limits on cell size with the interactions on CengageNow.

Section 3.3 Most cells are too small to see with the naked eye. Different types of microscopes use light or electrons to reveal different details of cells.

Section 3.4 A cell membrane is a mosaic of lipids (mainly phospholipids) and proteins. It functions as a selectively permeable barrier that separates an internal environment from an external one. The lipids are organized as a double layer in which the nonpolar tails of both layers are sandwiched between the polar heads.

The membranes of most cells can be described as a fluid mosaic. Proteins that are temporarily or permanently associated with a membrane carry out most membrane functions. All membranes have transport proteins. Plasma membranes also incorporate receptor proteins, adhesion proteins, enzymes, and recognition proteins.

CENGAGENOW Use the animations on CengageNow to learn about membrane structure and receptor proteins.

Section 3.5 Bacteria and archaeans are the prokaryotes. Prokaryotes have no nucleus, but many have a cell wall and one or more flagella or pili. Biofilms are shared living arrangements among bacteria and other microbial organisms.

CENGAGENOW View prokaryotic cell structure with the animation on CengageNow.

Section 3.6 Eukaryotic cells start out life with a nucleus and other membrane-enclosed organelles. Pores, receptors, and transport proteins in the nuclear envelope control the movement of molecules into and out of the nucleus.

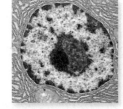

The endomembrane system includes rough and smooth endoplasmic reticulum, vesicles, and Golgi bodies. This set of organelles functions mainly to make and modify lipids and proteins; it also recycles molecules and particles such as worn-out cell parts, and inactivates toxins. Other eukaryotic organelles include mitochondria (which produce ATP), chloroplasts (which specialize in photosynthesis), peroxisomes, lysosomes, and vacuoles. Eukaryotic cells also have

a cytoskeleton that includes a mesh of microfilaments called the cell cortex. Motor proteins that are the basis of movement interact with microfilaments in pseudopods or microtubules in cilia and eukaryotic flagella.

CENGAGENOW Use the interaction and animations on CengageNow to survey the major types of eukaryotic organelles; learn more about cytoskeletal elements; view the nuclear envelope, endomembrane system, and a chloroplast; and study the structure of cell walls and junctions.

Section 3.7 Cells of most prokaryotes, protists, fungi, and all plant cells have a wall around the plasma membrane. Many eukaryotic cell types also secrete a cuticle. Cell junctions connect animal cells to one another and to extracellular matrix (ECM); plasmodesmata connect plant cells.

Self-Quiz   Answers in Appendix I

1. The _____ is the smallest unit of life.

2. Every cell is descended from another cell. This idea is called _____ .
   a. evolution
   b. the theory of relativity
   c. the cell theory
   d. cell biology

3. True or false? Some protists are prokaryotes.

4. Cell membranes consist mainly of a _____ .
   a. carbohydrate bilayer and proteins
   b. protein bilayer and phospholipids
   c. lipid bilayer and proteins

5. Unlike eukaryotic cells, prokaryotic cells _____ .
   a. have no plasma membrane   c. have no nucleus
   b. have RNA but not DNA   d. a and c

6. In a lipid bilayer, _____ of all the lipid molecules are sandwiched between all the _____ .
   a. hydrophilic tails; hydrophobic heads
   b. hydrophilic heads; hydrophilic tails
   c. hydrophobic tails; hydrophilic heads
   d. hydrophobic heads; hydrophilic tails

7. Enzymes contained in _____ break down worn-out organelles, bacteria, and other particles.

8. Put the following structures in order according to the pathway of a secreted protein:
   a. plasma membrane   c. endoplasmic reticulum
   b. Golgi bodies   d. post-Golgi vesicles

9. The main function of the endomembrane system is building and modifying _____ and _____ .

10. Is this statement true or false? The plasma membrane is the outermost component of all cells. Explain your answer.

Digging Into Data

Organelles and Cystic Fibrosis

CFTR is a transporter in the plasma membrane of epithelial cells. Sheets of these cells line the cavities and ducts of the lungs, liver, pancreas, intestines, reproductive system, and skin. The transporter pumps chloride ions out of these cells, and water follows the ions. A thin, watery film forms on the surface of the epithelial cell sheets. Mucus slides easily over the wet sheets of cells.

In some people, these epithelial cell membranes do not have enough working copies of the CFTR protein, and chloride ion transport is disrupted. Not enough chloride ions leave the cells, and so not enough water leaves them either. The result is thick, dry mucus that sticks to the epithelial cell sheets. In the respiratory tract, the mucus clogs airways to the lungs and makes breathing difficult. The mucus is too thick for the ciliated cells lining the airways to sweep out, and bacteria thrive in it. Low-grade infections occur and may persist for years.

These symptoms characterize cystic fibrosis (CF), the most common fatal genetic disorder in the United States. Even with a lung transplant, most CF patients live no longer than thirty years, at which time their lungs usually fail. There is no cure.

In most individuals with cystic fibrosis, the 508th amino acid of the CFTR protein (a phenylalanine) is missing (Figure 3.16**A**). A CFTR protein with this change is made correctly, and it can transport ions correctly, but it never reaches the plasma membrane to do its job. In 2000, Sergei Bannykh and his coworkers developed a way to measure the relative amounts of the CFTR protein localized in different areas of a cell. They compared the pattern of distribution of CFTR with and without the CF deletion (Figure 3.16**B**).

1. Which organelle contains the least amount of CFTR protein in normal cells? In cells with the deletion? Which contains the most?

2. In which organelle is the amount of CFTR protein most similar in both types of cells?

3. Where is the CFTR protein with the deletion getting held up?

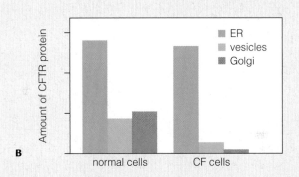

B

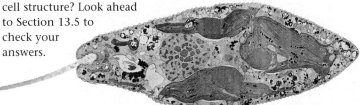
A

FIGURE 3.16 Changes in the CFTR protein affect intracellular transport. **A** Model of CFTR. The parts shown here are ATP-driven motors that widen or narrow a channel (*gray* arrow) across the plasma membrane. The tiny part of the protein that is missing in most people with cystic fibrosis is shown on the ribbon in *green*.

B Comparison of the amounts of CFTR protein associated with endoplasmic reticulum, vesicles traveling from ER to Golgi, and Golgi bodies. The patterns of CFTR distribution in normal cells, and cells with the deletion that causes cystic fibrosis, were compared.

---

11. Most membrane functions are carried out by _____ .
    a. proteins
    b. phospholipids
    c. nucleic acids
    d. hormones

12. No animal cell has a _____ .
    a. plasma membrane
    b. flagellum
    c. lysosome
    d. cell wall

13. _____ connect the cytoplasm of plant cells.
    a. Plasmodesmata
    b. Adhering junctions
    c. Tight junctions
    d. a and b

14. Match each cell component with its function.
    _____ mitochondrion
    _____ chloroplast
    _____ ribosome
    _____ smooth ER
    _____ Golgi body
    _____ rough ER

    a. protein synthesis
    b. associates with ribosomes
    c. ATP production
    d. sorts and ships
    e. assembles lipids; other tasks
    f. photosynthesis

Additional questions are available on CENGAGENOW™

Critical Thinking

1. In a classic episode of *Star Trek*, a titanic amoeba engulfs an entire starship. The crew of the ship blows the cell to bits before it reproduces. Think of at least one problem a biologist would have with this particular scenario.

2. A student is examining different samples with a transmission electron microscope. She discovers a single-celled organism (*below*) swimming in a freshwater pond. Which of this organism's structures can you identify? Is it a prokaryotic or eukaryotic cell? Can you be more specific about the type of cell based on what you know about cell structure? Look ahead to Section 13.5 to check your answers.

4 Energy and Metabolism

Ethanol is toxic: If you put more of it into your body than your enzymes can deal with, then you will die.

Impacts/Issues: A Toast to Alcohol Dehydrogenase

The next time someone asks you to have a drink, stop for a moment and think about the cells in your body that detoxify alcohol. It makes no difference whether you drink a bottle of beer, a glass of wine, or 1–1/2 ounces of vodka. Each holds the same amount of alcohol or, more precisely, ethanol. Ethanol molecules move quickly from the stomach and small intestine into the blood-stream. Almost all of the ethanol ends up in the liver, a large organ in the abdomen. The liver has impressive numbers of enzymes, and one of them, alcohol dehydrogenase, helps rid the body of ethanol and other toxins (Figure 4.1).

Ethanol is really hard on the liver. For one thing, breaking it down produces molecules that directly damage liver cells, so the more you drink, the fewer liver cells you have left to do the breaking down. Ethanol also interferes with normal processes of metabolism that keep the remaining liver cells—and the rest of the body—alive. For example, oxygen that would normally take part in breaking down fatty acids is diverted to breaking down ethanol, so fats tend to accumulate as large globules in the tissues of heavy drinkers. A common outcome of such processes is alcoholic hepatitis, a disease characterized by inflammation and destruction of liver tissue. Alcoholic cirrhosis, another possibility, leaves the liver permanently scarred. (The word cirrhosis is from the Greek word *kirros*, or orange-colored, after the abnormal skin color of people with the disease.)

Eventually, the liver of a heavy drinker just quits working, with dire health consequences. The liver is the largest gland in the human body, and it has many important functions. In addition to breaking down fats and toxins, it helps regulate the body's blood sugar level, and it makes blood proteins that are essential for blood clotting, immune function, and maintaining the solute balance of body fluids.

Binge drinking, a self-destructive behavior that involves consuming large amounts of alcohol in a brief period of time, is currently the most serious drug problem on college campuses in the United States. Tens of thousands of undergraduate students have been polled about their drinking habits in recent surveys. More than half of them reported that they regularly drink five or more alcoholic beverages within a two-hour period.

Binge drinking does far more than damage one's liver. Aside from the related 500,000 injuries from accidents, the 600,000 assaults by intoxicated students, 100,000 cases of date rape, and 400,000 incidences of (whoops) unprotected sex among students, binge drinking is responsible for killing or causing the death of more than 1,700 college students every year. Ethanol is toxic: If you put more of it into your body than your enzymes can deal with, then you will die. With this sobering example, we invite you to learn about how and why your cells break down organic compounds, including toxic molecules like ethanol.

FIGURE 4.1 Alcohol dehydrogenase helps the body break down ethanol and other toxic alcohols. This enzyme makes it possible for humans to drink beer, wine, and other alcoholic beverages.

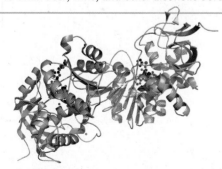

alcohol dehydrogenase

Life Runs on Energy

Energy, remember, is the capacity to do work, but this definition is not very satisfying. Even the brilliant physicists who study it cannot say what energy is, exactly. However, even without a perfect definition, we have an intuitive understanding of energy just by thinking about familiar forms of it, such as light, heat, electricity, and motion. We also understand intuitively that one form of energy can be converted to another. Think about how a lightbulb changes electricity into light, or how an automobile changes gasoline into motion.

Thermodynamics, which is the study of heat and other energy (*therm* is a Greek word for heat; *dynam* means energy), tells us that the total amount of energy before and after every conversion is always the same. In other words, energy cannot be created or destroyed—a phenomenon we call the **first law of thermodynamics**. Remember, a law of nature describes something that occurs without fail, but our explanation of why it occurs is incomplete (Section 1.6).

Thermodynamics also tells us that energy tends to disperse until no part of a system holds more than another part. For example, heat flows from a hot pan to air in a cool kitchen until the temperature of both is the same. We never see cool air raising the temperature of a hot pan. Again, this phenomenon is always true, but we do not know exactly why. Thus, we call the tendency of energy to disperse the **second law of thermodynamics**.

Work occurs as a result of an energy transfer. For example, making ATP (Section 2.10) is work, and it requires energy. A plant cell can do this work by transferring energy from the environment (light) to molecules that use the energy to build ATP. This particular transfer involves an energy conversion (light to chemical energy). Most cellular work occurs by transferring chemical energy from one molecule to another. For example, chemical energy transferred from ATP to other molecules is used for cellular work, such as making glucose.

Every time an energy transfer occurs, a bit of energy disperses, usually in the form of heat. For example, a typical incandescent lightbulb converts about 5 percent of the energy of electricity into light. The remaining 95 percent of the energy ends up as heat that radiates from the bulb. Dispersed heat is not very useful for doing work, and it is not easily converted to a more useful form of energy (such as electricity). Because some of the energy in every transfer disperses as heat, and heat is not useful for doing work, we can say that the total amount of energy available for doing work in the universe is always decreasing.

Is life an exception to this depressing flow? An organized body is hardly dispersed. Energy becomes concentrated in each new organism as the molecules of life organize into cells. Even so, living things constantly use energy to grow, to move, to acquire nutrients, to reproduce, and so on. Inevitable losses occur during the energy transfers that maintain life. Unless those losses are replenished with energy from another source, the complex organization of life will end.

Most of the energy that fuels life on Earth is energy from the sun. In our world, energy flows from the sun, through producers, then consumers (Figure 4.2). During this journey, the energy changes form and changes hands many times. Each time, some energy escapes as heat until, eventually, all of it is permanently dispersed. However, the second law of thermodynamics does not say how quickly the dispersal has to happen. Energy's spontaneous dispersal

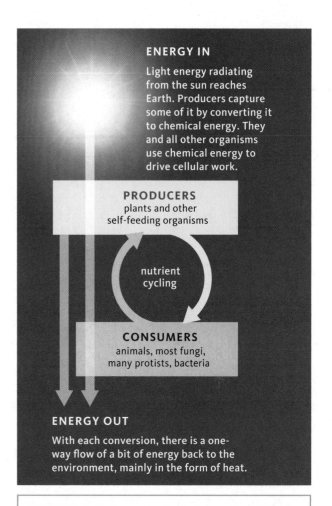

ENERGY IN

Light energy radiating from the sun reaches Earth. Producers capture some of it by converting it to chemical energy. They and all other organisms use chemical energy to drive cellular work.

PRODUCERS
plants and other self-feeding organisms

nutrient cycling

CONSUMERS
animals, most fungi, many protists, bacteria

ENERGY OUT

With each conversion, there is a one-way flow of a bit of energy back to the environment, mainly in the form of heat.

FIGURE 4.2 A one-way flow of energy into living organisms compensates for a one-way flow of energy out of them. Energy inputs drive a cycling of materials among producers and consumers.

energy The capacity to do work.
first law of thermodynamics Energy cannot be created or destroyed.
second law of thermodynamics Energy tends to disperse.

is resisted by chemical bonds. Think of all the bonds in the countless molecules that make up your skin, heart, liver, fluids, and other body parts. Those bonds hold the molecules, and you, together—at least for the time being.

Take-Home Message What is energy?

- Energy is the capacity to do work. It can be converted from one form to another, but it cannot be created or destroyed. It disperses spontaneously.

- Organisms can maintain their complex organization only as long as they replenish themselves with energy they harvest from someplace else.

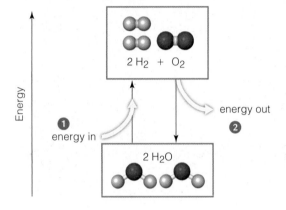

FIGURE 4.3 Energy inputs and outputs in chemical reactions.

❶ Some reactions convert molecules with lower energy to molecules with higher energy, so they require a net energy input to proceed.

❷ Other reactions convert molecules with higher energy to molecules with lower energy, so they end with a net energy output.

Figure It Out: Which law of thermodynamics explains energy inputs and outputs in chemical reactions?

Answer: The first law

activation energy Minimum amount of energy required to start a reaction.
ATP Adenosine triphosphate; the main energy carrier between reaction sites in cells.
phosphorylation Phosphate-group transfer.
product A molecule remaining at the end of a reaction.
reactant Molecule that enters a reaction.
reaction Process of chemical change.

4.3 Energy in the Molecules of Life

Cells store energy in chemical bonds, and access the energy stored in chemical bonds by breaking them. Both processes change molecules or ions into other molecules or ions (Section 2.3). A process by which a chemical change occurs is called a **reaction**. During a reaction, one or more **reactants** (molecules that enter a reaction) become one or more **products** (molecules that remain at the reaction's end). Intermediate molecules may form between reactants and products. We show a chemical reaction as an equation in which an arrow points from reactants to products:

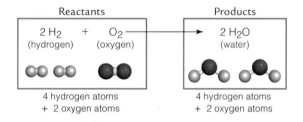

A number before a formula indicates the number of molecules. You can see that atoms shuffle around, but they never disappear: The same number of atoms that enter a reaction remain at the reaction's end.

Every chemical bond holds energy, and the amount of energy depends on which elements are taking part in the bond. For example, the covalent bond between an oxygen and a hydrogen atom in any water molecule holds a certain amount of energy, which differs from the amount of energy held by a covalent bond between two oxygen atoms in molecular oxygen (O_2). Thus, in most reactions, the energy of the reactants differs from the energy of the products (Figure 4.3). Reactions in which the reactants have less energy than the products require a net (or overall) energy input to proceed ❶.

Cells store energy in chemical bonds by running energy-requiring reactions. For example, light energy drives the overall reactions of photosynthesis, which convert carbon dioxide and water to glucose and oxygen. Unlike light, glucose can be stored in a cell. In other reactions, the reactants have greater energy than the products. Such reactions release energy ❷. Cells access the energy held in chemical bonds by running energy-releasing reactions. An example is the overall process of aerobic respiration, which releases the energy of glucose by breaking the bonds between its carbons.

Why the World Does Not Go Up in Flames The molecules of life release energy when they combine with oxygen. For example, think of how a spark ignites wood. Wood is mostly cellulose, a carbohydrate with long chains of repeating glucose units (Section 2.7). A spark starts a reaction that converts cellulose and oxygen to water and carbon dioxide. This reaction releases enough energy to start the same reaction with other cellulose and oxygen molecules. That is why wood keeps burning once it has been lit.

Earth is rich in oxygen—and in potential energy-releasing reactions. Why doesn't it burst into flames? Luckily, it takes energy to break the chemical bonds of reactants, even in an energy-releasing reaction. **Activation energy** is the minimum amount of energy that will get a chemical reaction started, a bit like a hill that reactants must climb before they can run down the other side to products (Figure 4.4).

Activation energy varies by the reaction. For example, guncotton, or nitrocellulose, is a highly explosive derivative of cellulose. Christian Schönbein accidentally discovered a way to make it when he used his wife's cotton apron to wipe up a nitric acid spill on his kitchen table, then hung it up to dry next to the oven. The apron promptly exploded. Being a chemist in the 1800s, Schönbein immediately made plans to market guncotton as a firearm explosive, but it turned out to be too unstable. So little activation energy is needed to make guncotton react with oxygen that it explodes spontaneously. After a few manufacturing plants burned down, guncotton was abandoned in favor of gunpowder, which has a higher activation energy for a reaction with oxygen.

ATP—The Cell's Energy Currency Cells pair reactions that require energy with reactions that release energy. ATP is part of that process for many reactions in cells. **ATP**, or adenosine triphosphate, is an energy carrier: It accepts energy released by energy-releasing reactions, and delivers energy to energy-requiring reactions. ATP is the main currency in a cell's energy economy, so we use a cartoon coin to symbolize it.

ATP is a nucleotide with three phosphate groups (Figure 4.5). The bonds that link those phosphate groups hold a lot of energy. When a phosphate group is transferred from ATP to another molecule, energy is transferred along with it. That energy contributes to the "energy in" part of an energy-requiring reaction. A phosphate-group transfer is called **phosphorylation**.

Cells constantly use up ATP to drive energy-requiring reactions, so they constantly replenish it. When ATP loses a phosphate, ADP (adenosine diphosphate) forms. ATP forms again when ADP binds phosphate:

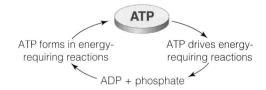

ATP forms in energy-requiring reactions

ATP drives energy-requiring reactions

ADP + phosphate

Take-Home Message **How do cells use energy?**

- Cells store and retrieve energy by making and breaking chemical bonds.
- Some reactions require a net input of energy. Others end with a net release of energy.
- ATP and other energy carriers couple energy-releasing reactions with energy-requiring ones.

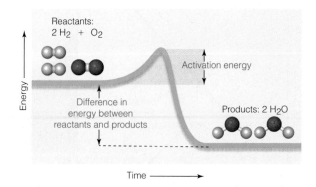

FIGURE 4.4 Animated! Activation energy. Most reactions will not proceed without an input of activation energy, which is shown here as a bump in an energy hill. In this example, the reactants have more energy than the products. Activation energy keeps such energy-releasing reactions from running spontaneously.

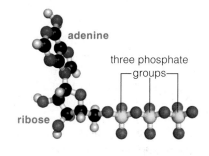

A Structure of ATP.

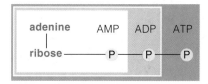

B After it loses one phosphate group, the nucleotide is ADP (adenosine diphosphate); after losing two phosphate groups, it is AMP (adenosine monophosphate).

FIGURE 4.5 Animated! ATP, the energy currency of all cells.

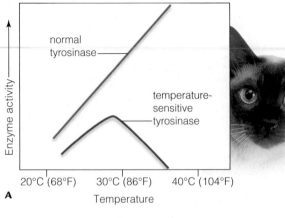

A

Temperature (x-axis): 20°C (68°F), 30°C (86°F), 40°C (104°F)
Enzyme activity (y-axis)

Labels: normal tyrosinase; temperature-sensitive tyrosinase

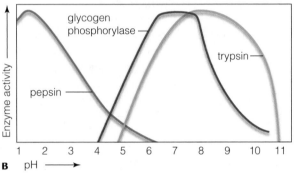

B

Enzyme activity (y-axis)
pH (x-axis): 1 2 3 4 5 6 7 8 9 10 11

Labels: glycogen phosphorylase; trypsin; pepsin

FIGURE 4.6 Enzymes, temperature, and pH.
A Tyrosinase is involved in the production of melanin, a black pigment in skin cells. The form of this enzyme in Siamese cats is inactive above about 30°C, so the warmer parts of the cat's body end up with less melanin, and lighter fur. **B** How pH values affect the activity of three enzymes in the human body.

active site Pocket in an enzyme where substrates bind and a reaction occurs.

allosteric Describes a region of an enzyme other than the active site that can bind regulatory molecules.

coenzyme An organic molecule that is a cofactor.

cofactor A metal ion or a coenzyme that associates with an enzyme and is necessary for its function.

enzyme Protein or RNA that speeds a reaction without being changed by it.

feedback inhibition Mechanism by which a change that results from some activity decreases or stops the activity.

metabolic pathway Series of enzyme-mediated reactions by which cells build, remodel, or break down an organic molecule.

substrate A reactant molecule that is specifically acted upon by an enzyme.

How Enzymes Work

Centuries might pass before sugar would break down to carbon dioxide and water on its own, yet that same conversion takes just moments inside a cell. Enzymes make the difference. **Enzymes** are molecules that make chemical reactions proceed much faster than they would on their own. Most are proteins, but some are RNAs. An enzyme is not consumed or changed by participating in a reaction; it can work again and again. Each kind recognizes and alters specific reactants called **substrates**. For instance, alcohol dehydrogenase removes two hydrogen atoms from an ethanol molecule (the enzyme's substrate), which results in a molecule of acetaldehyde.

Enzymes work because they have one or more **active sites**, which are pockets where substrates bind and where reactions proceed:

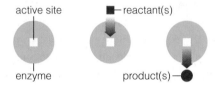

active site · reactant(s) · enzyme · product(s)

An active site is complementary in shape, size, polarity, and charge to the enzyme's substrate. That fit is the reason why each enzyme acts only on specific substrates. An enzyme's active site squeezes substrates close together, redistributes their charge, or causes some other change. The change reduces activation energy, so it lowers the barrier that prevents the reaction from proceeding.

Factors That Influence Enzyme Activity Adding heat boosts the energy of a system, which is one reason why molecular motion increases with temperature (Section 2.4). The greater the energy of reactants, the closer a reaction is to its activation energy. Thus, the rate of an enzymatic reaction typically increases with temperature, but only up to a point. Above a certain temperature, the hydrogen bonds that hold an enzyme in its characteristic shape break (Section 2.9). The enzyme stops working when it denatures, so the reaction rate falls sharply (Figure 4.6**A**). Body temperatures above 42°C (107.6°F) adversely affect many of your enzymes, which is why severe fevers are dangerous.

The pH tolerance of enzymes varies. Most enzymes in the human body work best at pH 6–8, but some work outside this typical range (Figure 4.6**B**). For example, pepsin functions only in stomach fluid, where it breaks down proteins in food. The fluid is very acidic, with a pH of about 2. An enzyme's activity is also influenced by the amount of salt in the surrounding fluid, because salt influences the hydrogen bonds that hold an enzyme in its three-dimensional shape.

Cofactors are atoms or molecules (other than proteins) that associate with enzymes and are necessary for their function. Some are metal ions. Organic molecules that are cofactors are called **coenzymes**. Some coenzymes must be tightly bound to an enzyme. Others, such as NAD^+ (nicotine adenine dinucleotide), participate in enzymatic reactions as separate molecules. Unlike enzymes, many coenzymes are modified by taking part in a reaction. They are typically regenerated in other reactions.

Organized, Enzyme-Mediated Reactions Metabolism, remember, refers to the enzyme-mediated chemical reactions by which cells acquire and use energy as they build and break down organic molecules (Section 2.6). Any series of enzyme-mediated reactions by which a cell builds, rearranges, or breaks down an organic substance is called a **metabolic pathway**. Some metabolic pathways build large molecules from smaller ones; others break down large molecules into smaller ones. A pathway may be linear, meaning that the reactions occur in a straight line from reactant to product:

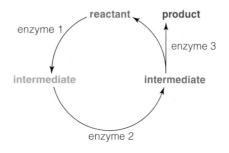

Other pathways are cyclic. In a cyclic pathway, the last step regenerates a reactant for the first step (*right*).

Cells conserve energy and resources by making what they need—no more, no less—at any given moment. How does a cell adjust the types and amounts of molecules it produces? Several mechanisms help a cell maintain, raise, or lower its production of thousands of different substances. For example, reactions do not only run from reactants to products. Many also run in reverse at the same time, with some of the products being converted back to reactants. The rates of the forward and reverse reactions often depend on the concentrations of reactants and products: A high concentration of reactants pushes the reaction in the forward direction, and a high concentration of products pushes it in the reverse direction.

Other mechanisms more actively regulate enzymes. Certain molecules in a cell influence how fast enzyme molecules are made, or influence the activity of enzymes that have already been built. For example, the end product of a series of enzymatic reactions may inhibit one of the enzymes in the series, an effect called **feedback inhibition** (Figure 4.7).

Some regulatory molecules activate or inhibit an enzyme by binding directly to its active site. Others bind to **allosteric** sites, which are regions of an enzyme (other than the active site) where regulatory molecules bind. *Allo*– means other, and *steric* means structure. Binding of an allosteric regulator alters the shape of the enzyme in a way that enhances or inhibits its function (Figure 4.8).

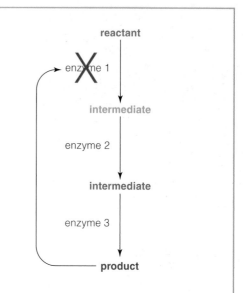

FIGURE 4.7 Animated! Feedback inhibition. In this example, three kinds of enzymes act in sequence to convert a substrate to a product, which inhibits the activity of the first enzyme.

Figure It Out: Is this an example of a cyclic or a linear metabolic pathway?

Answer: Linear

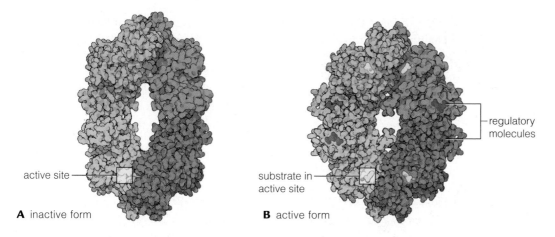

A inactive form   **B** active form

FIGURE 4.8 Animated! Allosteric effects. **A** Pyruvate kinase is an enzyme that consists of four identical polypeptide chains. Each chain has an active site and a binding site for a regulatory molecule. **B** When regulatory molecules (*magenta*) bind to the allosteric sites, the overall shape of the enzyme changes. The change makes the enzyme functional. Here, substrate molecules (*yellow*) and metal ion cofactors (*green*) are bound to the active sites.

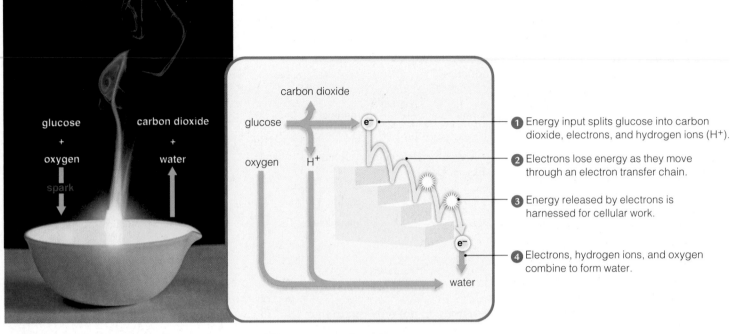

A Glucose and oxygen react (burn) when exposed to a spark. Energy is released all at once as light and heat when CO_2 and water form.

B The same overall reaction occurs in small steps with an electron transfer chain. Energy is released in amounts that cells can harness for cellular work.

1 Energy input splits glucose into carbon dioxide, electrons, and hydrogen ions (H^+).

2 Electrons lose energy as they move through an electron transfer chain.

3 Energy released by electrons is harnessed for cellular work.

4 Electrons, hydrogen ions, and oxygen combine to form water.

FIGURE 4.9 Animated! Comparing uncontrolled and controlled energy release.

Electron Transfers As you learned in Section 4.3, the bonds of organic molecules hold a lot of energy that can be released by combustion. Burning an organic substance in the presence of oxygen releases its energy all at once, explosively (Figure 4.9**A**). Cells use oxygen to break the bonds of organic molecules. However, they have no way to harvest an explosive burst of energy, so they break the molecules apart in several steps that release energy in small, manageable increments. Most of these steps are reactions in which a molecule accepts electrons from another molecule.

Coenzymes are among the many types of molecules that accept and donate electrons in electron transfer reactions. In the next chapter, you will learn about the importance of these reactions in electron transfer chains. An **electron transfer chain** is an organized series of reaction steps in which membrane-bound arrays of enzymes and other molecules give up and accept electrons in turn. Electrons are at a higher energy level (Section 2.2) when they enter a chain than when they leave. Electron transfer chains can harvest the energy given off by an electron as it drops to a lower energy level (Figure 4.9**B**).

concentration The number of molecules or ions per unit volume of a fluid.

concentration gradient Difference in concentration between adjoining regions of fluid.

diffusion Net movement of molecules or ions from a region where they are more concentrated to a region where they are less so.

electron transfer chain Array of enzymes and other molecules that accept and give up electrons in sequence, thus releasing the energy of the electrons in usable increments.

selective permeability Membrane property that allows some substances, but not others, to cross.

Take-Home Message What is a metabolic pathway?

■ Enzymes greatly enhance the rate of specific reactions. An enzyme's particular substrates bind at its active site, and it works best within certain ranges of temperature, pH, and salt concentration.

■ Control mechanisms enhance or inhibit the activity of many enzymes. The adjustments help cells produce only what they require in any given interval.

■ A metabolic pathway is a sequence of enzyme-mediated reactions that builds, breaks down, or remodels organic molecules.

■ Many metabolic pathways involve electron transfers.

Movement of Ions and Molecules

Metabolism only works because cells maintain the composition of their cytoplasm and internal fluids within certain ranges. If the composition of those fluids varied a lot, say in pH or the amount of salt, enzymes in the cell would stop working, and so would the cell.

Cells thrive even in environments that offer extreme conditions such as low pH and high salt. How does a cell keep its internal composition stable when that composition differs—often dramatically—from conditions outside of the cell? The answer begins with a property of membranes called **selective permeability**. Selectively permeable membranes allow some substances but not others to cross. This property helps a cell or membrane-enclosed organelle to control which substances and how much of each enter and leave it (Figure 4.10).

Such control is vital, because metabolism depends on the capacity to increase, decrease, and maintain concentrations of specific solutes. That capacity allows a cell or organelle to keep itself stocked with raw materials for reactions, to remove wastes, and to maintain internal fluid volume and pH within tolerable ranges.

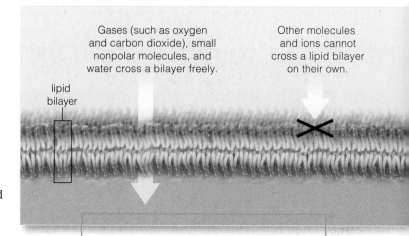

Gases (such as oxygen and carbon dioxide), small nonpolar molecules, and water cross a bilayer freely.

Other molecules and ions cannot cross a lipid bilayer on their own.

lipid bilayer

FIGURE 4.10 Animated! Selective permeability of cell membranes. Small, nonpolar molecules, gases, and water molecules freely cross the lipid bilayer. Polar molecules and ions only cross with the help of proteins that span the bilayer.

Diffusion Selective permeability begins with the behavior of fluids—specifically, the behavior of water and substances dissolved in it. A dissolved substance, remember, is a solute (Section 2.4). How much of it is dissolved in a given amount of fluid is the solute's **concentration**. A difference in solute concentration between two adjacent regions is called a **concentration gradient**. Solute molecules or ions tend to move "down" their concentration gradient, from a region of higher concentration to one of lower concentration. Why? Molecules and ions are always in motion; they collide at random and bounce off one another millions of times each second. The more crowded they are, the more often they collide. Thus, during any interval, more of them get knocked out of a region of higher concentration than get knocked into it.

Diffusion is the net (or overall) movement of molecules or ions in response to a concentration gradient. It is an essential way in which substances move into, through, and out of cells. In multicelled species, diffusion also moves substances between cells in different regions of the body or between cells and extracellular fluids. Any substance tends to diffuse in a direction set by its own concentration gradient, not by the gradients of any other solutes that may be sharing the same space in the solution. How quickly a particular solute diffuses through a particular solution depends on five factors:

1. Size. It takes less energy to move a smaller molecule, so smaller molecules diffuse more quickly than larger ones.

2. Temperature. Molecules move faster at higher temperature, so they collide more often. Thus, diffusion is faster at higher temperature.

3. Steepness of the concentration gradient. The rate of diffusion is higher with steeper gradients, because molecules collide more often in a region of greater concentration.

4. Charge. Opposite charges attract, and like charges repel. An overall charge can affect the rate and direction of diffusion of ions between two regions.

5. Pressure. Pressure squeezes molecules together, and molecules that are more crowded collide and rebound more frequently. Thus, diffusion occurs faster at higher pressures.

Osmosis and Tonicity Like any other substance, water molecules tend to diffuse in response to their own concentration gradient. How can water be more or less concentrated? Think of water's concentration in terms of relative numbers of water molecules and solute molecules in a solution. The concentration of water depends on the total number of molecules or ions dissolved in it. The higher the solute concentration, the lower the water concentration. For example, if you pour some sugar into a container that is partially filled with water, you increase the total volume of liquid. The number of water molecules is unchanged, but they are now dispersed in a larger total volume. As a result of the added solute (in this case, the sugar molecules), the number of water molecules per unit volume—the water concentration—has decreased:

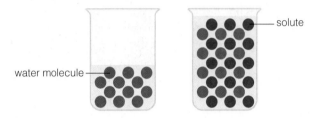

When water diffuses across a selectively permeable membrane, the diffusion is called **osmosis**. Turgor counters osmosis. **Turgor** is pressure that a volume of fluid exerts against a cell wall, membrane, tube, or other structure that holds it. The amount of turgor that can stop water from diffusing into cytoplasmic fluid or another solution is called **osmotic pressure**.

Think about growing plant cells. Cytoplasm typically contains more solutes than soil water does, so water tends to diffuse from soil water into these cells. The osmosis causes fluid pressure—turgor—to build up inside the cells. The turgor pushes against the cell walls from the inside, so it keeps the cells plump and the plant erect, just as high air pressure in a tire keeps it inflated. If the soil dries out, it loses water but not solutes, so its concentration of solutes increases. If the solutes in soil water become more concentrated than in cytoplasmic fluid, water will tend to diffuse out of the plant's cells and turgor in them will decrease. The plant wilts as the cytoplasm in its cells shrinks.

Tonicity refers to the relative concentrations of solutes in two fluids separated by a selectively permeable membrane. When the solute concentrations differ, the fluid with the lower concentration of solutes is said to be **hypotonic** (*hypo*–, under). The other one, with the higher solute concentration, is **hypertonic** (*hyper*–, over). Fluids that are **isotonic** have the same solute concentration.

Water tends to follow its gradient and diffuse from a hypotonic to a hypertonic fluid (Figures 4.11 and 4.12**A**). If a selectively permeable membrane separates two fluids that are not isotonic, water will cross the membrane, and move from the hypotonic fluid into the hypertonic one ❶. Osmosis will continue until the two fluids are isotonic ❷, or until some pressure against the hypertonic fluid counters the movement.

Cell membranes are selectively permeable, and water crosses them freely. Thus, if a cell's cytoplasm is hypertonic with respect to the fluid outside of its plasma membrane, water will diffuse into it. If its cytoplasm is hypotonic with respect to the fluid on the outside, water will diffuse out of it.

Most free-living cells have built-in mechanisms that compensate for differences in tonicity between cytoplasm and external fluid. Such cells can maintain their cytoplasm at a constant volume by pumping water one way or the other across their plasma membrane. Most cells of multicelled species have no such

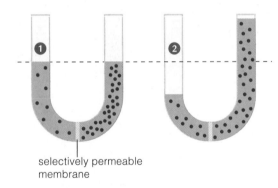

selectively permeable
membrane

FIGURE 4.11 Animated! Osmosis. Water moves across a selectively permeable membrane that separates two fluids of differing solute concentration.

❶ Initially, the volume of fluid is the same in the two compartments, but the solute concentration differs.

❷ The fluid volume in the two compartments changes as water follows its gradient and diffuses across the membrane.

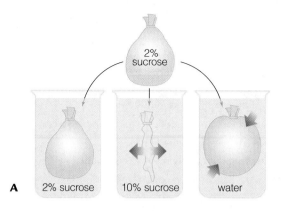

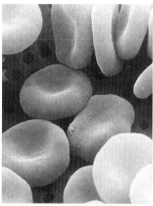

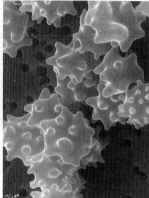

FIGURE 4.12 Animated! Tonicity.

A Tonicity experiment that shows what happens when a selectively permeable membrane bag is immersed in solutions of different tonicity.

B–D The micrographs show human red blood cells immersed in fluids of different tonicity.

Figure It Out: Which of the three solutions in **A** is hypertonic with respect to the fluid in the bag?

Answer: The 10% sucrose solution

B Red blood cells immersed in an isotonic solution do not change in volume. The fluid portion of blood is typically isotonic with cytoplasm.

C Red blood cells immersed in a hypertonic solution shrivel up because more water diffuses out of the cells than into them.

D Red blood cells immersed in a hypotonic solution swell up because more water diffuses into the cells than out of them.

mechanism, and their volume will change if the external fluid is not isotonic with the cytoplasm. Thus, in multicelled organisms, maintaining the tonicity of extracellular fluids is a critical part of homeostasis. Normally, body fluids are isotonic with fluid inside cells (Figure 4.12**B**). If extracellular fluid were to become hypertonic, the cells would lose water from their cytoplasm, and they would shrivel (Figure 4.12**C**). If the fluid were to become hypotonic, too much water would diffuse into the cells, and they would swell and burst (Figure 4.12**D**).

Take-Home Message **What influences the movement of ions and molecules?**

- Molecules or ions tend to diffuse into an adjoining region of fluid in which they are not as concentrated.
- Osmosis is a net diffusion of water between two fluids that differ in water concentration and are separated by a selectively permeable membrane.

4.6 Membrane-Crossing Mechanisms

Molecules that can cross a membrane on their own do so by diffusing directly across the lipid bilayer. Other molecules can cross only with the assistance of transport proteins (Section 3.4). Each type of transport protein can move a specific ion or molecule across a membrane. Glucose transporters only transport glucose; calcium pumps only pump calcium; and so on. The specificity of transport protein binding means that the amounts and types of substances that cross a membrane depend on which transport proteins are embedded in it.

hypertonic Describes a fluid with a high solute concentration relative to another fluid.

hypotonic Describes a fluid with a low solute concentration relative to another fluid.

isotonic Describes a fluid with the same solute concentration relative to another fluid.

osmosis The diffusion of water across a selectively permeable membrane.

osmotic pressure Amount of turgor that prevents osmosis into cytoplasm or other hypertonic fluid.

turgor Pressure that a fluid exerts against a wall, membrane, or some other structure that contains it.

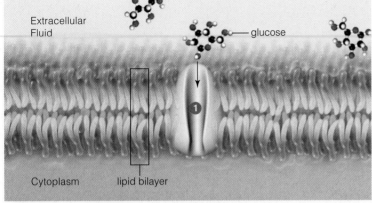

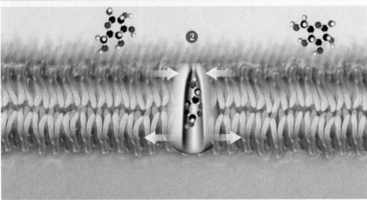

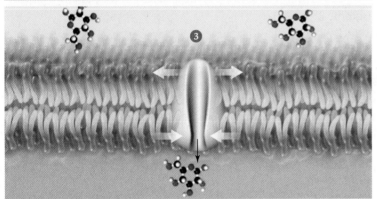

Extracellular Fluid

glucose

Cytoplasm   lipid bilayer

FIGURE 4.13 Animated! Passive transport. This model shows a glucose transporter. Glucose molecules can diffuse unassisted across a lipid bilayer, but the transporter increases the rate of diffusion by about 50,000 times.

❶ A glucose molecule (here, in extracellular fluid) binds to a glucose transporter in the plasma membrane.

❷ Binding causes the transport protein to change shape.

❸ The glucose molecule detaches from the transport protein on the other side of the membrane (here, in cytoplasm), and the protein resumes its original shape.

Figure It Out: Which fluid is hypotonic with respect to the other: the extracellular fluid or the cytoplasm?

Answer: The cytoplasm

For example, a glucose transporter in a plasma membrane can bind to a molecule of glucose, but not to a molecule of phosphorylated glucose. Enzymes in cytoplasm phosphorylate glucose as soon as it enters the cell. Phosphorylation prevents the molecule from moving back through the glucose transporter and leaving the cell.

Passive Transport In **passive transport**, the movement of a solute (and the direction of the movement) through a transport protein is driven entirely by the solute's concentration gradient. For this reason, passive transport is also called facilitated diffusion. The solute simply binds to the passive transport protein, and the protein releases it to the other side of the membrane (Figure 4.13).

A glucose transporter is an example of a passive transport protein ❶. This protein changes shape when it binds to a molecule of glucose ❷. The shape change moves the solute to the opposite side of the membrane, where it detaches. Then, the transporter reverts to its original shape ❸. Some passive transporters do not change shape; they form permanently open channels through a membrane. Others are gated, which means they open and close in response to a stimulus such as a shift in electric charge or binding to a signaling molecule.

Active Transport Solute concentrations shift constantly in the cytoplasm and in extracellular fluid. Maintaining a solute's concentration at a certain level often means transporting the solute against its gradient, to the side of a membrane where it is more concentrated. Pumping a solute against its gradient takes energy. In **active transport**, a transport protein uses energy to pump a solute against its gradient across a cell membrane. After a solute binds to an active transporter, an energy input (often in the form of a phosphate-group transfer from ATP) changes the shape of the protein. The change causes the transporter to release the solute to the other side of the membrane.

A calcium pump is an example of an active transporter. This protein moves calcium ions across muscle cell membranes (Figure 4.14). Muscle cells contract when the nervous system causes calcium ions to flood out from a special organelle, the sarcoplasmic reticulum, which is wrapped around the muscle fiber. The

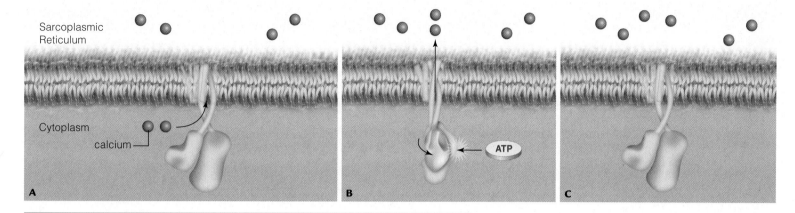

Sarcoplasmic
Reticulum

Cytoplasm

calcium

A

B

ATP

C

FIGURE 4.14 Animated! Active transport. This model shows a calcium transporter embedded in a muscle cell membrane. **A** Two calcium ions bind to the transport protein. **B** Energy in the form of a phosphate group is transferred from ATP to the protein. The transfer causes the protein to change shape so that it ejects the calcium ions to the opposite side of the membrane. **C** After it loses the calcium ions, the transport protein resumes its original shape.

flood clears out binding sites on motor proteins (Section 3.6) that make muscles contract. Contraction ends after calcium pumps have moved most of the calcium ions back into the sarcoplasmic reticulum, against their concentration gradient. Calcium pumps keep the concentration of calcium in that compartment 1,000 to 10,000 times higher than it is in muscle cell cytoplasm.

Cotransporters are active transport proteins that move two substances at the same time, in opposite directions across a membrane. Nearly all of the cells in your body have cotransporters called sodium–potassium pumps (Figure 4.15). Sodium ions (Na^+) in the cytoplasm diffuse into the pump's open channel and bind to its interior. A phosphate-group transfer from ATP causes the pump to change shape. Its channel opens to extracellular fluid, where it releases the Na^+. Then, potassium ions (K^+) from extracellular fluid diffuse into the channel and bind to its interior. The transporter releases the phosphate group and reverts to its original shape. The channel opens to the cytoplasm, where it releases the K^+.

Bear in mind, the membranes of all cells, not just those of animals, have active transporters. For example, active transporters in plant leaf cells pump sugars into tubes that distribute them throughout the plant body.

active transport Energy-requiring mechanism by which a transport protein pumps a solute across a cell membrane against its concentration gradient.

passive transport Mechanism by which a concentration gradient drives the movement of a solute across a cell membrane through a transport protein. Requires no energy input.

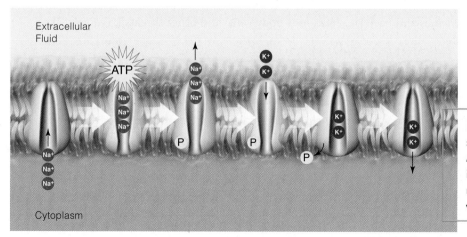

Extracellular
Fluid

ATP

Na+ Na+ Na+
Na+ Na+ Na+
Na+ Na+
Na+
Na+ Na+
P

K+
K+

P

K+
K+
P

K+
K+

Cytoplasm

FIGURE 4.15 Cotransport. This model shows how a sodium–potassium pump transports sodium ions (Na^+, *red*) from cytoplasm to extracellular fluid, and potassium ions (K^+, *purple*) in the other direction across the plasma membrane. A phosphate-group transfer from ATP provides energy for the transport.

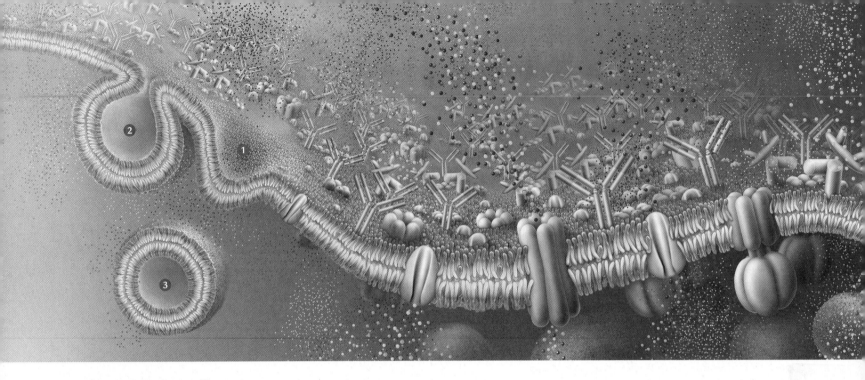

FIGURE 4.16 Animated! Membrane crossings.
A plasma membrane is a hub of activity: Molecules and
ions (colored balls) are constantly flowing into and out
of a cell via transport proteins embedded in its plasma
membrane. Vesicles are also taking in or expelling bulk
amounts of solutes and much larger particles.

1 By endocytosis, a small patch of plasma membrane
sinks inward.

2 As the membrane balloons into the cell, it wraps
itself around a small volume of extracellular fluid,
along with the solutes or particles it contains.

3 The balloon pinches off inside the cell as a vesicle,
which may deliver its contents to an organelle.

**FIGURE 4.17
Animated!**
Phagocytosis. Micro-
graph of a phagocytic
white blood cell (*top*)
engulfing a micro-
organism (*bottom*).

endocytosis Process by which a cell takes in a small
amount of extracellular fluid by the ballooning inward of its
plasma membrane.
exocytosis Process by which a cell expels a vesicle's contents to extracellular fluid.
phagocytosis Endocytic pathway by which a cell engulfs
particles such as microbes or cellular debris.

Membrane Trafficking When a lipid bilayer is disrupted, it seals itself,
because the disruption exposes the fatty acid tails of the phospholipids to their
watery surroundings. In watery fluids, phospholipids spontaneously rearrange
themselves so their nonpolar tails are together. A vesicle forms when a patch of
membrane bulges into the cytoplasm because the phospholipid tails of lipids in
the bilayer are repelled by water on both sides. The water "pushes" the phospholipid tails together, which helps round off the bulge as a vesicle, and also seals
the rupture in the membrane.

As part of vesicles, patches of membrane constantly move to and from
the cell surface (Figure 4.16). In a process called **endocytosis**, a cell takes up
substances near its outer surface. A small patch of plasma membrane balloons
inward, taking a small volume of extracellular fluid with it **1**. The balloon sinks
farther into the cytoplasm **2**, and then pinches off as a vesicle **3**. The vesicle
delivers its contents to an organelle or stores them in a cytoplasmic region. In
exocytosis, a vesicle moves to the cell surface, and its membrane fuses with
the plasma membrane. As the exocytic vesicle loses its identity, its contents are
released to the surroundings.

Phagocytosis ("cell eating") is an endocytic pathway. Phagocytic cells such
as amoebas engulf microorganisms, cellular debris, or other particles. Macrophages and other phagocytic white blood cells inside animal bodies engulf
viruses and bacteria, cancerous body cells, and other threats (Figure 4.17).

*Take-Home
Message* **How do molecules or ions that cannot diffuse through
a lipid bilayer cross a cell membrane?**

■ Transport proteins help specific molecules or ions to cross cell membranes.

■ In passive transport, a solute binds to a protein that releases it on the opposite
side of the membrane. The movement is driven by a concentration gradient.

■ In active transport, a protein pumps a solute across a membrane, against its
concentration gradient. The transporter requires an energy input, as from ATP.

■ Exocytosis, endocytosis, and phagocytosis move materials in bulk across plasma
membranes.

Impacts/Issues Revisited:
A Toast to Alcohol Dehydrogenase

In the human body, alcohol dehydrogenase (ADH) converts ethanol to acetaldehyde, an organic molecule even more toxic than ethanol and the most likely source of various hangover symptoms. A different enzyme, ALDH, very quickly converts acetaldehyde to nontoxic acetate. Thus, the overall pathway of ethanol metabolism in humans is:

$$\text{ethanol} \xrightarrow[\substack{NAD^+ \quad NADH}]{ADH} \text{acetaldehyde} \xrightarrow[\substack{NAD^+ \quad NADH}]{ALDH} \text{acetate}$$

In the average adult human body, this metabolic pathway can detoxify between 7 and 14 grams of ethanol per hour. The average alcoholic beverage contains between 10 and 20 grams of ethanol, which is why having more than one drink in any two-hour interval may result in a hangover.

Alcohol dehydrogenase detoxifies the tiny quantities of alcohols that form in some metabolic pathways. In animals, the enzyme also detoxifies alcohols made by gut-inhabiting bacteria, and those in foods such as ripe fruit.

Defects in ADH or ALDH can affect alcohol metabolism. For example, if ADH is overactive, acetaldehyde accumulates faster than ALDH can detoxify it:

$$\text{ethanol} \xrightarrow{ADH} \substack{\text{acetaldehyde} \\ \text{acetaldehyde} \\ \text{acetaldehyde}} \xrightarrow{ALDH} \text{acetate}$$

People with an overactive form of ADH become flushed and feel ill after drinking even a small amount of alcohol. The unpleasant experience may be part of the reason that these people are less likely to become alcoholic than others.

An underactive form of ALDH also causes acetaldehyde to accumulate:

$$\text{ethanol} \xrightarrow{ADH} \substack{\text{acetaldehyde} \\ \text{acetaldehyde} \\ \text{acetaldehyde}} \xrightarrow{\;\;X\;\;} \text{acetate}$$

Underactive ALDH is associated with the same effect—and the same protection from alcoholism—as overactive ADH. Both types of variant enzymes are common in people of Asian descent. For this reason, the alcohol flushing reaction is often called "Asian flush."

Having an underactive ADH enzyme has the opposite effect. It results in slowed alcohol metabolism, so people with an underactive ADH may not feel the ill effects of drinking alcoholic beverages as much as other people do. When these people drink alcohol, they have a tendency to become alcoholics. The study mentioned in the chapter opener showed that one-quarter of the undergraduate students who binged also had other signs of alcoholism.

Alcoholics will continue to drink despite the knowledge that doing so has tremendous negative consequences. In the United States, alcohol abuse is the leading cause of cirrhosis of the liver. The liver becomes so scarred, hardened, and filled with fat that it loses its function (Figure 4.18). It stops making the protein albumin, so the solute balance of body fluids is disrupted, and the legs and abdomen swell with watery fluid. It cannot remove drugs and other toxins from the blood, so they accumulate in the brain—which impairs mental functioning and alters personality. Restricted blood flow through the liver causes veins to enlarge and rupture, so internal bleeding is a risk. The damage to the body results in a heightened risk of diabetes and liver cancer. Once cirrhosis has been diagnosed, a person has about a 50 percent chance of death within 10 years.

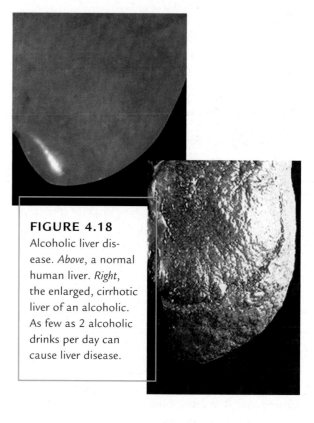

FIGURE 4.18
Alcoholic liver disease. *Above*, a normal human liver. *Right*, the enlarged, cirrhotic liver of an alcoholic. As few as 2 alcoholic drinks per day can cause liver disease.

How Would
YOU
☑ Vote ?

Some people have damaged their liver because they drank too much alcohol; others have had liver-damaging infections. Because there are not enough liver donors, should lifestyle be a factor in deciding who gets a liver transplant? **See CengageNow for details, then vote online (cengagenow.com).**

Summary

Section 4.1 Metabolic processes build and break down organic molecules such as ethanol and other toxins. Currently the most serious drug problem on college campuses is binge drinking.

Section 4.2 Energy is the capacity to do work. Energy cannot be created or destroyed (first law of thermodynamics), but it can be converted from one form to another and thus transferred between objects or systems. Energy tends to disperse spontaneously (second law of thermodynamics). A bit disperses at each energy transfer, usually in the form of heat.

Living things maintain their organization only as long as they harvest energy from someplace else. Energy flows in one direction through the biosphere, starting mainly from the sun, then into and out of ecosystems. Producers and then consumers use energy to assemble, rearrange, and break down organic molecules that cycle among organisms throughout ecosystems.

Section 4.3 Cells store and retrieve energy by making and breaking chemical bonds in metabolic reactions (Table 4.1). ATP is the main energy carrier in cells. Phosphate-group transfers (phosphorylations) to and from ATP couple energy-releasing reactions with energy-requiring ones.

CENGAGENOW Learn about chemical bookkeeping and investigate ATP with the animations on CengageNow.

Section 4.4 Enzymes are proteins or RNAs that greatly enhance the rate of a chemical reaction. Each type works best within a characteristic range of temperature, salt concentration, and pH. Controls over enzymes allow cells to conserve energy and resources by producing only what they require. Cells concentrate, convert, and dispose of most substances in enzyme-mediated reaction sequences called metabolic pathways. Electron transfer chains allow cells to harvest energy in manageable increments.

CENGAGENOW Investigate enzymes and activation energy, observe controls over enzymes, and compare the effects of controlled and uncontrolled energy release with the animations on CengageNow.

Section 4.5 A concentration gradient is a difference in the concentration of a substance between adjoining regions of fluid. Molecules or ions tend to follow their own gradient and move toward the region where they are less concentrated, a behavior called diffusion.

Table 4.1	Key Players in Metabolic Reactions
Reactant	Substance that enters a metabolic reaction; also called a substrate of an enzyme
Intermediate	Substance that forms in a reaction or pathway between the reactants and products
Product	Substance that remains at the end of a reaction
Enzyme	Protein or RNA that greatly enhances the rate of a reaction, but is not changed by participating in it
Cofactor	Molecule or ion that assists enzymes
Energy carrier	Mainly ATP; couples reactions that release energy with reactions that require energy

CENGAGENOW Investigate diffusion and osmosis with the animation on CengageNow.

Section 4.6 Gases, water, and small nonpolar molecules can diffuse across a lipid bilayer. Most other molecules and ions cross only with the help of transport proteins, which move specific solutes across membranes. Transporters allow a cell or membrane-enclosed organelle to control which substances enter and exit. The types of transport proteins in a membrane determine which substances cross it. Active transport proteins use energy, usually from ATP, to pump a solute against its concentration gradient. Passive transport proteins work without an energy input; solute movement is driven by the concentration gradient.

Osmosis is the diffusion of water across a selectively permeable membrane, from the region with a lower solute concentration (hypotonic) toward the region with a higher solute concentration (hypertonic).

CENGAGENOW Compare the processes of passive and active transport with the animation on CengageNow.

Self-Quiz Answers in Appendix I

1. _____ is life's primary source of energy.
   a. Food   b. Water   c. Sunlight   d. ATP

2. Energy _____ .
   a. cannot be created or destroyed
   b. can change from one form to another
   c. tends to disperse spontaneously
   d. all of the above

3. If we liken a chemical reaction to an energy hill, then an _____ reaction is an uphill run.
   a. energy-requiring   c. ATP-assisted
   b. energy-releasing   d. both a and c

4. _____ are always changed by participating in a reaction. (Choose all that are correct.)
   a. Enzymes   c. Reactants
   b. Cofactors   d. Intermediates

5. Enzymes _____ .
   a. are proteins, except for a few RNAs
   b. lower the activation energy of a reaction
   c. are changed by the reactions they catalyze
   d. a and b

6. Which of the following statements is not correct? A metabolic pathway _____ .
   a. is a sequence of enzyme-mediated reactions
   b. may build or break down molecules
   c. generates heat
   d. can include an electron transfer chain
   e. none of the above

7. Diffusion is the movement of ions or molecules from a region where they are _____ (more/less) concentrated to another where they are _____ (more/less) concentrated.

8. Name one molecule that can readily diffuse across a lipid bilayer.

9. Some sodium ions cross a cell membrane through transport proteins that first must be activated by an energy boost. This is an example of _____ .
   - a. passive transport
   - b. active transport
   - c. facilitated diffusion
   - d. a and c

10. Immerse a living cell in a hypotonic solution, and water will tend to _____ .
   - a. move into the cell
   - b. move out of the cell
   - c. show no net movement
   - d. move in by endocytosis

11. Fluid pressure against a wall or cell membrane is called _____ .

12. Vesicles form by _____ .
   - a. endocytosis
   - b. exocytosis
   - d. phagocytosis
   - e. all of the above

13. Match each term with its most suitable description.
   - _____ reactant
   - _____ enzyme
   - _____ first law
   - _____ product
   - _____ cofactor
   - _____ gradient
   - _____ phospholipid
   - _____ active transport
   - _____ phagocytosis
   - a. assists enzymes
   - b. there at reaction's end
   - c. enters a reaction
   - d. unchanged by participating in a reaction
   - e. energy cannot be created or destroyed
   - f. basis of diffusion
   - g. important in membranes
   - h. one cell engulfs another
   - i. requires energy boost

Additional questions are available on CENGAGENOW™

Critical Thinking

1. Beginning physics students are often taught the basic concepts of thermodynamics with two phrases: First, you can't win. Second, you can't break even. Explain.

2. Dixie Bee wanted to make Jell-O shots for her next party, but felt guilty about encouraging her guests to consume alcohol. She tried to compensate for the toxicity of the ethanol by adding healthy fresh pineapple to the shots, but then the Jell-O never set up. What do you think happened? *Hint:* Jell-O consists of sugar and gelatin, a protein manufactured from collagen. Heat dissolves Jell-O because it breaks the hydrogen bonds between its protein molecules. The hydrogen bonds form again and hold the protein molecules together as a solid when the Jell-O cools.

3. Seawater and extracellular fluid contain approximately the same kinds and proportions of salts, which is a legacy of life's probable origin in ocean water. However, those solutes are about four times more concentrated in seawater than in body fluids. On the outer surface of the body, a thick layer of dead, keratin-stuffed cells make skin waterproof. When you take a long bath, your skin wrinkles because the dry keratin molecules in the dead skin cells absorb water and swell up. Is skin wrinkling in water an example of osmosis or does it occur by a different mechanism? Would you predict that skin wrinkles faster in fresh water or salt water? Why?

Digging Into Data

Effects of Artichoke Extract on Hangovers

Ethanol is a toxin, so it makes sense that drinking it can cause various symptoms of poisoning—headache, stomachache, nausea, fatigue, impaired memory, dizziness, tremors, and diarrhea, among other ailments. All are symptoms of hangover, the common word for what happens as the body is recovering from a bout of heavy drinking.

The most effective treatment for a hangover is to avoid drinking in the first place. Folk remedies (such as aspirin, coffee, bananas, more alcohol, honey, barley grass, pizza, milkshakes, glutamine, raw eggs, charcoal tablets, or cabbage) abound, but few have been studied scientifically. In 2003, Max Pittler and his colleagues tested one of them. The researchers gave 15 participants an unmarked pill containing either artichoke extract or a placebo (an inactive substance) just before or after drinking enough alcohol to cause a hangover. The results are shown in Figure 4.19.

1. How many participants experienced a hangover that was worse with the placebo than with the artichoke extract?

2. How many participants experienced a worse hangover with the artichoke extract?

3. Calculate the numbers you counted in questions 1 and 2 as a percentage of the total number of participants. How much difference is there between the percentages?

4. Does these results support the hypothesis that artichoke extract is an effective hangover treatment? Why or why not?

Participant (Age, Gender)	Severity of Hangover	
	Artichoke Extract	Placebo
1 (34, F)	1.9	3.8
2 (48, F)	5.0	0.6
3 (25, F)	7.7	3.2
4 (57, F)	2.4	4.4
5 (34, F)	5.4	1.6
6 (30, F)	1.5	3.9
7 (33, F)	1.4	0.1
8 (37, F)	0.7	3.6
9 (62, M)	4.5	0.9
10 (36, M)	3.7	5.9
11 (54, M)	1.6	0.2
12 (37, M)	2.6	5.6
13 (53, M)	4.1	6.3
14 (48, F)	0.5	0.4
15 (32, F)	1.3	2.5

FIGURE 4.19 Results of a study that tested artichoke extract as a hangover preventive. All participants were tested once with the placebo and once with the extract, with a week interval between. Each rated the severity of 20 hangover symptoms on a scale of 0 (not experienced) to 10 ("as bad as can be imagined"). The 20 ratings were averaged as a single, overall rating, which is listed here.

5 Capturing and Releasing Energy

5.1 Impacts/Issues: Green Energy

Today, the expression "food is fuel" is not just about eating. With fossil fuel prices soaring, there is an increasing demand for biofuels, which are oils, gases, or alcohols made from organic matter that is not fossilized. Much of the material currently used for biofuel production consists of food crops—mainly corn, soybeans, and sugarcane. Growing these crops in large quantities is typically expensive and damaging to the environment, and using them to make biofuel competes with our food supply. The diversion of food crops to biofuels production is now contributing to a major increase in food prices worldwide.

How did we end up competing with our vehicles for food? We both run on the same fuel: energy that plants have stored in chemical bonds. Petroleum, coal, and natural gas are fossil fuels, which formed from the remains of ancient swamp forests that decayed and compacted over millions of years. They consist mainly of molecules originally assembled by ancient plants. Biofuels—and foods—consist mainly of molecules originally assembled by modern plants.

Plants use the energy of sunlight to assemble carbohydrates from carbon dioxide and water. That process is **photosynthesis**, and it is the way that plants make their own food. Directly or indirectly, photosynthesis also feeds most other life on Earth: We and almost all other organisms sustain ourselves by extracting energy stored in the organic products of photosynthesis. Animals and other **heterotrophs** get energy and carbon by breaking down organic molecules assembled by other organisms (*hetero*– means other; *–troph* refers to nourishment). Plants and other **autotrophs** harvest energy directly from the environment, and they get carbon from inorganic molecules (*auto*– means self).

A lot of energy is locked up in the chemical bonds of molecules made by plants. That energy can fuel heterotrophs, as when an animal cell powers ATP synthesis by breaking the bonds of sugars. It can also fuel our cars, which run on energy released by burning biofuels. Both processes are fundamentally the same: They release energy by breaking the bonds of organic molecules. Both use oxygen to break those bonds, and both produce carbon dioxide.

Plants such as corn and other food crops are typically rich in oils or sugars, molecules that are easily converted to biofuels. For example, we use heterotrophic bacteria to conveniently convert corn sugars to ethanol. Using other plant materials to make biofuels takes an extra step, because they contain a higher proportion of cellulose. Cellulose is a tough and insoluble carbohydrate, and bacteria cannot break it down. Breaking the bonds between cellulose's component sugars is an energy-intensive process that adds cost to the biofuel product. Thus, current biofuels research focuses on finding a cost-effective way to break down the abundant cellulose in fast-growing weeds (Figure 5.1) and agricultural wastes such as wood chips, wheat straw, cotton stalks, and rice hulls—all materials that we now dump in landfills or burn.

FIGURE 5.1 Switchgrass (*Panicum virgatum*), a fast-growing weed that grows wild in North American prairies. The abundant cellulose in its thick, strong stems holds enough energy to yield more than 1,100 gallons of ethanol biofuel per acre per year.

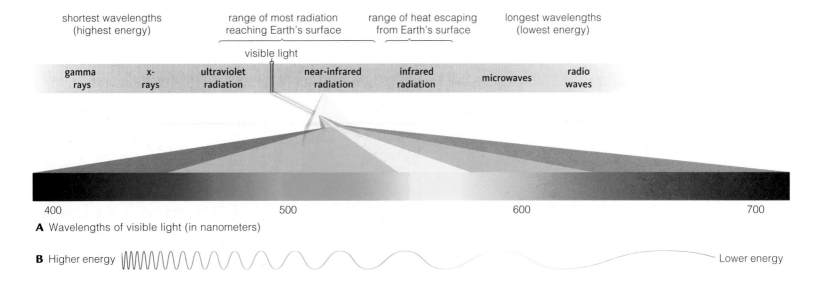

A Wavelengths of visible light (in nanometers)

B Higher energy ⋮⋮⋮⋮⋮⋮⋮⋮⋮⋮⋯⋯⋯⋯⋯⋯⋯ Lower energy

FIGURE 5.2 A Electromagnetic spectrum of radiant energy, which undulates across space as waves that are measured in nanometers. Visible light makes up a very small part of the electromagnetic spectrum. Raindrops or a prism can separate its different wavelengths, which are visible to us as different colors. **B** Light is organized as packets of energy called photons. The shorter a photon's wavelength, the greater its energy.

Capturing Rainbows

Harnessing the energy of sunlight for work is complicated business, or we would have been able to do it in an economically sustainable way by now. Plants do it by converting light energy to chemical energy, which they and most other organisms use to drive cellular work. The first step involves capturing light. In order to understand how that happens, you have to understand a little about the nature of light.

Most of the energy that reaches Earth's surface is in the form of visible light, so it should not be surprising that visible light is the energy that drives photosynthesis. Visible light is a very small part of a large spectrum of electromagnetic energy radiating from the sun. Light travels in waves, moving through space a bit like waves moving across an ocean. The distance between the crests of two successive waves is the light's **wavelength**, which we measure in nanometers (nm). Humans and many other organisms see particular wavelengths of light in the range of 380 to 750 nanometers as different colors (Figure 5.2**A**).

Light travels in waves, but it is also organized in packets of energy called photons. A photon's energy and its wavelength are related, so all photons traveling at the same wavelength carry the same amount of energy. Photons that carry the least energy travel in longer wavelengths; those that carry the most energy travel in shorter wavelengths (Figure 5.2**B**).

Photosynthesizers use pigments to capture light. A **pigment** is an organic molecule that selectively absorbs light of specific wavelengths. Wavelengths of light that are not absorbed are reflected, and that reflected light gives a pigment its characteristic color. For example, if a pigment absorbs violet, blue, and green light, it reflects the rest of the visible light spectrum, which consists of yellow, orange, and red light. This pigment would appear orange to us.

Chlorophyll *a* is by far the most common photosynthetic pigment in plants, and in photosynthetic protists and bacteria. Chlorophyll *a* absorbs purple and

autotroph Organism that makes its own food using carbon from inorganic molecules such as CO_2, and energy from the environment.

chlorophyll *a* Main photosynthetic pigment in plants.

heterotroph Organism that obtains energy and carbon from organic compounds assembled by other organisms.

photosynthesis Metabolic pathway by which most autotrophs capture light energy and use it to make sugars from CO_2 and water.

pigment An organic molecule that can absorb light of certain wavelengths.

wavelength Distance between the crests of two successive waves of light.

		Occurrence			
Table 5.1	**Some Examples of Photosynthetic Pigments**				
Pigment	Color	Plants	Protists	Bacteria	Archaeans
Chlorophyll *a*	green	●	●	●	
Other chlorophylls	green	●	●	●	
Phycobilins					
phycocyanobilin	blue		●	●	
phycoerythrobilin	red		●	●	
phycoviolobilin	purple		●	●	
Carotenoids					
beta-carotene	orange	●	●	●	
lycopene	red	●	●		
lutein	yellow	●	●	●	
zeaxanthin	yellow	●	●	●	
fucoxanthin	orange	●	●		
Anthocyanins	purple	●	●	●	
Retinal	purple				●

All life is sustained by inputs of energy, but not all forms of energy can sustain life.

red light, so it appears green to us. Accessory pigments, including other chlorophylls, absorb additional colors of light for photosynthesis. A few of the 600 or so known accessory pigments are listed in Table 5.1.

Most photosynthetic organisms use a combination of pigments for photosynthesis. In plants, chlorophylls are usually so abundant that they mask the colors of the other pigments, so leaves are typically green. Green leaves change color during autumn because they stop making pigments as the plant prepares for a period of dormancy. Chlorophyll breaks down faster than other pigments, so the leaves turn red, orange, yellow, or purple as their chlorophyll content declines and the accessory pigments show through. Accessory pigments also color many flowers, fruits, and roots red or orange. For example, the color of a tomato changes from green to red as it is ripening because it starts making carotenoids instead of chlorophylls.

In photosynthesis, pigment molecules are a bit like antennas that are specialized for receiving light energy. When a pigment absorbs a photon, its electrons become excited. The excited electrons return quickly to a lower energy level by emitting the extra energy (Section 2.2). As you will see, molecules in the membranes of photosynthetic cells can capture the energy emitted by an excited electron, and use it to drive ATP synthesis.

Take-Home Message How do photosynthetic organisms absorb light?

- Energy radiating from the sun travels through space in waves and is organized as packets called photons.

- The spectrum of radiant energy from the sun includes visible light. Humans perceive different wavelengths of visible light as different colors. The shorter the wavelength, the greater the energy.

- Pigments absorb light at specific wavelengths. Photosynthetic species use pigments such as chlorophyll *a* to harvest the energy of light for photosynthesis.

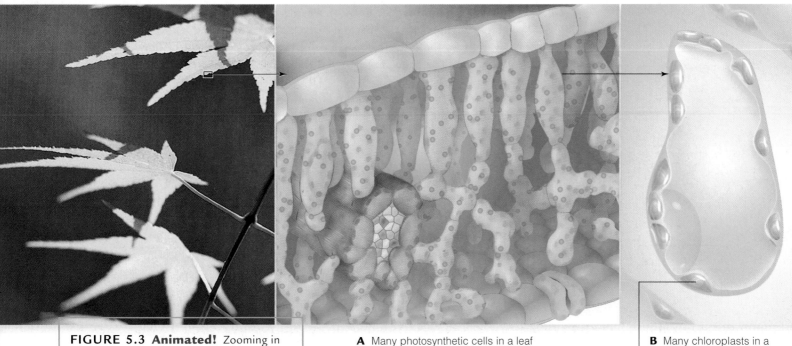

FIGURE 5.3 **Animated!** Zooming in on the sites of photosynthesis in a leaf.

A Many photosynthetic cells in a leaf

B Many chloroplasts in a photosynthetic cell

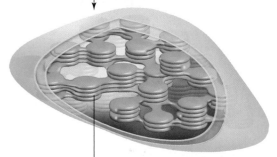

C Many thylakoids in a chloroplast

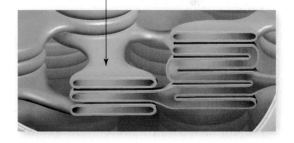

Storing Energy in Carbohydrates

All life is sustained by inputs of energy, but not all forms of energy can sustain life. Sunlight, for example, is abundant here on Earth, but it cannot be used to directly power protein synthesis or the other energy-requiring reactions that all organisms must run in order to stay alive. Photosynthesis converts the energy of light into the energy of chemical bonds. Unlike light, chemical energy can power the reactions of life, and it can be stored for use at a later time.

In plants, photosynthesis occurs in certain types of cells (Figure 5.3**A**). Photosynthetic cells contain **chloroplasts**, which are organelles specialized for photosynthesis (Figure 5.3**B**). Plant chloroplasts have two outer membranes, and they are filled with a semifluid substance called **stroma**. Suspended in the stroma are the chloroplast's DNA, some ribosomes, and a third, much-folded **thylakoid membrane**. The folds of a thylakoid membrane typically form stacks of interconnected disks called thylakoids (Figure 5.3**C**). The space enclosed by the thylakoid membrane forms one continuous compartment.

Many clusters of light-harvesting pigments are embedded in a thylakoid membrane. These clusters absorb photons of different energies. The membrane also incorporates **photosystems**, which are groups of hundreds of pigments and other molecules that work as a unit to begin the reactions of photosynthesis.

The overall process of photosynthesis can be summarized this way:

$$6CO_2 + 6H_2O \xrightarrow{\text{light energy}} C_6H_{12}O_6 + 6O_2$$

carbon dioxide   water   glucose   oxygen

The equation means that photosynthesis converts carbon dioxide and water to glucose and oxygen. However, photosynthesis is not one reaction. It is a

chloroplast Organelle of photosynthesis in plants and some protists.

photosystem Cluster of pigments and proteins that converts light energy to chemical energy in photosynthesis.

stroma Semifluid matrix between the thylakoid membrane and the two outer membranes of a chloroplast.

thylakoid membrane A chloroplast's highly folded inner membrane system; forms a continuous compartment in the stroma.

metabolic pathway (Section 4.4), a series of many reactions that occur in two stages (Figure 5.4). The first stage occurs at the thylakoid membrane. It is driven by light, so the collective reactions of this stage are called the **light-dependent reactions**. These reactions convert the energy of light to the chemical bond energy of ATP. The coenzyme NADP+ (nicotinamide adenine dinucleotide phosphate) accepts electrons and hydrogen ions. When this coenzyme is carrying electrons and hydrogen, we refer to it as NADPH. Water molecules are broken apart into oxygen and hydrogen atoms. Hydrogen ions stay in the chloroplast and participate in the process of ATP formation, and the oxygen atoms leave the cell as oxygen gas (O_2).

The reactions of the second stage of photosynthesis run in the stroma. They are collectively called the **light-independent reactions** because light does not power them. They run on energy delivered by molecules that formed in the first stage—ATP and NADPH. During the second-stage reactions, glucose and other carbohydrates are synthesized from carbon dioxide and water.

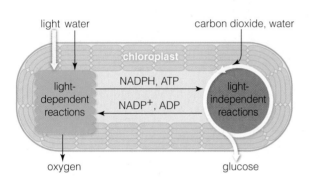

light water   carbon dioxide, water

chloroplast

light-dependent reactions →NADPH, ATP→ light-independent reactions
                           ←NADP+, ADP←

oxygen   glucose

FIGURE 5.4 Summary of the metabolic pathway of photosynthesis as it occurs in plant chloroplasts.

Take-Home Message   What is photosynthesis and where does it occur?

■ In the first stage of photosynthesis, light energy drives the formation of ATP and NADPH, and oxygen is released. These light-dependent reactions occur at the thylakoid membrane in plant chloroplasts.

■ In the second stage of photosynthesis, ATP and NADPH drive the synthesis of glucose and other carbohydrates from water and carbon dioxide. These light-independent reactions occur in the stroma.

5.4 The Light-Dependent Reactions

The light-dependent reactions of photosynthesis convert light energy to chemical bond energy in the form of ATP, and they also yield NADPH and O_2:

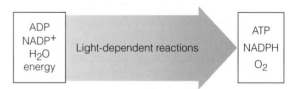

ADP
NADP+
H_2O
energy
→ Light-dependent reactions →
ATP
NADPH
O_2

The reactions begin when a pigment in a thylakoid membrane absorbs a photon, and the photon's energy boosts one of the pigment's electrons to a higher energy level (Section 2.2). The electron quickly emits the extra energy and drops back to its unexcited state. That energy would be lost to the environment if nothing else were to happen, but in a thylakoid membrane the energy of excited electrons is kept in play. The pigments in the membrane can hold onto energy by passing it back and forth, a bit like volleyball players hold onto a ball by passing it back and forth among team members.

The reactions of photosynthesis begin when energy being passed among photosynthetic pigments reaches a photosystem (Figure 5.5). At the center of each photosystem is a special pair of chlorophyll *a* molecules. Absorbing energy pops electrons right out of that special pair ❶. The electrons immediately enter an electron transfer chain (Section 4.4) in the thylakoid membrane.

electron transfer phosphorylation Metabolic pathway in which electron flow through electron transfer chains sets up a hydrogen ion gradient that drives ATP formation.
light-dependent reactions Metabolic pathway of photosynthesis that converts light energy to chemical energy of ATP and NADPH.
light-independent reactions Metabolic pathway of photosynthesis that uses ATP and NADPH to assemble sugars from water and CO_2.

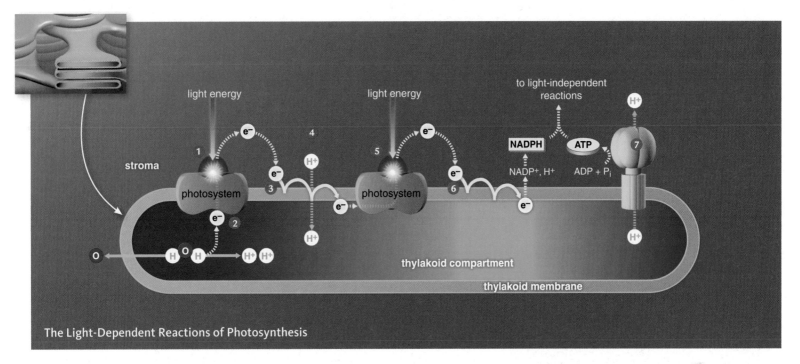

The Light-Dependent Reactions of Photosynthesis

A photosystem can donate only a few electrons to electron transfer chains before it must be restocked with more. Where do replacements come from? The photosystem gets more electrons by pulling them away from water molecules in the thylakoid compartment ❷. Losing electrons causes the water molecules to break apart into hydrogen ions and oxygen. The oxygen diffuses out of the cell as O₂ gas.

The actual conversion of light energy to chemical energy occurs when a photosystem donates electrons to an electron transfer chain ❸. Light does not take part in chemical reactions, but electrons do. In a series of reactions, electrons pass from one molecule of the chain to the next, and they release a bit of energy at each step.

The molecules of the electron transfer chain use the released energy to move hydrogen ions (H⁺) across the membrane, from the stroma to the thylakoid compartment ❹. Thus, the flow of electrons through electron transfer chains sets up and maintains a hydrogen ion gradient across the thylakoid membrane.

This gradient motivates hydrogen ions in the thylakoid compartment to move back into the stroma. However, ions cannot diffuse through lipid bilayers (Section 4.5). H⁺ leaves the thylakoid compartment only by flowing through membrane transport proteins called ATP synthases (Section 3.4).

Hydrogen ion flow through an ATP synthase causes this protein to attach a phosphate group to ADP ❼, so ATP forms in the stroma. The process by which the flow of electrons through electron transfer chains drives ATP formation is called **electron transfer phosphorylation**.

After the electrons have moved through the first electron transfer chain, they are accepted by another photosystem. This photosystem absorbs light energy, which causes electrons to pop out of its special pair of chlorophylls ❺. The electrons then enter a second, different electron transfer chain. At the end of this chain, NADP⁺ accepts the electrons along with H⁺, so NADPH forms ❻:

$$NADP^+ + 2e^- + H^+ \longrightarrow NADPH$$

ATP and NADPH continue to form as long as electrons continue to flow through transfer chains in the thylakoid membrane, and electrons flow through the

FIGURE 5.5 Animated! Light-dependent reactions of photosynthesis in the thylakoid membrane.

❶ Light energy ejects electrons from a photosystem.

❷ The photosystem pulls replacement electrons from water molecules, which break apart into oxygen and hydrogen ions. The oxygen leaves the cell as O₂.

❸ The electrons enter an electron transfer chain in the thylakoid membrane.

❹ Energy lost by the electrons as they move through the transfer chain causes hydrogen ions to be pumped from the stroma into the thylakoid compartment. A hydrogen ion gradient forms across the thylakoid membrane.

❺ Light energy pops electrons out of another photosystem. Replacement electrons come from an electron transfer chain.

❻ The electrons move through a second electron transfer chain, then combine with NADP⁺ and H⁺, so NADPH forms.

❼ Hydrogen ions in the thylakoid compartment are propelled through the interior of ATP synthases by their gradient across the thylakoid membrane. Hydrogen ion flow causes ATP synthases to attach phosphate to ADP, so ATP forms in the stroma.

chains as long as water and oxygen are plentiful. When water and oxygen are scarce, the flow of electrons slows, and so does ATP and NADPH production.

Take-Home Message What happens during the light-dependent reactions of photosynthesis?

- In the light-dependent reactions of photosynthesis, chlorophylls and other pigments in the thylakoid membrane transfer the energy of light to photosystems.
- Absorbing energy causes electrons to leave photosystems and enter electron transfer chains in the membrane. The flow of electrons through the transfer chains sets up hydrogen ion gradients that drive ATP formation.
- Oxygen is released and electrons end up in NADPH.

5.5 The Light-Independent Reactions

ATP and NADPH that form in the light-dependent reactions power the energy-requiring second stage of photosynthesis, which occurs in the stroma of chloroplasts. The light-independent reactions that make up this stage convert carbon dioxide to glucose and other carbohydrates:

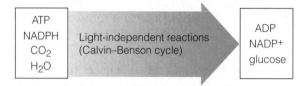

These reactions are also called the **Calvin–Benson cycle**, after the scientists who discovered them. Even though the Calvin–Benson cycle is often shown as a wheel or disk, it is not an object. It is a cyclic pathway, in which the product of the last reaction is the starting material of the first reaction (Section 4.4).

Extracting carbon atoms from an inorganic source and incorporating them into an organic molecule is a process called **carbon fixation**. In the Calvin–Benson cycle, an enzyme with a cumbersome name (ribulose-1,5-bisphosphate carboxylase, or **rubisco** for short) fixes carbon. Rubisco attaches a molecule of carbon dioxide to a molecule of ribulose bisphosphate, or RuBP. The molecule that forms splits at once into two 3-carbon intermediates called PGA, which continue in the cycle (Figure 5.6). It takes six cycles of Calvin–Benson reactions to fix the six carbon atoms necessary to make one glucose molecule.

Plants use the glucose they make in the light-independent reactions as building blocks for other organic molecules, or they break it down to access the energy in its bonds. However, most of the glucose is converted at once to sucrose or starch by other pathways that conclude the light-independent reactions. Excess glucose is stored as starch grains in the stroma of the chloroplast.

Rascally Rubisco Several adaptations allow plants to live where water is scarce or sporadically available. For example, a thin, waterproof coating called a cuticle prevents water loss by evaporation from aboveground plant parts. However, the cuticle also prevents gases from crossing through cells on the plant's surface. Gases play a critical role in photosynthesis, so photosynthetic parts are often studded with tiny, closable gaps called **stomata** (singular, stoma). When stomata are open, carbon dioxide for the light-independent reactions can diffuse

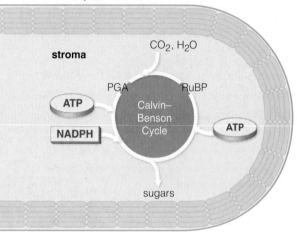

FIGURE 5.6 Animated! Light-independent reactions of photosynthesis which, in chloroplasts, occur in the stroma. The sketch is a summary of six cycles of the Calvin–Benson reactions and their product, one glucose molecule.

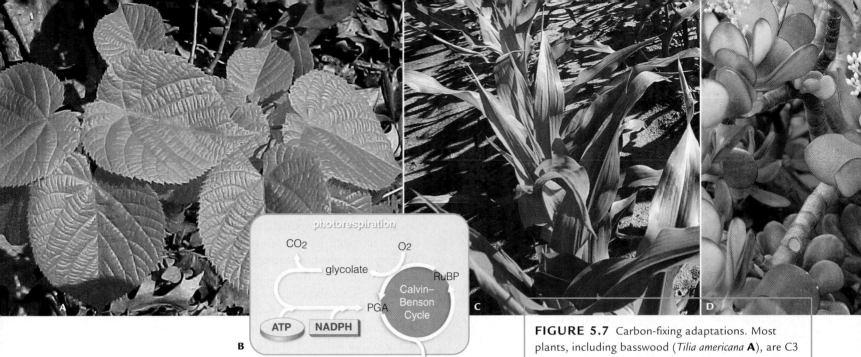

photorespiration

CO$_2$ O$_2$

glycolate   RuBP

                                    Calvin–
                                    Benson
                       PGA   Cycle

ATP   NADPH

B

sugars

A

C

D

from air into the plant's photosynthetic tissues, and oxygen produced by the light-dependent reactions can diffuse from photosynthetic cells into the air.

Plants that use only the Calvin–Benson cycle are called **C3 plants**, because 3-carbon PGA is the first stable intermediate to form in the light-independent reactions. C3 plants typically conserve water on dry days by closing their stomata. However, when stomata are closed, oxygen produced by the light-dependent reactions cannot escape from the plant. Oxygen that accumulates in photosynthetic tissues limits sugar production. Why? At high oxygen levels, rubisco attaches oxygen (instead of carbon) to RuBP. This pathway, **photorespiration**, produces carbon dioxide, so the plant loses carbon instead of fixing it (Figure 5.7**A,B**). In addition, ATP and NADPH are used to shunt the pathway's intermediates back to the Calvin–Benson cycle, so it takes extra energy to make sugars on dry days. C3 plants compensate for rubisco's inefficiency by making a lot of it: Rubisco is the most abundant protein on Earth.

Some plants have adaptations that help them minimize photorespiration. A 4-carbon molecule is the first stable intermediate that forms in the light-independent reactions of corn, bamboo, and other **C4 plants** (Figure 5.7**C**). Such plants fix carbon twice, in two kinds of cells. In the first kind of cell, carbon is fixed by an enzyme that does not use oxygen even at high oxygen levels. An intermediate is transported to another kind of cell and converted to carbon dioxide, which enters the Calvin–Benson cycle. The extra reactions keep the carbon dioxide level high near rubisco, thus minimizing photorespiration. In succulents, cactuses, and other plants, the extra reactions run at a different time rather than in different cells. The C4 reactions run during the day, and the Calvin–Benson cycle runs at night. Such **CAM plants** are named for *C*rassulacean *A*cid *M*etabolism, after the family of plants in which this pathway was first studied (Figure 5.7**D**).

Take-Home Message
What happens during the light-independent reactions of photosynthesis?

- Driven by the energy of ATP and electrons from NADPH, the light-independent reactions of photosynthesis use carbon and oxygen (from CO$_2$) to build sugars.

C3 plant Type of plant that uses only the Calvin–Benson cycle to fix carbon.

C4 plant Type of plant that minimizes photorespiration by fixing carbon twice, in two cell types.

Calvin–Benson cycle Light-independent reactions of photosynthesis; cyclic pathway that forms glucose from CO$_2$.

CAM plant Type of C4 plant that conserves water by fixing carbon twice, at different times of day.

carbon fixation Process by which carbon from an inorganic source such as carbon dioxide becomes incorporated into an organic molecule.

photorespiration Reaction in which rubisco attaches oxygen instead of carbon dioxide to ribulose bisphosphate.

rubisco Ribulose bisphosphate carboxylase. Carbon-fixing enzyme of the Calvin–Benson cycle.

stomata Gaps that open between guard cells on plant surfaces.

5.6 Photosynthesis and Aerobic Respiration: A Global Connection

The first cells on Earth did not tap into sunlight. Like some modern archaeans, these ancient organisms extracted energy and carbon from simple molecules such as methane and hydrogen sulfide—gases that were plentiful in the nasty brew that constituted Earth's early atmosphere (Figure 5.8 ❶).

The first photosynthetic autotrophs evolved about 3.2 billion years ago, probably in shallow ocean waters. Sunlight offered these **photoautotrophs** an essentially unlimited supply of energy, and they were very successful. Oxygen gas (O_2) released from uncountable numbers of water molecules began seeping out of uncountable numbers of photosynthesizers, and it accumulated in the ocean and the atmosphere. From that time on, the world of life would never be the same ❷.

Molecular oxygen had been a very small component of Earth's early atmosphere before photosynthesis evolved. The new abundance of atmospheric oxygen exerted tremendous selection pressure on all organisms. Why? Oxygen gas reacts easily with metals such as enzyme cofactors, and free radicals (molecules with unpaired electrons) form during those reactions. Free radicals damage biological molecules, so they are dangerous to life.

The ancient cells had no way to detoxify oxygen radicals, so most of them quickly died out. Only a few lineages persisted in deep water, muddy sediments, and other **anaerobic** (oxygen-free) habitats. New metabolic pathways that detoxified oxygen radicals evolved in the survivors. Organisms with such pathways were the first **aerobic** organisms—they could live in the presence of oxygen. As you will see, one of the pathways, aerobic respiration, put oxygen's reactive properties to good use. **Aerobic respiration** is one of several pathways by which organisms access the energy stored in carbohydrates. This pathway requires oxygen, and it produces carbon dioxide and water—the raw materials from which photosynthesizers make sugars. With this connection, the cycling of carbon, hydrogen, and oxygen through living things came full circle (*right*).

FIGURE 5.8 Then ❶ and now ❷, a view of how Earth's atmosphere was permanently altered by the evolution of photosynthesis. Photosynthesis is now the main pathway by which energy and carbon enter the web of life.

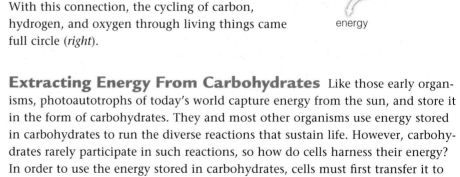

Extracting Energy From Carbohydrates Like those early organisms, photoautotrophs of today's world capture energy from the sun, and store it in the form of carbohydrates. They and most other organisms use energy stored in carbohydrates to run the diverse reactions that sustain life. However, carbohydrates rarely participate in such reactions, so how do cells harness their energy? In order to use the energy stored in carbohydrates, cells must first transfer it to ATP, which does participate in many of the energy-requiring reactions that a cell runs. The transfer occurs by breaking the bonds of carbohydrates, which releases

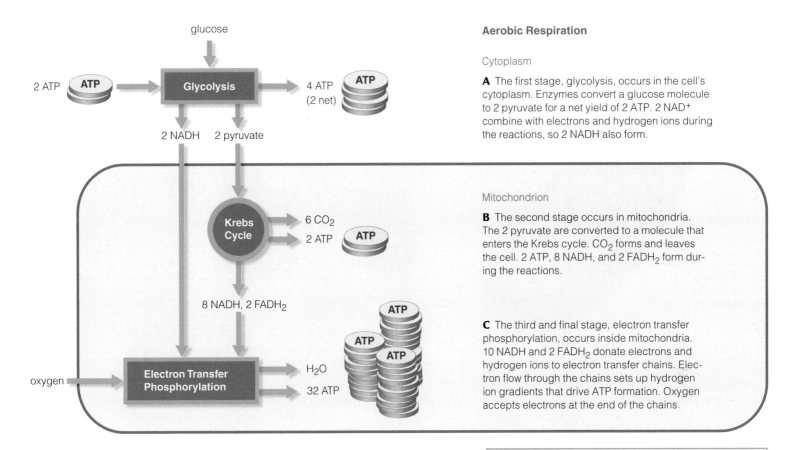

Cytoplasm

A The first stage, glycolysis, occurs in the cell's cytoplasm. Enzymes convert a glucose molecule to 2 pyruvate for a net yield of 2 ATP. 2 NAD$^+$ combine with electrons and hydrogen ions during the reactions, so 2 NADH also form.

Mitochondrion

B The second stage occurs in mitochondria. The 2 pyruvate are converted to a molecule that enters the Krebs cycle. CO_2 forms and leaves the cell. 2 ATP, 8 NADH, and 2 FADH$_2$ form during the reactions.

C The third and final stage, electron transfer phosphorylation, occurs inside mitochondria. 10 NADH and 2 FADH$_2$ donate electrons and hydrogen ions to electron transfer chains. Electron flow through the chains sets up hydrogen ion gradients that drive ATP formation. Oxygen accepts electrons at the end of the chains.

FIGURE 5.9 Animated! Overview of aerobic respiration. The reactions start in the cytoplasm and end in mitochondria.

Figure It Out: What is aerobic respiration's typical net yield of ATP? *Answer: 38 − 2 = 36 ATP per glucose*

energy that drives ATP synthesis. There are a few different kinds of carbohydrate-breakdown pathways, but aerobic respiration is the one that typical eukaryotic cells use at least most of the time. Aerobic respiration yields much more ATP than other carbohydrate breakdown pathways. You and other multicelled organisms could not live without its higher yield.

Aerobic Respiration Begins The reactions of aerobic respiration take place in three stages. The first stage, **glycolysis**, converts one 6-carbon molecule of glucose into two 3-carbon molecules of **pyruvate**:

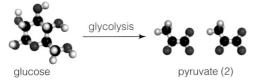

Glycolysis occurs in cytoplasm (Figure 5.9**A**). Two ATP are invested to begin the reactions, but four ATP form by the end. Thus, we say that the net (overall) yield of glycolysis is two ATP per glucose. Two NAD$^+$ (nicotinamide adenine dinucleotide) also combine with electrons and hydrogen ions during the reactions. When this coenzyme is carrying electrons and hydrogen, we refer to it as NADH.

After glycolysis, aerobic respiration continues in two more stages that run inside mitochondria (Figure 5.9**B,C**).

Aerobic Respiration Continues The second stage of aerobic respiration occurs inside mitochondria. It includes two sets of reactions, acetyl–CoA formation and the Krebs cycle, that break down the pyruvate from glycolysis. The reactions begin when the pyruvate enters the inner compartment of a mitochondrion, where it becomes converted to carbon dioxide and an intermediate,

aerobic Involving or occurring in the presence of oxygen.
aerobic respiration Aerobic pathway that breaks down carbohydrates to produce ATP.
anaerobic Occurring in the absence of oxygen.
glycolysis Set of reactions in which glucose or another sugar is broken down to two pyruvate for a net yield of two ATP.
photoautotroph Photosynthetic autotroph.
pyruvate Three-carbon end product of glycolysis.

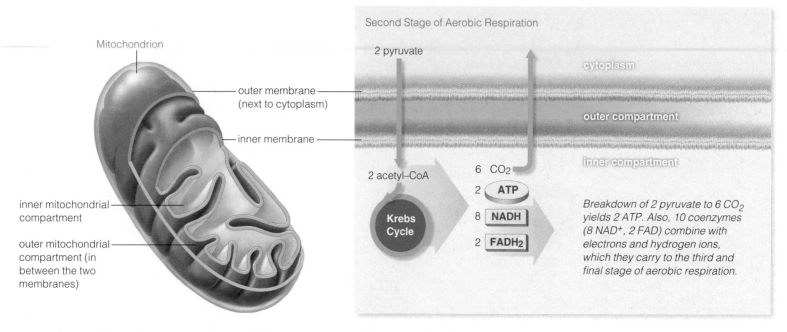

Mitochondrion

outer membrane
(next to cytoplasm)

inner membrane

inner mitochondrial
compartment

outer mitochondrial
compartment (in
between the two
membranes)

Second Stage of Aerobic Respiration

cytoplasm

outer compartment

inner compartment

2 pyruvate

2 acetyl–CoA

Krebs
Cycle

6 CO$_2$

2 ATP

8 NADH

2 FADH$_2$

Breakdown of 2 pyruvate to 6 CO$_2$ yields 2 ATP. Also, 10 coenzymes (8 NAD$^+$, 2 FAD) combine with electrons and hydrogen ions, which they carry to the third and final stage of aerobic respiration.

A An inner membrane divides a mitochondrion's interior into an inner compartment and an outer compartment. The second and third stages of aerobic respiration take place at the inner mitochondrial membrane.

B The second stage starts after membrane proteins transport pyruvate from the cytoplasm to the inner compartment. Six carbon atoms enter these reactions (in two molecules of pyruvate), and six leave (in six CO$_2$). Two ATP form and ten coenzymes accept electrons and hydrogen ions.

FIGURE 5.10 Animated! The second stage of aerobic respiration: acetyl–CoA formation and the Krebs cycle.

acetyl–CoA (Figure 5.10). The **Krebs cycle** then breaks down the acetyl–CoA to carbon dioxide. Like the Calvin–Benson cycle, the Krebs cycle is often depicted as a wheel or disk, but it is not a physical object. It is a cyclic pathway, in which the product of the last reaction is also a starting material of the first reaction.

The overall reactions of the second stage of aerobic respiration convert two molecules of pyruvate to six molecules of carbon dioxide. Two ATP form. Also, 8 NAD$^+$ and 2 FAD (flavin adenine dinucleotide, another coenzyme) combine with electrons and hydrogen ions, so 8 NADH and 2 FADH$_2$ form (Figure 5.10). At this point in aerobic respiration, one glucose molecule has been broken down completely; six carbon atoms have exited the cell, in six CO$_2$:

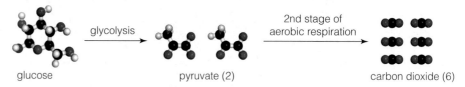

glucose → glycolysis → pyruvate (2) → 2nd stage of aerobic respiration → carbon dioxide (6)

A total of four ATP have formed by the end of the first two stages of aerobic respiration, but almost all of the energy released from the breakdown of the glucose is now carried by twelve coenzymes (eight NAD$^+$ and two FAD) that accepted electrons during the reactions. Energy carried by those electrons will drive the reactions of the third stage.

The Big Energy Payoff: Aerobic Respiration Ends The third and final stage of aerobic respiration, electron transfer phosphorylation, occurs in mitochondria (Figure 5.11). The reactions begin with NADH and FADH$_2$ that formed in the first two stages. These coenzymes now deliver their cargo of electrons and hydrogen ions to electron transfer chains in the inner mitochondrial membrane ❶.

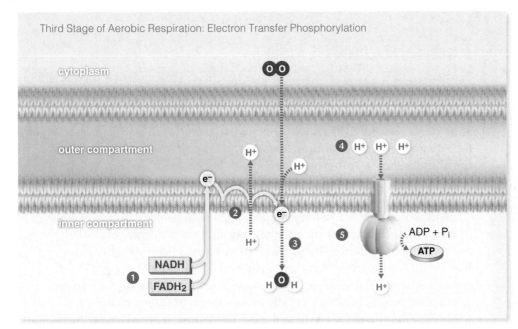

Third Stage of Aerobic Respiration: Electron Transfer Phosphorylation

cytoplasm

outer compartment

inner compartment

NADH

FADH₂

ADP + P$_i$

ATP

FIGURE 5.11 The final stage of aerobic respiration: electron transfer phosphorylation.

1 NADH and FADH₂ deliver electrons to electron transfer chains in the inner mitochondrial membrane.

2 Electron flow through the chains causes hydrogen ions (H^+) to be pumped from the inner to the outer compartment.

3 Oxygen (O_2) accepts electrons and hydrogen ions at the end of mitochondrial electron transfer chains, so water forms.

4 The activity of the electron transfer chains causes a hydrogen ion gradient to form across the inner mitochondrial membrane.

5 Hydrogen ions flow back to the inner compartment through ATP synthases. The flow drives the formation of ATP from ADP and phosphate (P_i).

Figure It Out: Which other pathway discussed in this chapter involves electron transfer phosphorylation?

Answer: Light-dependent reactions of photosynthesis

As the electrons pass through the electron transfer chains, they give up energy bit by bit. Some molecules of the chains use that energy to move hydrogen ions from the inner mitochondrial compartment to the outer one **2**.

At the end of the mitochondrial electron transfer chains, oxygen accepts electrons and combines with hydrogen ions, so water forms **3**. Aerobic respiration literally means "taking a breath of air," a name that refers to oxygen as the final electron acceptor in this pathway. Every breath you take provides your aerobically respiring cells with a fresh supply of oxygen.

As hydrogen ions are pumped from the inner to the outer compartment, a hydrogen ion concentration gradient forms across the inner mitochondrial membrane **4**. This gradient attracts hydrogen ions back toward the inner mitochondrial compartment. However, hydrogen ions cannot diffuse across a lipid bilayer without assistance. The ions can cross the inner mitochondrial membrane only by flowing through the interior of ATP synthases. The flow causes these transport proteins to attach a phosphate group to ADP, so ATP forms **5**. Thirty-two ATP typically form in the third stage of aerobic respiration. Add four ATP from the first and second stages, and the overall yield from the breakdown of one glucose molecule is thirty-six ATP:

$$C_6H_{12}O_6 \ + \ 6O_2 \ + \ 36 \ ADP \ \longrightarrow \ 6CO_2 \ + \ 6H_2O \ + \ 36 \ ATP$$
glucose   oxygen   carbon dioxide   water

Take-Home Message

How do cells typically access the chemical energy stored in carbohydrates?

■ Eukaryotic cells typically convert the chemical energy of carbohydrates to the chemical energy of ATP by the oxygen-requiring pathway of aerobic respiration.

■ Aerobic respiration occurs in three stages: glycolysis; acetyl–CoA formation and the Krebs cycle; and electron transfer phosphorylation.

■ Typically, thirty-six ATP form from the breakdown of one glucose molecule in aerobic respiration. Carbon dioxide and water also form.

Aerobic respiration literally means "taking a breath of air."

Krebs cycle Cyclic pathway that, along with acetyl–CoA formation, breaks down two pyruvate to carbon dioxide for a net yield of two ATP and many reduced coenzymes.

5.7 Fermentation

Fermentation is a type of anaerobic pathway that harvests energy from carbohydrates. Bacteria and single-celled protists that inhabit sea sediments, animal guts, improperly canned food, sewage treatment ponds, deep mud, and other anaerobic habitats use fermentation. Some of these organisms, including the bacteria that cause botulism, do not tolerate aerobic conditions, and will die when exposed to oxygen. Others, such as single-celled fungi called yeasts, can switch between fermentation and aerobic respiration. Animal muscle cells use both pathways.

Aerobic respiration and fermentation begin with precisely the same set of reactions in the cytoplasm: glycolysis. Again, two pyruvate, two NADH, and two ATP form during the reactions of glycolysis. However, after that, fermentation and aerobic respiration pathways differ. The final steps of fermentation occur in the cytoplasm. In these reactions, pyruvate is converted to other molecules, but it is not fully broken down to carbon dioxide and water. Electrons do not flow through transfer chains, so no more ATP forms. However, electrons are removed from NADH, so NAD$^+$ is regenerated. Regenerating this coenzyme allows glycolysis—along with the small ATP yield it offers—to continue. Thus, the net ATP yield of fermentation consists of the two ATP that form in glycolysis.

Alcoholic Fermentation

In **alcoholic fermentation**, the pyruvate from glycolysis is converted to ethyl alcohol, or ethanol. First, 3-carbon pyruvate is split into carbon dioxide and 2-carbon acetaldehyde. Then, electrons and hydrogen are transferred from NADH to the acetaldehyde, forming NAD$^+$ and ethanol (Figure 5.12). Bakers often use the alcoholic fermentation capabilities of one species of yeast, *Saccharomyces cerevisiae*, to make bread. These cells break down carbohydrates in bread dough, and release CO_2 by alcoholic fermentation. The dough expands (rises) as CO_2 forms bubbles in it. Some wild and cultivated strains of *Saccharomyces* are also used to produce wine. Crushed grapes are left in vats along with large populations of yeast cells, which convert the sugars in the juice to ethanol.

Lactate Fermentation

In **lactate fermentation**, the electrons and hydrogen ions carried by NADH are transferred directly to pyruvate. This reaction converts pyruvate to 3-carbon lactate (lactic acid), and also converts NADH to NAD$^+$:

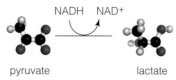

Some lactate fermenters spoil food, but we use others to preserve it. For instance, *Lactobacillus acidophilus* digests lactose in milk. We use this bacteria to produce dairy products such as buttermilk, cheese, and yogurt. Yeasts ferment and preserve pickles, corned beef, sauerkraut, and kimchi.

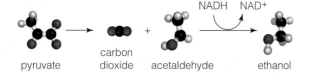

FIGURE 5.12 Animated! Alcoholic fermentation. *Top*, a vintner examines a fermentation product of a *Saccharomyces* yeast. *Bottom*, the last stage of alcoholic fermentation produces ethanol and NAD$^+$.

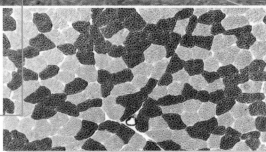

FIGURE 5.13 Lactate fermentation. *Right*, two types of fibers are visible in this cross-section through a human thigh muscle. Sprints (*above left*) and other intense bursts of activity are sustained by the light-colored fibers, which are fast-twitch fibers that make ATP by lactate fermentation. The dark-colored fibers, which are slow-twitch fibers that make ATP by aerobic respiration, contribute to endurance activities. Chickens (*above right*) spend most of their time walking. Their leg muscles consist mainly of slow-twitch fibers.

Animal skeletal muscles, which move bones, consist of cells fused as long fibers. The fibers differ in how they make ATP. Slow-twitch muscle fibers have many mitochondria and produce ATP by aerobic respiration. They dominate during prolonged activity, such as long runs. Slow-twitch fibers are red because they have an abundance of myoglobin, an oxygen-storing protein that is similar to hemoglobin.

Fast-twitch muscle fibers contain few mitochondria and no myoglobin, so they do not carry out a lot of aerobic respiration. Instead, they make most of their ATP by lactate fermentation. This pathway makes ATP quickly but not for long, so it is useful for quick, strenuous activities such as sprinting or weight lifting. It cannot support sustained activity. That is one reason chickens cannot fly very far: The flight muscles of a chicken are mostly fast-twitch fibers, which make up the "white" breast meat. Chickens fly only in short bursts. More often, a chicken walks or runs. Its leg muscles are mostly slow-twitch muscle, the "dark meat." Most human muscles consist of a mixture of fast-twitch and slow-twitch fibers (Figure 5.13), but the proportions vary among muscles and among individuals. Great sprinters tend to have more fast-twitch fibers. Great marathon runners tend to have more slow-twitch fibers. Section 20.4 offers a closer look at how skeletal muscles work.

Take-Home Message **What is fermentation?**

■ ATP can form in fermentation pathways, which are anaerobic (do not require oxygen). Fermentation reactions regenerate the coenzyme NAD^+, without which glycolysis (and ATP production) would stop.

■ The end product of lactate fermentation is lactate. The end product of alcoholic fermentation is ethanol. Both pathways have a net yield of two ATP per glucose molecule. The ATP forms during glycolysis.

alcoholic fermentation Anaerobic carbohydrate breakdown pathway that produces ATP and ethanol.
fermentation An anaerobic pathway by which cells harvest energy from carbohydrates.
lactate fermentation Anaerobic carbohydrate breakdown pathway that produces ATP and lactate.

Alternative Energy Sources in the Body

Excess dietary carbohydrate ends up as fat.

The Fate of Glucose at Mealtime As you eat, glucose and other breakdown products of digestion are absorbed across the gut lining, and blood transports these small organic molecules throughout the body. The concentration of glucose in the bloodstream rises, and in response the pancreas (an organ) increases its rate of insulin secretion. Insulin is a hormone (a type of signaling molecule) that causes cells to take up glucose, so an increase in insulin production causes cells to take up glucose faster. Once inside a cell, the glucose is immediately converted to an intermediate of glycolysis.

Thus, when a cell takes in more than enough glucose, its ATP-forming machinery goes into high gear. Unless the ATP is used quickly, its concentration rises in the cytoplasm. The high concentration of ATP causes the intermediate to be diverted away from glycolysis and into a pathway that forms glycogen, a polysaccharide (Section 2.7). Liver and muscle cells especially favor the conversion of glucose to glycogen, and these cells maintain the body's largest stores of glycogen.

What happens if you eat too many carbohydrates? When the blood level of glucose gets too high, acetyl–CoA is diverted away from the Krebs cycle and into a pathway that makes fatty acids. That is why excess dietary carbohydrate ends up as fat.

Between meals, the blood level of glucose declines. The pancreas responds to low glucose levels in the blood by secreting glucagon, a hormone that causes liver cells to convert stored glycogen to glucose. The cells release glucose into the bloodstream, so the blood glucose level rises. Thus, hormones control whether the body's cells use glucose as an energy source immediately or save it for use at a later time.

Glycogen makes up about 1 percent of an average adult's total energy reserves, which is the energy equivalent of about two cups of cooked pasta. Unless you eat regularly, you will completely deplete your liver's glycogen stores in less than twelve hours.

Energy From Fats Of the total energy reserves in a typical adult who eats well, about 78 percent (about 10,000 kilocalories) is stored in body fat, and 21 percent in proteins. How does a human body access its fat reservoir? A fat molecule, recall, has a glycerol head and one, two, or three fatty acid tails (Section 2.8). The body stores most fats as triglycerides, which have three fatty acid tails. Triglycerides accumulate in fat cells of adipose tissue. This tissue serves as an energy reservoir, and it also insulates and pads the buttocks and other strategic areas of the body.

When blood glucose level falls, triglycerides are tapped to provide energy (Figure 5.14). Enzymes in fat cells cut the bonds between glycerol and the fatty acids, and both are released into the

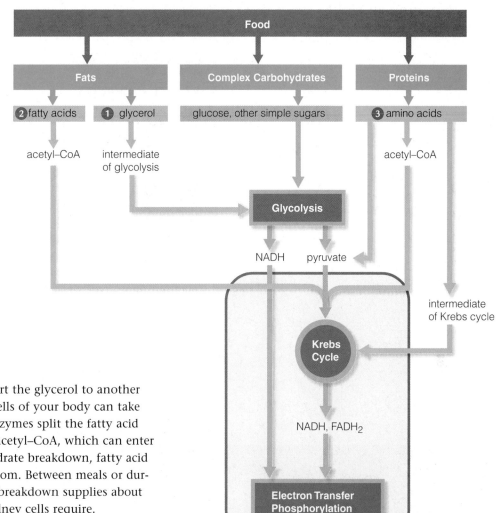

FIGURE 5.14 Animated!
Reaction sites where a variety of organic molecules enter aerobic respiration. Such molecules are energy sources for the body.

❶ Glycerol from fatty acid breakdown is converted to an intermediate of glycolysis.

❷ Fragments of fatty acids are converted to acetyl–CoA, which can enter the reactions of the Krebs cycle.

❸ Amino acids are converted to acetyl–CoA, pyruvate, or an intermediate of the Krebs cycle.

bloodstream. Enzymes in liver cells convert the glycerol to another intermediate of glycolysis ❶. Nearly all cells of your body can take up the free fatty acids. Inside the cells, enzymes split the fatty acid backbones and convert the fragments to acetyl–CoA, which can enter the Krebs cycle ❷. Compared to carbohydrate breakdown, fatty acid breakdown yields more ATP per carbon atom. Between meals or during steady, prolonged exercise, fatty acid breakdown supplies about half of the ATP that muscle, liver, and kidney cells require.

Diets that are extremely low in carbohydrates force the body to break down fats. The breakdown produces molecules called ketones, which most of your cells can use for energy instead of glucose. This metabolic state, which is normally associated with starvation, can raise LDL ("bad" cholesterol) and damage the kidneys and liver.

Energy From Proteins Some enzymes in your digestive system split dietary proteins into their amino acid subunits, which are then absorbed into the bloodstream. Cells use amino acids to build proteins or other molecules. Even so, when you eat more protein than your body needs, the amino acids become broken down further. The amino (NH_3^+) group is removed, and it becomes ammonia (NH_3), a waste product that the body eliminates in urine. Their carbon backbone is split, and acetyl–CoA, pyruvate, or an intermediate of the Krebs cycle forms, depending on the amino acid. Your cells can divert any of these organic molecules into the Krebs cycle ❸.

Take-Home Message How are molecules other than glucose metabolized?

■ In humans and other organisms, the entrance of organic compounds into an energy-releasing pathway depends on the kinds and proportions of carbohydrates, fats, and proteins in the diet.

Impacts/Issues Revisited:
Green Energy

Your body is about 9.5 percent carbon by weight, which means that you contain an enormous number of carbon atoms. Where did they all come from? You eat other organisms to get the carbon atoms your body uses for energy and for raw materials. Those atoms may have passed through other heterotrophs before you ate them, but at some point they were part of photoautotrophic organisms. Photoautotrophs strip carbon from carbon dioxide, then use the atoms to build organic compounds. Your carbon atoms—and those of most other organisms—were recently part of Earth's atmosphere, in molecules of carbon dioxide.

Photosynthesis removes carbon dioxide from the atmosphere, and locks its carbon atoms inside organic compounds. When photosynthesizers and other aerobic organisms break down the organic compounds for energy, carbon atoms are released in the form of CO_2, which then reenters the atmosphere.

Since photosynthesis evolved, these two processes have constituted a balanced cycle of the biosphere. You will learn more about the carbon cycle in Section 17.6. For now, know that the amount of carbon dioxide that photosynthesis removes from the atmosphere is roughly the same amount that organisms release back into it—at least it was, until humans came along.

FIGURE 5.15 Visible evidence of fossil fuel emissions in the atmosphere: the sky over New York City on a sunny day.

As early as 8,000 years ago, humans began burning forests to clear land for agriculture. When trees and other plants burn, most of the carbon locked in their tissues is released into the atmosphere as carbon dioxide. Fires that occur naturally release carbon dioxide the same way.

Today, we are burning a lot more than our ancestors ever did. In addition to wood, we are burning fossil fuels—coal, petroleum, and natural gas—to satisfy our greater and greater demands for energy (Figure 5.15). Fossil fuels are the organic remains of ancient organisms. When we burn these fuels, the carbon that has been locked in them for hundreds of millions of years is released back into the atmosphere—mainly as carbon dioxide.

Our activities have put Earth's atmospheric cycle of carbon dioxide out of balance. We are adding far more CO_2 to the atmosphere than photosynthetic organisms are removing from it. Today, we release about 28 billion tons of carbon dioxide into the atmosphere each year, more than ten times the amount we released in the year 1900. Most of it comes from burning fossil fuels. How do we know? Researchers can determine how long ago the carbon atoms in a sample of CO_2 were part of a living organism by measuring the ratio of different carbon isotopes in it (you will read more about radioisotope dating techniques in Section 11.4). These results are correlated with fossil fuel extraction, refining, and trade statistics.

Researchers find pockets of our ancient atmosphere in Antarctica. Snow and ice have been accumulating in layers there, year after year, for the last 15 million years. Air and dust trapped in each layer reveal the composition of the atmosphere that prevailed when the layer formed. Thus, we now know that the

FIGURE 5.16 Working toward a sustainable future. *Above*, biofuels researchers Ratna Sharma and Mari Chinn of North Carolina State University are trying to find an economically feasible way to manufacture biofuel from agricultural wastes and weeds such as switchgrass. *Right*, soils scientists Harry Pionke and Ron Schnabel examining a stand of switchgrass on a farm.

atmospheric CO_2 level had been relatively stable for about 10,000 years before the industrial revolution. Since 1850, the CO_2 level has been steadily rising. In 2008, it was higher than it had been in *24 million years*.

The increase in atmospheric carbon dioxide is having dramatic effects on climate. CO_2 contributes to global climate change. We are seeing a warming trend that mirrors the increase in CO_2 levels—Earth is now the warmest it has been for 12,000 years. The trend is affecting biological systems everywhere. Life cycles are changing: Birds are laying eggs earlier; plants are flowering earlier than usual; mammals are hibernating for shorter periods. Migration patterns and habitats are also changing. These changes may be too fast for many species, and the rate of extinctions is predicted to rise.

Under normal circumstances, extra carbon dioxide stimulates photosynthesis, which means extra carbon dioxide uptake. However, changes in temperature and moisture patterns as a result of global warming are offsetting this benefit because they are proving harmful to plants and other photosynthesizers.

Making biofuel production economically feasible is a high priority for today's energy researchers (Figure 5.16). Biofuels are a renewable source of energy: We can always make more of them simply by growing more biomass. Also, unlike fossil fuels, using plant matter for fuel recycles carbon that is already in the atmosphere, because plants remove carbon dioxide from the atmosphere as they grow. Other research efforts to minimize or reverse global climate change target photosynthesis. For example, scientists are trying to duplicate the function of photosystems in the laboratory. If successful, they will be able to harness the energy of light to split water into hydrogen, oxygen, and electrons—all of which can be used as clean, green sources of energy.

How Would

YOU

☑ **Vote ?**

Ethanol and other fuels manufactured from crops currently cost more than gasoline, but they are renewable energy sources and have fewer emissions. How much of a premium would you pay to drive a vehicle that runs on biofuels? **See CengageNow for details, then vote online (cengagenow.com).**

Summary

Section 5.1 Autotrophs make their own food using energy they get directly from the environment, and carbon from inorganic sources such as CO_2. By metabolic pathways of photosynthesis, plants and other autotrophs capture the energy of light and use it to build sugars from water and carbon dioxide. Heterotrophs get energy and carbon from molecules that other organisms have already assembled.

Section 5.2 Pigments such as chlorophyll *a* absorb visible light of particular wavelengths for photosynthesis. Light that is not absorbed is reflected as a pigment's characteristic color.

Section 5.3 In chloroplasts, the light-dependent reactions of photosynthesis occur at a much-folded thylakoid membrane. The membrane forms a continuous compartment in the chloroplast's stroma, where the light-independent reactions occur.

$$6CO_2 + 6H_2O \xrightarrow{\text{light energy}} C_6H_{12}O_6 + 6O_2$$
carbon dioxide   water   glucose   oxygen

CENGAGENOW Use the animation on CengageNow to view sites where photosynthesis takes place.

Section 5.4 In the light-dependent reactions, photosynthetic pigments in the thylakoid membrane absorb light energy and pass it to photosystems, which then release electrons. The photosystem replaces lost electrons by pulling them from water molecules, which then dissociate. The released oxygen atoms leave the cell as O_2. The electrons flow through electron transfer chains in the thylakoid membrane, and end up in NADPH. The activity of transfer chains cause a hydrogen ion gradient to form across the thylakoid membrane. The ions flow back across the membrane through ATP synthases. The flow causes these membrane transport proteins to produce ATP in the stroma.

CENGAGENOW Review the light-dependent reactions of photosynthesis with the animation on CengageNow.

Section 5.5 In the stroma of chloroplasts, the enzyme rubisco fixes carbon from CO_2 in the Calvin–Benson cycle. This cyclic pathway uses energy from ATP, carbon and oxygen from CO_2, and hydrogen and electrons from NADPH to make glucose and water. Alternative light-independent reactions minimize photorespiration in some types of plants.

CENGAGENOW See how glucose is produced in the light-independent reactions of photosynthesis with the animation on CengageNow.

Section 5.6 Most modern organisms convert the chemical energy of carbohydrates to the chemical energy of ATP by oxygen-requiring aerobic respiration. In eukaryotes, this pathway finishes in mitochondria. Coenzymes pick up electrons in the first two stages, acetyl–CoA formation and the Krebs cycle. The energy of those electrons drives ATP synthesis in the third stage, electron transfer phosphorylation. Oxygen combines with the electrons and H^+ at the end of electron transfer chains, so water forms.

$$C_6H_{12}O_6 + 6O_2 + 36\ ADP \longrightarrow 6CO_2 + 6H_2O + 36\ ATP$$
glucose   oxygen   carbon dioxide   water

CENGAGENOW Use the animation on CengageNow to see how each step in aerobic respiration contributes to a big energy harvest.

Section 5.7 Fermentation pathways finish in the cytoplasm. They do not use oxygen or electron transfer chains, so no ATP forms other than the small yield from glycolysis. The final steps regenerate NAD^+, which allows glycolysis to continue. The end product of lactate fermentation is lactate. The end product of alcoholic fermentation is ethyl alcohol, or ethanol.

CENGAGENOW Use the animation on CengageNow to compare alcoholic and lactate fermentation.

Section 5.8 In humans and other organisms, the simple sugars from carbohydrate breakdown, the glycerol and fatty acids from fat breakdown, and the carbon backbones of amino acids from protein breakdown may enter aerobic respiration at various reaction steps.

CENGAGENOW Use the interaction on CengageNow to follow the breakdown of different organic molecules.

Section 5.9 Human activities are disrupting the global cycling of carbon dioxide. We are adding more CO_2 to the atmosphere than photoautotrophs are removing from it, and the resulting imbalance fuels global warming.

Self-Quiz   Answers in Appendix I

1. A cat eats a bird, which ate a caterpillar that chewed on a weed. Which organisms are autotrophs? Heterotrophs?

2. Photosynthetic autotrophs use _____ from the air as a carbon source and _____ as their energy source.

3. Light-dependent reactions in plants occur in the _____ .
   a. thylakoid membrane   c. stroma
   b. plasma membrane   d. cytoplasm

4. In the light-dependent reactions, _____ .
   a. carbon dioxide is fixed   c. CO_2 accepts electrons
   b. ATP forms   d. sugars form

5. What accumulates inside the thylakoid compartment during the light-dependent reactions?
   a. glucose   b. RuBP   c. hydrogen ions   d. CO_2

6. Light-independent reactions proceed in the _____ .
   a. cytoplasm   b. plasma membrane   c. stroma

7. The Calvin–Benson cycle starts when _____ .
   a. light is available
   b. carbon dioxide is attached to RuBP
   c. electrons leave a photosystem

8. After photosynthesis evolved, its by-product, _____ , accumulated and changed the atmosphere.

Digging Into Data

Energy Efficiency of Biofuel Production From Corn, Soy, and Prairie Grasses

Most corn is grown intensively in vast swaths, which means farmers who grow it use fertilizers and pesticides, both of which are typically made from fossil fuels. Corn is an annual plant, and yearly harvests tend to cause runoff that depletes soil and pollutes rivers.

In 2006, David Tilman and his colleagues published the results of a 10-year study comparing the net energy output of various biofuels. The researchers grew a mixture of native perennial grasses without irrigation, fertilizer, pesticides, or herbicides, in sandy soil that was so depleted by intensive agriculture that it had been abandoned. They measured the usable energy in biofuels made from the grasses, from corn, and from soy. They also measured the energy it took to grow and produce each kind of biofuel (Figure 5.17).

1. About how much energy did ethanol produced from one hectare of corn yield? How much energy did it take to grow and produce that ethanol?

2. Which biofuel tested had the highest ratio of energy output to energy input?

3. Which of the three crops would require the least amount of land to produce a given amount of biofuel energy?

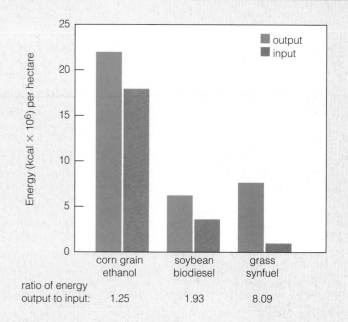

FIGURE 5.17 Energy inputs and outputs of biofuels from corn and soy grown on fertile farmland, and grassland plants grown in infertile soil. One hectare is about 2.5 acres.

---

9. Is the following statement true or false? Unlike animals, which make many ATP by aerobic respiration, plants make all of their ATP by photosynthesis.

10. Which molecule does not form during glycolysis?
   a. NADH   b. pyruvate   c. $FADH_2$ d. ATP

11. Glycolysis starts and ends in the _____ .
   a. nucleus   c. plasma membrane
   b. mitochondrion   d. cytoplasm

12. In eukaryotes, aerobic respiration is completed in the _____ .
   a. nucleus   c. plasma membrane
   b. mitochondrion   d. cytoplasm

13. In eukaryotes, fermentation is completed in the _____ .
   a. nucleus   c. plasma membrane
   b. mitochondrion   d. cytoplasm

14. In the third stage of aerobic respiration, _____ is the final acceptor of electrons from glucose.
   a. water   b. hydrogen   c. oxygen   d. NADH

15. Match the event with its most suitable description.
   _____ glycolysis   a. ATP, NADH, $FADH_2$,
   _____ fermentation   and CO_2 form
   _____ Krebs cycle   b. glucose to two pyruvates
   _____ electron transfer   c. NAD^+ regenerated, little ATP
            phosphorylation   d. H^+ flows through ATP
   _____ CO_2 fixation   synthases
                               e. rubisco function

Additional questions are available on CENGAGENOW™

Critical Thinking

1. About 200 years ago, Jan Baptista van Helmont wanted to know where growing plants get the materials necessary for increases in size. He planted a tree seedling weighing 5 pounds in a barrel filled with 200 pounds of soil and then watered the tree regularly. After five years, the tree weighed 169 pounds, 3 ounces, and the soil weighed 199 pounds, 14 ounces. Because the tree had gained so much weight and the soil had lost so little, he concluded that the tree had gained all of its additional weight by absorbing the water he had added to the barrel, but of course he was incorrect. What really happened?

2. At high altitudes, the oxygen levels are low. Climbers of very tall mountains risk altitude sickness, which is characterized by shortness of breath, weakness, dizziness, and confusion. The early symptoms of cyanide poisoning are the same as those for altitude sickness. Cyanide binds tightly to cytochrome c oxidase, a protein complex that is the last component of mitochondrial electron transfer chains. Cytochrome c oxidase with bound cyanide can no longer transfer electrons. Explain why cyanide poisoning starts with the same symptoms as altitude sickness.

3. As you learned, membranes impermeable to hydrogen ions are required for electron transfer phosphorylation. Membranes in mitochondria serve this purpose in eukaryotes. Prokaryotic cells do not have this organelle, but they do make ATP by electron transfer phosphorylation. How do you think they do it, given that they have no mitochondria?

6 DNA Structure and Function

6.1 Impacts/Issues: Here, Kitty, Kitty, Kitty, Kitty, Kitty

One of the most publicized demonstrations of the function of DNA occurred in 1997, when Scottish geneticist Ian Wilmut made the first genetic copy, or clone, of a fully grown animal. His team removed the nucleus (and the DNA it contained) from an unfertilized sheep egg. They replaced it with the nucleus of a cell taken from the udder of a different sheep. The hybrid egg became an embryo, and then a lamb. The lamb, whom the researchers named Dolly, was genetically identical to the sheep that had donated the udder cell.

At first, Dolly looked and acted like a normal sheep. But five years later, she was as fat and arthritic as a twelve-year-old sheep. The following year, Dolly contracted a lung disease that usually affects only old sheep, and she had to be euthanized. Dolly's telomeres hinted that she had developed health problems because she was a clone. Telomeres are short, repeated DNA sequences at the ends of chromosomes. They become shorter and shorter as an animal ages. When Dolly was only two years old, her telomeres were as short as those of a six-year-old sheep—the exact age of the adult animal that had been her genetic donor.

Since Dolly, many adult animals have been cloned, including mice, rats, rabbits, fish, pigs, cattle, goats, mules, deer, horses, cats, dogs, and a wolf. However, cloning mammals is still far from routine. It usually takes hundreds of attempts to produce one embryo, and most embryos that do form die before birth or shortly afterward. Many of the clones that survive have serious health problems.

What causes these problems? Even though all cells of an individual *inherit* the same DNA, an adult cell *uses* only a fraction of it compared to an embryonic cell. To make a clone from an adult cell, researchers must reprogram its DNA to function like the DNA of an egg. Even though we are getting better at doing that, we still have a lot to learn.

So why do we keep trying? The potential benefits are enormous. Already, cells of human–pig hybrid embryos are helping researchers unravel the molecular mechanisms of human genetic diseases. Cloned human cells may one day be induced to form replacement tissues or organs for people with incurable diseases. Endangered animals might be saved from extinction; extinct animals may be brought back. Livestock and pet animals are already being cloned commercially (Figure 6.1).

Perfecting the methods for cloning animals brings us closer to the possibility of cloning humans, both technically and ethically. For example, if cloning a lost cat for a grieving pet owner is acceptable, why would it not be acceptable to clone a lost child for a grieving parent? Different people have very different answers to such questions, so controversy over adult cloning continues to rage even as the techniques improve. Understanding the basis of heredity—what DNA is and how it works—will help inform your own opinions about the ethical issues surrounding cloning.

FIGURE 6.1 Adult cloning. Compare the markings on Tahini, a Bengal cat (*top*) with those of her clones, Tabouli and Baba Ganoush (*bottom*). Eye color changes as a Bengal cat matures; both of the kittens later developed the same eye color as Tahini's.

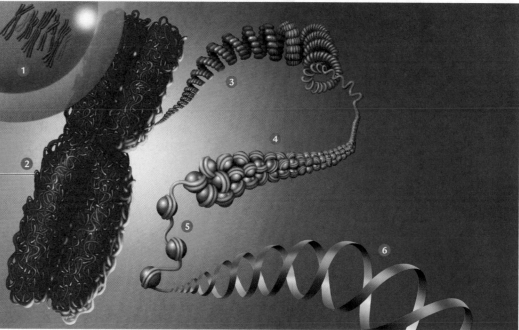

Chromosomes

All organisms pass DNA to offspring when they reproduce. Inside a cell, each DNA molecule is organized as a structure called a **chromosome** (Figure 6.2). Eukaryotic cells have a number of chromosomes ❶. During most of a cell's life, each of its chromosomes consists of one DNA molecule. As it prepares to divide, the cell duplicates its chromosomes, so that both of its offspring get a full set. After the chromosomes are duplicated, each consists of two DNA molecules, which are then called **sister chromatids**. Sister chromatids attach to each other at a constricted region called a **centromere**:

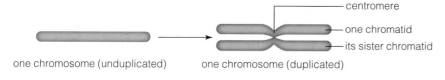

one chromosome (unduplicated)   one chromosome (duplicated)

If stretched out end to end, the 46 chromosomes in a human cell would be about 2 meters (6.5 feet) long. That is a lot of DNA to fit into a nucleus that is typically less than 10 micrometers in diameter! Interactions between a DNA molecule and the proteins that associate with it structurally organize a chromosome and help it pack tightly so it fits in a nucleus.

At its most condensed, a duplicated chromosome consists of two long, tangled filaments (the sister chromatids) bunched into a characteristic X shape ❷. A closer look reveals that each filament is actually a hollow tube formed by coils, like a phone cord ❸. The coils themselves consist of a twisted fiber of DNA and proteins ❹. The DNA molecule wraps twice at regular intervals around "spools" of proteins called **histones** ❺. In micrographs, the DNA–histone spools look like beads on a string. Each "bead" is called a **nucleosome**, the smallest unit of chromosomal organization in eukaryotes. As you will see in the next section, the DNA molecule consists of two strands twisted into a double helix ❻.

FIGURE 6.2 Animated! Zooming in on chromosome structure. Tight packing allows a lot of DNA to fit into a very small nucleus.

❶ The DNA inside the nucleus of a eukaryotic cell is typically divided up into a number of chromosomes. *Left*, a human chromosome.

❷ At its most condensed, a duplicated chromosome is packed tightly into an X shape.

❸ A chromosome unravels as a single fiber, a hollow cylinder formed by coiled coils.

❹ The coiled coils consist of a long molecule of DNA (*blue*) and the proteins that are associated with it (*purple*).

❺ At regular intervals, the DNA molecule is wrapped twice around a core of histone proteins. In this "beads-on-a-string" structure, the "string" is the DNA, and each "bead" is called a nucleosome.

❻ The DNA molecule itself has two strands that are twisted into a double helix.

centromere Constricted region in a eukaryotic chromosome where sister chromatids are attached.

chromosome A structure that consists of DNA and associated proteins; carries part or all of a cell's genetic information.

histone Type of protein that structurally organizes eukaryotic chromosomes.

nucleosome A length of DNA wound around a spool of histone proteins.

sister chromatid One of two attached members of a duplicated eukaryotic chromosome.

Table 6.1	Examples of Chromosome Number	
Species		**Chromosome Number**
Fruit fly		8
Amoeba		13
Garden pea		14
Frog		26
Cat		38
Human		46
Potato		48
Pineapple		50
Cow		60
Dog		78
Vizcacha rat		102
Horsetail		216
Adder's tongue fern		1200

Chromosome Number The genetic information of each eukaryotic species is distributed among some number of chromosomes, which differ in length and shape. The sum of all chromosomes in a cell of a given type is called the **chromosome number**. Each species has a characteristic chromosome number (Table 6.1). For example, the chromosome number of oak trees is 12, so the nucleus of a cell from an oak tree contains 12 chromosomes. The chromosome number of king crab cells is 208, so they have 208 chromosomes. That of human body cells is 46, so human body cells have 46 chromosomes.

Actually, human body cells have two of each type of chromosome, which means that their chromosome number is **diploid** (*2n*). The 23 pairs of chromosomes are like two sets of books numbered 1 to 23. There are two versions of each book: a pair. Except for a pairing of sex chromosomes (XY) in males, the two members of each pair have the same length and shape, and they hold information about the same traits. Think of them as two sets of books on how to build a house. Your father gave you one set. Your mother had her own ideas about wiring, plumbing, and so on. She gave you an alternate set that says slightly different things about many of those tasks.

Types of Chromosomes All except one pair of a diploid cell's chromosomes are **autosomes**, which are the same in both females and males. The two autosomes of a pair have the same length, shape, and centromere location. Members of a pair of **sex chromosomes** differ between females and males. The differences determine an individual's sex. The sex chromosomes of humans are called X and Y. Body cells of human females contain two X chromosomes (XX); those of human males contain one X and one Y chromosome (XY).

XX females and XY males are the rule among fruit flies, mammals, and many other animals, but there are other patterns. In butterflies, moths, birds, and certain fishes, males have two identical sex chromosomes, and females do not. Environmental factors (not sex chromosomes) determine sex in some species of invertebrates, turtles, and frogs. As an example, the temperature of the sand in which sea turtle eggs are buried determines the sex of the hatchlings.

In humans, a new individual inherits a combination of sex chromosomes that dictates whether it will become a male or a female. All eggs made by a human female have one X chromosome. One-half of the sperm cells made by a male carry an X chromosome; the other half carry a Y chromosome. If an X-bearing sperm fertilizes an X-bearing egg, the resulting individual will develop into a female. If the sperm carries a Y chromosome, the individual will develop into a male (Figure 6.3).

Karyotyping reveals an individual's diploid complement of chromosomes. With this technique, cells taken from an individual are treated to make their chromosomes condense, and stained so the chromosomes become visible under a microscope. The microscope reveals the chromosomes in every cell. A micro-

diploid reproductive cell in female

diploid reproductive cell in male

XX
X X

Xx
X Y

eggs

sperm

X X

X Y

X Y

	X	Y
X	XX	XY
X	XX	XY

union of sperm and egg at fertilization

FIGURE 6.3 Animated! Pattern of sex determination in humans. The grid shows how sex chromosome combinations result in female (*pink*) or male (*blue*) offspring.

Figure It Out: About what proportion of human newborns would you expect to be male?

Answer: About 50 percent

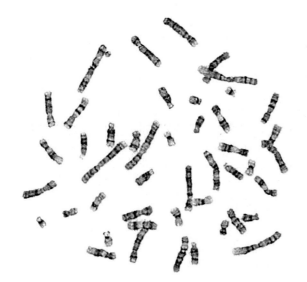

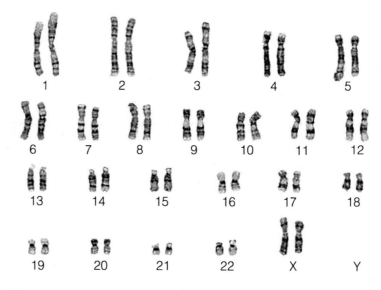

A The chromosomes of one body cell are isolated, then stained to reveal differences in banding patterns.

B The image is reassembled so that the chromosomes are paired by size, centromere position, and other characteristics.

FIGURE 6.4 Animated! A karyotype is an image of a single cell's diploid set of chromosomes. This human karyotype shows 22 pairs of autosomes and a pair of X chromosomes.

Figure It Out: Was this cell taken from a female or a male? *Answer: A female*

graph of a single cell is digitally rearranged so the images of the chromosomes are lined up by centromere location, and arranged according to size, shape, and length (Figure 6.4). The finished array constitutes the individual's **karyotype**, which is compared with a normal standard. A karyotype shows how many chromosomes are in the individual's cells. Comparison with a standard can also reveal any extra or missing chromosomes, and some structural abnormalities.

Take-Home Message **What are chromosomes?**

- A eukaryotic cell's DNA is divided among some characteristic number of chromosomes, which differ in length and shape.

- Members of a pair of sex chromosomes differ between males and females. Other chromosomes are autosomes—the same in males and females.

- Proteins that associate with DNA structurally organize chromosomes and allow them to pack tightly.

6.3 Fame and Glory

Each strand of DNA is a polymer of nucleotides that have been linked into a chain (Section 2.10). Even though a chain can be hundreds of millions of nucleotides long, only four kinds of nucleotides compose DNA. A DNA nucleotide has a five-carbon sugar, three phosphate groups, and one of four nitrogen-containing

autosome Any chromosome other than a sex chromosome.

chromosome number The sum of all chromosomes in a cell of a given type.

diploid Having two of each type of chromosome characteristic of the species ($2n$).

karyotype Image of an individual's complement of chromosomes arranged by size, length, shape, and centromere location.

sex chromosome Member of a pair of chromosomes that differs between males and females.

adenine (A)
deoxyadenosine triphosphate

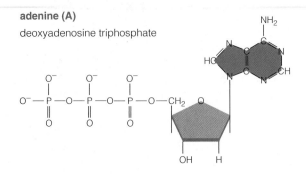

guanine (G)
deoxyguanosine triphosphate

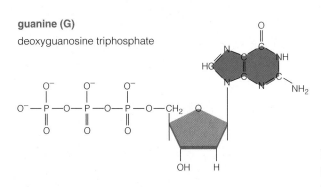

thymine (T)
deoxythymidine triphosphate

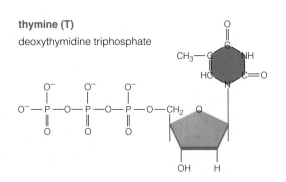

cytosine (C)
deoxycytidine triphosphate

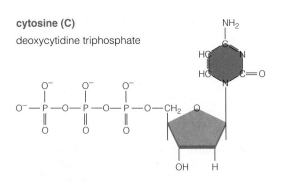

FIGURE 6.5 The four nucleotides in DNA. Each kind has three phosphate groups, a deoxyribose sugar (*orange*), and a nitrogen-containing base (*blue*) after which it is named.

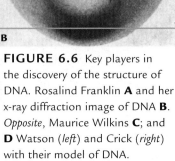

B

FIGURE 6.6 Key players in the discovery of the structure of DNA. Rosalind Franklin **A** and her x-ray diffraction image of DNA **B**. *Opposite*, Maurice Wilkins **C**; and **D** Watson (*left*) and Crick (*right*) with their model of DNA.

A

bases (Figure 6.5). Just how those four nucleotides—adenine (A), guanine (G), thymine (T), and cytosine (C)—are arranged in DNA was a puzzle that took over 50 years to solve. As molecules go, DNA is gigantic, and chromosomal DNA has a complex structural organization. Both factors made the molecule difficult to work with given the laboratory methods at the time.

In 1950, Erwin Chargaff, one of many researchers who had been trying to solve the structure of DNA, made two discoveries. First, the amounts of thymine and adenine in all DNA are the same, as are the amounts of cytosine and guanine. We call this discovery Chargaff's first rule:

$$A = T \text{ and } G = C$$

Chargaff's second discovery, or rule, is that the proportion of adenine and guanine differs among DNA of different species.

Meanwhile, American biologist James Watson and British biophysicist Francis Crick had been sharing their ideas about the structure of DNA. The helical pattern of secondary structure that occurs in many proteins (Section 2.9) had just been discovered, and Watson and Crick suspected that the DNA molecule was also a helix. They had spent many hours arguing about the size, shape, and bonding requirements of the four kinds of nucleotides that make up DNA. They had pestered chemists to help them identify bonds they might have overlooked. They had fiddled with cardboard cutouts, and made models from scraps of metal connected by suitably angled "bonds" of wire.

Rosalind Franklin (Figure 6.6**A**) had also been working on the structure of DNA. Like Crick, Franklin specialized in x-ray crystallography, a technique in which x-rays are directed through a purified and crystallized substance. Atoms in the substance's molecules scatter the x-rays in a pattern that can be captured as an image. Researchers use the pattern to calculate the size, shape, and spacing between any repeating elements of the molecules—all of which are details of molecular structure.

Franklin had already used x-ray crystallography to solve the complex and unorganized structure of coal. She had been told she would be the only one

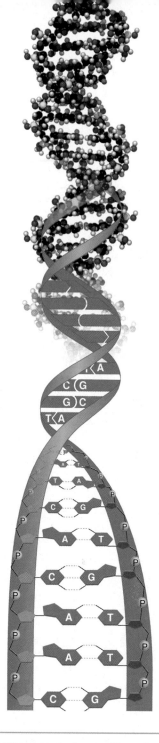

in her department working on the structure of DNA, so she did not know that Maurice Wilkins was already doing the same thing just down the hall.

Wilkins and Franklin had been given identical samples of carefully prepared DNA. Franklin's meticulous work with her sample yielded the first clear x-ray diffraction image of DNA as it occurs inside cells (Figure 6.6**B**), and she gave a presentation on this work in 1952. DNA, she said, had two chains twisted into a double helix, with a backbone of phosphate groups on the outside, and bases arranged in an unknown way on the inside. She had calculated DNA's diameter, the distance between its chains and between its bases, the pitch (angle) of the helix, and the number of bases in each coil. Crick, with his crystallography background, would have recognized the significance of the work—if he had been there. Watson was in the audience but he was not a crystallographer, and he did not understand the implications of Franklin's image or her calculations.

Franklin started to write a research paper on her findings. Meanwhile, and perhaps without her knowledge, Wilkins (Figure 6.6**C**) reviewed her x-ray diffraction image with Watson, and Watson and Crick read a report containing her unpublished data. Franklin's data provided Watson and Crick with the last piece of the DNA puzzle. In 1953, they put together all of the clues that had been accumulating for the last fifty years and built the first accurate model of the DNA molecule (Figure 6.6**D**). On April 25, 1953, Rosalind Franklin's work appeared third in a series of articles about the structure of DNA in the journal *Nature*. Wilkins's research paper was the second article in the series. The work of Franklin and Wilkins supported with experimental evidence Watson and Crick's theoretical model, which was presented in the first article.

The Double Helix Watson and Crick proposed that DNA's structure consists of two chains (or strands) of nucleotides, running in opposite directions and coiled into a double helix (Figure 6.7). Bonds between the sugar of one nucleotide and the phosphate of the next form the backbone of each chain. Hydrogen bonds between the internally positioned bases hold the two strands together. Only two kinds of base pairings form: A to T, and G to C, which explains the first of Chargaff's rules.

FIGURE 6.7 Animated! Structure of DNA, as illustrated by a composite of three different models. The two sugar–phosphate backbones coil in a helix around internally positioned bases. **Figure It Out:** What do the yellow balls represent? *Answer: Phosphate groups*

Variations in the base sequence of DNA are the foundation of life's diversity.

1 The two strands of a DNA molecule are complementary: their nucleotides match up according to base-pairing rules (G to C, T to A).

2 As replication starts, the two strands of DNA unwind at many sites along the length of the molecule.

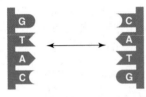

3 Each parent strand serves as a template for assembly of a new DNA strand from nucleotides, according to base-pairing rules.

4 DNA ligase seals any gaps that remain between bases of the "new" DNA, so a continuous strand forms. The base sequence of each half-old, half-new DNA molecule is identical to that of the parent.

FIGURE 6.8 Animated! DNA replication. Each strand of a DNA double helix is copied; two double-stranded DNA molecules result.

Most scientists had assumed (incorrectly) that the bases had to be on the outside of the helix, because they would be more accessible to DNA-copying enzymes that way. You will see in the next section how DNA replication enzymes access the bases on the inside of DNA's double helix.

Dozens of scientists contributed to the discovery of DNA's structure, but only three received recognition from the general public for their work. Rosalind Franklin died at age 37, of ovarian cancer probably caused by extensive exposure to x-rays. Because the Nobel Prize is not given posthumously, she did not share in the 1962 honor that went to Watson, Crick, and Wilkins for the discovery of the structure of DNA.

Patterns of Base Pairing Just two kinds of base pairings give rise to the incredible diversity of traits we see among living things. How? Even though DNA is composed of only four bases, the *order* in which one base pair follows the next—the **DNA sequence**—varies tremendously among species (which explains Chargaff's second rule). For example, a small piece of DNA from a tulip, a human, or any other organism might be:

one base pair

Notice how the two strands of DNA match up. They are complementary, which means each base on one strand is suitably paired with a partner base on the other. This bonding pattern (A to T, G to C) is the same in all molecules of DNA. However, which base pair follows the next differs among species, and among individuals of the same species. The information encoded by that sequence is the basis of visible traits that define species and distinguish individuals. Thus DNA, the molecule of inheritance in every cell, is the basis of life's unity. Variations in its base sequence are the foundation of life's diversity.

Take-Home Message What is the structure of DNA?

- A DNA molecule consists of two nucleotide chains (strands) running in opposite directions and coiled into a double helix. Internally positioned nucleotide bases hydrogen-bond between the two strands. A pairs with T, and G with C.
- The sequence of bases along a DNA strand is genetic information.
- DNA sequences vary among species and among individuals. This variation is the basis of life's diversity.

6.4 DNA Replication and Repair

During most of its life, a cell contains only one set of chromosomes. However, when the cell reproduces, it must have two sets of chromosomes: one for each of its future offspring. By a process called **DNA replication**, a cell copies its chromosomes before it divides (Figure 6.8).

Before replication begins, each chromosome consists of one molecule of DNA—one double helix **1**. During DNA replication, an enzyme breaks the hydrogen bonds that hold the double helix together, so the two DNA strands unwind **2**. An enzyme called **DNA polymerase** assembles a complementary strand of DNA on each of the parent strands.

The sequence of a cell's DNA is its genetic information, as you will see in the next chapter. Descendant cells must get an exact copy of it, or inheritance will go awry. As a DNA polymerase moves along a strand of DNA, it uses the sequence of bases as a template, or guide, to assemble a new strand of DNA from free nucleotides (Figure 6.9**A**). The base sequence of the new strand is complementary to that of the template, because DNA polymerase follows base-pairing rules. The polymerase adds a T to the end of the new DNA strand when it reaches an A in the parent DNA sequence; it adds a G when it reaches a C; and so on ❸. Phosphate-group transfers (Section 4.3) from the nucleotides provide energy for their own attachment to the end of a growing strand of DNA. Another enzyme, **DNA ligase**, seals any gaps, so a new, continuous strand of DNA results.

As each new DNA strand lengthens, it winds up with its "parent" strand into a double helix (Figure 6.9**B**). So, after replication, two double-stranded molecules of DNA have formed. One strand of each molecule is parental (old), and the other is new (Figure 6.9**C**). Because each new strand of DNA is complementary in sequence to a parent strand, both double-stranded molecules that result from DNA replication are duplicates of the parent ❹.

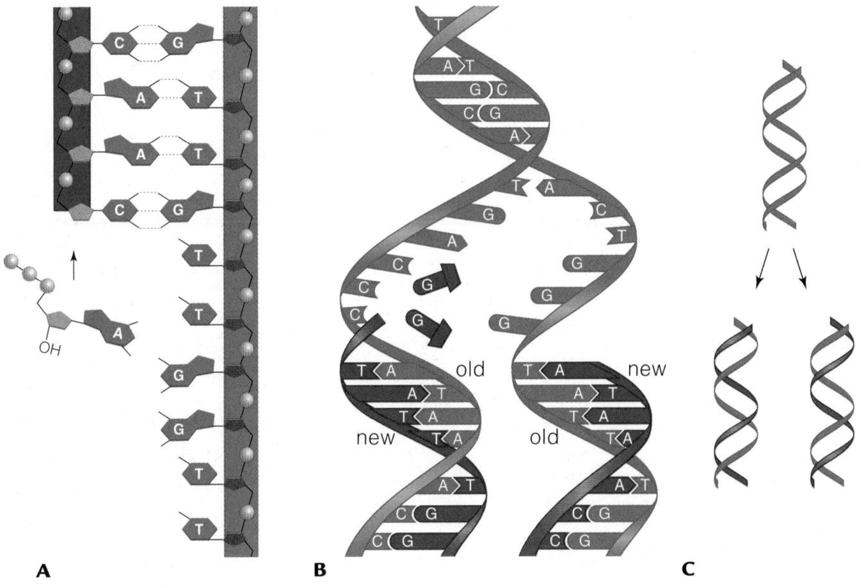

FIGURE 6.9 How a double-stranded molecule of DNA is replicated. **A** DNA polymerase uses a parental DNA strand as a template to assemble a new strand of DNA from nucleotides. **B** The two parental DNA strands (*blue*) stay intact. A new strand (*magenta*) is assembled on each of the parental (old) strands. **C** One strand of each DNA molecule that forms is new, and the other is parental.

Checking for Mistakes

DNA is not always replicated with perfect accuracy. Sometimes the wrong base is added to a growing DNA strand. At other times, bases get lost, or extra ones get added. Either way, the new DNA strand will no longer match up perfectly with its parent strand.

Some errors occur after the DNA becomes damaged by exposure to radiation or toxic chemicals, because DNA polymerases do not copy damaged DNA very well. In most cases, **DNA repair mechanisms** fix damaged DNA by removing and replacing any damaged or mismatched bases before replication begins.

Most DNA replication errors occur simply because DNA polymerases add nucleotides to a growing strand of DNA very quickly—up to 1,000 bases per second. Mistakes are inevitable, and some DNA polymerases make many of them. Luckily, most DNA polymerases proofread their own work. They correct any mismatches by immediately reversing the synthesis reaction to remove a mismatched nucleotide, and then resuming synthesis. If an error remains uncorrected, other mechanisms usually stop the cell from dividing.

When proofreading and repair mechanisms fail, an error becomes a **mutation**—a permanent change in the DNA sequence. Mutations can alter DNA's genetic instructions, and the outcome may be harmful or not. Mutations are the original source of variations in traits—the raw material of evolution, as you will see in later chapters.

Take-Home Message

How is DNA copied?

- A cell replicates its DNA before it divides. Each strand of the double helix serves as a template for synthesis of a new, complementary strand of DNA.

- DNA repair mechanisms and proofreading maintain the integrity of a cell's genetic information. Unrepaired errors may become mutations.

DNA ligase Enzyme that seals breaks or gaps in double-stranded DNA.

DNA polymerase DNA replication enzyme. Assembles a new strand of DNA based on the sequence of a DNA template.

DNA repair mechanism Any of several processes by which enzymes repair damaged DNA.

DNA replication Process by which a cell duplicates its DNA before it divides.

DNA sequence The order of nucleotide bases in a strand of DNA.

mutation Permanent change in DNA sequence.

Cloning Adult Animals

The word "cloning" means making an identical copy of something. In biology, cloning can refer to a laboratory method by which researchers copy DNA fragments (a technique we discuss in Chapter 10). It can also refer to interventions in reproduction that result in an exact genetic copy of an organism.

Genetically identical organisms occur all the time in nature. They arise mainly by the process of asexual reproduction, which we discuss in Chapter 8. Embryo splitting, another natural process, results in identical twins. The first few divisions of a fertilized egg form a ball of cells that sometimes splits spontaneously. If both halves continue to develop independently, identical twins result. Artificial embryo splitting has been routine in research and animal husbandry for decades. With this technique, a ball of cells is grown from a fertilized egg in a laboratory. The ball is teased apart into two halves, each of which goes on to develop as a separate embryo. The embryos are implanted in surrogate mothers, who give birth to identical twins. Artificial twinning and any other technology that yields genetically identical individuals is called **reproductive cloning**.

Twins get their DNA from two parents, which typically differ in their DNA sequence. Thus, although identical twins produced by embryo splitting are identical to one another, they are not identical to either parent. When animal breeders want an exact copy of a specific individual, they may turn to a cloning method that starts with a single cell taken from an adult organism. Such procedures present more of a technical challenge than embryo splitting. Unlike a fertilized egg, a body cell from an adult will not automatically start dividing. It must first be tricked into rewinding its developmental clock.

All cells descended from a fertilized egg inherit the same DNA. Thus, the DNA in each living cell of an individual is like a master blueprint that contains enough information to build an entirely new individual. As different cells in a developing embryo start using different subsets of their DNA, they differentiate, or become different in form and function. In animals, differentiation is usually a one-way path. Once a cell specializes, all of its descendant cells will be special-

egg

donor
cell

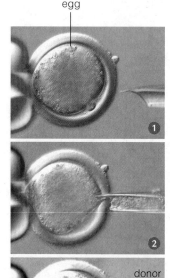

FIGURE 6.10 Animated! Somatic cell nuclear transfer, using cattle cells. This series of micrographs was taken by scientists at Cyagra, a commercial company that specializes in cloning livestock.

➊ A cow egg is held in place by suction through a hollow glass tube called a micropipette. The DNA in the cell is identified by a *purple* stain.

➋ Another micropipette punctures the egg and sucks out the DNA. All that remains inside the egg's plasma membrane is cytoplasm.

➌ A new micropipette prepares to enter the egg at the puncture site. The pipette contains a cell grown from the skin of a donor animal.

➍ The micropipette enters the egg and delivers the skin cell to a region between the cytoplasm and the plasma membrane.

➎ After the pipette is withdrawn, the donor's skin cell is visible next to the cytoplasm of the egg. The transfer is complete.

➏ The egg is exposed to an electric current. This treatment causes the foreign cell to fuse with and empty its nucleus into the cytoplasm of the egg. The egg begins to divide, and an embryo forms. After a few days, the embryo may be transplanted into a surrogate mother.

ized the same way. By the time a liver cell, muscle cell, or other specialized cell forms, most of its DNA has been turned off, and is no longer used.

To clone an adult, scientists must first transform one of its differentiated cells into an undifferentiated cell by turning its unused DNA back on. In **somatic cell nuclear transfer** (SCNT), a researcher removes the nucleus from an unfertilized egg, then inserts into the egg a nucleus from an adult animal cell (Figure 6.10). A somatic cell is a body cell, as opposed to a reproductive cell (*soma* is a Greek

word for body). If all goes well, the egg's cytoplasm reprograms the transplanted DNA to direct the development of an embryo, which is then implanted into a surrogate mother. The animal that is born to the surrogate is genetically identical with the donor of the nucleus (Figure 6.11). Dolly the sheep (pictured at *left*) and the other animals described in the chapter introduction were produced using SCNT.

Adult cloning is now a common practice among people who breed prized livestock. Among other benefits, many more offspring can be produced in a given time frame by cloning than by traditional breeding methods, and offspring can be produced after a donor animal is castrated or even dead.

The controversial issue with adult cloning is not necessarily about livestock. As the techniques become routine, cloning a human is no longer only within the realm of science fiction. Researchers are already using SCNT to produce human embryos for research, a practice called **therapeutic cloning**. The researchers harvest undifferentiated (stem) cells from the cloned human embryos. (We return to the topic of stem cells and their potential medical benefits in Chapter 19.) Reproductive cloning of humans is not the intent of such research, but somatic cell nuclear transfer would be the first step toward that end.

FIGURE 6.11 Liz the cow (*right*) and her clone. The clone was produced by somatic cell nuclear transfer, as in Figure 6.10.

Take-Home Message  What is cloning?

- Reproductive cloning technologies produce a clone: an exact genetic copy of an individual.

- Somatic cell nuclear transfer (SCNT) is a reproductive cloning method in which nuclear DNA of an adult donor is transferred to an egg with no nucleus. The hybrid cell develops into an embryo that is genetically identical to the donor.

- Therapeutic cloning uses SCNT to produce human embryos for research.

How Would YOU ☑ Vote?

Some view sickly or deformed clones as unfortunate but acceptable casualties of animal cloning research that also yields medical advances for human patients. Should animal cloning be banned? **See CengageNow for details, then vote online (cengagenow.com).**

6.6 Impacts/Issues Revisited:
Here, Kitty, Kitty, Kitty, Kitty, Kitty

Human eggs are difficult to come by. They also come with a hefty set of ethical dilemmas. Thus, researchers have started to make hybrid embryos using adult human cells and pig eggs. Cells from the resulting embryos are being used to study, among other things, how fatal diseases progress. For example, embryos created using cells from people with genetic heart defects will allow researchers to study how the defect causes developing heart cells to malfunction. Such research may ultimately lead to treatments for people who suffer from fatal diseases.

reproductive cloning Technology that produces genetically identical individuals.

somatic cell nuclear transfer (SCNT) Method of reproductive cloning in which genetic material is transferred from an adult somatic cell into an unfertilized, enucleated egg.

therapeutic cloning Using SCNT to produce human embryos for research.

Summary

Section 6.1 Making clones, or exact genetic copies, of adult animals is now a common practice in research and animal husbandry.

 Section 6.2 A eukaryotic chromosome is a molecule of DNA together with associated proteins. The proteins organize the DNA so it can pack into a small nucleus. Diploid cells have two of each type of chromosome. Chromosome number is the sum of all chromosomes in cells of a given type. A human body cell has twenty-three pairs of chromosomes. Members of a pair of sex chromosomes differ among males and females. All others are autosomes. Autosomes of a pair have the same length, shape, and centromere location, and they carry the same genes. Karyotyping reveals an individual's chromosomes.

CENGAGENOW Explore chromosomes, sex determination in humans, and karyotyping with the animations on CengageNow.

 Section 6.3 A DNA molecule consists of two strands of DNA coiled into a helix. Nucleotide monomers are joined to form each strand. A DNA nucleotide has a five-carbon sugar (deoxyribose), three phosphate groups, and one of four nitrogen-containing bases after which the nucleotide is named: adenine, thymine, guanine, or cytosine.

Bases of the two DNA strands in a double helix pair in a consistent way: adenine with thymine (A–T), and guanine with cytosine (G–C). The order of the bases (the DNA sequence) varies among species and among individuals. The DNA of each species has unique sequences that set it apart from the DNA of all other species.

CENGAGENOW Learn more about the structure of nucleotides and DNA with the animations on CengageNow.

Section 6.4 A cell replicates its DNA before it divides. Each double-stranded molecule of DNA results in two double-stranded DNA molecules identical to the parent. One strand of each molecule is new, and the other is parental.

During the replication process, the double helix unwinds. DNA polymerase uses each strand as a template to assemble new, complementary strands of DNA from free nucleotides. DNA ligase seals any gaps to form a continuous strand.

DNA repair mechanisms fix damaged DNA. Proofreading by DNA polymerases corrects most base-pairing errors. Uncorrected errors may become mutations.

CENGAGENOW See how a DNA molecule is replicated with the animation on CengageNow.

 Section 6.5 Reproductive cloning technologies produce genetically identical individuals (clones).

CENGAGENOW Observe procedures used to create clones of adult animals with the animation on CengageNow.

Self-Quiz Answers in Appendix I

1. The chromosome number _____ .
   a. refers to a particular chromosome pair in a cell
   b. is an identifiable feature of a species
   c. is like a set of books

2. What are the base-pairing rules for DNA?
   a. A–G, T–C   c. A–U, C–G
   b. A–C, T–G   d. A–T, G–C

3. One species' DNA differs from others in its _____ .
   a. sugars   c. base sequence
   b. phosphates   d. all of the above

4. When DNA replication begins, _____ .
   a. the two DNA strands unwind from each other
   b. the two DNA strands condense for base transfers
   c. two DNA molecules bond
   d. old strands move to find new strands

5. DNA replication requires _____ .
   a. template DNA   c. DNA polymerase
   b. free nucleotides   d. all of the above

6. Replication of a DNA molecule results in _____ .
   a. a single strand of DNA   c. two single strands of DNA
   b. a double-stranded DNA   d. two double strands of DNA

7. DNA polymerase adds nucleotides to _____ .
   a. double-stranded DNA   c. double-stranded RNA
   b. single-stranded DNA   d. single-stranded RNA

8. Show the complementary strand of DNA that forms on this template DNA fragment during replication: GGTTTCTTCAAGAGA

9. _____ is an example of reproductive cloning.
   a. Somatic cell nuclear transfer (SCNT)
   b. Multiple offspring from the same pregnancy
   c. Artificial embryo splitting
   d. a and c
   e. all of the above

10. Match the terms appropriately.
   _____ nucleotide   a. adds nucleotides to a
   _____ clone   growing DNA strand
   _____ DNA ligase   b. copy of an organism
   _____ DNA polymerase   c. fills in gaps, seals breaks
        in a DNA strand
        d. nitrogen-containing base,
        sugar, phosphate groups

Additional questions are available on **CENGAGENOW**

Critical Thinking

1. Mutations are permanent changes in a cell's DNA base sequence. They typically have negative consequences, but they are also the original source of genetic variation and the raw material of evolution. How can mutations accumulate, given that cells have repair systems that fix changes or breaks in DNA strands?

2. There may be millions of woolly mammoths frozen in the ice of Siberian glaciers. These huge elephant-like mammals have been extinct for about 10,000 years, but a few research groups are planning to resurrect one of them by cloning DNA isolated from frozen remains. What are some of the pros and cons, both technical and ethical, of cloning an extinct animal?

A Micrograph of three viruses injecting DNA into an *E. coli* cell.

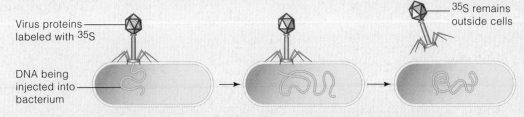

B In one experiment, bacteria were infected with virus particles labeled with a radioisotope of sulfur (^{35}S). The sulfur had labeled only viral proteins. The viruses were dislodged from the bacteria by whirling the mixture in a kitchen blender. Most of the radioactive sulfur was detected in the viruses, not in the bacterial cells. The viruses had not injected protein into the bacteria.

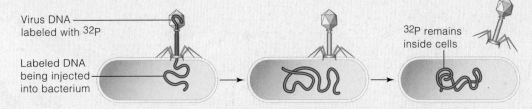

C In another experiment, bacteria were infected with virus particles labeled with a radioisotope of phosphorus (^{32}P). The phosphorus had labeled only viral DNA. When the viruses were dislodged from the bacteria, the radioactive phosphorus was detected mainly inside the bacterial cells. The viruses had injected DNA into the cells—evidence that DNA is the genetic material of this virus.

FIGURE 6.12 Animated! The Hershey–Chase experiments. Alfred Hershey and Martha Chase tested whether the genetic material injected by bacteriophage into bacteria is DNA, protein, or both. The experiments were based on the knowledge that proteins contain more sulfur (S) than phosphorus (P), and DNA contains more phosphorus than sulfur.

Digging Into Data

The Hershey–Chase Experiments

By 1950, researchers had discovered bacteriophage, a type of virus that infects bacteria. These infectious particles carry hereditary information about how to make new viruses. After a virus infects a cell, the cell starts making new virus particles. Bacteriophages inject genetic material into bacteria, but was that material DNA, protein, or both? Alfred Hershey and Martha Chase decided to find out by exploiting long-known properties of protein (high sulfur content) and of DNA (high phosphorus content). They cultured bacteria in a medium containing an isotope of sulfur, ^{35}S (Section 2.2). In this medium, the protein (but not the DNA) of bacteriophage that infected the bacteria became labeled with ^{35}S.

Hershey and Chase allowed the labeled viruses to infect bacteria. They knew from electron micrographs that phages attach to bacteria by their slender tails (Figure 6.12**A**). They reasoned it would be easy to break this precarious attachment, so they poured the mixture of virus and bacteria into a Waring blender and turned it on.

After blending, the researchers separated the bacteria from the virus-containing fluid, and measured the ^{35}S content of each separately. The fluid contained most of the ^{35}S. Thus, the viruses had not injected protein into the bacteria (Figure 6.12**B**).

Hershey and Chase repeated the experiment using an isotope of phosphorus, ^{32}P, which labeled the DNA (but not the proteins) of the bacteriophage. This time, they found that the bacteria contained most of the ^{32}P. The viruses had injected DNA into the bacteria (Figure 6.12**C**).

The graph in Figure 6.12**D** is reproduced from Alfred Hershey and Martha Chase's 1952 publication. "Infected bacteria" refers to the percentage of bacteria that survived the blender.

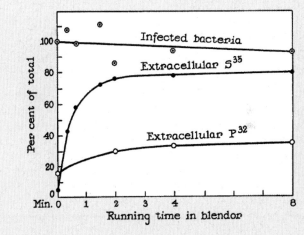

D

From the Journal of General Physiology, 36(1), September 20, 1952: "Independent Functions of Viral Protein and Nucleic Acid in Growth of Bacteriophage."

1. Why did the amount of radioactivity in each set of samples change over time?

2. After 4 minutes in the blender, what percentage of ^{35}S was outside the bacteria? What percentage of ^{32}P was outside the bacteria?

4. How did the researchers know that the radioisotopes in the fluid came from outside the bacterial cells (extracellular) and not from bacteria that had broken apart?

5. The extracellular concentration of which isotope, ^{35}S or ^{32}P, increased the most with blending? Why do these results imply that bacteriophage viruses inject DNA into bacteria?

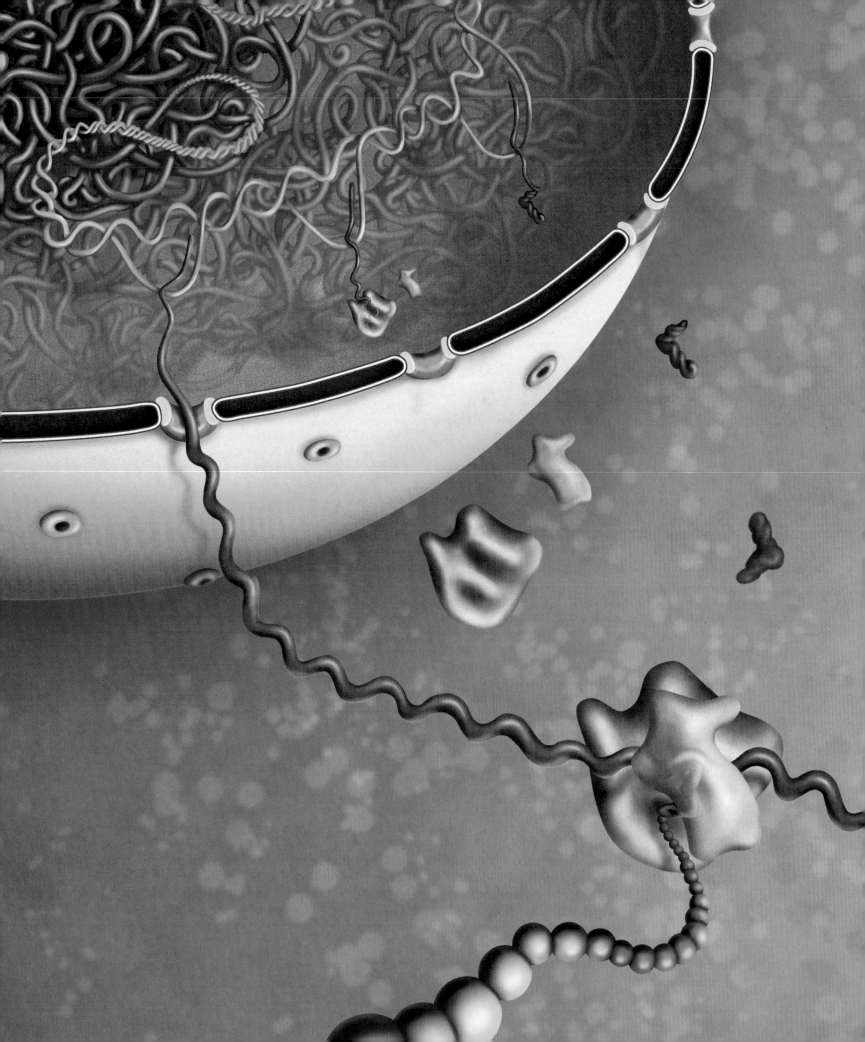

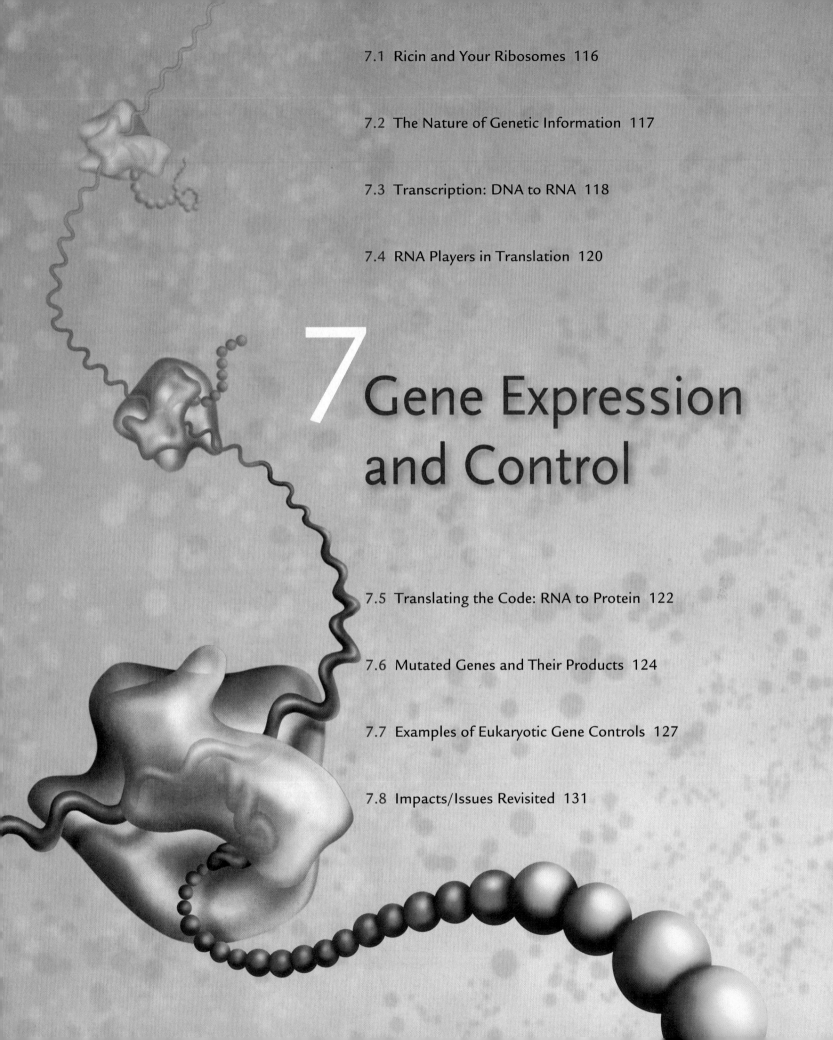

7 Gene Expression and Control

Proteins are critical to all life processes,
so cells that cannot make them die very quickly.

7.1 Impacts/Issues: Ricin and Your Ribosomes

Ricin is a highly toxic, naturally occurring protein: A dose as small as a few grains of salt can kill an adult. Only botulinum and tetanus toxins are more deadly, and there is no antidote. Ricin effectively deters many animals from eating any part of the castor-oil plant (*Ricinus communis*), which grows wild in tropical regions worldwide and is widely cultivated for its seeds (Figure 7.1). Castor-oil seeds are the source of castor oil, an ingredient in plastics, cosmetics, paints, soaps, polishes, and many other items. After the oil is extracted from the seeds, the ricin is typically discarded along with the leftover seed pulp.

The lethal effects of ricin were known as long ago as 1888, but using ricin as a weapon is now banned by most countries under the Geneva Protocol. However, controlling its production is impossible, because it takes no special skills or equipment to manufacture the toxin from easily obtained raw materials. Thus, ricin appears periodically in the news.

For example, at the height of the Cold War, the Bulgarian writer Georgi Markov had defected to England and was working as a journalist for the BBC. As he made his way to a bus stop on a London street, an assassin used the tip of a modified umbrella to jam a small, ricin-laced ball into Markov's leg. Markov died in agony three days later.

More recently, police in 2003 acted on an intelligence tip and stormed a London apartment, where they found laboratory glassware and castor-oil beans. Traces of ricin were found in a United States Senate mailroom and State Department building, and also in an envelope addressed to the White House in 2004. In 2005, the FBI arrested a man who had castor-oil beans and an assault rifle stashed in his Florida home. Jars of banana baby food laced with ground castor-oil beans also made the news in 2005. In 2006, police found pipe bombs and a baby food jar full of ricin in a Tennessee man's shed. In 2008, castor beans, firearms, and several vials of ricin were found in a Las Vegas motel room after its occupant was hospitalized for ricin exposure.

Ricin is toxic because it inactivates ribosomes, the organelles upon which amino acids are assembled into proteins in all cells (Sections 3.5 and 3.6). Proteins are critical to all life processes, so cells that cannot make them die very quickly. Someone who inhales ricin typically dies from low blood pressure and respiratory failure within a few days of exposure.

This chapter details how the information encoded by a gene becomes converted to the gene's product—an RNA or a protein. Even though it is extremely unlikely that your ribosomes will ever encounter ricin, protein synthesis is nevertheless worth appreciating for how it keeps you and all other organisms alive.

FIGURE 7.1 Beautiful and deadly: seeds of the castor-oil plant, source of ribosome-busting ricin. Eating eight of these seeds can kill an adult.

116 Unit Two Genetics

7.2 The Nature of Genetic Information

DNA is like a book, an encyclopedia that carries instructions for building a new individual. You already know the alphabet used to write the book: the four letters A, T, G, and C, for the four nucleotide bases adenine, thymine, guanine, and cytosine. A strand of DNA is a chain of those four kinds of nucleotides. The information it carries consists of which nucleotide follows the next along the strand—the DNA sequence (Section 6.3). Part of that information occurs in subsets called genes. A cell uses the sequence of a **gene** to build an RNA or protein product. Converting a gene's DNA sequence into its product starts with **transcription**: a process in which enzymes use the DNA sequence of a gene as a template to assemble a strand of RNA. Transcription makes an RNA copy of a gene. In other words, it *transcribes* the base sequence of a gene into a similar form: the base sequence of an RNA.

RNA is similar to a single strand of DNA in that both are chains of four kinds of nucleotides. However, the nucleotides that make up RNA differ slightly from those that make up DNA. The sugar in an RNA nucleotide is ribose, and that of a DNA nucleotide is a deoxyribose. Three of the bases (adenine, cytosine, and guanine) are the same in DNA and RNA nucleotides (Figure 7.2). The fourth base in RNA is uracil, not thymine.

Despite the seemingly small differences in structure, DNA and RNA have very different functions. DNA's important but only role is to store a cell's heritable information. By contrast, cells make several kinds of RNAs, each of which has a different function. Although all RNAs are encoded by DNA, **messenger RNA (mRNA)** is the only kind of RNA that carries a protein-building message. That message is encoded within the sequence of the mRNA itself by sets of three bases, "genetic words" that follow one another along the length of the mRNA. Like the words of a sentence, a series of genetic words can form a meaningful parcel of information—in this case, the sequence of amino acids of a protein. By the process of **translation**, the protein-building information in an mRNA is *translated* into a different language: a sequence of amino acids. The result is a polypeptide that twists and folds into a protein.

The processes of transcription and translation are part of **gene expression**, a multistep process by which information encoded in a gene becomes converted to an RNA or protein product. During gene expression, genetic information flows from DNA to RNA to protein:

$$DNA \xrightarrow{\text{transcription}} mRNA \xrightarrow{\text{translation}} protein$$

A cell's DNA sequence contains the information it needs to make the molecules of life. Each gene encodes an RNA, and RNAs interact to assemble proteins. Proteins (enzymes, in particular) assemble lipids and carbohydrates, replicate DNA, make RNA, and perform many other functions that keep the cell alive.

Take-Home Message What is the information carried by DNA?

- The nucleotide base sequence of a gene encodes instructions for building an RNA or protein product.
- A cell transcribes the base sequence of a gene into RNA.
- Messenger RNA (mRNA) carries a protein-building message.

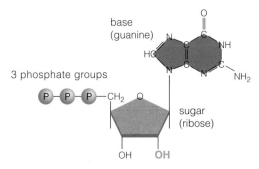

An RNA nucleotide: guanine (G), or guanosine triphosphate

A Guanine, one of the four nucleotides in RNA. The others (adenine, uracil, and cytosine) differ only in their component bases (*blue*). Three of the four bases in RNA nucleotides are identical to the bases in DNA nucleotides.

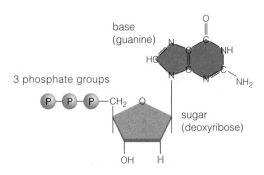

A DNA nucleotide: guanine (G), or deoxyguanosine triphosphate

B Compare the DNA nucleotide guanine. The only difference between the RNA and DNA versions of guanine (or adenine, or cytosine) is the hydrogen atom or hydroxyl group at one position on the sugar (shown in *green*).

FIGURE 7.2 Comparison between **A** an RNA nucleotide and **B** a DNA nucleotide.

gene Part of a DNA base sequence; specifies an RNA or protein product.

gene expression Process by which the information in a gene becomes converted to an RNA or protein product.

messenger RNA (mRNA) Type of RNA that has a protein-building message.

transcription Process by which an RNA is assembled from nucleotides using the base sequence of a gene as a template.

translation Process by which a polypeptide chain is assembled from amino acids in the order specified by an mRNA.

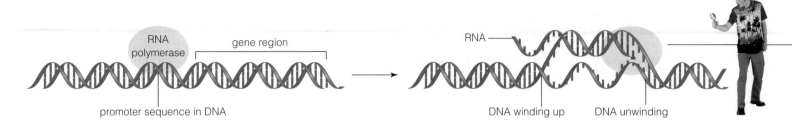

promoter sequence in DNA

RNA polymerase

gene region

RNA

DNA winding up   DNA unwinding

1 RNA polymerase binds to a promoter in the DNA. The binding positions the polymerase near a gene. In most cases, the base sequence of the gene occurs on only one of the two DNA strands. Only the DNA strand complementary to the gene sequence will be translated into RNA.

2 The polymerase begins to move along the DNA and unwind it. As it does, it links RNA nucleotides into a strand of RNA in the order specified by the base sequence of the DNA. The DNA winds up again after the polymerase passes. The structure of the "opened" DNA at the transcription site is called a transcription bubble, after its appearance.

FIGURE 7.3 Animated! Transcription assembles a strand of RNA from nucleotides using a gene region in DNA as a template.

7.3 Transcription: DNA to RNA

Remember that DNA replication begins with one DNA double helix and ends with two DNA double helices (Section 6.4). The two double helices are identical to the parent molecule because the process of DNA replication follows base-pairing rules. A nucleotide can be added to a growing strand of DNA only if it base-pairs with the corresponding nucleotide of the parent strand: G pairs with C, and A pairs with T (Section 6.3):

DNA
| A | T | G | T | A | C | T | C | A | G | T | G | T | A | C | A |
| T | A | C | A | T | G | A | G | T | C | A | C | A | T | G | T |
DNA

The same base-pairing rules also govern RNA synthesis in transcription. An RNA strand is structurally so similar to a DNA strand that the two can base-pair if their base sequences are complementary. In such hybrid molecules, G pairs with C, and A pairs with U (uracil):

RNA
| A | U | G | U | A | C | U | C | A | G | U | G | U | A | C | A |
| T | A | C | A | T | G | A | G | T | C | A | C | A | T | G | T |
DNA

During transcription, a strand of DNA acts as a template upon which a strand of RNA is assembled from RNA nucleotides. A nucleotide can be added to a growing strand of RNA only if it is complementary to the corresponding nucleotide of the parent strand of DNA: G pairs with C, and A pairs with U. Thus, each new RNA is complementary in sequence to the DNA strand that served as its template. As in DNA replication, each nucleotide provides the energy for its own attachment to the end of a growing strand.

Transcription is similar to DNA replication in that one strand of a nucleic acid serves as a template for synthesis of another. However, in contrast with DNA replication, only part of one DNA strand (a gene) serves as the template, not the whole molecule. The enzyme **RNA polymerase**, not DNA polymerase, adds nucleotides to the end of a growing strand. Also, transcription results in a single strand of RNA, not two DNA double helices.

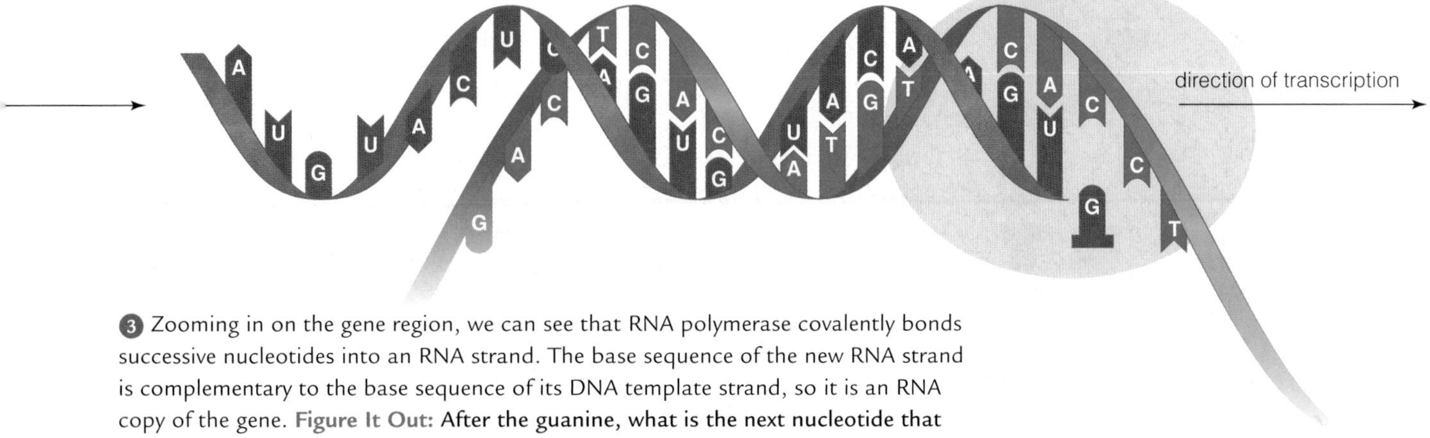

direction of transcription

3 Zooming in on the gene region, we can see that RNA polymerase covalently bonds successive nucleotides into an RNA strand. The base sequence of the new RNA strand is complementary to the base sequence of its DNA template strand, so it is an RNA copy of the gene. **Figure It Out:** After the guanine, what is the next nucleotide that will be added to this growing strand of RNA?

Answer: Another guanine (G)

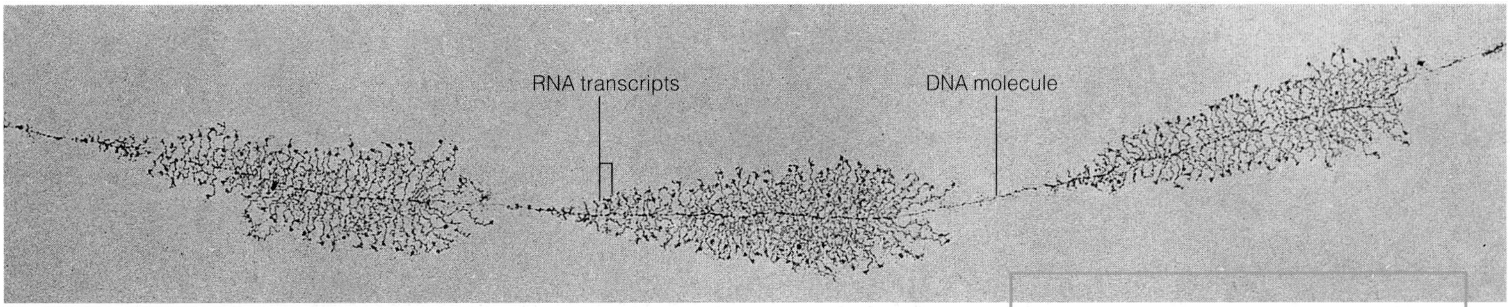

RNA transcripts

DNA molecule

FIGURE 7.4 Many RNA polymerases typically transcribe the same gene at the same time. The composite structure that forms is called a "Christmas tree" after its shape. Here, three genes next to one another on a chromosome are being transcribed by many polymerases.

Figure It Out: Are the polymerases transcribing this DNA molecule moving from left to right or from right to left?

Answer: Left to right (the transcripts get longer as the polymerases move along the DNA)

The Process of Transcription Transcription occurs in the nucleus of eukaryotic cells. It begins with a chromosome, which is a double helix molecule of DNA. The process gets under way when an RNA polymerase and several regulatory proteins attach to a specific binding site in the DNA called a **promoter** (Figure 7.3). The binding positions the polymerase at a transcription start site close to a gene **1**. The polymerase starts moving along the DNA, over the gene **2**. As it moves, the polymerase unwinds the double helix just a bit so it can "read" the base sequence of the noncoding DNA strand. As it reads, the polymerase joins free RNA nucleotides into a chain, in the order dictated by that DNA sequence. As in DNA replication, each nucleotide provides the energy for its own attachment to the end of a growing strand.

When the polymerase reaches the end of the gene, the DNA and the new RNA strand are released. The new RNA strand is complementary in base sequence to the DNA strand from which it was transcribed **3**. It is an RNA copy of a gene. Typically, many polymerases transcribe a particular gene region at the same time, so many new RNA strands can be produced very quickly (Figure 7.4).

Take-Home Message How is RNA assembled?

- In transcription, RNA polymerase uses the nucleotide base sequence of a gene region in a chromosome as a template to assemble a strand of RNA.
- The new strand of RNA is a copy of the gene from which it was transcribed.

promoter In DNA, a sequence to which RNA polymerase binds.

RNA polymerase Enzyme that carries out transcription.

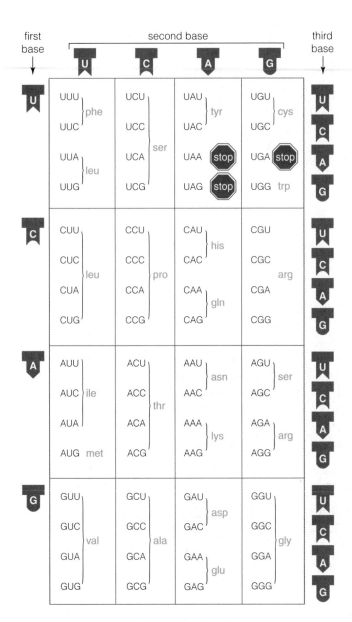

FIGURE 7.5 Animated! The genetic code. The left column lists a codon's first base, the top row lists the second, and the right column lists the third. Amino acid names are abbreviated. The list below gives the complete names and the one-letter abbreviations. **Figure It Out:** Which codons specify lysine (lys)? *Answer: AAA and AAG*

ala	alanine (A)	leu	leucine (L)
arg	arginine (R)	lys	lysine (K)
asn	asparagine (N)	met	methionine (M)
asp	aspartic acid (D)	phe	phenylalanine (F)
cys	cysteine (C)	pro	proline (P)
glu	glutamic acid (E)	ser	serine (S)
gln	glutamine (Q)	thr	threonine (T)
gly	glycine (G)	trp	tryptophan (W)
his	histidine (H)	tyr	tyrosine (Y)
ile	isoleucine (I)	val	valine (V)

7.4 RNA Players in Translation

mRNA and the Genetic Code An mRNA is a disposable copy of a gene. Its job is to carry DNA's protein-building information to the other two types of RNA for translation. That protein-building information consists of a linear sequence of genetic "words" spelled with an alphabet of the four bases A, C, G, and U.

Each of the genetic "words" carried by an mRNA is three bases long, and each is a code—a **codon**—for a particular amino acid. There are four possible bases in each of the three positions of a codon, so there are a total of sixty-four (or 4^3) mRNA codons. Collectively, the sixty-four codons constitute the **genetic code** (Figure 7.5). Which of the four nucleotides is in the first, second, and third position of a base triplet determines which amino acid the codon specifies. For instance, the codon AUG codes for the amino acid methionine (met), and UGG codes for tryptophan (trp).

One codon follows the next along the length of an mRNA, so the order of codons in an mRNA determines the order of amino acids in the polypeptide that will be translated from it. Thus, the base sequence of a gene is transcribed into the base sequence of an mRNA, which is in turn translated into an amino acid sequence:

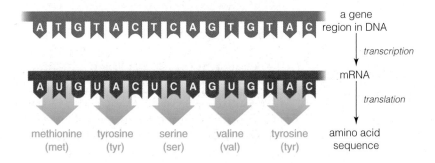

There are only twenty kinds of amino acids found in proteins. Sixty-four codons are more than are needed to specify twenty amino acids, so some amino acids are specified by more than one codon. For instance, GAA and GAG both code for glutamic acid. Some codons signal the beginning and end of a protein-coding sequence. For example, the first AUG in an mRNA is the signal to start translation in most species. AUG also happens to be the codon for methionine, so methionine is always the first amino acid in new polypeptides of such organisms. UAA, UAG, and UGA do not specify an amino acid. They are signals that stop translation, so they are called stop codons. A stop codon marks the end of the protein-coding sequence in an mRNA.

The genetic code is highly conserved, which means that many organisms use the same code and probably always have. Prokaryotes and some protists have a few codons that vary, as do mitochondria and chloroplasts. The variation was a clue that led to a theory of how organelles evolved (we will return to this topic in Section 13.3).

rRNA and tRNA—The Translators Ribosomes and transfer RNAs (tRNA) interact to translate an mRNA into a polypeptide. A ribosome has one large and one small subunit, both of which consist

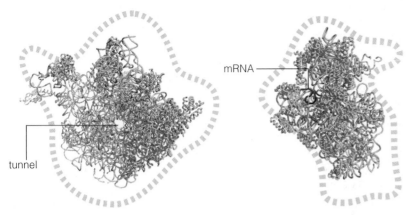

A large subunit of a ribosome   **B** small subunit of a ribosome

FIGURE 7.6 A ribosome (*right*) consists of **A** a large subunit and **B** a small subunit. Structural protein components of the subunits are shown in *green*.

Notice the tunnel through the interior of the large subunit. rRNA components of the ribosome (*tan*) assemble a polypeptide chain, which threads through this tunnel as it forms. We show an mRNA (in *red*) attached to the small subunit.

an intact ribosome

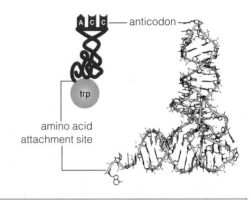

FIGURE 7.7 tRNA. *Above*, models of the tRNA that carries the amino acid tryptophan. Each tRNA's anticodon is complementary to an mRNA codon. Each also carries the amino acid specified by that codon.

Below, during translation, tRNAs dock at an intact ribosome. Here, three tRNAs are docked at the small ribosomal subunit (the large subunit is not shown, for clarity). The anticodons of the tRNAs line up with complementary codons in an mRNA (shown in *red*).

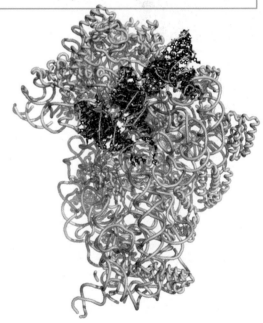

of structural proteins and rRNA. **Ribosomal RNA (rRNA)** is the main component of the subunits. During translation, a large and a small subunit converge as an intact ribosome on an mRNA (Figure 7.6). **Transfer RNAs (tRNAs)** deliver amino acids to ribosomes in the order specified by the mRNA. Each tRNA has two attachment sites: One is an **anticodon**, a triplet of nucleotides that base-pairs with an mRNA codon (Figure 7.7). The other binds to an amino acid—the one specified by the codon. tRNAs with different anticodons carry different amino acids. You will see in the next section how those tRNAs deliver amino acids, one after the next, to a ribosome–mRNA complex during translation.

rRNA is one of the few examples of RNA with enzymatic activity: The rRNA of a ribosome, not the protein, forms a peptide bond between amino acids. As the amino acids are delivered, the ribosome joins them via peptide bonds into a new polypeptide chain (Section 2.9). Thus, the order of codons in an mRNA— DNA's protein-building message—becomes translated into a new protein.

Take-Home Message **What roles do mRNA, tRNA, and rRNA play during translation?**

- mRNA carries protein-building information. The bases in mRNA are "read" in sets of three during protein synthesis. Most of these base triplets (codons) code for amino acids. The genetic code consists of all sixty-four codons.

- Ribosomes, which consist of two subunits of rRNA and proteins, assemble amino acids into polypeptide chains.

- A tRNA has an anticodon complementary to an mRNA codon, and it has a binding site for the amino acid specified by that codon. tRNAs deliver amino acids to ribosomes.

anticodon Set of three nucleotides in a tRNA; base-pairs with mRNA codon.

codon In mRNA, a nucleotide base triplet that codes for an amino acid or stop signal during translation.

genetic code Set of sixty-four mRNA codons, each of which specifies an amino acid or stop signal in translation.

ribosomal RNA (rRNA) A type of RNA that becomes part of ribosomes.

transfer RNA (tRNA) Type of RNA that delivers amino acids to a ribosome during translation.

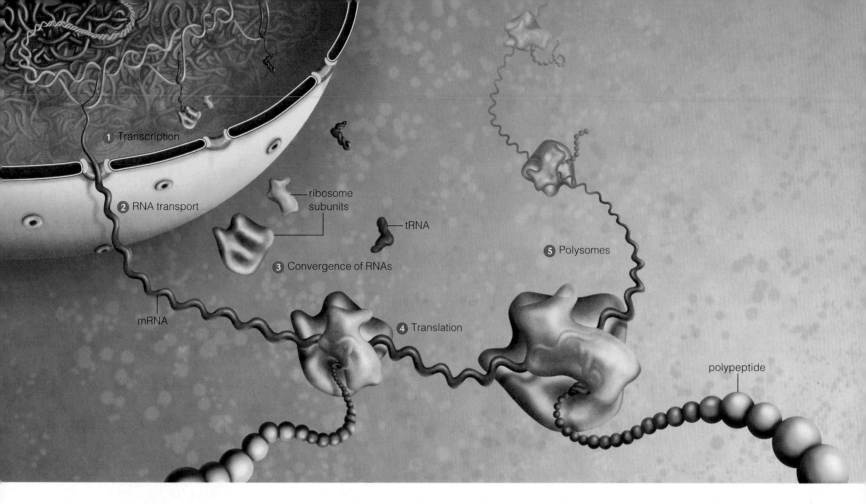

1 Transcription

2 RNA transport

ribosome subunits

tRNA

3 Convergence of RNAs

5 Polysomes

mRNA

4 Translation

polypeptide

FIGURE 7.8 Animated!

Translation in eukaryotes.

1 In eukaryotic cells, RNA is transcribed in the nucleus.

2 Finished RNA moves into the cytoplasm through nuclear pores.

3 Ribosomal subunits and tRNA converge on an mRNA.

4 A polypeptide chain forms as the ribosome moves along the mRNA, linking amino acids together in the order dictated by the mRNA codons.

5 Many ribosomes can translate an mRNA at the same time.

7.5 Translating the Code: RNA to Protein

Translation, the second part of protein synthesis, occurs in the cytoplasm of all cells (Figure 7.8). It has three stages: initiation, elongation, and termination. The initiation stage begins when a small ribosomal subunit binds to an mRNA. Next, the anticodon of a special tRNA called an initiator base-pairs with the first AUG codon of the mRNA. Then, a large ribosomal subunit joins the small subunit.

In the elongation stage, the ribosome assembles a polypeptide chain as it moves along the mRNA. The initiator tRNA carries the amino acid methionine, so the first amino acid of the new polypeptide chain is methionine. Another tRNA brings the second amino acid to the complex as its anticodons base-pair with the second codon in the mRNA. The ribosome joins the first two amino acids by way of a peptide bond (Section 2.9):

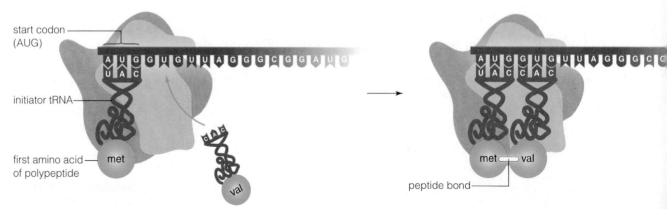

start codon (AUG)

initiator tRNA

first amino acid of polypeptide

met

val

met val

peptide bond

The first tRNA is released and the ribosome moves to the next codon. Another tRNA brings the third amino acid to the complex as its anticodons base-pair with the third codon of the mRNA. A peptide bond forms between the second and third amino acids:

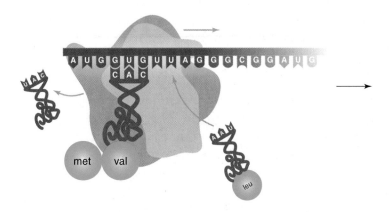
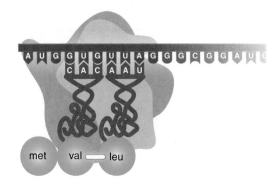

The second tRNA is released and the ribosome moves to the next codon. Another tRNA brings the fourth amino acid to the complex as its anticodons base-pair with the fourth codon of the mRNA. A peptide bond forms between the third and fourth amino acids:

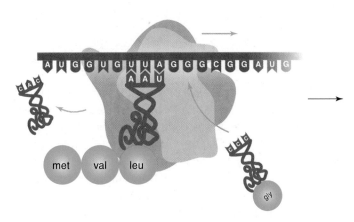
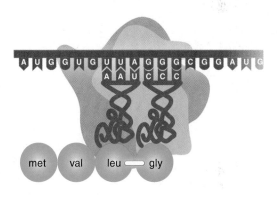

The new polypeptide chain grows as the ribosome causes peptide bonds to form between amino acids delivered by successive tRNAs.

Termination occurs when the ribosome reaches a stop codon in the mRNA. The mRNA and the polypeptide detach from the ribosome, and the ribosomal subunits separate from each other. Translation is now complete. The new polypeptide will join the pool of proteins in the cytoplasm, or it will enter rough ER of the endomembrane system (Section 3.6).

In cells that are making a lot of protein, many ribosomes may simultaneously translate the same mRNA, in which case they are called **polysomes**:

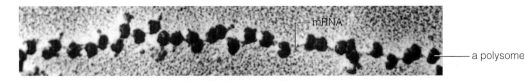

a polysome

Given that many polypeptides are translated from one mRNA, why would a cell also make many copies of an mRNA? Compared with DNA, RNA is not very stable. An mRNA may last only a few minutes before it is disassembled by enzymes in the cytoplasm. The fast turnover allows cells to adjust their protein synthesis quickly in response to changing needs.

polysome Cluster of ribosomes that are simultaneously translating an mRNA.

Transcription and translation both occur in the cytoplasm of prokaryotes, and these processes are closely linked in time and in space. Translation begins before transcription is done, so in these cells, a DNA transcription tree (Figure 7.4) may be decorated with polysome "balls."

Translation is an energy-intensive process. That energy is provided mainly in the form of phosphate-group transfers from the RNA nucleotide GTP (Figure 7.2**A**) to molecules involved in the process.

Take-Home Message How is mRNA translated into protein?

- Translation is an energy-requiring process that converts the protein-building information carried by an mRNA into a polypeptide.

- During initiation, an mRNA joins with an initiator tRNA and two ribosomal subunits.

- During elongation, amino acids are delivered to the complex by tRNAs in the order dictated by successive mRNA codons. As they arrive, the ribosome joins each to the end of the polypeptide chain.

- Termination occurs when the ribosome encounters a stop codon in the mRNA. The mRNA and the polypeptide are released, and the ribosome disassembles.

7.6 Mutated Genes and Their Products

Mutations are changes in the sequence of a cell's DNA. If a mutation changes the genetic instructions encoded in the DNA, an altered gene product may be the result. Remember, more than one codon can specify the same amino acid, so cells have a certain margin of safety. For example, a mutation that changes a UCU codon to UCC in an mRNA may not have further effects, because both codons specify serine. However, other mutations have drastic consequences.

The oxygen-binding properties of hemoglobin provide us with an example of how mutations can change the structure (and function) of a protein. A red blood cell carries oxygen molecules bound to hemoglobin molecules. As these cells circulate through the lungs, the hemoglobin proteins inside of them bind oxygen gas, then give it up in regions of the body with low oxygen levels. Then, the red blood cells return to the lungs and pick up more oxygen.

Hemoglobin can bind and release oxygen because of its structure. It consists of four polypeptides called globin chains, and each chain folds around a heme, a type of cofactor with an iron atom at its center. Oxygen binds to hemoglobin at those iron atoms. In adult humans, two alpha globin chains and two beta globin chains make up each hemoglobin molecule (Figure 7.9**A,B**). Defects in the chains can cause a condition called anemia, in which a person's blood is deficient in red blood cells or in hemoglobin. Either outcome limits the blood's ability to carry oxygen.

A **deletion** is a mutation in which one or more nucleotides are lost from DNA. The loss of a particular nucleotide from the beta globin gene causes beta thalassemia, a type of anemia. Like many other deletions, this one causes the reading frame of mRNA codons to shift. The shift garbles the genetic message, just as incorrectly grouping a series of letters garbles the meaning of a sentence:

> The cat ate the rat.
> T hec ata tet her at.
> Th eca tat eth era t.

If a mutation changes the genetic instructions encoded in DNA, an altered gene product may be the result.

---

base-pair substitution Type of mutation in which a single base-pair changes.

deletion Mutation in which one or more base pairs are lost.

insertion Mutation in which one or more base pairs become inserted into DNA.

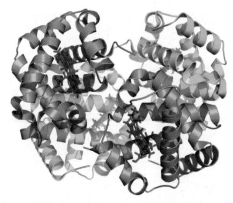

A Hemoglobin, an oxygen-transport protein in red blood cells. This protein consists of four globin chains: two alpha chains (*blue*) and two beta chains (*green*). Each globin chain folds up to form a pocket that cradles a type of cofactor called a heme (*red*). Oxygen binds to the iron atom at the center of each heme group.

FIGURE 7.9 Animated! Two common mutations of hemoglobin's beta globin chain.

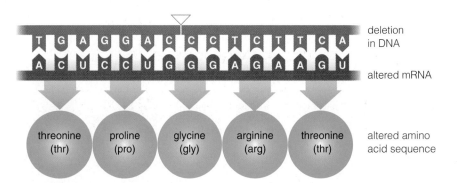

B Part of the DNA, mRNA, and amino acid sequence of the beta chain of a normal hemoglobin molecule.

C A single base-pair deletion causes the reading frame for the rest of the mRNA to shift, so a completely different protein product forms. This mutation results in a defective globin chain. The outcome is thalassemia, a genetic disorder in which a person has an abnormally low amount of hemoglobin.

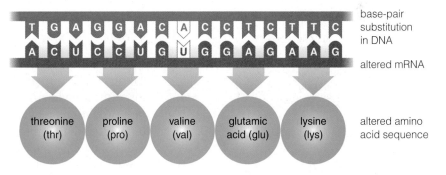

D A base-pair substitution in DNA replaces a thymine with an adenine. When the altered mRNA is translated, valine replaces glutamate as the sixth amino acid of the new polypeptide chain. Hemoglobin with this chain is called HbS, or sickle hemoglobin.

Expression of a beta globin gene with a thalassemia mutation results in a polypeptide that differs drastically from normal beta globin (Figure 7.9**C**). The result? Not enough hemoglobin molecules assemble, an outcome that is the source of the anemia.

Frameshifts may also be caused by **insertion** mutations, in which extra bases become inserted into DNA. Other mutations do not cause frameshifts. With a **base-pair substitution**, a nucleotide and its partner are replaced by a different base pair. A substitution may result in an amino acid change or a premature stop codon in a gene's protein product. Sickle-cell anemia, a type of anemia that is most common in people of African ancestry, arises because of a base-pair substitution. The substitution causes valine to be the sixth amino acid of the beta globin chain instead of glutamic acid (Figure 7.9**D**). Hemoglobin with this mutation in its beta chain is called sickle hemoglobin, or HbS.

Unlike glutamic acid, which carries a negative charge, valine carries no charge. As a result of that one substitution, a tiny patch of the beta globin chain changes from hydrophilic to hydrophobic, which in turn causes the hemoglobin's behavior to change slightly. HbS molecules stick together and form large, rodlike clumps under certain conditions. Red blood cells that contain the clumps become distorted into a crescent (sickle) shape. The sickled

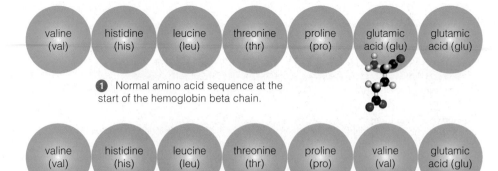

1 Normal amino acid sequence at the start of the hemoglobin beta chain.

2 One amino acid substitution results in the abnormal beta chain of sickle hemoglobin (HbS). The sixth amino acid in such chains is valine, not glutamic acid.

3 Glutamic acid carries an overall negative charge; valine carries no charge. This difference causes the protein to behave differently. At low oxygen levels, HbS molecules stick together and form rod-shaped clumps that distort normally round red blood cells into sickle shapes. (A sickle is a farm tool with a crescent-shaped blade.)

sickled cell

normal cell

4 Tionne "T-Boz" Watkins of the music group TLC is a celebrity spokesperson for the Sickle Cell Disease Association of America. She was diagnosed with sickle-cell anemia as a child.

FIGURE 7.10 Animated! How a single base-pair substitution causes sickle-cell anemia. *Below*, common symptoms of the disease.

Clumping of cells in bloodstream

↓

Circulatory problems, damage to brain, lungs, heart, skeletal muscles, gut, and kidneys

↓

Heart failure, paralysis, pneumonia, rheumatism, gut pain, kidney failure

Spleen concentrates sickle cells

↓

Enlargement of spleen; immune system function compromised

Rapid destruction of sickle cells

↓

Anemia, causing weakness, fatigue, impaired development, heart chamber dilation

↓

Impaired brain function, heart failure

cells clog tiny blood vessels, which in turn disrupts blood circulation throughout the body. Over time, repeated episodes of sickling can damage organs and even cause death (Figure 7.10).

What Causes Mutations? Insertion mutations are often caused by the activity of **transposable elements**, which are segments of DNA that can insert themselves anywhere in a chromosome. Certain kinds can move spontaneously from one place to another within the same chromosome, or to a different chromosome entirely. Transposable elements can be hundreds or thousands of base pairs long, so when one interrupts a gene sequence, it becomes a major insertion that changes the gene's product. These segments are common in the DNA of all species. For example, about 45 percent of human DNA consists of transposable elements or their remnants.

Many mutations occur spontaneously during DNA replication. That is not surprising, given the fast pace of replication (about twenty bases per second in humans, and one thousand bases per second in bacteria). DNA polymerases make mistakes at predictable rates, but most types fix errors as they occur (Section 6.4). Errors that remain uncorrected are mutations.

Harmful environmental agents also cause mutations. For example, some forms of energy such as x-rays can ionize atoms by knocking electrons right out of them. Such ionizing radiation can break chromosomes into pieces that may get lost during DNA replication (Figure 7.11**A**). Ionizing radiation also damages DNA indirectly when it penetrates living tissue, because it leaves a trail of destructive free radicals. Free radicals, remember, damage DNA (Section 5.6).

That is why doctors and dentists use the lowest possible doses of x-rays on their patients.

Nonionizing radiation boosts electrons to a higher energy level, but not enough to knock them out of an atom. DNA absorbs one kind, ultraviolet (UV) light. Exposure to UV light can cause two adjacent thymine bases to bond covalently to one another (Figure 7.11**B**). The resulting thymine dimer kinks the DNA. DNA polymerase may copy the kinked part incorrectly during replication, so a mutation becomes introduced into the DNA. Mutations that cause certain kinds of cancers begin with thymine dimers. Exposing unprotected skin to sunlight increases the risk of skin cancer because it causes thymine dimers to form in the DNA of skin cells.

Some natural or synthetic chemicals can also cause mutations. For instance, chemicals in cigarette smoke transfer small hydrocarbon groups to the bases in DNA. The altered bases mispair during replication, or stop replication entirely. Both events increase the chance of mutation.

Take-Home Message What is a mutation?

- A mutation is a permanent change in the nucleotide base sequence of DNA. A base-pair substitution, insertion, or deletion may alter a gene product.

- Most mutations arise during DNA replication as a result of unrepaired DNA polymerase errors. Some mutations occur as a result of transposable elements, or after exposure to harmful radiation or chemicals.

7.7 Eukaryotic Gene Controls

All of the cells in your body are descended from the same fertilized egg, so they all contain the same DNA—and the same genes. Some of the genes are transcribed by all cells; such genes affect structural features and metabolic pathways common to all cells. In other ways, however, nearly all of your body cells are specialized, or differentiated (Section 6.5). **Differentiation**, the process by which cells become specialized, occurs as different cell lineages begin to express different subsets of their genes. Which genes a cell uses determines the molecules it will produce, which in turn determines what kind of cell it will be. For example, most of your body cells express the genes that encode the enzymes of glycolysis, but only immature red blood cells use the genes that code for globin chains.

A cell rarely uses more than 10 percent of its genes at once. Which genes are expressed at any given time depends on many factors, such as conditions in the cytoplasm and extracellular fluid, and the type of cell. The factors affect controls governing all steps of gene expression, starting with transcription and ending with delivery of an RNA or protein product to its final destination. Such controls consist of processes that start, enhance, slow, or stop gene expression. For example, proteins called **transcription factors** affect whether and how fast certain genes are transcribed by binding directly to DNA. Some transcription factors prevent RNA polymerase from attaching to a promoter, which in turn prevents transcription of nearby genes. Others help RNA polymerase bind to a promoter, which speeds up transcription.

The remainder of this section introduces some specific examples and effects of gene controls in eukaryotes.

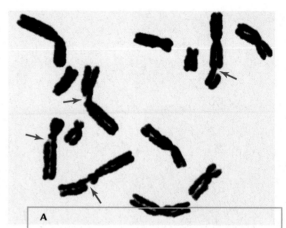

A

FIGURE 7.11 Two types of DNA damage that can lead to mutations. **A** Chromosomes from a human cell after exposure to gamma rays (a type of ionizing radiation). The broken pieces (*red* arrows) may get lost during DNA replication. **B** *Below*, a thymine dimer.

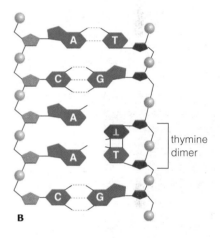

thymine dimer

B

differentiation The process by which cells become specialized; occurs as different cell lineages begin to express different subsets of their genes.

transcription factor Protein that influences transcription by binding to DNA.

transposable element Small segment of DNA that can spontaneously move to a new location in a chromosome.

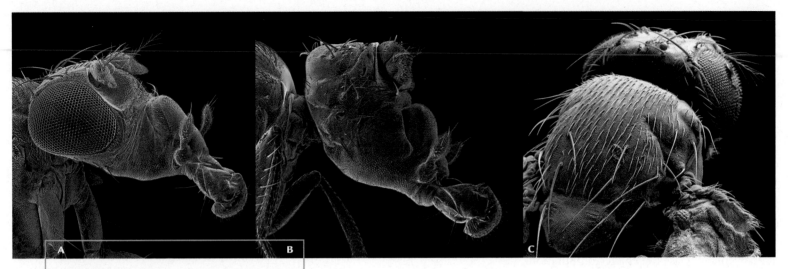

A B C

FIGURE 7.12 Eyes and *eyeless*. **A** A normal fruit fly has large, round eyes. **B** A fruit fly with a mutation in its *eyeless* gene develops with no eyes. **C** Eyes form wherever the *eyeless* gene is expressed in fly embryos—here, on a wing.

Humans, mice, squids, and other animals have a gene called *PAX6*, which is so similar to *eyeless* that it also triggers eye development when expressed in fly embryos. In humans, a *PAX6* mutation results in missing irises, a condition called aniridia **D**. Compare a normal iris **E**.

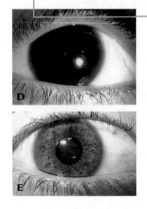

D

E

dosage compensation Theory that X chromosome inactivation equalizes gene expression between males and females.

gene knockout A gene that has been inactivated in an organism.

homeotic gene Type of master gene; its expression controls formation of specific body parts during development.

master gene Gene encoding a product that affects the expression of many other genes.

Homeotic Genes

Homeotic genes control the formation of specific body parts. All homeotic genes encode transcription factors with a homeodomain, a region of about sixty amino acids that can bind to a promoter or some other DNA sequence. Homeotic genes are one type of **master gene**, because their products affect expression of many other genes. The expression of a master gene causes other genes to be expressed, with the outcome being the completion of an intricate task such as the formation of an eye.

The function of many homeotic genes has been discovered by manipulating their expression, one at a time. With a **gene knockout**, researchers inactivate a gene by introducing a mutation or deleting it entirely. Then they observe how an organism that carries the knockout differs from normal individuals. The differences are clues to the function of the missing gene product.

Homeotic genes are expressed in animals during embryonic development. The process begins long before body parts develop, as various master genes are expressed in local areas of the early embryo. The master gene products form in concentration gradients that span the entire embryo. Depending on where they are located within the gradients, embryonic cells begin to transcribe different homeotic genes. Products of these homeotic genes form in specific areas of the embryo. The different products cause cells to differentiate into tissues that form specific structures such as wings or a head.

Researchers often name homeotic genes based on what happens in their absence. For instance, fruit flies with a mutated *eyeless* gene develop with no eyes (Figure 7.12**A,B**). *Dunce* is required for learning and memory. *Wingless*, *wrinkled*, and *minibrain* are self-explanatory. *Tinman* is necessary for development of a heart. Flies with a mutated *groucho* gene have too many bristles above their eyes. One gene was named *toll*, after what a German researcher exclaimed upon seeing the disastrous effects of its mutation (*toll* means "cool!" in German).

Homeotic genes control development by the same mechanisms in all eukaryotes, and many are interchangeable among different species. Thus, we can infer that they evolved in the most ancient eukaryotic cells. Homeodomains often differ among species only in conservative substitutions—one amino acid has replaced another with similar chemical properties. Consider the *eyeless* gene. Eyes form in embryonic fruit flies wherever this gene is expressed: typically in the head, but also in wings, legs, or other parts (Figure 7.12**C**). Humans, squids, mice, fish, and many other animals have a gene called *PAX6*, which is similar

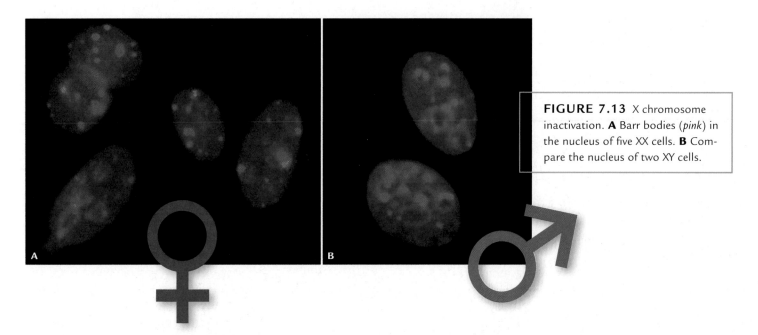

FIGURE 7.13 X chromosome inactivation. **A** Barr bodies (*pink*) in the nucleus of five XX cells. **B** Compare the nucleus of two XY cells.

to *eyeless*. In humans, *PAX6* mutations cause eye disorders such as aniridia, in which a person's irises are underdeveloped or missing (Figure 7.12**D,E**). *PAX6* works across different species. For example, if the *PAX6* gene from a human or mouse is inserted into an *eyeless* mutant fly, it has the same effect as the *eyeless* gene: An eye forms wherever it is expressed. Such studies are evidence of a shared ancestor among these evolutionarily distant animals.

Sex Chromosome Genes In humans and other mammals, a female's cells each contain two X chromosomes, one inherited from her mother, the other one from her father. One X chromosome is always tightly condensed (Figure 7.13). We call the condensed X chromosomes "Barr bodies," after Murray Barr, who discovered them. RNA polymerase cannot access most of the genes on the condensed chromosome. According to the theory of **dosage compensation**, the inactivation equalizes expression of X chromosome genes between the sexes. The body cells of male mammals (XY) have one set of X chromosome genes. The body cells of female mammals (XX) have two sets, but only one is expressed. Normal development of female embryos depends on this control.

The human X chromosome carries 1,336 genes. Some of those genes are associated with sexual traits, such as the distribution of body fat and hair. However, most of the genes on the X chromosome govern nonsexual traits such as blood clotting and color perception. Such genes are expressed in both males and females. Males, remember, also inherit one X chromosome.

The human Y chromosome carries only 307 genes, but one of them is the *SRY* gene—the master gene for male sex determination. Its expression in XY embryos triggers the formation of testes, which are male gonads (Figure 7.14). Some of the cells in these primary male reproductive organs make testosterone, a sex hormone that controls the emergence of male secondary sexual traits such as facial hair, increased musculature, and a deep voice. We know that *SRY* is the master gene that controls emergence of male sexual traits because mutations in this gene cause XY individuals to develop external genitalia that appear female. An XX embryo has no Y chromosome, no *SRY* gene, and much less testosterone, so primary female reproductive organs (ovaries) form instead of testes. Ovaries make estrogens and other sex hormones that will govern the development of female secondary sexual traits, such as enlarged, functional breasts, and fat deposits around the hips and thighs.

Structures that will give rise to external genitalia appear at seven weeks

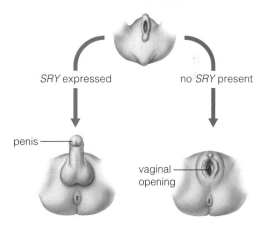

SRY expressed   no *SRY* present

penis

vaginal opening

birth approaching

FIGURE 7.14 Development of reproductive organs in human embryos. An early human embryo appears neither male nor female. Gene expression determines what reproductive organs form.

In an XY embryo, the *SRY* gene product triggers the formation of testes, male gonads that secrete testosterone. This hormone initiates development of other male traits. In an XX embryo, ovaries form in the absence of the Y chromosome and its *SRY* gene.

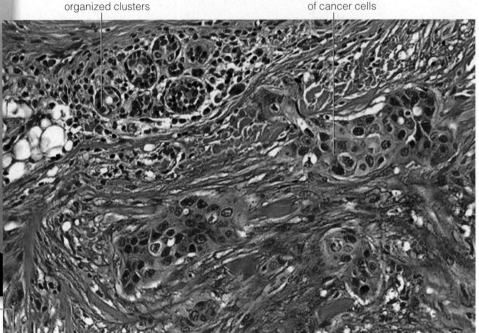

normal cells in
organized clusters

irregular clusters
of cancer cells

FIGURE 7.15 Breast cancer: an outcome of failed gene expression controls. *Right*, this micrograph shows irregular clusters of cancer cells that infiltrated milk ducts in breast tissue. *Above*, Robin Shoulla. Diagnostic tests revealed abnormal cells such as these in her body when she was seventeen years old.

cancer Disease that occurs when a malignant neoplasm physically and metabolically disrupts body tissues.
tumor Abnormally growing and dividing mass of cells.

Cancer: Gene Expression Out of Control Every second, millions of cells in your skin, bone marrow, gut lining, liver, and elsewhere are dividing and replacing their worn-out, dead, and dying predecessors. They do not divide at random. Many gene expression controls regulate cell growth and division. When those controls fail, cancer is the outcome.

Cancer is a multistep process in which abnormally growing and dividing cells disrupt body tissues. Gene expression controls that usually keep cells from getting overcrowded in tissues are lost, so cancer cell populations may reach extremely high densities. Unless chemotherapy, surgery, or another procedure eradicates them, cancer cells can put an individual on a painful road to death. Each year, cancers cause 15 to 20 percent of all human deaths in developed countries alone.

Cancer typically begins with a mutation in a gene whose product is part of the controls over cell growth and division. The mutation may be inherited, or it may be a new one, as when DNA becomes damaged by environmental agents (Section 7.6). If the mutation alters the gene's protein product so that it no longer works properly, one level of control over the cell's growth and division has been lost. Such genes typically have at least two backups: other genes whose products regulate the same process. If the backup genes also mutate, the cell begins to divide over and over, and its descendants form an abnormal mass called a **tumor**. Tumor cells sometimes lose the membrane recognition proteins that keep them in their home tissue, in which case they break free and establish themselves in other parts of the body—a process of cancer called metastasis.

Mutations in some genes predispose individuals to develop certain kinds of cancer. Tumor suppressor genes are named because tumors are more likely to occur when these genes mutate. Two examples are *BRCA1* and *BRCA2*. A mutated version of one or both of these genes is often found in breast and ovarian cancer cells. If a *BRCA* gene mutates in one of three especially dangerous ways, a woman has an 80 percent chance of developing breast cancer before the age of seventy.

BRCA proteins promote transcription of genes that encode DNA repair enzymes (Section 6.4). Any mutations that alter this function also alter a cell's capacity to repair damaged DNA. Other mutations are likely to accumulate, and that sets the stage for cancer.

BRCA proteins also bind to receptors for the hormones estrogen and progesterone, which are abundant in breast and ovarian tissues. Binding regulates the transcription of growth factor genes. Among other things, growth factors stimulate cells to divide during normal, cyclic renewals of breast and ovarian tissues. When a mutation results in a *BRCA* protein that cannot bind to hormone receptors, the growth factors are overproduced. Cell division goes out of control, and tissue growth becomes disorganized. In other words, cancer develops.

Because mutations in genes such as *BRCA* can be inherited, cancer is not only a disease of the elderly. Robin Shoulla was only seventeen when she was diagnosed with metastatic breast cancer (Figure 7.15). At an age when most young women are thinking about school, parties, and potential careers, she was dealing with radical mastectomy: the removal of a breast, all lymph nodes under the arm, and skeletal muscles in the chest wall under the breast. She was pleading with her oncologist not to use her jugular vein for chemotherapy and wondering if she would survive to see the next year. Robin's ordeal was one of more than 200,000 new cases of breast cancer in the United States each year. About 5,700 of those cases occur in women and men under thirty-four years of age.

Robin Shoulla survived. Although radical mastectomy is rarely performed today (a modified procedure is less disfiguring), it is the only option when cancer cells invade muscles under the breast. It was Robin's only option. Now, seventeen years later, she has what she calls a normal life—career, husband, children. Her goal as a cancer survivor: "To grow very old with gray hair and spreading hips, smiling."

Take-Home Message What is gene expression control?

- Cells of multicelled organisms differentiate when they start expressing a unique subset of their genes, an outcome of controls that govern gene expression.

- All steps between transcription and delivery of gene product are regulated.

- Controlling gene expression is critical for proper development and functioning of a eukaryotic body. Mutations that disrupt the normal system of controls can cause developmental problems, or cancer in an adult.

7.8 Impacts/Issues Revisited:
Ricin and Your Ribosomes

 One of ricin's two polypeptide chains is an enzyme that removes a specific adenine base from one of the rRNA chains in the large ribosomal subunit. Once that happens, the ribosome stops working. Protein synthesis grinds to a halt as the ricin inactivates the cell's ribosomes, and the cell quickly dies. A modified form of ricin is currently being tested as a treatment for some kinds of cancer. The ricin is attached to an antibody that can find cancer cells in a person's body. Researchers hope that the attached ricin will kill the cancer cells without harming normal cells.

How Would **YOU** ☑ **Vote?** Some women who are genetically at high risk of developing breast cancer opt for preventive surgical removal of their breasts before cancer develops. Statistically, many of those women never would have developed cancer. Should surgery be restricted to cancer treatment? See CengageNow for details, then vote online (cengagenow.com).

Summary

Section 7.1 The ability to make proteins is critical to all life processes.

Section 7.2 The genetic information in DNA consists of its base sequence. Genes are subunits of that sequence. A cell uses the information in a gene to make an RNA or protein product. The process of gene expression involves transcription (DNA to RNA), and translation (mRNA, or messenger RNA, to protein). Translation requires the participation of tRNA (transfer RNA) and rRNA (ribosomal RNA).

$$\text{DNA} \xrightarrow{\text{transcription}} \textbf{mRNA} \xrightarrow{\text{translation}} \text{protein}$$

Section 7.3 In transcription, RNA polymerase binds to a promoter in the DNA near a gene. It assembles a strand of RNA by linking RNA nucleotides in the order dictated by the base sequence of the gene. Thus, the new RNA is a copy of the gene from which it was transcribed.

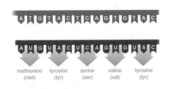

CENGAGENOW™ Explore transcription with the animation on CengageNow.

Section 7.4 mRNA carries DNA's protein-building information. The information consists of a series of codons, sets of three nucleotides. Sixty-four codons, most of which specify amino acids, constitute the genetic code. Each tRNA has an anticodon that can base-pair with a codon, and it binds to the kind of amino acid specified by the codon. rRNA and proteins make up the two subunits of ribosomes.

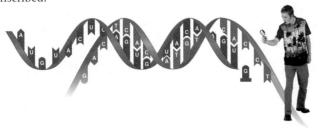

CENGAGENOW™ Learn about the genetic code with the interaction on CengageNow.

Section 7.5 Genetic information carried by an mRNA directs the synthesis of a polypeptide during translation. First, an mRNA, an initiator tRNA, and two ribosomal subunits converge. The intact ribosome then joins successive amino acids, which are delivered by tRNAs in the order specified by the codons in the mRNA. Translation ends when the polymerase encounters a stop codon.

CENGAGENOW™ See the translation of an mRNA with the animation on CengageNow.

Section 7.6 Insertions, deletions, and base-pair substitutions are mutations that arise by replication error, transposable element activity, or exposure to environmental hazards. A mutation that changes a gene's product may have harmful effects. Sickle-cell anemia, which is caused by a base-pair substitution in the hemoglobin beta chain gene, is one example.

CENGAGENOW™ Investigate the effects of mutation with the animation on CengageNow.

Section 7.7 Controls over gene expression are the basis of differentiation, the process by which cells in a multicelled body become specialized. Such controls operate at every step between gene and gene product, and are a critical part of the embryonic development and normal functioning of a multicelled body.

Self-Quiz   Answers in Appendix I

1. A chromosome contains many genes that are transcribed into different _____ .
   - a. proteins
   - b. polypeptides
   - c. RNAs
   - d. a and b

2. A binding site for RNA polymerase is called a _____ .
   - a. homeotic gene
   - b. promoter
   - c. transcription factor
   - d. b and c

3. Energy that drives transcription is provided mainly by _____ .
   - a. ATP
   - b. RNA nucleotides
   - c. GTP
   - d. all are correct

4. An RNA molecule is typically _____ ; a DNA molecule is typically _____ .
   - a. single-stranded; double-stranded
   - b. double-stranded; single-stranded
   - c. both are single-stranded
   - d. both are double-stranded

5. RNAs form by _____ ; proteins form by _____ .
   - a. replication; translation
   - b. translation; transcription
   - c. transcription; translation
   - d. replication; transcription

6. Up to how many amino acids can be encoded by a gene that consists of 45 nucleotides?

7. Most codons specify a(n) _____ .
   - a. protein
   - b. polypeptide
   - c. amino acid
   - d. mRNA

8. Anticodons pair with _____ .
   - a. mRNA codons
   - b. DNA codons
   - c. RNA anticodons
   - d. amino acids

9. Energy that drives translation is provided mainly by _____ .
   - a. ATP
   - b. RNA nucleotides
   - c. GTP
   - d. all are correct

10. Where does transcription take place in a eukaryotic cell?
    - a. nucleus
    - b. ribosome
    - c. cytoplasm
    - d. b and c are correct

11. Where does translation take place in a eukaryotic cell?
    - a. nucleus
    - b. ribosome
    - c. cytoplasm
    - d. b and c are correct

Digging Into Data

BRCA Mutations in Women Diagnosed With Breast Cancer

Investigating a correlation between specific cancer-causing mutations and the risk of mortality in humans is challenging, in part because each cancer patient is given the best treatment available at the time. There are no "untreated control" cancer patients, and ideas about which treatments are best change quickly as new drugs become available and new discoveries are made.

Figure 7.16 shows results from a 2007 study by Pal Moller and his colleagues. The researchers looked for *BRCA* mutations in 442 women who had been diagnosed with breast cancer, and followed their treatments and progress over several years.

All of the women in the study had at least two affected close relatives, so their risk of developing breast cancer due to an inherited factor (such as a *BRCA* mutation) was estimated to be greater than that of the general population.

1. According to this study, what is a woman's risk of dying of cancer if two of her close relatives have breast cancer?

2. What is her risk of dying of cancer if she carries a mutated *BRCA1* gene?

3. According to these results, is a *BRCA1* or *BRCA2* mutation more dangerous in breast cancer cases?

4. What other data would you have to see in order to make a conclusion about the effectiveness of preventive mastectomy or oophorectomy?

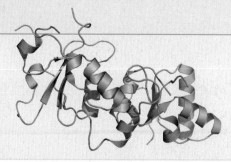

BRCA Mutations in Women Diagnosed With Breast Cancer				
	BRCA1	BRCA2	No *BRCA* Mutation	Total
Total number of patients	89	35	318	442
Avg. age at diagnosis	43.9	46.2	50.4	
Preventive mastectomy	6	3	14	23
Preventive oophorectomy	38	7	22	67
Number of deaths	16	1	21	38
Percent died	18.0	2.8	6.9	8.6

FIGURE 7.16 Results from a 2007 study investigating *BRCA* mutations in women diagnosed with breast cancer. All women in the study had a family history of the disease.

Some of the women underwent preventive mastectomy (removal of the noncancerous breast) during their course of treatment. Others had preventive oophorectomy (surgical removal of the ovaries) to prevent the possibility of getting ovarian cancer.

Top, model of the unmutated *BRCA1* protein.

12. Use Figure 7.5 to translate this nucleotide base sequence into an amino acid sequence, starting at the first base:

GGUUUCUUGAAGAGA

13. Translate the sequence of bases in the previous question, starting at the second base.

14. Homeotic gene products _____ .
   a. map out the overall body plan in embryos
   b. control the formation of specific body parts

15. Which of the following includes all of the others?
   a. homeotic genes   c. *SRY* gene
   b. master genes   d. *PAX 6*

16. A cell with a Barr body is _____ .
   a. prokaryotic   c. from a female mammal
   b. a sex cell   d. infected by Barr virus

17. Match each term with the most suitable description.
   _____ genetic message   a. cells become specialized
   _____ sequence   b. gets around
   _____ polysome   c. read as base triplets
   _____ *eyeless*   d. linear order of bases
   _____ genetic code   e. occurs only in groups
   _____ differentiation   f. set of 64 codons
   _____ transposable element   g. homeotic

Additional questions are available on CENGAGENOW™

Critical Thinking

1. An anticodon has the sequence GCG. What amino acid does this tRNA carry? What would be the effect of a mutation that changed the C of the anticodon to a G?

2. Each position of a codon can be occupied by one of four (4) nucleotides. What is the minimum number of nucleotides per codon necessary to specify all 20 of the amino acids that are found in proteins?

3. Unlike most rodents, guinea pigs are well developed at the time of birth. Suppose a breeder decides to separate baby guinea pigs from their mothers three weeks after they were born. He wants to raise the males and the females in different cages. However, he has trouble identifying the sex of young guinea pigs. Suggest how a quick look through a microscope can help him identify the females.

4. Cigarette smoke contains at least fifty-five different chemicals identified as carcinogenic (cancer-causing) by the International Agency for Research on Cancer (IARC). When these carcinogens enter the bloodstream, enzymes convert them to a series of chemical intermediates that are easier to excrete. Some of the intermediates bind irreversibly to DNA. Propose a hypothesis about why such binding causes cancer.

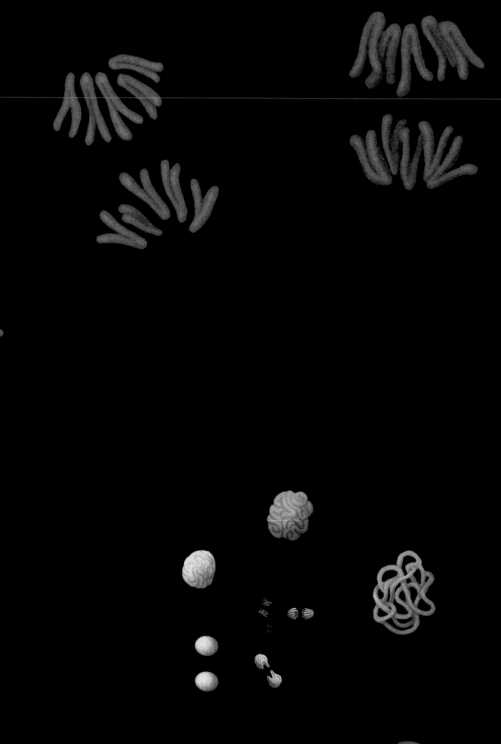

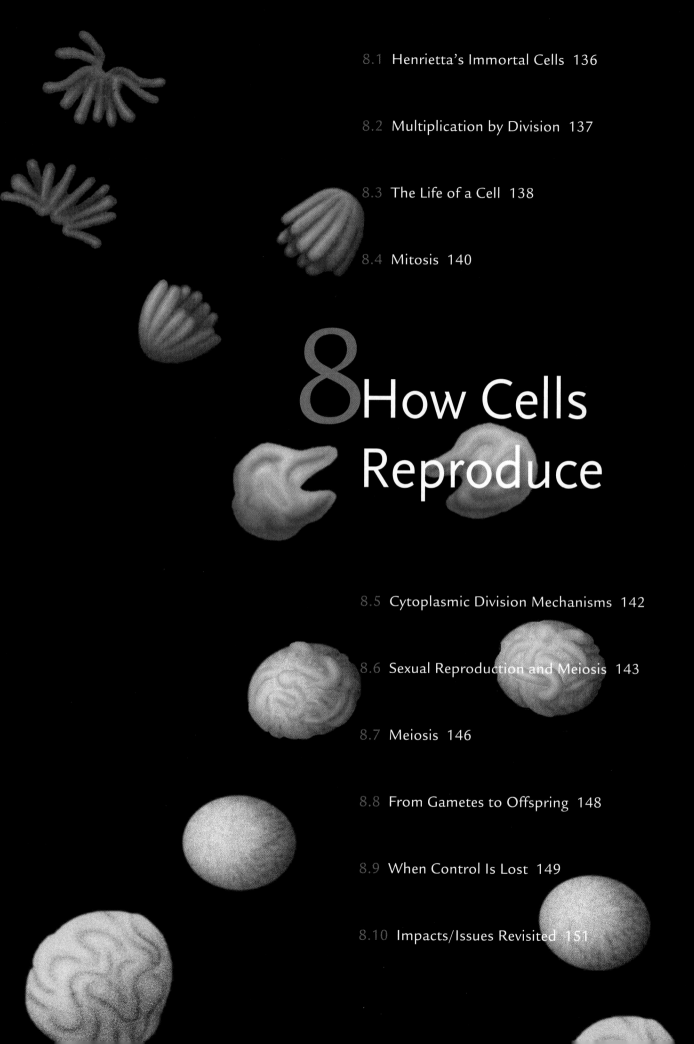

8 How Cells Reproduce

8.1 Impacts/Issues: Henrietta's Immortal Cells

Each human starts out as a fertilized egg. By the time of birth, the human body consists of about a trillion cells, all descended from that single cell. Even in an adult, billions of cells divide every day as new cells replace worn-out ones. However, human cells tend to divide a few times and die within weeks when grown in the laboratory.

Researchers started trying to coax human cells to become immortal—to keep dividing outside of the body—in the mid-1800s. Why? Many human diseases occur only in human cells. Immortal cell lineages, or cell lines, would allow researchers to study human diseases (and potential cures for them) without experimenting on people.

At Johns Hopkins University, George and Margaret Gey were among the researchers trying to culture human cells. They had been working on the problem for almost thirty years when, in 1951, their assistant Mary Kubicek prepared a sample of human cancer cells. Mary named the cells *HeLa*, after the first and last names of the patient from whom the cells had been taken.

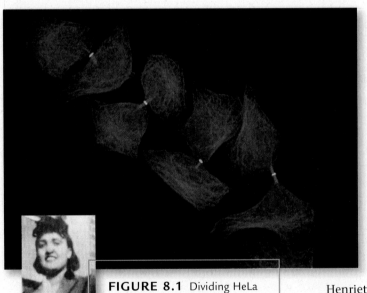

The HeLa cells began to divide, again and again. The cells were astonishingly vigorous. They quickly coated the inside of their test tube and consumed the nutrient broth in which they were bathed. Four days later, there were so many cells that the researchers had to transfer them to more tubes. The cell populations increased at a phenomenal rate. The cells were dividing every twenty-four hours and coating the inside of the tubes within days.

Sadly, cancer cells in the patient were dividing just as fast. Just six months after she had been diagnosed with cervical cancer, malignant cells had invaded tissues throughout her body. Two months after that, Henrietta Lacks, a young African American woman from Baltimore, was dead.

Although Henrietta passed away, her cells lived on in the Geys' laboratory (Figure 8.1). The Geys were able to grow poliovirus in HeLa cells, a practice that enabled them to find out which strains of the virus cause polio. That work was a critical step in the development of polio vaccines, which have since saved millions of lives.

FIGURE 8.1 Dividing HeLa cells (*above*), a cellular legacy of Henrietta Lacks (*left*), who was a young casualty of cancer.

Henrietta Lacks's cells, frozen away in tiny tubes and packed in Styrofoam boxes, continue to be shipped among laboratories all over the world. Researchers use those cells to investigate cancer, viral growth, protein synthesis, and the effects of radiation. They helped several researchers win Nobel Prizes for research in medicine and chemistry. Some HeLa cells even traveled into space for experiments on the *Discoverer XVII* satellite.

Henrietta Lacks was thirty-one, a wife and mother of five, when runaway cell divisions killed her. Her legacy continues to help people, through her cells that are still dividing, again and again, more than fifty years after she died. Understanding why cancer cells are immortal—and why we are not—begins with understanding the structures and mechanisms that cells use to divide.

Table 8.1	Comparison of Division Mechanisms		
		Mechanism	
Function	Mitosis	Meiosis	Prokaryotic Fission
Increases in body size during growth	X		
Replacement of dead or worn-out cells	X		
Repair of damaged tissues	X		
Asexual reproduction	X		X
Sexual reproduction		X	

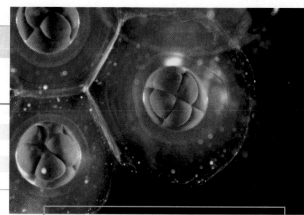

FIGURE 8.2 A multicelled animal develops by repeated divisions of a fertilized egg. This photo shows early frog embryos, each a product of three divisions of one fertilized egg. **Figure It Out:** Each of these embryos consists of how many cells? *Answer: Eight*

8.2 Multiplication by Division

A cell reproduces by dividing in two. Each of its cellular offspring must inherit a complete set of chromosomes, or it will not grow or function properly. Each descendant cell must also inherit some of the parent's cytoplasm, which contains all the enzymes, organelles, and other metabolic machinery necessary for life. A descendant cell that inherits a blob of cytoplasm is getting startup metabolic machinery that will keep it running until it can make its own.

In general, a eukaryotic cell cannot simply split in two, because only one of its descendant cells would get the nucleus (and the DNA). Thus, a cell's cytoplasm splits only after its DNA has been packaged into more than one nucleus by way of mitosis or meiosis. These two processes have much in common, but their outcomes differ (Table 8.1).

Mitosis is a nuclear division mechanism that maintains the chromosome number. Mitosis and cytoplasmic division are the basis of increases in body size during development (Figure 8.2), and ongoing replacements of damaged or dead cells. Many species of plants, animals, fungi, and protists also make copies of themselves, or reproduce asexually, by mitosis and cytoplasmic division. Prokaryotes (bacteria and archaeans) also reproduce asexually, but they do it by prokaryotic fission, a mechanism that we consider in Section 13.4.

Meiosis, a nuclear division mechanism that halves the chromosome number, is the basis of sexual reproduction. In humans and all other mammals, meiosis in reproductive cells results in sperm and eggs. Spores, which protect and disperse new generations, form by meiosis during the life cycle of fungi, plants, and many kinds of protists.

Take-Home Message   What is cell division and why does it happen?

- When a cell divides, each descendant cell receives a set of chromosomes and some cytoplasm. In eukaryotic cells, the nucleus divides first, then cytoplasm.

- Mitosis is a nuclear division mechanism that maintains the chromosome number. It is the basis of body size increases, cell replacements, and tissue repair in multicelled eukaryotes; and asexual reproduction in single-celled and some multicelled eukaryotes.

- In eukaryotes, a nuclear division mechanism called meiosis precedes the formation of gametes or spores. It is the basis of sexual reproduction.

meiosis Nuclear division process that halves the chromosome number. Basis of sexual reproduction.

mitosis Nuclear division mechanism that maintains the chromosome number. Basis of body growth, tissue repair and replacement in multicelled eukaryotes; also asexual reproduction in some plants, animals, fungi, and protists.

The Life of a Cell

Just as we describe the series of events in an animal's life as a life cycle, we can describe the events that occur from the time a cell forms until the time it divides as a **cell cycle** (Figure 8.3). There are three phases to the cell cycle: mitosis and cytoplasmic division (topics of the following sections), and interphase. During **interphase**, a cell increases its mass, roughly doubles the number of its cytoplasmic components, and replicates its DNA. Interphase is typically the longest phase, and it consists of three stages:

❶ G1, the first interval (or gap) of cell growth before DNA replication

❷ S, the time of synthesis (DNA replication)

❸ G2, the second interval (or gap) when the cell prepares to divide

Gap intervals were named because outwardly they seem to be periods of inactivity. Actually, most cells going about their metabolic business are in G1. Cells preparing to divide enter S, when they copy their DNA. During G2, they make the proteins that will drive mitosis. Once S begins, DNA replication usually proceeds at a predictable rate and ends before the cell divides.

Different types of cells proceed through the cell cycle at different rates. For example, the stem cells in your red bone marrow divide every 12 hours. Their descendants become red blood cells that replace 2 to 3 million worn-out ones in your blood each second. Cells in the tips of a bean plant root divide every 19 hours. The cells in a fruit fly embryo divide every 10 minutes.

Gene expression controls (Section 7.7) affect the rate of the cell division. Some of the controls function as built-in brakes on the cell cycle. Apply the brakes that work in G1, and the cycle stalls in G1. Lift the brakes, and the cycle runs again. Sometimes the brakes are not lifted. For example, the nerve cells of adult humans are stuck in G1 of interphase. Because the cell cycle of these cells cannot proceed, they do not divide. Thus, damaged nerve cells cannot be replaced by new ones.

If such controls over the cell cycle stop working, the body may be endangered. As you will see shortly, cancer begins this way. Crucial controls are lost, and the cell cycle spins out of control.

FIGURE 8.3 Animated! The eukaryotic cell cycle. The length of the intervals differs among cells. G1, S, and G2 are part of interphase.

❶ G1 is the interval of growth before DNA replication. The cell's chromosomes are unduplicated during this stage.

❷ S is the time of synthesis. The name refers to DNA synthesis, because the cell copies its DNA during this stage.

❸ G2 is the interval after DNA replication and before mitosis. The cell prepares to divide during this phase.

❹ Mitosis begins as interphase ends.

❺ The nucleus divides during mitosis.

❻ At the end of mitosis, the cytoplasm typically divides, and the cycle begins anew in interphase for each descendant cell.

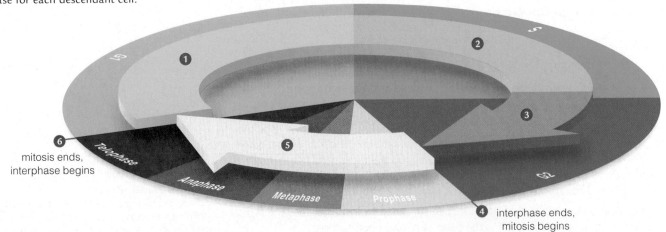

mitosis ends, interphase begins

interphase ends, mitosis begins

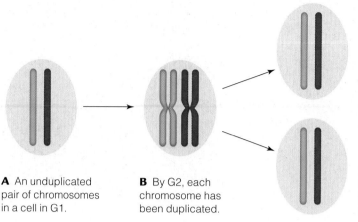

A An unduplicated pair of chromosomes in a cell in G1.

B By G2, each chromosome has been duplicated.

C Mitosis and cytoplasmic division package one copy of each chromosome into each of two new cells.

FIGURE 8.4 How mitosis maintains the chromosome number.

Chromosomes During the Cell Cycle Remember from Section 6.2 that human body cells have 46 chromosomes—two of each type. Except for a pairing of sex chromosomes (XY) in males, the chromosomes of each pair are **homologous**, which means they have the same length, shape, and collection of genes (*hom*– means alike). Typically, each member of a chromosome pair was inherited from one of two parents.

With mitosis followed by cytoplasmic division, a diploid parent cell produces two diploid descendant cells. Both offspring have the same number and kind of chromosomes as the parent. Thus, mitosis maintains the chromosome number. However, it is not just that each new cell gets twelve or forty-eight or two hundred and eight chromosomes. If only the total mattered, then one of the cell's offspring might get, say, two pairs of chromosome 22 and no pairs whatsoever of chromosome 9. A cell cannot function properly without a full complement of DNA, which means it needs to have one copy of each chromosome.

When a cell is in G1, each of its chromosomes consists of one double-stranded DNA molecule (Figure 8.4**A**). A cell replicates its DNA in S. By G2, each chromosome consists of *two* double-stranded DNA molecules (Figure 8.4**B**). The two molecules stay attached to one another at the centromere until mitosis is almost over. Until they separate, the two DNA molecules are called sister chromatids. By the time a cell enters mitosis (Figure 8.3 ❹), its chromosomes have already been duplicated. As you will see in the next section, mitosis parcels one copy of each chromosome (one sister chromatid) into each of two new nuclei ❺. When the cytoplasm divides, the two nuclei are packaged into separate cells (Figure 8.4**C**). Each new cell has a full complement of unduplicated chromosomes, and each starts the cell cycle over again in G1 of interphase ❻.

Take-Home Message

What is a cell cycle?

■ A cell cycle is a sequence of stages (interphase, mitosis, and cytoplasmic division) through which a cell passes during its lifetime.

■ During interphase, a new cell increases its mass, doubles the number of its cytoplasmic components, and duplicates its chromosomes. The cycle ends after the cell undergoes mitosis and then divides its cytoplasm.

cell cycle A series of events from the time a cell forms until its cytoplasm divides.

homologous Refers to the two members of a pair of chromosomes with the same length, shape, and genes.

interphase In a eukaryotic cell cycle, the interval between mitotic divisions when a cell grows, roughly doubles the number of its cytoplasmic components, and replicates its DNA.

FIGURE 8.5 Animated! Mitosis. Micrographs here and opposite show plant cells (onion root, *left*), and animal cells (whitefish embryo, *right*). *This page*, interphase cells are shown for comparison, but interphase is not part of mitosis.

Opposite page, the stages of mitosis. The drawings show a diploid (2*n*) animal cell. For clarity, only two pairs of chromosomes are illustrated, but nearly all eukaryotic cells have more than two. The two chromosomes of the pair inherited from one parent are *pink*; the two chromosomes from the other parent are *blue*.

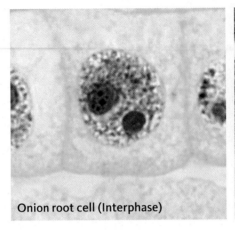

Onion root cell (Interphase)

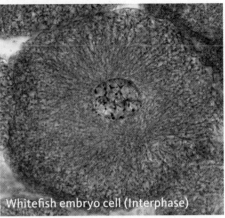

Whitefish embryo cell (Interphase)

8.4 Mitosis

A cell's chromosomes are loosened during interphase to allow for transcription and DNA replication. In preparation for nuclear division, they begin to condense tightly (Figure 8.5 ❶). Transcription and DNA replication stop as the chromosomes pack into their more easily transported compact form.

We can tell that a cell is in **prophase**, the first stage of mitosis, when its chromosomes have condensed so much that they are visible under a light microscope ❷. "Mitosis" is from the Greek word for thread, *mitos*, after the threadlike appearance of chromosomes during nuclear division.

Most animal cells have a centrosome, a region near the nucleus that organizes microtubules while they are forming. The centrosome usually includes two barrel-shaped centrioles, and it is duplicated just before prophase begins. During prophase, one of the two centrosomes (along with its pair of centrioles) moves to the opposite side of the cell, and microtubules that will form a spindle begin to grow from both centrosomes. A **spindle** is a dynamic network of microtubules that attaches to and moves chromosomes during nuclear division. Motor proteins (Section 3.6) traveling along the microtubules help the spindle grow in the appropriate directions.

As prophase ends, the nuclear envelope breaks up and spindle microtubules penetrate the nuclear region ❸. Some microtubules from each spindle pole stop growing after they overlap in the middle of the cell. Others continue to grow until they reach a chromosome and attach to it. At that point, one chromatid of each chromosome is attached to microtubules extending from one spindle pole, and its sister chromatid is attached to microtubules extending from the other spindle pole. The opposing sets of microtubules begin a tug-of-war by adding and losing tubulin subunits. As the microtubules grow and shrink, they push and pull the chromosomes. When all the microtubules are the same length, the chromosomes are aligned midway between the spindle poles ❹. The alignment marks **metaphase** (from *meta*, the ancient Greek word for between).

During **anaphase**, microtubules of the spindle separate the sister chromatids of each duplicated chromosome, and move them toward opposite spindle poles ❺. Each DNA molecule has now become a separate chromosome. Anaphase ends as the chromosomes are heading toward opposite spindle poles.

Telophase begins when the two clusters of chromosomes reach the spindle poles ❻. Each cluster consists of the parental complement of chromosomes—two of each, if the parent cell was diploid. A new nucleus forms around each

anaphase Stage of mitosis in which sister chromatids separate and move to opposite spindle poles.

metaphase Stage of mitosis during which the cell's chromosomes align midway between poles of the spindle.

prophase Stage of mitosis in which chromosomes condense and become attached to a newly forming spindle.

spindle Dynamically assembled and disassembled array of microtubules that moves chromosomes during nuclear division.

telophase Stage of mitosis during which chromosomes arrive at the spindle poles and decondense, and new nuclei form.

centrosome

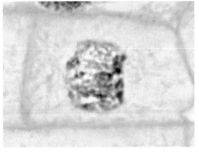
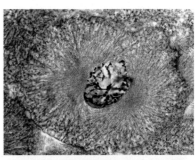

① Early Prophase

Mitosis begins. In the nucleus, the DNA begins to appear grainy as it organizes and condenses. The centrosome is duplicated.

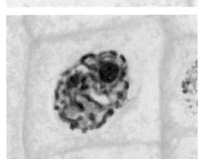

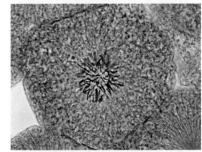

② Prophase

The chromosomes become visible as distinct structures as they condense further. Microtubules assemble and move one of the two centrosomes to the opposite side of the nucleus, and the nuclear envelope breaks up.

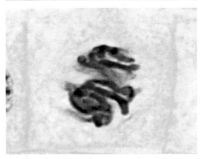
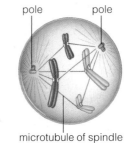
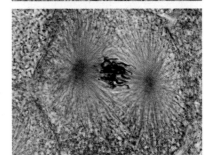

pole   pole

③ Transition to Metaphase

The nuclear envelope is gone, and the chromosomes are at their most condensed. Spindle microtubules assemble and attach sister chromatids to opposite spindle poles.

microtubule of spindle

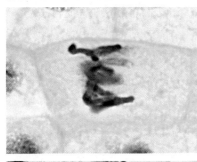

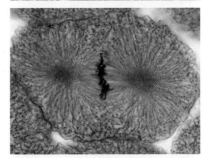

④ Metaphase

All of the chromosomes are aligned midway between the spindle poles. Microtubules attach each chromatid to one of the spindle poles, and its sister to the opposite pole.

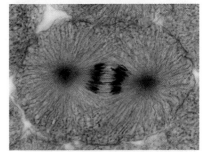

⑤ Anaphase

Motor proteins moving along spindle microtubules drag the chromatids toward the spindle poles, and the sister chromatids separate. Each sister chromatid is now a separate chromosome.

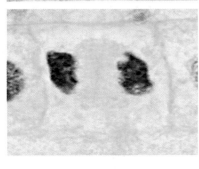

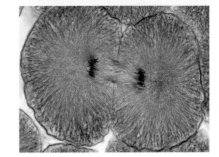

⑥ Telophase

The chromosomes reach the spindle poles and decondense. A nuclear envelope forms around each cluster. Mitosis is over.

cluster as the chromosomes loosen up again. Thus, two nuclei form. The parent cell in our example was diploid, so each new nucleus is diploid too. Once two nuclei have formed, telophase is over, and so is mitosis.

Take-Home Message

What happens during mitosis?

- Each chromosome in a cell's nucleus was duplicated before mitosis begins, so each consists of two DNA molecules (sister chromatids).

- In prophase, the chromosomes condense and a spindle forms. The nuclear envelope breaks up. Spindle microtubules attach to the chromosomes.

- At metaphase, the chromosomes (still duplicated) are aligned midway between the spindle poles.

- In anaphase, microtubules separate the sister chromatids of each chromosome, and pull them toward opposite spindle poles. Each DNA molecule is now a separate chromosome.

- In telophase, two clusters of chromosomes reach the spindle poles. A new nuclear envelope forms around each cluster.

- Two new nuclei form at the end of mitosis. Each one has the same chromosome number as the parent cell's nucleus.

8.5 Cytoplasmic Division Mechanisms

A cell's cytoplasm usually divides after mitosis, so two cells form, each with their own nucleus. The process of cytoplasmic division differs among eukaryotes.

Typical animal cells divide simply by pinching in two. How? The cell cortex, which is the mesh of cytoskeletal elements just under the plasma membrane, includes a band of actin and myosin filaments that wraps around the cell's midsection. The band of filaments is called a **contractile ring** because it contracts when its component proteins are energized by ATP. When the ring contracts, it shrinks, and it drags the plasma membrane inward as it does. The sinking plasma membrane is visible on the outside of the cell as an indentation between the former spindle poles (Figure 8.6**A,B**). The indentation is called a **cleavage furrow**. The furrow advances around the cell, deepening as it does, until the cytoplasm (and the cell) is pinched in two (Figure 8.6**C,D**). Two new cells form this way. Each has a nucleus and some of the parent cell's cytoplasm, and each is enclosed in its own plasma membrane.

FIGURE 8.6 Animated! Cytoplasmic division of an animal cell.

 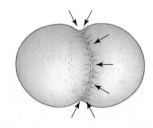 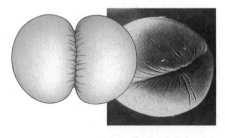

A After mitosis is completed, the spindle begins to disassemble.

B At the midpoint of the former spindle, a ring of actin and myosin filaments attached to the plasma membrane contracts.

C This contractile ring pulls the cell surface inward as it shrinks.

D The ring contracts until it pinches the cell in two.

 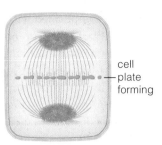 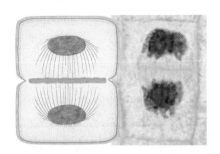 

cell
plate
forming

A The plane of division was established before mitosis began. Vesicles cluster here when mitosis ends.

B As the vesicles fuse with each other, they form a cell plate along the plane of division.

C The cell plate expands outward along the plane of division. When it reaches the plasma membrane, it attaches to the membrane and partitions the cytoplasm.

D The cell plate matures as two new cell walls that join with the parent cell wall, so each descendant cell becomes enclosed by its own cell wall.

FIGURE 8.7 Animated! Cytoplasmic division of a plant cell.

Dividing plant cells face a particular challenge because they have cell walls that are stiffened with cellulose (Section 2.7). The mechanism that pinches an animal cell in two would not be strong enough to pinch a plant cell wall. Accordingly, plant cells have an extra step in their cytoplasmic division process, which begins before mitosis is finished. By the end of anaphase in a plant cell, a set of short microtubules has formed on either side of the division plane. These microtubules now guide vesicles from Golgi bodies and the cell surface to the division plane (Figure 8.7**A**). There, the vesicles and their wall-building contents start to fuse into a disk-shaped **cell plate** (Figure 8.7**B**). The plate grows outward until its edges reach the plasma membrane (Figure 8.7**C**). It attaches to the membrane, and so partitions the cytoplasm. In time, the cell plate will develop into a primary cell wall that merges with the parent cell's wall. Thus, by the end of division, each of the descendant cells will be enclosed by its own plasma membrane and its own cell wall (Figure 8.7**D**).

Take-Home Message **How do cells divide?**

- After mitosis, the cytoplasm of the parent cell typically is partitioned into two descendant cells, each with its own nucleus. The process of cytoplasmic division differs between plants and animals.

- In animal cells, a contractile ring pinches the cytoplasm in two. In plant cells, a cell plate that forms midway between the spindle poles partitions the cytoplasm when it reaches and connects to the parent cell wall.

8.6 Sexual Reproduction and Meiosis

Introducing Alleles An individual's genes collectively contain the information necessary to make a new individual. With **asexual reproduction**, one parent produces offspring, so all of its offspring inherit the same number and kinds of genes. Mutations aside, then, all offspring of asexual reproduction are genetically identical copies of the parent, or **clones**.

asexual reproduction Reproductive mode by which offspring arise from one parent and inherit that parent's genes only.

cell plate After nuclear division in a plant cell, a disk-shaped structure that forms a cross-wall between the two new nuclei.

cleavage furrow In a dividing animal cell, the indentation where cytoplasmic division will occur.

clone A genetically identical copy of an organism.

contractile ring A thin band of actin and myosin filaments that wraps around the midsection of an animal cell. During cytoplasmic division, the band contracts and pinches the cytoplasm in two.

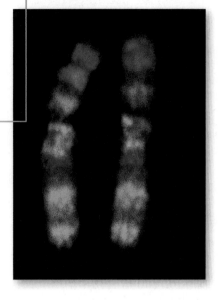

FIGURE 8.8
A chromosome pair. The two appear identical in this micrograph, but any gene might differ in DNA sequence from its partner on the other chromosome.

Inheritance is far more complicated with **sexual reproduction**, a process in which two parents contribute genes to offspring. So why sex? Variation in forms and combinations of heritable traits is typical of sexually reproducing populations. Some forms of traits help an organism survive under conditions that prevail in its environment. If those conditions change, some of the diverse offspring of sexual reproducers may have forms of traits or combinations of them that help these individuals to survive the change. All offspring of asexual reproducers are adapted the same way to the environment—and equally vulnerable to changes in it. Sexual reproduction mixes up the genetic information of two individuals that have different forms of heritable traits. It brings together adaptive traits, and separates adaptive traits from maladaptive ones, in far fewer generations than does mutation alone.

The body cells of multicelled organisms that reproduce sexually contain pairs of chromosomes. Typically, one chromosome of each pair is maternal and the other is paternal (Figure 8.8). Except for the pair of nonidentical sex chromosomes, the two chromosomes of every pair carry the same set of genes. If the information in every gene pair were identical, then sexual reproduction would produce clones, just like asexual reproduction. Just imagine: The entire human population might consist of clones, in which case everybody would look alike.

But the two genes of a pair are often not identical. Why not? Mutations that inevitably accumulate in DNA change its sequence. Thus, the genes of a pair might encode slightly different forms of the gene's product.

Different forms of the same gene are called **alleles**. Alleles influence differences in thousands of traits. For example, the mutated genes that cause sickle-cell anemia and thalassemia (Section 7.6) are two of over 700 known alleles of the human beta globin gene. Alleles are one reason that the individuals of a sexually reproducing species do not all look exactly the same. The offspring of sexual reproducers inherit new combinations of alleles, which is the basis of new combinations of traits.

Meiosis Halves the Chromosome Number Sexual reproduction involves the fusion of reproductive cells from two parents. The process begins with meiosis, a nuclear division process that halves the chromosome number. Meiosis occurs in immature reproductive cells (germ cells) of multicelled eukaryotes that reproduce sexually. In animals, meiosis in germ cells results in mature reproductive cells called **gametes**. A sperm cell is an example of a male gamete. An egg is a female gamete. Gametes usually form inside special reproductive structures or organs (Figure 8.9).

Gametes have a single set of chromosomes, so they are **haploid** (n): Their chromosome number is half of the diploid ($2n$) number. Human body cells are diploid, with 23 pairs of homologous chromosomes. Meiosis of a human germ cell ($2n$) normally produces gametes with 23 chromosomes: one of each pair (n). The diploid chromosome number is restored at **fertilization**, when two haploid gametes (one egg and one sperm, for example) fuse to form a **zygote**, which is the first cell of a new individual.

The process of meiosis is similar to mitosis in certain respects. A cell duplicates its DNA before the division process starts. The two DNA molecules and associated proteins stay attached at the centromere. For as long as they remain attached, they are sister chromatids. As in mitosis, the microtubules of a spindle move the chromosomes to opposite poles of the cell. However, meiosis sorts the chromosomes into new nuclei twice. Two consecutive nuclear divisions form

Reproductive organs of a human male

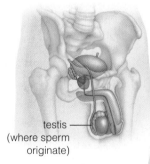

testis (where sperm originate)

Reproductive organs of a human female

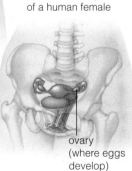

ovary (where eggs develop)

FIGURE 8.9 Animated!
Meiosis of germ cells in reproductive organs gives rise to gametes.

four haploid nuclei. There is typically no interphase between the two divisions, which are called meiosis I and II:

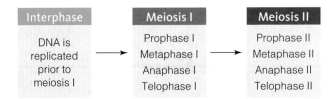

Interphase	Meiosis I	Meiosis II
DNA is replicated prior to meiosis I	Prophase I Metaphase I Anaphase I Telophase I	Prophase II Metaphase II Anaphase II Telophase II

In meoisis I, every duplicated chromosome aligns with its partner. After they are sorted and arranged this way, each homologous chromosome is pulled away from its partner:

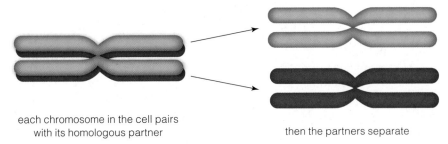

each chromosome in the cell pairs with its homologous partner

then the partners separate

After the homologous chromosomes are pulled apart, each ends up in one of two new nuclei. At this stage, the chromosomes are still duplicated (sister chromatids are still attached to one another). During meiosis II, the sister chromatids of each chromosome are pulled apart, so each becomes an individual, unduplicated chromosome:

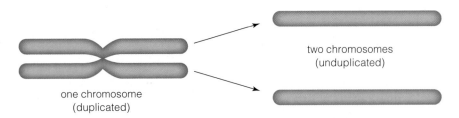

one chromosome
(duplicated)

two chromosomes
(unduplicated)

Meiosis distributes the duplicated chromosomes of a diploid nucleus (*2n*) into four new nuclei. Each new nucleus is haploid (*n*), with one unduplicated version of each chromosome.

Sexual repro-
duction mixes
up the genetic
information
of two indi-
viduals that
have different
forms of heri-
table traits.

Take-Home Message How does sexual reproduction introduce variation in heritable traits?

■ The cells of sexually reproducing organisms have pairs of homologous chromosomes, with pairs of genes. Paired genes may vary in sequence as alleles.

■ Alleles are the basis of traits. Sexual reproduction mixes up alleles from two parents. It results in new combinations of alleles—thus new combinations of traits—in offspring.

■ Meiosis occurs in immature reproductive cells of sexually reproducing eukaryotes. It halves the diploid (*2n*) chromosome number, to the haploid number (*n*), for forthcoming gametes.

■ When two gametes fuse at fertilization, the chromosome number is restored. The new individual has two sets of chromosomes, one from each parent.

alleles Forms of a gene that encode slightly different versions of the gene's product.

fertilization Fusion of a sperm nucleus and an egg nucleus, the result being a single-celled zygote.

gamete Mature, haploid reproductive cell.

haploid Having one of each type of chromosome characteristic of the species.

sexual reproduction Reproductive mode by which offspring arise from two parents and inherit genes from both.

zygote Cell formed by fusion of gametes; first cell of a new individual.

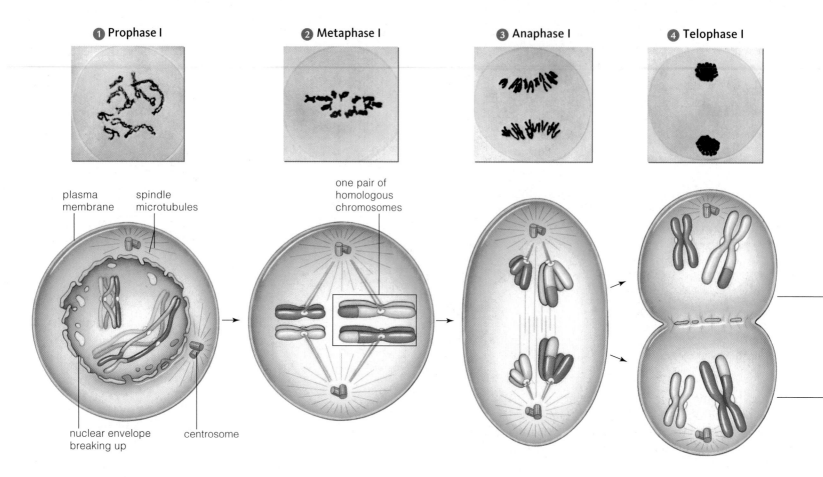

❶ Prophase I **❷ Metaphase I** **❸ Anaphase I** **❹ Telophase I**

plasma membrane spindle microtubules

one pair of homologous chromosomes

nuclear envelope breaking up centrosome

FIGURE 8.10 Animated! Meiosis. Two pairs of chromosomes are illustrated in a diploid (2*n*) animal cell. Homologous chromosomes are indicated in *blue* and *pink*. Micrographs show meiosis in a lily plant cell (*Lilium regale*).

❶ Prophase I. The (duplicated) chromosomes condense, and spindle microtubules attach to them as the nuclear envelope breaks up.

❷ Metaphase I. The (duplicated) chromosomes are aligned midway between spindle poles.

❸ Anaphase I. Homologous chromosomes separate.

❹ Telophase I. Two clusters of chromosomes reach the spindle poles. A new nuclear envelope encloses each cluster, so two haploid (*n*) nuclei form.

❺ Prophase II. The (still duplicated) chromosomes condense, and spindle microtubules attach to each sister chromatid as the nuclear envelope breaks up.

❻ Metaphase II. The chromosomes are aligned midway between the spindle poles.

❼ Anaphase II. Sister chromatids separate and become individual chromosomes (unduplicated).

❽ Telophase II. A cluster of (unduplicated) chromosomes reaches each spindle pole. A new nuclear envelope encloses each cluster, so four haploid (*n*) nuclei form.

8.7 Meiosis

A cell's chromosomes are duplicated prior to meiosis. As meiosis I begins, each chromosome consists of two sister chromatids. The nucleus is diploid (2*n*): It contains two sets of chromosomes, one from each parent. During the first stage of meiosis, prophase I, the chromosomes condense, and homologous chromosomes align tightly and swap segments (Figure 8.10). The centrosome (with its two centrioles) is duplicated, and one centriole pair moves to the opposite side of the cell as the nuclear envelope breaks up. Spindle microtubules begin to extend from the centrosomes ❶. By the end of prophase I, spindle microtubules connect the chromosomes to the spindle poles. Each chromosome is now attached to one spindle pole, and the homologous chromosome of the pair is attached to the other. The microtubules grow and shrink, pushing and pulling the chromosomes as they do. At metaphase I, all of the microtubules are the same length, and the chromosomes are aligned midway between the poles of the spindle ❷.

In anaphase I, the spindle microtubules separate homologous chromosomes and pull them toward opposite spindle poles ❸. During telophase I, the chromosomes reach the spindle poles ❹. New nuclear envelopes form around the two clusters of chromosomes as the DNA loosens up. Each of the two haploid (*n*) nuclei that form contains one set of (duplicated) chromosomes. The cytoplasm may divide at this point to form two haploid cells, but the DNA is not replicated before meiosis II begins.

During prophase II, the chromosomes condense as a new spindle forms. One centriole moves to the opposite side of each new nucleus, and the nuclear envelopes break up. By the end of prophase II, microtubules connect the chro-

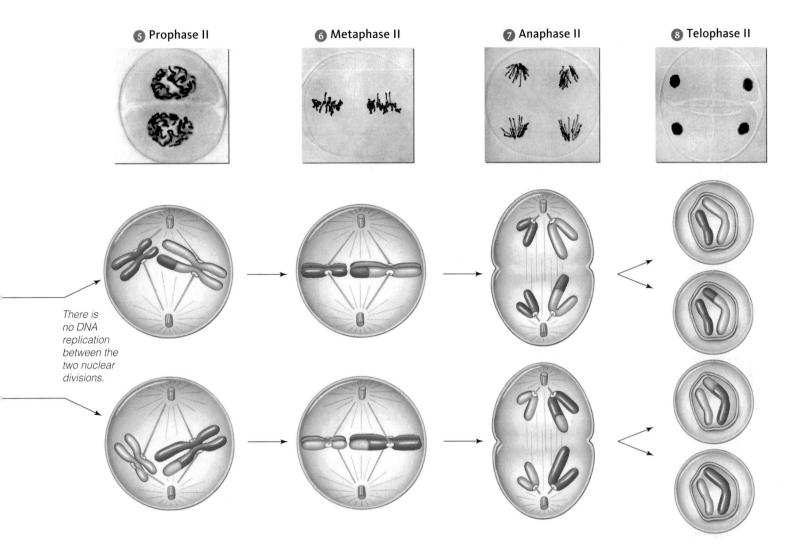

⑤ Prophase II **⑥ Metaphase II** **⑦ Anaphase II** **⑧ Telophase II**

There is no DNA replication between the two nuclear divisions.

mosomes to the spindle poles. Each chromatid is now attached to one spindle pole, and its sister is attached to the other ⑤. The microtubules grow and shrink, pushing and pulling the chromosomes as they do. At metaphase II, all of the microtubules are the same length, and the chromosomes are aligned midway between the spindle poles ⑥.

In anaphase II, the spindle microtubules pull the sister chromatids apart ⑦. Each chromosome now consists of one molecule of DNA. During telophase II, the chromosomes (unduplicated) reach the spindle poles ⑧. New nuclear envelopes form around the four clusters of chromosomes as the DNA loosens up. Each of the four haploid (*n*) nuclei that form contains one set of (unduplicated) chromosomes. The cytoplasm may divide to form four haploid cells.

Crossing Over Early in prophase I of meiosis, all chromosomes in a germ cell condense. When they do, each is drawn close to its homologue. The chromatids of one homologous chromosome become tightly aligned with the chromatids of the other along their length:

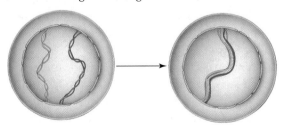

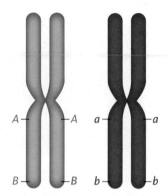

A Here, we focus on only two genes. One gene has alleles *A* and *a*; the other has alleles *B* and *b*.

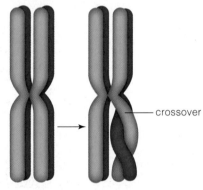

— crossover

B Close contact between the homologous chromosomes promotes crossing over between nonsister chromatids, so paternal and maternal chromatids exchange segments.

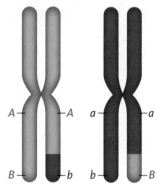

C Crossing over mixes up paternal and maternal alleles on homologous chromosomes.

FIGURE 8.11 Animated! Crossing over. *Blue* signifies a paternal chromosome, and *pink*, its maternal homologue. For clarity, we show only one pair of homologous chromosomes and one crossover, but more than one crossover may occur in each chromosome pair.

This tight, parallel orientation favors **crossing over**, a process by which a chromosome and its homologous partner exchange corresponding pieces of DNA (Figure 8.11). Crossing over is a normal and frequent process in meiosis, but the rate of crossing over varies among species and among chromosomes. In humans, between 46 and 95 crossovers occur per meiosis, so each chromosome crosses over at least once. Each crossover event is an opportunity for homologous chromosomes to exchange alleles. Thus, crossing over introduces novel combinations of alleles in both members of a pair of homologous chromosomes, which results in novel combinations of traits among offspring.

Take-Home Message   What happens during meiosis?

■ During meiosis, the nucleus of a diploid (2*n*) cell divides twice. Four haploid (*n*) nuclei form, each with a full set of chromosomes—one of each type.

■ Crossing over is recombination between nonsister chromatids of homologous chromosomes during prophase I. It makes new combinations of parental alleles.

8.8 From Gametes to Offspring

All gametes are typically haploid, but they differ in their details. For example, human male gametes—**sperm**—have one flagellum. Opossum sperm have two, and roundworm sperm have none. A flowering plant's male gamete is simply a sperm nucleus. We leave most of the details of sexual reproduction for later chapters, but you will need to know a few concepts before you get there.

In plants, two kinds of multicelled bodies form. **Sporophytes** are typically diploid, and spores form by meiosis in their specialized parts (Figure 8.12**A**). Spores consist of one or a few haploid cells. These cells undergo mitosis and give rise to a **gametophyte**, a multicelled haploid body inside which one or more gametes form. As an example, sequoia trees are sporophytes. Male and female gametophytes develop inside different types of cones that form on each tree. In flowering plants, gametophytes form in flowers.

The gametes of animals arise by mitosis of diploid germ cells (Figure 8.12**B**). In male animals, a germ cell develops into a primary spermatocyte. This large cell divides by meiosis, producing four haploid cells that develop into spermatids. Each spermatid matures as a sperm.

In female animals, a germ cell becomes a primary oocyte, which is an immature egg. This cell undergoes meiosis and division, as occurs with a primary spermatocyte. However, the cytoplasm of a primary oocyte divides unequally, so the cells that result differ in size and function. Two haploid cells form when the primary oocyte divides after meiosis I. One of the cells is called a first polar body. The other cell, the secondary oocyte, is much larger because it gets nearly all of the parent cell's cytoplasm. This cell undergoes meiosis II and cytoplasmic division. One of the two cells that forms is a second polar body. The other cell gets most of the cytoplasm and matures into a female gamete, which is called an ovum (plural, ova), or **egg**.

Polar bodies are not nutrient-rich or plump with cytoplasm, and generally do not function as gametes. In time they degenerate. Their formation simply ensures that the egg will have a haploid chromosome number, and also will get enough metabolic machinery to support early divisions of the new individual.

Fertilization At fertilization, the fusion of two gametes produces a zygote. Fertilization restores the parental chromosome number. If meiosis did not precede fertilization, the chromosome number would double with every generation. If the chromosome number changes, so does the individual's set of genetic instructions. This set is like a fine-tuned blueprint that must be followed exactly, page by page, in order to build a body that functions normally. Changes in the blueprint can have lethal consequences, particularly in animals.

Fertilization also contributes to the variation that we see among offspring of sexual reproducers. Think about it in terms of human reproduction. Cells that give rise to human gametes have twenty-three pairs of homologous chromosomes. Each time a human germ cell undergoes meiosis, the four gametes that form end up with one of 8,388,608 (or 2^{23}) possible combinations of homologous chromosomes. On top of that, any number of genes may occur as different alleles on the maternal and paternal chromosomes, and crossing over makes mosaics of that genetic information. Then, out of all the male and female gametes that form, which two actually get together at fertilization is a matter of chance. Are you getting an idea of why such fascinating combinations of traits show up among the generations of your own family tree?

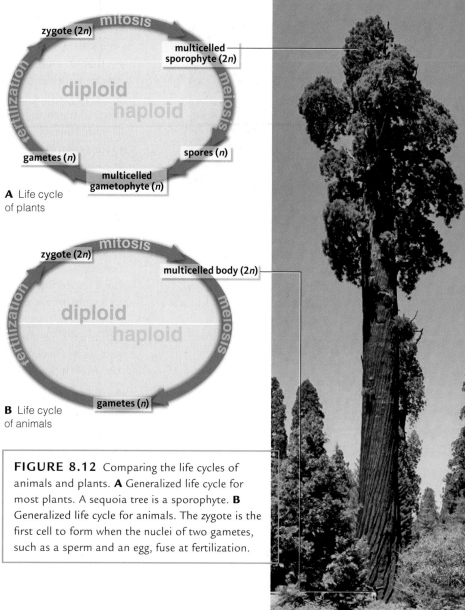

A Life cycle of plants

B Life cycle of animals

FIGURE 8.12 Comparing the life cycles of animals and plants. **A** Generalized life cycle for most plants. A sequoia tree is a sporophyte. **B** Generalized life cycle for animals. The zygote is the first cell to form when the nuclei of two gametes, such as a sperm and an egg, fuse at fertilization.

Take-Home Message Where does meiosis fit into the life cycle of plants and animals?

- Meiosis and cytoplasmic division precede the development of haploid gametes in animals and spores in plants.

- The union of two haploid gametes at fertilization results in a diploid zygote.

8.9 When Control Is Lost

Checkpoint Genes and Tumors What happens when something goes wrong during mitosis or meiosis? Suppose sister chromatids do not separate as they should during mitosis. As a result, one descendant cell ends up with too many chromosomes and the other with too few. Or suppose DNA gets damaged by free radicals, chemicals, or environmental assaults such as ultraviolet radiation. Such problems are frequent and inevitable, so the cell cycle has built-in checkpoints that allow problems to be corrected before the cycle advances.

crossing over Process in which homologous chromosomes exchange corresponding segments during meiosis.

egg Mature female gamete, or ovum.

gametophyte A haploid, multicelled body in which gametes form during the life cycle of plants.

sperm Mature male gamete.

sporophyte Diploid, spore-producing body of a plant.

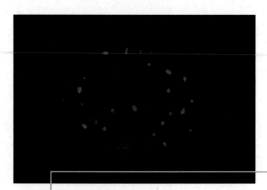

FIGURE 8.13 Checkpoint genes put the brakes on cancer. Exposure to ionizing radiation damaged the DNA inside this breast cell nucleus. *Red* dots pinpoint the location of BRCA1 protein, which has clustered around breaks in the cell's chromosomes. The combined action of BRCA1 and other checkpoint gene products blocks mitosis until the breaks are fixed.

FIGURE 8.14 Animated! Metastasis.

❶ Benign cancer cells grow slowly and stay in their home tissue.

❷ Malignant cancer cells can break away from their home tissue.

❸ The metastasizing cells become attached to the wall of a blood vessel or lymph vessel. They release digestive enzymes that create an opening in the wall, then enter the vessel.

❹ The cells creep or tumble along inside blood vessels, then leave the bloodstream the same way they got in. They start new tumors in new tissues.

Certain proteins, the products of "checkpoint" genes, can monitor whether a cell's DNA has been copied completely, whether it is damaged, and even whether enough nutrients to support cell growth are available. These proteins interact to advance, delay, or stop the cell cycle. Some checkpoint gene products activate a cascade of signaling events that ultimately causes the cell to die (you will read more about this cellular self-destruct mechanism in Section 26.2).

Sometimes a mutation modifies a checkpoint gene so that its protein product no longer works properly. The result may be that the cell skips interphase, and division occurs over and over with no resting period. Or, damaged DNA may be replicated, a process that frequently introduces new mutations (Section 7.6). In still other cases, a mutation alters the signals that make an abnormal cell commit suicide. When checkpoint mechanisms fail, a cell loses control over its cell cycle, and its descendants form a tumor.

Checkpoint genes whose products inhibit mitosis are called tumor suppressors, because tumors form when they are missing. The *BRCA* genes that you read about in Section 7.7 are examples. BRCA proteins regulate, among other things, the expression of enzymes that repair broken DNA (Figure 8.13). That is why mutations in these genes are often found in many types of cancer cells. Viruses such as HPV (human papillomavirus) cause a cell to make proteins that interfere with some checkpoint gene products. Infection with HPV causes noncancerous skin growths called warts, and some kinds are associated with cervical cancers.

Cancer Moles and other tumors are neoplasms: abnormal masses of cells that lost control over how they grow and divide (Figure 8.14). Ordinary skin moles are among the noncancerous, or benign, neoplasms. They grow very slowly, and their cells retain the surface recognition proteins that keep them in their home tissue ❶. Unless a benign neoplasm grows too large or becomes irritating, it poses no threat to the body.

A malignant neoplasm is one that is dangerous to health. Cancer occurs when the abnormally dividing cells of a malignant neoplasm disrupt body tissues, physically and metabolically (Figure 8.15). These typically disfigured cells can break loose from home tissues ❷, slip into and out of blood vessels and lymph vessels ❸, and invade other tissues where they do not belong ❹.

Cancer cells typically display the following three characteristics:

• First, cancer cells grow and divide abnormally.

• Second, cancer cells often have an altered plasma membrane and cytoplasm. The membrane may be leaky and have altered or missing proteins. The cytoskeleton may be shrunken, disorganized, or both. The balance of metabolism is often shifted, as in an amplified reliance on ATP formation by fermentation rather than by aerobic respiration.

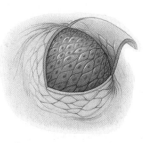

❶ benign tumor

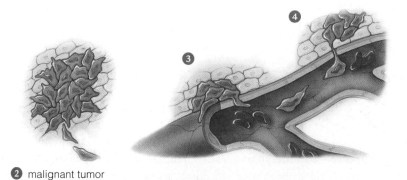

❷ malignant tumor

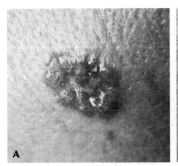 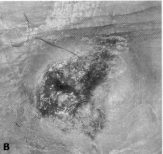 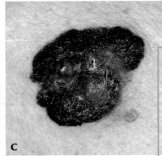

FIGURE 8.15 Skin cancer: one outcome of cell cycle checkpoint failure. **A** A basal cell carcinoma is the most common type. This slow-growing, raised lump is typically uncolored, reddish-brown, or black.

B The second most common form of skin cancer is a squamous cell carcinoma. This pink growth, firm to the touch, grows under the surface of skin.

C Malignant melanoma spreads fastest. Cells form dark, encrusted lumps that may itch or bleed easily.

• Third, cancer cells often have a weakened capacity for adhesion. Because their recognition proteins are altered or lost, they do not stick together the way normal cells do, and may break away form their home tissue to establish colonies elsewhere in the body. Metastasis is the name for this process of abnormal cell migration and tissue invasion.

Researchers already know about many mutations that contribute to cancer. They are working to identify drugs that target and destroy cancer cells or stop them from dividing. For example, HeLa cells were used in early tests of taxol, a drug that keeps microtubules from disassembling. If spindle fibers cannot disassemble, mitosis stops. Frequent divisions of cancer cells make them more vulnerable to this poison than normal cells. Such research may yield drugs that can put the brakes on cancer.

Take-Home Message How does cancer start?

- Built-in checkpoints are gene expression controls that advance, delay, or block the cell cycle in response to internal and external conditions.
- The failure of cell cycle checkpoints results in the uncontrolled cell divisions that characterize cancer.

8.10 Impacts/Issues Revisited:
Henrietta's Immortal Cells

These days, physicians and researchers are required to obtain a signed consent form before they take tissue samples from a patient, but not in the 1950s. It was common at that time for doctors to experiment on patients without their knowledge or consent. Thus, the young resident who was treating Henrietta Lacks's cancerous cervix probably never thought twice about asking permission before he took a sample of it. That sample was the one that the Geys used to establish the HeLa cell line. No one in Henrietta's family knew about the cells until 25 years after she died, and they have not received any compensation to date. It is now illegal to sell one's own organs or tissues, but no one is under a legal obligation to share profits reaped from research on a patient's tissues—or cell lines derived from them—with the donor.

How Would YOU Vote?

No one asked Henrietta Lacks's permission to use her cells. Her family did not find out about them until twenty-five years after she died. HeLa cells are still being sold worldwide. Should the family of Henrietta Lacks share in the profits? See CengageNow for details, then vote online (cengagenow.com).

Summary

 Sections 8.1 and 8.2 A cell reproduces by dividing in two. Each descendant cell receives a full set of chromosomes and some cytoplasm. Nuclear division mechanisms partition chromosomes of a parent cell into new nuclei. The cytoplasm divides by a separate mechanism. Mitosis and cytoplasmic division are the basis of growth, cell replacements, and tissue repair in multicelled species, and asexual reproduction in many species (Figure 8.16). Meiosis, the basis of sexual reproduction in eukaryotes, precedes the formation of gametes or sexual spores.

Section 8.3 A cell cycle starts when a new cell forms, and ends when the cell reproduces. Most of a cell's activities, including DNA replication, occur in interphase. Mitosis maintains the chromosome number.

CENGAGENOW™ Investigate the stages of the cell cycle with the interaction on CengageNow.

 Section 8.4 During mitosis, duplicated homologous chromosomes line up at the spindle equator, then sister chromatids move apart to opposite spindle poles. Nuclear envelopes form around the two clusters of chromosomes, forming two new nuclei with the parental chromosome number.

CENGAGENOW™ See mitosis proceed with the animation on CengageNow.

 Section 8.5 Mechanisms of cytoplasmic division differ. In animal cells, a contractile ring pinches the cytoplasm in two. In plant cells, a cross-wall forms.

CENGAGENOW™ Compare the cytoplasmic division of plant and animal cells with the animation on CengageNow.

Section 8.6 Offspring of asexual reproduction are clones. Offspring of sexual reproduction vary in shared traits. In sexual reproduc-

 tion, meiosis in germ cells forms haploid gametes that fuse at fertilization. Offspring of most sexual reproducers inherit pairs of chromosomes, one of each pair from the mother and the other from the father. Except for a pair of nonidentical sex chromosomes, the members of a chromosome pair have the same length, shape, and set of genes. Paired genes on homologous chromosomes often vary slightly in DNA sequence, in which case they are called alleles.

Section 8.7 Meiosis shuffles parental combinations of alleles. In meiosis, two nuclear divisions halve the parental chromosome number. In the first nuclear division, duplicated homologous chromosomes line up and cross over, then move apart, toward opposite spindle poles. Two new nuclear envelopes form around the two clusters of still-duplicated chromosomes. The second nuclear division separates sister chromatids. Four haploid nuclei typically form. Each has one complete set of unduplicated chromosomes.

CENGAGENOW™ Explore what happens in the stages of meiosis with the animation on CengageNow.

Section 8.8 Haploid gametes form by meiosis during the life cycles of plants and animals. Familiar plants are diploid sporophytes that make haploid spores. Germ cells in the reproductive organs of animals give rise to sperm or eggs. The fusion of two haploid gamete nuclei during fertilization restores the parental chromosome number in the zygote, the first cell of the new individual.

CENGAGENOW™ See how eggs and sperm form with the animation on CengageNow.

Section 8.9 Checkpoint gene products are gene expression controls that regulate the cell cycle. Disruption of checkpoint gene products, such as by mutations or

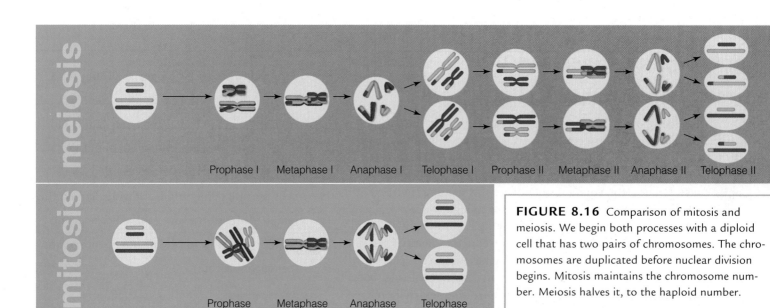

Prophase I · Metaphase I · Anaphase I · Telophase I · Prophase II · Metaphase II · Anaphase II · Telophase II

meiosis

mitosis

Prophase · Metaphase · Anaphase · Telophase

FIGURE 8.16 Comparison of mitosis and meiosis. We begin both processes with a diploid cell that has two pairs of chromosomes. The chromosomes are duplicated before nuclear division begins. Mitosis maintains the chromosome number. Meiosis halves it, to the haploid number.

Digging Into Data

HeLa Cells Are a Genetic Mess

 HeLa cells (*left*) continue to be an extremely useful tool in cancer research. One early finding was that HeLa cells vary in chromosome number. The panel of chromosomes in Figure 8.17, originally published in 1989 by Nicholas Popescu and Joseph DiPaolo, shows all of the chromosomes in a single metaphase HeLa cell.

1. What is the chromosome number of this HeLa cell?

2. How many extra chromosomes does this cell have, compared to a normal human body cell?

3. Can you tell that this cell came from a female? How?

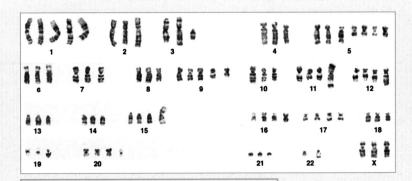

FIGURE 8.17 Chromosomes in a HeLa cell.

viruses, causes tumors that may end up as cancer. Cancer cells may metastasize—break loose and colonize distant tissues.

CENGAGENOW™ See how cancers spread through the body with the animation on CengageNow.

Self-Quiz   Answers in Appendix I

1. How many chromatids does a duplicated chromosome have?

2. Mitosis and cytoplasmic division function in _____ .
   a. asexual reproduction of single-celled eukaryotes
   b. growth and tissue repair in multicelled species
   c. gamete formation in prokaryotes
   d. both a and b

3. _____ maintains the chromosome number; _____ halves it.
   a. mitosis; meiosis   b. meiosis; mitosis

4. Generally, a pair of homologous chromosomes _____ .
   a. carry the same genes   c. are the same length, shape
   b. interact at meiosis   d. all of the above

5. A cell with two of each type of chromosome has a(n) _____ chromosome number.
   a. diploid   b. haploid   c. tetraploid   d. abnormal

6. Interphase is the part of the cell cycle when _____ .
   a. a cell ceases to function
   b. a cell forms its spindle apparatus
   c. a cell grows and duplicates its DNA
   d. mitosis proceeds

7. After mitosis, the chromosome number of the two new cells is _____ the parent cell's.
   a. the same as   c. rearranged compared to
   b. one-half of   d. doubled compared to

8. Only _____ is not a stage of mitosis.
   a. prophase   b. interphase   c. metaphase   d. anaphase

9. Meiosis and cytoplasmic division function in _____ .
   a. asexual reproduction of single-celled eukaryotes
   b. growth and tissue repair
   c. sexual reproduction
   d. both a and b

10. The cell shown at *right* is in anaphase II. I know this because _____ .

11. Sexual reproduction always requires _____ .
    a. mitosis   c. spore formation
    b. fertilization   d. a and b

12. What is the name for alternative forms of the same gene?

13. Match each term with the best description.
    ____ prophase   a. homologous chromosomes
    ____ telophase   aligned at the spindle equator
    ____ metaphase I   b. maybe none between meiosis I, II
    ____ metaphase II   c. unduplicated chromosomes aligned
    ____ interphase   at the spindle equator
    ____ anaphase II   d. chromosomes start to condense
    ____ prophase I   e. new nuclei form
    ____ anaphase I   f. homologous chromosomes move apart
                          g. sister chromatids move apart
                          h. homologues swap segments

Additional questions are available on CENGAGENOW™

Critical Thinking

1. The anticancer drug taxol was first isolated from Pacific yews (*Taxus brevifolia*), which are slow-growing trees. Bark from about six yew trees provided enough taxol to treat one patient, but removing the bark killed the trees. Fortunately, taxol is now produced using plant cells that grow in big vats rather than in trees. What challenges do you think had to be overcome to get plant cells to grow and divide in laboratories?

2. Suppose you have a way to measure the amount of DNA in one cell during the cell cycle. You first measure the amount at the G1 phase. At what points in the rest of the cycle will you see a change in the amount of DNA per cell?

3. Why do you think that sexual reproduction tends to give rise to greater genetic diversity among offspring in fewer generations than asexual reproduction does?

4. Which nuclear division, meiosis I or meiosis II, is most like mitosis? Why?

9 Patterns of Inheritance

Most CF patients live no longer than thirty years, at which time their tormented lungs usually fail.

9.1 Impacts/Issues: Menacing Mucus

In 1988, researchers discovered a gene that, when mutated, causes cystic fibrosis (CF). Cystic fibrosis is the most common fatal genetic disorder in the United States. The gene in question, *CFTR*, encodes a protein that moves chloride ions out of epithelial cells. Sheets of these cells line the passageways and ducts of the lungs, liver, pancreas, intestines, reproductive system, and skin. When the CFTR protein pumps chloride ions out of these cells, water follows the ions by osmosis. This two-step process maintains a thin film of water on the surface of the epithelial sheets. Mucus slides easily over the wet sheets of cells.

The most common CF mutation is a deletion of one codon that specifies the 508th amino acid of the CFTR protein, a phenylalanine. The deletion disrupts membrane trafficking of CFTR so that newly assembled polypeptides are stranded in the endoplasmic reticulum. The protein itself can function properly, but it never reaches the cell surface to do its job.

One outcome is that the transport of chloride ions out of epithelial cells is disrupted. If not enough chloride ions leave the cells, not enough water leaves them either, so the surfaces of epithelial cell sheets are not as wet as they should be. Mucus that normally slips and slides through the body's tubes sticks to them instead. Thick, sticky globs of mucus accumulate and clog the passageways of the lungs, making breathing difficult.

The CFTR protein also functions as a receptor that alerts the body to the presence of bacteria. Bacteria bind to CFTR. The binding triggers endocytosis, which speeds the immune system's defensive responses. Without the CFTR protein on the surface of epithelial cell linings, disease-causing bacteria that enter the ducts and passageways of the body can persist there. Thus, chronic bacterial infections of the intestine and lungs are hallmarks of cystic fibrosis.

Daily routines of posture changes and thumps on the chest and back help clear the lungs of some of the thick mucus, and antibiotics help control infections, but there is no cure. Even with a lung transplant, most CF patients live no longer than thirty years, at which time their tormented lungs usually fail (Figure 9.1).

More than 10 million people carry the CF mutation in one of their two copies of the *CFTR* gene, but most of them do not realize it. Cystic fibrosis only occurs when a person inherits two mutated genes, one from each parent. This unlucky event occurs in about 1 of 3,300 births. Why is it that people with one copy of a mutated CFTR gene are healthy, but people with two copies are ill with cystic fibrosis? Why is the mutation that causes CF so common if its effects are so devastating? You will begin to find the answers to such questions in this chapter, in which we introduce principles of inheritance—how new individuals are put together in the image of their parents.

Cody, 23   Jeff, 21   Lindsay, 22
Ben, 23   Savannah, 19   Brandon, 18

FIGURE 9.1 A few of the many victims of cystic fibrosis. At least one young person dies every day in the United States from complications of this disease.

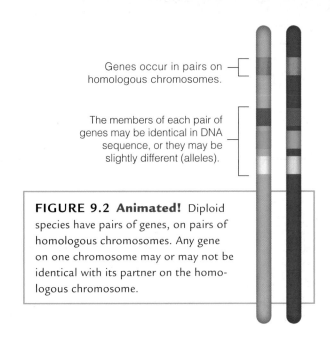

Genes occur in pairs on homologous chromosomes.

The members of each pair of genes may be identical in DNA sequence, or they may be slightly different (alleles).

FIGURE 9.2 Animated! Diploid species have pairs of genes, on pairs of homologous chromosomes. Any gene on one chromosome may or may not be identical with its partner on the homologous chromosome.

FIGURE 9.3 Animated!
Breeding garden pea plants.

1 In this example, pollen from a plant that breeds true for purple flowers is brushed onto the flower of a plant that breeds true for white flowers.

2 Seeds develop inside pods of the cross-fertilized plant. The seeds grow into new plants.

3 Every plant that arises from this cross has purple flowers. Predictable patterns such as this offer evidence of how inheritance works.

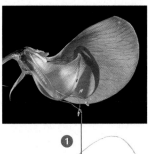
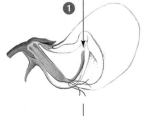

9.2 Tracking Traits

Alleles and Phenotype By the mid-1800s, most people had an idea that parents contribute hereditary material to their offspring. At the time, no one knew about genes, meiosis, or chromosomes. Some thought that hereditary material must be fluid, with fluids from both parents blending at fertilization like milk into coffee. However, the idea of "blending inheritance" failed to explain the obvious. For example, if parental fluids blended, then children of the same parents would all have the same eye color, which would be a blended shade of their parents' eye colors. But children of the same parents often differ

in eye color. Meanwhile, an Austrian monk named Gregor Mendel (*left*) had been collecting evidence of how inheritance really works. He was carefully documenting how traits are passed from generation to generation of garden pea plants. This plant is self-fertilizing, which means that its flowers make both male and female gametes. Pea plants can be easily cross-fertilized by transferring pollen from the flower of one plant to the flower of another.

Mendel knew that some pea plants "breed true" for traits such as white flowers. Breeding true means that offspring have the same form of the trait as the parent(s), generation after generation. For example, all offspring of pea plants that breed true for white flowers also have white flowers.

Flower color is an example of a pea plant's **phenotype**, its set of observable traits. Mendel did not know that phenotype is governed by pairs of genes on pairs of homologous chromosomes (Figure 9.2). He also did not know that variation in traits occurs because genes of a pair often vary slightly in sequence, as alleles (Section 8.6). A pea plant breeds true for a trait such as white flowers because it has identical alleles that govern that trait. An individual with identical alleles is said to be **homozygous** for the gene (*homo*– means the same; *zygotos* is a Greek word that means linked).

Mendel crossed pea plants that bred true for different forms of a trait, and discovered that the phenotypes of the resulting **hybrid** plants occur in predictable patterns (Figure 9.3). The patterns he saw were clues about the plants'

homozygous Having identical alleles of a gene.
hybrid The offspring of a cross between two individuals that breed true for different forms of a trait.
phenotype An individual's observable traits.

FIGURE 9.4 Gene segregation. Homologous chromosomes separate during meiosis, so the pairs of genes they carry separate too. Each of the resulting gametes carries only one of the two members of each gene pair.

For clarity, we show nuclei of reproductive cells that carry only one chromosome.

❶ All gametes produced by a parent homozygous for a dominant allele carry the dominant allele.

❷ All gametes produced by a parent homozygous for a recessive allele carry the recessive allele.

❸ If these two parents are crossed, the union of any of their gametes at fertilization produces a zygote with both alleles. Thus, all of the offspring of this cross will be heterozygous.

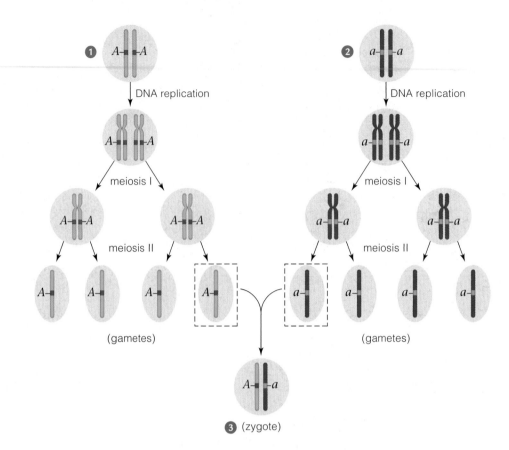

genotype, which refers to the particular alleles that an individual carries. The patterns occur because the effect of some alleles masks the effect of others. We say that an allele is **dominant** if its effect masks that of a **recessive** allele paired with it. Uppercase italic letters (*A*) usually refer to dominant alleles and lowercase italic letters (*a*) refer to their recessive partners.

Consider a gene that specifies flower color in pea plants. For this gene, the allele that specifies purple is dominant over the allele that specifies white. Let's designate the purple color allele *A*, and the white color allele *a*. So, a pea plant with two *A* alleles (*AA*) has purple flowers, and one with two *a* alleles (*aa*) has white flowers. Mendel crossed plants that bred true for purple flowers (*AA*) with plants that bred true for white flowers (*aa*). All of the offspring of this cross had purple flowers. Why? Homologous chromosomes separate during meiosis, so each gamete that forms carries only one of the two alleles (Figure 9.4 ❶, ❷). If plants homozygous for different alleles are crossed (*AA* × *aa*), only one outcome is possible: All of the offspring will have one of each allele (*Aa*) ❸. All of them carry the dominant allele *A*, so all of them will have purple flowers. Individuals that have nonidentical alleles are **heterozygous** for the gene (*hetero–*, mixed).

Mendelian Inheritance Patterns Dominance relationships among alleles can sometimes be determined by breeding heterozygous individuals together. The pattern in which traits show up among offspring of such crosses may indicate whether one allele is dominant over the other. For example, a **monohybrid cross** is used to check for a dominance relationship between alleles of one gene. In a monohybrid cross, individuals that are identically heterozygous at one gene are bred together (*Aa* × *Aa*), and the traits that show up among the offspring are observed. Mendel made hundreds of monohybrid crosses, and recorded the traits of thousands of their offspring.

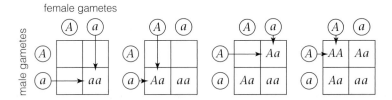

A Making a Punnett square. The possible parental gametes are listed on the top and left sides of the grid (in circles). Each square is filled in with the combination of alleles that would result if the gametes in the corresponding row and column met up.

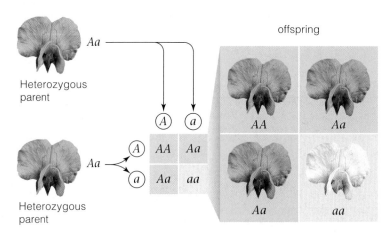

B A Punnett square shows that the ratio of dominant-to-recessive phenotypes among offspring of a monohybrid cross is about 3:1 (3 purple to 1 white).

FIGURE 9.5 Animated! Example of a monohybrid cross.

Figure It Out: How many genotypes are possible among the off-spring of this cross?

Answer: Three: AA, Aa, and aa

He discovered that about three of every four of those offspring had one trait (such as purple flowers), and about one of every four had the other trait (white flowers). The phenotypic ratio of 3:1 indicates that the traits Mendel studied were specified by alleles with a clear dominant–recessive relationship. To understand why, we have to revisit probability—the chance that a particular outcome will occur (Section 1.7). That chance depends on the total number of possible outcomes. Grids called **Punnett squares** allow us to determine the probability of genetic outcomes of a cross. Figure 9.5**A** shows how to construct a Punnett square, using a monohybrid cross between our purple-flowered heterozygotes as an example (*Aa* × *Aa*). With respect to the flower color gene, only two types of gametes (*A* and *a*) can form in each *Aa* plant. So, in a monohybrid cross between two *Aa* plants, those two types of gametes can meet up in four possible ways at fertilization:

Possible Event	Outcome
Sperm *A* meets egg *A*	offspring is *AA*
Sperm *A* meets egg *a*	offspring is *Aa*
Sperm *a* meets egg *A*	offspring is *Aa*
Sperm *a* meets egg *a*	offspring is *aa*

In our example, three out of those four possible outcomes include at least one copy of the dominant allele *A*. Each time fertilization occurs, there are 3 chances in 4 that the resulting offspring will inherit an *A* allele, and have purple flowers. There is 1 chance in 4 that it will inherit two recessive *a* alleles, and have white flowers. Thus, the probability that a particular offspring of this cross will have purple or white flowers is 3 purple to 1 white, which we represent as a ratio of 3:1 (Figure 9.5**B**).

If the probability of one individual inheriting a particular genotype is difficult to imagine, think about probability in terms of the phenotypes of many offspring of our monohybrid cross. In this example, there will be roughly three purple-flowered plants for every white-flowered one. A phenotypic ratio of 3:1 is typical of the offspring of a monohybrid cross with dominant and recessive alleles, which is the pattern that Mendel saw.

A monohybrid cross tracks alleles of one gene pair. What about alleles of two gene pairs? How two gene pairs get sorted into gametes depends partly on whether the two genes are on the same chromosome. Remember, chromosomes line up and then separate during meiosis (Section 8.7). Homologous chromo-

dominant Refers to an allele that masks the effect of a recessive allele paired with it.

genotype The particular alleles carried by an individual.

heterozygous Having two different alleles of a gene.

monohybrid cross Experiment in which individuals with different alleles of a gene are crossed.

Punnett square Diagram used to predict the outcome of a cross.

recessive Refers to an allele with an effect that is masked by a dominant allele on the homologous chromosome.

A This example shows just two pairs of homologous chromosomes in the nucleus of a diploid (2n) reproductive cell. Maternal and paternal chromosomes, shown in *pink* and *blue*, have already been duplicated.

B Either chromosome of a pair may get attached to either spindle pole during meiosis I. With two pairs of homologous chromosomes, there are two different ways that the maternal and paternal homologues can get attached to opposite spindle poles.

C Two nuclei form with each scenario, so there are a total of four possible combinations of parental chromosomes in the nuclei that form after meiosis I.

D Thus, when sister chromatids separate during meiosis II, the gametes that result have one of four possible combinations of maternal and paternal chromosomes.

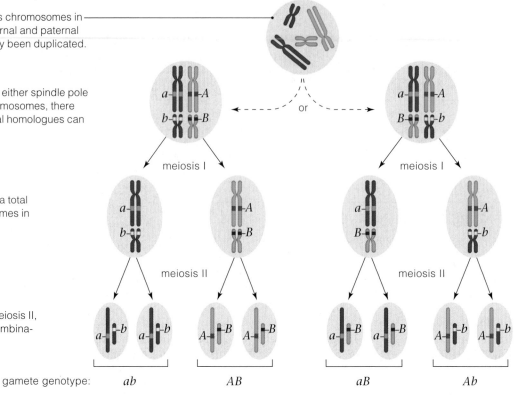

gamete genotype: *ab*   *AB*   *aB*   *Ab*

FIGURE 9.6 Independent assortment during meiosis. Here, we track how alleles of two gene pairs on two chromosomes are distributed into gametes.

somes become attached to opposite spindle poles during meiosis I, but which homologue gets attached to which pole is entirely random. Microfilaments that extend from one spindle pole attach to the first chromosome of a pair they contact, and the homologous chromosome gets attached to the other spindle pole. Thus, when homologous chromosomes separate, either homologue can end up in a particular nucleus. The random sorting of chromosomes during meiosis means that gene pairs on one chromosome become sorted into gametes independently of gene pairs on other chromosomes (Figure 9.6).

Punnett squares are particularly useful when predicting inheritance patterns of two or more genes simultaneously, such as with a dihybrid cross. A **dihybrid cross** tests for dominance relationships between alleles of two genes. In a typical dihybrid cross, individuals identically heterozygous for two genes (dihybrids) are bred together (*AaBb* × *AaBb*), and the traits that show up among the offspring are observed.

To make a dihybrid cross, we would start with individuals that breed true for two different traits. Let's use genes for flower color (*A*, purple; *a*, white) and height (*B*, tall; *b*, short), and assume that these genes occur on separate chromosomes. Figure 9.7 shows a dihybrid cross starting with one parent plant that breeds true for purple flowers and tall stems (*AABB*), and another that breeds true for white flowers and short stems (*aabb*). Each homozygous plant makes only one type of gamete ❶. So, all of the offspring from a cross between these parent plants (*AABB* × *aabb*) will be dihybrids (*AaBb*) and have purple flowers and tall stems ❷. Thus, four combinations of alleles are possible in the gametes of *AaBb* dihybrids: *AB*, *Ab*, *aB*, and *ab* ❸.

If two of the resulting *AaBb* dihybrids are crossed (a dihybrid cross, *AaBb* × *AaBb*), the four types of gametes that form can combine in sixteen possible ways at fertilization (4 × 4 = 16). In this example, those sixteen genotypes would result in four different phenotypes ❹. Nine would be tall with purple

dihybrid cross Experiment in which individuals with different alleles of two genes are crossed.

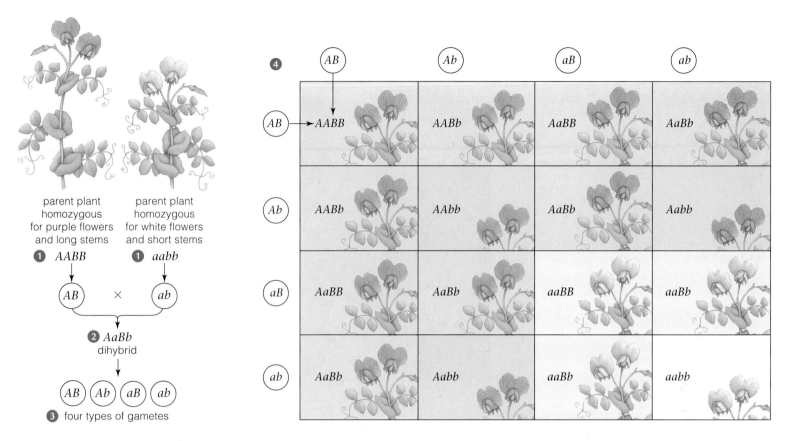

parent plant
homozygous
for purple flowers
and long stems

parent plant
homozygous
for white flowers
and short stems

❶ *AABB*

❶ *aabb*

AB × ab

❷ *AaBb*
dihybrid

AB Ab aB ab

❸ four types of gametes

FIGURE 9.7 Animated! A dihybrid cross between plants that differ in flower color and plant height. *A* and *a* stand for dominant and recessive alleles for flower color. *B* and *b* stand for dominant and recessive alleles for height.

❶ Meiosis in a homozygous individual results in one type of gamete.

❷ A cross between two homozygous individuals yields one type of gamete. All offspring that form in this example are dihybrids (heterozygous for two genes).

❸ Meiosis in dihybrid individuals results in four kinds of gametes.

❹ If two dihybrid individuals are crossed (a dihybrid cross), the four types of gametes can meet up in 16 possible ways. Out of 16 possible genotypes of the offspring, 9 will result in plants that are purple-flowered and tall; 3, purple-flowered and short; 3, white-flowered and tall; and 1, white-flowered and short. The ratio of phenotypes in a dihybrid cross is 9:3:3:1.

flowers, three would be short with purple flowers, three would be tall with white flowers, and one would be short with white flowers. The ratio of these phenotypes is 9:3:3:1. Mendel discovered the 9:3:3:1 ratio of phenotypes among the offspring of dihybrid crosses, although he had no idea what it meant. He published his results in 1866, but apparently his work was read by few and understood by no one. In 1871 he became abbot of his monastery, and his pioneering experiments ended. He died in 1884, never to know that his experiments would be the starting point for modern genetics.

The Contribution of Crossovers Alleles of genes on different chromosomes assort independently into gametes. What about genes on the same chromosome? Mendel studied seven genes in pea plants, which have seven pairs of chromosomes. Was he lucky enough to choose one gene on each of those seven chromosome pairs? As it turns out, some of the genes Mendel studied are on the same chromosome. The genes are far enough apart that crossing over occurs between them very frequently—so frequently that they tend to assort into gametes independently, just as if they were on different chromosomes. By contrast, genes that are very close together on a chromosome do not tend to assort independently, because crossing over does not happen very often between them. Thus, gametes usually receive parental combinations of alleles of such genes. Human gene linkages were identified by tracking inheritance in families over several generations. One thing became clear: Crossovers are not at all rare, and are often a required step in order for meiosis to run to completion.

A Human Example: Skin Color Certain types of skin cells have melanosomes, which are organelles that make red or brownish-black pigments called melanins. Most people have about the same number of melanosomes in their skin cells. Variations in skin color begin with variations in the kinds and amounts of melanins made by the melanosomes.

FIGURE 9.8 Variation in skin color and in most other human traits begins with alleles. Fraternal twin girls Kian and Remee inherited different sets of alleles that govern skin color from their parents Kylie (*left*) and Remi (*right*).

The twins' grandmothers are both of European descent, and have pale skin. Their grandfathers are both of African descent, and have dark skin.

All humans share the genetic legacy of a common ancestry.

Like most other human traits, skin color has a genetic basis. The products of more than 100 genes affect how melanins are made and deposited in skin cells. Different alleles of those genes contribute to variations of human skin color. For example, one gene encodes a transport protein in melanosome membranes. Nearly all people of native African, American, or east Asian descent have the same allele of this gene. By contrast, nearly all people of native European descent carry a particular mutation in that gene, and a slightly different gene product is the result. The European allele results in less melanin, and lighter skin color, than the unmutated version of the gene.

Occasionally, skin color offers a vivid demonstration of how alleles separate during meiosis. In 2006, two young parents made the news. Both have dark skin, as does their daughter Kian. However, Kian's fraternal twin Remee was born with fair skin (Figure 9.8). Both of the children's grandmothers are of European descent, and have pale skin. Both of their grandfathers are of African descent, and have dark skin. The twins inherited different alleles of some of the genes that affect skin color from their parents, who, given the appearance of their children, must be heterozygous for the alleles.

Skin color is only one of many human traits that can vary because of mutations in single genes. The small scale of such differences is a reminder that all humans share the genetic legacy of a common ancestry.

Take-Home Message How do alleles contribute to traits?

- Diploid cells have pairs of genes, on pairs of homologous chromosomes. The two genes of a pair (which may be identical or not) separate during meiosis, so they end up in different gametes.

- Dominant alleles mask the effects of recessive ones. Which alleles an individual carries (its genotype) is the basis of its unique traits (its phenotype).

- A gene tends to be distributed into gametes independently of how other genes are distributed.

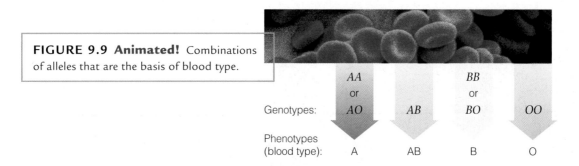

Genotypes: AA or AO   AB   BB or BO   OO

Phenotypes (blood type): A   AB   B   O

Beyond Simple Dominance

The inheritance patterns in the last section are examples of simple dominance, in which a dominant allele fully masks the expression of a recessive one. However, there are many other examples in which both genes of a pair are expressed at the same time. The one gene = one trait equation does not always apply, either. In many cases, several gene products influence the same trait. In other cases, the expression of a single gene influences multiple traits.

Codominance Two nonidentical alleles of a gene may be **codominant**, which means that both alleles are fully expressed in heterozygous individuals. The three alleles of the *ABO* gene offer an example. The enzyme encoded by the *ABO* gene modifies a carbohydrate that imparts an identity to human red blood cell membranes. The *A* and *B* alleles encode slightly different versions of the enzyme, which in turn modify the carbohydrate differently. The *O* allele has a mutation that prevents its enzyme product from becoming active at all.

The two alleles you carry for the *ABO* gene determine the form of the carbohydrate on your blood cells, and that carbohydrate is the basis of your blood type (Figure 9.9). The *A* and the *B* allele are codominant when paired. If your genotype is *AB*, then you have both versions of the enzyme, and your blood is type AB. The *O* allele is recessive when paired with either *A* or *B*. If you are *AA* or *AO*, your blood is type A. If you are *BB* or *BO*, it is type B. If you are *OO*, it is type O.

Receiving incompatible blood cells in a transfusion is very dangerous, because the immune system usually attacks red blood cells bearing molecules that do not occur in one's own body. The attack can cause the blood cells to clump or burst, a transfusion reaction with potentially fatal consequences. People with type O blood can donate blood to anyone else, so they are called universal donors. However, they can receive transfusions of type O blood only. People with type AB blood can receive a transfusion of any blood type, so they are called universal recipients.

Incomplete Dominance With **incomplete dominance**, one allele of a gene pair is not fully dominant over the other, so the heterozygous phenotype is between the two homozygous phenotypes. A gene that influences flower color in snapdragon plants is an example. A cell has one copy of this gene on each homologous chromosome, and both copies are expressed. One allele (*R*) encodes an enzyme that makes a red pigment. Another allele (*r*) carries a mutation, so the enzyme it encodes cannot make any pigment. Plants homozygous for the *R* allele (*RR*) make a lot of red pigment, so they have red flowers. Plants homozygous for the *r* allele (*rr*) do not make any pigment at all, so their flowers are white. Heterozygous plants (*Rr*) make only enough red pigment to color their flowers pink (Figure 9.10**A**). A cross between pink-flowered heterozygous plants yields red-, pink-, and white-flowered offspring in a 1:2:1 ratio (Figure 9.10**B**).

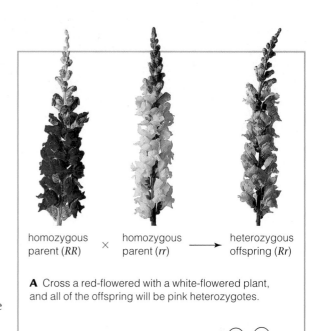

homozygous parent (*RR*)   ×   homozygous parent (*rr*)   ⟶   heterozygous offspring (*Rr*)

A Cross a red-flowered with a white-flowered plant, and all of the offspring will be pink heterozygotes.

B If two of the heterozygotes are crossed, the phenotypes of the resulting offspring will occur in a 1:2:1 ratio.

	R	*r*
R	*RR*	*Rr*
r	*Rr*	*rr*

FIGURE 9.10 Incomplete dominance of flower color alleles in snapdragon plants. **Figure It Out: Is the experiment in B a monohybrid or dihybrid cross?**

Answer: A monohybrid cross

codominant Refers to two alleles that are both fully expressed in heterozygous individuals.

incomplete dominance Condition in which one allele is not fully dominant over another, so the heterozygous phenotype is between the two homozygous phenotypes.

	EB	Eb	eB	eb
EB	EEBB	EEBb	EeBB	EeBb
Eb	EEBb	EEbb	EeBb	Eebb
eB	EeBB	EeBb	eeBB	eeBb
eb	EeBb	Eebb	eeBb	eebb

FIGURE 9.11 Epistasis. In this example, the products of two genes influence coat color in Labrador retrievers. Alleles *B* and *b* influence melanin synthesis. Alleles *E* and *e* govern how much of the melanin is deposited in a dog's fur.

❶ Dogs with a dominant *E* allele and a dominant *B* allele have black fur.

❷ Dogs with a dominant *E* allele and two recessive *b* alleles have brown fur.

❸ Dogs that are homozygous for the recessive *e* allele have yellow fur.

Epistasis

Epistasis Some traits are affected by multiple gene products, an effect called polygenic inheritance or **epistasis**. Human skin color, which is a result of interactions among several gene products, is an example. Similar genes affect Labrador retriever coat color, which can be black, yellow, or brown (Figure 9.11). One gene is involved in the synthesis of the pigment melanin. A dominant allele of the gene (*B*) specifies black fur, and its recessive partner (*b*) specifies brown fur. A dominant allele of a different gene (*E*) causes melanin to be deposited in fur, and its recessive partner (*e*) reduces melanin deposition. Thus, a dog that carries an *E* and a *B* allele has black fur ❶. One that carries an *E* allele and is homozygous for the *b* allele has brown fur ❷. A dog homozygous for the *e* allele has yellow fur regardless of its *B* or *b* alleles ❸.

Pleiotropy

Pleiotropy A **pleiotropic** gene is one that influences multiple traits. Mutations in such genes are associated with complex genetic disorders such as sickle-cell anemia (Section 7.6), cystic fibrosis (Section 9.1), and Marfan syndrome. Marfan syndrome can be very difficult to diagnose. Affected people are often tall, thin, and loose-jointed, but there are plenty of tall, thin, loose-jointed people without Marfan syndrome. Symptoms may not be apparent, so many people die suddenly and early without ever knowing they had the disorder. 21-year-old basketball star Haris Charalambous (*right*) was one. Charalambous died when his aorta burst during warmup exercises in 2006.

The pleiotropic gene involved in Marfan syndrome encodes fibrillin. Long fibers of this protein impart elasticity to tissues of the heart, skin, blood vessels, tendons, and other body parts. In Marfan

bell curve Bell-shaped curve; typically results from graphing frequency versus distribution for a trait that varies continuously in a population.

continuous variation A range of small differences in a shared trait.

epistasis Effect in which a trait is influenced by the products of multiple genes.

pleiotropic Refers to a gene whose product influences multiple traits.

syndrome, mutations in the fibrillin gene result in tissues that form with defective fibrillin or none at all. The largest blood vessel leading from the heart, the aorta, is particularly affected. In Marfan syndrome, muscle cells in the aorta's thick wall do not work very well, and the wall itself is not as elastic as it should be. The aorta expands under pressure, so the lack of elasticity eventually makes it thin and leaky. Calcium deposits accumulate inside. Inflamed, thinned, and weakened, the aorta can rupture abruptly during exercise.

Take-Home Message **Are all alleles clearly dominant or recessive?**

- An allele may be fully dominant, incompletely dominant, or codominant with its partner on a homologous chromosome.

- In epistasis, two or more gene products influence a trait.

- The product of a pleiotropic gene influences two or more traits.

9.4 Complex Variations in Traits

Most organic molecules are made in metabolic pathways that involve many enzymes. Genes encoding those enzymes can mutate in any number of ways, so their products may function within a spectrum of activity that ranges from excessive to not at all. Thus, the end product of a metabolic pathway can be produced within a range of concentration and activity. Environmental factors often add further variations on top of that. In the end, phenotype often results from complex interactions among gene products and the environment.

Continuous Variation The individuals of a species typically vary in many of their shared traits. Some of those traits appear in two or three distinct forms. Others occur in a range of small differences that is called **continuous variation**. The more genes and environmental factors that influence a trait, the more continuous is its variation.

How do we determine whether a trait varies continuously? First, we divide the total range of phenotypes into measurable categories, such as inches of height. The number of individuals that fall into each category gives the relative frequencies of phenotypes across our range of measurable values. Finally, we plot the data as a bar chart (Figure 9.12). A graph line around the top of the bars shows the distribution of values for the trait. If the line is a bell-shaped curve, or **bell curve**, the trait varies continuously.

Human eye color is another example of a trait that varies continuously. The colored part is the iris, a doughnut-shaped, pigmented structure. Iris color, like skin color, is the result of epistasis among gene products that make and distribute melanins. The more melanin deposited in the iris, the less light is reflected from it. Dark irises have dense melanin deposits that absorb almost all light, and reflect almost none. Melanin deposits are not as extensive in brown eyes, which reflect some incident light. Green and blue eyes have the least amount of melanin, so they reflect the most light.

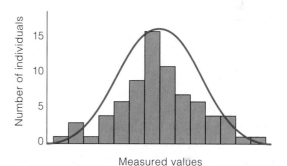

5/3 5/4 5/5 5/6 5/7 5/8 5/9 5/10 5/11 6/0 6/1 6/2 6/3 6/4 6/5

Height (feet/inches)

FIGURE 9.12 Animated! Continuous variation in height among male biology students at the University of Florida. The students were divided into categories of one-inch increments in height and counted. The resulting bell-shaped curve indicates that height varies continuously.

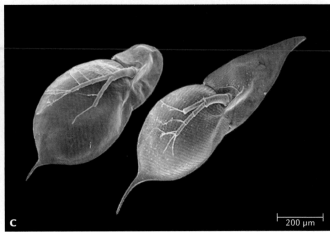

A

B

C

200 µm

FIGURE 9.13 Environmental effects on animal phenotype. The color of the snowshoe hare's fur varies by season: **A** summer, and **B** winter. Both forms offer seasonally appropriate camouflage from predators. **C** The body form of the *Daphnia* at the *left* develops in environments with few predators. A longer tail spine and a pointy head (*right*) develop in response to chemicals emitted by predatory insects.

Height (centimeters)

60   60   60

0   0   0

A Plant grown at high elevation (3,060 meters above sea level)

B Plant grown at mid-elevation (1,400 meters above sea level)

C Plant grown at low elevation (30 meters above sea level)

FIGURE 9.14 Environmental effects on plant phenotype. Cuttings from the same yarrow plant grow to different heights at three different elevations.

Environmental Effects on Phenotype Variations in traits are not always the result of differences in alleles. Environmental factors often affect gene expression, which in turn affects phenotype. For example, seasonal changes in temperature and the length of day affect the production of melanin and other pigments that color the skin and fur of many animals. These animals have different color phases in different seasons (Figure 9.13**A,B**). Other factors, such as the presence of predators, can influence phenotype. Consider *Daphnia*, a microscopic freshwater relative of shrimp. Aquatic insects prey on them. *Daphnia* living in ponds with few predators have rounded heads, but those in ponds with many predators have pointy heads (Figure 9.13**C**). *Daphnia*'s predators emit chemicals that trigger the different phenotype.

Yarrow plants offer another example of how environment influences phenotype. Yarrow is useful for experiments because it grows from cuttings. All cuttings of a plant are genetically identical, so experimenters know that genes are not the basis for any phenotypic differences among them. In one study, genetically identical yarrow plants had different phenotypes when grown at different altitudes (Figure 9.14).

The environment also affects human genes. One of our genes encodes a protein that transports serotonin across the membrane of brain cells. Serotonin lowers anxiety and depression during traumatic times. Some mutations in the serotonin transporter gene can reduce the ability to cope with stress. It is as if some of us are bicycling through life without an emotional helmet. Only when we crash does the mutation's phenotypic effect—depression—appear. Other human genes affect emotional state, but mutations in this one reduce our capacity to recover from emotional setbacks.

Take-Home Message **How does phenotype vary?**

- Some traits have a range of small differences, or continuous variation. The more genes and other factors that influence a trait, the more continuous the distribution of phenotype.

- Enzymes and other gene products control steps of most metabolic pathways. Mutations, interactions among genes, and environmental conditions can affect one or more steps, and thus contribute to variation in phenotypes.

Human Genetic Analysis

Some organisms, including pea plants and fruit flies, are ideal for genetic analysis. They have only a few chromosomes, they reproduce quickly under controlled conditions, and breeding them poses few ethical problems. It does not take long to track a trait through many generations. Humans, however, are a different story. Unlike flies grown in laboratories, we humans live under variable conditions, in different places, and we live as long as the geneticists who study us. Most of us select our own mates and reproduce if and when we want to. Most human families are not large, which means that there are not enough offspring to clarify inheritance patterns.

Thus, inheritance patterns in humans are typically studied by tracking observable genetic disorders that crop up in families. Geneticists gather information from multiple generations in order to minimize sampling error (Section 1.7), and graph their results as standardized charts of genetic connections called **pedigrees** (Figure 9.15). Such studies can show whether genetic abnormalities and disorders are caused by a dominant or recessive allele, and whether the allele is on an autosome or a sex chromosome. They also allow geneticists to determine the probability that a genetic disorder will recur in future generations of a family or a population.

The next sections of this chapter offer examples of the inheritance patterns of genetic abnormalities and disorders. When considering them, keep in mind that a genetic abnormality is simply a rare or uncommon version of a trait, such as when a person is born with six digits on each hand or foot instead of the usual five. Abnormalities are not inherently life-threatening, and how you view them is a matter of opinion. By contrast, a genetic disorder sooner or later causes medical problems that may be severe. A genetic disorder is characterized by a specific set of symptoms called a **syndrome**. We reserve the term "disease" for an illness caused by infection or other environmental factor.

Alleles that give rise to severe genetic disorders are generally rare in populations, because they compromise the health and reproductive ability of their bearers. Why don't they disappear entirely? Mutations periodically reintroduce them. In some cases, a normal allele in heterozygotes masks the effects of a harmful one. In others, a codominant allele offers a survival advantage in a particularly hazardous environment.

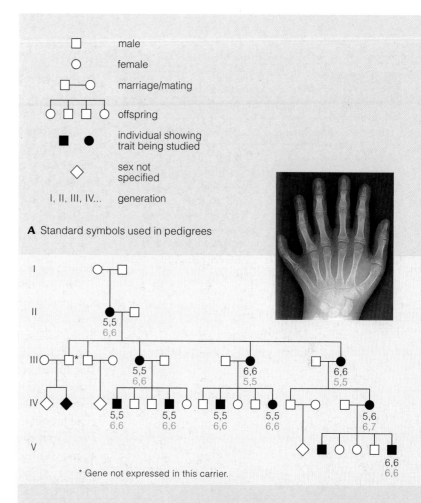

A Standard symbols used in pedigrees

* Gene not expressed in this carrier.

B A pedigree for *polydactyly*, which is characterized by extra fingers, toes, or both. The *black* numbers signify the number of fingers on each hand; the *blue* numbers signify the number of toes on each foot.

FIGURE 9.15 **Animated!** Pedigrees.

Take-Home Message How do we study human inheritance patterns?

- A genetic abnormality is a rare version of an inherited trait. A genetic disorder is an inherited condition that causes medical problems.

- Inheritance patterns in humans are often studied by tracking inheritance patterns of genetic disorders through generations.

pedigree Chart showing the pattern of inheritance of a trait in a family.

syndrome The set of symptoms that characterize a genetic disorder.

Human Genetic Disorders

Table 9.1	Some Autosomal Dominant Disorders or Abnormalities
Disorder or Abnormality	**Main Symptoms**
Achondroplasia	One form of dwarfism
Camptodactyly	Rigid, bent fingers
Familial hyper-cholesterolemia	High cholesterol levels in blood, eventually causing clogged arteries
Huntington's disease	Nervous system degenerates progressively, irreversibly
Marfan syndrome	Abnormal connective tissue
Polydactyly	Extra fingers, toes, or both
Progeria	Drastic premature aging
Neurofibromatosis	Tumors of nervous system and skin

Autosomal Dominant Disorders Some examples of genetic disorders caused by autosomal dominant alleles are listed in Table 9.1. A dominant allele on an autosome (an autosomal dominant allele) is expressed in homozygotes and heterozygotes, so any trait it specifies tends to appear in every generation. When one parent is heterozygous, and the other is homozygous for the recessive allele, each of their children has a 50 percent chance of inheriting the dominant allele—and displaying the trait associated with it (Figure 9.16**A**).

Achondroplasia is an autosomal dominant disorder that affects about 1 out of 10,000 people. Adult heterozygotes average about four feet, four inches tall, and have abnormally short arms and legs relative to other body parts. While they were still embryos, the cartilage model on which a skeleton is constructed did not form properly. Most homozygotes die before or not long after birth.

Huntingon's disease is also caused by an autosomal dominant allele. With this genetic disorder, involuntary muscle movements increase as the nervous system slowly deteriorates. Typically, symptoms do not start until after age thirty; affected people die during their forties or fifties. The mutation that causes Huntington's alters a protein necessary for brain cell development. It is an expansion mutation, in which three nucleotides are duplicated many times. Hundreds of thousands of other expansion repeats occur harmlessly in and between genes on human chromosomes. This one alters the function of a critical gene product.

Hutchinson–Gilford progeria, a genetic disorder that causes accelerated aging, is inherited in an autosomal dominant pattern. Progeria arises by spontaneous mutation of a gene for a protein that normally makes up intermediate filaments in the nucleus (Section 3.6). The altered protein is not processed properly. It builds up on the inner nuclear membrane and distorts the nucleus. How this buildup causes the symptoms of progeria is not yet known. Those symptoms start before the age of two. Skin that should be plump and resilient starts

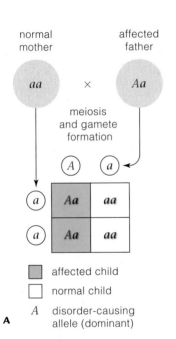

normal mother

affected father

meiosis and gamete formation

A affected child

normal child

A disorder-causing allele (dominant)

FIGURE 9.16 Animated! Autosomal dominance.

A Autosomal dominant inheritance pattern. In this example, the father is heterozygous for the dominant allele (*A*).

B Progeria. Mickey (*left*) and Fransie (*right*) met at a gathering of progeriacs at Disneyland, California. They were not yet ten years old. Fransie was seventeen when he died. Mickey was twenty.

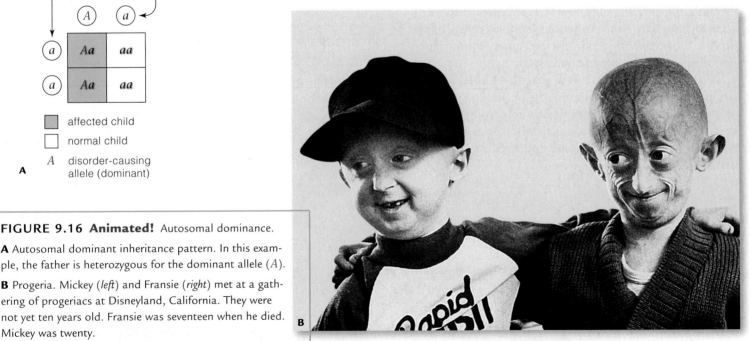

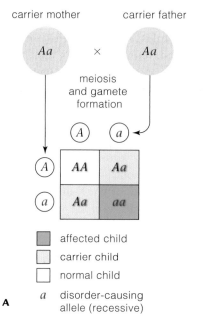

carrier mother   carrier father

Aa   ×   Aa

meiosis and gamete formation

(A)   (a)

	(A)	(a)
(A)	AA	Aa
(a)	Aa	aa

- ▨ affected child
- ▢ carrier child
- ▢ normal child

A   *a*   disorder-causing allele (recessive)

B

FIGURE 9.17 Animated! Autosomal recessive inheritance pattern. **A** In this example, both parents are carriers: they are heterozygous for the recessive allele (*a*), but do not have the disorder.

B Most mutations in the human gene for tyrosinase (Section 4.4) disable the enzyme, and melanocytes produce no melanin in the absence of tyrosinase. The resulting phenotype, albinism, is inherited in an autosomal recessive pattern.

to thin. Skeletal muscles weaken. Limb bones that should lengthen and grow stronger soften. Premature baldness is inevitable. Most progeriacs can expect to die in their early teens as a result of strokes or heart attacks. These final insults are brought on by a hardening of artery walls, a condition typical of advanced age (Figure 9.16**B**).

Progeria does not run in families because affected people do not usually live long enough to reproduce. Other dominant alleles that cause severe problems can persist if their expression does not interfere with reproduction. The one associated with achondroplasia is an example. With Huntington's and other late-onset disorders, people tend to reproduce before symptoms appear, so the allele may be passed unknowingly to children.

Autosomal Recessive Disorders A recessive allele on an autosome (an autosomal recessive allele) is expressed only in homozygotes, so traits associated with it may skip generations. Heterozygotes are carriers, which means that they have the allele but not the trait. Any child of two carriers has a 25 percent chance of inheriting the allele from both parents and being a homozygote with the trait (Figure 9.17). All children of homozygous parents will be homozygous. Some examples of genetic disorders caused by autosomal recessive alleles are listed in Table 9.2, including cystic fibrosis (Section 9.1).

Galactosemia is a heritable metabolic disorder that affects about 1 in every 50,000 newborns. This case of autosomal recessive inheritance involves an allele for an enzyme that helps digest the lactose in milk or in milk products. The body normally converts lactose to glucose and galactose. Then, a series of three enzymes converts the galactose to an intermediate that can enter glycolysis (Section 5.6) or be converted to glycogen. People with galactosemia do not make one of these three enzymes; they are homozygous for a mutated recessive allele. An intermediate that accumulates to toxic levels in their body can be detected in the urine. The condition leads to malnutrition, diarrhea, vomiting, and damage to the eyes, liver, and brain. When they do not receive treatment, galactosemics typically die young. If they are quickly placed on a diet that excludes all dairy products, the symptoms may not be as severe.

Table 9.2	Some Autosomal Recessive Disorders or Abnormalities	
Disorder or Abnormality		**Main Symptoms**
Albinism		Absence of pigmentation
Hereditary methemoglobinemia		Blue skin coloration
Cystic fibrosis		Abnormal secretions leading to tissue, organ damage
Ellis–van Creveld syndrome		Dwarfism, heart defects, polydactyly
Fanconi anemia		Physical abnormalities, bone marrow failure
Galactosemia		Brain, liver, eye damage
Phenylketonuria (PKU)		Mental impairment
Sickle-cell anemia		Adverse pleiotropic effects on organs throughout body

Table 9.3	Some X-Linked Recessive Disorders or Abnormalities	
Disorder or Abnormality	**Main Symptoms**	
Androgen insensitivity syndrome	Male with female traits; sterility	
Red–green color blindness	Inability to distinguish red and green	
Fragile X syndrome	Mental impairment	
Hemophilia	Impaired blood clotting	
Muscular dystrophies	Progressive loss of muscle function	
X-linked anhidrotic dysplasia	Abnormal hair, skin, nails, teeth	

X-Linked Recessive Disorders

The X chromosome carries more than 6 percent of all human genes. Mutations on this sex chromosome are known to cause or contribute to over 300 genetic disorders (Table 9.3).

A recessive allele on an X chromosome (an X-linked recessive allele) leaves certain clues when it causes a genetic disorder. First, more males than females are affected by such X-linked recessive disorders because heterozygous females have a second X chromosome that carries the dominant allele, which masks the effects of the recessive one. Heterozygous males have only one X chromosome, so they are not similarly protected (Figure 9.18). Second, an affected father cannot pass his X-linked recessive allele to a son because all children who inherit their father's X chromosome are female. Thus, a heterozygous female is the bridge between an affected male and his affected grandson.

X-linked dominant alleles that cause disorders are rarer than X-linked recessive ones, probably because they tend to be lethal in male embryos. Females, with two X chromosomes, most often have one functional allele that can dampen the effects of a mutated one.

The pattern of X-linked recessive inheritance shows up among individuals who have some degree of color blindness (Figure 9.18**A–D**). The term refers to a range of conditions in which an individual cannot distinguish among some or all colors in the spectrum of visible light. Mutated genes result in altered function of the light-sensitive receptors in the eyes. Normally, humans can sense the differences among 150 colors. A person who is red–green color blind sees fewer than 25 colors because receptors that respond to red and green wavelengths are weakened or absent. Some color blind people confuse red and green colors. Others see green as shades of gray, but perceive blues and yellows quite well.

Duchenne muscular dystrophy (DMD) is one of several X-linked recessive disorders that is characterized by muscle degeneration. DMD affects about 1 in 3,500 people, almost all of them boys. A gene on the X chromosome encodes dystrophin, which is a protein that structurally supports the fused cells in muscle fibers by anchoring the cell cortex to the plasma membrane. When dystrophin is abnormal or absent, the cell cortex weakens and muscle cells die. The cell debris that remains in the tissues triggers chronic inflammation. DMD is typically diagnosed in boys between the ages of three and seven. The rapid progression of this disorder cannot be stopped. When an affected boy is about twelve years old, he will begin to use a wheelchair. His heart muscles will start to break down. Even with the best of care, he will probably die before he is thirty years old, from a heart disorder or from respiratory failure (suffocation).

carrier mother   normal father

XX   ×   XY

meiosis and gamete formation

X   Y

	X	Y
X	XX	XY
X	XX	XY

☐ normal daughter or son
▨ carrier daughter
▨ affected son

X recessive allele on X chromosome

FIGURE 9.18 Animated! X-linked recessive inheritance. *Above*, in this example, the mother carries a recessive allele on one of her X chromosomes (*red*). *Right*, color blindness in inherited in an X-linked recessive pattern. **A** View with red–green color blindness. The perception of blues and yellows is normal, but red and green appear similar. **B** Compare what a person with normal vision sees.

Far right, two Ishihara plates, which are standardized tests for color blindness. **C** You may have one form of red–green color blindness if you see the number 7 instead of 29 in this circle. **D** You may have another form if you see a 3 instead of an 8.

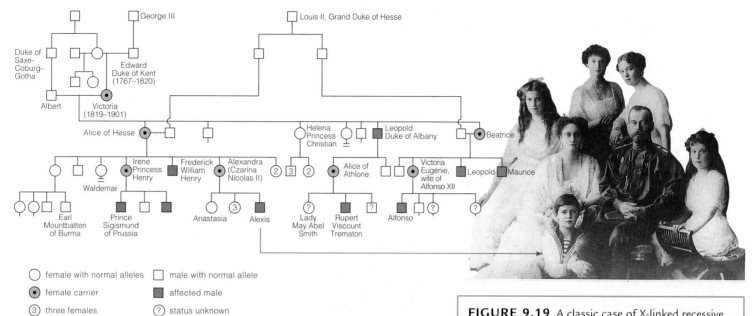

FIGURE 9.19 A classic case of X-linked recessive inheritance: a partial pedigree of the descendants of Queen Victoria of England. At one time, the recessive X-linked allele that resulted in hemophilia was present in eighteen of Victoria's sixty-nine descendants, who sometimes intermarried.

Of the Russian royal family members shown, the mother (Alexandra Czarina Nicolas II) was a carrier.

Figure It Out: How many of Alexis's siblings were affected by hemophilia A?

Answer: None

Pedigree legend:
- ○ female with normal alleles
- ◉ female carrier
- ③ three females
- □ male with normal allele
- ■ affected male
- ⑦ status unknown

Hemophilia A is an X-linked recessive disorder that interferes with blood clotting. Most of us have a blood clotting mechanism that quickly stops bleeding from minor injuries. That mechanism involves protein products of genes on the X chromosome. Bleeding can be prolonged in males who carry a mutated form of one of these X-linked genes, or in females who are homozygous for a mutation (clotting time is close to normal in heterozygous females). Affected people tend to bruise very easily, and internal bleeding causes problems in their muscles and joints.

In the nineteenth century, the incidence of hemophilia A was relatively high in royal families of Europe and Russia, probably because the common practice of inbreeding kept the harmful allele circulating in their family trees (Figure 9.19). Today, about 1 in 7,500 people is affected, but that number may be rising because the disorder is now treatable. More affected people are living long enough to transmit the allele to children.

Take-Home Message
What causes genetic disorders?

- Many genetic disorders are linked to dominant or recessive alleles on autosomes.

- Recessive alleles on sex chromosomes cause or contribute to more than 300 genetic disorders.

9.7 Changes in Chromosome Number

Rarely, offspring end up with a chromosome number that differs from their parents. In humans, such major changes to the genetic blueprint can have serious effects on an individual's health and fertility (Section 8.8 and Table 9.4).

Table 9.4	Some Disorders Associated With Chromosome Number Changes
Disorder or Abnormality	**Main Symptoms**
Down syndrome	Mental impairment; heart defects
Turner syndrome (XO)	Sterility; abnormal ovaries, abnormal sexual traits
Klinefelter syndrome	Sterility, mental impairment
XXX syndrome	Minimal abnormalities
XYY condition	Mild mental impairment

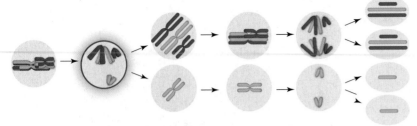

Metaphase I   Anaphase I   Telophase I   Metaphase II   Anaphase II   Telophase II

A

> **FIGURE 9.20** Nondisjunction during meiosis. **A** *Above*, of the two pairs of homologous chromosomes shown in this example, one fails to separate during anaphase I of meiosis. *Below*, one outcome of nondisjunction, Down syndrome **B**, occurs when an individual has three copies of chromosome 21 **C**.

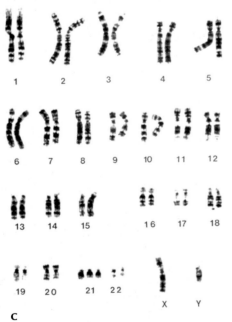

Chromosome number changes often arise through **nondisjunction**, in which one or more pairs of chromosomes do not separate properly during mitosis or meiosis (Figure 9.20**A**). Nondisjunction affects the chromosome number at fertilization. For example, suppose that a normal gamete fuses with an $n+1$ gamete (one that has an extra chromosome). The new individual will be trisomic ($2n+1$), having three of one type of chromosome and two of every other type. As another example, if an $n-1$ gamete fuses with a normal n gamete, the new individual will be $2n-1$, or monosomic.

Trisomy and monosomy are types of **aneuploidy**, a condition in which cells have too many or too few copies of a chromosome. Autosomal aneuploidy is usually fatal in humans, and it causes many miscarriages. However, about seventy percent of flowering plant species, and some insects, fishes, and other animals are **polyploid**, which means that their cells have three or more of each type of chromosome.

Autosomal Change and Down Syndrome

A few trisomic humans are born alive, but only those that have trisomy 21 will reach adulthood. A newborn with three chromosomes 21 will develop Down syndrome. This autosomal disorder is the most common type of aneuploidy in humans; it occurs once in 800 to 1,000 births and affects more than 350,000 people in the United States alone. Figure 9.20**C** shows a karyotype for a trisomic 21 male. The affected individuals have upward-slanting eyes, a fold of skin that starts at the inner corner of each eye, a deep crease across the sole of each palm and foot, one (instead of two) horizontal furrows on their fifth fingers, slightly flattened facial features, and other symptoms (Figure 9.20**B**).

Not all of the outward symptoms develop in every individual. That said, trisomic 21 individuals tend to have moderate to severe mental impairment and heart problems. Their skeleton grows and develops abnormally, so older children have short body parts, loose joints, and misaligned bones of the fingers, toes, and hips. The muscles and reflexes are weak, and motor skills such as speech develop slowly. With medical care, trisomy 21 individuals live about fifty-five years. Early training can help affected individuals learn to care for themselves and to take part in normal activities.

The incidence of nondisjunction generally rises with the increasing age of the mother. Nondisjunction may occur in the father, although far less frequently. Trisomy 21 is one of the hundreds of conditions that can be detected easily through prenatal diagnosis (Section 9.8).

Change in the Sex Chromosome Number

Nondisjunction also causes alterations in the number of X and Y chromosomes, with a frequency of about 1 in 400 live births. Most often, such alterations lead to difficulties in learning and impaired motor skills such as a speech delay, but problems may be so subtle that the underlying cause is never diagnosed.

Female Sex Chromosome Abnormalities

Individuals with Turner syndrome have an X chromosome and no corresponding X or Y chromosome (XO). About 1 in 2,500 to 10,000 newborn girls are XO (Figure 9.21). At least 98 percent of XO embryos will spontaneously abort early in pregnancy, so there are fewer cases compared with other sex chromosome abnormalities. XO individuals are not as disadvantaged as other aneuploids. They grow up well proportioned but short (with an average height of four feet, eight inches). Most do not have functional ovaries, so they do not make enough sex hormones to become sexually mature. The development of secondary sexual traits such as breasts is also affected.

A female may inherit three, four, or five X chromosomes. The resulting XXX syndrome occurs in about 1 of 1,000 births. Only one X chromosome is typically active in female cells (Section 7.7), so having extra X chromosomes usually does not result in physical or medical problems.

Male Sex Chromosome Abnormalities About 1 out of every 500 males has an extra X chromosome (XXY). Most cases are an outcome of nondisjunction during meiosis. The resulting disorder, Klinefelter syndrome, develops at puberty. XXY males tend to be overweight, tall, and within a normal range of intelligence. They make more estrogen and less testosterone than normal males, and this hormone imbalance has feminizing effects. Affected men tend to have small testes and prostate glands, low sperm counts, sparse facial and body hair, high-pitched voices, and enlarged breasts. Testosterone injections during puberty can reverse these traits.

About 1 in 500 to 1,000 males has an extra Y chromosome (XYY). Adults tend to be taller than average and have mild mental impairment, but most are otherwise normal. XYY men were once thought to be predisposed to a life of crime. This misguided view was based on sampling error (too few cases in narrowly chosen groups such as prison inmates) and bias (the researchers who gathered the karyotypes also took the personal histories of the participants).

FIGURE 9.21 A 6-year-old with Turner's syndrome. Affected girls tend to be shorter than average, but daily hormone injections can help them reach normal height.

Take-Home Message   **What are the effects of chromosome number changes?**

- Nondisjunction can change the number of autosomes or sex chromosomes in gametes. Such changes usually cause genetic disorders in offspring.
- Sex chromosome abnormalities are usually associated with learning difficulties, speech delays, and motor skill impairment.

9.8 Some Prospects in Human Genetics

With the first news of pregnancy, parents-to-be typically wonder if their baby will be healthy. Quite naturally, they want their baby to be free of genetic disorders, and most babies are. Many prospective parents have difficulty coming to terms with the possibility that a child of theirs might develop a severe genetic disorder, but sometimes that happens. What are their options?

Genetic Counseling Genetic counseling starts with diagnosis of parental genotypes, pedigrees, and genetic testing for known disorders. Using information gained from these tests, genetic counselors can predict a couple's probability of having a child with a genetic disorder.

Parents-to-be commonly ask genetic counselors to compare the risks associated with diagnostic procedures against the likelihood that their future child will be affected by a severe genetic disorder. At the time of counseling, they also should consider the small overall risk (3 percent) that complications during the

aneuploidy A chromosome abnormality in which there are too many or too few copies of a particular chromosome.

nondisjunction Failure of sister chromatids or homologous chromosomes to separate during meiosis or mitosis.

polyploid Having three or more of each type of chromosome characteristic of the species.

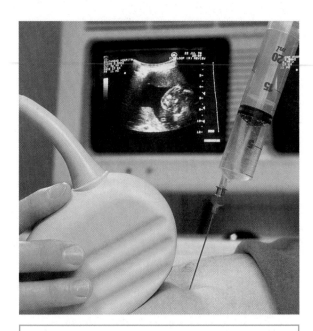

FIGURE 9.22 Animated! Amniocentesis. A pregnant woman's doctor draws a sample of amniotic fluid into a syringe. The path of the needle is monitored by an ultrasound device. Amniotic fluid contains fetal cells and wastes that can be analyzed for genetic disorders.

birth process can affect any child. They should talk about their age, because the risk of having a child with a genetic disorder increases with the age of the parents.

As a case in point, suppose a first child or a close relative has a severe disorder. A genetic counselor will evaluate the pedigrees of the parents, and the results of any genetic tests. Using this information, counselors can predict risks for disorders in future children. The same risk will apply to each pregnancy.

Prenatal Diagnosis Doctors commonly use methods of prenatal diagnosis to determine the sex of embryos or fetuses and to screen for more than 100 known genetic problems. Prenatal means before birth. Embryo is a term that applies until eight weeks after fertilization, after which fetus is appropriate.

Suppose a forty-five-year-old woman becomes pregnant and worries about Down syndrome. Between fifteen and twenty weeks after conception, she might opt for amniocentesis (Figure 9.22). In this diagnostic procedure, a physician uses a syringe to withdraw a small sample of fluid from the amniotic cavity. The "cavity" is a fluid-filled sac, enclosed by a membrane—the amnion—that in turn encloses the fetus. The fetus normally sheds some cells into the fluid. Cells suspended in the fluid sample can be analyzed for many genetic disorders, including Down syndrome, cystic fibrosis, and sickle-cell anemia.

Chorionic villi sampling (CVS) is a diagnostic procedure similar to amniocentesis. A physician withdraws a few cells from the chorion, which is a membrane that surrounds the amnion and helps form the placenta. The placenta is an organ that keeps the blood of mother and embryo separate, while allowing substances to be exchanged between them. Unlike amniocentesis, CVS can be performed as early as eight weeks into pregnancy.

It is now possible to see a live, developing fetus with a procedure called fetoscopy. In fetoscopy, a fiber-optic device called an endoscope is used to directly visualize and photograph the fetus, umbilical cord, and placenta in high resolution (Figure 9.23). Characteristic physical effects of certain genetic abnormalities or disorders can be diagnosed by fetoscopy.

All three prenatal diagnosis procedures are associated with risks to the fetus. Risks include punctures, infections, and loss of too much amniotic fluid (if the amnion does not reseal itself quickly). Amniocentesis increases the risk of miscarriage by 1 to 2 percent. CVS occasionally disrupts the placenta's development and thus causes underdeveloped or missing fingers and toes in 0.3 percent of newborns. Fetoscopy raises the miscarriage risk by 2 to 10 percent.

Preimplantation Diagnosis Preimplantation diagnosis is a procedure associated with *in vitro* fertilization. Sperm and eggs from prospective parents are mixed so that an egg becomes fertilized. Then, mitotic cell divisions turn the fertilized egg into a ball of eight cells within forty-eight hours.

 All of the cells in the tiny, free-floating ball (*left*) have the same genes, but they are not yet committed to being specialized. Doctors can remove one of these undifferentiated cells and analyze its genes. If it has no detectable genetic defects, the ball may be inserted into the mother's uterus. The withdrawn cell will not be missed, and the ball can develop into an embryo. Many of the resulting "test-tube babies" are born in good health. Some couples who are at risk of passing on alleles associated with cystic fibrosis, muscular dystrophy, or another genetic disorder have opted for this procedure.

Phenotypic Treatments Surgery, prescription drugs, hormone replacement therapy, and often dietary controls can minimize and in some cases eliminate the symptoms of many genetic disorders. For instance, strict dietary controls work in cases of phenylketonuria, or PKU. Individuals affected by this genetic disorder are homozygous for a recessive allele on an autosome. They cannot make a functional form of an enzyme that catalyzes the conversion of one amino acid (phenylalanine) to another (tyrosine). Because the conversion is blocked, phenylalanine accumulates and is diverted into other metabolic pathways. The outcome is an impairment of brain function. Affected people who restrict phenylalanine intake can lead essentially normal lives. They must avoid diet soft drinks and other products that are sweetened with aspartame, a compound that contains phenylalanine.

Genetic Screening Genetic screening is the widespread, routine testing for alleles associated with genetic disorders. It provides information on reproductive risks, and helps families that are already affected by a genetic disorder. In some cases, early phenotypic treatments can minimize the effects of a genetic disorder. Hospitals routinely screen newborns for certain genetic disorders such as PKU. Affected infants receive early treatment, so we now see fewer individuals with symptoms of the disorder. Besides helping individuals, the information from genetic screening can help us estimate the prevalence and distribution of harmful alleles in populations. However, knowledge of personal genetic information comes with social risks. What would happen if you were labeled as someone who carries a "bad" allele? If you decided to become a parent even though you know you have a "bad" allele, how would you feel if your child ends up with a genetic disorder? There are no easy answers.

Take-Home Message **How do we use human genetic information?**

- Genetic testing can provide prospective parents with information about the health of their future children.

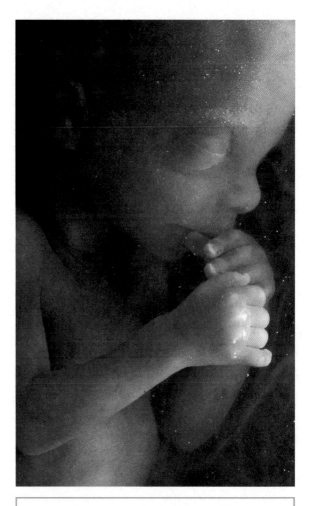

FIGURE 9.23 Fetoscopy yields high-resolution images of a fetus.

9.9 Impacts/Issues Revisited:
Menacing Mucus

Compared with other alleles that cause lethal genetic diseases, the one associated with cystic fibrosis is very common. The mutation itself is estimated to be at least 50,000 years old, and up to 1 in 25 people carry it in some populations. Why does it persist at such high frequency if it is so dangerous? The CF allele may be the lesser of two evils: It is lethal in homozygotes, but it offers heterozygotes a survival advantage against certain deadly infectious diseases.

The CFTR protein triggers endocytosis when it binds to bacteria. This process is an essential part of the body's immune response to bacteria in the respiratory tract. However, the same function of CFTR allows bacteria to enter  cells of the gastrointestinal tract, where they can be deadly. For example, endocytosis of *Salmonella typhi* (*left*) into epithelial cells lining the gut results in a dangerous infection called typhoid fever. The CF mutation prevents endocytosis of bacteria into intestinal cells, so people that carry it may have a decreased susceptibility to typhoid fever and other bacterial diseases that begin in the intestinal tract.

How Would

YOU

☑ **Vote ?**

Tests for predisposition to genetic disorders will be available soon. Do you support the new legislation preventing genetic discrimination? **See CengageNow for details, then vote online (cengagenow.com).**

Summary

Section 9.1 Cystic fibrosis, the most common fatal genetic disorder in the United States, is caused by a deletion in the *CFTR* gene. The allele persists at high frequency despite its devastating effects. Only those homozygous for the CF allele have the disorder.

Section 9.2 Diploid cells have homologous chromosomes (usually one inherited from each of two parents) with pairs of genes. The genes of a pair may or may not be identical. Individuals that carry two identical alleles are homozygous for the gene. Hybrids, or heterozygotes, have two nonidentical alleles. A dominant allele masks the effect of a recessive allele partnered with it. An individual's alleles constitute genotype. Gene expression results in phenotype, which refers to an individual's observable traits.

A cross, or mating, between heterozygous individuals can reveal dominance relationships among the alleles under study. We use Punnett squares to calculate the probability of the genotype and phenotype of the offspring of crosses.

During meiosis, the genes of each pair separate, so each gamete gets one or the other gene. Meiosis assorts gene pairs on homologous chromosomes independently of other gene pairs on the other chromosomes. The random attachment of homologous chromosomes to opposite spindle poles during prophase I is the basis of this outcome.

CENGAGENOW™ Carry out monohybrid crosses, observe the results of a dihybrid cross, and learn how Mendel crossed garden pea plants with the interactions on CengageNow.

Section 9.3 With incomplete dominance, an allele is not fully dominant over its partner on a homologous chromosome, and both are expressed. The combination of alleles gives rise to an intermediate phenotype. Codominant alleles are fully expressed in heterozygotes. In epistasis, interacting products of one or more genes often affect the same trait. A pleiotropic gene affects two or more traits.

	Ⓡ	ⓡ
Ⓡ	RR	Rr
ⓡ	Rr	rr

CENGAGENOW™ Explore patterns of non-Mendelian inheritance with the interaction on CengageNow.

Section 9.4 A trait that is influenced by the products of multiple genes often occurs in a range of small increments of phenotype (continuous variation). An individual's phenotype may be influenced by environmental factors.

CENGAGENOW™ Graph variation in height with the interaction on CengageNow.

Section 9.5 Inheritance patterns in humans are studied by following inherited genetic disorders in families. A genetic abnormality is an uncommon version of a heritable trait that does not result in medical problems. A genetic disorder is a heritable condition that results in a syndrome of mild or severe medical problems.

CENGAGENOW™ Learn about pedigrees with the animation on CengageNow.

Section 9.6 Some dominant or recessive alleles on autosomes or the X chromosome are associated with genetic abnormalities or

disorders. Males cannot transmit a recessive X-linked allele to their sons. A female passes such alleles to male offspring.

CENGAGENOW™ Investigate autosomal and X-linked inheritance with the interactions on CengageNow.

Section 9.7 The chromosome number of a cell can change permanently. Most often, such a change is an outcome of nondisjunction, which is the failure of duplicated chromosomes to separate during meiosis. In aneuploidy, cells have too many or too few copies of a chromosome. The most common aneuploidy in humans, trisomy 21, causes Down syndrome.

Section 9.8 Geneticists estimate the chance that a couple's offspring will inherit a genetic abnormality or disorder. Potential parents who may be at risk of transmitting a harmful allele to offspring have screening or treatment options.

CENGAGENOW™ Observe amniocentesis with the animation on CengageNow.

Self-Quiz Answers in Appendix I

1. A heterozygote has a _____ for a trait being studied.
   a. pair of identical alleles
   b. pair of nonidentical alleles
   c. haploid condition, in genetic terms

2. The observable traits of an organism constitute its _____ .
   a. phenotype   c. genotype
   b. sociobiology   d. pedigree

3. The offspring of the cross *AA* × *aa* are _____ .
   a. all *AA*   c. all *Aa*
   b. all *aa*   d. 1/2 *AA* and 1/2 *aa*

4. Assuming alleles have a clear dominance relationship, a dihybrid cross leads to a phenotypic ratio in offspring that is close to _____ .
   a. 1:2:1   b. 3:1   c. 1:1:1:1   d. 9:3:3:1

5. The probability of a crossover occurring between two genes on the same chromosome is _____ .
   a. unrelated to the distance between them
   b. increased if they are close together
   c. increased if they are far apart

6. If one parent is heterozygous for a dominant autosomal allele and the other parent does not carry the allele, a child of theirs has a _____ chance of being heterozygous.
   a. 25 percent   c. 75 percent
   b. 50 percent   d. no chance; it will die

7. A bell curve indicates _____ in a trait.
   a. pleiotropy   c. continuous variation
   b. crossing over   d. aneuploidy

8. Nondisjunction at meiosis can result in _____ .
   a. pleiotropy   c. continuous variation
   b. crossing over   d. aneuploidy

Digging Into Data

Cystic Fibrosis and Typhoid Fever

The CF mutation disables the receptor function of the CFTR protein, so it inhibits the endocytosis of bacteria into epithelial cells. Endocytosis is an important part of the respiratory tract's immune defenses against common *Pseudomonas* bacteria, which is why *Pseudomonas* infections are a chronic problem in CF patients.

The CF mutation also inhibits endocytosis of *Salmonella typhi* into cells of the gastrointestinal tract, where internalization of this bacteria can cause typhoid fever. Typhoid fever is a common worldwide disease. Its symptoms include extreme fever and diarrhea, and the resulting dehydration causes delirium that may last several weeks. If untreated, it kills up to 30 percent of those infected. Around 600,000 people die annually from typhoid fever. Most of them are children.

In 1998, Gerald Pier and his colleagues compared *S. typhi* uptake by epithelial cells homozygous for the normal allele with cells heterozygous for the CF mutation. (Cells homozygous for the mutation do not take up *S. typhi* bacteria.) Some of their results are shown in Figure 9.24.

1. Regarding the Ty2 strain of *S. typhi*, about how many more bacteria were able to enter cells expressing unmutated CFTR than cells expressing the CF-mutated protein?

2. Which strain of bacteria entered normal epithelial cells most easily?

3. The CF mutation inhibited the entry of all three *S. typhi* strains into epithelial cells. Can you tell which strain was most inhibited?

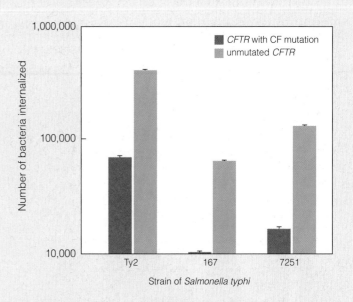

FIGURE 9.24 Effect of the CF mutation in epithelial cells on the uptake of three different strains of *Salmonella typhi* bacteria.

---

9. Color blindness is a case of _____ inheritance.
   a. autosomal dominant   c. X-linked dominant
   b. autosomal recessive   d. X-linked recessive

10. Klinefelter syndrome (XXY) is easily diagnosed by _____ .
    a. pedigree analysis   c. karyotyping
    b. aneuploidy   d. phenotypic treatment

11. Is this statement true or false? A son can inherit an X-linked recessive allele from his father.

12. Is this statement true or false? Body cells may inherit three or more of each type of chromosome characteristic of the species, a condition called polyploidy.

13. A recognized set of symptoms that characterize a specific disorder is a _____ .
    a. syndrome   b. disease   c. pedigree

14. Match each example with the most suitable description.
    _____ dihybrid cross   a. *bb*
    _____ monohybrid cross   b. *AaBb* × *AaBb*
    _____ homozygous condition   c. *Aa*
    _____ heterozygous condition   d. *Aa* × *Aa*

15. Match the terms appropriately.
    _____ polyploidy   a. symptoms that characterize
    _____ syndrome   a genetic disorder
    _____ aneuploidy   b. extra sets of chromosomes
    _____ nondisjunction   c. gametes with the wrong
           during meiosis   chromosome number
    _____ epistasis   d. one gene affects another
                               e. one extra chromosome

Additional questions are available on CENGAGENOW™

Genetics Problems   Answers in Appendix II

1. Mendel crossed a true-breeding pea plant with green pods and a true-breeding pea plant with yellow pods. All offspring had green pods. Which color is recessive?

2. Assuming that independent assortment occurs during meiosis, what type(s) of gametes will form in individuals with the following genotypes?
   a. *AABB*   b. *AaBB*   c. *Aabb*   d. *AaBb*

3. Refer to problem 2. Determine the frequencies of each genotype among offspring from the following matings:
   a. *AABB* × *aaBB*   c. *AaBb* × *aabb*
   b. *AaBB* × *AABb*   d. *AaBb* × *AaBb*

4. Suppose you identify a new gene in mice. One of its alleles specifies white fur, another specifies brown. You want to see if the two interact in simple or incomplete dominance. What sorts of tests would give you the answer?

5. Certain genes are vital for development. When mutated, they are lethal in homozygous recessives. Even so, heterozygotes can perpetuate recessive, lethal alleles. The allele *Manx* (M^L) in cats is an example. Homozygous cats ($M^L M^L$) die before birth. In heterozygotes ($M^L M$), the spine develops abnormally, and the cats end up with no tail. Two $M^L M$ cats mate. What is the probability that any one of their surviving kittens will be heterozygous?

6. As you read earlier, Duchenne muscular dystrophy arises through the expression of a recessive X-linked allele. Usually, symptoms start to appear in childhood. Gradual, progressive loss of muscle function leads to death, usually by age twenty or so. Unlike color blindness, the disorder is nearly always restricted to males. Suggest why.

10 Biotechnology

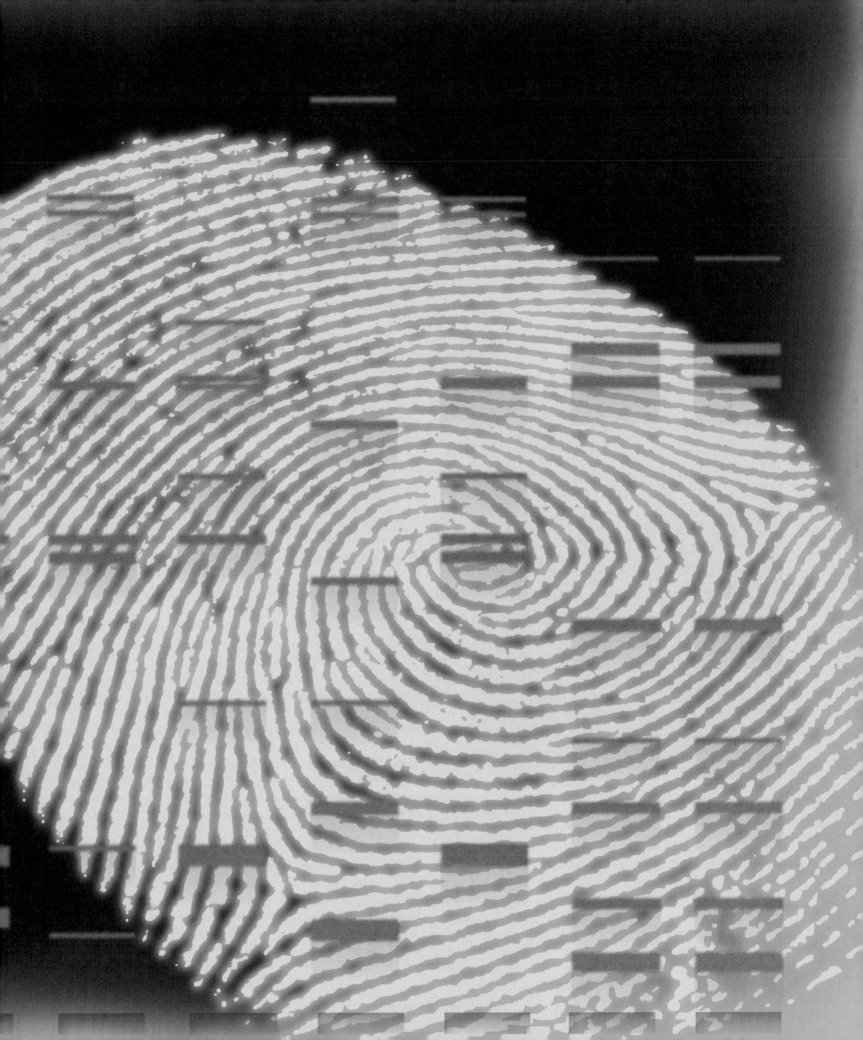

10.1 Impacts/Issues: Golden Rice or Frankenfood?

Vitamin A is necessary for good vision, growth, and immune system function. A small child can get enough of it just by eating a carrot every few days. Nonetheless, around 140 million children under the age of six become seriously ill from vitamin A deficiency every year. Vitamin A–deficient children do not grow as they should, and they succumb easily to infection. As many as 500,000 of them become blind, and of those half die within a year of losing their sight. These are some harsh statistics for such an easily prevented deficiency.

It is no coincidence that populations with the highest incidence of vitamin A deficiency also are the poorest. Most people in impoverished populations tend to eat few animal products, vegetables, or fruits—all foods that are rich sources of vitamin A. Correcting and preventing vitamin A deficiency can be as simple as supplementing the diet with these foods, but changes in dietary habits are often limited by cultural traditions and poverty. Political and economic issues have hampered long-term vitamin supplementation programs.

Geneticists Ingo Potrykus and Peter Beyer wanted to help impoverished people by improving the nutritional value of rice. Why rice? Rice is the dietary staple for 3 billion people in developing countries around the world. Economies, traditions, and cuisines are based on growing and eating rice. So, growing and eating rice that happens to contain enough vitamin A to prevent deficiency would be compatible with traditional methods of agriculture and dietary preferences.

The body can easily convert beta-carotene, an orange photosynthetic pigment, into vitamin A. However, getting rice grains to make beta-carotene is beyond the scope of conventional plant breeding methods. For example, corn seeds (the kernels) make and store beta-carotene, but even the best gardener cannot cross-breed rice plants with corn plants.

Potrykus and Beyer genetically modified rice plants to make beta-carotene in their seeds—in the grains of Golden Rice (Figure 10.1). Like many other **genetically modified organisms** (GMOs), Golden Rice is **transgenic**, which means it carries genes from a different species. GMOs are made in laboratories, not on farms, but they are an extension of breeding practices used for many thousands of years to coax new plants and new breeds of animals from wild species.

No one wants children to suffer or die. However, many people are opposed to any GMO. Some worry that our ability to tinker with genetics has surpassed our ability to understand the impact of the tinkering. Should we be more cautious? Two people created a way to keep millions of children from dying. How much of a risk should we as a society take to help those children?

At this time, geneticists hold molecular keys to the kingdom of inheritance. As you will see, what they are unlocking is already having an impact on life in the biosphere.

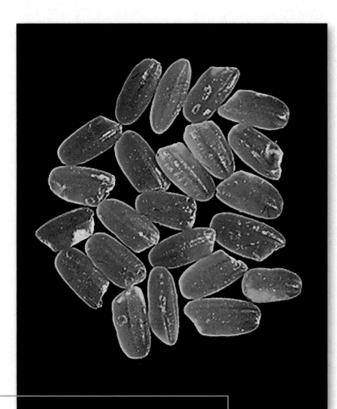

FIGURE 10.1 Golden Rice: miracle of modern science or Frankenfood? Rice plants with artificially inserted genes make and store the orange pigment beta-carotene in their seeds, or rice grains.

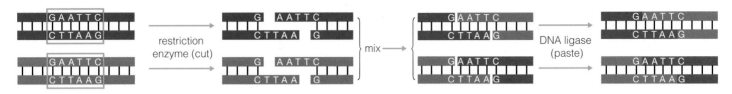

1 A restriction enzyme recognizes a specific base sequence in DNA (*green* boxes) from two sources.

2 The enzyme cuts DNA from both sources into fragments that have sticky ends.

3 The DNA fragments from the two sources are mixed together. The matching sticky ends base-pair with each other.

4 DNA ligase joins the fragments of DNA where they overlap. Molecules of recombinant DNA are the result.

FIGURE 10.2 Making recombinant DNA.

10.2 Finding Needles in Haystacks

Cut and Paste In the 1950s, excitement over the discovery of DNA's structure gave way to frustration: No one could determine the order of nucleotides in a molecule of DNA. Identifying a single base among thousands or millions of others turned out to be a huge technical challenge.

A seemingly unrelated discovery offered a solution. Some types of bacteria resist infection by viruses that inject their DNA into bacterial cells. Werner Arber, Hamilton Smith, and their coworkers discovered that special enzymes inside these bacteria chop up any injected viral DNA before it has a chance to integrate into the bacterial chromosome. The enzymes restrict viral growth; hence their name, restriction enzymes. A **restriction enzyme** can cut DNA wherever a specific nucleotide sequence occurs, regardless of the source of the DNA. For example, the enzyme *Eco*RI (named after the organism from which it was isolated, *E. coli*) cuts DNA only at the sequence GAATTC.

The discovery allowed researchers to cut huge molecules of chromosomal DNA into manageable fragments. It also allowed them to combine DNA fragments from different organisms to make **recombinant DNA** (Figure 10.2). Making recombinant DNA typically involves cutting DNA from multiple sources with the same restriction enzyme **1**. Many restriction enzymes leave single-stranded tails ("sticky ends") on DNA fragments **2**. When the DNA fragments from different sources are mixed together, their matching sticky ends base-pair together **3**. DNA ligase is used to join the fragments as hybrid molecules **4**.

Making recombinant DNA is the first step in **DNA cloning**, a set of laboratory methods that uses living cells to make many copies of specific DNA fragments. For example, researchers often insert DNA fragments into **plasmids**, small circles of bacterial DNA independent of the chromosome. Before a bacterium divides, it copies any plasmids it carries along with its chromosome, so both descendant cells get one of each. If a plasmid carries a fragment of foreign DNA, that fragment gets copied and distributed to descendant cells along with the plasmid. Thus, plasmids can be used as **cloning vectors**, which are molecules that carry foreign DNA into host cells (Figure 10.3). A host cell that takes up a cloning vector can be grown in the laboratory to yield a huge population of genetically identical cells called **clones**. Each clone contains a copy of the vector and the fragment of foreign DNA it carries. Researchers can harvest the DNA fragment from the clones in large quantities.

Cloning isolates a small fragment of DNA away from a larger molecule. The development of this laboratory technique was a critical step that allowed researchers to unravel the genetic information that DNA carries (Section 7.2).

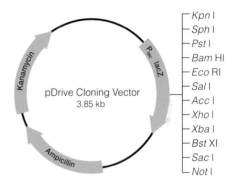

FIGURE 10.3 A commercial plasmid cloning vector. Restriction enzyme recognition sequences are indicated on the *right* by the name of the enzyme that cuts them. Researchers insert foreign DNA into the vector at these sites.

Bacterial genes (*green*) help researchers identify host cells that take up a vector with inserted DNA.

clone A genetically identical copy of DNA, a cell, or an organism.

cloning vector A DNA molecule that can accept foreign DNA, be transferred to a host cell, and get replicated in it.

DNA cloning Set of procedures that uses living cells to make many identical copies of a DNA fragment.

genetically modified organism (GMO) An organism whose genome has been deliberately modified.

plasmid A small, circular DNA molecule in bacteria, replicated independently of the chromosome.

recombinant DNA A DNA molecule that contains genetic material from more than one organism.

restriction enzyme Type of enzyme that cuts specific base sequences in DNA.

transgenic Refers to an organism that has been genetically modified to carry a gene from a different species.

A The enzyme reverse transcriptase transcribes mRNA into DNA.

mRNA

cDNA

B DNA polymerase replicates the DNA strand.

cDNA

cDNA

*Eco*RI recognition site

C The result is a double-stranded molecule of DNA that can be cut and pasted into a cloning vector.

FIGURE 10.4 Making cDNA. mRNA must be transcribed into cDNA for cloning.

Thirty PCR cycles may amplify the number of template DNA molecules a billionfold.

cDNA Cloning Researchers who study eukaryotic genes and their expression work with messenger RNA. mRNA cannot be cloned directly, because restriction enzymes and DNA ligase cut and paste only double-stranded DNA. However, mRNA can be used as a template to make double-stranded DNA in a test tube. **Reverse transcriptase**, a replication enzyme from certain types of viruses, assembles a strand of complementary DNA, or **cDNA**, on an mRNA template (Figure 10.4**A,B**). DNA polymerase added to the mixture strips the RNA from the hybrid molecule as it copies the cDNA into a second strand of DNA. The outcome is a double-stranded DNA copy of the original mRNA (Figure 10.4**C**). Like any other DNA, double-stranded cDNA may be cut with restriction enzymes and pasted into a cloning vector using DNA ligase.

Libraries The entire set of genetic material—the **genome**—of most organisms consists of thousands of genes. To study or manipulate a single gene, researchers must first separate it from all of the others. Researchers can isolate a gene by cutting an organism's DNA into pieces, and then cloning all the pieces. The result is a genomic library, a set of clones that collectively contain all of the DNA in an organism's genome.

In genomic or cDNA libraries, a cell that contains a particular DNA fragment of interest may be mixed up with thousands or millions of others that do not. All the cells look the same, so researchers get tricky to find that one clone among all of the others—the needle in the haystack. Using a probe is one trick. A **probe** is a fragment of DNA labeled with a tracer (Section 2.2). Researchers design probes to match a targeted DNA sequence. For example, they might synthesize a short chain of nucleotides based on a known DNA sequence, then attach a radioactive phosphate group to it.

The nucleotide sequence of a probe is complementary to that of the targeted gene, so the probe can base-pair with the gene. Base pairing between DNA (or DNA and RNA) from more than one source is called **nucleic acid hybridization**. A probe mixed with DNA from a library base-pairs with (hybridizes to) the targeted gene. Researchers pinpoint a clone that hosts the gene by detecting the label on the probe. The clone is then cultured, and the DNA fragment of interest is extracted in bulk from the cultured cells.

PCR The polymerase chain reaction (**PCR**) is a technique used to mass-produce copies of a particular DNA fragment without having to clone it. The reaction can transform a needle in a haystack—that one-in-a-million DNA fragment—into a huge stack of needles with a little hay in it (Figure 10.5). The starting material for PCR is any sample of DNA with at least one molecule of a target sequence. It might be DNA from a mixture of 10 million different clones, a sperm, a hair left at a crime scene, or a mummy. Essentially any sample that has DNA in it can be used for PCR.

First, the starting material is mixed with DNA polymerase, nucleotides, and primers ❶. **Primers** are short single strands of DNA that base-pair with a certain sequence—here, on either end of the DNA to be amplified (or mass-produced). Researchers expose the reaction mixture to repeated cycles of high and low temperature. High temperature disrupts the hydrogen bonds that hold the two strands of a DNA double helix together (Section 6.3). During a high-temperature cycle, every molecule of double-stranded DNA unwinds and becomes single-stranded ❷. During a low-temperature cycle, the single DNA strands hybridize with complementary partner strands, and double-stranded DNA forms again.

Most DNA polymerases denature at the high temperatures required to separate DNA strands. The kind that is used in PCR reactions, *Taq* polymerase, is from *Thermus aquaticus*. This bacterial species lives in hot springs and hydrothermal vents, so its DNA polymerase is necessarily heat-tolerant.

Taq polymerase recognizes hybridized primers as places to start DNA synthesis. During a low-temperature cycle, it starts synthesizing DNA where primers have hybridized with template ❸. Synthesis proceeds along the template strand until the temperature rises and the DNA separates into single strands ❹. The newly synthesized DNA is a copy of the target.

When the mixture cools, the primers rehybridize, and DNA synthesis begins again. The number of copies of the target DNA can double with each cycle of heating and cooling ❺. Thirty PCR cycles may amplify the number of template DNA molecules a billionfold.

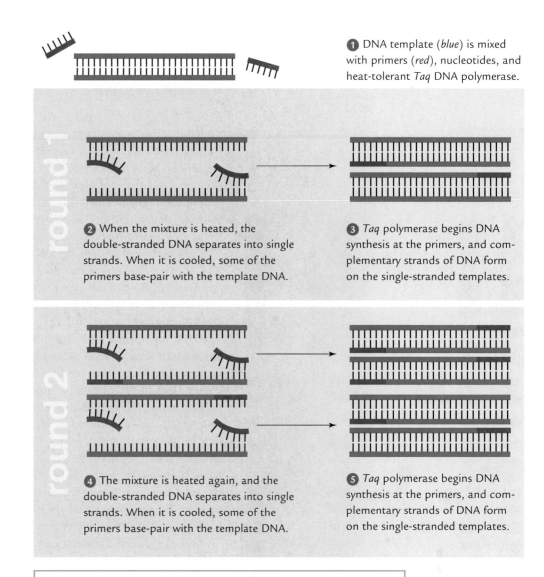

❶ DNA template (*blue*) is mixed with primers (*red*), nucleotides, and heat-tolerant *Taq* DNA polymerase.

round 1

❷ When the mixture is heated, the double-stranded DNA separates into single strands. When it is cooled, some of the primers base-pair with the template DNA.

❸ *Taq* polymerase begins DNA synthesis at the primers, and complementary strands of DNA form on the single-stranded templates.

round 2

❹ The mixture is heated again, and the double-stranded DNA separates into single strands. When it is cooled, some of the primers base-pair with the template DNA.

❺ *Taq* polymerase begins DNA synthesis at the primers, and complementary strands of DNA form on the single-stranded templates.

FIGURE 10.5 Animated! Two rounds of PCR. Each cycle of this reaction can double the number of target DNA molecules. Thirty cycles can amplify a template DNA a billionfold.

Take-Home Message

What techniques allow researchers to study DNA?

- DNA cloning uses living cells to mass-produce DNA fragments. Restriction enzymes cut DNA into pieces that are inserted into cloning vectors. The resulting recombinant DNA molecules are introduced into host cells, which copy the DNA as they divide.

- Researchers make DNA libraries or use PCR to isolate one gene from the many other genes in a genome.

- Probes are used to identify one clone that hosts a DNA fragment of interest among many other clones in a DNA library.

- PCR, the polymerase chain reaction, quickly mass-produces copies of a particular DNA fragment.

cDNA DNA synthesized from an RNA template by the enzyme reverse transcriptase.

genome An organism's complete set of genetic material.

nucleic acid hybridization Base-pairing between DNA or RNA from different sources.

PCR Polymerase chain reaction. Method that rapidly generates many copies of a specific DNA fragment.

primer Short, single strand of DNA designed to hybridize with a DNA fragment.

probe Short fragment of DNA labeled with a tracer; designed to hybridize with a nucleotide sequence of interest.

reverse transcriptase A viral enzyme that uses mRNA as a template to make a strand of cDNA.

Studying DNA

DNA Fingerprinting Each human has a unique set of fingerprints. Like members of other sexually reproducing species, each also has a **DNA fingerprint**, a unique array of DNA sequences. More than 99 percent of the DNA in all humans is the same, but the other fraction of 1 percent is unique to an individual. Some of the unique sequences are sprinkled throughout the genome as short tandem repeats, which are many copies of the same 2- to 10-base-pair sequences, positioned one after the next along the length of a chromosome.

For example, one person's DNA might contain fifteen repeats of the bases TTTTC in a certain location. Another person's DNA might have TTTTC repeated two times in the same location. One person might have ten repeats of CGG; another might have fifty. These repetitive sequences slip spontaneously into DNA during replication, and their numbers grow or shrink over generations.

DNA fingerprinting reveals differences in the number of tandem repeats among individuals. With this technique, PCR is used to copy a region of a chromosome known to have tandem repeats of 4 or 5 nucleotides. The size of the copied DNA fragment differs among most individuals, because the number of tandem repeats in that region also differs. Thus, the genetic differences between individuals can be detected by **electrophoresis**, a technique in which an electric field pulls DNA fragments through a semisolid gel. DNA fragments of different sizes move through the gel at different rates. The shorter the fragment, the faster it moves, because shorter fragments slip through the tangled molecules of the gel faster than longer fragments do. All fragments of the same length move through the gel at the same speed. Thus, DNA fragments gather into discrete bands according to length as they migrate through the gel (Figure 10.6).

Several regions of chromosomal DNA are amplified by PCR and subjected to electrophoresis. The banding patterns on the resulting gel constitute an individual's DNA fingerprint, which, for all practical purposes, is unique. Unless two people are identical twins, the chances that they have identical tandem repeats in even three regions of DNA is 1 in 1,000,000,000,000,000,000—or one in a quintillion—which is far more than the number of people alive on Earth.

A standard set of thirteen short tandem repeat regions is typically tested to make a DNA fingerprint usable as evidence in a U.S. court. DNA fingerprints are now routinely used to resolve paternity disputes and criminal cases. A few drops of blood, semen, or cells from a hair follicle at a crime scene or on a suspect's clothing yield enough DNA to amplify with PCR for a DNA fingerprint. The technique is used not only to convict the guilty, but also to exonerate the innocent: As of this writing, DNA fingerprinting evidence has helped release more than 160 innocent people from prison.

DNA fingerprinting has many applications. For example, it was used to identify the remains of many individuals who died in the World Trade Center on September 11, 2001. It confirmed that human bones exhumed from a shallow pit in Siberia belonged to five individuals of the Russian imperial family, all shot to death in secrecy in 1918. Researchers also use DNA fingerprinting to study population dispersal in humans and other animals. Because only a tiny amount of DNA is necessary, such studies are not limited to living populations. Short tandem repeats on the Y chromosome are also used to determine genetic relationships among male relatives and descendants, and to trace an individual's ethnic heritage.

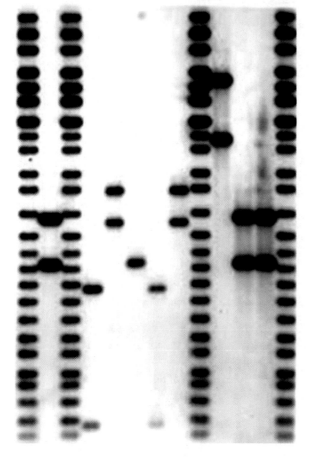

Evidence from Crime Scene

Size Reference | Control DNA | Size Reference | Victim | Suspect 1 | Suspect 2 | Female Cells | Semen | Size Reference | Boyfriend | Control DNA | Control DNA | Size Reference

FIGURE 10.6 DNA fingerprinting in an actual investigation of sexual assault. A short tandem repeat region was amplified from evidence at the crime scene: the perpetrator's semen and the victim's cells. The two samples were compared with the same tandem repeat region amplified from DNA of the victim, her boyfriend, and two suspects (1 and 2).

Note the three samples of control DNA (to confirm that the PCR was working correctly), and the four size reference samples.

The photo shows an x-ray film image of an electrophoresis gel from a forensics laboratory. The bands represent DNA fragments labeled with a radioactive tracer. **Figure It Out: Which of the two suspects was found to be guilty?** *Answer: Suspect 1*

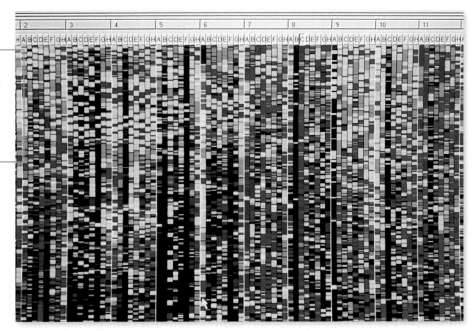

FIGURE 10.7 Sequencing, a method of determining the sequence of a fragment of DNA. On this computer screen, the four bases (adenine, thymine, guanine, and cytosine) are color-coded green, red, yellow, and blue, respectively. The order of colors represents the sequence of the DNA fragment.

The Human Genome Project

Around 1986, people were arguing about sequencing the human genome. **Sequencing** is a method of determining the order of nucleotide bases in an isolated DNA fragment (Figure 10.7). Many insisted that deciphering the sequence of the human genome would have enormous payoffs for medicine and research. Others said the project would divert funds from work that was more urgent—and that also had a better chance of success. At that time, sequencing 3 billion bases was a daunting task: It would take at least 6 million sequencing reactions, all done by hand. Given the techniques available, the work would have taken more than fifty years to complete.

But techniques kept getting better, so more bases could be sequenced in less time. Automated (robotic) DNA sequencing and PCR had just been invented. Both of these techniques were still cumbersome and expensive, but many researchers sensed their potential. Waiting for faster technologies seemed the most efficient way to sequence 3 billion bases, but how fast did they need to be before the project could begin?

Private companies started to sequence the human genome in 1987. Walter Gilbert, one of the early inventors of DNA sequencing, started a company that intended to sequence and patent the human genome. This development provoked widespread outrage, but it also spurred commitments in the public sector. In 1988, the National Institutes of Health (NIH) effectively annexed the project by hiring James Watson (of DNA structure fame) to head the official Human Genome Project, and providing $200 million per year to fund it. A consortium formed between the NIH and international institutions that were sequencing different parts of the genome. Watson set aside 3 percent of the funding for studies of ethical and social issues arising from the research. He later resigned over a patent disagreement, and geneticist Francis Collins took his place.

Amid ongoing squabbles over patent issues, Celera Genomics formed in 1998. With biologist Craig Venter at its helm, the company intended to commercialize genetic information. Celera started to sequence the genome using new, faster techniques, because the first to have the complete sequence had a legal basis for patenting it. The competition motivated the public consortium to move its efforts into high gear.

Then, in 2000, U.S. President Bill Clinton and British Prime Minister Tony Blair jointly declared that the sequence of the human genome could not be patented. Celera kept on sequencing anyway. Celera and the public consortium separately published about 90 percent of the sequence in 2001. By 2003, fifty years after the discovery of the structure of DNA, the sequence of the human genome was officially completed. At this writing, about 99 percent of its coding regions have been identified. Researchers have not discovered what all of the

DNA fingerprint An individual's unique array of short tandem repeats.

electrophoresis Technique that separates DNA fragments by size.

sequencing Method of determining the order of nucleotides in DNA.

Table 10.1	A Few Sequenced Genomes		
Organism	Complete	Size (×10⁶ bases)	Protein-Coding Genes
Yeast	1997	12	6,532
Roundworm	1999	101	20,176
Arabidopsis	2000	120	18,641
Fruit fly	2000	120	14,141
Human	2001	2,858	20,067
Mouse	2002	2,562	22,010
Rat	2004	2,705	19,684
Zebrafish	2005	1,533	18,957
Chimpanzee	2005	2,836	23,186
Dog	2005	2,385	15,914
Chicken	2007	1,013	12,351
Horse	2007	2,429	15,567
Cow	2007	2,623	16,954
Macaque	2007	2,778	18,379
Duckbill platypus	2008	1,840	11,201

genes encode, only where they are in the genome. What do we do with this vast amount of data? The next step is to find out what the sequence means.

Genomics Investigations into the genomes of humans and other species have converged into the new research field of **genomics**. Structural genomics focuses on determining the three-dimensional structure of the proteins encoded by a genome. Comparative genomics compares genomes of different species; similarities and differences reflect evolutionary relationships.

The human genome sequence is a massive collection of seemingly cryptic data. One way we are able to decipher it is by comparing it to genomes of other organisms (Table 10.1), the premise being that all organisms are descended from shared ancestors, so all genomes are related to some extent. We see evidence of such genetic relationships just by comparing the sequence data. For example, the human and mouse sequences are about 78 percent identical; the human and banana sequences are about 50 percent identical.

Intriguing as these percentages might be, gene-by-gene comparisons offer more practical benefits. We have learned about the function of many human genes by studying their counterpart genes in other species. For example, researchers studying a human gene might disable expression of the same gene in mice. The effects of the gene's absence on mice are clues to its function in humans. These types of knockout experiments are revealing the function of many human genes. For example, researchers comparing the human and mouse genomes discovered a human version of the mouse gene *APOA5*. This gene encodes a protein that, as part of HDL lipoproteins, helps transport dietary fats in the blood. Mice with an *APOA5* knockout have four times the normal level of triglycerides in their blood. The researchers then looked for—and found—a correlation between *APOA5* mutations and high triglyceride levels in humans. High triglycerides are a risk factor for coronary artery disease.

Take-Home Message
What are some of the ways researchers study DNA, and how do we use what they discover?

- Electrophoresis is used to separate different sized fragments of DNA.
- A DNA fingerprint is an individual's unique array of short tandem repeats. DNA fingerprinting is used in forensics, as court evidence, and in other applications.
- Analysis of the human genome sequence is yielding new information about human genes and how they work. The information has practical applications in medicine, research, and other fields.

10.4 Genetic Engineering

Traditional cross-breeding methods can alter genomes, but only if individuals with the desired traits will interbreed. **Genetic engineering**, a laboratory process by which deliberate changes are introduced into an individual's genome, takes gene-swapping to an entirely new level. A gene may be transferred to the

genetic engineering Process by which deliberate changes are introduced into an individual's genome.
genomics The study of genomes.

genome of another species (or even another genus) to produce a transgenic organism. Or, a gene may be altered and reinserted into an individual of the same species. Both methods result in genetically modified organisms (GMOs).

Genetically Modified Microorganisms

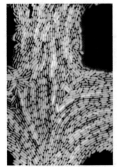

The most common GMOs are bacteria and yeast. These cells have the metabolic machinery to make complex organic molecules, and they are easily modified. For example, the *E. coli* on the *left* have been modified to produce a fluorescent protein from jellyfish. The cells are genetically identical, so the visible variation in fluorescence among them reveals differences in gene expression. Such differences may help us discover why some bacteria of a population become dangerously resistant to antibiotics, and others do not.

Bacteria and yeast have been modified to produce medically important proteins. Diabetics were among the first beneficiaries of such organisms. Insulin for their injections was once extracted from animals, but it provoked an allergic reaction in some people. Human insulin, which does not provoke allergic reactions, has been produced by transgenic *E. coli* since 1982. Slight modifications of the gene have also yielded fast-acting and slow-release human insulin.

Engineered microorganisms also produce proteins used in food manufacturing. For example, cheese is traditionally made with an extract of calf stomachs, which contain the enzyme chymotrypsin. Most cheese manufacturers now use chymotrypsin made by genetically modified bacteria. Other examples are GMO-made enzymes that improve the taste and clarity of beer and fruit juice, slow bread staling, or modify fats.

Designer Plants *Agrobacterium tumefaciens* is a species of bacteria that infects many plants, including peas, beans, potatoes, and other important crops. It carries a plasmid with genes that cause tumors to form on infected plants; hence the name Ti plasmid (for *Tumor-inducing*). Researchers use the Ti plasmid as a vector to transfer foreign or modified genes into plants. They remove the tumor-inducing genes from the plasmid, then insert desired genes. Whole plants can be grown from plant cells that take up the modified plasmid (Figure 10.8).

Modified *A. tumefaciens* bacteria are used to deliver genes into some food crop plants, including soybeans, squash, and potatoes. Researchers also transfer genes into plants by way of electric or chemical shocks, or by blasting them with DNA-coated pellets.

As crop production expands to keep pace with human population growth, it places unavoidable pressure on ecosystems everywhere. Irrigation leaves mineral and salt residues in soils. Tilled soil erodes, taking topsoil with it. Runoff clogs rivers, and fertilizer in it causes algae to grow so fast that fish suffocate. Pesticides can harm humans, other animals, and beneficial insects.

Pressured to produce more food at lower cost and with less damage to the environment, many farmers have begun to rely on genetically modified crop plants. Some of these modified plants carry genes that impart resistance to devastating plant diseases. Others offer improved yields, such as a strain of transgenic wheat that has double the yield of unmodified wheat. GMO crops such as *Bt* corn and soy help farmers use smaller amounts of toxic pesticides. Organic farmers often spray their crops with spores of *Bt* (*Bacillus thuringiensis*), a bacterial species that makes a protein toxic only to insect larvae. Researchers transferred the gene encoding the *Bt* protein into plants. The engineered plants produce the *Bt* protein, but otherwise they are identical to unmodified plants.

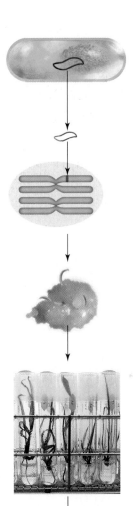

❶ An *A. tumefaciens* bacterium has been engineered to contain a Ti plasmid. The plasmid carries a foreign gene.

❷ The bacterium infects a plant cell and transfers the Ti plasmid into it. The plasmid DNA becomes integrated into one of the cell's chromosomes.

❸ The plant cell divides, and its descendants form an embryo.

❹ The embryo develops into a transgenic plant.

❺ The transgenic plant expresses the foreign gene. This tobacco plant is expressing a gene from a firefly.

FIGURE 10.8 Animated! Using the Ti plasmid to make a transgenic plant.

FIGURE 10.9 Genetically modified crops can help farmers use less pesticide. *Left*, the *Bt* gene conferred insect resistance to the genetically modified plants that produced this corn. *Right*, corn produced by unmodified plants is more vulnerable to insect pests.

Insect larvae die shortly after eating their first (and only) GMO meal. Farmers can use much less pesticide on crops that make their own (Figure 10.9).

Transgenic crop plants are also being developed for regions that are affected by severe droughts, such as Africa. Genes that confer drought tolerance and insect resistance are being transferred into crop plants such as corn, beans, sugarcane, cassava, cowpeas, banana, and wheat. Such crops may help people that rely on agriculture for food and income in drought-stricken, impoverished regions of the world.

The USDA Animal and Plant Health Inspection Service (APHIS) regulates the introduction of GMOs into the environment. At this writing, the agency has deregulated seventy-four genetically modified crop plants, which means the plants are approved for unregulated use in the United States.

The most widely planted GMO crops include corn, sorghum, cotton, soy, canola, and alfalfa engineered for resistance to glyphosate. Rather than tilling the soil to control weeds, farmers can spray their fields with this herbicide, which kills the weeds but not the engineered crops. However, weeds are becoming resistant to glyphosate, so spraying it no longer kills the weeds in glyphosate-resistant crop fields. The engineered gene is also appearing in wild plants and in nonengineered crops, which means that transgenes can (and do) escape into the environment. The genes are probably being transferred from transgenic plants to nontransgenic ones via pollen carried by wind or insects.

Controversy raised by such GMO use invites you to read the research and form your own opinions. The alternative is to be swayed by media hype (the term "Frankenfood," for instance), or by reports from possibly biased sources (such as herbicide manufacturers).

Biotech Barnyards Traditional cross-breeding has produced animals so unusual that transgenic animals may seem a bit mundane in comparison (Figure 10.10**A**). Cross-breeding is also a form of genetic manipulation, but many transgenic animals would probably never occur in nature (Figure 10.10**B,C**).

The first genetically modified animals were mice. Today, such mice are commonplace, and they are invaluable in research. We have discovered the function of many human genes by inactivating their counterparts in mice. *APOA5* (discussed in the previous section) is an example. Genetically modified mice are also used as models of many human diseases. For example, researchers inactivated the molecules involved in the control of glucose metabolism, one by one, in

xenotransplantation Transplant of an organ from one species into another.

mice. Studying the effects of the knockouts has resulted in much of our current understanding of how diabetes works in humans. Genetically modified animals such as these mice are allowing researchers to study human diseases (and their potential cures) without experimenting on humans.

Genetically modified animals also make proteins that have medical and industrial applications. Various transgenic goats produce proteins used to treat cystic fibrosis, heart attacks, blood clotting disorders, and even nerve gas exposure. Milk from goats transgenic for lysozyme, an antibacterial protein in human milk, may protect infants and children in developing countries from acute diarrheal disease. Goats transgenic for a spider silk gene produce the protein in their milk. Once researchers figure out how to spin it like spiders do, the silk may be used to manufacture fashionable fabrics, bulletproof vests, sports equipment, and biodegradable medical supplies. Rabbits make human interleukin-2, a protein that triggers divisions of immune cells. Genetic engineering has also given us dairy goats with heart-healthy milk, low-fat pigs, pigs with environmentally friendly low-phosphate manure, extra-large sheep, and cows that are resistant to mad cow disease.

Many people view transgenic animal research as unconscionable. Many others see it as simply an extension of thousands of years of acceptable animal husbandry practices: The techniques have changed, but not the intent. We humans still have a vested interest in improving our livestock.

Knockout Cells and Organ Factories Millions of people suffer with organs or tissues that are damaged beyond repair. In any given year, more than 80,000 of them are on waiting lists for an organ transplant in the United States alone. Human donors are in such short supply that illegal organ trafficking is now a common problem.

Pigs are a potential source of organs for transplantation, because pig and human organs are about the same in both size and function. However, the human immune system battles anything it recognizes as nonself. It rejects a pig organ at once, because it recognizes a particular protein on the plasma membrane of pig cells. Within a few hours, blood coagulates inside the organ's vessels and dooms the transplant. Drugs can suppress the immune response, but they also render organ recipients particularly vulnerable to infection.

Researchers have produced genetically modified pigs that lack the offending protein on their cells. The human immune system may not reject tissues or organs transplanted from these pigs. Transferring an organ from one species into another is called **xenotransplantation**. Critics of xenotransplantation are concerned that, among other things, pig-to-human transplants would invite pig viruses to cross the species barrier and infect humans, perhaps catastrophically. Their concerns are not unfounded. Evidence suggests that some of the worst pandemics arose when animal viruses adapted to new hosts: humans.

Take-Home Message **What is genetic engineering?**

- Genetic engineering is the directed alteration of an individual's genome, and it results in a genetically modified organism (GMO).
- A transgenic organism carries a gene from a different species. Transgenic organisms, including bacteria and yeast, are used in research, medicine, and industry.
- Plants with modified or foreign genes are now common farm crops.
- Animals that would be impossible to produce by traditional breeding methods are being created by genetic engineering.

FIGURE 10.10 Modified livestock.

A Featherless chicken developed by traditional cross-breeding methods. Such chickens survive in deserts where cooling systems are not an option.

B A goat transgenic for a human protein that inhibits blood clotting. **C** *Left*, a pig transgenic for a jellyfish protein that fluoresces yellow. Its nontransgenic littermate is on the *right*.

Genetically Modified Humans

Getting Better We know of more than 15,000 serious genetic disorders. Collectively, they cause 20 to 30 percent of infant deaths each year, and account for half of all mentally impaired patients and a fourth of all hospital admissions. They also contribute to many age-related disorders, including cancer, Parkinson's disease, and diabetes. Drugs and other treatments can minimize the symptoms of some genetic disorders, but gene therapy is the only cure. **Gene therapy** is the transfer of recombinant DNA into an individual's body cells, with the intent to correct a genetic defect or treat a disease. The transfer, which occurs by way of viral vectors or lipid clusters, inserts an unmutated gene into an individual's chromosomes.

Human gene therapy is a compelling reason to embrace genetic engineering research. It is now being tested as a treatment for cystic fibrosis, hemophilia A, several types of cancer, and inherited diseases of the eye, the ear, and the immune system. The results are encouraging. For example, little Rhys Evans (*left*) was born with a severe immune disorder, SCID-X1. SCID-X1 stems from mutations in the *IL2RG* gene, which encodes a receptor for an immune signaling molecule. Children affected by this disorder can survive only in germ-free isolation tents, because they cannot fight infections. In 1998, a viral vector was used to insert unmutated copies of *IL2RG* into cells taken from the bone marrow of eleven boys with SCID-X1. Each child's modified cells were infused back into his bone marrow. Months later, ten of the boys left their isolation tents for good. Their immune systems had been repaired by the gene therapy. Since then, gene therapy has freed many other SCID-X1 patients from life in an isolation tent. Rhys is one of them.

Getting Worse Manipulating a gene within the context of a living individual is unpredictable even when we know its sequence and where it is within the genome. No one, for example, can predict where a virus-injected gene will insert into chromosomes. Its insertion might disrupt other genes. If it interrupts a gene that is part of the controls over cell division, then cancer might be the outcome. Three boys from the 1998 SCID-X1 clinical trial have since developed a type of bone marrow cancer called leukemia, and one of them died. The researchers had wrongly predicted that cancer related to the gene therapy would be rare. Research now implicates the very gene targeted for repair, especially when combined with the viral vector that delivered it.

Other unanticipated problems sometimes occur with gene therapy. Jesse Gelsinger had a rare genetic deficiency of a liver enzyme that helps the body rid itself of ammonia, a toxic by-product of protein breakdown. Jesse's health was fairly stable while he was on a low-protein diet, but he had to take a lot of medication. In 1999, Jesse volunteered to be in a clinical trial of a gene therapy. He had a severe allergic reaction to the viral vector, and four days after receiving the treatment, his organs shut down and he died. He was 18. Our understanding of how the human genome works clearly lags behind our ability to modify it.

Getting Perfect The idea of selecting the most desirable human traits, **eugenics**, is an old one. It has been used as a justification for some of the most horrific episodes in human history, including the genocide of 6 million Jews during World War II. Thus, it continues to be a hotly debated social issue. For example, using gene therapy to cure human genetic disorders seems like a socially acceptable goal to most people. However, imagine taking this idea a bit

eugenics Idea of deliberately improving the genetic qualities of the human race.
gene therapy The transfer of a normal or modified gene into an individual with the goal of treating a genetic defect or disorder.

further. Would it also be acceptable to engineer the genome of an individual who is within a normal range of phenotype in order to modify a particular trait? Researchers have already produced mice that have improved memory, enhanced learning ability, bigger muscles, and longer lives. Why not people?

Given the pace of genetic research, the eugenics debate is no longer about how we would engineer desirable traits, but how we would choose the traits that are desirable. Realistically, cures for many severe but rare genetic disorders will not be found, because the financial return will not even cover the cost of the research. Eugenics, however, might just turn a profit. How much would potential parents pay to be sure that their child will be tall or blue-eyed? Would it be okay to engineer "superhumans" with breathtaking strength or intelligence? How about a treatment that can help you lose that extra weight, and keep it off permanently? The gray area between interesting and abhorrent can be very different depending on who is asked. In a survey conducted in the United States, more than 40 percent of those interviewed said it would be fine to use gene therapy to make smarter and cuter babies. In one poll of British parents, 18 percent would be willing to use it to keep a child from being aggressive, and 10 percent would use it to keep a child from growing up to be homosexual.

Getting There Some people are adamant that we must never alter the DNA of anything. The concern is that gene therapy puts us on a slippery slope that may result in irreversible damage to ourselves and to the biosphere. We as a society may not have the wisdom to know how to stop once we set foot on that slope. One is reminded of our peculiar human tendency to leap before we look. And yet, something about the human experience allows us to dream of such things as wings of our own making, a capacity that carried us into space. In this brave new world, the questions before you are these: What do we stand to lose if serious risks are not taken? And, do we have the right to impose the consequences of taking such risks on those who would choose not to take them?

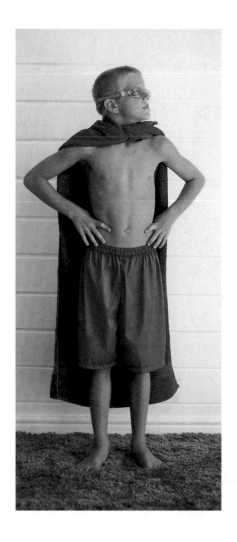

Take-Home Message **Do we genetically modify people?**

- Genes can be transferred into a person's cells to correct a genetic defect or treat a disease. However, the outcome of altering a person's genome remains unpredictable given our current understanding of how the genome works.

10.6

Impacts/Issues Revisited:
Golden Rice or Frankenfood?

Beta-carotene is an orange photosynthetic pigment (Section 5.2) that is remodeled by cells of the small intestine into vitamin A. Potrykus and Beyer transferred two genes in the beta-carotene synthesis pathway into rice plants—one gene from corn and one from bacteria. Both genes are under the control of a promoter that works only in seeds. The transgenic rice plants make beta-carotene in their seeds—in the grains of Golden Rice. One cup of Golden Rice has enough beta-carotene to satisfy a child's daily recommended amount of vitamin A. The rice was ready in 2005, but is still not available for human consumption. The biosafety experiments required by regulatory agencies are far too expensive for a humanitarian agency in the public sector. Most of the transgenic organisms used for food today were carried through the deregulation process by private companies.

How Would YOU ☑ Vote? Packaged food in the United States must have a nutrition label, but there is no requirement that genetically modified foods be labeled as such. Should food producers and distributors be required to identify products made from GMO plants or livestock? **See CengageNow for details, then vote online (CengageNow.com).**

Summary

 Section 10.1 Genes from one species may be inserted into an individual of a different species to make a transgenic organism, or a gene may be modified and reinserted into an individual of the same species. The result of either process is a genetically modified organism (GMO).

Section 10.2 Recombinant DNA consists of the fused DNA of different organisms. In DNA cloning, restriction enzymes cut DNA into pieces, then DNA ligase splices the pieces into plasmids or other cloning vectors. The resulting hybrid molecules are inserted into host cells such as bacteria. When a host cell divides, it forms huge populations of genetically identical descendant cells, or clones. Each clone has a copy of the foreign DNA.

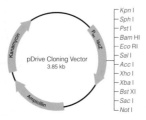

RNA cannot be cloned directly. Reverse transcriptase, a viral enzyme, is used to copy single-stranded RNA into cDNA for cloning.

A library is a collection of cells that host different fragments of DNA, often representing an organism's entire genome. Researchers can use probes to identify cells that host a specific fragment of DNA. Base pairing between nucleic acids from different sources is called nucleic acid hybridization.

The polymerase chain reaction (PCR) uses primers and a heat-resistant DNA polymerase to rapidly increase the number of molecules of a DNA fragment.

CENGAGENOW Learn how researchers isolate and copy genes, and survey their tools with the interaction and animation on CengageNow.

Section 10.3 Short tandem repeats are multiple copies of a short DNA sequence that follow one another along a chromosome. The number and distribution of short tandem repeats, unique in each 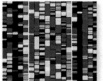 individual, is revealed by electrophoresis as a DNA fingerprint.

The genomes of several organisms have been sequenced. Genomics, or the study of genomes, is providing insights into the function of the human genome.

CENGAGENOW Observe the process of DNA fingerprinting with the animations on CengageNow.

 **Section 10.4** Recombinant DNA technology and genome analysis are the basis of genetic engineering, which is directed modification of an organism's genetic makeup with the intent to modify its phenotype.

Genetically modified bacteria and yeast produce medically valuable proteins. Transgenic crop plants are helping farmers produce food more efficiently. Genetically modified animals produce human proteins, and may one day provide a source of organs and tissues for transplantation into humans.

CENGAGENOW See how the Ti plasmid is used to genetically modify plants with the animation on CengageNow.

 Section 10.5 With gene therapy, a gene is transferred into body cells to correct a genetic defect or treat a disease. As with any new technology, the potential benefits of genetically modifying humans must be weighed against the potential risks, including social implications.

Self-Quiz   Answers in Appendix I

1. _____ cut(s) DNA molecules at specific sites.
   a. DNA polymerase   c. Restriction enzymes
   b. DNA probes   d. Reverse transcriptase

2. A _____ is a small circle of bacterial DNA that contains only a few genes and is separate from the bacterial chromosome.
   a. plasmid   c. nucleus
   b. chromosome   d. double helix

3. By reverse transcription, _____ is assembled on a(n) _____ template.
   a. mRNA; DNA   c. DNA; ribosome
   b. cDNA; mRNA   d. protein; mRNA

4. For each species, all _____ in the complete set of chromosomes is the _____ .
   a. genomes; phenotype   c. mRNA; start of cDNA
   b. DNA; genome   d. cDNA; start of mRNA

5. A set of cells that host various DNA fragments collectively representing an organism's entire set of genetic information is a _____ .
   a. genome   c. genomic library
   b. clone   d. GMO

6. PCR can be used to _____ .
   a. increase the number of specific DNA fragments
   b. check DNA fingerprints
   c. modify a human genome
   d. a and b are correct

7. Fragments of DNA can be separated by electrophoresis according to _____ .
   a. sequence   b. length   c. species

8. Sequencing determines the order of bases in _____ .
   a. DNA   c. electrophoresis
   b. PCR   d. RNA

9. Which of the following can be used to carry foreign DNA into host cells? Choose all correct answers.
   a. RNA   e. lipid clusters
   b. viruses   f. blasts of pellets
   c. PCR   g. xenotransplantation
   d. plasmids   h. sequencing

10. A transgenic organism _____ .
   a. carries a gene from another species
   b. has been genetically modified
   c. both a and b

11. _____ can be used to correct a genetic defect.
   a. Xenotransplantation   d. Cloning vectors
   b. Gene therapy   e. a and b
   c. Cloning   f. all of the above

Digging Into Data

Enhanced Spatial Learning Ability in Mice with Autism Mutation

Autism is a neurobiological disorder with a range of symptoms that include impaired social interactions, stereotyped patterns of behavior such as hand-flapping or rocking, and, occasionally, greatly enhanced intellectual abilities.

Autism may have a genetic basis. Some autistic people have a mutation in neuroligin 3, a type of cell adhesion protein (Section 3.4) that connects brain cells to one another. One mutation changes amino acid 451 from arginine to cysteine.

Mouse and human neuroligin 3 are very similar. In 2007, Katsuhiko Tabuchi and his colleagues genetically modified mice to carry the same arginine-to-cysteine substitution in their neuroligin 3. The mutation caused an increase in transmission of some types of signals between brain cells. Mice with the mutation had impaired social behavior, and, unexpectedly, enhanced spatial learning ability. Figure 10.11 shows the results from some of their tests.

1. In the first test, how many days did unmodified mice need to learn to find the location of a hidden platform within 10 seconds?

2. Did the modified or the unmodified mice learn the location of the platform faster in the first test?

3. Which mice learned faster the second time around?

4. Which mice showed the greatest improvement in memory between the first and the second test?

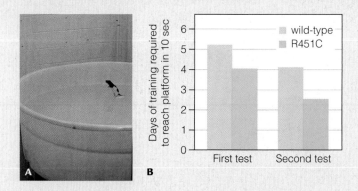

FIGURE 10.11 Enhanced spatial learning ability in mice with a mutation in neuroligin 3 (R451C), compared with unmodified (wild-type) mice. **A** The mice were tested in a water maze, in which a platform is submerged a few millimeters below the surface of a deep pool of warm water. The platform is not visible to swimming mice.

Mice do not particularly enjoy swimming, so they locate a hidden platform as fast as they can. When tested again, they can remember its location by checking visual cues around the edge of the pool. **B** How quickly they remember the platform's location is a measure of spatial learning ability. The platform was moved and the experiment was repeated for the second test.

---

12. Match the terms with the best description.

    _____ DNA fingerprint   a. having a foreign gene

    _____ Ti plasmid   b. cuts DNA at a specific sequence

    _____ nucleic acid hybridization   c. a person's unique collection of short tandem repeats

    _____ eugenics

    _____ restriction enzyme   d. base pairing of DNA or DNA and RNA from different sources

    _____ transgenic

    _____ GMO   e. selecting "desirable" traits

     f. genetically modified

     g. used in some gene transfers

Additional questions are available on CENGAGENOW™

Critical Thinking

1. The *FOXP2* gene encodes a transcription factor associated with vocal learning in mice, bats, birds, and humans. Mutations in *FOXP2* result in altered vocalizations in mice, and severe language disorders in humans. The chimpanzee, gorilla, and rhesus *FOXP2* proteins are identical; the human version differs in two of 715 amino acids. The change of two amino acids may have contributed to the development of language in humans. Would it be okay to transfer the human *FOXP2* gene into a nonhuman primate?

Spoken language is a complex, epistatic trait (Section 9.3). Thus, biologists do not anticipate that the transfer of the *FOXP2* gene alone would confer speaking ability upon a nonhuman recipient.

What do you think might happen if this prediction is incorrect, and an animal transgenic for *FOXP2* learned how to speak?

2. Animal viruses can mutate so that they infect humans, occasionally with disastrous results. In 1918, an influenza pandemic that apparently originated with a strain of avian flu killed 50 million people worldwide. Researchers isolated samples of that virus, which was called influenza A(H1N1) strain, from bodies of infected people preserved in Alaskan permafrost since 1918. From the samples, the researchers reconstructed the DNA sequence of the viral genome, then reconstructed the virus. Being 39,000 times more infectious than modern influenza strains, the reconstructed A(H1N1) virus proved to be 100 percent lethal in mice.

The researchers reconstructed the virus because understanding how the A(H1N1) strain works may help us defend ourselves against similar influenza strains. For example, they are using the reconstructed virus to discover which of its mutations made it so infectious and deadly in humans. Their work is urgent. A dangerous new strain of avian influenza in Asia shares some mutations with the A(H1N1) strain. Even now, researchers are working to test the effectiveness of antiviral drugs and vaccines on the reconstructed virus, and to develop new ones.

Critics of the A(H1N1) reconstruction are concerned. If the virus escapes its high-security containment facilities (even though it has not done so yet), it might cause another pandemic. Worse, terrorists could use the published DNA sequence and methods to make the virus for horrific purposes. Do you think this research makes our society more or less safe?

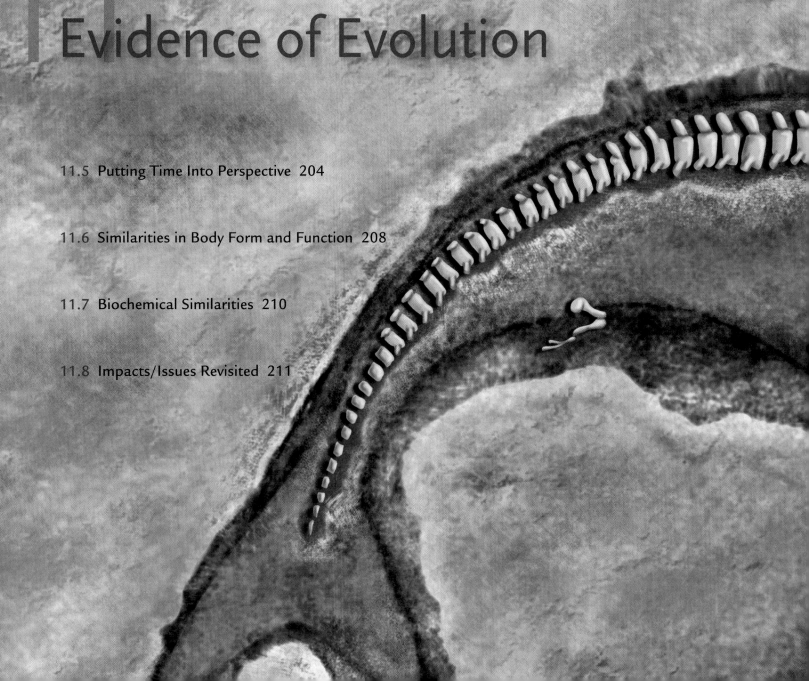

11 Evidence of Evolution

Natural phenomena in the past can be explained by the same physical, chemical, and biological processes that operate today.

11.1

Impacts/Issues: Reflections of a Distant Past

How do you think about time? Is your understanding limited to your own generation? Perhaps you can conceive of a few hundred years of human events, but how about a few million? Envisioning the distant past requires an intellectual leap from the familiar to the unknown.

One way to make that leap involves, surprisingly, asteroids. Asteroids are small planets hurtling through space. They range in size from 1 to 1,500 kilometers (roughly 0.5 to 1,000 miles) wide. Millions of them orbit around the sun between Mars and Jupiter—cold, stony leftovers from the formation of our solar system. Asteroids are difficult to see even with the best telescopes, because they do not emit light. Many cross Earth's orbit, but most of those pass us by before we know about them. Some have passed too close for comfort.

The mile-wide Barringer Crater in Arizona is difficult to miss (Figure 11.1). A 300,000-ton asteroid made this impressive pockmark in the desert sandstone when it slammed into Earth 50,000 years ago. The impact was 150 times more powerful than the bomb that leveled Hiroshima.

No human could have witnessed the impact, so how can we know anything about what happened? We often find physical evidence of events that occurred before we were around to see them. Geologists were able to infer the most probable cause of the Barringer Crater by analyzing tons of meteorites, melted sand, and other rocky clues at the site.

Similar evidence points to even larger asteroid impacts in the more distant past. For example, a **mass extinction**, or permanent loss of major groups of organisms, occurred 65.5 million years ago. It is marked by an unusual, worldwide layer of rock called the K–T boundary layer. There are plenty of dinosaur fossils below the K–T boundary layer. Above it, in rock layers that were deposited more recently, there are no dinosaur fossils anywhere. An impact crater off the coast of what is now the Yucatán Peninsula dates to about 65.5 million years ago. Coincidence? Many scientists say no. They have inferred from the evidence that the impact of an asteroid about 20 km (12 miles) wide caused a global catastrophe that wiped out the dinosaurs.

You are about to make an intellectual leap through time, to places that were not even known about a few centuries ago. We invite you to launch yourself from this premise: Natural phenomena that occurred in the past can be explained by the same physical, chemical, and biological processes that operate today. That premise is the foundation for scientific research into the history of life. The research represents a shift from experience to inference—from the known to what can only be surmised—and it gives us astonishing glimpses into the distant past.

FIGURE 11.1 From evidence to inference. What made the Barringer Crater? Rocky evidence points to a 300,000-ton asteroid that collided with Earth 50,000 years ago.

Early Beliefs, Confusing Discoveries

The seeds of biological inquiry were taking hold in the Western world more than 2,000 years ago. Aristotle, the Greek philosopher, was making connections between observations in an attempt to explain the order of the natural world. Like few others of his time, Aristotle viewed nature as a continuum of organization, from lifeless matter through complex plants and animals. Aristotle was one of the first **naturalists**, people who observe life from a scientific perspective.

By the fourteenth century, Aristotle's earlier ideas about nature had been transformed into a rigid view of life, in which a "great chain of being" extended from the lowest form (snakes), to humans, to spiritual beings. Each link in the chain was a species, and each was said to have been forged at the same time in a perfect state. A species would never change after it had been created, because any variation from its ideal form and function would constitute imperfection. Once every species had been discovered, the meaning of life would be revealed.

European naturalists that embarked on globe-spanning survey expeditions brought back tens of thousands of plants and animals from Asia, Africa, North and South America, and the Pacific Islands. Each newly discovered species was carefully catalogued as another link in the chain of being.

By the late 1800s, Alfred Wallace and a few other naturalists were seeing patterns in where species live and how they might be related, and had started to think about the natural forces that shape life. These naturalists were pioneers in **biogeography**, the study of patterns in the geographic distribution of species. Some of the patterns raised questions that could not be answered within the framework of prevailing belief systems. For example, globe-trotting explorers had discovered plants or animals living in extremely isolated places. The isolated species looked suspiciously similar to species living across vast expanses of open ocean, or on the other side of impassable mountain ranges. Could different species be related? If so, how could the related species end up geographically isolated from one another?

The birds in Figure 11.2, for example, share very similar features, although each lives on a different continent. All three flightless birds sprint about on long, muscular legs in flat, open grasslands about the same distance from the equator. All raise their long necks to watch for predators. Wallace thought that the unique set of similarities might mean that the three types of birds are descendants of an ancient common ancestor (and he was correct), but he had no idea how they could have ended up on different continents.

FIGURE 11.2 Similar species that are native to distant geographic realms.

Above, **A** South American rhea, **B** Australian emu, and **C** African ostrich. All three types of ratite birds live in similar habitats. These related birds are unlike most other birds in several traits, including their long, muscular legs and their inability to fly.

Below, similar-looking, unrelated plants: a spiny cactus native to the hot deserts of the American Southwest **D**, and a spiny spurge native to southwestern Africa **E**.

biogeography Study of patterns in the geographic distribution of species and communities.

mass extinction Simultaneous loss of many lineages from Earth.

naturalist Person who observes life from a scientific perspective.

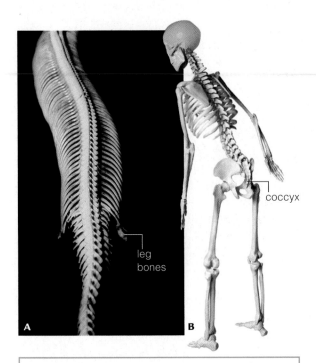

FIGURE 11.3 Vestigial body parts. **A** Pythons and boa constrictors have tiny leg bones, but snakes do not walk. **B** We humans use our legs, but not our coccyx (tail) bones.

leg bones

coccyx

Naturalists of the time also had trouble classifying organisms that are very similar in some features, but different in others. For example, the plants in Figure 11.2**D,E** are native to different continents. Both live in hot deserts where water is seasonally scarce. Both have rows of sharp spines that deter herbivores, and both store water in their thick, fleshy stems. However, their reproductive parts are very different, so these plants cannot be as closely related as their outward appearance might suggest.

Comparative morphology, the study of body plans and structures among groups of organisms, added to the confusion when naturalists discovered body parts with no apparent function. According to prevailing beliefs, each species had been created in a perfect state. If that were so, then why were there useless parts such as leg bones in snakes (which do not walk), or the vestiges of a tail in humans (Figure 11.3)?

Fossils were puzzling too. Geologists mapping rock formations exposed by erosion or quarrying had discovered identical sequences of rock layers in different parts of the world. Deep layers held fossils of simple marine life. Layers above them held similar but more intricate fossils. In higher layers, fossils that were similar but even more intricate looked like modern species. The photos on the *right* show one such series, ten fossilized foraminiferan shells from ten stacked layers of sedimentary rock. What did these fossil sequences mean? Fossils of gigantic animals that had no living representatives were also being unearthed. If the animals had been perfect at the time of creation, why were they now extinct?

Taken as a whole, findings from biogeography, comparative morphology, and geology did not fit with prevailing beliefs of the nineteenth century. If species had not been created in a perfect state (and fossil sequences and "useless" body parts implied they had not), then perhaps species had changed over time.

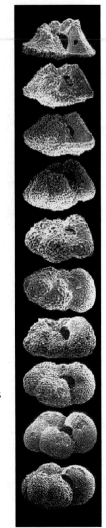

Take-Home Message How have observations of the natural world changed our thinking?

- Increasingly extensive observations of nature in the nineteenth century did not fit with prevailing belief systems.

- The cumulative findings from biogeography, comparative morphology, and geology led to new ways of thinking about the natural world.

11.3 A Flurry of New Theories

Squeezing New Evidence Into Old Beliefs In the nineteenth century, naturalists were faced with increasing evidence that life on Earth, and even Earth itself, had changed over time. Around 1800, Georges Cuvier, an expert in zoology and paleontology, was trying to make sense of the new information. He had observed abrupt changes in the fossil record, and knew that many fossil species seemed to have no living counterparts. Given this evidence,

catastrophism Now-abandoned hypothesis that catastrophic geologic forces unlike those of the present day shaped Earth's surface.

comparative morphology Scientific study of body plans and structures among groups of organisms.

evolution Change in a line of descent.

theory of uniformity Idea that gradual, repetitive processes occurring over long time spans shaped Earth's surface.

he proposed a startling idea: Many species that had once existed were now extinct. Cuvier also knew about evidence that Earth's surface had changed. For example, he had seen fossilized seashells on mountainsides far from modern seas. Like most others of his time, he assumed Earth's age to be in the thousands, not millions, of years. He reasoned that geologic forces unlike those known today would have been necessary to raise sea floors to mountaintops in this short time span. Catastrophic geological events would have caused extinctions, after which surviving species repopulated the planet. Cuvier's idea came to be known as **catastrophism**. We now know it is incorrect; geologic processes have not changed over time.

Another scholar, Jean-Baptiste Lamarck, was thinking about processes that drive **evolution**, or change in a line of descent. Lamarck thought that a species gradually improved over generations because of an inherent drive toward perfection, up the chain of being. This drive directed an unknown "fluida" into body parts needing change. By Lamarck's hypothesis, environmental pressures and internal needs cause changes in an individual's body, and offspring inherit the changes. Try using Lamarck's hypothesis to explain why the giraffe's neck is very long. We might predict that some short-necked ancestor of the modern giraffe stretched its neck to browse on leaves beyond the reach of other animals. The stretches may have even made its neck a bit longer. By Lamarck's hypothesis, that animal's offspring would inherit a longer neck, and after many generations strained to reach ever loftier leaves, the modern giraffe would have been the result. Lamarck was correct in thinking that environmental factors affect a species' traits, but incorrect about how inheritance works.

Voyage of the Beagle

In 1831, the twenty-two-year-old Charles Darwin was wondering what to do with his life. Ever since he was eight, he had wanted to hunt, fish, collect shells, or watch insects and birds—anything but sit in school. After attempting to study medicine in college, he earned a degree in theology from Cambridge. All through school, however, Darwin spent most of his time with faculty members who embraced natural history. Botanist John Henslow arranged for Darwin to become a naturalist aboard the *Beagle*, a ship about to embark on a survey expedition to South America.

The *Beagle* set sail for South America in December, 1831 (Figure 11.4). The young man who had hated school and had no formal training in science quickly became an enthusiastic naturalist. During the *Beagle*'s five-year voyage, Darwin found many unusual fossils. He saw diverse species living in environments that ranged from the sandy shores of remote islands to the plains high in the Andes. Along the way, he read the first volume of a new and popular book, Charles Lyell's *Principles of Geology*. Lyell was a proponent of what became known as the **theory of uniformity**, the idea that gradual, repetitive change had shaped Earth. For many years, geologists had been chipping away at the sandstones, limestones, and other types of rocks that form from accumulated sediments in lakebeds, river bottoms, and ocean floors. These rocks held evidence that gradual processes of geologic change operating in the present were the same ones that operated in the distant past.

The theory of uniformity held that strange catastrophes were not necessary to explain Earth's surface. Over great spans of time, gradual, everyday geologic processes such as erosion could have sculpted Earth's current landscape. The theory challenged the prevailing belief that Earth was 6,000 years old. According to traditional scholars, people had recorded everything that happened in those 6,000 years—and in all that time, no one had mentioned seeing a species evolve. However, by Lyell's calculations, it must have taken millions of years to sculpt Earth's surface. Darwin's exposure to Lyell's ideas gave him insights into the

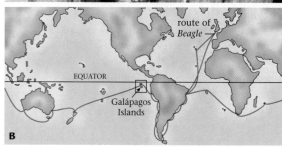

FIGURE 11.4 A Charles Darwin. **B** Voyage of the *Beagle*, during which Darwin discovered many puzzling fossils and species. In the Andes, for example, he found fossils of marine organisms in rock layers 3.6 kilometers (2.2 miles) above sea level.

FIGURE 11.5 Ancient relatives? **A** A modern armadillo, about a foot long. **B** Fossil of a glyptodon, an automobile-sized mammal that lived between about 2 million and 15,000 years ago.

Glyptodons and armadillos are widely separated in time, but they share a restricted distribution and unusual traits, including a shell and helmet of keratin-covered bony plates—a material similar to crocodile and lizard skin. (The fossil in **B** is missing its helmet.) Their unique shared traits were a clue that helped Darwin develop a theory of evolution by natural selection.

geological history of the regions he would encounter on his journey. Was million of years enough time for species evolve? Darwin thought that it was.

Natural Selection Darwin sent to England the thousands of specimens he had collected on his voyage. Among them were fossil glyptodons from Argentina. These armored mammals are extinct, but they have many traits in common with modern armadillos (Figure 11.5). For example, armadillos live only in places where glyptodons once lived. Like glyptodons, armadillos have helmets and protective shells that consist of unusual bony plates. Could the odd shared traits mean that glyptodons were ancient relatives of armadillos? If so, perhaps traits of their common ancestor had changed in the line of descent that led to armadillos. But why would such changes occur?

Back in England, Darwin pondered his notes and fossils. He also read an essay by one of his contemporaries, economist Thomas Malthus. Malthus had correlated increases in human population size with famine, disease, and war. He proposed that humans run out of food, living space, and other resources because they tend to reproduce beyond the capacity of their environment to sustain them. When that happens, the individuals of a population must either compete with one another for the limited resources, or develop technology to increase their productivity. Darwin realized that Malthus's ideas had wider application: All populations, not just human ones, have the capacity to produce more individuals than their environment can support.

Darwin also thought about species he had observed during his voyage. He knew that individuals of a species were not always identical. They had many traits in common, but they might vary in size, color, or other traits. It dawned on Darwin that having a particular version of a variable trait might give an individual an advantage over competing members of its species. The trait might enhance the individual's ability to survive and reproduce in its particular environment. Darwin realized that in any population, some individuals have traits that make them better suited to their environment than others. In other words, individuals of a natural population vary in **fitness**. We define fitness as the degree of adaptation to a specific environment, and measure it as relative genetic contribution to future generations. A trait that enhances an individual's fitness is called an evolutionary **adaptation**, or **adaptive trait**.

Over many generations, individuals with the most adaptive traits tend to survive longer and reproduce more than their less fit rivals. Darwin understood that this process, which he called **natural selection**, could be a driving force of evolution. If an individual has a form of a trait that makes it better suited to an environment, then it is better able to survive. If an individual is better able to

adaptation (**adaptive trait**) A heritable trait that enhances an individual's fitness.

fitness The degree of adaptation to an environment, as measured by an individual's relative genetic contribution to future generations.

natural selection A process of evolution in which individuals of a population who vary in the details of heritable traits survive and reproduce with differing success.

FIGURE 11.6 Alfred Wallace, codiscoverer of natural selection.

survive, then it has a better chance of living long enough to produce offspring. If individuals that bear an adaptive, heritable trait produce more offspring than those that do not, then the frequency of that trait will tend to increase in the population over successive generations. Table 11.1 summarizes this reasoning.

Great Minds Think Alike Darwin wrote out his ideas about natural selection, but let ten years pass without publishing them. In the meantime, Alfred Wallace, who had been studying wildlife in the Amazon basin and the Malay Archipelago, wrote an essay and sent it to Darwin for advice. Wallace's essay had outlined Darwin's theory! Wallace had written earlier letters to Lyell and Darwin about patterns in the geographic distribution of species; he too had connected the dots. Wallace is now called the father of biogeography (Figure 11.6).

In 1858, just weeks after Darwin received Wallace's essay, their similar theories were presented jointly at a scientific meeting. Wallace was still in the field and knew nothing about the meeting, which Darwin did not attend. The next year, Darwin published *On the Origin of Species*, which laid out detailed evidence in support of his theory. Many scholars readily accepted the idea of descent with modification, or evolution. However, there was a fierce debate over the idea that evolution occurs by natural selection. Decades would pass before experimental evidence from the field of genetics led to its widespread acceptance in the scientific community.

As you will see in the remainder of this chapter, Darwin's theory is supported by and helps explain the fossil record as well as similarities in the form, function, and biochemistry of living things.

Table 11.1	Principles of Natural Selection

Observations about populations

- Natural populations have an inherent reproductive capacity to increase in size over time.

- As a population expands, resources that are used by its individuals (such as food and living space) eventually become limited.

- When resources are limited, the individuals of a population compete for them.

Observations about genetics

- Individuals of a species share certain traits.

- Individuals of a natural population vary in the details of those shared traits.

- Shared traits have a heritable basis, in genes. Alleles (slightly different forms of a gene) arise by mutation.

Inferences

- A certain form of a shared trait may make its bearer better able to survive.

- The individuals of a population that are better able to survive tend to leave more offspring.

- Thus, an allele associated with an adaptive trait tends to become more common in a population over time.

Take-Home Message How did new evidence change the way people in the nineteenth century thought about the history of life?

- In the 1800s, fossils and other evidence led some naturalists to propose that Earth and the species on it had changed over time. The naturalists also began to reconsider the age of Earth.

- These ideas set the stage for Darwin's theory of evolution by natural selection. Natural selection is the differential survival and reproduction among individuals of a population that vary in the details of shared, inherited traits.

- Traits favored by natural selection are said to be adaptive. An adaptive trait increases the chances that an individual bearing it will survive and reproduce.

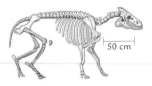

A A 30-million-year-old fossil of *Elomeryx*. This small terrestrial mammal was a member of the same artiodactyl group that gave rise to hippopotamuses, pigs, deer, sheep, cows, and whales.

B *Rodhocetus*, an ancient whale, lived about 47 million years ago. Its distinctive ankle bones point to a close evolutionary connection to artiodactyls. Inset: compare a *Rodhocetus* ankle bone (*left*) with that of a modern artiodactyl, a pronghorn antelope (*right*).

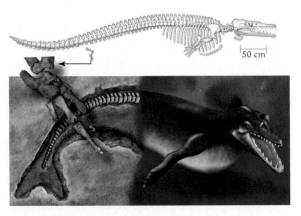

C *Dorudon atrox*, an ancient whale that lived about 37 million years ago. Its artiodactyl-like ankle bones (*left*) were much too small to have supported the weight of its huge body on land, so this mammal had to be fully aquatic.

FIGURE 11.7 New links in the ancient lineage of whales. With their artiodactyl-like ankle bones, *Rodhocetus* and *Dorudon* were probably offshoots of the ancient artiodactyl-to-modern-whale lineage as it transitioned back to life in water. Modern cetaceans do not have even a remnant of an ankle bone.

About Fossils

Even before Darwin's time, **fossils** were recognized as stone-hard evidence of earlier forms of life. Most fossils are mineralized bones, teeth, shells, seeds, spores, or other hard body parts. Trace fossils such as footprints and other impressions, nests, burrows, trails, eggshells, or feces (*left*) are evidence of an organism's activities.

The process of fossilization begins when an organism or its traces become covered by sediments or volcanic ash. Water seeps into the remains, and metal ions and other inorganic compounds dissolved in the water gradually replace the minerals in the bones and other hard tissues. Sediments that accumulate on top of the remains exert increasing pressure on them. After a very long time, the pressure and mineralization process transform the remains into rock.

Most fossils are found in layers of sedimentary rock such as mudstone, sandstone, and shale. These rocks form as rivers wash silt, sand, volcanic ash, and other particles from land to sea. The particles settle on sea floors in horizontal layers that vary in thickness and in composition. After hundreds of millions of years, the layers of sediments became compacted into layers of rock.

We study layers of sedimentary rock in order to understand the historical context of fossils we find in them. Usually, the deeper layers in a stack were the first to form, and those closest to the surface formed most recently. Thus, the deeper the layer of sedimentary rock, the older the fossils it contains. A layer's composition and thickness relative to other layers is also a clue about local or global events that were occurring as it formed. For instance, layers of sedimentary rock deposited during ice ages are thinner than other layers. Why? Tremendous volumes of water froze and became locked in glaciers during the ice ages. Rivers dried up, and sedimentation slowed. When the glaciers melted, sedimentation resumed and the layers became thicker.

The Fossil Record We have fossils for more than 250,000 known species. Considering the current range of biodiversity, there must have been many millions more, but we will never know all of them. Why not?

The odds are against finding evidence of an extinct species, because fossils are relatively rare. Most of the time, an organism's remains are quickly obliterated by scavengers or decay. Organic materials decompose in the presence of oxygen, so remains endure only if they are encased in an air-excluding material such as sap, tar, ice, or mud. Remains that do become fossilized are often deformed, crushed, or scattered by erosion and other geologic assaults. For us to find a fossil of an extinct species, at least one specimen had to be buried before it decomposed or something ate it. The burial site had to escape destructive geologic events, and it had to end up where we might be able to find it.

However, despite these challenges, the fossil record is substantial enough to help us reconstruct patterns in the history of life. We have been able to find fossil evidence of the evolutionary history of many species, as illustrated by the following example.

Evolutionary biologists had thought for a long time that the ancestors of whales probably walked on land. The skull and lower jaw of cetaceans—which include whales, dolphins, and porpoises—have distinctive features that are also characteristic of ancient carnivorous land animals. DNA sequence comparisons suggested that those animals were probably artiodactyls, hooved mammals with two or four toes on each foot. Modern representatives of the line of descent—the **lineage**—include hippopotamuses, camels, pigs, deer, sheep, and cows.

We have fossil skeletons representing gradual changes in thirty genera between the ancient artiodactyls and modern whales. However, until recently the skeletons of these transitional forms were incomplete, so the story of the transition remained speculative. Then, in 2000, archaeologists found two of the missing pieces of the puzzle in Pakistan when they collected complete fossil skeletons of two ancient whales. Intact, sheeplike ankle bones and ancient whalelike skull bones present in the same skeletons provide clear evidence of the transition from terrestrial to aquatic life (Figure 11.7).

Radiometric Dating

A radioisotope is a form of an element with an unstable nucleus (Section 2.2). Atoms of a radioisotope become atoms of other elements as their nucleus disintegrates. Such decay is not influenced by temperature, pressure, chemical bonding state, or moisture; it is influenced only by time. Like the ticking of a perfect clock, each type of radioisotope decays at a constant rate into predictable products called daughter elements. For example, radioactive uranium 238 decays into thorium 234, which decays into something else, and so on until it becomes lead 206. The time it takes for half of a radioisotope's atoms to decay into a product is called **half-life** (Figure 11.8A). The half-life of uranium 238 to lead 206 decay is 4.5 billion years.

The predictability of radioactive decay can be used to find the age of a volcanic rock—the date it cooled. Rock deep inside Earth is hot and molten, so atoms swirl and mix in it. Rock that reaches the surface cools and hardens. As the rock cools, minerals crystallize in

zircon

it. Each kind of mineral has a characteristic structure and composition. For example, the mineral zircon consists primarily of ordered arrays of zircon silicate molecules ($ZrSiO_4$). Some of the molecules in a zircon crystal have uranium atoms substituted for zirconium atoms, but never lead atoms. Thus, new zircon crystals that form as molten rock cools contain no lead. However, uranium decays into lead at a predictable rate. Thus, over time, uranium atoms disappear from a zircon crystal, and lead atoms accumulate in it. The ratio of uranium atoms to lead atoms in a zircon crystal can be measured precisely. That ratio can be used to calculate how long ago the crystal formed (its age).

We have just described **radiometric dating**, a method that can reveal the age of a material by measuring its content of a radioisotope and daughter elements. The oldest known terrestrial rock, a tiny zircon crystal from Australia, is 4.404 billion years old.

Recent fossils that still contain carbon can be dated by measuring their carbon 14 content (Figure 11.8B–D). Most of the ^{14}C in a fossil will have decayed after about 60,000 years. The age of fossils older than that can be estimated only by dating volcanic rocks in lava flows above and below the fossil-containing rock.

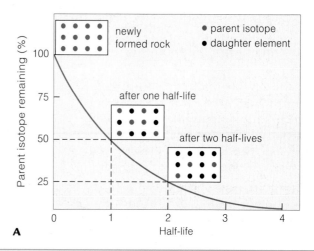
newly formed rock
• parent isotope
• daughter element
after one half-life
after two half-lives

A

FIGURE 11.8 Animated! Radiometric dating. **A** Half-life, the time it takes for half of the atoms in a sample of radioisotope to decay. **B–D** Using radiometric dating to find the age of a fossil. Carbon 14 (^{14}C) is a radioisotope of carbon. It forms in the atmosphere and combines with oxygen to become CO_2, which enters food chains by way of photosynthesis.

Figure It Out: How much of any radioisotope remains after two of its half-lives have passed?

Answer: 25 percent

B Long ago, trace amounts of ^{14}C and a lot more ^{12}C were incorporated into the tissues of a nautilus. The carbon atoms were part of organic molecules in the nautilus's food. As long as it was alive, the nautilus gained carbon atoms from its food. Thus, the proportion of ^{14}C to ^{12}C in its tissues stayed the same.

C When the nautilus died, it stopped eating, so its body stopped gaining carbon. The ^{14}C in its body continued to decay, so the amount of ^{14}C decreased relative to the amount of ^{12}C. Half of the ^{14}C decayed in 5,370 years, half of what was left decayed after another 5,370 years, and so on.

D Fossil hunters discover the fossil. They measure its ^{14}C to ^{12}C ratio and use it to calculate the number of half-life reductions since the organism died. For example, if the ^{14}C to ^{12}C ratio is one-eighth of the ratio in living organisms, then this nautilus died about 16,000 years ago.

Take-Home Message
What are fossils?

■ Fossils are evidence of organisms that lived in the remote past, a stone-hard historical record of life.

■ Researchers use the predictability of radioisotope decay to estimate the age of rocks and fossils.

fossil Physical evidence of an organism that lived in the ancient past.

half-life Characteristic time it takes for half of a quantity of a radioisotope to decay.

lineage Line of descent.

radiometric dating Method of estimating the age of a rock or fossil by measuring the content and proportions of a radioisotope and its daughter elements.

Putting Time Into Perspective

Radiometric dating and fossils allow us to recognize similar sequences of sedimentary rock layers around the world. Transitions between some of the layers mark boundaries of great intervals of time in the **geologic time scale**, or chronology of Earth's history (Figure 11.9). Each layer holds clues to life on Earth during the period of time that it formed. For example, the Grand Canyon's

FIGURE 11.9 Animated! The geologic time scale correlated with some of the sedimentary rock layers of the Grand Canyon.

Eon	Era	Period	Epoch	mya	Major Geologic and Biologic Events
PHANEROZOIC	CENOZOIC	QUATERNARY	Recent	0.01	Modern humans evolve. Major extinction event is now under way.
			Pleistocene	1.8	
		TERTIARY	Pliocene	5.3	Tropics, subtropics extend poleward. Climate cools; dry woodlands and grasslands emerge. Adaptive radiations of mammals, insects, birds.
			Miocene	23.0	
			Oligocene	33.9	
			Eocene	55.8	
			Paleocene	65.5 ◄	**Major extinction event**, perhaps precipitated by asteroid impact. Mass extinction of all dinosaurs and many marine organisms.
	MESOZOIC	CRETACEOUS	Late		
				99.6	
			Early		Climate very warm. Dinosaurs continue to dominate. Important modern insect groups appear (bees, butterflies, termites, ants, and herbivorous insects including aphids and grasshoppers). Flowering plants originate and become dominant land plants.
				145.5	
		JURASSIC			Age of dinosaurs. Lush vegetation; abundant gymnosperms and ferns. Birds appear. Pangea breaks up.
				199.6 ◄	**Major extinction event**
		TRIASSIC			Recovery from the major extinction at end of Permian. Many new groups appear, including turtles, dinosaurs, pterosaurs, and mammals.
				251 ◄	**Major extinction event**
	PALEOZOIC	PERMIAN			Supercontinent Pangea and world ocean form. Adaptive radiation of conifers. Cycads and ginkgos appear. Relatively dry climate leads to drought-adapted gymnosperms and insects such as beetles and flies.
				299	
		CARBONIFEROUS			High atmospheric oxygen level fosters giant arthropods. Spore-releasing plants dominate. Age of great lycophyte trees; vast coal forests form. Ears evolve in amphibians; penises evolve in early reptiles (vaginas evolve later, in mammals only).
				359 ◄	**Major extinction event**
		DEVONIAN			Land tetrapods appear. Explosion of plant diversity leads to tree forms, forests, and many new plant groups including lycophytes, ferns with complex leaves, and seed plants.
				416	
		SILURIAN			Radiations of marine invertebrates. First appearances of land fungi, vascular plants, bony fish, and perhaps terrestrial animals (millipedes, spiders).
				443 ◄	**Major extinction event**
		ORDOVICIAN			Major period for first appearances. The first land plants, fish, and reef-forming corals appear. Gondwana moves toward the South Pole and becomes frigid.
				488	
		CAMBRIAN			Earth thaws. Explosion of animal diversity. Most major groups of animals appear (in the oceans). Trilobites and shelled organisms evolve.
				542	
PROTEROZOIC					Oxygen accumulates in atmosphere. Origin of aerobic metabolism. Origin of eukaryotic cells, then protists, fungi, plants, animals. Evidence that Earth mostly freezes over in a series of global ice ages between 750 and 600 mya.
				2,500	
ARCHAEAN AND EARLIER					3,800–2,500 mya. Origin of prokaryotes.
					4,600–3,800 mya. Origin of Earth's crust, first atmosphere, first seas. Chemical, molecular evolution leads to origin of life (from protocells to anaerobic prokaryotes).

A Transitions in sedimentary rock layers mark great time spans in Earth's history (not to the same scale). mya: millions of years ago.

Dates from International Commission on Stratigraphy data, 2007.

B We can reconstruct some of the events in the history of life by studying rocky clues in the layers. Here, the *red* triangles mark times of great mass extinctions. "First appearance" refers to appearance in the fossil record, not necessarily the first appearance on Earth; we often discover fossils that are significantly older than previously discovered specimens.

Kaibab Limestone

Toroweap Formation

Coconino Sandstone

Hermit Shale

Esplanade Sandstone

Wescogame Formation

Manakacha Formation

Watahomigi Formation

Redwall Limestone

Temple Butte Formation

Muav Limestone

Bright Angel Shale

Tapeats Sandstone*

Sixtymile Formation*

Chuar Group*

Nankoweap Formation*

Unkar Group*

Vishnu Basement Rocks*

C Sedimentary rock layers exposed by erosion in the Grand Canyon. Each layer has a characteristic composition and set of fossils (some are shown) that reflect events during its deposition.
For example, Coconino Sandstone, which stretches from California to Montana, consists mainly
of greatly weathered sand. Ripple marks and reptile tracks are the only fossils in it. Many think it
is the remains of a vast sand desert, like the Sahara is today. *Layers not visible in this view.

geologic time scale Chronology of Earth history.

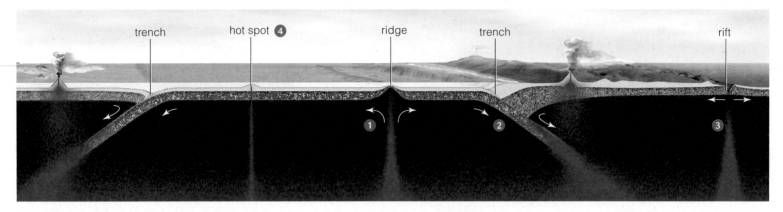

trench   hot spot ④   ridge   trench   rift

FIGURE 11.10 Plate tectonics. Huge pieces of Earth's outer layer of rock slowly drift apart and collide. As the plates move, they raft continents around the globe.

① At oceanic ridges, huge plumes of molten rock welling up from Earth's interior drive the movement of tectonic plates. New crust spreads outward as it forms on the surface, forcing adjacent tectonic plates away from the ridge and into trenches elsewhere.

② At trenches, the advancing edge of one plate plows under an adjacent plate and buckles it.

③ Faults are ruptures in Earth's crust where plates meet. Plates move apart at rifts. The aerial photo *below* shows about 4.2 kilometers (2.6 miles) of the San Andreas Fault, which extends 1,300 km (800 miles) through California. This fault is a boundary between two tectonic plates slipping by one another.

④ Plumes of molten rock rupture a tectonic plate at what are called "hot spots." The Hawaiian Islands have been forming this way.

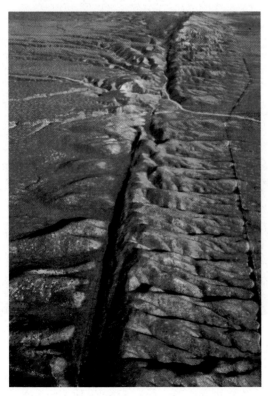

uppermost layer, Kaibab limestone, is its youngest. The layer consists mainly of limestone, a sandy sedimentary rock that forms at the bottom of shallow oceans. It also contains fossils of marine organisms. If the Kaibab limestone layer was originally deposited at the bottom of an ocean, a lot more rock must have formed on top of it in the last 250 million years. What happened to that rock? Part of the answer can be found in a neighboring park, Utah's Bryce Canyon, where about 12,000 feet (3,650 meters) of sedimentary rock layers occur on top of Kaibab limestone. These younger layers no longer exist in the Grand Canyon because they have eroded away.

Drifting Continents, Changing Seas Wind, water, and other natural forces continuously sculpt the surface of Earth, but they are only part of a bigger picture of geologic change. Earth itself also changes dramatically. For instance, the Atlantic coasts of South America and Africa seem to "fit" like jigsaw puzzle pieces. By one model, all continents were once part of a bigger supercontinent—**Pangea**—that had split into fragments and drifted apart. The idea explained why the same types of fossils occur in sedimentary rock on both sides of the vast Atlantic Ocean.

At first, most scientists did not accept this model, which was called continental drift. Continents drifting about Earth seemed to be an outrageous idea, and no one knew what would drive such movement. However, evidence that supported the model kept turning up. For instance, molten rock deep inside Earth wells up and solidifies on the surface. Some iron-rich minerals become magnetic as they solidify, and their magnetic poles align with Earth's poles when they do. If continents never moved, then all of these ancient rocky magnets would be aligned north-to-south, like compass needles. Indeed, the magnetic poles of rock formations on different continents are aligned—but not north-to-south. They point in many different directions. Either Earth's magnetic poles veer dramatically from their north–south axis, or the continents wander.

Then, deep-sea explorers discovered that ocean floors are not as static and featureless as had been assumed. Immense ridges stretch thousands of kilometers across the sea floor (Figure 11.10). Molten rock spewing from the ridges pushes old sea floor outward in both directions **①**, then cools and hardens into new sea floor. Elsewhere, older sea floor plunges into deep trenches **②**.

Such discoveries swayed the skeptics. Finally, there was a plausible mechanism for continental drift. By this mechanism, Earth's relatively thin outer layer of rock is cracked into immense plates, like a gigantic cracked eggshell. The plates grow from undersea ridges and continental rifts **③**, and they sink into trenches. As the plates move, they act like colossal conveyer belts, rafting continents on top of them to new locations. The movement is no more than about

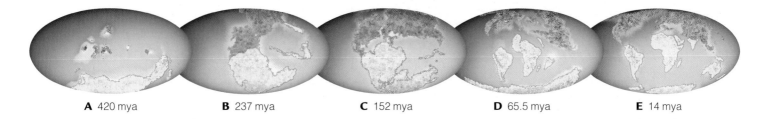

A 420 mya   **B** 237 mya   **C** 152 mya   **D** 65.5 mya   **E** 14 mya

10 centimeters (4 inches) a year—about half as fast as your toenails grow—but that is enough to carry a continent all the way around the world after 40 million years or so. Evidence of the movement is all around us, in various geological features of our landscapes ❹.

Researchers soon applied the theory of continental movement, which is now called **plate tectonics**, to some long-standing puzzles. For example, fossils of a type of fern (*Glossopteris*) and of an early reptile (*Lystrosaurus*) had been found in an unusual, 260 million-year-old geologic formation that occurs in Africa, India, South America, and Australia. The formation consists of a sequence of rock layers so complex that the identical sequence is quite unlikely to have formed in more than one place. However, the formation also predates Pangea, which started to form 237 million years ago.

The fern's seeds were too heavy to float or to be wind-blown over an ocean, and the reptile was not built to be able to swim between continents. Researchers suspected that both organisms evolved on a supercontinent that had broken up by the time Pangea formed. The older supercontinent, **Gondwana**, included most of the land masses now in the Southern Hemisphere as well as India and Arabia (Figure 11.11). Based on this hypothesis, the researchers predicted that *Glossopteris* and *Lystrosaurus* fossils and the same geologic formation would also occur in Antarctica, which was mostly unexplored at the time. Later expeditions found the formations and the fossils. The discovery supported the prediction— and the plate tectonics theory.

We now know that many modern species, including the ratite birds pictured in Figure 11.2**A–C**, live only in places that were once part of Gondwana. After it formed, Gondwana drifted south, across the South Pole, then north until it merged with other continents to form Pangea.

At least five times since Earth's outer layer of rock solidified 4.55 billion years ago, a single supercontinent with one ocean lapping at its coastline formed and then split up again. The resulting changes in Earth's surface, atmosphere, and waters have had a profound impact on the course of life's evolution. A continent's climate changes—often dramatically so—with its position on Earth. Colliding continents physically separate organisms living in oceans, and bring together those that had been living apart on land. Continents that break up separate organisms living on land, and bring together aquatic ones. Such changes are a major driving force of evolution, as you will see in the next chapter. Lineages that cannot adapt die out, and new evolutionary opportunities open up for the survivors.

FIGURE 11.11 Animated! A series of reconstructions of the drifting continents. **A** The supercontinent Gondwana (*yellow*) had begun to break up by the Silurian. **B** The supercontinent Pangea formed during the Triassic, then **C** began to break up in the Jurassic. **D** K–T boundary. **E** The continents reached their modern configuration in the Miocene.

Take-Home Message
What events have influenced life's history on Earth?

■ Slow movements of Earth's crust have changed the land, atmosphere, and oceans, and so have influenced the course of life's evolution.

Gondwana Supercontinent that formed more than 500 million years ago.

Pangea Supercontinent that formed about 237 million years ago and broke up about 152 million years ago.

plate tectonics Theory that Earth's outer layer of rock is cracked into plates, the slow movement of which rafts continents to new locations over geologic time.

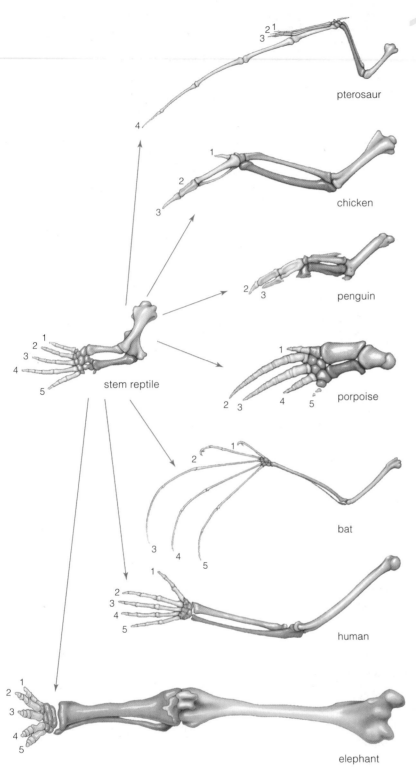

Similarities in Body Form and Function

How do we know about evolution that occurred in the ancient past? Like asteroid impacts, evolution also leaves evidence that it occurred. Fossils are one example, but organisms that are alive today provide others. To a biologist, remember, evolution means change in a line of descent. Clues about the history of a lineage may be revealed by comparative morphology, because similarities in the structure of body parts are often evidence of a common ancestor. Similar body parts that reflect shared ancestry are called **homologous structures** (*hom*– means the same). Such structures may be used for different purposes in different groups, but the same genes direct their development.

Morphological Divergence A body part that appears very different in different lineages may be quite similar in some underlying aspect of form. For example, even though vertebrate forelimbs are not the same in size, shape, or function from one group to the next, they clearly are alike in the structure and positioning of bony elements. They also are alike in the patterns of nerves, blood vessels, and muscles that develop inside of them. In addition, comparisons of the early embryos of different vertebrates reveal strong resemblances in patterns of bone development. Such similarities are evidence of shared ancestry.

Populations that are not interbreeding tend to diverge genetically, and in time they diverge morphologically. Change from the body form of a common ancestor is an evolutionary pattern called **morphological divergence**. We have evidence from fossilized limb bones that all modern land vertebrates share an ancestor, a "stem reptile" that crouched low to the ground on four legs. Descendants of this ancestor diversified into many new habitats on land, and gave rise to the groups we call reptiles, birds, and mammals. A few lineages that had become adapted to walking on land even returned to life in the seas.

The stem reptile's five-toed limbs were evolutionary clay. Over millions of years, they became molded into limbs with very different functions across many lineages (Figure 11.12). In extinct reptiles called pterosaurs, most birds, and bats, they have been modified for flight. In penguins and porpoises, the limbs are now flippers useful for swimming. In humans, five-toed limbs became modified into the arms and hands, in which the thumb evolved in opposition to the fingers. An opposable thumb was the basis of more precise motions and a firmer grip. Among elephants, the limbs are now strong and pillarlike, capable of supporting a great deal of weight. Limbs degenerated to nubs in pythons and boa constrictors, and to nothing at all in other snakes.

FIGURE 11.12 Morphological divergence among vertebrate forelimbs, starting with the bones of a stem reptile. The number and position of many skeletal elements were preserved when these diverse forms evolved; notice the bones of the forearms. Certain bones were lost over time in some of the lineages (compare the digits numbered 1 through 5). The drawings are not to the same scale.

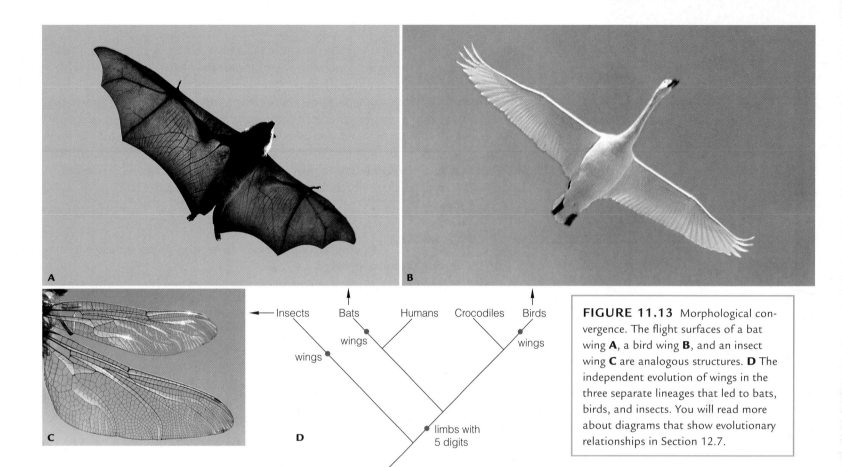

FIGURE 11.13 Morphological convergence. The flight surfaces of a bat wing **A**, a bird wing **B**, and an insect wing **C** are analogous structures. **D** The independent evolution of wings in the three separate lineages that led to bats, birds, and insects. You will read more about diagrams that show evolutionary relationships in Section 12.7.

Morphological Convergence

Similar body parts are not always homologous. Similar structures may evolve independently in separate lineages as adaptations to the same environmental pressures, in which case they are called analogous structures. **Analogous structures** look alike in different lineages but did not evolve in a shared ancestor; they evolved independently after the lineages diverged. Evolution of similar body parts in different lineages is known as **morphological convergence**.

We can sometimes identify analogous structures by studying their underlying form. For example, bird, bat, and insect wings all perform the same function: flight. However, several clues tell us that the flight surfaces of these wings are not homologous. The wing surfaces are adapted to the same physical constraints that govern flight, but the adaptations are different. In the case of birds and bats, the limbs themselves are homologous, but the adaptations that make those limbs useful for flight differ. The surface of a bat wing is a thin, membranous extension of the animal's skin. By contrast, the surface of a bird wing is a sweep of feathers, which are specialized structures derived from skin. Insect wings differ even more. An insect wing forms as a saclike extension of the body wall. Except at forked veins, the sac flattens and fuses into a thin membrane. The veins are reinforced with chitin, which structurally support the wing. The unique adaptations for flight are evidence that wing surfaces of birds, bats, and insects are analogous structures—they evolved after the ancestors of these modern groups diverged (Figure 11.13).

Comparative Embryology

The development of an embryo into the body of a plant or animal is orchestrated by layer after layer of master gene expression (Section 7.7). The failure of any single master gene to participate in this symphony of expression can result in a drastically altered body plan, typically with devastating consequences.

analogous structures Similar structures that evolved separately in different lineages.

homologous structures Similar body parts that reflect shared ancestry among lineages.

morphological convergence Evolutionary pattern in which similar body parts evolve separately in different lineages.

morphological divergence Evolutionary pattern in which a body part of an ancestor changes in its descendants.

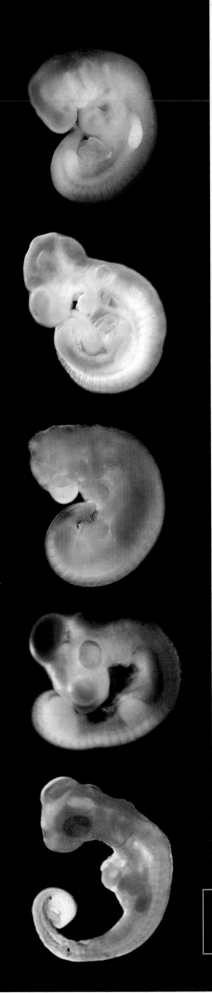

Because a mutation in a master gene typically unravels development, these genes tend to be highly conserved, which means they have changed very little or not at all over evolutionary time. Thus, a master gene with a similar sequence and function across different lineages is evidence that those lineages are related.

For example, master genes called homeotic genes guide formation of specific body parts during development. A mutation in one homeotic gene can disrupt details of the body's form, so these genes (and the patterns of development they govern) are often conserved in related lineages. When the same master genes direct development, embryos develop in similar ways (Figure 11.14).

Take-Home Message Do similarities in body form and parts indicate an evolutionary relationship?

- In morphological divergence, a body part inherited from a common ancestor becomes modified differently in different lines of descent. Such parts are called homologous structures.

- In morphological convergence, body parts that appear alike evolved independently in different lineages, not in a common ancestor. Such parts are called analogous structures.

- Similarities in patterns of embryonic development are the result of master genes that have been conserved over evolutionary time.

11.7 Biochemical Similarities

Each lineage has a unique set of traits that is a mixture of ancestral and novel characteristics, including biochemical features such as the nucleotide sequence of its DNA. Inevitable mutations change that sequence over time. The process of mutation is random, so the changes can occur anywhere in a chromosome.

We can use our knowledge of mutations to investigate how species are related. Most mutations are neutral, which means they have no effect on fitness and are not affected by natural selection. Think about what happens when one species splits into two. After the split occurs, each species begins to accumulate different neutral mutations independently. Mutations occur at a predictable rate, so the number of accumulated differences between species indicates how long ago they shared an ancestor. The more recent the divergence, the less time there has been for unique mutations to accumulate. That is why the DNA sequences of closely related species are more similar than those of distantly related ones.

Comparing the DNA of different species is now very fast and accurate, thanks to DNA sequencing and fingerprinting (Section 10.3). Comparative genomics studies have shown us (for example) that about 30 percent of the 6,609 yeast genes have homologous genes in the human genome. So do 50 percent of the 30,971 fruit fly genes and 40 percent of the 19,023 roundworm genes.

FIGURE 11.14 Comparative embryology. All vertebrates go through an embryonic stage in which they have four limb buds and a tail. From *top* to *bottom*: human, mouse, bat, chicken, and alligator embryos.

```
       honeycreeper ...CRDVQFGWLIRNLHANGASFFFICIYLHIGRGIYYGSYLNK--ETWNIGVILLLTLMATAFVGYVLPWGQMSFWG...
  Gough Island finch ...CRDVQFGWLIRNIHANGASFFFICIYLHIGRGLYYGSYLYK--ETWNVGVILLLTLMATAFVGYVLPWGQMSFWG...
        song sparrow ...CRDVQFGWLIHANGASFFFICIYLHIGRGIYYGSYLNK--ETWNVGIILLLALMATAFVGYVLPWGQMSFWG...
          deer mouse ...CRDVNYGWLIRYMHANGASMFFICLFLHVGRGMYYGSYTFT--ETWNIGIVLLFAVMATAFMGYVLPWGQMSFWG...
   Asiatic black bear ...CRDVHYGWIIRYMHANGASMFFICLFMHVGRGLYYGSYLLS--ETWNIGIILLFTVMATAFMGYVLPWGQMSFWG...
       bogue (a fish) ...CRDVNYGWLIRNLHANGASFFFICIYLHVGRGLYYGSYLYK--ETWNIGVVLLLLVMGTAFVGYVLPWGQMSFWG...
               human ...TRDVNYGWIIRYLHANGASMFFICLFLHIGRGLYYGSFLYS--ETWNIGIILLLATMATAFMGYVLPWGQMSFWG...
 thale cress (a plant) ...MRDVEGGWLLRYMHANGASMFLIVVYLHIFRGLYHASYSSPREFVWCLGVVIFLLMIVTAFIGYVLPWGQMSFWG...
       baboon louse ...ETDVMNGWMVRSIHANGASWFFIMLYSHIFRGLWVSSFTQP--LVWLSGVIILFLSMATAFLGYVLPWGQMSFWG...
        baker's yeast ...MRDVHNGYILRYLHANGASFFFMVMFMHMAKGLYYGSYRSPRVTLWNVGVIIFTLTIATAFLGYCCVYGQMSHWG...
```

Amino acid sequence comparisons are also used to determine relationships among species. Two species with many identical proteins are likely to be close relatives. Two species with few similar proteins probably have not shared an ancestor for a long time because many mutations have accumulated in the DNA of their separate lineages.

Some essential genes have evolved very little; they are highly conserved across diverse species. One such gene encodes cytochrome *b* (Figure 11.15). This protein is an important component of electron transfer chains in mitochondria. In humans, its primary structure consists of 378 amino acids.

In amino acid sequence comparisons, similarities among species are a legacy of shared ancestry. The amino acids that differ also provide clues about evolution. For example, when one amino acid replaces another of similar charge, size, and polarity, the function of the resulting protein may be unchanged. However, when one amino acid is replaced by another with very different properties, the change in the protein can be dramatic. Such nonconservative substitutions often affect phenotype, and most mutations that affect phenotype are selected against. Thus, we are more likely to see nonconservative substitutions in lineages that diverged long ago. Nonconservative substitutions, deletions, and insertions are often at the root of differences in phenotype among divergent lineages.

FIGURE 11.15 Alignment of part of the amino acid sequence of mitochondrial cytochrome *b*. This protein is a crucial component of mitochondrial electron transfer chains.

One-letter abbreviations for the amino acids are spelled out in Figure 7.5 (Section 7.4). Amino acids that differ from the honeycreeper sequence are shown in *red*. Gaps are indicated by dashes.

Figure It Out: Of the amino acid sequences shown here, which most closely matches the honeycreeper sequence?   *Answer: The Gough Island finch sequence*

Take-Home Message **How does biochemistry reflect evolutionary history?**

- DNA and amino acid sequence differences are greatest among lineages that diverged long ago, and less among lineages that diverged more recently.

11.8 Impacts/Issues Revisited:
Reflections of a Distant Past

The K–T boundary layer (*left*) consists of an unusual clay that formed 65 million years ago, worldwide (the red pocketknife is shown for scale). The clay is rich in iridium, an element rare on Earth's surface but common in asteroids. After finding the iridium, researchers looked for evidence of an asteroid big enough to cover the entire Earth with its debris. They found a crater that is about 65 million years old, buried under sediments off the coast of Mexico's Yucatán Peninsula. It is so big—273.6 kilometers (170 miles) across and 1 kilometer (3,000 feet) deep—that no one had realized it was a crater. This crater is evidence of an asteroid impact 40 million times more powerful than the one that made the Barringer Crater, certainly big enough to have influenced life on Earth in a big way.

How Would YOU Vote? Many theories and hypotheses about events in the ancient past are necessarily based on traces left by those events, not on data collected by direct observations. Is indirect evidence ever enough to prove a theory about a past event? **See CengageNow for details, then vote online (cengagenow.com).**

Summary

Section 11.1 Events of the ancient past can be explained by the same physical, chemical, and biological processes that operate in today's world.

Section 11.2 By the 19th century, naturalists on globe-spanning survey expeditions were bringing back increasingly detailed observations of nature. Studies of biogeography and comparative morphology led to new ways of thinking about the natural world.

Section 11.3 Prevailing belief systems may influence interpretation of the underlying cause of a natural event. Nineteenth-century naturalists tried to explain the accumulating evidence of evolution. Charles Darwin and Alfred Wallace came up with a theory of how species evolve:

A population tends to grow until it begins to exhaust environmental resources—food, shelter from predators, and so on. As that happens, competition for those resources intensifies among individuals of the population.

Individuals with forms of traits that make them more competitive tend to produce more offspring.

Adaptive traits (adaptations) that impart greater fitness to an individual become more common in a population over generations, compared with less competitive forms.

Differential survival and reproduction of individuals of a population that vary in the details of shared traits is called natural selection. It is one of the processes that drives evolution.

Section 11.4 Many fossils are found in stacked layers of sedimentary rock. Younger fossils usually occur in more recently deposited layers, on top of older fossils in older layers. Fossils are relatively scarce, so the fossil record will always be incomplete. Even so, it helps us to reconstruct the history of life. The characteristic half-life of a radioisotope allows us to determine the age of rocks and fossils using a technique called radiometric dating.

CENGAGENOW Learn more about fossil formation and half-life with the animation and interaction on CengageNow.

Section 11.5 Transitions in the fossil record became boundaries for great intervals of the geologic time scale. The scale correlates geologic and evolutionary events. Earth's tectonic plates raft land masses to new positions over geologic time. The movements have had profound impacts on evolution.

CENGAGENOW Investigate the geologic time scale and learn more about the drifting continents with the interaction and animation on CengageNow.

Section 11.6 Homologous structures are evidence of a common ancestor. Analogous structures are body parts that look alike in different lineages but did not evolve in a common ancestor. Similarities in patterns of embryonic development often reflect shared ancestry.

Section 11.7 We can discover and clarify evolutionary relationships through comparisons of nucleic acid and protein sequences. Neu-

tral mutations tend to accumulate in DNA at a predictable rate, so lineages that diverged recently have more nucleotide or amino acid sequences in common than ones that diverged long ago.

CENGAGENOW Use the interaction on CengageNow to learn more about amino acid sequence comparisons.

Self-Quiz   Answers in Appendix I

1. Which of the following is not part of natural selection?
   a. environmental resources   c. neutral mutations
   b. differences in traits   d. competition among individuals

2. The number of species on an island depends on the size of the island and its distance from a mainland. This statement would most likely be made by _____ .
   a. an explorer   c. a geologist
   b. a biogeographer   d. a philosopher

3. The bones of a bird's wing are similar to the bones in a bat's wing. This observation is an example of _____ .
   a. uniformity   c. comparative morphology
   b. evolution   d. a lineage

4. Evolution is _____ .
   a. natural selection   c driven by natural selection
   b. change in a line of descent d. b and c are correct

5. If the half-life of a radioisotope is 20,000 years, then a sample in which three-quarters of that radioisotope has decayed is _____ years old.
   a. 15,000   c. 30,000
   b. 26,667   d. 40,000

6. _____ has/have influenced the fossil record.
   a. Sedimentation and compaction
   b. Tectonic plate movements
   c. Prevailing belief systems
   d. a and b

7. Evidence suggests that life originated in the _____ .
   a. Archaean   c. Phanerozoic
   b. Proterozoic   d. Cambrian

8. The Cretaceous ended _____ million years ago.

9. Forces of geologic change include _____ (select all that are correct).
   a. erosion   e. tectonic plate movements
   b. fossilization   f. wind
   c. volcanoes   g. asteroid impacts
   d. evolution   h. hot spots

10. Through _____ , a body part of an ancestor is modified differently in different lines of descent.
    a. analogous structures   c. morphological convergence
    b. homologous structures   d. morphological divergence

11. Homologous structures among major groups of organisms may differ in _____ .
    a. size   c. function
    b. shape   d. all of the above

12. Some mutations are neutral because they do not affect _____ .
    a. amino acid sequence
    b. nucleotide sequence
    c. the chances of survival
    d. body form

13. By altering steps in the program by which embryos develop, a mutation in a _____ may lead to major differences in body form between related lineages.
    a. coprolite
    b. homeotic gene
    c. homologous structure
    d. all of the above

14. Match the terms with the most suitable description.
    _____ fitness
    _____ fossils
    _____ natural selection
    _____ half-life
    _____ catastrophism
    _____ uniformity
    _____ analogous structures
    _____ sedimentary rock
    _____ homologous structures

    a. measured by reproductive success
    b. geologic change occurs continuously
    c. geologic change occurs in sudden major events
    d. good for finding fossils
    e. survival of the fittest
    f. characteristic of a radioisotope
    g. insect wing and bird wing
    h. human arm and bird wing
    i. evidence of life in distant past

Additional questions are available on CENGAGENOW™

Critical Thinking

1. Radiometric dating does not measure the age of an individual atom. It is a measure of the age of a quantity of atoms—a statistic. As with any statistical measure, its values may deviate around an average (sampling error, Section 1.7). Imagine that one sample of rock is dated ten different ways. Nine of the tests yield an age close to 225,000 years. One test yields an age of 3.2 million years. Do the nine consistent results imply that the one that deviates is incorrect, or does the one odd result invalidate the nine that are consistent?

2. If you think of geologic time spans as minutes, life's history might be plotted on a clock such as the one shown on the right. According to this clock, the most recent epoch started in the last 0.1 second before noon. Where does that put you?

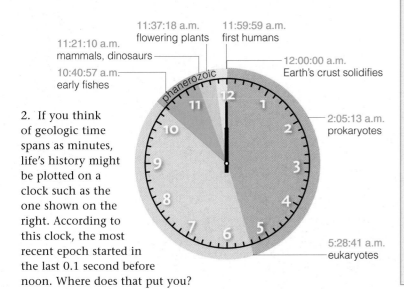

11:37:18 a.m. flowering plants
11:59:59 a.m. first humans
11:21:10 a.m. mammals, dinosaurs
10:40:57 a.m. early fishes
12:00:00 a.m. Earth's crust solidifies
phanerozoic
2:05:13 a.m. prokaryotes
5:28:41 a.m. eukaryotes

Digging Into Data

Abundance of Iridium in the K–T Boundary Layer

In the late 1970s, geologist Walter Alvarez was investigating the composition of the 1-centimeter-thick layer of clay that marks the Cretaceous–Tertiary (K–T) boundary all over the world. He asked his father, Nobel Prize–winning physicist Luis Alvarez, to help him analyze the elemental composition of the layer. The photo shows Luis and Walter Alvarez with a section of the K–T boundary layer.

The Alvarezes and their colleagues tested the layer in Italy and in Denmark. The researchers discovered that the K–T boundary layer had a much higher iridium content than the surrounding rock layers. Some of their results are shown in the table in Figure 11.16.

Iridium belongs to a group of elements (Appendix VI) that are much more abundant in asteroids and other solar system materials than they are in Earth's crust. The Alvarez group concluded that the K–T boundary layer must have originated with extraterrestrial material. They calculated that an asteroid 14 kilometers (8.7 miles) in diameter would contain enough iridium to account for the iridium in the K–T boundary layer.

1. What was the iridium content of the K–T boundary layer?

2. How much higher was the iridium content of the boundary layer than the sample taken 0.7 meters above the layer?

Sample Depth	Average Abundance of Iridium (ppb)
+ 2.7 m	< 0.3
+ 1.2 m	< 0.3
+ 0.7 m	0.36
boundary layer	41.6
− 0.5 m	0.25
− 5.4 m	0.30

FIGURE 11.16 Abundance of iridium in and near the K–T boundary layer in Stevns Klint, Denmark. Many rock samples taken from above, below, and at the boundary layer were tested for iridium content. Depths are given as meters above or below the boundary layer.

The iridium content of an average Earth rock is 0.4 parts per billion (ppb) of iridium. An average meteorite contains about 550 parts per billion of iridium.

12 Processes of Evolution

12.1

Impacts/Issues: Rise of the Super Rats

Slipping in and out of the pages of human history are rats—*Rattus*—the most notorious of mammalian pests. Rats thrive in urban centers, where garbage is plentiful and natural predators are not (Figure 12.1). The average city in the United States sustains about one rat for every ten people. Part of their success stems from an ability to reproduce very quickly. Rat populations can expand within weeks to match the amount of garbage available for them to eat. Unfortunately for us, rats carry pathogens and parasites that cause bubonic plague, typhus, and other infectious diseases. They chew their way through walls and wires, and eat or foul 20 to 30 percent of our total food production. Rats cost us about $19 billion per year.

FIGURE 12.1 Rats thrive wherever people do. Dousing buildings and soil with poisons does not usually exterminate rat populations, which recover quickly. The practice selects for rats that are resistant to the poisons.

For years, people have been fighting back with poisons, including arsenic and cyanide. Baits laced with warfarin, an organic compound that interferes with blood clotting, were popular in the 1950s. Rats that ate the poisoned baits died within days after bleeding internally or losing blood through cuts or scrapes. Warfarin was very effective, and compared to other rat poisons, it had much less impact on harmless species. It quickly became the rodenticide of choice. In 1958, however, a Scottish researcher reported that warfarin was not working against some rats. Similar reports from other European countries followed. About twenty years later, about 10 percent of rats caught in urban areas of the United States were resistant to warfarin. What happened?

To find out, researchers compared warfarin-resistant rats with still-vulnerable rats. They traced the difference to a gene on one of the rat chromosomes. Certain mutations in the gene were common among warfarin-resistant rat populations but rare among vulnerable ones. Warfarin inhibits the gene's product, an enzyme that recycles vitamin K after it has been used to activate blood clotting factors. The mutations made the enzyme insensitive to warfarin.

"What happened" was evolution by natural selection. As warfarin exerted pressure on rat populations, the populations changed. The previously rare mutations became adaptive. Rats that had an unmutated gene died after eating warfarin. The lucky ones that had one of the mutations survived and passed it to their offspring. The rat populations recovered quickly, and a higher proportion of rats in the next generation carried the mutations. With each onslaught of warfarin, the frequency of the mutation in rat populations increased.

Selection pressures can and often do change. When warfarin resistance increased in rat populations, people stopped using warfarin. The frequency of the mutation in rat populations declined, probably because rats with the mutation are not as healthy as normal rats. Now, savvy exterminators in urban areas know that the best way to control a rat infestation is to exert another kind of selection pressure: Remove their source of food, which is usually garbage. Then the rats will eat each other.

FIGURE 12.2 Sampling of phenotypic variation in humans. Variation in shared traits among individuals is an outcome of variations in genes that influence the traits.

12.2 Making Waves in the Gene Pool

Remember from Section 1.2 that a population is a group of individuals of the same species in some specified area. The individuals of a species—and a population—share certain features. For example, giraffes normally have very long necks, brown spots on white coats, and so on. These are examples of morphological traits (*morpho–* means form). Individuals of a species also share physiological traits, such as metabolic activities. They also respond the same way to certain stimuli, as when hungry giraffes feed on tree leaves. These are behavioral traits. Genotype gives rise to phenotype, so individuals of a population have the same traits because they have the same genes. Together, the genes of a population comprise a pool of genetic resources called a **gene pool**.

Almost every shared trait varies a bit among individuals of a population (Figure 12.2). Alleles of shared genes are the main source of this variation. Many traits have two or more distinct forms, or morphs. A trait with only two forms is dimorphic (*di–* means two). Purple and white flower color in the pea plants that Gregor Mendel studied is an example of a dimorphic trait (Section 9.2). Dimorphic flower color occurs in this case because the interaction of two alleles with a clear dominance relationship gives rise to the trait. Traits with more than two distinct forms are polymorphic (*poly–*, many). Human blood type, which is determined by the codominant *ABO* alleles, is one example (Section 9.3). Traits that vary continuously among the individuals of a population often arise by interactions among alleles of several genes, and may be influenced by environmental factors (Section 9.4).

In earlier chapters, you learned about the processes that introduce variation in traits among individuals of a species. Table 12.1 summarizes the key events involved. Mutation is the source of new alleles. Other events shuffle the alleles into different combinations, but what a shuffle that is! There are $10^{116,446,000}$ possible combinations of human alleles. Not even 10^{10} people are living today. Unless you have an identical twin, it is unlikely that another person with your precise genetic makeup has ever lived, or ever will.

Mutation Revisited Being the original source of new alleles, mutations are worth another look—this time within the context of their impact on populations. We cannot predict when or in which individual a particular gene will mutate. We can, however, predict the average mutation rate of a species, which is the probability that a mutation will occur in a given interval. In humans, that rate is about 175 mutations per person per generation.

Many mutations give rise to structural, functional, or behavioral alterations that reduce an individual's chances of surviving and reproducing. Even one

Table 12.1	Sources of Variation in Traits Among Individuals of a Species
Genetic Event	Effect
Mutation	Source of new alleles
Crossing over at meiosis I	Introduces new combinations of alleles into chromosomes
Independent assortment at meiosis I	Mixes maternal and paternal chromosomes
Fertilization	Combines alleles from two parents
Changes in chromosome number or structure	Transposition, duplication, or loss of chromosomes

gene pool All of the genes in a population.

No pattern of evolution is purposeful. Evolution simply fills the nooks and crannies of opportunity.

biochemical change may be devastating. For instance, the skin, bones, tendons, lungs, blood vessels, and other vertebrate organs incorporate the protein collagen. If the collagen gene mutates in a way that changes the protein's function, the entire body may be affected. A mutation that drastically changes phenotype is called a **lethal mutation** because it results in death.

A **neutral mutation** changes the base sequence in DNA, but the alteration has no effect on survival or reproduction. It neither helps nor hurts the individual. For instance, if you carry a mutation that keeps your earlobes attached to your head instead of swinging freely, attached earlobes should not in itself stop you from surviving and reproducing as well as anybody else. So, natural selection does not affect the frequency of this particular mutation in a population.

Occasionally, a change in the environment favors a mutation that had previously been neutral or even somewhat harmful. The warfarin resistance gene in rats is an example. Even if a beneficial mutation bestows only a slight advantage, natural selection can increase its frequency in a population over time because it operates on traits with a genetic basis. Natural selection, remember, is the differential survival and reproduction among individuals of a population that vary in the details of their shared traits (Section 11.3).

Mutations have been altering genomes for billions of years, and they are still at it. Cumulatively, they have given rise to Earth's staggering biodiversity. Think about it. The reason you do not look like an avocado or an earthworm or even your next-door neighbor began with mutations that occurred in different lines of descent.

Allele Frequencies **Allele frequency** refers to an allele's abundance among individuals of a population. Change in allele frequency is the same thing as change in a line of descent—evolution. Evolution within a population or species is called **microevolution**.

A theoretical reference point, **genetic equilibrium**, occurs when the allele frequencies of a population do not change (in other words, the population is not evolving). Genetic equilibrium can only occur if every one of the following five conditions are met: (1) mutations never occur; (2) the population is infinitely large; (3) the population is isolated from all other populations of the species (no gene flow); (4) mating is random; and (5) all individuals survive and produce the same number of offspring. As you can imagine, all five conditions are never met in nature, so natural populations are never in equilibrium. Microevolution is always occurring in natural populations, because the processes that drive it are always in play. The remaining sections of this chapter explore those processes—mutation, natural selection, genetic drift, and gene flow—and their effects. Remember, even though we can recognize patterns of evolution, none of them are purposeful. Evolution simply fills the nooks and crannies of opportunity.

allele frequency Abundance of a particular allele among members of a population.

directional selection Mode of natural selection in which phenotypes at one end of a range of variation are favored.

genetic equilibrium Theoretical state in which a population is not evolving.

lethal mutation Mutation that drastically alters phenotype; causes death.

microevolution Change in allele frequencies in a population or species.

neutral mutation A mutation that has no effect on survival or reproduction.

Take-Home Message What mechanisms drive evolution?

- We partly characterize a natural population by shared morphological, physiological, and behavioral traits. Most of those traits have a heritable basis.

- Different alleles are the basis of differences in the details of a population's shared traits.

- Alleles of all individuals in a population comprise the population's gene pool.

- Natural populations are always evolving, which means that allele frequencies in their gene pool are always changing over generations.

- Microevolution is evolutionary change in a population or species.

Natural Selection Revisited

12.3

With natural selection (Section 11.3), differential survival and reproduction alters allele frequencies of a population over many generations. We observe different patterns of natural selection depending on the selection pressures and organisms involved. For example, individuals with a trait at one extreme of a range of variation may be selected against, and those at the other extreme favored. In other cases, midrange forms are favored, and the extremes are selected against. Or, forms at the extremes of the range of variation are favored, and intermediate forms are selected against.

Directional Selection With **directional selection**, allele frequencies shift in a consistent direction, so forms at one end of a range of phenotypic variation become more common over time (Figure 12.3). The following examples show how field observations provide evidence of directional selection.

Peppered moths feed and mate at night, and rest motionless on trees during the day. Their behavior and coloration camouflage them from day-flying, moth-eating birds. Light-colored moths were the most common form in preindustrial England. A dominant allele that resulted in the dark color was rare. The air was clean, and light-gray lichens grew on the trunks and branches of most trees. Light moths were camouflaged when they rested on the lichens (Figure 12.4**A**), but dark moths were not (Figure 12.4**B**).

By the 1850s, the dark moths were appearing more frequently. Why? The industrial revolution had begun, and smoke from coal-burning factories was beginning to change the environment. Air pollution was killing the lichens. Researchers hypothesized that dark moths were now better camouflaged from predators on the soot-darkened trees than light moths (Figure 12.4**C,D**).

In the 1950s, H. B. Kettlewell bred both moth forms in captivity and marked hundreds so that they could be easily identified after being released in the wild. He released them near highly industrialized areas around Birmingham and near an unpolluted part of Dorset. His team recaptured more of the dark moths in the polluted area and more light

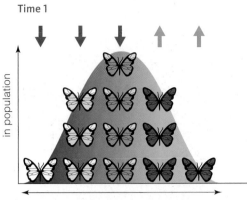

Time 1

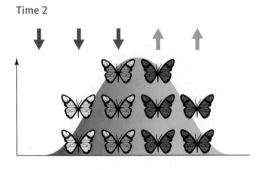

Time 2

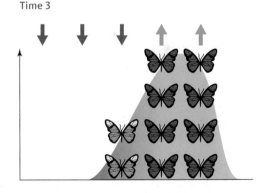

Time 3

FIGURE 12.3 Animated! Directional selection. The bell curves indicate continuous variation in a butterfly wing-color trait. *Red* arrows indicate which forms are being selected against; *green*, forms that are being favored.

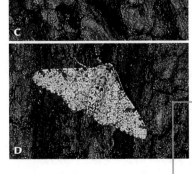

FIGURE 12.4 Directional selection of two forms of a trait in different settings. **A** Light peppered moths on a nonsooty tree trunk are hidden from predators. **B** Dark ones stand out. In places where soot darkens tree trunks, the dark color **C** is more adaptive than **D** the light color.

FIGURE 12.5 Directional selection in populations of rock pocket mice.

1 Mice with dark fur are more common in areas with dark basalt rock.

2 Mice with light fur are more common in areas with light-colored granite.

3 Mice with coat colors that do not match their surroundings are more easily seen by predators, so they are preferentially eliminated from the populations.

stabilizing selection Mode of natural selection in which intermediate phenotypes are favored over extremes.

ones near Dorset. They also observed predatory birds eating more light moths in Birmingham, and more dark moths in Dorset.

Pollution controls went into effect in 1952. Tree trunks became free of soot, and lichens made a comeback. Moth phenotypes shifted in the reverse direction: Wherever pollution decreased, the frequency of dark moths decreased as well. Many other researchers since Kettlewell have confirmed the rise and fall of the dark-colored form of the peppered moth.

Directional selection also affects the color of rock pocket mice in Arizona's Sonoran Desert. Rock pocket mice are small mammals that spend the day sleeping in underground burrows, emerging at night to forage for seeds (Figure 12.5). The Sonoran Desert is dominated by outcroppings of light brown granite. There are also patches of dark basalt rock, the remains of ancient lava flows. Most of the mice in populations that inhabit the dark rock have dark gray coats **1**. Most of the mice in populations that inhabit the light brown granite have light brown coats **2**. Why the difference? Individuals that match the rock color in each habitat are camouflaged from their natural predators. Night-flying owls more easily see mice that do not match the rocks **3**, and they preferentially eliminate easily seen mice from each population. Predation by owls causes a directional shift in the frequency of alleles that affect coat color.

Researchers analyzed genes of mice from granite and basalt populations to find out which genes predation is affecting. They found that allele frequencies differed at a gene that influences deposition of the dark pigment melanin. Compared to granite-dwelling populations, populations living on basalt have a much higher frequency of four mutations that cause increased melanin deposition.

Our attempts to control the environment can result in directional selection, as is the case with the warfarin-resistant rats. The use of antibiotics is another example. When your grandparents were still young, scarlet fever, tuberculosis, and pneumonia caused one-fourth of the annual deaths in the United States alone. Since the 1940s, we have been relying on antibiotics such as penicillin to fight these and other dangerous bacterial diseases. We also use them in other, less dire circumstances. Antibiotics are used preventively, both in humans and in livestock. They are part of the daily rations of millions of cattle, pigs, chickens, fish, and other animals that are raised on factory farms.

Bacteria evolve at an accelerated rate compared with humans, in part because they reproduce very quickly. For example, the common intestinal bacteria *E. coli* can divide every 17 minutes. Each new generation is an opportunity for mutation, so the gene pool of a bacterial population varies greatly. Thus, it is very likely that some cells will have alleles that allow them to survive an antibiotic treatment. Then, natural selection takes over. A typical two-week course of antibiotics can potentially exert selection pressure on over a thousand generations of bacteria, and antibiotic-resistant strains may be the outcome.

Antibiotic-resistant bacteria have plagued hospitals for many years, and now they are becoming similarly common in schools. Even as researchers scramble to find new antibiotics, this trend is bad news for the millions of people each year who contract cholera, tuberculosis, or another dangerous bacterial disease.

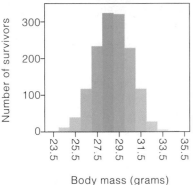

FIGURE 12.7 Stabilizing selection in sociable weavers (*top*). Graph shows the number of birds (out of 977) that survived a breeding season.

Figure It Out: What is the optimal weight of a sociable weaver bird?　　*Answer: About 29 grams*

Stabilizing Selection With **stabilizing selection**, an intermediate form of a trait is favored, and extreme forms are not. This mode of selection tends to preserve the midrange phenotypes in a population (Figure 12.6). For example, the body weight of sociable weaver birds is subject to stabilizing selection. Weaver birds build large communal nests in the African savanna. Between 1993 and 2000, Rita Covas and her colleagues captured, tagged, weighed, and released birds living in communal nests before the breeding season began. The researchers then recaptured and weighed the surviving birds after the breeding season was over.

Covas's field studies indicated that body weight in sociable weavers is a trade-off between the risks of starvation and predation (Figure 12.7). Foraging is not easy in the sparse habitat of an African savanna, and leaner birds do not store enough fat to avoid starvation. A meager food supply selects against birds with low body weight. Fatter birds may be more attractive to predators, and not as agile when escaping. Predators select against birds of high body weight. Thus, birds of intermediate weight have the selective advantage, and make up the bulk of sociable weaver populations.

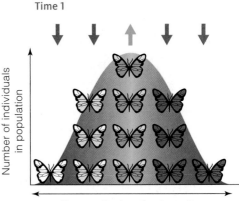

Time 1

Range of values for the trait

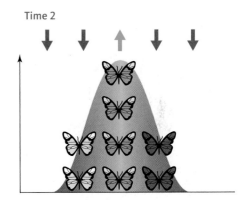

Time 2

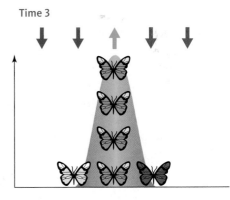

Time 3

FIGURE 12.6 Animated! Stabilizing selection eliminates extreme forms of a trait, and maintains the predominance of an intermediate phenotype in a population. *Red* arrows indicate which forms are being selected against; *green*, forms that are being favored. Compare the data set from a field experiment shown in Figure 12.7.

Time 1

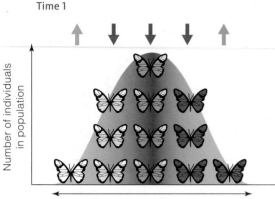

Number of individuals in population

Range of values for the trait

Time 2

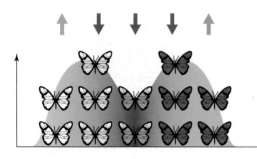

Time 3

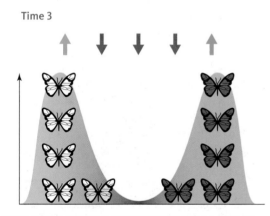

FIGURE 12.8 Animated! Disruptive selection eliminates midrange forms of a trait, and maintains extreme forms.

Disruptive Selection With **disruptive selection**, forms of a trait at both ends of a range of variation are favored, and intermediate forms are selected against (Figure 12.8). Consider the black-bellied seedcracker native to Cameroon, Africa. These finches are dimorphic for bill size. They tend to have either a large bill or a small one—but no sizes between (Figure 12.9). It is as if every human adult were four feet or six feet tall, with no one of intermediate height. Large-billed and small-billed birds live side by side, and mate with one another without regard to bill size.

Factors that affect seedcracker feeding performance maintain the dimorphism in bill size. The finches feed mainly on the seeds of two types of sedge, which is a grasslike plant. One sedge produces hard seeds; the other, soft seeds. Small-billed birds are better at opening the soft seeds, and large-billed birds are better at cracking the hard ones. During Cameroon's wet seasons, all seeds are abundant, and all seedcrackers feed on both types of seeds. During the region's dry seasons, sedge seeds are scarce, and each bird focuses on eating the seeds that it opens most efficiently. Small-billed birds feed mainly on soft seeds, and large-billed birds feed mainly on hard seeds. Birds with intermediate-sized bills cannot open either type of seed as efficiently as the other birds, so they are less likely to survive the dry seasons.

lower bill 12 mm wide

lower bill 15 mm wide

FIGURE 12.9
Disruptive selection in African seedcrackers. Birds with bills that are about 12 *or* 15 millimeters wide are favored. The difference is a result of competition for scarce food during dry seasons.

Take-Home Message **How does natural selection drive evolution?**

■ With directional selection, allele frequencies underlying a range of variation shift in a consistent direction in response to selection pressure.

■ With stabilizing selection, an intermediate phenotype is favored, and extreme forms are eliminated.

■ With disruptive selection, an intermediate form of a trait is selected against, and extreme phenotypes are favored.

12.4 Factors That Affect Variation in Traits

Selection pressures that operate on natural populations are typically not as clear-cut as the examples in the previous section might suggest. Non-ecological pressures, including random genetic change, are part of natural selection. Complex interactions among genes and the environment also play a role.

Nonrandom Mating Not all natural selection occurs as a result of interactions between a species and its environment. Competition within a species also drives evolution. Consider how the individuals of many sexually reproducing species have a distinct male or female phenotype—a sexual dimorphism. Individuals of one sex (often males) tend to be more colorful, larger, and more

aggressive than individuals of the other sex. These traits seem puzzling because they take energy and time away from an individual's survival activities. Some are probably maladaptive because they attract predators. Why do they persist?

The answer is **sexual selection**, in which the genetic winners outreproduce others of a population because they are better at securing mates. In sexual selection, the most adaptive forms of a trait are those that help individuals defeat same-sex rivals for mates, or are the ones most attractive to the opposite sex.

By choosing among mates, a male or female acts as a selective agent on its own species. For example, the females of some species will shop for a mate among males that vary slightly in appearance and courtship behavior. Females choose males that display species-specific cues that often include flashy body parts or behaviors (Figure 12.10**A**). Such traits can be a physical hindrance and they may attract predators. However, a flashy male's survival despite his obvious handicap implies health and vigor, two traits that are likely to improve a female's chances of bearing healthy, vigorous offspring. The selected males pass the alleles for their attractive traits to the next generation of males. Females pass alleles that influence mate preference to the next generation of females.

Gerald Wilkinson and his colleagues demonstrated female preference for an exaggerated male trait in the stalk-eyed fly. The eyes of this Malaysian species form on long, horizontal eyestalks that provide no obvious adaptive advantage to their bearers, other than perhaps provoking sexual interest in other flies (Figure 12.10**B**). The researchers predicted that if eyestalk length is a sexually selected trait, then males with longer eyestalks would be more attractive to female flies than males with shorter eyestalks. They bred males with extra-long eyestalks, and found those males were indeed preferred by the female flies. Such experiments show how exaggerated traits can arise by sexual selection in nature.

The females of some species cluster in defendable groups when they are sexually receptive. When they do, males are likely to compete for access to clusters. Competition for ready-made harems favors combative males (Figure 12.10**C**).

FIGURE 12.10 Sexual selection in action.

A This male bird of paradise (*Paradisaea raggi ana*) is engaged in a flashy courtship display. He caught the eye (and, perhaps, the sexual interest) of the smaller, less colorful female. The males of this species of bird compete fiercely for females, which are selective. (Why do you suppose the females are not as flashy as the males?)

B Ooh, what sexy eyestalks! Stalk-eyed flies cluster on aerial roots to mate, and females cluster preferentially around males with the longest eyestalks. This photo, taken in Kuala Lumpur, Malaysia, shows a male with very long eyestalks (*top*) that has captured the interest of the three females below him.

C Male elephant seals fight for sexual access to a cluster of females.

disruptive selection Mode of natural selection that favors two forms in a range of variation; intermediate forms are selected against.

sexual selection Mode of natural selection in which some individuals outreproduce others of a population because they are better at securing mates.

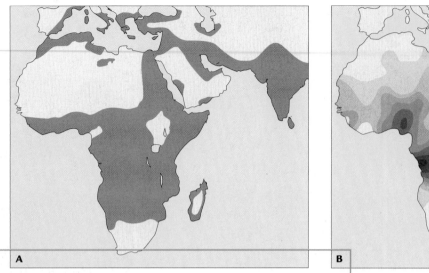

FIGURE 12.11 Malaria and sickle-cell anemia. **A** Distribution of malaria cases (*orange*) reported in Africa, Asia, and the Middle East in the 1920s, before the start of programs to control mosquitoes, which transmit *Plasmodium*. **B** Distribution (by percentage) of people that carry the sickle-cell allele. Notice the correlation between the maps. **C** Physician searching for mosquito larvae in Southeast Asia.

Legend for map B:
- 0%–2%
- 2%–4%
- 4%–6%
- 6%–8%
- 8%–10%
- 10%–12%
- 12%–14%
- more than 14%

balanced polymorphism Maintenance of two or more alleles for a trait in some populations, as a result of natural selection against homozygotes.

bottleneck Reduction in population size so severe that it reduces genetic diversity.

fixed Refers to an allele for which all members of a population are homozygous.

founder effect Change in allele frequencies that occurs after a small number of individuals establish a population.

genetic drift Change in allele frequencies in a population due to chance alone.

Balanced Polymorphism Any mode of natural selection may maintain two or more alleles of a gene in a population. This state, which is called **balanced polymorphism**, occurs when environmental conditions favor heterozygotes (individuals with nonidentical alleles of a gene). For example, certain environments select for balanced polymorphism of the human *Hb* gene. This gene encodes the beta globin chain of hemoglobin, the oxygen-transporting protein in blood. Hb^A is the normal allele, and the Hb^S allele carries a mutation that causes homozygotes to develop sickle-cell anemia (Section 7.6). Individuals homozygous for the Hb^S allele often die in their teens or early twenties.

Despite being so harmful, the Hb^S allele persists at very high frequency among the human populations in tropical and subtropical regions of Asia and Africa. Why? Populations with the highest frequency of the Hb^S allele also have the highest incidence of malaria (Figure 12.11). Mosquitoes transmit the parasitic agent of malaria, *Plasmodium*, to human hosts. The protozoan multiplies in the liver and then in red blood cells. The cells rupture and release new parasites during severe, recurring bouts of illness.

It turns out that Hb^A/Hb^S heterozygotes are more likely to survive malaria than people who make only normal hemoglobin. Several mechanisms are possible. For example, infected cells of heterozygotes take on a sickle shape. The abnormal shape brings infected cells to the attention of the immune system, which destroys them—along with the parasites they harbor. Infected cells of Hb^A/Hb^A homozygotes do not sickle, and the parasite may remain hidden from the immune system.

The persistence of the Hb^S allele may be a matter of relative evils. Malaria and sickle-cell anemia are both potentially deadly. In areas where malaria is common, Hb^A/Hb^S heterozygotes are more likely to survive and reproduce than Hb^A/Hb^A homozygotes. Heterozygotes are not completely healthy, but they do make enough normal hemoglobin to survive. Malaria or not, they are more likely to live long enough to reproduce than Hb^S/Hb^S homozygotes. The result is that nearly one-third of individuals that live in the most malaria-ridden regions of the world are Hb^A/Hb^S heterozygotes.

Genetic Drift

Genetic drift is a random change in allele frequencies over time, brought about by chance alone. We explain genetic drift in terms of probability—the chance that some event will occur (Section 1.7). The probability of an event occurring is expressed as a percentage. For instance, if 10 million people enter a drawing, each has the same probability of winning: 1 in 10 million, or a very improbable 0.00001 percent.

Remember, sample size is important in probability. For example, every time you flip a coin, there is a 50 percent chance it will land heads up. With 10 flips, the proportion of times heads actually land up may be very far from 50 percent. With 1,000 flips, that proportion is more likely to be near 50 percent. We can apply the same rule to populations. Because population sizes are not infinite, there will be random changes in allele frequencies. These random changes have a minor impact on large populations. However, such changes can lead to dramatic shifts in the allele frequencies of small populations.

Imagine two populations. Population I has 10 individuals, and population II has 100. Say an allele *b* occurs in both populations at a 10 percent frequency. Only one person carries the allele in population I. If that person does not reproduce, allele *b* will be lost from population I. However, ten people in population II carry the allele. All ten would have to die without reproducing for the allele to be lost from population II. Thus, the chance that population I will lose allele *b* is greater than that for population II. Steven Rich and his colleagues demonstrated this effect in populations of flour beetles (Figure 12.12).

Random change in allele frequencies can lead to the loss of genetic diversity and to the homozygous condition. These outcomes of genetic drift are possible in all populations, but they are more likely to occur in small ones. When all individuals of a population are homozygous for one allele, we say that the allele has become **fixed**. Once an allele is fixed, its frequency will not change again unless mutation or gene flow introduces new alleles.

Bottlenecks and the Founder Effect

Genetic drift can be dramatic when a few individuals rebuild a population or start a new one, such as occurs after a bottleneck. A **bottleneck** is a drastic reduction in population size brought about by severe pressure. If a contagious disease, habitat loss, or overhunting nearly wipes out a population, the allele frequencies in the survivors will have been altered at random. For example, northern elephant seals (shown in Figure 12.10**C**) were once on the brink of extinction, with only twenty known survivors in the world in the 1890s. Hunting restrictions implemented since then have allowed the population to recover to about 170,000 individuals. Every seal is homozygous for all of the genes analyzed to date.

In one type of bottleneck, a small group of individuals founds a new population. The group is not representative of the original population in terms of allele frequencies, so the new population is not representative of it either. This outcome is called the **founder effect**. If a founding group is very small, the new population's genetic diversity may be quite reduced. For instance, imagine a patch of pink and yellow lily flowers on a mainland. By chance, a seabird lands on a yellow lily, and a few seeds stick to its feathers. The bird flies to a remote island and drops the seeds. Most of the seeds have the allele for yellow flowers. The seeds sprout, and a small, isolated population of lily plants establishes itself;

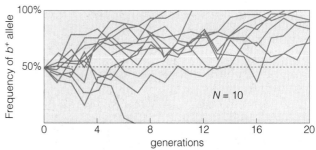

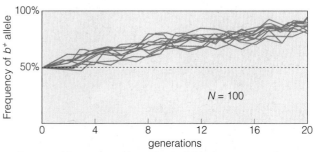

A   *N* = number of breeding individuals per generation

B   *N* = number of breeding individuals per generation

FIGURE 12.12 Animated! Genetic drift in flour beetles (*Tribolium castaneum*, shown *below* on a flake of cereal). Randomly selected beetles heterozygous for alleles *b*⁺ and *b* were maintained in populations of **A** 10 or **B** 100 individuals for 20 generations.

Graph lines in **B** are smoother than in **A**, indicating that drift was greatest in the sets of 10 beetles and least in the sets of 100. Allele *b*⁺ was lost in one population (one graph line ends at 0). Notice that the average frequency of allele *b*⁺ rose at the same rate in both groups, an indication that natural selection was at work too: Allele *b*⁺ was weakly favored.

Figure It Out: In how many populations did allele *b*⁺ become fixed?

Answer: Six

10 mm

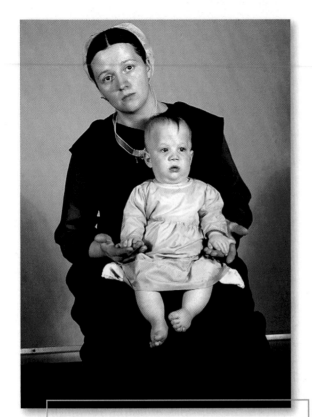

FIGURE 12.13 An Amish child with Ellis–van Creveld syndrome. The syndrome is characterized by dwarfism, polydactyly, and heart defects, among other symptoms. The recessive allele that causes it is unusually common in the Old Order Amish of Lancaster County, an outcome of the founder effect and moderate inbreeding.

most of the plants have yellow flowers. In the absence of gene flow or natural selection for flower color, genetic drift may fix the allele for yellow flowers.

Genetic drift particularly affects allele frequency in inbred populations. **Inbreeding** is nonrandom breeding or mating between close relatives, which share a large number of alleles. Inbreeding lowers the genetic diversity of a population. Loss of diversity tends to be a bad thing, because more people become homozygous for recessive alleles with harmful effects. This is one reason that most societies discourage or forbid incest, which is mating between parents and children or between siblings. However, more distant relatives such as cousins do often mate in some societies.

The Old Order Amish in Lancaster County, Pennsylvania, offer an example. Amish people marry only within their community. Intermarriage with other groups is not permitted, and no "outsiders" are allowed to join the community. As a result, Amish populations are moderately inbred, and many of their individuals are homozygous for harmful alleles. The Lancaster population has an unusually high frequency of a recessive allele associated with Ellis–van Creveld syndrome (Figure 12.13). The particular allele that affects the Lancaster population has been traced to one man and his wife, who were among a group of 400 Amish who immigrated to the United States in the mid-1700s. As a result of the founder effect and inbreeding since then, about 1 of 8 people in the population is now heterozygous for the allele, and 1 in 200 is homozygous.

Gene Flow Individuals tend to mate or breed most frequently with other members of their own population. However, most populations of a species are not completely isolated from one another, and there may be intermating among adjacent populations. Also, individuals sometimes leave one population and join another. **Gene flow**, the movement of alleles among populations, occurs in both cases. Gene flow tends to keep allele frequencies stable, so it counters the effects of mutation, natural selection, and genetic drift. It is most pronounced among populations of animals, which tend to be more mobile, but it also occurs in plant populations.

Many opponents of genetic engineering are concerned about gene flow from transgenic organisms into wild populations. The flow is already occurring: The *bt* gene and herbicide-resistance genes (Section 10.4) have been found in weeds and unmodified crop plants adjacent to test fields of transgenic plants. The long-term effects of this gene flow are currently unknown.

gene flow The movement of alleles into and out of a population, as by individuals that immigrate or emigrate.

inbreeding Nonrandom mating among close relatives.

reproductive isolation Absence of gene flow between populations; part of speciation.

speciation Process by which new species arise from existing species.

Take-Home Message **What mechanisms maintain or reduce phenotypic variation in natural populations?**

- With sexual selection, some version of a trait gives an individual an advantage over others in securing mates. Sexual dimorphism is one outcome of sexual selection.

- Environmental pressures may maintain diversity by favoring balanced polymorphism (persistence of two or more alleles at high frequency in a population).

- Genetic drift is a random change in allele frequencies over generations. The magnitude of its effect is greatest in small populations, such as one that endures a bottleneck.

- Gene flow is the physical movement of alleles into and out of a population. It tends to counter the evolutionary effects of mutation, natural selection, and genetic drift on a population.

FIGURE 12.14 Four butterflies, two species: Which are which? Two forms of the species *Heliconius melpomene* are on the *top* row; two of *H. erato* are on the *bottom* row. These two species never cross-breed. Their alternate but similar patterns of coloration evolved as a shared warning signal to local predators that the butterflies taste terrible.

12.5 Speciation

Mutation, natural selection, and genetic drift operate on all natural populations, and they do so independently in populations that are not interbreeding. When gene flow does not keep populations alike, genetic changes accumulate in each one independently. Sooner or later, the genetic divergences lead to new species. The evolutionary process by which new species arise is called **speciation**.

There are tremendous differences between species such as petunias and whales, beetles and emus, and so on. Such organisms look very different, so it is easy to tell them apart. Their separate lineages probably diverged so long ago that many changes accumulated in them. Species that share a more recent ancestor may be much more difficult to tell apart (Figure 12.14).

Evolutionary biologist Ernst Mayr defined a species as one or more groups of individuals that potentially can interbreed, produce fertile offspring, and do not interbreed with other groups. This "biological species concept" is useful but it is not universally applicable. For example, not all populations of a species actually continue to interbreed. In many cases, we may never know whether populations separated by a great distance could interbreed successfully even if they did get together. Also, populations often continue to interbreed even as they diverge into separate species. Thus, it is useful to remember that a "species" is a convenient but artificial construct of the human mind.

Every species is an outcome of its own unique history and interactions with the environment, so speciation is not a predictable process. The details differ with every speciation event, but we can identify some recurring patterns. For example, speciation is often triggered by a physical barrier that arises and cuts off gene flow between populations. However speciation happens, reproductive isolation is always part of the process.

Reproductive Isolation **Reproductive isolation** refers to the end of gene flow between populations. It is part of the process by which sexually reproducing species attain and maintain their separate identities. By preventing successful interbreeding, mechanisms of reproductive isolation reinforce differences between diverging populations. Several prezygotic mechanisms are barriers to pollination or mating, and postzygotic mechanisms result in weak or infertile hybrids (Figure 12.15).

Some populations cannot interbreed because the timing of their reproduction differs. Periodical cicadas (*left*) are examples. Cicadas feed on roots as they mature underground, then emerge to reproduce every 17 years. There are three species of cicada. Each has a sibling species with nearly identical form and behavior, except that the siblings emerge on a 13-year cycle instead of a 17-year cycle. Sibling species have the potential to interbreed, but they can only get together every 221 years!

In other cases, the size or shape of an individual's reproductive parts prevent it from mating with members of another population. For example, black sage

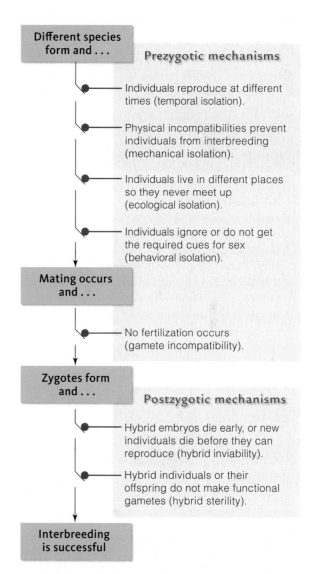

Different species form and . . .

Prezygotic mechanisms

- Individuals reproduce at different times (temporal isolation).
- Physical incompatibilities prevent individuals from interbreeding (mechanical isolation).
- Individuals live in different places so they never meet up (ecological isolation).
- Individuals ignore or do not get the required cues for sex (behavioral isolation).

Mating occurs and . . .

- No fertilization occurs (gamete incompatibility).

Zygotes form and . . .

Postzygotic mechanisms

- Hybrid embryos die early, or new individuals die before they can reproduce (hybrid inviability).
- Hybrid individuals or their offspring do not make functional gametes (hybrid sterility).

Interbreeding is successful

FIGURE 12.15 Animated! How reproductive isolating mechanisms prevent interbreeding. The mechanisms include barriers to meeting, mating, or pollination; barriers to fertilization; and reduced fitness or fertility of hybrid offspring.

anther

FIGURE 12.16 Mechanical isolation in sage.
A The pollen-bearing anthers of white sage flowers are at the tips of filaments that project high above the petals. **B** Honeybees that land on white sage flowers are too small to touch the anthers, so only larger insects such as bumblebees **C** can pollinate this species. **D** Black sage flowers are too small to support bumblebees. This species is pollinated mainly by smaller bees, including honeybees.

FIGURE 12.17 **Animated!** Behavioral isolation. Species-specific courtship displays precede sex among many birds, including these albatrosses.

(*Salvia mellifera*) and white sage (*S. apiana*) grow in the same areas, but hybrids rarely form because the flowers of the two species have become specialized for different pollinators. The pollen-bearing parts (anthers) of white sage flowers are at the tips of long filaments that extend far above two large petals (Figure 12.16**A**). The anthers are too far above the petals to reach the back of a small bee, so honeybees and other small bees cannot pollinate white sage (Figure 12.16**B**). Thus, this species is pollinated mainly by bumblebees and other bees that are big enough to brush the anthers (Figure 12.16**C**). Large bees have difficulty finding footing on the tiny flowers of black sage, so this species is pollinated mainly by smaller bees (Figure 12.16**D**).

Populations adapted to different microenvironments in the same region may be ecologically isolated. For example, two species of manzanita (a plant) native to the Sierra Nevada mountain range rarely hybridize. One species that is better adapted for conserving water inhabits dry, rocky hillsides high in the foothills. The other lives on lower slopes where water stress is not as intense. The physical separation makes cross-pollination unlikely.

In animals, behavioral differences can stop gene flow between related species. For instance, males and females of some bird species engage in courtship displays before sex (Figure 12.17). A female recognizes the vocalizations and movements of a male of her species as an overture to sex, but females of different species do not.

Even if gametes of different species do meet up, they often have molecular incompatibilities that prevent them from fusing. Gamete incompatibility may be the primary speciation route of animals that fertilize their eggs by releasing free-swimming sperm in water.

As populations begin to diverge, so do their genes. Even chromosomes of species that diverged recently may have major differences. Thus, a hybrid zygote may have extra or missing genes, or genes with incompatible products. In either case, its development probably will not proceed correctly and the resulting embryo will die prematurely. Hybrid offspring that survive may have reduced fitness, such as occurs with ligers and tigons (offspring of lions and tigers), which have more health problems and a shorter life expectancy than individuals of either parent species.

Some interspecies crosses produce robust but sterile offspring. For example, the offspring of a female horse (64 chromosomes) mated with a male donkey (62 chromosomes) is a mule. The mule's 63 chromosomes cannot pair up evenly during meiosis, so this animal makes few viable gametes. If hybrids are fertile, their offspring usually have lower fitness with each successive generation. A mismatch between nuclear and mitochondrial DNA may be the cause (mitochondrial DNA is inherited from the mother only).

Allopatric Speciation Genetic changes that lead to a new species usually begin with physical separation between populations, so **allopatric speciation** may be the most common way that new species form (*allo–* means different; *patria*, fatherland). By this speciation mode, a physical barrier separates two population and ends gene flow between them. Then, reproductive isolating mechanisms arise, so even if the populations meet up again their individuals could not interbreed.

Whether a geographic barrier can block gene flow depends on an organism's means of travel or the way its gametes disperse. Populations of most species are separated by distance, and gene flow between them is usually intermittent. The Great Wall of China is an example of a barrier that arose abruptly. As it was being built, the wall cut off gene flow among nearby populations of wind-pollinated plants. DNA sequences show that trees, shrubs, and herbs on either side of the wall are now diverging genetically.

Geographic isolation can occur slowly, such as when plate tectonics (Section 11.5) caused the Isthmus of Panama to form. This land bridge between continents cut off the flow of water—and gene flow among populations of aquatic organisms—as it separated one large ocean into what are now the Pacific and the Atlantic oceans (Figure 12.18).

Sympatric Speciation In **sympatric speciation**, populations inhabiting the same geographic region speciate in the absence of a physical barrier between them (*sym–* means together). Sympatric speciation can occur in an instant with a change in chromosome number. Nondisjunction during meiosis can change the chromosome number at fertilization (Section 9.7). Or, the chromosome number doubles when body cells duplicate their chromosomes but fail to divide during mitosis. In plants, the resulting polyploid cells can proliferate to form shoots and flowers. If the flowers self-fertilize, a new species may be the result.

Sometimes a new species arises by chromosome multiplication in one parent. In other cases, a species originates after related species hybridize, and then the chromosome number multiplies in the offspring. Common bread wheat is an example (Figure 12.19).

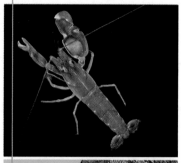

FIGURE 12.18 Allopatric speciation in snapping shrimp (*right*). As the Isthmus of Panama (*below*) formed about 3.5 million years ago, it cut off gene flow among populations of these aquatic shrimp. Individuals from opposite sides of the isthmus are so similar that they might interbreed, but they are now behaviorally isolated: Instead of mating when they are brought together, they snap their claws at one another aggressively.

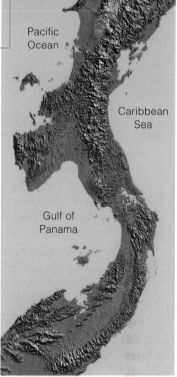

Pacific Ocean

Caribbean Sea

Gulf of Panama

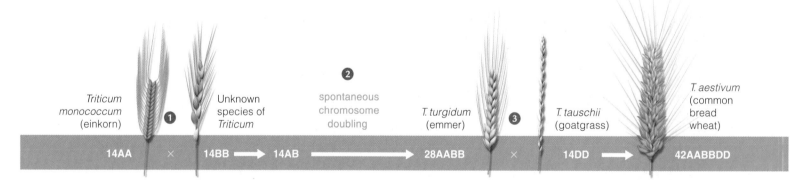

Triticum monococcum (einkorn)	❶	Unknown species of Triticum		❷ spontaneous chromosome doubling	T. turgidum (emmer)	❸	T. tauschii (goatgrass)	T. aestivum (common bread wheat)
14AA	×	14BB ➜ 14AB			28AABB	×	14DD ➜	42AABBDD

FIGURE 12.19 Animated! Sympatric speciation in wheat. ❶ Einkorn has a diploid chromosome number of 14 (two sets of 7, shown here as 14 AA). Wild einkorn probably hybridized with another wild species having the same chromosome number (14 BB) about 11,000 years ago. The resulting hybrid was diploid (14 AB).

❷ About 8,000 years ago, the chromosome number of an AB hybrid plant spontaneously doubled. The resulting species, emmer, is tetraploid: it has two sets of 14 chromosomes (28 AABB). ❸ Emmer probably hybridized with a wild goatgrass having a diploid chromosome number of 14 (two sets of 7 DD). The resulting common bread wheat has six sets of 7 chromosomes (42 AABBDD).

allopatric speciation Speciation pattern in which a physical barrier that separates members of a population ends gene flow between them.

sympatric speciation Pattern in which speciation occurs in the absence of a physical barrier.

FIGURE 12.20 Red fish, blue fish: sympatric speciation in cichlids. *Above*, Lake Victoria, Africa. *Below*, five of the hundreds of cichlids that speciated in this lake. Mutations in genes that affect females' perception of color in particular regions of the lake also affect their choice of mates.

Figure It Out: What form of natural selection has been driving sympatric speciation in Lake Victoria cichlids?

Answer: Sexual selection

Polyploidy occurs more often in plants than in animals. About 95 percent of ferns originated by polyploidy; 70 percent of flowering plants are now polyploid, as well as a few conifers, insects and other arthropods, mollusks, fishes, amphibians, and reptiles.

Plant breeders often artificially induce polyploidy in plants by treating buds or seeds with colchicine. This microtubule poison prevents assembly of a spindle, so chromosomes cannot separate during mitosis or meiosis. Polyploid plants tend to be larger and more robust than diploids. Backcrossing a polyploid plant with its parent species usually produces infertile offspring, because the mismatched chromosomes pair abnormally during meiosis. This type of cross is used to produce sterile offspring that are valued for agriculture, such as seedless watermelons and citrus fruits.

Sympatric speciation also occurs with no chromosome number change. For example, more than 500 extremely diverse species of cichlid, a freshwater fish, arose with no physical barrier to gene flow in the shallow waters of Lake Victoria. This large freshwater lake sits isolated from river inflow on an elevated plain in Africa's Great Rift Valley. Since it formed about 400,000 years ago, the lake has dried up completely three times.

DNA sequence comparisons indicate that almost all of the cichlid species in Lake Victoria arose there since the last time the lake went dry, within the last 12,400 years. How could more than 500 species of fish arise so quickly in a lake? The answer begins with differences in the color of ambient light and water clarity in different parts of the lake. Water absorbs blue light, so the deeper it is, the less blue light penetrates it. The light in shallower, clear water is mainly blue; the light that penetrates deeper, muddier water is mainly red.

Cichlids vary in color and in patterning (Figure 12.20). Females rarely mate with males of other species outside of captivity, preferring instead to mate with brightly colored males of their own species. Their preference has a molecular basis in genes that encode light-sensitive pigments of the eye. The pigments made by species that live mainly in shallower, clear water are more sensitive to blue light. The males of these species are also the bluest. Variant pigments made by species that live mainly in deeper, murkier water are more sensitive to red light. The males of these species are redder. In other words, the colors that a female cichlid sees best are the same colors displayed by males of its species. Thus, mutations in genes that affect color perception are likely to affect the choice of mates and of habitats. Such mutations are probably the way sympatric speciation occurs in these fish.

Take-Home Message How do species attain and maintain separate identities?

- Speciation is an evolutionary process by which new species form. It varies in its details and duration, but reproductive isolation is always part of the process.

- With allopatric speciation, a physical barrier that intervenes between populations or subpopulations of a species prevents gene flow among them. As gene flow ends, genetic divergences give rise to new species.

- A new species may arise from a population in the absence of a physical barrier, by sympatric speciation.

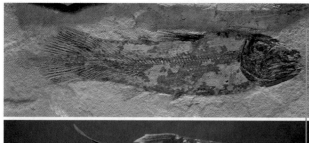

FIGURE 12.21 The coelacanth, a living fossil. *Top*, 320-million-year-old fossil found in Montana. *Bottom*, a live individual caught off the waters of Sulawesi in 1998. The coelacanth lineage has changed very little over evolutionary time.

proboscis

nectar tube

10 cm

FIGURE 12.22 Coevolved species. The orchid *Angraecum sesquipedale*, discovered in Madagascar in 1852, stores its nectar at the base of a floral tube 30 centimeters (12 inches) long. Charles Darwin predicted that someone would eventually discover an insect in Madagascar with a proboscis (mouthpart) long enough to reach the nectar and pollinate the flower. Decades later, the hawkmoth *Xanthopan morgani praedicta* was discovered in Madagascar. Its proboscis is 30–35 cm long.

12.6 Macroevolution

Microevolution is change in allele frequencies within a single species or population. **Macroevolution** is our name for evolutionary patterns on a larger scale. Flowering plants evolved from seed plants, animals with four legs (tetrapods) evolved from fish, birds evolved from dinosaurs—all of these are examples of macroevolution that occurred over millions of years.

Stasis With the simplest macroevolutionary pattern, **stasis**, lineages persist for millions of years with little or no change. For example, a type of ancient lobe-finned fish, the coelacanth, had been assumed extinct for at least 70 million years until a fisherman caught one in 1938. The modern coelacanth is very similar to fossil specimens hundreds of millions of years old (Figure 12.21).

Coevolution The process by which close ecological interactions between two species cause them to evolve jointly is **coevolution**. Each species acts as an agent of selection on the other, and each adapts to changes in the other. Over evolutionary time, the two species may become so interdependent that they can no longer survive without one another. We know of many coevolved species of predator and prey, host and parasite, pollinator and flower (Figure 12.22). Later chapters detail specific examples.

Exaptation A major evolutionary novelty typically stems from the adaptation of an existing structure for a completely different purpose. This macroevolutionary pattern is called preadaptation or **exaptation**. Some traits serve a very different purpose in modern species than they did when they first evolved. For example, feathers that allow modern birds to fly are derived from feathers that first evolved in some dinosaurs. Those dinosaurs could not have used their feathers for flight, but they probably did use them for insulation. Flight feathers in birds are an exaptation of insulating feathers in dinosaurs.

Extinction By current estimates, more than 99 percent of all species that ever lived are now **extinct**, or irrevocably lost from Earth. In addition to continuing small-scale extinctions, the fossil record indicates that there have been more than twenty **mass extinctions**, which are simultaneous losses of many lineages. These include five catastrophic events in which the majority of species on Earth disappeared (Section 11.5).

coevolution The joint evolution of two closely interacting species; each species is a selective agent for traits of the other.

exaptation Adaptation of an existing structure for a completely different purpose; a major evolutionary novelty.

extinct Refers to a species that has been permanently lost.

macroevolution Patterns of evolution that occur above the species level.

mass extinction Simultaneous extinction of many lineages.

stasis Macroevolutionary pattern in which a lineage persists with little or no change over evolutionary time.

FIGURE 12.23 Hawaiian honeycreeper diversity, as represented by a small sampling of the 71 remaining species. The bills of these birds are adapted to feed on insects, seeds, fruits, nectar in floral cups, and other foods. Honeycreepers descended by adaptive radiations of a common ancestor that probably resembled the housefinch (*Carpodacus*) shown at *right*.

Adaptive Radiation In an evolutionary pattern called **adaptive radiation**, a lineage rapidly diversifies into several new species. Adaptive radiation occurs when a lineage encounters a set of new niches. Think of a niche as a certain way of life, such as "burrowing in seafloor sediments" or "catching winged insects in the air at night." A lineage that encounters a new set of niches tends to diversify over time. Genetic divergences then give rise to many new species that fill the niches.

A lineage may encounter new niches when genetic changes allow some of its individuals to exploit new opportunities within their existing habitat. A **key innovation** is a structural or functional modification that bestows upon its bearer the opportunity to exploit a habitat more efficiently or in a novel way. New niches also open up when geologic or climatic events change an existing habitat so that it becomes very different, or when individuals gain physical access to a new habitat.

The adaptive radiation of finches that migrated to the Hawaiian Islands more than 4 million years ago gave rise to hundreds of species of Hawaiian honeycreepers (Figure 12.23). No predators had preceded the birds, but tasty insects and plants that bore tender leaves, nectar, seeds, and fruits were already there. The finches thrived. Populations of their descendants radiated into habitats along the coasts, through dry lowland forests, and into highland rain forests. Over many generations, unique forms and behaviors evolved in many different lineages of birds. Such traits allowed the birds to exploit special opportunities presented by their island habitats.

Evolutionary Theory Biologists do not doubt that macroevolution occurs, but many disagree about how it occurs. However we choose to categorize evolutionary processes, the very same genetic change may be at the root of all evolution—fast or slow, large-scale or small-scale. Dramatic jumps in morphology, if they are not artifacts of gaps in the fossil record, may be the result of mutations in homeotic or other regulatory genes. Macroevolution may include more processes than microevolution, or it may not. It may be an accumulation of many microevolutionary events, or it may be an entirely different process. Evolutionary biologists may disagree about these and other hypotheses, but all of them are trying to explain the same thing: how all species are related by descent from common ancestors.

Take-Home Message What is macroevolution?

■ Macroevolution refers to large-scale patterns of evolutionary change such as adaptive radiations, the origin of major groups, and loss through extinction.

DOMAIN	Eukarya	Eukarya	Eukarya	Eukarya	Eukarya
KINGDOM	Plantae	Plantae	Plantae	Plantae	Plantae
PHYLUM	Magnoliophyta	Magnoliophyta	Magnoliophyta	Magnoliophyta	Magnoliophyta
CLASS	Magnoliopsida	Magnoliopsida	Magnoliopsida	Magnoliopsida	Magnoliopsida
ORDER	Apiales	Rosales	Rosales	Rosales	Rosales
FAMILY	Apiaceae	Cannabaceae	Rosaceae	Rosaceae	Rosaceae
GENUS	*Daucus*	*Cannabis*	*Malus*	*Rosa*	*Rosa*
SPECIES	*carota*	*sativa*	*domesticus*	*acicularis*	*canina*
COMMON NAME	carrot	marijuana	apple	arctic rose	dog rose

FIGURE 12.24 Linnaean classification of five species that are related at different levels. Each species has been assigned to ever more inclusive groups, or taxa: in this case, from genus to domain. **Figure It Out:** Which of the plants shown here are in the same order?

Answer: Marijuana, apple, arctic rose, and dog rose are all in the order Rosales.

12.7 Organizing Information About Species

A Rose by Any Other Name . . . **Taxonomy**, the science of naming and classifying species, began thousands of years ago. However, naming species in a consistent way became a priority in the eighteenth century. At the time, European explorers discovering the scope of life's diversity were having more and more trouble communicating about species, which often had multiple names. For example, one plant species native to Europe, Africa, and Asia was alternately known as the dog rose, briar rose, witch's briar, herb patience, sweet briar, dog berry, briar hip, eglantine gall, hep tree, hip fruit, hip rose, hip tree, hop fruit, and hogseed—and those are only the English names! In addition, each species often had been given multiple scientific names. The scientific names were in Latin, and were descriptive but cumbersome. For example, the scientific name of the dog rose was *Rosa sylvestris inodora seu canina* (odorless woodland dog rose), and also *Rosa sylvestris alba cum rubore, folio glabro* (pinkish white woodland rose with smooth leaves).

An eighteenth-century naturalist, Carolus Linnaeus, came up with a much simpler naming system that we still use. By the Linnaean system, every species is given a unique two-part scientific name: The first part is the genus name, and together with the second part it designates the species. Thus, the dog rose now has one official scientific name: *Rosa canina*. Linnaeus ranked species into ever more inclusive categories. Each category, or **taxon** (plural, taxa) is a group of organisms. The categories above species—genus, family, order, class, phylum, kingdom, and domain—are the higher taxa (Figure 12.24). Each higher taxon consists of a group of the next lower taxon.

adaptive radiation A burst of genetic divergences from a lineage gives rise to many new species.

key innovation An evolutionary adaptation that gives its bearer the opportunity to exploit a particular environment more efficiently or in a new way.

taxon (**taxa**) A grouping of organisms.

taxonomy Science of naming and classifying species.

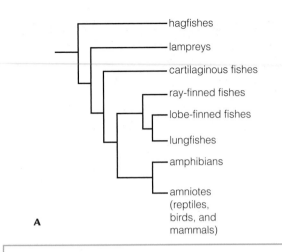

A

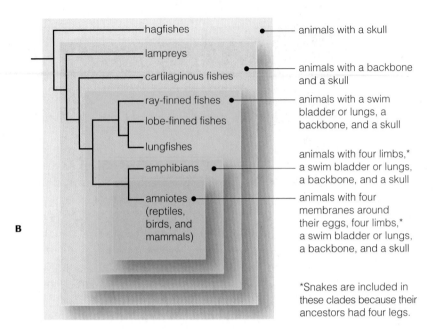

B

animals with a skull

animals with a backbone and a skull

animals with a swim bladder or lungs, a backbone, and a skull

animals with four limbs,* a swim bladder or lungs, a backbone, and a skull

animals with four membranes around their eggs, four limbs,* a swim bladder or lungs, a backbone, and a skull

*Snakes are included in these clades because their ancestors had four legs.

FIGURE 12.25 Cladograms. **A** This example shows evolutionary relationships among some of the major groups of animals. **B** We can visualize the same cladogram as "sets within sets" of characters.

> **Every living thing is related if you just go back far enough in time.**

character Quantifiable, heritable characteristic or trait.

clade A group of species that share a set of characters.

cladistics Method of determining evolutionary relationships by grouping species into clades.

cladogram Evolutionary tree diagram that shows a network of evolutionary relationships among clades.

evolutionary tree Type of diagram that summarizes evolutionary relationships among a group of species.

monophyletic group An ancestor and all of its descendants.

phylogeny Evolutionary history of a species or groups of species.

sister groups The two lineages that emerge from a node on a cladogram.

A species is usually assigned to higher taxa based on traits it shares with other species. That assignment may change as we discover more about the species and the traits involved. For example, Linnaeus grouped plants by the number and arrangement of reproductive parts, a scheme that resulted in odd pairings such as castor-oil plants with pine trees. Today we place these plants in separate phyla.

Ranking Versus Grouping Linnaeus devised his system of taxonomy before anyone knew about evolution. As we now know, evolution is a dynamic, extravagant, messy, and ongoing process that can be challenging for those who like their categories neat. For example, speciation does not usually occur at a precise moment in time: Individuals often continue to interbreed even as populations are diverging, and populations that have already diverged may come together and interbreed again.

Linnaean taxonomy can be problematic when species boundaries are fuzzy, but the bigger problem is that the rankings do not necessarily reflect evolutionary relationships. Our increasing understanding of evolution is prompting a major, ongoing overhaul of the way biologists view life's diversity. Instead of trying to divide that tremendous diversity into a series of ranks, most biologists are now focusing on evolutionary connections. Each species is viewed not as a member or representative of a rank in a hierarchy, but rather as part of a bigger picture of evolution.

Phylogeny is the evolutionary history of a species or group of them, a kind of genealogy that follows a lineage's evolutionary relationships through time. The central question of phylogeny is, "Who is related to whom?" Thus, methods of finding the answer to that question are an important part of phylogenetic classification systems. One method, **cladistics**, groups species on the basis of shared **characters**, which are quantifiable, heritable characteristics. A character can be a physical, behavioral, physiological, or molecular trait of an organism. Because each species has many characters, cladistic groupings may differ depending on which characters are used.

The result of a cladistic analysis is a **cladogram**, a diagram that shows a network of evolutionary relationships (Figure 12.25). Each line in a cladogram represents a lineage, which may branch into two lineages at a node. The node

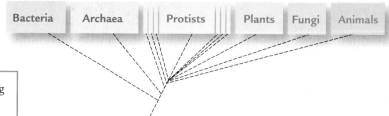

FIGURE 12.26 Animated! Different ways to see the big picture of evolutionary connections.

A A six-kingdom classification system assigns all prokaryotes to kingdoms Bacteria and Archaea. Kingdom Protista includes the most ancient multicelled and all single-celled eukaryotes. Plants, fungi, and animals have their own kingdoms.

B A three-domain system sorts all life into three domains: Bacteria, Archaea, and Eukarya.

A This tree represents all life classified into six kingdoms. We have discovered that the kingdom of protists is not monophyletic, so some biologists now divide it up into a number of new kingdoms.

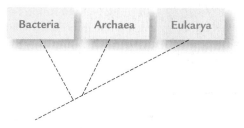

B This tree represents all life classified into three domains. The kingdoms protists, plants, fungi, and animals are subsumed into domain Eukarya.

represents a common ancestor of the two lineages. Every branch ends with a **clade** (from *klados*, a Greek word for twig or branch), a group of species that share a set of characters. Ideally, a clade is a **monophyletic group** that comprises an ancestor and all of its descendants. Cladograms and other types of **evolutionary trees** summarize our best data-supported hypotheses about how a group of species evolved. We use these diagrams to visualize evolutionary trends and patterns. For instance, the two lineages that emerge from a node on a cladogram are called **sister groups**. Sister groups are, by default, the same age. We may not know what that age is, but we can compare sister groups on a cladogram and say something about their relative rates of evolution.

Like other hypotheses, evolutionary tree diagrams get revised as new information is gathered. However, the diagrams are based on a solid premise: All species are interconnected by shared ancestry. Every living thing is related if you just go back far enough in time (Figure 12.26). Evolutionary biologists' job is to figure out where the connections are.

12.8 Impacts/Issues Revisited: Rise of the Super Rats

The gene most commonly involved in warfarin resistance in rats is inherited in a simple dominance pattern. The gene's normal allele is recessive when paired with the mutated allele that confers resistance. In the presence of warfarin, the dominant allele is clearly adaptive. Rats with this allele require a lot of extra vitamin K, but being vitamin K–deficient is not so bad when compared with being dead from rat poison. However, in the absence of warfarin, individuals that have the warfarin resistance allele are at a serious disadvantage compared with those who do not. Rats with the allele cannot easily obtain enough vitamin K from their diet to sustain normal blood clotting and bone formation. Thus, the allele is adaptive when warfarin is present, and maladaptive when it is not, so periodic exposure to warfarin maintains a balanced polymorphism of the resistance gene in rat populations.

How Would YOU ☑ Vote? Antibiotic-resistant strains of bacteria are now widespread. One standard animal husbandry practice includes continually dosing healthy livestock with the same antibiotics prescribed for people. Should this practice stop? See CengageNow for details, then vote online (CengageNow.com).

Summary

 Sections 12.1 and 12.2 Individuals of a population share a gene pool. Mutations are the original source of alleles, but many are lethal or neutral. Microevolution, or change in the allele frequencies of a population, always occurs in natural populations. Evolution is an opportunistic process.

CENGAGENOW™ Investigate allele frequency with the interaction on CengageNow.

Section 12.3 Natural selection occurs in different patterns depending on the species and selection pressures involved. Directional selection shifts the range of variation in traits in one direction. Stabilizing selection favors intermediate forms of a trait. Disruptive selection favors extreme forms.

CENGAGENOW™ See how directional selection, disruptive selection, and stabilizing selection work with the animations on CengageNow.

Section 12.4 Sexual selection leads to forms of traits that enhance reproductive success. Sexual dimorphism is one outcome. In balanced polymorphism, selection against homozygotes maintains nonidentical alleles for a trait in a population.

Genetic drift can lead to loss of genetic diversity or fixation. It is pronounced in small or inbreeding populations, such as those that have been through an evolutionary bottleneck. A bottleneck can result in the founder effect. Gene flow counters the effects of mutation, natural selection, and genetic drift.

CENGAGENOW™ Explore genetic drift with the interaction on CengageNow.

Section 12.5 Individuals of sexually reproducing species can interbreed successfully under natural conditions, produce fertile offspring, and are reproductively isolated from other species. Reproductive isolation typically occurs after gene flow stops. Divergences then lead to speciation (Table 12.2).

In allopatric speciation, a geographic barrier interrupts gene flow between populations. Genetic divergences then give rise to new species. With sympatric speciation, populations in physical contact speciate. Polyploid species of many plants (and a few animals) originated by chromosome doublings and hybridizations.

CENGAGENOW™ Explore how species become reproductively isolated, learn more about speciation, and see sympatric speciation in wheat with the animations on CengageNow

Section 12.6 Macroevolution refers to patterns of evolution above the species level. With stasis, a lineage changes very little over evolutionary time. In exaptation, a lineage uses a structure for a different purpose than its ancestor did. An adaptive radiation is rapid diversification into new species that occupy novel niches. Coevolution occurs when two species act as agents of selection upon one another. Permanent loss of a lineage is extinction.

Section 12.7 Taxonomy is the science of naming and classifying species. In traditional taxonomy systems, species are organized into a series of ranks (taxa) based on their traits. Such  systems do not necessarily reflect evolutionary relationships.

Cladistics is a set of methods that allow us to reconstruct evolutionary history (phylogeny). The result of a cladistic analysis is an evolutionary tree diagram, in which a line represents a lineage. In evolutionary trees called cladograms, a line (lineage) can branch into two sister groups at a node, which represents a shared ancestor. A six-kingdom classification system and a three-domain classification system are two different ways to organize life's diversity.

CENGAGENOW™ Review different classification systems with the animation on CengageNow.

Self-Quiz   Answers in Appendix I

1. Biologists define evolution as _____ .
   a. purposeful change in a lineage
   b. heritable change in a line of descent
   c. acquiring traits during the individual's lifetime

2. Evolution can only occur in a population when _____ .
   a. mating is random
   b. there is selection pressure
   c. neither is necessary

3. Stabilizing selection tends to _____ (select all that apply).
   a. eliminate extreme forms of a trait
   b. favor extreme forms of a trait
   c. eliminate intermediate forms of a trait
   d. favor intermediate forms of a trait

4. Disruptive selection tends to _____ (select all that apply).
   a. eliminate extreme forms of a trait
   b. favor extreme forms of a trait
   c. eliminate intermediate forms of a trait
   d. favor intermediate forms of a trait
   e. shift allele frequencies in one direction

Table 12.2	Different Speciation Models	
	Allopatric	**Sympatric**
Original population	⬤	⬤
Initiating event	◖◗ barrier arises	⬤ genetic change
Reproductive isolation	◖◗ in isolation	⬤ within population
New species	⬤⬤	⬤⬤

5. Directional selection tends to _____ (select all that apply).
   a. eliminate extreme forms of a trait
   b. favor extreme forms of a trait
   c. eliminate intermediate forms of a trait
   d. favor intermediate forms of a trait
   e. shift allele frequencies in one direction

6. Sexual selection, such as competition between males for access to fertile females, frequently influences aspects of body form and can lead to _____ .
   a. male aggression   c. sexual dimorphism
   b. sexual reproduction   d. both a and c

7. The persistence of the sickle allele at high frequency in a population is a case of _____ .

8. _____ tends to keep different populations of a species similar to one another.

9. A fire devastates all trees in a wide swath of forest. Populations of a species of tree-dwelling frog on either side of the burned area diverge to become separate species. This is an example of _____ .

10. Cladistics is based on _____ .
    a. reconstructing evolutionary relationships
    b. grouping species on the basis of shared characters
    c. both a and b

11. In evolutionary trees, each node represents a(n) _____ .
    a. single lineage   c. divergence
    b. extinction   d. adaptive radiation

12. In cladograms, sister groups are _____ .
    a. inbred   c. represented by nodes
    b. the same age   d. members of the same family

13. Match the evolution concepts.
    _____ gene flow   a. can lead to interdependent species
    _____ natural   b. changes in a population's allele
          selection   frequencies due to chance alone
    _____ mutation   c. alleles enter and leave a population
    _____ genetic   d. evolutionary history
          drift   e. occurs in different patterns
    _____ adaptive   f. burst of divergences from one
          radiation   lineage into a set of niches
    _____ coevolution   g. source of new alleles
    _____ phylogeny   h. diagram of sets within sets
    _____ cladogram

Additional questions are available on CENGAGENOW™

Critical Thinking

1. Rama the cama, a llama–camel hybrid, was born in 1997. The idea was to breed an animal that has the camel's strength and endurance and the llama's gentle disposition. However, instead of being large, strong, and sweet, Rama is smaller than expected and has a camel's short temper. He has his eye on Kamilah, a female cama born in early 2002. The breeders wonder whether any offspring from such a match would be fertile, and what the offspring would look like.

Digging Into Data

Resistance to Rodenticides in Wild Rat Populations

Beginning in 1990, rat infestations in northwestern Germany started to intensify despite continuing use of rodenticides. In 2000, Michael H. Kohn and his colleagues analyzed the genetics of wild rat populations around Munich. For part of their research, they trapped wild rats in five towns, and tested those rats for resistance to warfarin and the second-generation anticoagulant bromadiolone. The results are shown in Figure 12.27.

1. In which of the five towns were most of the rats susceptible to warfarin?

2. Which town had the highest percentage of anticoagulant-resistant wild rats?

3. What percentage of rats in Olfen were resistant to warfarin?

4. In which town do you think the application of bromadiolone was most intensive?

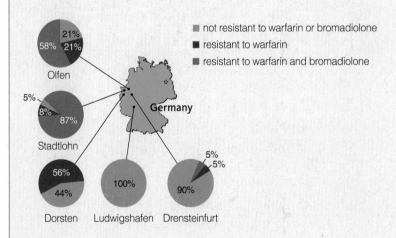

FIGURE 12.27 Resistance to anticoagulant rodenticides in wild populations of rats in Germany, 2000.

What does Rama's story tell you about the genetic changes required for reproductive isolation in nature? Explain why a biologist might not view Rama as evidence that llamas and camels are the same species.

2. Some theorists have hypothesized that many of our uniquely human traits arose by sexual selection. Over many thousands of years, women attracted to charming, witty men perhaps prompted the development of human intellect far beyond what was necessary for mere survival. Men attracted to women with juvenile features may have shifted the species as a whole to be less hairy and softer featured than any of our simian relatives. Can you think of a way to test these hypotheses?

3. Some people refer to different species as "primitive" or "advanced." For example, they may say that mosses are primitive and flowering plants are advanced, or that crocodiles are primitive and mammals are advanced. Why is it incorrect to refer to a modern taxon as primitive?

13 Early Life Forms and the Viruses

> If numbers and distribution are measures of success, then microorganisms are Earth's most successful inhabitants by far.

13.1 Impacts/Issues: Microbial Menaces

By the time Chedo Gowero (Figure 13.1) was thirteen years old, she had lost both her parents to AIDS, left school, and was working to support herself and her ten-year-old brother. Stories like Chedo's are common in Zimbabwe, a country where one-fifth of the population has AIDS.

AIDS is short for *Acquired Immune Deficiency Syndrome*, and it is caused by a virus called HIV, for *Human Immunodeficiency Virus*. The virus infects certain white blood cells and destroys the body's ability to defend itself. AIDS occurs when other diseases run rampant in the undefended body. Worldwide, more than 20 million people have died from AIDS and about 39 million are infected with HIV. In developed countries, antiviral drugs help slow the onset and progression of AIDS in most affected people. However, in less developed countries, only a small percentage of the population has access to such drugs.

Less developed countries also suffer greatly from malaria, which is mainly a tropical disease. Mosquitoes carry the protist that causes malaria from one human to another. Malaria destroys oxygen-carrying red blood cells, causing weakness. Affected people have bouts of chills and fever. They often become jaundiced as waste materials accumulate in their body and tint them yellow. When infected blood cells get into the brain, they can cause blindness, seizures, coma, and even death. Children are especially susceptible to malaria-induced brain damage.

Malaria was common in the United States, especially the South, until an aggressive campaign during the 1940s eliminated its sources. Swamps and ponds where mosquitoes bred were drained, and the insecticide DDT was sprayed inside millions of homes. Today, nearly all cases of malaria in the United States occur in people who became infected outside the country.

Malaria remains common in parts of Mexico, South and Central America, and Asia and the Pacific Islands. However, it takes its greatest toll in Africa, where it has been a potent selective force on humans. As Section 12.4 explained, the allele responsible for sickle-cell anemia also reduces mortality from malaria.

This chapter focuses on how life first began and on the prokaryotes, protists, and viruses, groups that evolved long before the plants, fungi, and animals. If we think of these mainly microscopic groups at all, it is usually because some of them cause disease. However, most microorganisms are not harmful and many are beneficial. They are food for larger organisms and their activities keep nutrients cycling through ecosystems.

Moreover, although microorganisms are smaller and structurally simpler than the more familiar organisms, all have been shaped by the same biological processes. Microorganisms respond to selective pressures and show remarkable adaptations to their extraordinarily diverse environments. As you will see, these ancient lineages survive today in some of Earth's most challenging habitats. If numbers and distribution are measures of success, then microorganisms are Earth's most successful inhabitants by far.

FIGURE 13.1 Animated! Chedo Gowero, one of the more than 12 million African children orphaned by AIDS. The smaller photo shows HIV, the virus that causes AIDS.

13.2
Before There Were Cells

Conditions on the Early Earth Our solar system began assembling when asteroids (large rocks) orbiting the sun collided, forming bigger rocky objects. The heavier these pre-planets became, the more gravitational pull they had, and the more material they gathered. By about 4.6 billion years ago, this process had formed Earth and the other planets that orbit our sun.

Planet formation did not clear out all debris from the orbit around the sun, so the early Earth received a constant hail of meteorites upon its still molten surface. More molten rock and gases spewed from volcanoes. Gases released by such volcanoes were the main source of the early atmosphere.

Studies of volcanic eruptions, meteorites, ancient rocks, and other planets provide clues about Earth's early atmosphere. They suggest that the air held water vapor, carbon dioxide, and gaseous hydrogen and nitrogen. The geologic record provides evidence that there was little oxygen or no oxygen early on. Iron in rocks rusts when exposed to oxygen, and the oldest rocks with rust date to about 2.3 billion years ago.

Relatively low levels of oxygen gas probably made the origin of life possible, because oxygen gas is reactive. If it had been present, complex compounds would not have been able to form and persist. Oxygen gas would have reacted with and destroyed these compounds as soon as they formed.

Although the air had little oxygen, oxygen was present in combination with hydrogen as water. At first, any water falling on Earth's molten surface evaporated immediately. As the surface cooled, rocks formed. Later, rains washed mineral salts out of these rocks and the salty runoff pooled in early seas. It was in these seas that life first began.

Origin of the Building Blocks of Life Until the early 1800s, chemists thought that organic molecules possessed a special "vital force" and could only be made inside living organisms. Then, in 1825, a German chemist synthesized urea, a molecule that is abundant in urine. Later, another chemist made alanine, an amino acid. These synthetic reactions demonstrated that nonliving mechanisms could yield organic molecules.

Could chemical reactions on the early Earth have yielded the types of organic molecules that are now the building blocks of life? In the 1950s Stanley Miller became the first to test this hypothesis. He put gases chosen to simulate Earth's early atmosphere into a reaction chamber (Figure 13.2). He kept the mixture circulating and zapped it with sparks to simulate lightning. Within a week, amino acids, sugars, and other organic compounds had formed.

Since Miller's experiment, researchers have revised their ideas about which gases were present in Earth's early atmosphere. Miller and others have repeated his experiment using different gases and adding various ingredients to the mix. Amino acids form easily under some conditions, but not under others. Scientists continue to debate which conditions are most like those on the early Earth.

The presence of amino acids, sugars, and nucleotide bases in meteorites that fell to Earth suggests an alternative origin for life's building blocks. Perhaps these molecules formed in interstellar clouds of ice, dust, and gases and were delivered to Earth by meteorites. Keep in mind that during Earth's early years, meteorites fell to Earth thousands of times more frequently than they do today.

A

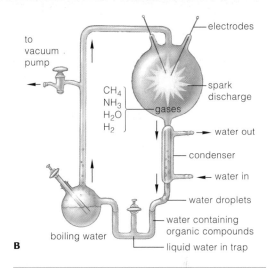

B

to vacuum pump

electrodes

CH_4
NH_3
H_2O
H_2

gases

spark discharge

water out

condenser

water in

water droplets

water containing organic compounds

boiling water

liquid water in trap

FIGURE 13.2 Animated! Conditions on early Earth. **A** Artist's depiction of the early Earth. **B** Diagram of the apparatus that Stanley Miller used to test whether organic building blocks of life could have formed under conditions that prevailed on early Earth. Miller circulated water vapor, hydrogen gas (H_2), methane (CH_4), and ammonia (NH_3) in a glass chamber to simulate the first atmosphere. Sparks provided by an electrode simulated lightning.

Figure It Out: Which compound in Miller's mixture provided the nitrogen for the amino group in the amino acids? *Answer: Ammonia*

FIGURE 13.3 Where did complex organic compounds first form? Two candidate locations: **A** Tidal flats rich in negatively charged clay particles, and **B** iron-sulfide rocks at hydrothermal vents on the ocean floor. Rocks at these vents are covered with cell-sized chambers.

Assembly of Complex Organic Compounds Modern cells take up organic subunits, concentrate them, and assemble them into complex organic compounds. The first step toward such metabolism may have been a nonbiological process that concentrated organic subunits in one place, thus increasing the chance that the subunits would combine into larger molecules.

By one hypothesis, such a process may have occurred in shallow pools on clay-rich tidal flats (Figure 13.3**A**). Clay particles have a slight negative charge, so positively charged organic subunits dissolved in seawater tend to stick to them. If this happened on ancient tidal flats, evaporation would have concentrated the subunits even more, and energy from the sun could have caused them to bond together. Simulation of this process has resulted in short chains of amino acids.

By another hypothesis, metabolic reactions began at **hydrothermal vents**, underwater openings from which hot, mineral-rich water flows. Minerals such as iron sulfide settle out of the water and build up as rocky deposits near the vents (Figure 13.3**B**). These mineral-rich rocks are pocked with tiny chambers about the size of cells. In simulations of this environment, iron sulfide in the chamber walls donated hydrogen and electrons to dissolved carbon dioxide, forming organic molecules that accumulated inside the chambers.

Origin of Genetic Material DNA is the genetic material in all modern cells. Cells pass copies of their DNA to descendant cells, which use instructions encoded in DNA to build proteins. Some of these proteins aid synthesis of new DNA, which is passed along to descendant cells, and so on. Protein synthesis depends on DNA, which is built by proteins. How did this cycle begin?

In the 1960s, Francis Crick and Leslie Orgel addressed this dilemma. They suggested that RNA may have been the first informational molecule. Since then, evidence for an early **RNA world**—a time when RNA both stored genetic information and functioned like an enzyme in protein synthesis—has accumulated. For example, some RNAs still serve as enzymes in living cells. An rRNA in ribosomes speeds formation of peptide bonds during protein synthesis (Section 7.4). Also, RNAs called ribozymes cut noncoding bits out of newly formed RNAs. In the laboratory, researchers made synthetic, self-replicating ribozymes that copy

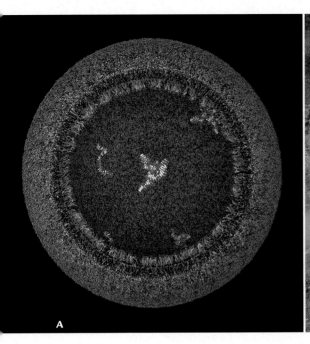

FIGURE 13.4 Testing hypotheses about protocells.

A Illustration of one type of protocell that has been synthesized in the laboratory. It has a fatty acid membrane that encloses strands of RNA.

B David Deamer tests the ability of a mix of organic subunits and phosphates to form protocells in a hot, acidic pool in Russia. No complex organic molecules or membrane-like structures formed in this experiment. The negative result led Deamer to revise his hypothesis about the conditions under which protocells can form in nature.

themselves by assembling free nucleotides. Whether and how ribozymes could have formed on early Earth remains an open question.

Formation of Protocells In cells, a selectively permeable plasma membrane composed of lipids and proteins encloses metabolic machinery and genetic information. An intermediate stage may have been **protocells**, simple membranous sacs that enclosed interacting organic molecules.

Researchers have combined organic molecules in the laboratory to produce synthetic protocells that have some of the properties we associate with life. For example, they have made vesicle-like spheres of fatty acid that hold RNA (Figure 13.4**A**). The spheres "grow" by incorporating fatty acids and nucleotides from their surroundings. Mechanical force causes the spheres to divide. Researchers are also carrying out field tests to determine what type of conditions favor or interfere with protocell formation (Figure 13.4**B**).

Simulations and experiments cannot prove how life or cells began, but they do show us what is plausible. A variety of investigations by many researchers tell us this: Chemical and physical processes that operate today can produce simple organic compounds, concentrate them, and assemble them into protocells. Billions of years ago, the same processes operating in the same way may have opened the way to the first life.

> **Simulations and experiments cannot prove how life or cells began, but they can show us what is plausible.**

Take-Home Message What do scientific studies tell us about events that could have preceded life?

- Small organic subunits could have formed on the early Earth, or formed in space and fallen to Earth on meteorites.
- Complex organic molecules could have self-assembled from simpler ones.
- The first genetic material may have been RNA rather than DNA.
- Protocells—chemical-filled membranous sacs that grow and divide—may have been the ancestors of the first cells.

hydrothermal vent Submerged opening where hot, mineral-rich water streams out.

protocells Membranous sacs that contain interacting organic molecules.

RNA world Hypothetical early interval when RNA served as the material of inheritance.

Life's Early Evolution

The Age of Prokaryotes As you might imagine, it is difficult to find fossils of individual cells. However, some early cells that lived in shallow seas formed dense mats that tended to trap sediments. Over thousands of years, as sediments built up and new cell layers grew old, the mats formed large dome-shaped structures called **stromatolites** (Figure 13.5). The oldest fossil stromatolites date back about 3.5 billion years, so we can assume that cells arose sometime before that.

What were the first cells like? Data from gene comparisons among modern species, as well as the size and structure of fossil cells, indicate that the first cells were **prokaryotes**, cells that do not have a nucleus (Section 3.5). Oxygen was scarce, so the earliest cells probably were anaerobic. Like some modern prokaryotes, they may have used dissolved carbon as their carbon source and mineral ions as their energy source.

There are currently two prokaryotic domains: Bacteria and Archaea, and their lineages diverged very early in the history of life. Shortly after this divergence, some members of both groups began to capture and use light energy in photosynthetic pathways that did not produce oxygen. Many modern prokaryotes still carry out photosynthesis by such pathways.

Then, 2.7 billion years ago, one lineage of bacteria began to carry out photosynthesis by the oxygen-releasing pathway. As populations of these cells increased in number, oxygen began to accumulate in the seas and air.

Rising oxygen levels had three important consequences:

1. Oxygen prevents the self-assembly of complex organic compounds, so life could no longer arise from nonliving materials.

2. Oxygen served as a selective pressure, putting organisms that thrived in higher oxygen conditions at an advantage. The pathway of aerobic respiration (Section 5.6) evolved and became widespread. This pathway uses oxygen, and it is far more efficient at releasing energy than other pathways. It would later meet the high energy needs of multicelled eukaryotes.

3. As oxygen enriched the atmosphere, an **ozone layer** formed. This layer keeps much of the sun's ultraviolet (UV) radiation from reaching Earth's surface. UV radiation can damage DNA and other biological molecules. Without the ozone layer to protect it, life could not have moved onto land.

Origin of Eukaryotes and Their Organelles
Eukaryotes, cells in which the DNA is enclosed inside a nucleus, first appear in the fossil record about 1.8 million years ago. The outer membrane of the eukaryotic nucleus (the nuclear envelope) is continuous with the membrane of the endoplasmic reticulum (ER). Both membranes probably evolved from plasma membrane that folded inward (Figure 13.6). Some modern prokaryotes have such infoldings.

At first, membrane infoldings may have been advantageous because they increased the surface area for membrane-related reactions. Infoldings also would have divided the cytoplasm into compartments where specific reactions could take place. Later, membrane folds that extended around the DNA, and thus formed a nuclear envelope, may have been favored because they protected the cell's genetic material.

FIGURE 13.5
Early prokaryotic life.
A Artist's depiction of stromatolites in an ancient sea. Each mound stood about 30 centimeters high and consisted of prokaryotic cells and sediments.

B,C Fossil cells that lived in stromatolites 850 million years ago.

In terms of the genes inside their nucleus, eukaryotes are more similar to archaeans than to bacteria. However, eukaryotes are thought to have a composite ancestry, with some organelles descended from bacteria. Mitochondria and chloroplasts resemble bacteria in their structure and genome and these organelles probably evolved by **endosymbiosis**. By this process, one species enters another, then lives and replicates inside it. (*Endo*– means within; *symbiosis* means living together.) Endosymbionts inside a cell can be passed along to the cell's descendents when the cell divides.

Mitochondria resemble some modern bacteria that are aerobic heterotrophs, so biologists infer that mitochondria share an ancestor with these bacteria. Presumably, the bacteria invaded or were engulfed by an early eukaryote and began living inside it. Over generations, the bacteria came to rely on their eukaryotic host for raw materials, while the host used ATP produced by its bacterial guests. Over time, genes that specified the same or similar proteins in both the host and its symbiont were free to mutate. If a gene lost its function in one partner, a gene from the other could take up the slack. Eventually, the host and symbiont both became incapable of living independently.

Similarly, genes of chloroplasts resemble those of modern oxygen-producing photosynthetic bacteria. As a result, biologists infer that chloroplasts evolved by endosymbiosis from relatives of these bacteria. The bacteria would have provided their host with sugars and received shelter and carbon dioxide in return.

Eukaryotic Diversification The first eukaryotes were also the first protists. The earliest protist fossil that we can assign to any modern group is *Bangiomorpha pubescens*, a red alga that lived about 1.2 billion years ago (Figure 13.7). It was multicelled and there was a cellular division of labor. Some of the cells held the alga in place, and others produced two types of sexual spores. Apparently, *B. pubescens* was one of the earliest sexually reproducing organisms.

Stromatolites dominated the world's oceans for billions of years, but they began to decline about 750 million years ago. Newly evolved algal competitors and bacteria-eating protists probably played a role in their decline.

So far, the earliest animal fossils known date back to 570 million years ago. These early animals were less than a millimeter across. They shared the oceans with bacteria, archaeans, fungi, and protists, including the lineage of green algae that would later give rise to land plants. Animal diversity increased tremendously during a great adaptive radiation in the Cambrian, 543 million years ago. When that period finally ended, all of the major animal lineages, including vertebrates (animals with backbones), were represented in the seas.

On the next two pages, Figure 13.8 summarizes the evolutionary events discussed in this section. We then devote the remainder of this chapter to discussions of prokaryotes, protists, and viruses. The next two chapters discuss the evolution and diversity of the more familiar plants, fungi, and animals.

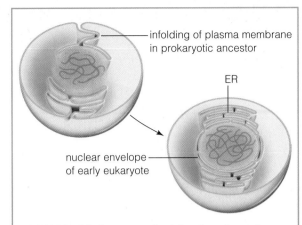

FIGURE 13.6 Proposed origin of a eukaryotic nuclear envelope and ER membrane by infolding of the plasma membrane of a prokaryotic ancestor.

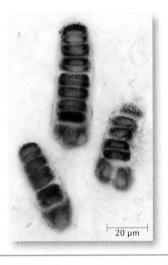

FIGURE 13.7 A eukaryotic fossil, a multicelled red alga that lived 1.2 billion years ago.

Take-Home Message **What was early life like and how did it change Earth?**

■ Life arose by 3–4 billion years ago; it was initially anaerobic and prokaryotic.

■ An early divergence separated ancestors of modern bacteria from the lineage that would lead to archaeans and eukaryotic cells.

■ Oxygen accumulation in the air and seas halted spontaneous formation of the molecules of life, formed a protective ozone layer, and favored organisms that carried out the highly efficient pathway of aerobic respiration.

endosymbiosis One species lives inside another.

eukaryote Organism that encloses its DNA in a nucleus; a protist, plant, fungus, or animal.

ozone layer Atmospheric layer with a high concentration of ozone that prevents much UV radiation from reaching Earth's surface.

prokaryote Single-celled organism in which the DNA resides in the cytoplasm; a bacterium or archaean.

stromatolites Dome-shaped structures composed of layers of prokaryotic cells and sediments; form in shallow seas.

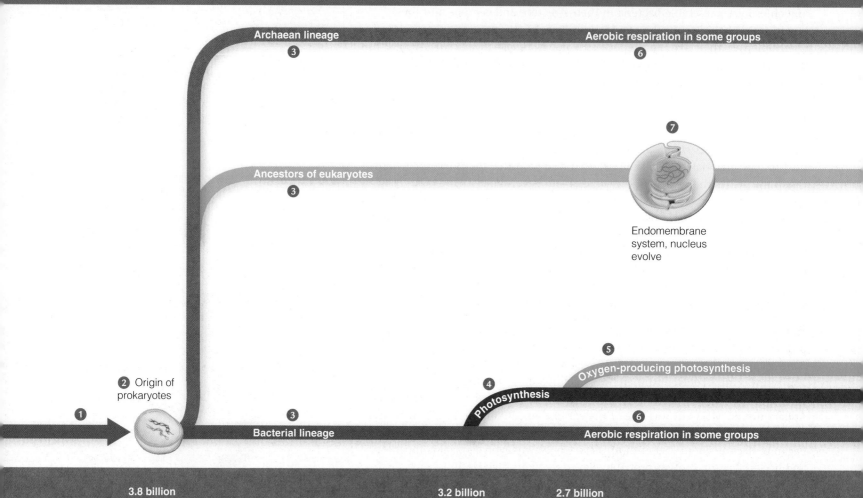

Archaean lineage
③

Aerobic respiration in some groups
⑥

⑦

Ancestors of eukaryotes
③

Endomembrane
system, nucleus
evolve

⑤

Oxygen-producing photosynthesis

② Origin of
prokaryotes

④

Photosynthesis

①

③

⑥

Bacterial lineage

Aerobic respiration in some groups

3.8 billion
years ago

3.2 billion
years ago

2.7 billion
years ago

Steps Preceding Cells

① Between 5 billion and 3.8 billion years ago, lipids, proteins, nucleic acids, and complex carbohydrates formed from the simple organic compounds present on early Earth.

Prokaryotic Cells

② The first cells evolved by 3.5 billion years ago. They were prokaryotic—they did not have a nucleus or other organelles. Oxygen was scarce, so the first cells made ATP by anaerobic pathways.

Three Domains of Life

③ The first major divergence gave rise to bacteria and a common ancestor of archaeans and eukaryotic cells. Not long after that, the archaeans and eukaryotic cells parted ways.

Photosynthesis and Aerobic Respiration

④ Photosynthetic pathways that did not produce oxygen evolved in some bacterial lineages.

⑤ Oxygen-producing photosynthesis evolved in a branch from this lineage, and oxygen began to accumulate.

⑥ Aerobic respiration became the predominant metabolic pathway in some bacteria and archaea.

Endomembrane System, Nucleus

⑦ Cell sizes and the amount of genetic information continued to expand in ancestors of what would become the eukaryotic cells. The endomembrane system, including the nuclear envelope, arose through the modification of cell membranes between 3 and 2 billion years ago.

FIGURE 13.8 Animated! Milestones in the history of life, based on the most widely accepted hypotheses. This figure also shows the evolutionary connections among all groups of organisms. The time line is not to scale.

Figure It Out: Which organelle evolved first, mitochondria or chloroplasts?

Answer: Mitochondria

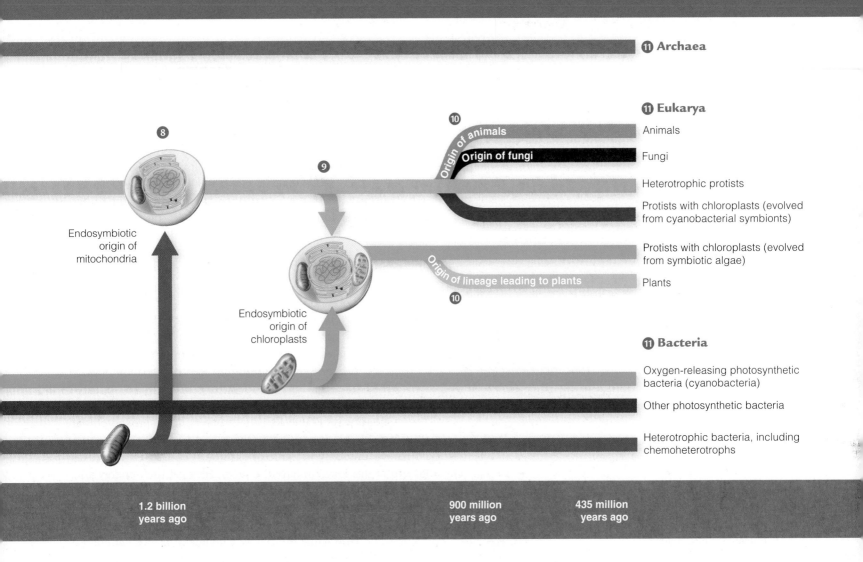

11 Archaea

11 Eukarya

8

10

Origin of animals

Origin of fungi

9

Animals

Fungi

Heterotrophic protists

Protists with chloroplasts (evolved from cyanobacterial symbionts)

Endosymbiotic origin of mitochondria

Endosymbiotic origin of chloroplasts

Origin of lineage leading to plants

10

Protists with chloroplasts (evolved from symbiotic algae)

Plants

11 Bacteria

Oxygen-releasing photosynthetic bacteria (cyanobacteria)

Other photosynthetic bacteria

Heterotrophic bacteria, including chemoheterotrophs

1.2 billion years ago

900 million years ago

435 million years ago

Endosymbiotic Origin of Mitochondria

8 An aerobic bacterium entered a eukaryotic cell and, over generations, descendants of the bacterial cell became mitochondria.

Endosymbiotic Origin of Chloroplasts

9 An oxygen-producing, photosynthetic bacterial cell entered a eukaryotic cell. Over generations, bacterial descendants evolved into chloroplasts.

Plants, Fungi, and Animals

10 By 900 million years ago, representatives of all major lineages—including fungi, animals, and the algae that would give rise to plants—had evolved in the seas.

Lineages That Have Endured to the Present

11 Modern organisms are related by descent and share certain traits. However, each lineage also has characteristic traits that evolved in response to the unique selective pressures it experienced.

13.4 The Prokaryotes

Enduring, Abundant, and Widespread The earliest cells had no nucleus; they were the first prokaryotes. Modern prokaryotes also lack a nucleus and, compared to eukaryotes, they are structurally simple. This simplicity has served them well. Prokaryotes existed for at least 2 billion years before eukaryotes evolved, and they have coexisted with eukaryotes for more than a billion years. Many prokaryotes survive by feeding on larger, more complex eukaryotes. From an evolutionary viewpoint, prokaryotes and eukaryotes are both successful.

In numbers prokaryotes are unparalleled. Biologists at the University of Georgia recently estimated that 5,000,000,000,000,000,000,000,000,000,000 bacterial cells are at this moment alive on Earth. Prokaryotes are abundant in and on our bodies, in our food and water, and in the air we breathe. They endure in places where little or no other life exists. Some make their home in scalding hot water near hydrothermal vents on the sea floor. Others live in glacial ice that never thaws. Prokaryotes have been found in acidic springs, alkaline pools, and Earth's driest desert. One species, *Bacillus infernus,* lives at 75°C (167°F) on rocks 3 kilometers (a little less than 2 miles) beneath the soil surface in Virginia. Its name means "bacterium from hell." In short, you will find prokaryotes just about any where there is a source of energy and carbon.

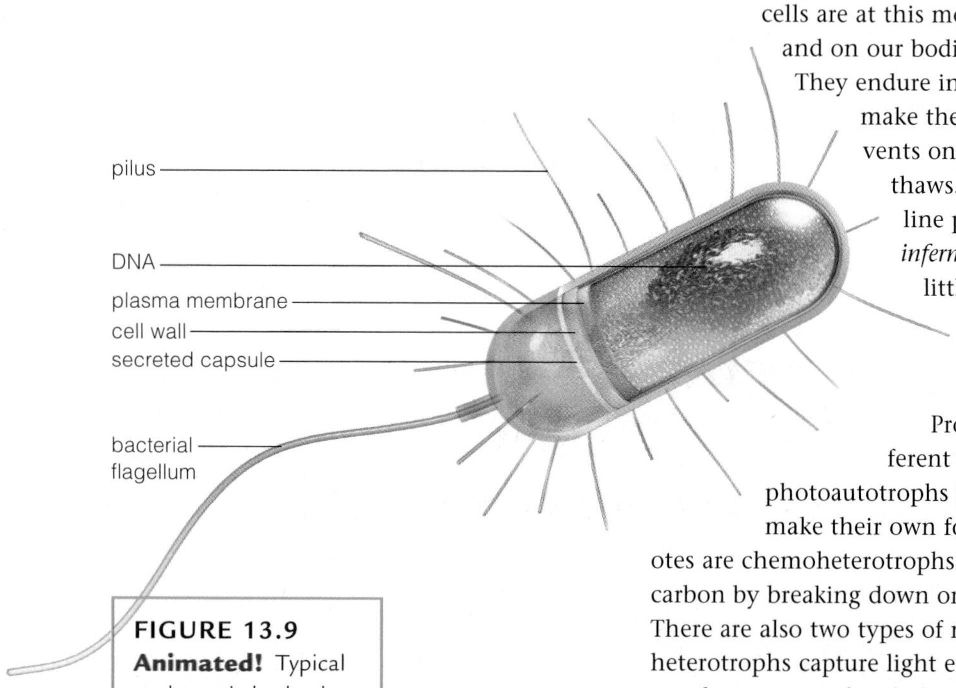

Prokaryotes meet their nutritional needs in four different ways (Table 13.1). Like plants, some prokaryotes are photoautotrophs that use sunlight energy and inorganic carbon to make their own food by the process of photosynthesis. Other prokaryotes are chemoheterotrophs. Like animals and fungi, they get their energy and carbon by breaking down organic compounds assembled by other organisms. There are also two types of nutrition that are unique to prokaryotes. Photoheterotrophs capture light energy, but rather than making their own food, they use the energy to break down organic molecules. Chemoautotrophs get energy by removing electrons from inorganic substances such as minerals in rocks, and use that energy to build organic compounds from carbon dioxide and water.

Prokaryotic Structure and Function Prokaryotes are small and, with rare exceptions, cannot be seen without a light microscope. Section 3.5 described the structure of prokaryotic cells, which we review briefly here. Figure 13.9 shows a typical prokaryotic cell. There is no nucleus or other membrane-enclosed organelles. The single prokaryotic chromosome (a ring of DNA) lies in the cytoplasm, as do the ribosomes.

Nearly all prokaryotes have a porous cell wall around their plasma membrane. The wall gives the cell its shape, which may be spherical, spiral, or rod-shaped. We describe a spherical prokaryotic cell as a coccus, a spiral-shaped one as a spirillum, and a rod-shaped one as a bacillus. The cell depicted in Figure 13.9 is a bacillus. Like many prokaryotes, it has a capsule made of secreted material around its cell wall.

Most prokaryotes can move from place to place. Some have one or more bacterial flagella that rotate like a propeller. Other bacteria glide along surfaces by using thin protein filaments called pili (singular, pilus) as grappling hooks. The pilus is

pilus
DNA
plasma membrane
cell wall
secreted capsule
bacterial flagellum

FIGURE 13.9 Animated! Typical prokaryotic body plan.

Table 13.1	How Organisms Get Energy and Carbon	
Nutritional Type	Energy Source	Carbon Source
Photoautotroph	Light	Carbon dioxide
Chemoheterotroph	Organic compounds	Organic compounds
*Photoheterotroph	Light	Organic compounds
*Chemoautotroph	Inorganic compounds	Carbon dioxide

* Prokaryotes only

extended out to a surface, sticks to it, then shortens, drawing the cell forward. Pili can also be used to lock a cell in place, or to draw cells together prior to the exchange of genetic material, as described below.

Reproduction and Gene Transfers
Prokaryotes have staggering reproductive potential. Some can divide every twenty minutes. One cell becomes two, then two become four, then four become eight, and so on. In some species, a daughter cell buds from a parent. More commonly, a cell reproduces by **prokaryotic fission**, a process that yields two equal-sized, genetically identical descendant cells (Figure 13.10**A**).

Prokaryotes do not reproduce sexually, but they do transfer genetic material among existing individuals. In the process of **prokaryotic conjugation**, one cell gives a small circle of DNA, called a **plasmid**, to another. A plasmid is separate from the larger bacterial chromosome and usually has only a few genes. Cells get together for conjugation when one cell extends a sex pilus out to a prospective partner and reels it in (Figure 13.10**B**). Once the cells are close together, the cell that made the sex pilus passes a copy of its plasmid to its partner. The cells then separate. Afterwards, each cell has a copy of the plasmid. Each will duplicate the plasmid and pass it along to its descendants. Each cell can also transfer a copy of the plasmid to other cells by conjugation.

Prokaryotic genomes also change by two other methods. First, a cell can take up DNA from its environment. Second, viruses that infect prokaryotes sometimes move genes from one cell to another.

The ability of prokaryotes to acquire new genetic information by conjugation or other methods has important implications. Suppose a gene for antibiotic resistance arises in one bacterial cell. Not only can this gene be passed on to that cell's descendants, but it can also be transferred to other existing cells. Gene transfers speed the rate at which a gene spreads through a population, thus accelerating the response to selective pressures.

Phylogeny and Classification
Until recently, reconstructing the evolutionary history of the prokaryotes seemed an impossible task. With the exception of stromatolites, there are very few prokaryotic fossils. Today, comparisons of gene sequences are helping to clarify relationships. For example, gene comparison studies opened the way for the division of all organisms into three domains, Archaea, Bacteria, and Eukarya:

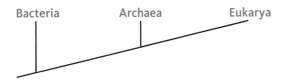

Classification of prokaryotes into species is complicated by the fact that many prokaryotes will not grow in the lab, which makes them difficult to characterize. Also, gene exchanges between species are common in prokaryotes. Prokaryotes even swap genes with members of the other domain. As a result, when defining prokaryotic species, microbiologists focus on shared ancestry (as revealed by gene sequence studies) and shared phenotypic traits, rather than the potential to interbreed.

In addition to naming species, microbiologists identify **strains**, subgroups within a species that are characterized by a specific trait or traits. For example, most strains of *Escherichia coli* cells that live in the mammalian gut are harmless or beneficial. However, one strain, *E. coli* O157:H7, makes a toxin that can taint food and cause a potentially deadly food poisoning.

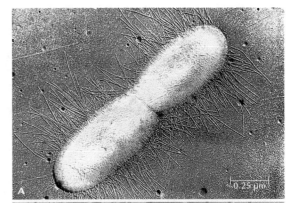

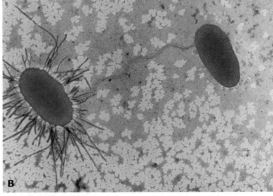

FIGURE 13.10 Animated! Prokaryotic reproduction and gene transfer. **A** One bacterial cell undergoing prokaryotic fission to form two cells. **B** A sex pilus from one cell contacts another cell prior to conjugation, which is a process that transfers genes.

plasmid Of a prokaryote, a small ring of nonchromosomal DNA with a few genes.

prokaryotic conjugation One prokaryotic cell transfers a plasmid to another.

prokaryotic fission Method of asexual reproduction in which one prokaryotic cell divides and forms two identical descendant cells.

strain A subgroup within a species that has a characteristic trait or traits.

FIGURE 13.11 Some archaean habitats.
A Heat-loving archaeans live in Yellowstone
National Park's hot springs. **B** Evaporation
ponds in Utah's Great Salt Lake are home to
some salt-loving archaeans. **C** Archaeans that
produce methane live inside the gut of cattle
and other animals.

Archaean Traits and Ecology **Archaea** is the more recently
discovered and less known prokaryotic domain. Gene comparisons indicate
that archaeans are the prokaryote group most closely related to eukaryotes.
Some can withstand conditions as harsh as those on the early Earth, so they
may resemble the first cells. Their group name, *archae–*, means "ancient."

Some of the archaea are **extreme thermophiles**, organisms that live in
very hot places. For example, archaeans have been found in scalding hot
water near deep-sea hydrothermal vents and in thermal pools such as those
in Yellowstone National Park (Figure 13.11**A**).

Other archaeans are **extreme halophiles**, organisms that live in highly
salty water, as in Utah's Great Salt Lake (Figure 13.11**B**). Some salt-loving
archaeans use purple pigments to capture light energy and make ATP. These
photoheterotrophs use the ATP to break down organic compounds, rather
than making their own food from carbon dioxide and water.

Still other archaeans are **methanogens**, or methane makers. These
chemoautotrophs release methane gas as a by-product of their metabolic
reactions. Methanogens cannot tolerate oxygen; it kills them. They live
in swamp sediments and the guts of many animals, including termites,
humans, and cattle (Figure 13.11**C**). Methane produced by gut-dwelling
methanogens exits animal bodies when the animals belch or "pass gas."

As biologists continue to explore archaean diversity, they are finding
that archaeans are not restricted to extreme environments. Archaeans live
alongside bacteria nearly everywhere. They are particularly abundant in the
ocean's deep waters. So far, scientists have not found any archaeans that
pose a major threat to human health, although some that live in the mouth
may encourage gum disease.

Bacterial Traits and Ecology **Bacteria** is the more diverse and
better-studied domain of prokaryotes. The bacterial plasma membrane, cell
wall, and flagella are built somewhat differently than those of archaeans
and the two groups are genetically distinct.

Many bacteria have important roles in chemical cycles. For example,
some photosynthetic bacteria put oxygen into the atmosphere. Photosynthesis evolved in many bacterial lineages, but only the cyanobacteria
(Figure 13.12**A**) use a pathway that produces oxygen. Biologists infer that
ancient cyanobacteria were the ancestors of modern chloroplasts. Thus, we
have cyanobacteria and their chloroplast relatives to thank for nearly all of
the oxygen that we breathe.

Cyanobacteria are also among the bacteria that make nitrogen available
to plants and photosynthetic protists. There is a lot of nitrogen gas in the
air, but eukaryotes cannot use it. Of all organisms, only bacteria can carry
out **nitrogen fixation**. In this process, bacteria make ammonia by combining hydrogen with nitrogen atoms they get from nitrogen gas. Plants and
other photosynthetic eukaryotes can then take up the ammonia and use it
to build essential molecules, such as amino acids.

Some nitrogen-fixing bacteria live as free cells in the soil. Others enter
and live inside the roots of legumes, a group of plants that includes peas,
alfalfa, and clover. The plants benefit from the presence of the bacteria,
which provide them with ammonia. The bacteria benefit by living in the
shelter of the roots and receiving sugar from the plant.

Bacteria also help cycle nutrients when they serve as **decomposers**,
organisms that break organic material down into its inorganic subunits.
Together with fungi, bacteria ensure that nutrients in wastes and remains of
organisms return to the soil in a form that plants can take up and use.

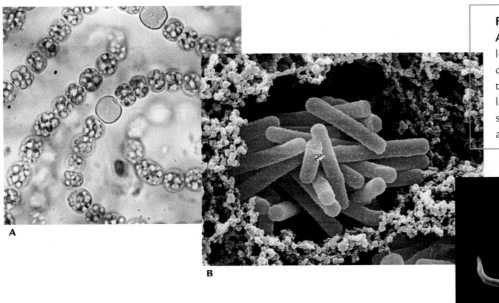

FIGURE 13.12 Examples of bacteria.
A Aquatic cyanobacteria. This species forms long chains of cells. Some members of the chain specialize in nitrogen fixation and share the nitrogen with the others. **B** Cluster of lactate-fermenting bacteria in yogurt. **C** A spirochete. This species causes Lyme disease, and ticks **D** serve as its vector.

Bacteria that carry out lactate fermentation are among the decomposers. Sometimes lactate-fermenting bacteria get into our food and spoil it, as when they cause milk to go sour. On the other hand, we use some lactate fermenters to make sauerkraut, pickles, cheese, and yogurt (Figure 13.12**B**).

Lactate fermenters are also among the estimated 100 trillion bacterial cells in a healthy human intestine. The acidity of the lactate these bacteria produce helps keep disease-causing organisms from taking hold in our gut. Other intestinal bacteria benefit us by producing essential vitamins or by breaking down materials we could not otherwise digest.

Beneficial prokaryotes and other microorganisms that normally live in or on our body are referred to as our **normal flora**. Their importance can become apparent when we take antibiotics. These drugs kill beneficial bacteria as well as harmful ones, so they can disrupt the balance among microorganisms. When this happens in the intestine, it can cause diarrhea. When it happens in the vagina, the result is an overgrowth of yeast (fungal) cells that cause vaginitis.

Despite being small and structurally simple, some bacteria show remarkably complex behavior. Myxobacteria are tiny predators that glide about as a cohesive swarm eating other soil bacteria. When food dwindles, thousands of cells form a multicelled structure called a fruiting body. Cells that will become part of the fruiting body's branches move into place and then die. Other cells become resting spores in capsules at the tips of the branches. Researchers are studying how bacterial cells communicate with one another and coordinate their movements and development as they form a fruiting body.

Disease-Causing Bacteria

Some bacteria are **pathogens**, organisms that infect other species and cause disease. Pathogenic bacteria enter the human body in a variety of ways. For example, people develop Lyme disease after a bite from an infected tick introduces a spiral-shaped bacteria into their body (Figure 13.12**C**,**D**). We call an animal that transmits a disease between hosts a **vector**. Ticks are the vector for the bacterium that causes Lyme disease.

Pathogenic bacteria that live in the human reproductive tract do not need a vector. Instead, human sexual behavior delivers them to new hosts. Gonorrhea, syphilis, and chlamydia are examples of sexually transmitted bacterial diseases.

Bacteria also enter our body when we eat or drink tainted food or water. In the United States, the rod-shaped bacterium *Salmonella* is a commonly reported

Archaea Prokaryotic domain most closely related to eukaryotes; many members live in extreme environments.

Bacteria Most diverse prokaryotic domain.

decomposer Organism that breaks organic material down into its inorganic subunits.

extreme halophile Organism that lives where the salt concentration is high.

extreme thermophile Organism that lives where the temperature is very high.

methanogen Organism that produces methane gas as a metabolic by-product.

nitrogen fixation Process of combining nitrogen gas with hydrogen to form ammonia.

normal flora Collection of microorganisms that normally live in or on a healthy animal or person.

pathogen Organism that infects another organism and causes disease.

vector Animal that transmits a pathogen between its hosts.

cause of foodborne disease. Infection by *Salmonella* causes nausea, abdominal cramps, and diarrhea, but is rarely life-threatening. *E. coli* O157:H7 is a rarer but more dangerous foodborne pathogen. It causes about 60 deaths per year in the United States. Worldwide, waterborne bacterial diseases such as cholera are major killers, responsible for an estimated 2 million deaths per year.

Some bacteria form resting structures called endospores that can survive boiling, irradiation, and drying out. When conditions improve, an endospore germinates and a bacterium emerges. Endospores that germinate in the human body can be lethal. In 2001, *Bacillus anthracis* endospores sent in the mail caused fatal cases of anthrax in five people who inhaled them. *Clostridium tetani* endospores from the soil sometimes gets into wounds. The result is a disease called tetanus. It locks muscles in contraction, an effect referred to as "lockjaw."

Take-Home Message

What are the modern prokaryotes like?

- Nearly all prokaryotes are microscopically small. They reproduce by prokaryotic fission or budding, and they swap genes by conjugation and other processes.

- Collectively, prokaryotes are the most metabolically diverse organisms. They live almost anywhere there is carbon and energy.

- Archaeans are the prokaryotes most closely related to eukaryotes. Many live in extremely hot or salty habitats.

- Bacteria, better known and more diverse, include species that play important roles in nutrient cycles. Some live in or on the human body, either as normal flora or as pathogens.

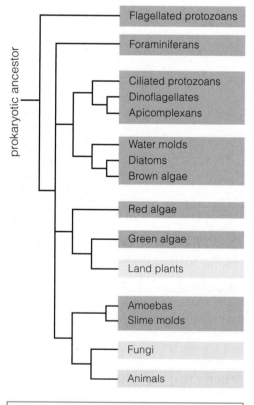

FIGURE 13.13 Evolutionary tree for the eukaryotes. *Orange* boxes indicate the lineages that are considered protists.

13.5 The Protists

Of all existing species, **protists** are the most like the first eukaryotic cells. Unlike prokaryotes, protist cells have a nucleus and a cytoskeleton that includes microtubules. Most also have mitochondria, endoplasmic reticulum, and Golgi bodies. All protists have multiple chromosomes, each consisting of DNA with many proteins attached. Protists reproduce asexually, sexually, or both.

Most protists are single-celled species. But there are also colonial forms and multicelled species. Many protists are photoautotrophs that have chloroplasts. Others are predators, parasites, or decomposers.

Until recently, protists were defined largely by what they are *not*—not prokaryotes, not plants, not fungi, not animals. They were lumped together in a kingdom between prokaryotes and the "higher" forms of life. Thanks to improved comparative methods, protists are now being assigned to groups that reflect evolutionary relationships. Figure 13.13 shows where the protists that we cover in this book fit in the eukaryote family tree. There are many other protist lineages, but this sampling will demonstrate their diversity and importance. Notice that protists are not a single lineage. Some of the protists are actually more closely related to plants, animals, or fungi than they are to other protists.

Flagellated Protozoans "Protozoans" is the general term for heterotrophic protists that live as single cells. **Flagellated protozoans** are single, unwalled cells with one or more flagella (structures described in Section 3.6). All groups are entirely or mostly heterotrophic. A **pellicle**, a layer of elastic proteins just beneath the plasma membrane, helps the cells retain their shape.

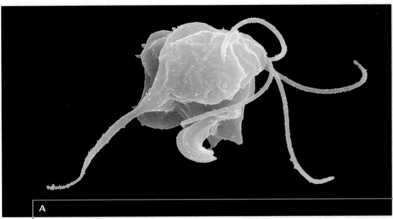

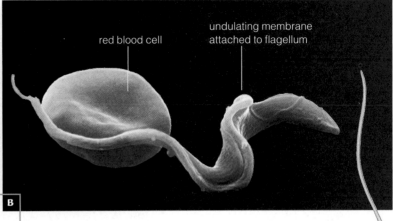

red blood cell

undulating membrane attached to flagellum

A

B

FIGURE 13.14 **Animated!** Flagellated protozoans. **A** *Trichomonas*, a sexually transmitted parasite of humans that has many flagella. **B** The trypanosome that causes African sleeping sickness. **C** Structure of a photosynthetic euglenoid.

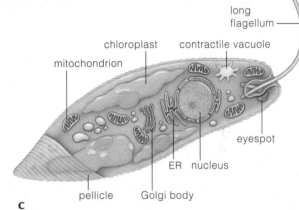

long flagellum

chloroplast   contractile vacuole

mitochondrion

eyespot

ER   nucleus

pellicle   Golgi body

C

Some protozoans have multiple flagella that they use to swim through the body fluids of animals. For example, *Trichomonas* (Figure 13.14**A**) infects the human reproductive tract. It causes trichomoniasis, one of the most common sexually transmitted diseases among college students. Another multiflagellated protozoan, *Giardia*, lives as an intestinal parasite inside humans, cattle, and some wild animals. *Giardia* cells can survive outside the host body in a resting stage called a cyst. If you drink water from a stream contaminated with *Giardia* cysts, you could end up with the disease giardiasis. Symptoms range from mild cramps to severe diarrhea that sometimes persists for weeks.

Trypanosomes are another group of parasitic flagellates. These long, tapered cells have a single flagellum that runs the length of the body inside a membrane. Biting insects act as vectors for trypanosomes that parasitize humans. In sub-Saharan Africa, tsetse flies spread the trypanosome that causes African sleeping sickness (Figure 13.14**B**). This fatal disease affects the nervous system. Infected people become drowsy during the daytime and often cannot sleep at night.

Euglenoids have a single long flagella and most live in ponds and lakes. As in other freshwater protists, one or more **contractile vacuoles** counter the tendency of water to diffuse into the cell. Excess water collects in the contractile vacuoles, which contract and expel it to the outside (Figure 13.14**C**). Euglenoids are related to trypanosomes, and the early ones were heterotrophs. Many still are, but others such as the one depicted in Figure 13.14**C** have chloroplasts that evolved by endosymbiosis.

Foraminiferans **Foraminiferans**, or forams, are single-celled predators that secrete a shell containing calcium carbonate. Threadlike cytoplasmic extensions protrude through openings in the cell. Most forams live on the sea floor, where they probe the water and sediments for prey. Others are part of the marine **plankton**, the mostly microscopic organisms that drift or swim in the open sea. The planktonic forams often have photosynthetic protists that live in their cytoplasm (Figure 13.15). Long ago, the calcium-rich remains of forams and other protists with calcium carbonate shells began to accumulate on the sea floor. Over great spans of time, these deposits became limestone and chalk.

FIGURE 13.15 A planktonic foraminiferan. *Gold* dots on its cytoplasmic extensions are algal cells.

contractile vacuole In freshwater protists, an organelle that collects and expels excess water.

flagellated protozoan Member of a heterotrophic lineage of protists that have one or more flagella.

foraminiferan Heterotrophic protist that secretes a calcium carbonate shell.

pellicle Layer of proteins that gives shape to many unwalled, single-celled protists.

plankton Community of tiny drifting or swimming organisms.

protist A eukaryote that is not a fungus, plant, or animal.

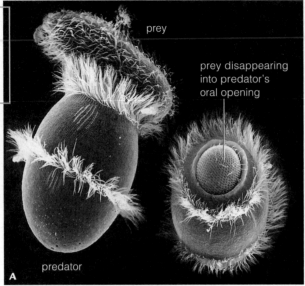

FIGURE 13.16 Aquatic protists. **A** A barrel-shaped ciliate with tufts of cilia (*Didinium*), catching and engulfing a *Paramecium*, another ciliate. **B** Two dinoflagellate cells.

prey

prey disappearing into predator's oral opening

predator

A

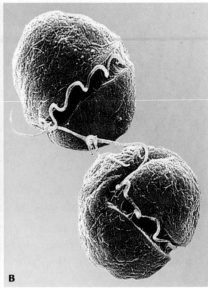

B

FIGURE 13.17 Dinoflagellate bioluminescence. Tropical waters shimmer with light when countless dinoflagellates emit light in response to a disturbance.

The Ciliates Ciliated protozoans, or ciliates, are close relatives of dinoflagellates and apicomplexans. **Ciliates** are unwalled cells that have many cilia (structures described in Section 3.6). Most ciliates are predators in seawater or fresh water. They feed on bacteria, algae, and one another (Figure 13.16**A**). Ciliates also are among the microorganisms that live in the gut of cattle, sheep, and other mammalian grazers. Like some bacteria, these ciliates help their host animals digest plant material. Only one ciliate is known to cause disease in humans. It also infects pigs, and people usually become infected when pig feces with a resting form of the ciliate gets into their drinking water. Infection causes nausea and diarrhea.

Dinoflagellates The name **dinoflagellate** means "whirling flagellate." These single-celled protists typically have two flagella, one at the cell's tip and the other running in a groove around its middle like a belt (Figure 13.16**B**). Combined action of the two flagella causes the cell to rotate as it moves forward. Most dinoflagellates deposit cellulose just beneath their plasma membrane, and these deposits form thick protective plates.

Dinoflagellates live in fresh water and the oceans. Some prey on bacteria, others are parasites of animals, and still others have chloroplasts that evolved from algal cells. A few photosynthetic species live inside cells of reef-building corals. They supply their coral host, which is an invertebrate animal, with essential sugars.

Some marine dinoflagellates are **bioluminescent**. Like fireflies, they can convert ATP energy into light (Figure 13.17). Emitting light may protect a cell by startling a predator that was about to eat it. By another hypothesis, the flash of light acts like a car alarm. It attracts the attention of other organisms, including predators that pursue would-be dinoflagellate eaters.

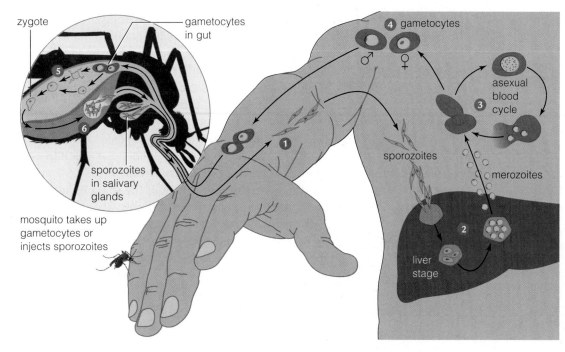

zygote

gametocytes in gut

④ gametocytes

③ asexual blood cycle

② sporozoites

merozoites

liver stage

sporozoites in salivary glands

mosquito takes up gametocytes or injects sporozoites

FIGURE 13.18 Animated! Life cycle of the apicomplexan that causes malaria.

❶ Infected mosquito bites a human. Sporozoites enter the blood, which carries them to the liver.

❷ Sporozoites reproduce asexually in liver cells, then mature into merozoites. Merozoites leave the liver and enter the bloodstream, where they infect red blood cells.

❸ Inside some red blood cells, merozoites reproduce asexually. These cells burst and release more merozoites into the bloodstream.

❹ Inside other red blood cells, merozoites develop into male and female gametocytes.

❺ A female mosquito bites and sucks blood from the infected person. Gametocytes in red blood cells enter her gut and mature into gametes, which fuse to form zygotes.

❻ The zygotes develop into sporozoites that migrate to the mosquito's salivary glands.

In nutrient-enriched water, free-living photosynthetic dinoflagellates or other aquatic protists sometimes undergo great increases in population size, a phenomenon known as an **algal bloom**. Algal blooms can harm other organisms. When the cells die, aerobic bacteria that feed on their remains use up all the oxygen in the water, so aquatic animals suffocate. Also, some dinoflagellates produce toxins that can kill aquatic organisms directly and sicken people.

Apicomplexans

Apicomplexans are parasitic protists that spend part of their life inside cells of their hosts. Their name refers to a complex of microtubules at their apical (top) end that allows them to enter a host cell. They are also sometimes called sporozoans.

Apicomplexans infect a variety of animals, from worms and insects to humans. Their life cycle often involves more than one host species. For example, Figure 13.18 shows the life cycle of *Plasmodium*, the apicomplexan that causes malaria. Saliva of a female *Anopheles* mosquito transmits an infective stage (called the sporozoite) to a human host when she bites ❶. A sporozoite travels in blood vessels to liver cells, where it reproduces asexually ❷. Some offspring, called merozoites, enter red blood cells, where they produce more merozoites ❸. Other merozoites enter red blood cells and develop into immature gametes, or gametocytes ❹. When a mosquito bites an infected person, gametocytes are taken up with blood and mature in the mosquito gut. Gametes fuse and form zygotes ❺. Zygotes develop into new sporozoites that migrate to the insect's salivary glands, where they await transfer to a new vertebrate host ❻. Malaria symptoms usually start a week or two after a bite, when the infected liver cells rupture and release merozoites and cellular debris into blood (art at *right*). After the initial episode, symptoms may subside, but continued infection damages the body and is eventually fatal. Malaria kills more than a million people each year.

Water Molds, Diatoms, and Brown Algae

Water molds are heterotrophs known to scientists as oomycotes. The term means "egg fungus," and these organisms were once mistakenly grouped with the fungi. Like fungi, the oomycotes form a mesh of nutrient-absorbing filaments, but the two groups

algal bloom Population explosion of single-celled aquatic organisms such as dinoflagellates.

apicomplexan Parasitic protist that enters and lives inside the cells of its host.

bioluminescent Able to use ATP to produce light.

ciliate Single-celled, heterotrophic protist with many cilia.

dinoflagellates Single-celled, aquatic protist typically with cellulose plates and two flagella; may be heterotrophic or photosynthetic.

water mold Heterotrophic protist that forms a mesh of nutrient-absorbing filaments.

differ in many structural traits and are genetically distinct. Most oomycotes help decompose organic debris and dead organisms in aquatic habitats, but a few are parasites that have significant economic effects. Some grow as fuzzy white patches on fish in fish farms and aquariums. Others infect land plants, destroying crops and forests. Members of the genus *Phytophthora* are especially notorious. Their name means "plant destroyer" and they cause an estimated $5 billion dollars in crop losses each year. In the mid-1800s, one species destroyed Irish potato crops, causing a famine that killed and displaced millions of people. Today, another species is causing an epidemic of sudden oak death in Oregon, Washington, and California. Millions of oaks have already died.

The closest relatives of the water molds are two photosynthetic groups: diatoms and brown algae (Figure 13.19). Both have chloroplasts that include a brownish accessory pigment (fucoxanthin) that tints them olive, golden, or dark brown.

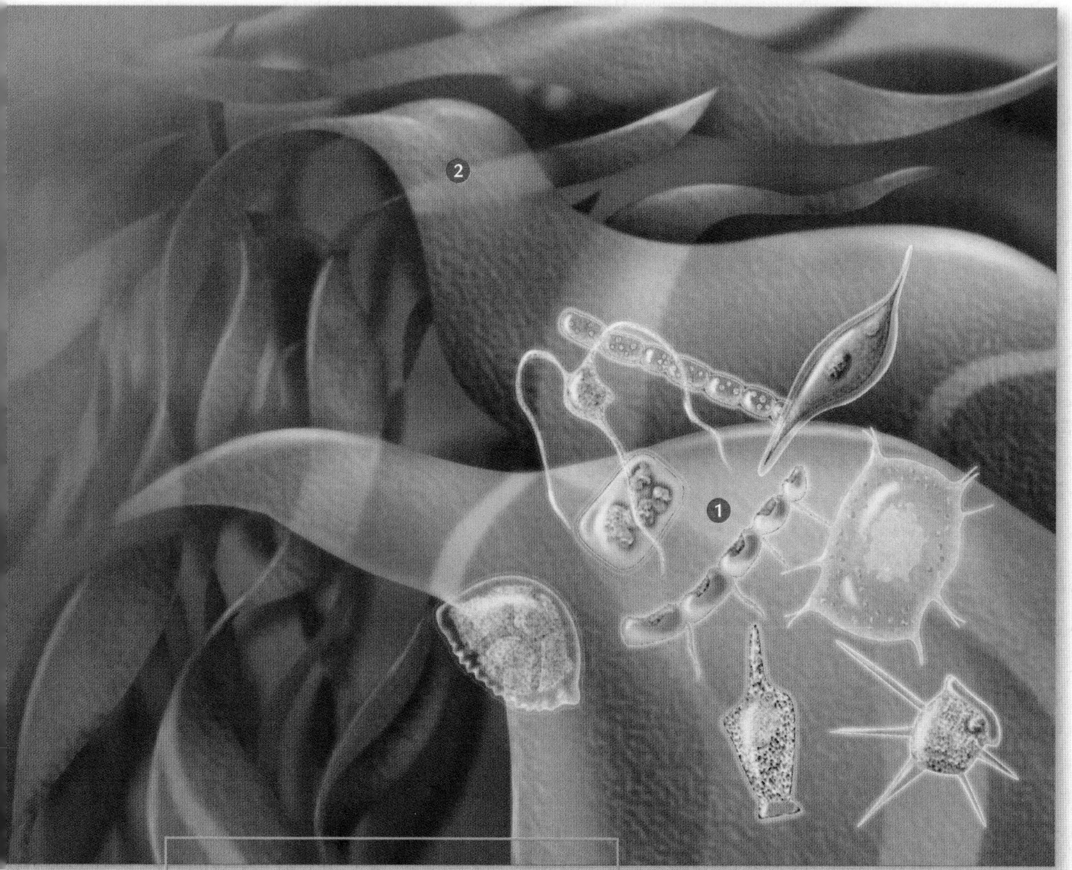

Diatoms have a two-part silica shell, with upper and lower parts that fit together like a shoe box. Some cells live individually, and others form chains ❶. Most diatoms float near the surface of seas or lakes, but some live in moist soil or in water droplets that cling to mosses in damp habitats.

Marine diatoms have lived and died in the oceans for many millions of years, and their remains form vast deposits on the sea floor. In places where these deposits have been uplifted and are now on land, silica-rich diatomaceous earth is quarried for use in filters, abrasive cleaners, and as an insecticide that is not harmful to vertebrates.

Brown algae are multicelled inhabitants of temperate or cool seas. In size, they range from microscopic filaments to giant kelps that stand 30 meters (100 feet) tall. Giant kelps form forestlike stands in coastal waters of the Pacific Northwest ❷. Like trees in a forest, kelps shelter a wide variety of other organisms. The Sargasso Sea in the North Atlantic Ocean is named for its abundance of *Sargassum*. This kelp forms vast, floating mats up to 9 meters (30 feet) thick. The mats provide food and shelter to fish, sea turtles, and invertebrates.

Sargassum and other brown algae have commercial uses. Alginic acid from the cell walls of brown algae is used to produce aligns, which serve as thickeners, emulsifiers, and suspension agents. Algins are used to manufacture ice cream, pudding, jelly beans, toothpaste, cosmetics, and other products.

FIGURE 13.19 Artist's depiction of two related protist groups common in California's coastal waters, ❶ microscopic diatoms with a silica shell and ❷ a large multicelled kelp (a brown alga).

Red Algae and Green Algae Some **red algae** are single cells, but most are multicelled forms that live in tropical seas. Most commonly they have a branching structure (Figure 13.20**A**), but some form thin sheets. Coralline algae (red algae with cell walls hardened by calcium carbonate) are a component of tropical coral reefs. Red algae are tinted red to black by accessory pigments called phycobilins. These pigments absorb the blue-green light that penetrates deep into water. The presence of phycobilins allows red algae to carry out photosynthesis at greater depths than other algae.

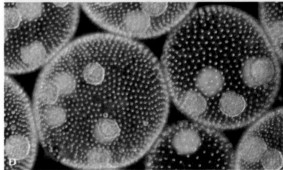

FIGURE 13.20 Red algae and green algae. **A** A marine red alga with the typical branching body plan. **B** A multicelled marine green alga with spongy branches. **C** A single-celled freshwater green alga that has just divided to form two cells. The larger part of each cell is a remnant of the parent. The smaller portion is new. **D** *Volvox*, a colonial green freshwater alga. Each sphere is a colony made up of flagellated cells linked by thin cytoplasmic strands. New colonies (smaller spheres) form inside the parent colony.

Red algae have many commercial uses. Nori, the sheets of seaweed used to wrap some sushi, is a red alga that is grown commercially. Agar and carrageenan are valuable products extracted from the cell walls of other red algae. Agar keeps baked goods and cosmetics moist, helps jellies set, and is used to make capsules that hold medicines. Carrageenan is added to soy milk, dairy foods, and the fluid that is sprayed on airplanes to prevent ice formation.

Green algae include single-celled, colonial, and multicelled forms (Figure 13.20**B–D**). Most live in fresh water, some are marine, and some grow on soil, trees, or other damp surfaces. A few partner with a fungi to form lichens.

Some multicelled green algae are harvested from the sea as food, especially in Asia. The single-celled freshwater species *Chlorella* is grown commercially and sold in dried form as a food supplement.

Shared traits of red algae, green algae, and land plants include the presence of chloroplasts, a particular type of chlorophyll, and a cell wall of cellulose. These similarities are evidence that these three groups share a common ancestor. Both algal groups evolved from an ancestral protist that had chloroplasts descended from cyanobacteria. After the red algal and green algal lineages split, land plants evolved from one lineage of green algae.

Amoebozoans We began this section by describing how the lineages previously lumped together as protists are now being sorted into smaller groups. The new group Amoebozoa is an outcome of this research. Most amoebozoans do not have a cell wall, shell, or constrictive pellicle and so can alter their shape. A cell may be a compact blob at one moment, then long and narrow the next. Cells engulf prey and move about by extending lobes of cytoplasm called pseudopods, as described in Section 3.6.

brown alga Multicelled, photosynthetic protist with brown accessory pigments.

diatom Single-celled photosynthetic protist with brown accessory pigments and a two-part silica shell.

green alga Single-celled, colonial, or multicelled photosynthetic protist belonging to the group most closely related to land plants.

red alga Single-celled or multicelled photosynthetic protist with red accessory pigment.

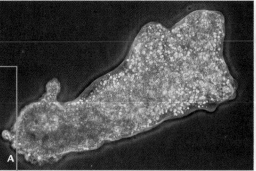

FIGURE 13.21 Animated!
Examples of amoebozoans. **A** One of the freshwater amoebas. **B** Plasmodial slime mold streaming across a log. **C** Life cycle of a cellular slime mold (*Dictyostelium*).

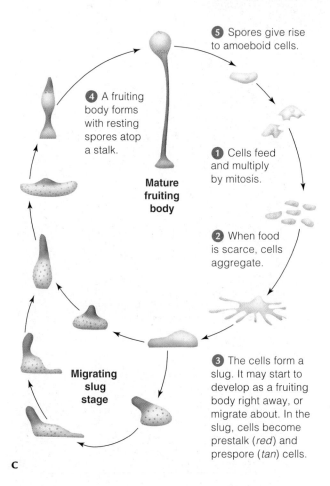

5 Spores give rise to amoeboid cells.

4 A fruiting body forms with resting spores atop a stalk.

Mature fruiting body

1 Cells feed and multiply by mitosis.

2 When food is scarce, cells aggregate.

3 The cells form a slug. It may start to develop as a fruiting body right away, or migrate about. In the slug, cells become prestalk (*red*) and prespore (*tan*) cells.

Migrating slug stage

C

The **amoebas** live as single cells. Figure 13.21**A** shows one example. Like most amoebas, this one is a predator in freshwater habitats. Amoebas can also live in the animal gut, and some of these cause disease. Each year, about 50 million people suffer from amebic dysentery after drinking water contaminated with pathogenic amoebas.

Slime molds are sometimes called "social amoebas." There are two kinds. **Plasmodial slime molds** spend most of their life cycle as a flat, slimy mass that streams over logs and along the forest floor (Figure 13.21**B**). This mass is usually a few centimeters across, large enough to be seen with the naked eye. It is essentially a giant cell with hundreds of nuclei. The mass feeds on bacteria and when food supplies dwindle, it develops into one or more spore-bearing structures, or fruiting bodies.

In contrast, **cellular slime molds** spend the bulk of their existence as individual amoeba-like cells (Figure 13.21**C**). Each cell eats bacteria and reproduces by mitosis **1**. When the food runs out, thousands of cells come together **2**. Often they form a "slug" that migrates in response to light and heat **3**. When the slug reaches a suitable spot, its cells differentiate and form a fruiting body **4**. A stalk forms and lengthens, then spores form at its tip. Germination of a spore releases a diploid amoeboid cell that starts the cycle anew **5**.

Cellular slime molds are often used in research. They provide clues as to how signaling pathways of multicelled animals evolved. Coordinated behavior, which is an ability to respond to stimuli as a unit, requires cell-to-cell communication, which may have originated in an amoeboid ancestor.

amoeba Single-celled heterotrophic protist that moves and feeds by extending pseudopods.

bacteriophage Virus that infects prokaryotes.

cellular slime mold Heterotrophic protist that usually lives as a single-celled predator. When conditions are unfavorable, cells aggregate into a cohesive group that can form a fruiting body.

plasmodial slime mold Heterotrophic protist that moves and feeds as a multinucleated mass; forms a fruiting body when conditions are unfavorable.

virus Infectious particle that consists of protein and nucleic acid and replicates only inside a living cell.

Take-Home Message
What are the protists?

- The protists are not a natural group, but rather a diverse collection of eukaryotic lineages. Some are only distantly related to one another.

- Most protists live as single cells, but there are colonial and multicelled species.

- We find protistan photosynthesizers, predators, and decomposers in lakes, seas, and damp places on land. Protists also live inside the bodies of other eukaryotes, including humans. Some of these protists are helpful, but others are parasites and pathogens.

13.6

The Viruses

Viral Characteristics and Diversity A **virus** is a noncellular infectious agent. It consists of a protein coat wrapped around genetic material (either RNA or DNA) and a few viral enzymes. A virus is far smaller than any cell and does not have ribosomes or other metabolic machinery. To replicate, the virus must insert its genetic material into a cell of a specific organism, which we call its host. A viral infection is like a cellular hijacking. Viral genes take over a host cell's machinery and direct it to synthesize viral proteins and nucleic acids. These components then self-assemble as new viral particles.

Each kind of virus multiplies only in certain hosts. **Bacteriophages** are viruses that infect prokaryotes. One well-studied group of bacteriophages has a complex structure (Figure 13.22**A**). Their DNA is encased in a protein "head." Attached to the head is a rodlike "tail" with fibers that help the virus attach to its host. As Figure 6.12 illustrated, experiments using bacteriophages helped reveal DNA's function as genetic material.

Plant cells have a thick wall, so plant viruses usually infect a plant only after insects or pruning create a wound that allows the virus into a cell. The tobacco mosaic virus is rod-shaped, with coat proteins arranged in a helix around a strand of RNA (Figure 13.22**B**). It was the first virus ever discovered.

Adenoviruses are naked (non-enveloped) viruses that infect animals. Their 20-sided protein coat has a distinctive protein spike at each corner (Figure 13.22**C**). Human upper respiratory ailments are usually caused by adenoviruses. Other naked viruses cause hepatitis, polio, common colds, and warts.

More often, animal viruses have an envelope made of membrane derived from the host cell in which the virus self-assembled. For example, herpesviruses are DNA viruses with a 20-sided coat beneath their envelope (Figure 13.22**D**). Enveloped RNA viruses cause AIDS, rabies, rubella (German measles), bronchitis, mumps, measles, yellow fever, and West Nile encephalitis.

> A viral infection is like a cellular hijacking. Viral genes take over a host's cellular machinery.

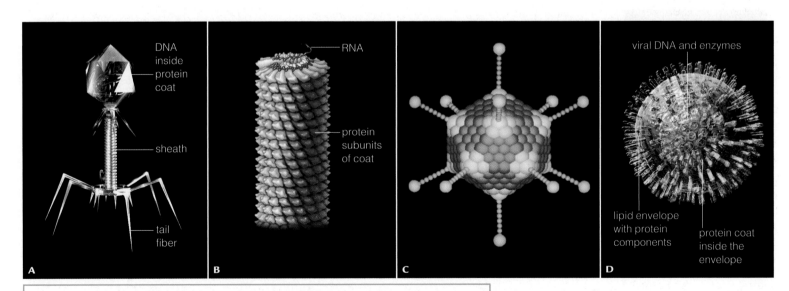

FIGURE 13.22 Examples of viral structure. **A** A bacteriophage with a complex coat structure. **B** Protein subunits bound to nucleic acid form a spiral in this tobacco mosaic virus. **C** An adenovirus has a coat with a 20-sided shape. **D** The envelope of a herpesvirus consists of membrane from the cell in which the virus formed.

Viral Multiplication Cycles Viral multiplication cycles are varied, but nearly all go through the following steps. The virus first attaches to an appropriate host cell by binding to a specific protein or proteins in the host's plasma membrane. A virus cannot seek out its host, but rather relies on a chance encounter. The virus, or just its genetic material, enters the cell. Viral genes direct the cell to replicate the viral DNA or RNA and to build viral proteins. These components self-assemble to form new viral particles. The new viruses bud from the infected host cell or are released when the host bursts.

Figure 13.23 shows two pathways that are common among bacteriophages. In the **lytic pathway**, the virus attaches to the host cell and injects its DNA. Viral genes direct the host to make viral DNA and proteins, which assemble as viral particles. Under direction of viral genes, the host now produces an enzyme that initiates **lysis**: The enzyme breaks up the plasma membrane, killing the cell and allowing new viral particles to escape into the environment.

In the **lysogenic pathway**, a virus enters a latent state that extends the multiplication cycle. Viral DNA becomes integrated into the host chromosome. The viral DNA is copied along with host DNA, and is passed along to all descendants of the host cell. Like tiny time bombs, the viral DNA inside these descendant cells awaits a signal to enter the lytic pathway. When the cell does enter this pathway, copies of some bacterial genes may become packaged into new viruses, along with the viral genes. This is how bacteriophages can transfer genes from one prokaryotic cell to another.

As another example of replication, consider the human immunodeficiency virus, or HIV. This virus causes AIDS (acquired immune deficiency syndrome). We delve into the symptoms of AIDS in depth when we discuss the immune system in Chapter 22. Here, we look at the steps by which HIV infects and multiplies inside human cells.

HIV is an enveloped RNA virus, and Figure 13.24 shows the structure of a free viral particle, or viroid. Notice the protein spikes that extend out through the envelope. Replication begins when one of these spikes bonds with a protein

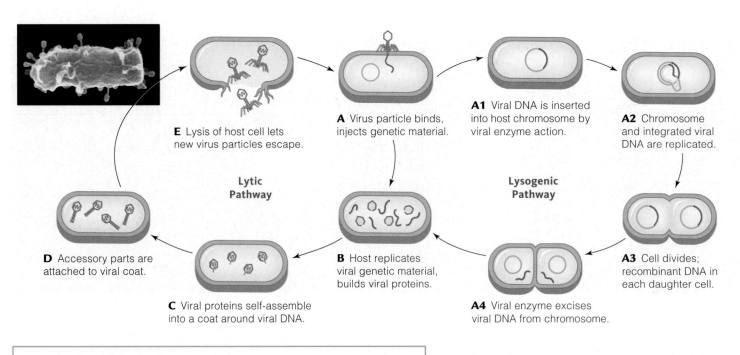

E Lysis of host cell lets new virus particles escape.

A Virus particle binds, injects genetic material.

A1 Viral DNA is inserted into host chromosome by viral enzyme action.

A2 Chromosome and integrated viral DNA are replicated.

Lytic Pathway

Lysogenic Pathway

D Accessory parts are attached to viral coat.

B Host replicates viral genetic material, builds viral proteins.

A3 Cell divides; recombinant DNA in each daughter cell.

C Viral proteins self-assemble into a coat around viral DNA.

A4 Viral enzyme excises viral DNA from chromosome.

FIGURE 13.23 Animated! The two bacteriophage replication pathways.

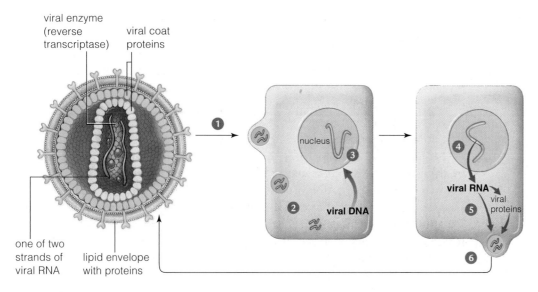

viral enzyme (reverse transcriptase)

viral coat proteins

nucleus

viral DNA

viral RNA

viral proteins

one of two strands of viral RNA

lipid envelope with proteins

FIGURE 13.24 Animated!
Multiplication cycle of HIV, the retrovirus that causes AIDS.

❶ Viral RNA and protein enter the host cell.

❷ Viral reverse transcriptase uses viral RNA to make double-stranded viral DNA.

❸ Viral DNA gets integrated into the host's chromosome.

❹ Viral DNA gets transcribed along with the host's genes.

❺ Some RNA transcripts are new viral RNA. Others are translated into viral proteins. RNA and proteins assemble as new virus particles.

❻ Viral particles bud from the infected cell. *Below*, electron micrographs show HIV budding from the plasma membrane of a cell.

at the surface of a specific type of white blood cell. The viral envelope fuses with the cell's plasma membrane and the viral contents enter the host cell ❶. These contents include a few enzymes and the viral RNA. A viral enzyme called reverse transcriptase converts viral RNA into double-stranded DNA ❷. This conversion puts viral genes into a form that the cell's genetic machinery can read.

The viral DNA enters the host cell's nucleus, where it becomes integrated into the host's chromosome ❸. A viral enzyme assists in this process. Once integrated, the viral DNA is transcribed along with the host's genes ❹. Some of the resulting RNA will become the genetic material of new HIV particles. Other transcripts encode viral proteins. The transcripts leave the nucleus and are translated on the host's ribosomes. Viral genetic material and viral proteins in the cytoplasm self-assemble as new viral particles ❺. In the final step, these particles move to the cell's plasma membrane and acquire their envelope as they bud from the cell ❻. Each viral particle can now infect other white blood cells. Once a cell has been infected, diversion of cell resources to virus production and the ongoing loss of membrane to budding virus will eventually kill it.

Drugs designed to fight HIV take aim at steps in viral replication. Some interfere with binding of HIV to a host cell. Others such as AZT slow reverse transcription of RNA. Integrase inhibitors prevent viral DNA from integrating into a human chromosome. Protease inhibitors prevent the processing of newly translated polypeptides into mature viral proteins.

Viral Origins and Ecology
How did viruses originate, and how are they related to cellular organisms? There are three main hypotheses. The first is that viruses are descendants of cells that were parasites inside other cells. Over time, most functions of these parasitic cells were delegated to the host, leaving the virus unable to survive on its own. A second hypothesis is that viruses are genetic elements that escaped from cells. The fact that some viral genes are similar to genes in cellular organisms lends support to these two hypotheses. But as noted earlier, viruses sometimes pick up genes from their hosts. The third hypothesis is that viruses represent a separate evolutionary branch, that they arose independently from the replicating molecules that preceded the origin of cellular life. This hypothesis would explain why most viral proteins are unlike any found in cellular organisms.

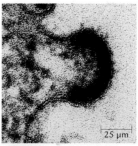

25 μm

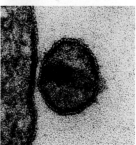

lysogenic pathway Bacteriophage replication pathway in which the virus becomes integrated into the host's chromosome and is passed to its descendants.

lysis Breaking of a cell's plasma membrane; results in death of the cell.

lytic pathway Bacteriophage replication pathway in which the virus replicates in its host and quickly kills it.

Viruses infect organisms in all three domains of life. Generally, a viral infection decreases a host's ability to survive and reproduce. This is bad news for the infected organism, but it can benefit other organisms. For example, the ocean is filled with viruses. One sample of seawater from near the coast of Texas had a million viral particles per milliliter of water. Most of the viruses, which infected marine cyanobacteria, were relatives of herpesviruses. Actions of the viruses help prevent overgrowth of these bacteria. The lysis of bacteria also frees up nutrients that other marine producers can take up and use.

Viruses benefit humans when they target bacteria that cause human disease or an insect that eats our crops. As the saying goes, "The enemy of my enemy is my friend." One application of this principle is the prevention of food poisoning by a new food additive that contains bacteriophages. When sprayed on meat or fish, the product inhibits growth of bacterial populations. As another example, insect-attacking baculoviruses are available in sprays for use on crops. Compared to chemical pesticides, virus-based ones are safer because viruses attack only specific host species, and virus-based pesticides do not persist in the environment.

Take-Home Message What are viruses and how do they affect us?

- Viruses are noncellular particles that consist of genetic material wrapped in a protein coat.

- Viruses can replicate only inside living cells, and they infect members of all three domains of life.

- Some viruses harm us by causing disease, but others help us by killing human pathogens. They also keep nutrients flowing in ecosystems.

13.7

Evolution and Infectious Disease

The Nature of Disease An infection occurs when a pathogen gets past the body's surface barriers and multiplies. **Disease** follows when the body's defenses cannot be called into action quickly enough to keep a pathogen's activities from interfering with normal body functions. Infectious diseases are spread by contact with tiny amounts of mucus, blood, or other body fluid that can contain a pathogen. Table 13.2 lists the deadliest infectious diseases.

In an **epidemic**, a disease spreads quickly through part of a population, then subsides. **Sporadic diseases**, such as whooping cough, occur irregularly and affect few people. **Endemic diseases** occur more or less continually but do not spread far in large populations. Tuberculosis is like this. So is impetigo, a highly contagious bacterial infection that typically spreads no further than a single day-care center, or a similarly limited location.

In a **pandemic**, a disease breaks out and spreads worldwide. AIDS is a pandemic that has no end in sight. A 2002–2003 outbreak of SARS (severe acute respiratory syndrome) was a brief pandemic (Figure 13.25). It began in China, and travelers quickly carried it to thirty countries around the world. Before government-ordered quarantines (isolation of those infected) halted its spread,

Table 13.2	Deaths From Infectious Diseases*	
Disease	Type of Pathogen	Deaths per Year
Acute respiratory infections	Bacteria, viruses	4 million
AIDS	Virus (HIV)	2.7 million
Diarrheas	Bacteria, viruses, protists	1.8 million
Tuberculosis	Bacteria	1.6 million
Malaria	Protists	1.3 million
Measles	Viruses	600,000
Whooping cough	Bacteria	294,000
Tetanus	Bacteria	204,000
Meningitis	Bacteria, viruses	173,000
Syphilis	Bacteria	157,000

* Deaths worldwide, based on *World Health Report 2004*.

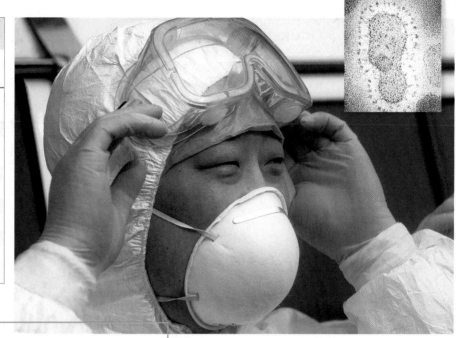

FIGURE 13.25 A health care worker in China dons protective gear during the SARS pandemic. The SARS virus, shown in the inset, is a coronavirus, like some viruses that cause common colds. Doctors and nurses who cared for SARS patients were among the casualties of the SARS pandemic.

about 8,000 people were sickened. About 10 percent of them died. Researchers quickly determined that a previously unknown coronavirus causes SARS. Other coronaviruses cause less dire respiratory infections such as colds.

There have been no reported cases of SARS since 2003. Is the pathogen gone for good? Only time will tell. Diseases sometimes disappear for years, then break out again without warning.

From the Pathogen's Perspective Consider disease in terms of a pathogen's prospects for survival. A pathogen stays around only for as long as it has access to outside sources of energy and raw materials. To a microscopic organism or virus, a human is a treasure trove of both. With bountiful resources, a pathogen can reach amazing population sizes. Evolutionarily speaking, the pathogens that leave the most descendants win.

Two barriers prevent pathogens from killing off all of their hosts. First, any species that has a history of being attacked by a specific pathogen has coevolved with it. Coevolution is a type of natural selection in which two species that interact closely over many generations exert selective pressure on one another. Selection favors hosts that can ward off pathogens, so organisms have evolved defenses against common pathogens. Second, if a pathogen kills its host too fast, the pathogen might disappear along with the host. Having a less-than-fatal effect increases a pathogen's reproductive success. Think of an infected host as a factory that makes and distributes pathogens. Killing the host would shut down this facility. The longer the infected individual survives, the more copies of the pathogen can be made and spread around.

Pathogens most often kill hosts who are weakened by age or by the presence of other pathogens. Death may also occur after a pathogen infects a new kind of host that has not evolved any defenses against it.

disease Condition that arises when a pathogen interferes with an organism's normal body functions.

endemic disease A disease that remains present at low levels in a population.

epidemic A disease spreads rapidly through a population.

pandemic A disease spreads worldwide.

sporadic disease A disease that breaks out occasionally and affects only a small portion of a population.

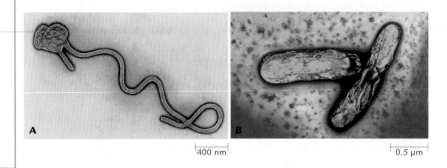

FIGURE 13.26 Which is deadlier?
A The Ebola virus turns blood vessels to mush and kills up to 90 percent of those infected. So far, outbreaks have been infrequent and restricted to small areas in Africa. **B** *Mycobacterium tuberculosis* causes tuberculosis (TB). Without treatment, the virus kills about 50 percent of those infected, and it infects about 300 million worldwide.

A

B

400 nm

0.5 µm

Emerging Diseases As human populations have soared, people have moved farther and farther into jungles and other previously uninhabited or sparsely inhabited regions. In such places, they eat the local animals and sometimes meet up with pathogens that have not coevolved with humans. Sometimes one of these pathogens infects people, with disastrous results.

Close contact with wild animals probably started the SARS epidemic. Bats are hosts for a virus that has genes nearly identical to those of the SARS virus. The bats suffer little or no harm, making them a reservoir, from which the virus can spread to hosts that it does sicken. The SARS virus also has genes that it picked up from birds. Researchers suspect that SARS may have been so deadly because mammalian genes and bird genes combined in the virus. The resulting hybrid genes encoded proteins that the human immune system was unfamiliar with and had no coevolved defenses against.

In Africa, bats serve as a reservoir for the highly deadly Ebola virus (Figure 13.26**A**), which often kills between 50 and 90 percent of those infected. The first symptoms are high fever and flu-like aches. Within a few days vomiting and diarrhea begin. Blood vessels are destroyed. Blood seeps into surrounding tissues and leaks out through all the body's orifices. Contact with body fluids of infected people can spread the disease. Understandably, at the start of an Ebola outbreak, government agencies throughout the world are notified. During 2007, independent outbreaks in Uganda and the Congo infected hundreds of people and killed more than one hundred.

World health officials are currently keeping a close eye on the H5N1 strain of avian influenza (bird flu). The first humans infected by this strain turned up in Hong Kong during 1997. Since then, human infections have been reported in other parts of Asia, Africa, the Pacific, Europe, and the Middle East. Half of those infected die. So far, there seems to be little or no human-to-human transmission of H5N1. Nearly all people have become infected by direct contact with infected birds or bird feces. The virus has spread rapidly among birds and is expected to eventually affect them worldwide. With many infected birds, cases of bird-to-human transfer will no doubt rise. However, an even greater concern is the risk that a mutation will allow human-to-human transfer. A 1918 flu pandemic that infected one-third of the world population and killed 50 million people was caused by a distantly related strain of bird flu.

The Threat of Drug Resistance As explained in Section 12.3, use of medications to treat infectious disease results in directional selection. In a population of pathogens, those least affected by the drug are at a selective advantage. Drug-resistant individuals survive and reproduce, while their drug-susceptible competitors die. As a result, the frequency of drug-resistant individuals increases over generations, which, for most pathogens, are short.

For example, *Streptococcus pneumoniae*, common among children at day-care centers, causes pneumonia, meningitis, and chronic ear infections. Penicillin-resistant strains of *S. pneumoniae* first appeared in 1967. Today, about half of the known strains are resistant. Penicillin-resistant strains of bacteria can arise by mutations or result from gene transfers. In the case of *S. pneumoniae*, genetic comparisons showed that the genes for antibiotic resistance were transferred to *S. pneumoniae* from a related species of bacteria.

Multidrug resistance is also a problem. Some strains of the bacterium that causes tuberculosis are now resistant to nearly all available drugs (Figure 13.26).

Viruses are not cells, but they do have genes that can mutate, so they can evolve by natural selection. For example, use of antiviral drugs has selected for HIV strains with mutations that make them resistant to these drugs.

Take-Home Message   How do infectious organisms evolve?

■ An infectious organism that encounters a new host with no coevolved defenses may prove deadly. A pathogen will become less harmful over time, if individual pathogens that allow their host to live have higher reproductive success.

■ Drugs used to treat infection exert selective pressure on pathogens and favor drug-resistant members of the pathogen population.

13.8 Impacts/Issues Revisited:
Microbial Menaces

To be successful, a pathogen must have a way to get from one host to another. For example, many viruses cause their hosts to cough and sneeze. This action

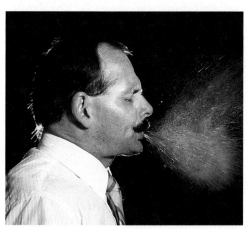

disperses viral particles into the environment, where they may encounter a new host.

Other pathogens have evolved ways of increasing the odds of successful transfer by manipulating their current host. For example, when an animal such as a dog has the disease rabies, viral particles accumulate in its saliva. Rabies virus also affects areas of the animal's brain that control its emotions, causing it to become less fearful and more aggressive. The changes in behavior increase the odds that the dog will bite another animal, thus passing along the rabies-causing virus.

Plasmodium, the protist that causes malaria, alters the feeding behavior of the mosquitoes that carry it in a way that helps spread the disease. A mosquito that is carrying sporozoites (the infectious form of the protist) bites a greater number of people per night than a mosquito without sporozoites.

How Would YOU ☑ Vote? Developing drugs that fight AIDS is expensive, and so are the drugs, which are protected by patents. Should patents be waived so that poor countries can produce their own supply of drugs at a lower cost? **See CengageNow for details, then vote online (cengagenow.com).**

Summary

Section 13.1   We share the Earth with countless microorganisms that affect our health and have important roles in ecological processes. Like their larger counterparts, microorganisms evolve in response to natural selection and their traits are adaptations to their environments.

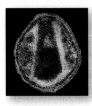

Section 13.2   Laboratory simulations provide indirect evidence that organic subunits self-assemble under certain conditions. They also show how complex organic compounds and protocells may have formed on the early Earth.

CENGAGENOW˜   See organic molecules form with animation on CengageNow.

Section 13.3   Fossils of early life forms date back more than 3 billion years. The earliest forms were prokaryotes and their lineage branched early, with one branch giving rise to bacteria, and the other to archaeans and the ancestors of eukaryotes. Eukaryotes have organelles (mitochondria and chloroplasts) that evolved from bacteria by the process of endosymbiosis.

CENGAGENOW˜   Review the history of life with animation on CengageNow.

Section 13.4   Archaea and bacteria are the only prokaryotes. They are metabolically diverse, reproduce by prokaryotic fission, and can swap genes. Many archaea live in places too hostile for other organisms. Bacteria are the most common prokaryotes. Many are decomposers and photosynthesizers. Some can cause disease.

CENGAGENOW˜   Examine a typical prokaryotic cell and observe cell division and conjugation with the animation on CengageNow.

Section 13.5   The "protist kingdom" is now being reorganized to reflect new understanding of evolutionary relationships. Most protists are single-celled, but some lineages include multicelled species. Nearly all live in water or moist habitats, including host tissues.

Flagellated protozoans are single cells that are typically heterotrophs, although some of the euglenoids have chloroplasts that evolved from algal cells.

Foraminiferans are single-celled heterotrophs in secreted calcium carbonate shells. Their remains contribute to limestone and chalk.

Ciliates, dinoflagellates, and apicomplexans are close relatives. Ciliates are single-celled heterotrophs that use cilia to move and feed. Dinoflagellates are single cells that move with a whirling motion and can be heterotrophs or photosynthetic. Some emit light. Apicomplexans, such as the species that cause malaria, are parasites that spend part of their life in cells of their host.

Water molds are heterotrophs that grow as filaments. Some are pathogens of fish or plants. The related diatoms are single-celled, silica-shelled, aquatic producers. The also related brown algae include the giant kelps, which are the largest protists.

Red algae have accessory pigments that allow them to live at greater depths than other algae. Red algae share a common ancestor

with the green algae. The land plants evolved from one lineage of green algae.

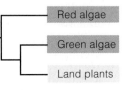
Red algae
Green algae
Land plants

Amoebas extend cytoplasmic lobes to feed and move. They live in aquatic habitats and animal bodies. The related slime molds spend part of their life as a single cell and part in a cohesive group that can migrate and differentiate to form a spore-bearing body. The amoebozoans share a common ancestor with fungi and animals.

CENGAGENOW˜   Explore the body plan of a euglenoid and the life cycles of an apicomplexan and a slime mold with the animation on CengageNow.

Section 13.6   A virus is a noncellular infectious particle that can multiply only by pirating the metabolic machinery of a host cell. It consists of DNA or RNA inside a protein coat. Some replication cycles of viruses include a latent phase, when viral genetic material is integrated into the host genome.

CENGAGENOW˜   Use the animation on CengageNow to explore viral structure and multiplication pathways.

Section 13.7   Evolution shapes the characteristics of pathogens, as it does all organisms. Pathogens and their hosts have coevolved. Most hosts have evolved defenses against common pathogens, and most pathogens usually do not kill their hosts. Use of antibiotics favors the spread of drug resistance among pathogens.

Self-Quiz   Answers in Appendix I

1. An abundance of _____ in the atmosphere would have prevented the assembly of complex organic compounds.

2. Stanley Miller's experiment demonstrated _____ .
   a. the great age of Earth
   b. that amino acids can assemble under some conditions
   c. that oxygen is necessary for life
   d. all of the above

3. The evolution of _____ resulted in an increase in the amount of oxygen in the atmosphere.
   a. prokaryotic fission   c. aerobic respiration
   b. sexual reproduction   d. photosynthesis

4. Mitochondria are most likely descendants of _____ .
   a. archaeans   c. cyanobacteria
   b. aerobic bacteria   d. anaerobic bacteria

5. Bacteria transfer genes to a partner cell by _____ .
   a. prokaryotic fission   c. conjugation
   b. the lytic pathway   d. endospore formation

6. The first eukaryotes were _____ .
   a. bacteria   b. protists   c. fungi   d. animals

7. True or false? Some protists are more related to plants than to other protists.

Digging Into Data

How *Plasmodium* Summons Mosquitoes

Dr. Jacob Koella and his associates hypothesized that *Plasmodium* might benefit by making its human host more attractive to hungry mosquitoes when gametocytes were available in the host's blood. Gametocytes can be taken up by the mosquito and mature into gametes inside its gut (Figure 13.18).

To test their hypothesis, the researchers recorded the response of mosquitoes to the odor of *Plasmodium*-infected children and uninfected children over the course of 12 trials on 12 separate days. Figure 13.27 shows their results.

1. On average, which group of children was the most attractive to mosquitoes?

2. Which group of children attracted the fewest mosquitoes on average?

3. What percentage of the total number of mosquitoes were attracted to the most attractive group?

4. Did the data support Dr. Koella's hypothesis?

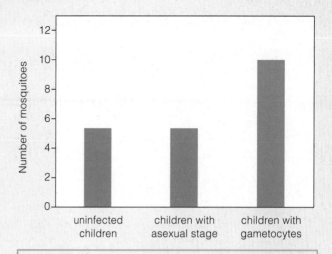

FIGURE 13.27 Number of mosquitoes (out of 100) attracted to uninfected children, children with the asexual stage of *Plasmodium*, and children with gametocytes in their blood. The bars show the average number of mosquitoes attracted to that category of child over the course of 12 separate trials.

---

8. Freshwater protists have a _____ that expels excess water.

9. The _____ are parasitic eukaryotes that live in other cells.
   a. viruses   c. euglenoids   e. both a and b
   b. apicomplexans   d. slime molds   f. all are correct

10. Remains of _____ form chalk and limestone deposits.
    a. ciliates   c. foraminiferans
    b. diatoms   d. dinoflagellates

11. Some of the _____ are human pathogens.
    a. slime molds   c. flagellated protozoans
    b. archaea   d. both a and c

12. Green algae are most closely related to _____ algae.

13. Silica-rich remains of _____ are used as an insecticide.
    a. dinoflagellates   c. foraminiferans
    b. diatoms   d. apicomplexans

14. The genetic material of a _____ may be DNA or RNA.
    a. bacteria   b. ciliate   c. dinoflagellate   d. virus

15. Match these terms with the appropriate definition.
    _____ green algae   a. protist population explosion
    _____ virus   b. social amoeba
    _____ bacteria   c. most diverse prokaryotes
    _____ brown algae   d. noncellular infectious agent
    _____ endospore   e. include the largest protists
    _____ euglenoid   f. flagellate with chloroplasts
    _____ algal bloom   g. closest relative of plants
    _____ dinoflagellate   h. layered prokaryotes and sediment
    _____ slime mold   i. resistant resting stage
    _____ stromatolite   j. whirling cell

Additional questions are available on **CENGAGENOW**

Critical Thinking

1. Researchers looking for fossils of the earliest life forms face many hurdles. For example, few sedimentary rocks date back more than 3 billion years. Review what you learned about plate tectonics (Section 11.5). Explain why so few remaining samples of these early rocks remain.

2. Craig Venter and Claire Fraser are working to create a "minimal organism." They are starting with *Mycoplasma genitalium*, a bacterium that has 517 genes. By disabling its genes one at a time, they discover that 265–350 of them code for essential proteins. The scientists are synthesizing the essential genes and inserting them, one by one, into an engineered cell consisting only of a plasma membrane and cytoplasm. They want to see how few genes it takes to build a new life form. What properties would such a cell have to exhibit for you to conclude that it was alive?

3. Viruses that do not have a lipid envelope tend to remain infectious outside the body longer than enveloped viruses. "Naked" viruses are also less likely to be rendered harmless by soap and water. Can you explain why?

4. When planting bean seeds, gardeners are advised to sprinkle powder that contains nitrogen-fixing *Rhizobium* cells into the soil around the seed. The bacterial will later infect the plant's roots. How could the presence of these bacteria benefit the plants?

14
Plants and Fungi

In less than four decades, humans have destroyed
nearly half of the world's tropical forests.

Impacts/Issues: Speaking for the Trees

The world's great forests influence life in profound ways. They take up carbon dioxide from the air and release oxygen into it. At the same time, the forests act like giant sponges absorbing water from the rains, then releasing it slowly. By sucking up water and holding soil in place, forests prevent erosion, flooding, and sedimentation that can disrupt rivers, lakes, and reservoirs. Strip forests away, and exposed soil can wash away or lose nutrients.

Developing nations of Asia, Africa, and Latin America have fast-growing populations and high demands for food, fuel, and lumber. To meet their own needs and create products for sale in the global market, these nations turn to their tropical forests. These forests have been home to at least 50 percent of all land-dwelling species for at least 10,000 years. In less than four decades, humans have destroyed more than half of tropical forests by logging. The pace of logging is the equivalent of leveling thirty-four city blocks every minute.

Deforestation affects evaporation rates, runoff, and regional patterns of rainfall. In tropical forests, most of the water vapor in the air is released from trees. In logged-over regions, annual rainfall declines, and rain swiftly drains away from the exposed, nutrient-poor soil. The region gets hotter and drier, and soil fertility and moisture decline.

Like many environmental issues, this ever-accelerating deforestation can seem like an overwhelming problem. Wangari Maathai, a Kenyan biologist, suggested a simple solution: Plant new trees (Figure 14.1).

Maathai founded the Green Belt Movement, which helps women in Kenya and elsewhere plant and care for trees. Maathai believes that if a woman plants and cares for a tree seedling and watches it grow, she will begin to value it. As many women plant trees, they see the resulting improved environment and understand that they can be a force for positive change.

FIGURE 14.1 Wangari Maathai holding one of the many trees that were planted through her efforts.

Maathai began her tree-planting campaign in 1977. By 2007, members of her organization had planted 40 million trees in Africa. These trees provide shade, reduce erosion, and have begun to restore the diverse forests that had been disappearing. Just as importantly, Maathai's practical solution to a seemingly overwhelming problem has inspired others to do what they can to stem the tide of environmental degradation. In 2004, the Nobel Committee honored Maathai's efforts with the Nobel Peace Prize.

As Maathai recognized, we are dependent on plants for our well-being. This chapter explains how these green partners of ours evolved and diversified. It also discusses the fungi, a group that interacts with plants and without which most plants could not thrive.

14.2 Plant Characteristics and Evolution

Plants are multicelled eukaryotes and almost all are photosynthetic. They evolved from a lineage of green algae, and share many traits with this group. Both have the same kind of chlorophyll and both store starch. The defining trait that sets plants apart is a multicelled embryo that forms, develops within, and is nourished by tissues of the parental plant.

Structural Adaptations to Life on Land

Plants have many unique traits that adapt them to life on land. A multicelled green algae is surrounded by water, so it can absorb water and dissolved nutrients across its body surface. Water also buoys its parts, thus helping it stand upright. In contrast, land plants face the threat of drying out and must hold themselves upright.

The oldest plant lineages are the **bryophytes**, with mosses being the most familiar group. We know from fossils that some bryophytes were living on land by 475 million years ago. Like algae, bryophytes absorb water and any dissolved minerals that they require across their surface. Unlike most algae, bryophytes can revive after being dried out. Also, some bryophytes secrete a waxy **cuticle** that reduces evaporative water loss. Tiny pores called **stomata** (singular, stoma) allow the plant to balance water conservation with the need to obtain carbon dioxide for photosynthesis. The stomata open or close as needed.

Bryophytes have structures that hold them in place, but these structures are not true roots; those evolved later. Roots not only anchor plants, they also take up water with dissolved mineral ions from the soil. Plants with true roots have a vascular system. This system consists of internal pipelines in which water and nutrients move through the plant body. **Xylem** is the vascular tissue that distributes water and mineral ions. **Phloem** is the vascular tissue that distributes sugars made by photosynthetic cells. Of 295,000 or so modern plant species, more than 90 percent have xylem and phloem and are known as **vascular plants**.

A variety of adaptations contribute to the success of vascular plants. For example, their vascular tissues are reinforced by **lignin**, an organic compound that stiffens them and provides structural support. Vascular tissues with lignin not only distribute materials, they also help plants stand upright and allow them to branch. The most diverse vascular lineages evolved leaves, which increase a plant's surface area for intercepting sunlight and for exchanging gases. Figure 14.2 shows a cross-section through a vascular plant leaf.

The Plant Life Cycle

Life cycle changes also adapted vascular plants to life in drier habitats. In the animal life cycle, a multicelled diploid individual produces haploid cells (eggs or sperm) that combine to produce a new diploid individual. In contrast, plants go through two multicelled stages in their life cycle, an alternation of generations (Figure 14.3). The diploid form of a plant is a **sporophyte**, which makes spores, not gametes. A plant spore is a nonmotile cell that divides and develops into a multicelled, haploid **gametophyte**. The gametophyte makes eggs and sperm that combine and give rise to a new sporophyte.

The gametophyte is the largest and longest-lived part of the bryophyte life cycle, but the sporophyte dominates the life cycle of vascular plants. Spores withstand drying out better than gametes do.

Reproduction and Dispersal

Both bryophytes and the oldest vascular plant lineages require water for fertilization. Both have flagellated sperm that swim to eggs through droplets of water that cling to the plant. These plants disperse their new generation by releasing spores.

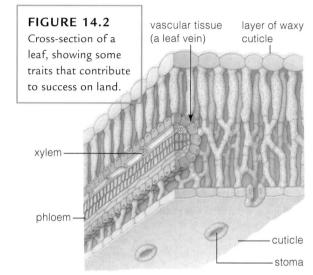

FIGURE 14.2 Cross-section of a leaf, showing some traits that contribute to success on land.

vascular tissue (a leaf vein)
layer of waxy cuticle
xylem
phloem
cuticle
stoma

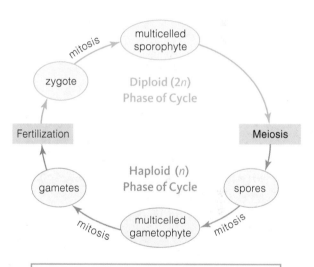

FIGURE 14.3 Generalized plant life cycle.

multicelled sporophyte
zygote
mitosis
Diploid (2*n*) Phase of Cycle
Meiosis
Fertilization
Haploid (*n*) Phase of Cycle
gametes
spores
mitosis
multicelled gametophyte
mitosis

bryophyte Member of an early evolving plant lineage that does not have vascular tissue; for example, a moss.

cuticle Secreted covering at a body surface.

gametophyte Haploid gamete-forming stage in a plant life cycle.

lignin Material that stiffens cell walls of vascular plants.

phloem Vascular tissue that distributes dissolved sugars.

plant Multicelled, photosynthetic organism; develops from an embryo that forms on the parent and is nourished by it.

sporophyte Diploid spore-forming stage in a plant life cycle.

stomata Adjustable pores in a plant cuticle.

vascular plant A plant that has xylem and phloem.

xylem Vascular tissue that distributes water and dissolved mineral ions.

angiosperms Largest seed plant lineage. Its members make flowers and fruits.

gymnosperms Seed plant lineage that does not make flowers or fruits.

pollen grain Male gametophyte of a seed plant.

seed Embryo sporophyte of a seed plant packaged with nutritive tissue inside a protective coat.

One group of vascular plants, the seed plants, evolved an ability to reproduce even in dry places by releasing pollen. A **pollen grain** is a walled, immature male gametophyte that can be delivered to eggs by wind or animals.

Seed plants also have another advantage. Their female gametophytes form within a protective chamber on the sporophyte body. In seed plants, gamete formation and fertilization produces a **seed**, which is an embryo sporophyte and tissues that nourish it, all encased in a protective coat. The embryo can lie dormant during seasons that do not favor its growth. When conditions become more favorable, the seed germinates and a new sporophyte develops.

There are two lineages of seed plants. The **gymnosperms** were the first to evolve. A pine tree is a familiar gymnosperm. **Angiosperms**, the flowering plants, branched off from a gymnosperm lineage and are now the most diverse group of plants. They alone make flowers and release their seeds inside a fruit. Figure 14.4 summarizes the relationships among the major plant groups and the traits of each group.

Take-Home Message

What adaptive traits allow plants to live on land and in dry places?

■ Plants are multicelled, typically photosynthetic species that protect and nourish their multicelled embryos. Plants evolved from a lineage of green algae.

■ Adaptations to life on land include a waxy cuticle with stomata, true roots, and vascular tissues that distribute materials and provide structural support.

■ The plant life cycle includes two multicelled forms: a haploid gametophyte and a diploid sporophyte.

■ The most recently evolved plant lineages produce pollen grains and seeds. These adaptations allowed the seed plants to spread into diverse habitats.

FIGURE 14.4 Traits of the four plant groups and the relationships among them.

Bryophytes
- No xylem or phloem
- Gametophyte predominant
- Water required for fertilization
- Seedless

liverworts hornworts mosses

Seedless vascular plants
- Vascular tissue present
- Sporophyte predominant
- Water required for fertilization
- Seedless

club mosses,   whisk ferns,
spike mosses   horsetails,
                ferns

Gymnosperms
- Vascular tissue present
- Sporophyte predominant
- Pollen grains; water not required for fertilization
- "Naked" seeds

gnetophytes, ginkgos, conifers, cycads

Angiosperms
- Vascular tissue present
- Sporophyte predominant
- Pollen grains; water not required for fertilization
- Seeds form inside a fruit; flowers

monocots, eudicots, and relatives

ancestral alga

14.3

Bryophytes

Modern bryophytes include 24,000 species that belong to three lineages: mosses, hornworts, and liverworts. Some mosses have tubes that conduct water and sugar, but none of the bryophytes have the lignin-stiffened vascular pipelines that vascular plants do. As a result, few bryophytes stand more than 20 centimeters (8 inches) tall.

The gametophyte is the largest, most conspicuous phase of a bryophyte life cycle. Eggs or sperm form in multicellular chambers (gametangia) in or on the gametophyte's surface. The sperm have flagella and swim to eggs. Sperm may also be transported by crawling mites and insects that happen to pick the sperm up on their bodies.

Sporophytes are unbranched and remain attached to the gametophyte even when mature. They produce wind-dispersed spores that withstand drying out, making bryophytes important colonists of rocky places.

Figure 14.5 shows the life cycle of a common moss. The gametophyte has a stemlike axis with tiny leaflike parts arranged around it. Rootlike threads hold the gametophyte in place, but they do not take up water or nutrients. These resources are absorbed across the leafy surfaces.

The sporophyte consists of a capsule and a stalk ❶. It is attached to the gametophyte and depends on it for photosynthetic sugars. Spores form inside the capsule by meiosis and are dispersed by the wind ❷. Each spore grows into a male or female gametophyte that makes sperm or eggs in a chamber at its tip ❸. Rain causes these chambers to open, and flagellated sperm swim through a film of water to reach eggs ❹. Fertilization occurs in the egg chamber of a female gametophyte ❺. The zygote forms inside that chamber and then develops into a new sporophyte ❻.

The moss in this example has either male or female sexual organs. In some other moss species, each plant makes both eggs and sperm. Mosses and other bryophytes can also reproduce asexually, as when new gametophytes grow from pieces that break off the parent.

Mosses are the most diverse group of bryophytes, with 14,000 or so species. Among these, 350 or so species of peat mosses (*Sphagnum*) are of great ecological and commercial importance. Peat mosses are the main plants in peat bogs that cover more than 350 million acres in Europe, Northern Asia, and North America. Many of these bogs have persisted for thousands of years, and layer upon layer of plant remains have become compressed as peat. Peat is harvested for use as a fuel, especially in Ireland, where it is even burned in power plants to generate electricity. Compared to coal, peat produces far fewer air pollutants when burned.

Freshly harvested peat moss is also an important commercial product. The moss is dried and added to planting mixes to help soil retain moisture. Dead strands of peat moss absorb water, then release it slowly over time so plant roots are kept moist.

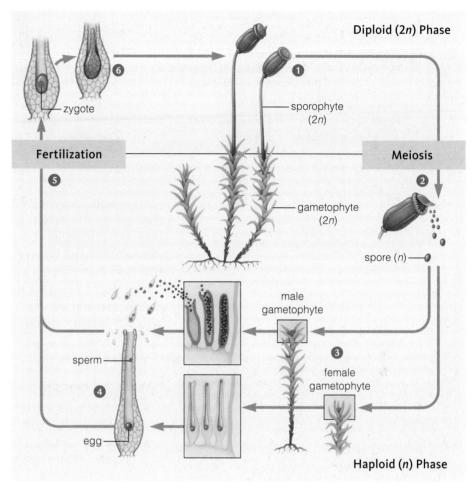

FIGURE 14.5 Animated! Life cycle of the moss *Polytrichum*.

❶ A mature sporophyte (a stalk and capsule) is attached to a gametophyte even when mature.

❷ Spores form by meiosis and are released.

❸ Spores develop into male and female gametophytes, which produce eggs or sperm in chambers at their tips.

❹ Sperm released from tips of male gametophytes swim through water to eggs at tips of female gametophytes.

❺ Fertilization occurs in the egg chamber of the female gametophyte and produces a zygote.

❻ The zygote grows and develops into a sporophyte while remaining attached to and nourished by its female parent.

FIGURE 14.6 Less familiar bryophyte lineages. **A** One of the hornworts. Its "horn" makes the spores. **B** Female liverwort. Eggs form in cells on the umbrella-like structures which are part of the gametophyte.

Figure It Out: Are cells in the gametophyte of a hornwort haploid or diploid?

Answer: Gametophytes are always haploid.

sporophyte

A gametophyte

Liverworts and hornworts often live with mosses in damp, shaded places. A hornwort's gametophyte is ribbonlike. The attached hornlike sporophyte can be several centimeters tall (Figure 14.6**A**). Many liverworts also have a ribbon-like gametophyte. When conditions favor sexual reproduction, umbrella-shaped structures that hold the gamete-making cells grow from the gametophyte's surface (Figure 14.6**B**). A few liverwort species commonly grow in commercial greenhouses and are important pests. Liverwort infestations can be difficult to get rid of because the tiny spores persist even after the plants are killed.

Take-Home Message
What are bryophytes?

- Bryophytes include three lineages of low-growing plants (mosses, hornworts, and liverworts) that have flagellated sperm and disperse by releasing spores.

- Bryophytes are the only modern plants in which a gametophyte dominates the life cycle and the sporophyte is dependent upon it.

14.4 Seedless Vascular Plants

Ferns, club mosses, and horsetails are examples of seedless vascular plants. They evolved from bryophytes and share some traits with these lineages. Like bryo-phytes, they have flagellated sperm that require a film of water to swim to eggs. Also like bryophytes, they disperse a new generation by releasing spores.

However, seedless vascular plants differ from bryophytes in other aspects of their life cycle and structure. The gametophyte is reduced in size and is relatively short lived. Although the sporophyte forms on the gametophyte, it can live independently after the gametophyte dies. Lignin stiffens a sporophyte's body, and a system of vascular tissue distributes water, sugars, and minerals through it. These innovations in support and plumbing allow seedless vascular sporophytes to be large and structurally complex, with roots, stems, and leaves.

Ferns We begin our survey of the seedless vascular plants with the largest and most familiar lineage, the ferns. Figure 14.7 shows the life cycle of a typical fern. The leafy plant we envision when we think of a fern is a sporophyte ❶. In most ferns, roots and fronds (leaves) grow from **rhizomes**, stems that grow along or just below the ground. Spores form by meiosis in capsules on the underside of leaves ❷. They are released and dispersed by wind when the capsule pops open.

The spore germinates and grows into a tiny gametophyte a few centimeters across. Eggs and sperm form in chambers on its underside ❸. The sperm swim to eggs and fertilize them ❹. The resulting zygote develops into a new sporo-phyte, and its parental gametophyte dies ❺.

epiphyte Plant that grows on the trunk or branches of another plant but does not harm it.

rhizome Stem that grows horizontally along the ground.

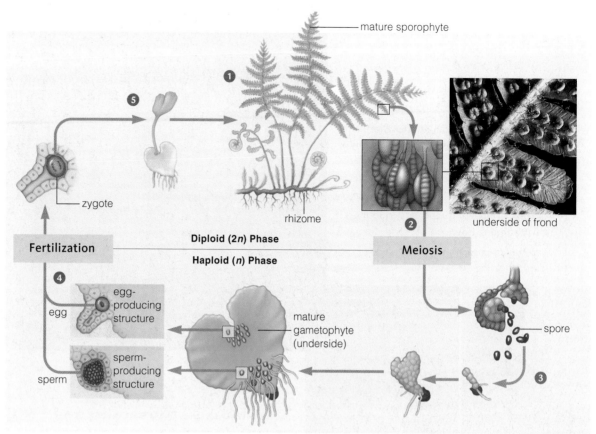

FIGURE 14.7 Animated!
Life cycle of a fern (*Woodwardia*).

1 The familiar leafy form is the diploid sporophyte.

2 Meiosis in cells on the underside of fronds produces haploid spores.

3 After their release, the spores germinate and grow into tiny gametophytes that produce eggs and sperm.

4 Sperm swim to eggs and fertilize them, forming a zygote.

5 The sporophyte begins its development attached to the gametophyte, but it continues to grow and live independently after the gametophyte dies.

mature sporophyte

zygote

Diploid (2*n*) Phase

Haploid (*n*) Phase

Fertilization

Meiosis

rhizome

underside of frond

egg

egg-producing structure

sperm

sperm-producing structure

mature gametophyte (underside)

spore

In many ferns, asexual reproduction occurs more frequently than sexual reproduction. As a rhizome grows through the soil, new shoots and roots grow from it. Then the connection to the parental plant breaks, and that segment of rhizome becomes an independent plant. Unlike sexual reproduction, asexual reproduction can take place even when there is little water present.

Ferns are the most diverse group of seedless vascular plants, and their sporophytes show enormous variation in size and form. Some float on freshwater ponds and have fronds less than 1 centimeter wide (Figure 14.8**A**). Many tropical ferns are **epiphytes**, plants that live attached to the trunk or branches of another plant but do not withdraw nutrients from it (Figure 14.8**B**). The largest ferns are tree ferns that can be 25 meters (80 feet) tall (Figure 14.8**C**).

Compared to seed plants, ferns are of lesser commercial importance. Tree ferns and other ferns are sold as potted plants, and ferns also are grown for use in flower arrangements. The trunks of some tree ferns are shredded to produce a potting medium used to grow orchids. Young, tightly coiled fronds of other ferns are harvested from the wild as food. The young fronds are called fiddleheads, because they resemble the coil at the top of a violin.

FIGURE 14.8 A sampling of fern diversity. **A** The floating fern *Azolla pinnata*. The whole plant is not as wide as a finger. Chambers in the leaves shelter nitrogen-fixing cyanobacteria. Southeast Asian farmers grow this species in rice fields as a natural alternative to chemical fertilizers. **B** Bird's nest fern (*Asplenium nidus*), one of the epiphytes. **C** Forest of tree ferns in Australia's Tarra-Bulga National Park.

FIGURE 14.9 Then and now. *Right*, an artist's depiction of a swamp forest during the Carboniferous period. An understory of ferns ❶ is shaded by tree-sized lycophytes ❷ and horsetails ❸.

Below left, a modern lycophyte, a club moss, growing among leaves on the forest floor.

Below right, a modern horsetail growing in a damp clearing.

Club Mosses, Horsetails, and the Coal Forests Ferns are currently the most diverse seedless vascular plants, but relatives of the modern club mosses and horsetails were the dominant plants in swamp forests during the Carboniferous period from 360 to 300 million years ago (Figure 14.9). The remains of these lush forests became compacted and were transformed over time to coal. When we burn coal, we are releasing the energy of sunlight captured by these plants hundreds of millions of years ago.

Modern club mosses and horsetails are far smaller than their Carboniferous counterparts. Club mosses are common on the floor of temperate forests. Their branching form makes them look like tiny pine trees and they are often collected for use in holiday wreaths. Spores form in a conelike reproductive structure or at the tips of branches. The spores have a waxy coating that makes them ignite easily. They were used to create flashes for early photography and to produce special effects in theaters.

Horsetails thrive along streams and roadsides, and in disturbed areas. Their stems contain silica. Before the invention of modern abrasive cleansers, the silica-rich stems of some *Equisetum* species were used to scrub pots and polish metals. These species are commonly referred to as scouring rushes. Their high silica content is an adaptation that helps them resist insect predators.

Take-Home Message **What are seedless vascular plants?**

- Seedless vascular plants include ferns, club mosses, and horsetails.

- These plants disperse by releasing spores, and the life cycle is dominated by a sporophyte that has vascular tissue and lignin. The gametophyte is small and relatively short-lived.

- Like bryophytes, seedless vascular plants have flagellated sperm that must swim through a film of water to reach eggs.

Rise of the Seed Plants

Seed plants evolved from a lineage of seedless vascular plants about 400 million years ago during the Devonian. They survived alongside bryophytes and seedless nonvascular plants until the late Carboniferous, then rose to dominance as the climate became drier. As you will see, unique traits of seed plants give them a competitive advantage over seedless plants when water is scarce.

Like some bryophytes and seedless vascular plants, seed plants have separate male and female gametophytes. Each type of gametophyte arises from a different type of spore. **Microspores** develop into male gametophytes, and **megaspores** develop into female gametophytes (Figure 14.10).

Other plants release their spores into the environment, but seed plants hold onto them. The male and female gametophytes are reduced in size, and they develop on the sporophyte. In gymnosperms, spores form and develop into gametophytes in cones or other special reproductive structures. In angiosperms, or flowering plants, these events occur in flowers.

The male gametophytes of seed plants are pollen grains and they form inside pollen sacs. The wind or animals can deliver pollen released by a seed plant to the female parts of another plant, a process called **pollination**. After pollination, a tube grows out from the pollen grain and delivers sperm to the egg. Because their sperm do not need to swim through a film of water to reach an egg, seed plants can reproduce even during dry times.

The female gametophytes of seed plants form inside an **ovule**. After pollination and fertilization of the egg, the ovule develops into a seed. As noted earlier, a seed is an embryo sporophyte and some nutritive tissue inside a protective coat. Dispersing seeds, rather than spores, puts seed plants at an advantage over seedless plants because it increases the odds that the new generation will survive. With its waterproof protective coat and a source of food, the embryo sporophyte inside a seed can stay dormant until conditions favor growth. In contrast, spores released by bryophytes and seedless vascular plants are single cells without stored food.

Seed plants also evolved structural traits that gave them an advantage over seedless lineages. Many seed plants undergo secondary growth that produces woody tissue. **Secondary growth** is growth in diameter, rather than height. **Wood** is the tissue produced by adding new lignin-stiffened xylem. Only seed plants produce true wood. Although some now-extinct nonvascular plants that lived in ancient forests did undergo secondary growth and became quite large, their trunks were softer and more flexible.

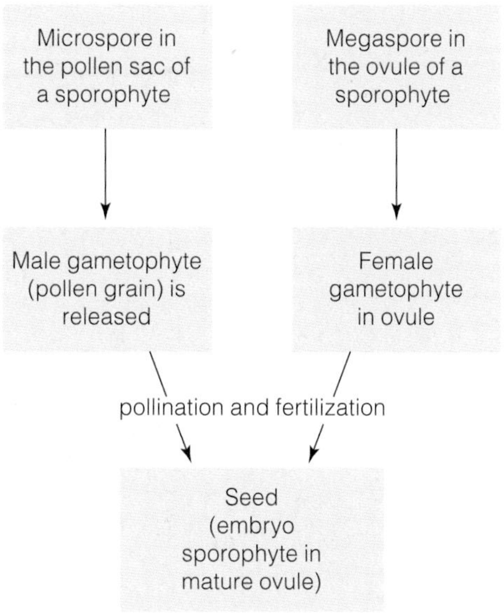

FIGURE 14.10 How seeds form.

Unique traits put seed plants at an advantage where water is scarce.

Take-Home Message What adaptive traits allowed seed plants to move into drier habitats?

- Seed plants package their male gametophytes as pollen grains that wind or animals can carry to the female parts of plants. As a result, these plants can reproduce even under dry conditions.

- The female gametophyte of a seed plant forms inside an ovule that becomes a seed after fertilization. Dispersing seeds rather than spores increases the reproductive success of seed plants.

megaspore In seed plants, a haploid cell that gives rise to a female gametophyte.

microspore In seed plants, a haploid cell that gives rise to a male gametophyte (pollen grain).

ovule Of seed plants, chamber inside which the female gametophyte forms.

pollination Delivery of pollen to female part of a plant.

secondary growth Increase in diameter of a plant part.

wood Lignin-stiffened secondary growth of some seed plants.

14.6 Gymnosperms

Gymnosperms are vascular seed plants that produce seeds on the surface of ovules. Seeds are said to be "naked," because unlike those of angiosperms they are not inside a fruit. (*Gymnos* means naked and *sperma* is taken to mean seed.) However, many gymnosperms enclose their seeds in a fleshy or papery covering.

Conifers **Conifers** are the most diverse gymnosperms. Most are woody trees or shrubs with needlelike or scalelike leaves. Most conifers are evergreen, meaning they do not shed their leaves all at once. A few are deciduous. A deciduous plant sheds all of its leaves at the same time, usually before a dry or cold season.

A pine tree is a conifer sporophyte. Pine gametophytes form in its cones. Each spring, the wind disperses pollen from male cones (Figure 14.11**A**). A pine's female cone has clusters of papery or woody scales with ovules on their upper surface (Figure 14.11**B**). Pollination occurs when a pollen grain lands on an exposed ovule of a female cone scale. The pollen grain germinates and a pollen tube begins growing toward the egg. The pollen tube grows so slowly that it takes about a year to reach the egg. When it does, a nonmotile sperm delivered by the tube fertilizes the egg. The resulting zygote develops into an embryo sporophyte. The sporophyte and surrounding ovule tissue become a seed.

Conifers include the tallest trees in the Northern Hemisphere (redwoods), as well as the most abundant (pines). They also include the amazingly long-lived bristlecone pines (Figure 14.11**C**). One bristlecone pine that was cut down was estimated to be 4,900 years old. The oldest living specimen is 4,600 years old.

Conifers are of tremendous ecological and economic importance. They are the main plants in cool Northern Hemisphere forests. Conifers, especially pines, provide lumber for building homes, and we use conifer bark as mulch in our gardens. Some pines make a sticky resin that deters insects from boring into them. We use this resin to make turpentine. We use oils from cedar in cleaning products, and we eat the seeds, or "pine nuts," of some pines.

FIGURE 14.11 Conifers. **A** Pollen wafting from the male cone of a pine. **B** Cross-section through a female pine cone, showing the ovule. **C** A bristlecone pine in the Sierra Nevada.

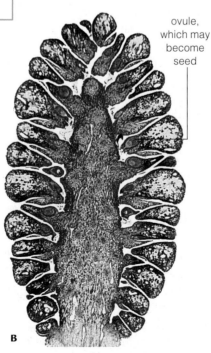

ovule, which may become seed

FIGURE 14.12 Beyond conifers: a sampling of other gymnosperms. **A** An Australian cycad with its seeds. *Ginkgo biloba*: **B** fan-shaped leaves, **C** fleshy seeds. **D** *Ephedra*, the source of an herbal stimulant.

A Sample of Gymnosperm Diversity Cycads and ginkgos were diverse in dinosaur times. We see their evolutionary connection to seedless plants in one reproductive trait. They are the only modern gymnosperms with flagellated sperm. In both cycads and ginkgos, a male plant produces pollen that is carried by the wind to a female plant. The ovule of the female plant secretes fluid that traps pollen grains. Sperm emerge from pollen grains, then swim through the secreted fluid to eggs in the plant's ovule.

About 130 species of cycads survived to the present, mainly in tropical and subtropical regions. Cycads look like palms or ferns but are not close relatives of either (Figure 14.12**A**). The "sago palms" commonly used in landscaping and as houseplants are actually cycads.

The only living ginkgo species is *Ginkgo biloba*, the maidenhair tree (Figure 14.12**B,C**). Ginkgos are native to China, but their attractive fan-shaped leaves and resistance to air pollution make them popular street trees in American cities. Usually male trees are planted because the fleshy seeds dropped by female trees have a strong odor and make a mess. Dietary supplements made from ginkgo may help slow memory loss in people who have Alzheimer's disease.

The shrubby plant *Ephedra* belongs to another gymnosperm lineage (Figure 14.12**D**). The plant makes a compound that has a stimulating effect on the nervous system, and extracts are sold as an herbal stimulant and weight-loss aid. Misuse of such supplements can be dangerous. A few people, including the professional baseball player Steve Bechler, have died while using them.

Take-Home Message **What are gymnosperms?**

- Gymnosperms are one of the two lineages of seed-bearing vascular plants. Seeds form on the surface of cones or other spore-producing structures.

- Conifers, the most familiar and largest gymnosperm group, are of great ecological and economic importance.

conifers Cone-bearing gymnosperms.

Angiosperms—The Flowering Plants

Angiosperms are vascular seed plants, and the only plants that make flowers and fruits. Their name refers to **ovaries**, the chambers that enclose one or more egg-producing ovules. (*Angio*– means enclosed chamber, and *sperma*, seed.) After fertilization, an ovule matures into a seed and the ovary becomes the **fruit**.

Keys to Angiosperm Diversity Nearly 90 percent of all modern plant species are flowering plants. They survive in nearly every land habitat. What accounts for angiosperm diversity? For one thing, they tend to grow faster than gymnosperms. Think of how a plant like a dandelion or a grass can grow from a seed and produce seeds of its own within a few months. In contrast, most gymnosperms take years to mature and produce their first seeds.

The **flower**, a specialized reproductive shoot (Figure 14.13**A**), is another important angiosperm adaptation. After pollen-producing plants evolved, some insects began feeding on the plants' highly nutritious pollen. Plants gave up some pollen but gained a reproductive edge. How? Insects move pollen from male parts of one flower to female parts of another.

Many flowering plants evolved traits that attracted specific **pollinators**, animals that move pollen of one plant species onto female reproductive structures of the same species. Insects are the most common pollinators (Figure 14.13**B**), but birds, bats, and other animals also serve in this role. Brightly colored petals, sugary nectar, or a strong fragrance can attract pollinators. The wind-pollinated plants tend to have small, pale, unscented flowers that lack nectar. For example, you have probably never noticed the small, pale flowers of grasses, which are wind pollinated.

Over time, plants coevolved with their animal pollinators. Again, coevolution refers to two or more species jointly evolving as a result of a close ecological interaction. Inherited changes in one species exert selective pressure on the other species, which also evolves.

A variety of fruit structures helped angiosperms disperse their seeds. Some fruits float in water, ride the winds, stick to animal fur, or survive a trip through an animal's gut. Gymnosperm seeds have less diverse dispersal mechanisms.

Major Groups The vast majority of flowering plants belong to one of two lineages. The 80,000 or so **monocots** include orchids, palms, lilies, and grasses, such as rye, wheat, corn, rice, sugarcane, and other valued plants. The **eudicots** include most herbaceous (nonwoody) plants such as tomatoes, cabbages, roses, daisies, most flowering shrubs and trees, and cacti. Monocots and eudicots evolved different structural details such as the arrangement of their vascular tissues and the number of flower petals. Also, some eudicots put on secondary growth and become woody. No monocots produce true wood. Section 27.2 describes the differences between eudicots and monocots in detail.

A Flowering Plant Life Cycle Figure 14.14 shows the life cycle of a lily, one of the monocots. A lily plant is the sporophyte ❶. Pollen forms in a flower's male parts, the **stamens** ❷. A thin filament holds the upper part of the stamen (the anther) aloft. The anther contains pollen sacs where microspores develop into pollen grains, the male gametophytes. Female gametophytes form

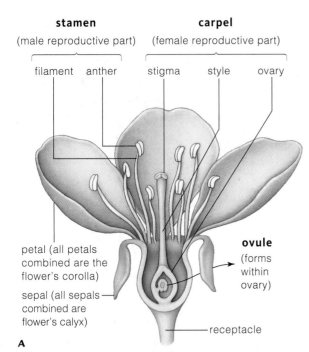

stamen
(male reproductive part)

filament   anther

carpel
(female reproductive part)

stigma   style   ovary

petal (all petals combined are the flower's corolla)

sepal (all sepals combined are flower's calyx)

ovule
(forms within ovary)

receptacle

A

B

FIGURE 14.13 Flowers. **A** Floral parts. Not all flowers have carpels and stamens. **B** A bee pollinates flowers as it gathers pollen and sips nectar.

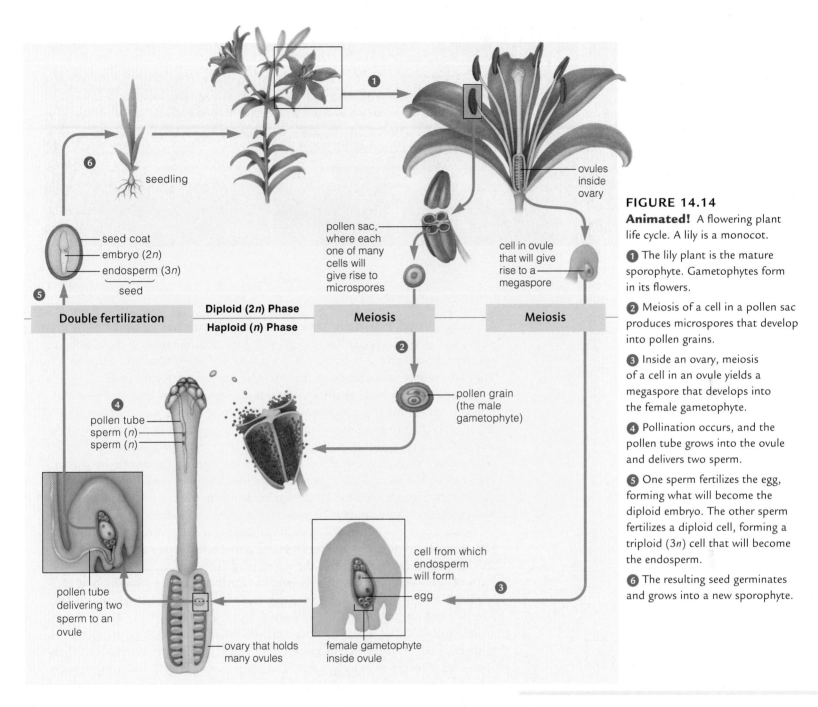

FIGURE 14.14

Animated! A flowering plant life cycle. A lily is a monocot.

❶ The lily plant is the mature sporophyte. Gametophytes form in its flowers.

❷ Meiosis of a cell in a pollen sac produces microspores that develop into pollen grains.

❸ Inside an ovary, meiosis of a cell in an ovule yields a megaspore that develops into the female gametophyte.

❹ Pollination occurs, and the pollen tube grows into the ovule and delivers two sperm.

❺ One sperm fertilizes the egg, forming what will become the diploid embryo. The other sperm fertilizes a diploid cell, forming a triploid (3n) cell that will become the endosperm.

❻ The resulting seed germinates and grows into a new sporophyte.

Labels within figure:

seedling

seed coat
embryo (2n)
endosperm (3n)
seed

ovules inside ovary

pollen sac, where each one of many cells will give rise to microspores

cell in ovule that will give rise to a megaspore

Double fertilization

Diploid (2n) Phase
Haploid (n) Phase

Meiosis

Meiosis

pollen grain (the male gametophyte)

pollen tube
sperm (n)
sperm (n)

pollen tube delivering two sperm to an ovule

cell from which endosperm will form

egg

ovary that holds many ovules

female gametophyte inside ovule

---

in a flower's female parts, or **carpels**. A carpel consists of a slender projection where pollen will land (the stigma) and a part that contains the ovaries (the style). An ovary has many ovules in which female gametophytes form from megaspores ❸. Pollination occurs when a pollen grain lands on a receptive stigma. In lilies, an insect usually serves as the pollinator.

After pollination, a pollen tube forms ❹. The tube grows down through the style and into the ovary. It then delivers two sperm to the ovule, where **double fertilization** occurs ❺. One sperm fertilizes the egg, forming what will become the embryo. The other sperm fertilizes a cell that has two nuclei, forming a triploid cell. This cell will divide and become the **endosperm**, a nutrient-rich tissue unique to angiosperms. Ovary tissue matures into a fruit that encloses the seed, which grows into a new sporophyte ❻.

carpel Female part of a flower.

double fertilization In flowering plants, one sperm fertilizes the egg, forming the zygote, and another fertilizes a diploid cell, forming what will become endosperm.

endosperm Nutritive tissue in an angiosperm seed.

eudicots Largest lineage of angiosperms; includes herbaceous plants, woody trees, and cacti.

flower Specialized reproductive shoot of a flowering plant.

fruit Mature ovary tissue that encloses a seed or seeds.

monocots Lineage of angiosperms that includes grasses, orchids, and palms.

ovary Of angiosperms only, a floral chamber that holds one or more ovules.

pollinator Animal that moves pollen from one plant to another, thus facilitating pollination.

stamen Male part of a flower.

FIGURE 14.15 Growing angiosperms for food, fabric, or drugs. *Left*, Mechanized harvesting of wheat. *Middle*, Field of cotton. *Right*, Marijuana discovered growing illegally in Linn County, Oregon.

Ecology and Human Uses of Angiosperms It would be nearly impossible to overestimate the importance of the angiosperms. As the dominant plants in most land habitats, they provide food and shelter for land animals. They also supply many products that meet human needs (Figure 14.15).

Angiosperms provide nearly all of our food, directly or as feed for livestock. Cereal crops are the most widely planted. The United States devotes more acreage to corn than to any other plant, and in 2006 American farmers harvested 282 million metric tons of it. Worldwide, rice feeds more people than any other crop. Wheat, barley, and sorghum are other widely grown grains. All are grasses. Legumes are the second most important source of human food. They can be paired with grains to provide all the amino acids the human body needs to build proteins. Soybeans, lentils, peas, and peanuts are examples of legumes.

Name a part of a plant, and humans probably eat some version of it. In addition to the seeds of grains and legumes, we dine on the leaves of lettuce and spinach, the stems of asparagus, the developing flowers of broccoli, the modified roots of potatoes, carrots, and beets, and the fruits of tomatoes, apples, and blueberries. The stamens of crocus flowers provide the spice saffron, and we grate the bark of a tropical tree to make cinnamon.

The fibers used to make clothing come from two main sources, petroleum and plants. Plant fibers include cotton, flax, ramie, and hemp. We also use fibers from seed plants to weave rugs, and in many places to thatch roofs. Angiosperm woods are "hardwoods," as opposed to gymnosperm "softwoods." Furniture and flooring are often made from oak or other hardwoods.

We extract medicines and psychoactive drugs from plants. Aspirin is derived from a compound discovered in willows. Digitalis from foxglove strengthens a weak heartbeat. Coffee, tea, and tobacco are widely used stimulants. Marijuana, illegal in the United States, is one of the country's most valuable cash crops. Worldwide, cultivation of opium poppies (the source of heroin) and coca (the source of cocaine) have wide-reaching health, economic, and political effects.

Take-Home Message **What are angiosperms?**

- Angiosperms are plants in which seeds develop in ovaries that become fruit.
- Angiosperms are the most diverse plants. Adaptations that contributed to their success include short life cycles, coevolution with insect pollinators, and a variety of fruit structures that aid in dispersal of seeds.

14.8 Fungal Traits and Diversity

Yeasts, Molds, and Mushrooms With this section, we begin our survey of another major group of eukaryotes, the fungi. Although a mushroom may seem more similar to a plant than to an animal, fungi are more closely related to animals than to plants:

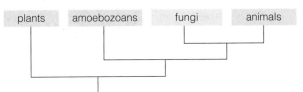

Fungi are more closely related to animals than to plants.

Fungi are spore-producing heterotrophs that have cell walls made of chitin, a polysaccharide also found in the external skeletons of insects and crabs.

Some fungi live as single cells and are commonly called yeasts (Figure 14.16**A**). The yeasts sold for baking bread are fungi, as are the organisms that cause yeast infections of the mouth or vagina.

Mushrooms and molds are multicelled fungi (Figure 14.16**B,C**). You may have seen mushrooms growing in the wild, and you probably have encountered them at salad bars and on pizzas. Molds are familiar to anyone who has left produce or other food sitting around a bit too long.

A multicelled fungus grows as a mesh of threadlike branching filaments collectively called the **mycelium** (plural, mycelia). Each filament is one **hypha** (plural, hyphae) and consists of cells arranged end to end (Figure 14.16**D**). Fungi do not have any type of specialized vascular tissue, but the cell walls between cells in a hypha are porous, so materials can flow among cells. As a result, nutrients or water taken up in one part of the mycelium are shared with cells in other regions of the fungal body.

Some fungal bodies are enormous. The largest organism we know about is a soil fungus in Oregon. Its hyphae extend through the soil over an area of about 2,200 acres, and it is still growing. Given its growth rate and size, researchers estimate that this individual has been alive for 8,000 years.

All fungi obtain nutrients by extracellular digestion and absorption. As a fungus grows in or over organic matter, it secretes digestive enzymes. The fungus then absorbs the resulting breakdown products. Most fungi are free-living **saprobes**: organisms that feed on and decompose organic wastes and remains. In this role, they help keep nutrients cycling in ecosystems. Other fungi live in or on other organisms. Some of these are parasites. Other fungi benefit their host or have no effect. Most plants have beneficial fungi growing in or on their roots. Fungi also form partnerships with green algae or cyanobacteria, forming a composite organism that we call a lichen. We return to relationships between fungi and other species in Section 14.9.

one cell (part of a hypha in a mycelium)

FIGURE 14.16 Examples of fungi. **A** One type of yeast. Cells reproduce by budding. **B** Green mold on grapefruit. **C** Scarlet hood mushroom.

Molds and mushrooms are examples of a mycelium, a multicelled body made of interwoven hyphae **D**. Material flows easily among the cells of a hypha.

fungus Spore-producing heterotroph with cell walls of chitin that feeds by extracellular digestion and absorption.
hypha A single filament in a fungal mycelium.
mycelium Mass of threadlike filaments (hyphae) that make up the body of a multicelled fungus.
saprobe Organism that feeds on wastes and remains.

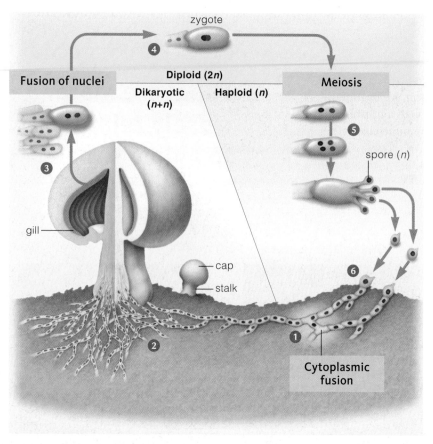

zygote

Fusion of nuclei

Diploid (2n)

Dikaryotic (n+n) **Haploid (n)**

Meiosis

spore (n)

gill

cap

stalk

Cytoplasmic fusion

FIGURE 14.17 Animated!
Generalized life cycle of many club fungi. Hyphae of different mating strains often grow through the same patch of soil.

❶ Two haploid hyphal cells meet and their cytoplasm fuses, forming a dikaryotic (n+n) cell.

❷ Mitotic cell divisions form a mycelium that produces a mushroom.

❸ Spore-making cells form at the edges of the mushroom's gills.

❹ Inside these dikaryotic cells, nuclei fuse, making the cells diploid (2n).

❺ The diploid cells undergo meiosis, forming haploid (n) spores.

❻ Spores are released and give rise to a new haploid mycelium.

Figure It Out: Are cells that make up the stalk of a mushroom haploid, diploid, or dikaryotic? *Answer: dikaryotic*

Fungal Diversity

About 56,000 species of fungi have been named, and there may be a million more. Fungi have been diversifying for a long time. Some fossil fungi date to 500 million years ago. The fossils resemble chytrids, the only modern fungi that are mainly aquatic and have a flagellated stage in their life cycle. The fossils and the existence of the aquatic, flagellated chytrids are taken as evidence that fungi, like plants and animals, evolved from some type of aquatic protist.

Three major lineages of fungi became established by 300 million years ago. They are the club fungi, zygote fungi, and sac fungi. We focus on these groups for the remainder of this chapter.

Fungal Life Cycles

Multicelled club fungi spend most of their life cycle growing as a dikaryotic mycelium. Dikaryotic means "having two nuclei." (*Di-* means two and *karyon* means nucleus, as in eukaryotic.) We represent the dikaryotic state as *n+n*.

Mushrooms are the reproductive parts, or fruiting bodies, of club fungi. They appear only briefly, when conditions favor sexual reproduction. Figure 14.17 shows the life cycle of a club fungus. Haploid hyphae of two different mating strains of club fungus (represented in the figure by cells with different colored nuclei) sometimes meet in the soil. If they do, their cytoplasm fuses and a dikaryotic cell forms ❶. Mitotic divisions produce a dikaryotic mycelium that grows through the soil ❷. When conditions favor reproduction, a sudden burst of hyphal growth produces a mushroom that emerges above the ground.

Typically the mushroom has a stalk and a cap, with thin sheets of tissue called gills on its underside ❸. Inside cells at the edges of the gills, nuclear fusion forms diploid zygotes ❹. The diploid stage is brief. Meiosis yields haploid spores ❺ that germinate and divide by mitosis, forming a haploid mycelium ❻.

The sexual stage differs a bit in the other groups of fungi. For example, bread molds are zygote fungi, which do not make large fruiting bodies. Instead, hyphae fuse and produce a diploid spore called a zygospore. The zygospore can survive unfavorable conditions, then undergo meiosis and release haploid spores. Like many fungi, bread molds also produce spores asexually. Specialized hyphae form and produce spores by mitosis at their tips (Figure 14.18**A**).

Truffles and morels are fruiting bodies of sac fungi (Figure 14.18**B,C**). Both are highly valued as food. Morels form above ground like the mushrooms of club fungi. A truffle forms underground. When truffle spores mature, the fungus gives off a scent like that of a male pig seeking a mate. Female pigs that catch a whiff disperse truffle spores as they root through the soil in search of this seemingly subterranean suitor. Dogs can also be trained to snuffle out truffles. The search can be worthwhile. In 2006, a single Italian truffle weighing 1.5 kilograms (about 3 pounds) sold for $160,000.

Human Uses of Fungi

As the truffle example illustrates, some fungi are important as food crops. Others are used in the production of foods or beverages. Baking yeast holds spores of a sac fungus (*Saccharomyces cerevisiae*). When bread dough is set out to rise in a warm place, the spores germinate, releasing yeast cells. Fermentation reactions carried out by the yeast produce the carbon

FIGURE 14.18 Spore-producing structures. **A** Black bread mold, a zygote fungus. The tiny black dots are spore sacs at the tips of special hyphae, as shown in the inset photo. Truffles **B** and morels **C** are highly prized edible fungi. Both are fruiting bodies of sac fungi. **D** The death cap mushroom (*Amanita phalloides*). The stalk, cap, and spores of this club fungus are toxic. Even with treatment, about a third of poisonings are fatal. Worldwide, *Amanita* species cause about 90 percent of mushroom poisonings.

dioxide that causes dough to expand (rise). Fermentation by fungi also helps produce beer, wine, and soy sauce. Other sac fungi produce the citric acid used to preserve and flavor soft drinks. Still others are responsible for the tangy blue veins in cheeses such as Roquefort.

Fungi cannot run away from predators, but they have evolved other defenses. Some defend themselves by producing toxins. Each year thousands of people are sickened after eating poisonous mushrooms that they mistook for an edible species. In some cases, mushroom poisoning is fatal (Figure 14.18**D**).

Some mushroom toxins alter mood and cause hallucinations. The drug LSD was isolated from such a mushroom, and psilocybin-containing mushrooms are eaten for their mind-altering effects. Both LSD and psilocybin-containing mushrooms are illegal in the United States.

Defensive compounds made by mushrooms are also sources of beneficial medicines. Most famously, the initial source of the antibiotic penicillin was a soil fungus, *Penicillium*. Fungi have also yielded drugs used to lower blood pressure or to prevent rejection of transplanted organs.

Take-Home Message What are fungi?

- Fungi are heterotrophs that absorb nutrients from their environment. They live as single cells or as a multicelled mycelium and disperse by producing spores.
- Most fungi produce spores both sexually and asexually. Mushrooms are the spore-producing fruiting bodies of club fungi.

Fungal Ecology

Fungi as Partners Nearly all plants form mutually beneficial relationships, or **mutualisms**, with fungi. Many soil fungi, including truffles, live in or on tree roots in a partnership known as a **mycorrhiza** (plural, mycorrhizae). In some cases, fungal hyphae form a dense net around roots but do not penetrate them (Figure 14.19**A**). In other cases, hyphae grow inside root cell walls.

Hyphae of both kinds of mycorrhizae grow through soil and serve to increase the absorptive surface area of their plant partner. The fungus shares minerals it takes up with the plant, and the plant gives up sugars to the fungus. It is a beneficial trade; many plants do poorly without a fungal partner.

Lichens are composite organisms consisting of a fungus and a single-celled photosynthetic species, either a green alga or a cyanobacterium. The fungus makes up most of the lichen's mass and shelters the photosynthetic species, which shares nutrients with the fungus.

Lichens grow on many exposed surfaces (Figure 14.19**B**,**C**). They are ecologically important as colonizers in places that are too hostile for other organisms. By releasing acids and retaining water that freezes and thaws, lichens help break down rocks and form soil. Formation of soil allows plants to move in and take root. Millions of years ago, lichens may have preceded plants onto the land.

Lichens take up water and nutrients across their surface, but they also absorb toxic air pollutants. Where air pollution levels are high, lichens can become poisoned and die. For this reason, researchers sometimes monitor the health of lichen populations as an indicator of air quality.

Fungi as Decomposers and Pathogens Fungi are important decomposers. They secrete digestive enzymes onto wastes and remains, then absorb some of the breakdown products. Because not all nutrients are absorbed, some remain in the soil to nourish other organisms. Club fungi play an especially important role as decomposers in forests. They are the only fungi that can break down lignin, which is abundant in wood.

Some fungi do not wait until an organism is dead to feed on it. Such parasitic fungi can be important plant pathogens. For example, airborne spores spread wheat stem rust among wheat plants (Figure 14.20**A**). Infection by this club fungus can reduce crop yield by up to 70 percent. Another fungus caused the demise of the American chestnut. This tree species was abundant in eastern forests until the early 1900s. Then, a pathogenic fungus introduced from Asia caused mature trees to die back to the ground. Today, a few chestnuts still sprout from old root systems, but they cannot reach mature size.

Fungi also feed on living humans. Most frequently, they infect skin, causing flaking, redness, and itching. For example, several species of fungus can take up residence in the thin skin between your toes, causing what is commonly called "athlete's foot" (Figure 14.20**B**). Such infections can usually be cured with over-the-counter medications. To prevent athlete's foot, do not go barefoot in public showers or other places where infected people may have walked and shed fungal spores. Also, keep your feet dry. Skin fungi grow best in damp places.

FIGURE 14.19 Fungal partnerships. **A** Mycorrhizal fungus growing on roots of a young hemlock tree. **B** Leaflike lichen on a birch tree. **C** Encrusting lichens on granite.

Low numbers of single-celled fungi normally occur in the vagina, but sometimes their population explodes and causes fungal vaginitis (a vaginal yeast infection). Symptoms usually include itching or burning sensations and a thick, odorless, whitish vaginal discharge. Intercourse is often painful. Disrupting the normal populations of bacteria in the vagina by douching or using antibiotics increases the risk of fungal vaginitis, as does use of oral contraceptives. Nonprescription medications placed into the vagina will usually control the infection. If this treatment is not effective, a woman should see a doctor.

Histoplasmosis is a common fungal disease in the midwestern and south central United States, where the soil holds spores of *Histoplasma capsulatum*. Most people who inhale spores of this species experience a brief episode of coughing or no symptoms at all. However, in immune-suppressed or elderly people, the fungus may spread from the lungs, through the blood, and into other organs, with fatal results.

Similarly, soils in the American Southwest hold spores of *Coccidioides*, which can cause coccidioidomycosis, or valley fever. Like histoplasmosis, this fungal disease can be fatal to the elderly or to those with an impaired immune system.

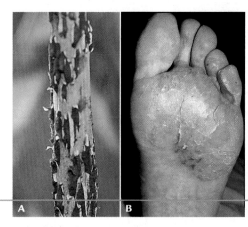

FIGURE 14.20 Effects of pathogenic fungi. **A** Wheat stalk infected by wheat stem rust. **B** Cracked, red, itchy skin caused by a fungal infection. The fungus feeds by secreting enzymes that dissolve keratin—the main skin protein— and other components of skin.

Take-Home Message How do fungi interact with other species?

- Mycorrhizal fungi live in or on plant roots in a mutually beneficial relationship. Fungi also live with single-celled photosynthetic cells as lichens.

- Fungi benefit other organisms when they feed on wastes and remains, releasing nutrients into the soil. They harm other organisms, including humans, when they infect them and feed on their tissues.

lichen Composite organism consisting of a fungus and a single-celled photosynthetic alga or bacterium.
mycorrhiza Fungus–plant root partnership.
mutualism Species interaction that benefits both species.

14.10

Impacts/Issues Revisited:
Speaking for the Trees

When we walk through a forest, we glimpse only a tiny fraction of the life it holds. There is a world beneath our feet. By one estimate, if all the fungal hyphae in a gram of forest soil were laid out end to end they would extend for 200 meters, more than two football fields.

Most forest plants benefit by the nutrients their roots receive from this fungal web, but some species take it to an extreme. The photo to the *right* shows an unusual plant commonly known as Indian pipe. It grows on the forest floor and does not have chlorophyll. Instead of making its own food, it taps into pipelines of mycorrhizae in the soil.

Logging not only removes trees, it also disrupts relationships between plants and soil fungi. Newly planted stands of trees have far fewer types of fungi than older forests. Damage to soil fungi affects the plants that depend on them. In studies of forests that had been clear-cut (had all trees removed), researchers found that populations of some plants, including Indian pipe, had not recovered more than 50 years after the logging took place.

How Would
YOU
☑ **Vote ?**

Demand for paper is a big factor in deforestation, but using recycled paper can add to the cost of a product. Would you be willing to pay more for papers, books, and magazines that are printed on recycled paper? **See CengageNow for details, then vote online (cengagenow.com).**

Summary

Section 14.1 Plants provide important eco-logical services, so deforestation is a threat to human well-being. Planting trees helps raise awareness of the essential role of plants.

Section 14.2 Plants evolved from green algae. Nearly all existing plants are photosynthetic. Key adaptations that allowed plants to move into dry habitats include a waterproof cuticle with stomata, and internal pipelines of vascular tissue (xylem and phloem).

Plant life cycles include two multicelled forms, the haploid gametophyte and the diploid sporophyte. Seeds and male gametophytes that can be dispersed without water (pollen grains) were important adaptations of seed plants.

CENGAGENOW˜ Compare the relative contributions of haploid and diploid stages to different plant life cycles with the interaction on CengageNow.

Section 14.3 Mosses, liverworts, and hornworts are bryophytes. These nonvascular plants lack xylem and phloem. Their flagellated sperm reach eggs by swimming through films or droplets of water that cling to the plant. The sporophyte begins development inside gametophyte tissues. It remains attached to and often dependent upon the gametophyte even when mature.

sporophyte —

gametophyte —

CENGAGENOW˜ Observe the life cycle of a typical moss with the animation on CengageNow.

Section 14.4 Ferns, club mosses, and horse-tails are among the seedless vascular plants. Their sporophytes have vascular tissues and are the larger, longer-lived phase of the life cycle. Like bryophytes, they have flagellated sperm and require water for fertilization.

CENGAGENOW˜ Observe the life cycle of a typical fern with the animation on CengageNow.

Section 14.5 Gymnosperms and flowering plants (angiosperms) are seed-bearing vascular plants. They produce two types of spores. Microspores give rise to pollen grains, the sperm-producing male gametophytes. Even in the absence of water, pollen grains can deliver sperm to egg cells. Megaspores form in ovules, and give rise to egg-producing female gametophytes. The seed is a mature ovule, with an embryo sporophyte and some nutritive tissue inside it.

Section 14.6 Conifers, cycads, and ginkgos are among the gymnosperms. They are adapted to dry climates and bear seeds on exposed surfaces of spore-bearing structures. In conifers, these spore-bearing structures are distinctive cones.

CENGAGENOW˜ Learn about the stages in the life cycle of a typical gymno-sperm with the animation on CengageNow.

Section 14.7 Angiosperms are the dominant land plants. They alone have flowers. Most flowering plants coevolved with pollinators. After pol-lination, the flower's ovary becomes a fruit that contains one or more seeds. A flowering plant

seed includes an embryo sporophyte and endosperm, a nutritious tissue. The vast majority of crop plants are angiosperms.

CENGAGENOW˜ Explore the angiosperm life cycle with the animation on CengageNow.

Section 14.8 Fungi are single-celled and multicelled heterotrophs. They secrete enzymes onto organic mate-rial and then absorb the resulting breakdown products. Multicelled fungi grow as a mycelium composed of many hyphal filaments. Fungi produce spores both sex-ually and asexually. A mushroom is a spore-producing body of a club fungus.

CENGAGENOW˜ Explore the life cycle of a mushroom and a bread mold with the animation on CengageNow.

Section 14.9 Most plants rely on mycorrhizal fungi that live in or on their roots and help them take up food. Fungi also partner with photosynthetic cells and form lichens. Fungi are important decomposers, espe-cially of wood. Some fungi that are parasites of plants or animals cause disease.

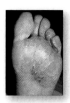

Section 14.10 Logging reduces a community's diversity of fungi as well as plants.

Self-Quiz Answers in Appendix I

1. Which of the following statements is not correct?
   a. Gymnosperms are the simplest vascular plants.
   b. Bryophytes are nonvascular plants.
   c. Ferns and angiosperms are vascular plants.
   d. Only angiosperms produce flowers.

2. Which does *not* apply to gymnosperms or angiosperms?
   a. vascular tissues   c. single spore type
   b. diploid dominance   d. all of the above

3. Bryophytes have independent _____ and attached, dependent _____ .
   a. sporophytes; gametophytes
   b. gametophytes; sporophytes

4. Ferns are classified as _____ plants.
   a. multicelled aquatic   c. seedless vascular
   b. nonvascular seed   d. seed-bearing vascular

5. The _____ produce flagellated sperm.
   a. ferns   c. monocots
   b. conifers   d. a and c

6. The _____ produced in the male cones of a conifer develop into pollen grains.
   a. ovules   c. megaspores
   b. ovaries   d. microspores

7. A seed is _____ .
   a. a female gametophyte   c. a mature pollen tube
   b. a mature ovule   d. an immature spore

8. The _____ are the most diverse seedless vascular plants.

9. Match the terms appropriately.

_____ gymnosperm	a. gamete-producing body
_____ sporophyte	b. help control water loss
_____ horsetail	c. "naked" seeds
_____ bryophyte	d. spore-producing body
_____ gametophyte	e. nonvascular land plant
_____ stomata	f. seedless vascular plant
_____ angiosperm	g. flowering plant

10. All fungi _____ .
a. are multicelled   c. are heterotrophs
b. form flagellated spores   d. all of the above

11. Saprobic fungi derive nutrients from _____ .
a. nonliving organic matter   c. living animals
b. living plants   d. photosynthesis

12. A mushroom is _____ .
a. the food-absorbing part of a fungus
b. the only part of the fungal body not made of hyphae
c. a reproductive structure that releases sexual spores
d. the longest-lived part of the fungal life cycle

13. A _____ is a fungus-plant root partnership.

14. A _____ is a composite organism composed of a fungus and a single-celled photosynthetic species.

15. True or false? Fungal skin infections are often fatal.

Additional questions are available on CENGAGENOW™

Critical Thinking

1. Early botanists admired ferns but found their life cycle perplexing. In the 1700s, they learned to propagate them by sowing what appeared to be tiny dustlike "seeds" from the undersides of fronds. Despite many attempts, the scientists could not find the pollen source, which they assumed must stimulate the "seeds" to develop. Imagine you could write to one of these botanists. Compose a note that would clear up their confusion.

2. Today, the tallest bryophytes reach a maximum height of 20 centimeters (8 inches) or so. So far as we know from fossils, there were no giants among their ancestors. Lignin and vascular tissue first evolved in relatives of club moss, and some extinct species stood 40 meters (130 feet) high. Among modern seed plants, *Sequoia* (a gymnosperm) and *Eucalyptus* (an angiosperm) can be more than 100 meters (330 feet) high. Explain why evolution of vascular tissues and lignin would have allowed such a dramatic increase in plant height. How might being tall give one plant species a competitive advantage over another?

3. Some poisonous mushrooms such as those shown at the *right* have bright, distinctive colors that mushroom-eating animals learn to recognize. Once sickened, an animal avoids these species. Other toxic mushrooms are as dull-looking as edible ones, but they have an unusually strong odor. Some scientists hypothesize that the strong odors aid in defense against mushroom-eating animals that are active at night. Devise an experiment to test this hypothesis.

Digging Into Data

Removing Stumps to Save Trees

The club fungus *Armillaria ostoyae* infects living trees and acts as a parasite, withdrawing nutrients from them. When the tree dies, the fungus continues to dine on its remains. Fungal hyphae grow out from the roots of infected trees and roots of dead stumps. If these hyphae contact roots of a healthy tree, they can invade and cause a new infection.

Canadian forest pathologists hypothesized that removing stumps after logging could help prevent tree deaths. To test this hypothesis, they carried out an experiment. In half of a forest, they removed stumps after logging. In a control area, they left stumps behind. For more than 20 years, they recorded tree deaths and whether *A. ostoyae* caused them. Figure 14.21 shows the results.

1. Which tree species was most often killed by *A. ostoyae* in control forests? Which was least affected by the fungus?

2. For the species most affected, what percentage of deaths did *A. ostoyae* cause in control and in experimental forests?

3. Do the overall data support the hypothesis that stump removal helps protect living trees from infection by *A. ostoyae*?

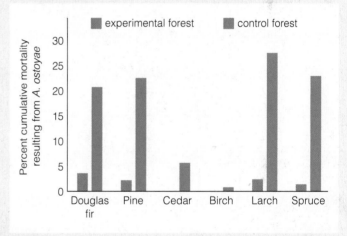

FIGURE 14.21 Results of a long-term study of how logging practices affect tree deaths caused by the fungus *A. ostoyae*. In the experimental forest, whole trees—including stumps—were removed (*brown* bars). The control half of the forest was logged conventionally, with stumps left behind (*blue* bars).

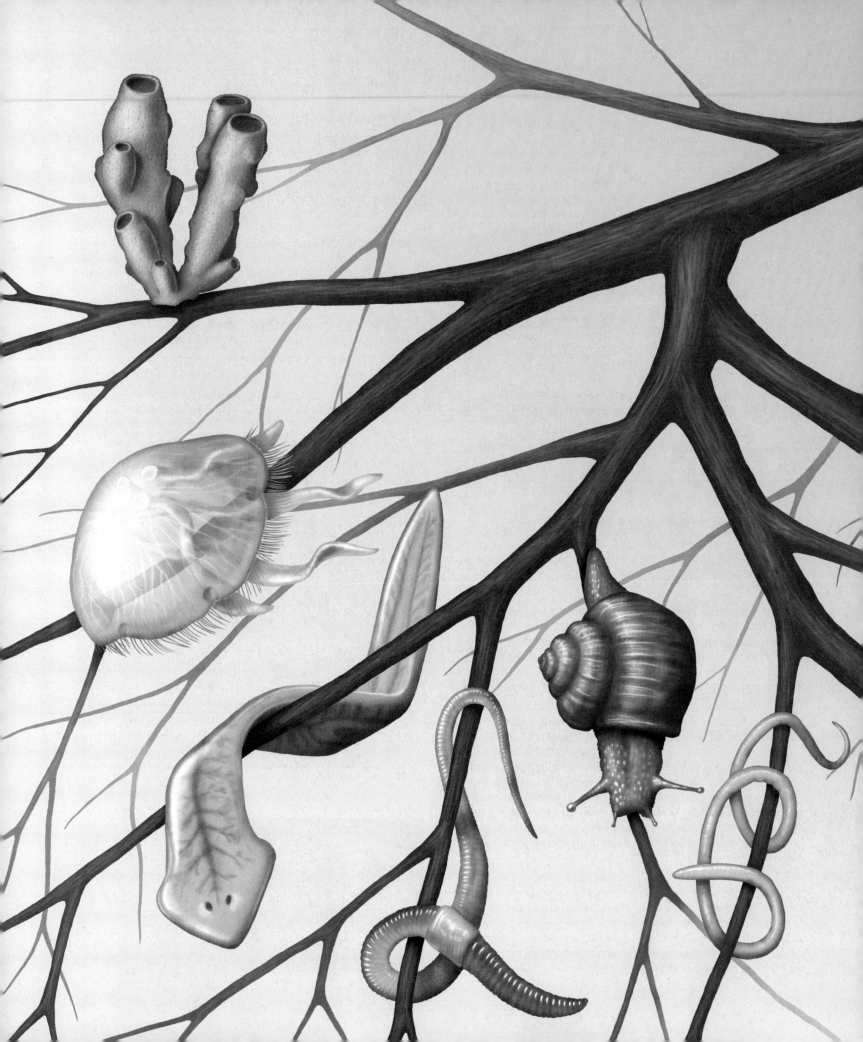

15 Animal Evolution

15.1 Impacts/Issues: Windows on the Past

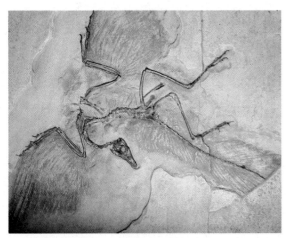

Archaeopteryx

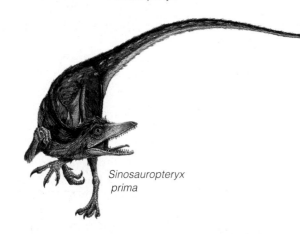

Sinosauropteryx prima

Confuciusornis sanctus

In Darwin's time, the presumed absence of transitional fossils was an obstacle to acceptance of his new theory of evolution. Skeptics wondered, if new species evolve from existing ones, then where are the fossils that document these transitions? Where are the fossils that bridge major groups? In fact, one of these "missing links" was unearthed by workers at a limestone quarry in Germany just one year after Darwin's *On the Origin of Species* was published.

That fossil, about the size of a large crow, looked like a small meat-eating dinosaur. It had a long, bony tail, three clawed fingers on each forelimb, and a heavy jaw with short, spiky teeth. It also had feathers. The new fossil species was named *Archaeopteryx* (ancient winged one). Eventually, eight *Archaeopteryx* fossils were unearthed. One is shown at the *upper left*. Radiometric dating (Section 11.4) indicates that *Archaeopteryx* lived 150 million years ago.

Archaeopteryx is the most widely known transitional fossil in the bird lineage, but there are others. In 1994, a farmer in China discovered a fossil dinosaur with small forelimbs and a long tail. Unlike most dinosaurs, this one (shown in the reconstruction at the *middle left*) had downy feathers like those that cover modern chicks. Researchers named the farmer's fuzzy find *Sinosauropteryx prima*, which means first Chinese feathered dragon. Given its shape and lack of long feathers, *S. prima* was certainly flightless. The feathers probably helped the animal stay warm.

Both *S. prima* and *Archaeopteryx* had a long tail and a mouth full of sharp teeth. In contrast, modern birds have a small tailbone and their toothless jaws are covered with hard layers of protein that form a beak. Another fossil from China, *Confuciusornis sanctus* (sacred Confucius bird), is the earliest known bird with a beak. As shown at the *lower left*, it had a short birdlike tail with long feathers. Even so, its dinosaur ancestry remains apparent. Unlike modern bird wings, the wings of *C. sanctus* had grasping digits with claws at their tips.

As these examples show, fossils are an important source of information about the past. They provide evidence of the relationships among animal lineages and information about transitions between major groups. High-quality fossils of birds, mammals, and other animals with a backbone are especially rare and scientifically valuable. For this reason, scientists are concerned about sale of these specimens. Once a fossil is privately owned, there is no guarantee that it will be properly cared for or made available for study. Also, a thriving market in fossils encourages people to poach fossils from public land. United States federal laws protect important archaeological sites and regulate the sale of important archaeological artifacts. The Society of Vertebrate Paleontologists argues that fossil-rich sites and fossils should receive similar protections.

With this introduction, we turn to the story that fossils and other data tell about animal evolution. This is the tale of how the unique traits that characterize modern animal lineages arose. It is also the tale of our own ancestry, as it stretches backs millions of years.

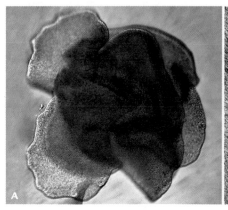

15.2 Animal Origins, Traits, and Trends

Animals are multicelled heterotrophs that take food into their body, where they digest it and absorb the released nutrients. An animal has a few to hundreds of types of unwalled cells. Cells become specialized as an animal develops from an embryo (an early developmental stage) to an adult. Most animals reproduce sexually, some reproduce asexually, and some do both. Nearly all animals are motile (can move from place to place), during part or all of their life cycle.

Animal Origins and Early Evolution Animals probably evolved from a heterotrophic protist that formed colonies. At first, all cells in the colony were similar. Each could reproduce and carry out all other essential tasks. Later, mutations produced cells that specialized in some tasks and did not carry out others. Perhaps these cells captured food more efficiently, but did not make gametes. Colonies that had interdependent cells and a division of labor were at a selective advantage, and new specialized cell types evolved. Eventually this process produced the first animal.

Studies of the oldest animal lineages give us an idea of what the earliest animals may have been like. For example, the **placozoans** represent an early branch on the animal family tree. They have the fewest genes and simplest body plan of any living animal. A placozoan's flattened body is about 2 millimeters across and consists of four cell types (Figure 15.1**A**). All other animals have more kinds of cells. Ciliated cells on the placozoan's lower surface allow it to move about. A placozoan eats bacteria and single-celled algae. Gland cells, also on its lower surface, secrete enzymes onto food, then take up breakdown products.

The first animals may have evolved in the seas as early as 1 billion years ago. However, fossils show that most animal lineages arose during an explosion of diversity in the Cambrian (544–505 million years ago). During this time, the oxygen concentration in seawater increased dramatically. All animals carry out aerobic respiration, so abundant oxygen would have allowed larger, more active animals to evolve. At the same time, early supercontinents were breaking up. Movement of land masses cut off gene flow among populations, encouraging speciation (Section 12.5). Species interactions may have also encouraged evolutionary innovations. For example, once the first predators arose, mutations that produced protective hard parts were favored (Figure 15.1**B**). Changes in genes that regulate body plans (homeotic genes, Section 7.7) may have sped things along. Mutations in these genes would have allowed adaptive changes to body form in response to predation or other selective forces.

animal A eukaryotic heterotroph that is made up of unwalled cells and develops through a series of stages. Most ingest food, reproduce sexually, and move.

placozoan Structurally simplest animal known, with only four types of cells and a small genome.

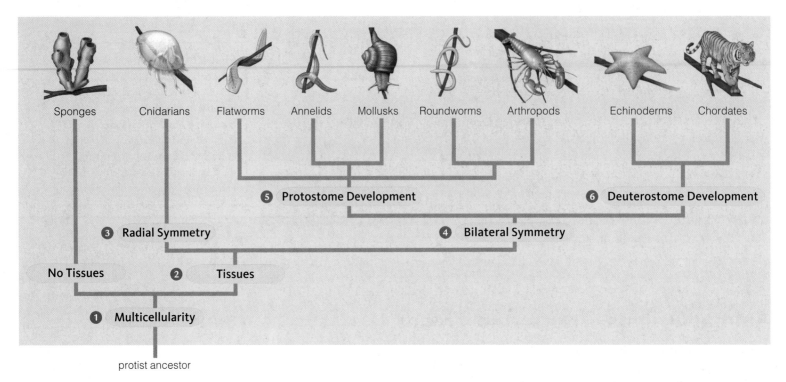

FIGURE 15.2 Family tree for the animals. The vertebrates, animals with a backbone, are a subgroup of the chordates.

Of about 2 million named animals, only about 50,000 are vertebrates.

Major Animal Groups and Evolutionary Trends Figure 15.2 shows relationships among the major animal groups covered in this book. All animals are descended from a common multicelled ancestor ❶. The earliest ones were aggregations of cells, and placozoans and sponges still show this level of organization. However, most animals have tissues ❷. A **tissue** consists of one or more types of cells that are organized in a specific pattern and that carry out a particular task. In the early animal lineages, embryos had two tissue layers: an outer ectoderm and an inner endoderm. In later lineages, cells began to rearrange themselves and form a middle embryonic layer called mesoderm. The evolution of a three-layer embryo allowed an increase in complexity. Most animal groups have organs derived from mesoderm.

The structurally simplest animals such as sponges are asymmetrical—you cannot divide their body into halves that are mirror images. Jellies and other cnidarians have **radial symmetry**: Body parts are repeated around a central axis, like the spokes of a wheel ❸. Radial animals have no front or back end. They attach to an underwater surface or drift along, so their food can arrive from any direction. Most animals have **bilateral symmetry**. They have a right and left half, with body parts repeated on either side of the body ❹. Bilateral animals have a distinctive "head end" that has a concentration of nerve cells.

The head end of a bilateral animal also has an opening for taking in food. In flatworms, food enters a saclike digestive cavity, then wastes leave through the same body opening. However, most bilateral animals have a tubular gut, with a mouth at one end and an anus at the other. A tubular gut, known as a complete digestive system, has advantages. Parts of the tube can be specialized for taking in food, digesting food, absorbing nutrients, or compacting the wastes. Unlike a saclike cavity, a complete gut can carry out all of these tasks simultaneously.

Two lineages of bilateral animals differ in their embryonic development. In protostomes, the first opening that appears on the embryo becomes the mouth ❺. *Proto–* means first and *stoma* means opening. In deuterostomes, the first embryonic opening becomes an anus and the second becomes the mouth ❻.

A flatworm's digestive cavity is enclosed by a more or less solid mass of tissues and organs (Figure 15.3**A**). However, in most animals, a fluid-filled body cavity surrounds the gut (Figure 15.3**B,C**). If this cavity has a lining of tissue

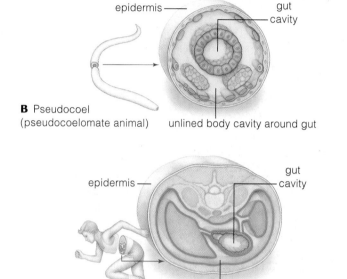

FIGURE 15.3 Types of body cavity. **A** A flatworm has no internal body cavity—its gut is surrounded by tissues. **B** A roundworm has a pseudocoel, a cavity around the gut that is not fully lined with mesoderm. **C** Like all vertebrates, humans have a lined body cavity, or coelom.

A No coelom (acoelomate animal)

epidermis — gut cavity

organs packed between gut and body wall

B Pseudocoel (pseudocoelomate animal)

epidermis — gut cavity

unlined body cavity around gut

C Coelom (coelomate animal)

epidermis — gut cavity

body cavity with a lining (*dark blue*) derived from mesoderm

derived from mesoderm, it is called a **coelom** (Figure 15.3**C**). A cavity that is only partially lined by mesodermal tissue is a **pseudocoel**, which means false coelom.

A fluid-filled coelom or pseudocoel provided three advantages. First, materials could diffuse through the fluid to body cells. Second, muscles could redistribute the fluid to alter the body shape and aid locomotion. Finally, organs were not hemmed in by a mass of tissue, so they could grow larger and move more freely.

Gases and nutrients diffuse quickly through the body of a small animal. However, diffusion alone cannot move substances through the body of a big animal fast enough to keep it alive. In most animals, a circulatory system speeds the distribution of substances through the body. In a closed circulatory system, a heart or hearts propel blood through a continuous system of vessels. Materials carried by the blood diffuse out of vessels and into cells, and vice versa. A closed system allows for faster blood flow than an open one. In an open circulatory system, blood is pumped out of vessels into internal spaces, from which it is taken up again by a heart.

Most bilaterally symmetrical animals have some degree of **segmentation**, a division of a body into interconnecting units that are repeated one after the other along the main body axis. We clearly see body segments in annelids such as earthworms. We also find clues to our own origins in the segmented body of an early human embryo. Segmentation opened the way to evolutionary innovations in body form. When many segments have organs that carry out the same function, some segments can become modified without endangering the animal's survival.

Of about 2 million named animals, only about 50,000 are **vertebrates**, animals with a backbone. These include fishes, amphibians, reptiles, birds, and mammals. The vast majority of animals are **invertebrates**, animals that do not have a backbone, and they are the starting point for our survey of animal diversity beginning in the next section.

Take-Home Message **What are animals?**

- Animals are multicelled heterotrophs that typically ingest food. Their cells are unwalled.

- Animals reproduce sexually and, in many cases, asexually. They go through a period of embryonic development, and most move about during at least part of the life cycle.

- Animal bodies differ in their symmetry, type of digestive cavity, type of body cavity (if any), and degree of segmentation.

- Most animals are invertebrates, animals that do not have a backbone.

bilateral symmetry Having right and left halves with similar parts, and a front and back that differ.
coelom A body cavity with a complete lining of tissue derived from mesoderm.
invertebrate Animal without a backbone.
pseudocoel Body cavity not fully lined with mesoderm.
radial symmetry Having parts arranged around a central axis, like spokes around a wheel.
segmentation Having a body composed of units that repeat along its length.
tissue One or more types of cells that are organized in a specific pattern and that carry out a particular task.
vertebrate Animal with a backbone.

Invertebrate Diversity

water out

central cavity

glasslike structural elements

amoeboid cell

pore

semifluid matrix

flattened surface cells

collar cell

water in

A

B

FIGURE 15.4 Animated! A Body plan of a glass sponge. Arrows indicate direction of water flow. **B** Natural bath sponge.

Sponges

A **sponge** (phylum Porifera) is an aquatic animal that typically has no symmetry, tissues, or organs. The phylum name, Porifera, means pore-bearing and refers to the body's many tiny pores.

Adult sponges do not move about. They attach to a surface and eat bacteria that they filter from the water. A sponge has two cell layers, an outer layer of flattened cells and an inner layer of flagellated collar cells (Figure 15.4**A**). A jellylike matrix lies between the cell layers. Waving of the collar cells' flagella keeps water flowing into the body through pores, then out through one or more larger openings. The collar cells trap bacteria in the water, engulf them by phagocytosis (Section 4.6), and digest them. Amoeba-like cells move throughout the matrix, receiving food from collar cells and distributing it to other cells in the body.

In many species, cells in the matrix also secrete fibrous proteins or glassy spikes that structurally support the body and discourage predators. Some of the protein-rich sponges are commercially important. The sponges are harvested from the sea, dried, cleaned, and bleached, then sold for bathing and cleaning (Figure 15.4**B**). About $40 million worth of sponges are harvested each year.

A typical sponge is a **hermaphrodite**: an individual that produces both eggs and sperm. After fertilization, a zygote forms and develops into a ciliated larva. A **larva** (plural, larvae) is a free-living, sexually immature form in an animal life cycle. Sponge larvae exit the parental body, swim briefly, then settle and develop into adults.

Cnidarians

Like sponges, adult **cnidarians** (phylum Cnidaria) have a two-layer body with a jellylike matrix between the layers. However, cnidarians evolved a more complex, radially symmetrical body plan, with special cells that give them some abilities sponges do not have. For example, unlike sponges, cnidarians have nerve cells and contractile cells. Interactions among these cells allow cnidarians to detect stimuli and respond by moving their body or its parts. The nerve cells interconnect to form a nerve net, but there is no central information processing region that functions like a brain.

outer tissue layer

gastrovascular cavity

jellylike matrix

inner tissue layer

gastrovascular cavity

A   **Medusa**

Polyp

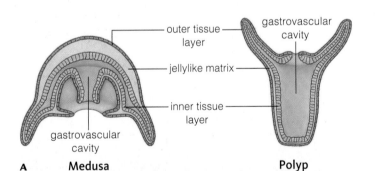

B   **C**   **D**

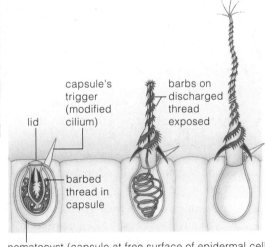

capsule's trigger (modified cilium)

lid

barbs on discharged thread exposed

barbed thread in capsule

E   nematocyst (capsule at free surface of epidermal cell)

FIGURE 15.5 Animated! Cnidarians. **A** There are two body plans: medusa and polyp. The *Obelia* life cycle includes both free-swimming medusae **B** and polyps. Sea anemones **C** and corals **D** live only as polyps.

E Cnidarian tentacles contain unique stinging cells with specialized organelles called nematocysts that discharge barbed threads when a trigger is touched.

FIGURE 15.6 Animated!
Life cycle of the beef tapeworm.

1 A person becomes infected by eating undercooked beef that contains the resting stage of the tapeworm.

2 In the human intestine, the tapeworm uses its barbed scolex to attach to the intestinal wall. The worm grows by adding new body units called proglottids. Over time, it can become many meters long.

3 Each proglottid produces both eggs and sperm, which combine. Proglottids with fertilized eggs leave the body in feces.

4 Cattle eat grass contaminated with proglottids or early larvae.

5 The larval tapeworm forms a cyst in the cattle muscle.

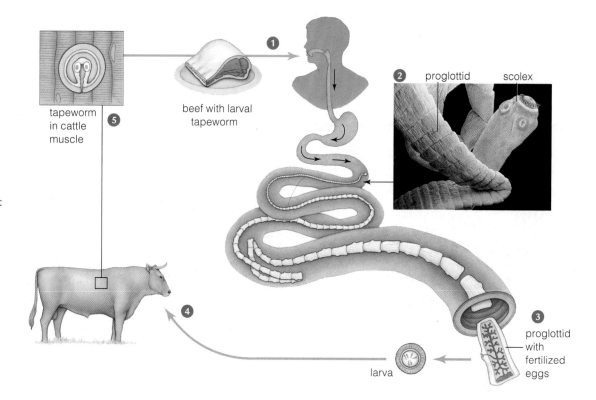

There are two cnidarian body plans, and one or both may occur in the life cycle (Figure 15.5**A**–**D**). **Medusae** look like a bell or umbrella. They swim or drift about. Jellies (also called jellyfish) are medusae. **Polyps**, such as sea anemones and corals, are tubular, and one end usually attaches to a surface. In both polyps and medusae, a tentacle-ringed opening leads to a gastrovascular cavity that takes in and digests food, expels wastes, and also functions in gas exchange.

The phylum name Cnidaria is from *cnidos*, the Greek word for nettle, a kind of stinging plant. It refers to a unique cnidarian trait, specialized stinging cells that have organelles called **nematocysts**. When something brushes against a cnidarian's tentacles, nematocysts in cells on the tentacle discharge a barb with venom into it (Figure 15.5**E**). Human swimmers who brush up against cnidarians can trigger this response and receive painful, and occasionally deadly, stings. More typically, the tentacles snag tiny invertebrates or fish. The captured prey is pulled through the mouth into the gastrovascular cavity, where gland cells secrete enzymes that digest it.

Flatworms With **flatworms** (phylum Platyhelminthes) we begin our survey of the protostomes, one of the two lineages of bilateral animals. The phylum name comes from Greek; *platy–* means flat, and *helminth* means worm. As shown in Figure 15.3**A**, flatworms have no body cavity. Their gut is surrounded by tissue and organs. Flatworms are typically hermaphrodites.

Flukes and tapeworms are flatworms that spend part of their life cycle as parasites of vertebrates. Figure 15.6 shows the life cycle of the beef tapeworm, which can infect humans who eat undercooked beef.

Many flatworms live in the ocean and a few live in damp places on land. The planarians are free-living, freshwater flatworms (Figure 15.7). They have a branching gut with a single opening. A muscular tube (the pharynx) sucks in food or lets out waste. A pair of nerve cords runs the length of the body. Clusters of nerve cell bodies in the head serve as a simple brain. The head also has chemical receptors and light-detecting eyespots.

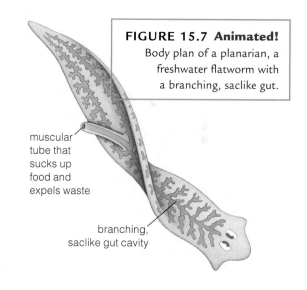

FIGURE 15.7 Animated!
Body plan of a planarian, a freshwater flatworm with a branching, saclike gut.

muscular tube that sucks up food and expels waste

branching, saclike gut cavity

cnidarian Radially symmetrical invertebrate with two tissue layers; uses tentacles with stinging cells to capture food.

flatworm Bilaterally symmetrical invertebrate with organs but no body cavity; for example, a planarian or tapeworm.

hermaphrodite Animal that makes both eggs and sperm.

larva Preadult stage in some animal life cycles.

medusa Bell-shaped, free-swimming cnidarian body form.

nematocyst Touch-sensitive stinging organelle unique to cnidarians.

polyp Tubular cnidarian body form that usually attaches to some surface.

sponge Aquatic invertebrate that has no tissues or organs and filters food from the water.

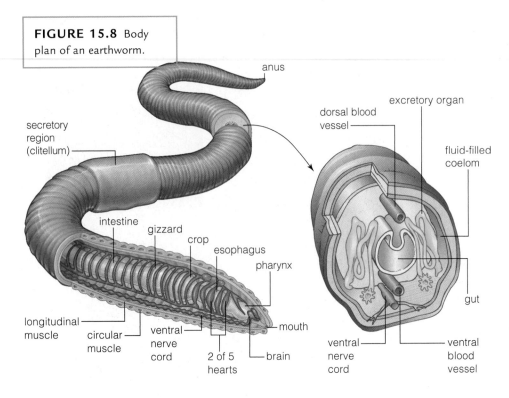

FIGURE 15.8 Body plan of an earthworm.

anus

secretory region (clitellum)

intestine gizzard crop esophagus pharynx

longitudinal muscle circular muscle ventral nerve cord 2 of 5 hearts brain mouth

dorsal blood vessel

excretory organ

fluid-filled coelom

gut

ventral nerve cord

ventral blood vessel

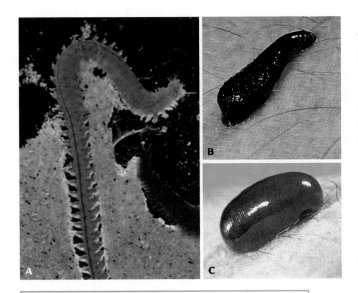

FIGURE 15.9 Annelid diversity. **A** A sandworm, a predatory marine polychaete. **B** A leech beginning to feed, and **C** fully engorged after a blood meal.

Annelids Annelids (phylum Annelida) are worms that evolved a segmented body, a coelom, a complete digestive system, and a closed circulatory system. The phylum name is from the Greek word *annulus*, meaning ringed.

Earthworms, the most familiar annelids, live on land and most have more than one hundred segments (Figure 15.8). An earthworm eats its way through the soil, digesting any organic material that it takes in. A few chitin-stiffened bristles on each segment help the animal burrow.

Each segment has a fluid-filled coelomic chamber with paired excretory organs that remove waste from the fluid. A digestive tract with specialized regions runs through the coelom. A nerve cord extends the length of the body and connects to a simple brain. Multiple hearts pump blood through vessels.

Earthworms are hermaphrodites. A secretory region, the clitellum, makes mucus that glues worms together while they swap sperm. Later, the clitellum secretes a silky case for the fertilized eggs the worm deposits in the soil.

Polychaetes, a lineage of mostly marine annelids, typically have many bristles per segment. (*Poly*– means many and *chaete* means bristle.) Some, including the sandworms often sold as bait, are active predators (Figure 15.9). Others have appendages specialized for filtering food from currents.

Leeches are another annelid lineage. They typically live in fresh water, although some are found in damp places on land. Most are scavengers and predators of invertebrates. The few that suck blood from vertebrates evolved a protein in their saliva that keeps blood from clotting while the leech feeds (Figure 15.9**B,C**). Doctors who reattach a severed finger or ear sometimes apply leeches to the reattached part. The leeches prevent unwanted clots from forming in the reattached part's blood vessels.

Mollusks Mollusks (phylum Mollusca) have a small coelom and a soft, unsegmented body. *Mulluscus* is Latin for soft. Mollusks probably evolved from a wormlike ancestor, but the body has been highly modified. One unique mollusk trait is the **mantle**. This skirtlike extension of the upper body wall drapes over the body mass, forming a mantle cavity (Figure 15.10**A**). In most mollusks, the mantle secretes a hard, calcium-rich shell.

Among animals, mollusks are second only to arthropods in diversity. Most live in the seas, but some live in fresh water or on land. With about 60,000 species of snails and slugs, the **gastropods** are the largest subgroup. Their name means "belly foot." Most gastropods glide about on a broad muscular foot that makes up most of their lower body mass. A typical gastropod uses its **radula**, a tonguelike organ with bits of hard chitin, to scrape up algae. However, some snails have adapted to a predatory life-style. For example, a cone snail has a harpoonlike radula that injects paralyzing venom into prey such as small fish (Figure 15.10**B**).

The only land-dwelling mollusks, land snails and slugs, are gastropods. They have a lung instead of gills and can breathe air. Both land slugs and sea slugs do not have a shell. Most defend themselves by secreting

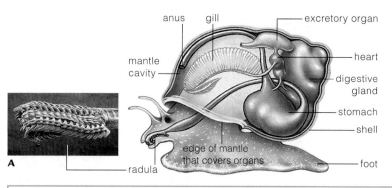

anus — gill — excretory organ
mantle cavity
heart
digestive gland
stomach
shell
edge of mantle that covers organs
foot
radula

FIGURE 15.10 Mollusk diversity. **A** Body plan of an aquatic snail, a gastropod. The photo shows a snail radula, a tonguelike organ that the snail uses to scrape up algae. **B** A cone snail, a predatory gastropod, eats a fish paralyzed by venom injected by the snail's modified radula. Studies of cone snail venom have led to new drugs. One molecule from the venom of the snail shown here may be useful as a treatment for epilepsy. **C** Spanish shawl nudibranchs (sea slugs) have no shell, but nematocysts stored in the feathery outgrowths on their back deter predators.

D A bivalve (scallop) with a two-part hinged shell and many eyes (*blue* dots) around the edge of its mantle. **E** A squid, a speedy cephalopod with no shell.

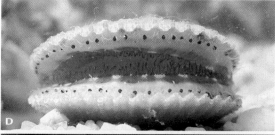

substances that predators find distasteful. The sea slugs shown in Figure 15.10**C** have a different defense. They evolved the ability to eat cnidarians and store undischarged nematocysts from their prey in the feathery outgrowths on their backs. A predator that attacks them gets stung by secondhand nematocysts.

Most mollusks that end up on our dinner plates, such as mussels, oysters, clams, and scallops, are **bivalves** (Figure 15.10**D**). About 15,000 species of bivalves live in fresh water and the seas. Their hinged, two-part shell encloses a body that does not have a distinct head or a radula. Bivalves typically attach to a surface or burrow into sediment. They feed by drawing water into the mantle cavity and filtering out bits of food. Being filter-feeders, bivalves occasionally take in toxins or pathogens that can sicken people who eat them. Risk of illness is greatest when bivalves are eaten raw or only partially cooked.

Cephalopods include about 700 species of squids, octopuses, cuttlefish, and nautiluses (Figure 15.10**E**). All are predators and most have beaklike, biting mouthparts in addition to a radula. Cephalopods move by jet propulsion. They draw water into the mantle cavity, then force it out through a funnel-shaped siphon that is an evolutionary modification of the foot.

Five hundred million years ago, cephalopods with a long conelike shell were the top predators in the seas. Some had shells 5 meters (15 feet) long. Today, nautiluses have a coiled external shell, but other cephalopods have a highly reduced shell or none at all. Most likely, evolution of jawed fishes 400 million years ago was a factor in this shift in body form. Once jawed fishes began to compete with cephalopods for prey, fast-moving, agile cephalopods were at an advantage. Reduction or loss of the shell improved speed and agility.

A speedy life-style required other changes as well. Of all mollusks, only cephalopods have a closed circulatory system. Competition with fishes also favored improved eyesight. Like vertebrates, cephalopods have eyes with lenses that focuses light. Because mollusks and vertebrates are not closely related, this type of eye is assumed to have evolved independently in these groups.

Cephalopods include the fastest (squids), biggest (giant squid), and smartest (octopuses) invertebrates. Of all invertebrates, octopuses have the largest brain relative to body size, and show the most complex behavior.

annelid Segmented worm with a coelom, complete digestive system, and closed circulatory system.

bivalve Mollusk with a hinged two-part shell.

cephalopod Predatory mollusk with a closed circulatory system; moves by jet propulsion.

gastropod Mollusk that moves about on its enlarged foot.

mantle Skirtlike extension of tissue in mollusks; covers the mantle cavity and secretes the shell in species that have a shell.

mollusk Invertebrate with a reduced coelom and a mantle.

radula Tonguelike organ of many mollusks.

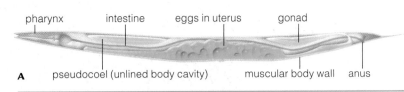

pharynx   intestine   eggs in uterus   gonad

A   pseudocoel (unlined body cavity)   muscular body wall   anus

FIGURE 15.11 Roundworms. **A** Body plan of a typical roundworm. **B** Doctor examining a man who has lymphatic filariasis. Enlargement of the man's left leg is the result of a chronic roundworm infection.

Figure It Out: Does the roundworm shown *above* have a complete digestive system?

Answer: Yes

Roundworms

Roundworms, or nematodes (phylum Nematoda), are more closely related to the arthropods such as insects and crabs than to other worms. At least 12,000 species are known. Most are microscopic or only a few millimeters long. The unsegmented body is cylindrical and bilaterally symmetrical. There is a complete gut and a pseudocoel filled with reproductive organs (Figure 15.11**A**). The roundworm cuticle (secreted body covering) is molted repeatedly as the animal grows. **Molting** refers to a periodic casting off of an outer covering.

Most roundworms are free-living species that are decomposers. They are remarkably abundant in soil. Many play vital roles in the cycling of nutrients through ecosystems. *Caenorhabditis elegans* is a favorite experimental organism among biologists. It has the same tissue types as more complex animals, but it is tiny and transparent. Researchers can easily monitor the developmental fate of each of its 950 or so body cells. In addition, this worm has a very small genome, and hermaphroditic forms can be self-bred to yield worms that are homozygous for specific desired traits.

Some roundworms are pests. Soil-dwelling roundworms infect crop plant roots and harm them. Roundworms also parasitize animals. For example, mosquitoes transmit roundworms called heartworms from dog to dog.

Children in the United States frequently become infected by pinworms (*Enterobius*), a roundworm about the size of a staple. The worms live in the rectum and females venture out at night to lay eggs on the skin around the anus. Their migration causes the area to itch. Scratching transfers the eggs onto fingers, particularly under the fingernails. Eggs can survive outside the body for about two weeks. Swallowing them causes a new infection.

Roundworms in tropical and subtropical regions cause lymphatic filariasis, or elephantiasis (Figure 15.11**B**). Mosquitoes carry larval roundworms to a human host. Once in the body, the worms reproduce inside lymph vessels, damaging them. As a result, fluid that should be returned to the blood instead pools in a body part, most commonly a leg or, in men, the scrotum. The resulting swelling stretches and breaks the skin, encouraging bacterial infections. Chronic infection causes a disfiguring, irreversible enlargement of the affected area. Worldwide, an estimated 40 million people have lymphatic filariasis.

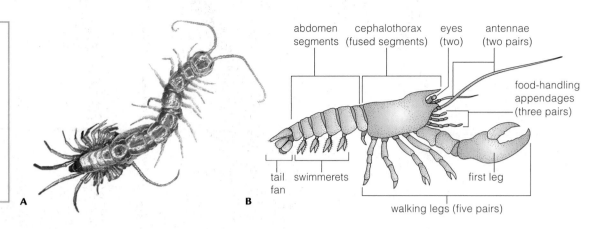

FIGURE 15.12 Arthropod traits. **A** Red centipede molting its old exoskeleton.
B Lobster, with fused segments, specialized appendages, and sensory structures.
C Fruit fly development. A larva becomes a pupa, which is remodeled into an adult during metamorphosis.

abdomen segments   cephalothorax (fused segments)   eyes (two)   antennae (two pairs)

food-handling appendages (three pairs)

tail fan   swimmerets   first leg

walking legs (five pairs)

A   **B**

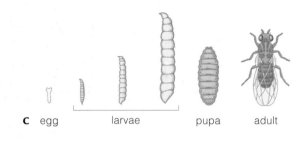

C   egg   larvae   pupa   adult

Arthropod Traits

Arthropods (phylum Arthropoda) are invertebrates with jointed legs. The trilobites mentioned in Section 15.2 are an extinct arthropod lineage. Living groups include centipedes and millipedes, horseshoe crabs, spiders and their relatives, crustaceans such as lobsters, and insects. Researchers have identified more than a million living arthropod species. Six evolutionary adaptations contribute to the great success of this lineage.

1. *A hardened exoskeleton.* Arthropod cuticle is hardened with chitin, the same material in the mollusk radula. The cuticle serves as an external skeleton, or an **exoskeleton**. It helps fend off predators, and muscles that attach to it move body parts. Among land arthropods, the exoskeleton also helps conserve water and support an animal's weight.

A hardened exoskeleton does not restrict growth, because, like roundworms, arthropods molt their cuticle after each growth spurt. A new cuticle forms under the old one, which is shed (Figure 15.12**A**).

2. *Jointed appendages.* If an arthropod's cuticle were uniformly hard like a plaster cast, it would prevent movement. However, arthropod cuticle thins at joints, regions where two hard body parts meet. "Arthropod" means jointed leg. Body parts move when the muscles that attach to the exoskeleton on either side of a joint contract.

3. *Specialized segments and appendages.* In early arthropods, body segments were distinct and all appendages were alike. In many of their descendants, segments fused into structural units such as a head, a thorax (midsection), and an abdomen (hind section). Specialized appendages developed. For example, lobsters have claws (Figure 15.12**B**) and some insects have wings.

4. *Respiratory structures.* Many freshwater and marine arthropods use gills as respiratory organs. Land-dwelling groups have air-conducting tubes that start at pores on the body surface. The tubes deliver air deep inside the body, to fluid near tissues. Diffusion of oxygen into this fluid supports the high rates of aerobic respiration essential for flight and other energy-demanding activities.

5. *Specialized sensory structures.* Most arthropods have one or more pairs of eyes. Insects and crustaceans have **compound eyes** that consist of many individual units, each with its own lens. Such eyes are highly sensitive to movement. Most arthropods also have one or two pairs of **antennae**, sensory structures on the head that detect touch, odor, and vibrations.

6. *Specialized developmental stages.* The body plan of most arthropods changes during the life cycle. Individuals typically undergo **metamorphosis**: The body plan is dramatically remodeled as larvae develop into adults. For instance, wingless fruit fly larvae metamorphose into winged adults that disperse and find mates (Figure 15.12**C**). Each stage has a body plan that adapts it to its tasks. Having different bodies also prevents adults and juveniles from competing for the same resources.

antenna Of some arthropods, sensory structure on the head that detects touch and odors.

arthropod Invertebrate with jointed legs and a hardened exoskeleton that is periodically molted.

compound eye Eye that consists of many individual units, each with its own lens.

exoskeleton External skeleton.

metamorphosis Dramatic remodeling of body form during the transition from larva to adult.

molting Periodic shedding of an outer body layer or part.

roundworm Worm with a pseudocoel and a cuticle that is molted as the animal grows.

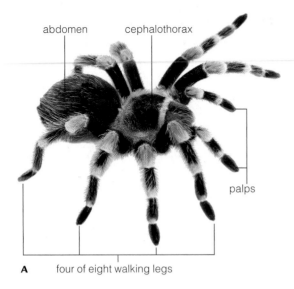

abdomen   cephalothorax

palps

A   four of eight walking legs

stinger

palps

B

rudderlike spine

C

FIGURE 15.13
Body plan variations.
A A tarantula has
two body regions, a
cephalothorax and
an abdomen, with
a "waist" between
them. It grasps prey
with armlike palps.
B Scorpion with no
waist between the
cephalothorax and
abdomen, clawlike
palps, and a venom-
injecting stinger.
C Horseshoe crab
with a shield over its
cephalothorax and a
rudderlike spine on its
abdomen.

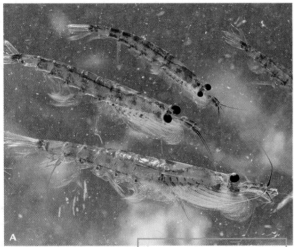

A

B

FIGURE 15.14
Crustaceans. **A** Krill,
tiny shrimplike animals,
use their many legs to
swim in the sea. **B** Goose
barnacles. Adults cement
themselves to one spot
head down, and filter
food from the water with
their feathery legs.

Arthropod Diversity and Ecology Spiders, scorpions, ticks, and mites are **arachnids**. Nearly all live on land. They have four pairs of walking legs, a pair of touch-sensitive palps, and no antennae. Spiders are predators with fanglike mouthparts that deliver venom. Their head and thoracic segments are fused as a cephalothorax, and the abdomen has silk-making glands (Figure 15.13**A**). Scorpions are predators with clawlike palps, and a venom-producing stinger on the final abdominal segment (Figure 15.13**B**).

Horseshoe crabs (Figure 15.13**C**) are the closest living relatives of the extinct trilobites, as well as close relatives of the arachnids. Horseshoe crabs live in the seas, where they eat clams and worms. The last segment of the abdomen has evolved into a spine that helps the animal steer as it swims or right itself after a wave flips it over. Horseshoe crabs do not make venom. There are only four species of horseshoe crab, but they remain ecologically important. The billions of eggs they lay onshore each spring are essential food for some species of migratory birds.

Crustaceans are a group of mostly marine arthropods with two pairs of antennae. Shrimps, crabs, and lobsters are crustaceans that humans harvest as food. Smaller crustaceans are an important food source for marine animals. For example, krill (euphausiids) have a shrimplike body a few centimeters long and swim in upper ocean waters (Figure 15.14**A**). Krill are so plentiful and nutritious that a blue whale weighing more than 100 tons can subsist almost entirely on the krill that it filters from seawater.

Barnacles are marine crustaceans and the only arthropods that secrete a calcium-rich external shell. The shell closes up tight to protect the animal from predators, drying winds, and surf. Barnacle larvae swim, then settle and develop into adults that attach by their heads to rocks, piers, boats, and even whales. They feed by filtering food from seawater with their feathery feet (Figure 15.14**B**). Given that a barnacle is glued in place at its head end, you might think finding a sexual partner could be challenging. But barnacles tend to cluster in groups and most are hermaphrodites. A barnacle extends its penis, which can be several times the body length, out to its neighbors.

abdomen

thorax with six legs

head with two eyes, and two antennae

A

B

D

E

FIGURE 15.15 Insects. **A** Wingless bedbug illustrating the basic insect body plan: a head with a pair of antennae, a thorax with three pairs of legs, and an abdomen. The bug is 7 millimeters (0.25 inches) long and feeds on human blood.

Members of the four most diverse insect lineages. **B** Mediterranean fruit fly. Larvae of this insect destroy citrus fruit and other crops. **C** Ladybird beetles with a distinctive red and black spotted wing cover. **D** Bald-faced hornet, a type of wasp. This fertile female, or queen, lives in a papery nest with her many offspring. **E** A butterfly.

Insects, the most diverse arthropods, are mainly land-dwellers. They have a three-part body: (1) a head with a single pair of antennae, (2) a thorax with three pairs of legs, and (3) an abdomen. Some insect groups such as fleas, lice, and true bugs are wingless (Figure 15.15**A**), but the four most diverse subgroups all have two pairs of wings on the thorax. There are approximately 150,000 species of flies, or dipterans (Figure 15.15**B**), and at least as many beetles, or coleopterans (Figure 15.15**C**). The wasp in Figure 15.15**D** is one of about 130,000 hymenopterans, a group that also includes the bees and ants. Lepidopterans, the moths and butterflies, weigh in with about 120,000 species (Figure 15.15**E**). As a comparison, consider that there are about 4,500 species of mammals.

Most flowering plants are pollinated by a member of one of the four highly diverse insect groups. It is likely that close interactions between these pollinators and flowering plants contributed to an increased rate of speciation in both.

Insects have other important ecological roles. For example, they serve as food for many kinds of wildlife. Larval moths and butterflies, commonly called caterpillars or inchworms, feed songbird nestlings. Aquatic larvae of some insects such as dragonflies serve as food for fish. Most amphibians and reptiles feed mainly on insects. In addition, insects dispose of wastes and remains. Flies and beetles are quick to discover an animal corpse or a pile of feces. They lay their eggs in or on this organic material, and the larvae that hatch devour it.

On the other hand, some insects are pests. Insects are our main competitors for crop plants. It is estimated that about a quarter to a third of all crops grown in the United States are lost to insects. Insects are also vectors for dangerous diseases. For example, mosquitoes can transmit malaria, viruses, or parasitic roundworms to people. Fleas that bite rats and then bite humans can transmit bubonic plague. Body lice can transmit typhus.

There are about 150,000 species of flies, compared to 4,500 species of mammals.

arachnids Land-dwelling arthropods with four pairs of walking legs and no antennae; for example, a spider, scorpion, or tick.

crustaceans Mostly marine arthropods with two pairs of antennae; for example, a shrimp, crab, lobster, or barnacle.

insects Land-dwelling arthropods with a pair of antennae, three pairs of legs, and—in the most diverse groups—wings.

FIGURE 15.16 Echinoderms.

A Sea star, and a close-up of its arm with little tube feet.

B Sea urchins, which move about on spines and a few tube feet.

C Sea cucumber, with rows of tube feet along its elongated body. The hard parts have been reduced to microscopic plates embedded in a soft body.

chordates Animal phylum characterized by a notochord, dorsal nerve cord, pharyngeal gill slits, and a tail that extends beyond the anus. Includes invertebrate and vertebrate groups.

echinoderms Invertebrates with a water–vascular system and an endoskeleton made of hardened plates and spines.

endoskeleton Internal skeleton.

lancelets Invertebrate chordates that have a fishlike shape and retain their defining chordate traits into adulthood.

notochord Stiff rod of connective tissue that runs the length of the body in chordate larvae or embryos.

tunicates Invertebrate chordates that lose their defining chordate traits during the transition to adulthood.

water–vascular system Of echinoderms, a system of fluid-filled tubes and tube feet that function in locomotion.

Echinoderms So far, all of the animals with organs that we have discussed have been members of the protostome lineage. We turn now to another major animal lineage, the deuterostomes. It includes echinoderms and chordates.

Echinoderms (phylum Echinodermata) include about 6,000 marine invertebrates such as sea stars, sea urchins, and sea cucumbers (Figure 15.16). Their phylum name means spiny-skinned and refers to interlocking spines and plates of calcium carbonate embedded in their skin. The plates form an internal skeleton called an **endoskeleton**. Adults have a radially symmetrical body plan.

Sea stars (sometimes called starfish) are the most familiar echinoderms. Sea stars do not have a brain, but they do have a decentralized nervous system. Eye spots at the tips of their arms detect light and movement. A typical sea star is an active predator that moves about on tiny, fluid-filled tube feet (Figure 15.16**A**). Tube feet are part of a **water–vascular system**, a system of fluid-filled tubes that is unique to echinoderms.

Most sea stars eat bivalve mollusks. They slide their stomach out through their mouth and into a bivalve's shell. The stomach secretes acid and enzymes that kill the mollusk and begin to digest it. Partially digested food enters the stomach and digestion is completed with the aid of digestive glands in the arms.

Sexes are separate. Reproductive organs are in the arms, and eggs and sperm are released into the water. Fertilization produces an embryo that develops into a ciliated, bilateral larva. The larva swims about and develops into the adult. The bilateral larva, along with evidence from genetic studies, indicates that the ancestor of echinoderms was a bilateral animal.

Take-Home Message What are the traits of the main invertebrate groups?

- Sponges are filter-feeders that have specialized cells but no tissues. Cnidarians have two tissue layers and a radial body plan. Most other invertebrates have a bilateral body plan with three tissue layers that form organ systems.

- Three groups of "worms" differ in their traits and are not closely related. Flatworms do not have a coelom. Annelids have a coelom and a segmented body. Roundworms have a pseudocoel and are the only worms that molt.

- Mollusks and arthropods are the two most diverse invertebrate phyla. Mollusks are soft-bodied, although many make a shell. Arthropods have a hardened exoskeleton and jointed legs. Insects, an arthropod subgroup, are the most diverse invertebrates and the only winged ones.

- Echinoderms are radial animals with bilateral larvae. They are on the same branch of the animal family tree as the chordates.

Introducing the Chordates

Chordate Traits The majority of deuterostomes are **chordates** (phylum Chordata). Chordate embryos have four defining traits: (1) A **notochord**, a rod of stiff but flexible connective tissue, extends the length of the body and provides support. (2) A dorsal, hollow nerve cord parallels the notochord. (3) Gill slits open across the wall of the pharynx (throat region). (4) A muscular tail extends beyond the anus. Depending on the chordate group, some, all, or none of these traits persist in the adult.

The Invertebrate Chordates There are two groups of invertebrate chordates, tunicates and lancelets. Both are marine and both filter food from currents of water that pass through gill slits in their pharynx.

In **tunicates**, larvae have typical chordate traits, but adults retain only the pharynx with gill slits (Figure 15.17). Larvae swim about briefly, then undergo metamorphosis. The tail breaks down and other parts are rearranged into the adult body form.

A secreted carbohydrate-rich covering or "tunic" encloses the adult body and gives tunicates their common name. Most adult tunicates attach to an undersea surface and filter food from the water. As water flows in an oral opening and past gill slits, bits of food stick to mucus on the gills and are then sent to a gut. Water leaves through another body opening.

Lancelets are fish-shaped chordates 3 to 7 centimeters long (Figure 15.18). They retain all characteristic chordate traits as adults. The dorsal nerve cord extends into the head. A single eyespot at the end of the nerve cord detects light, but the head has no brain or paired sensory organs like those of fishes.

Until recently, lancelets were considered to be the closest invertebrate relatives of vertebrates. An adult lancelet certainly looks more like a fish than an adult tunicate does, but such superficial similarities are sometimes deceiving. New studies of developmental processes and gene sequences suggest that tunicates are the closest invertebrate relatives of vertebrates.

Keep in mind that neither tunicates nor lancelets are ancestors of vertebrates. These groups share a recent common relative, but each has unique traits that put it on a separate branch of the animal family tree.

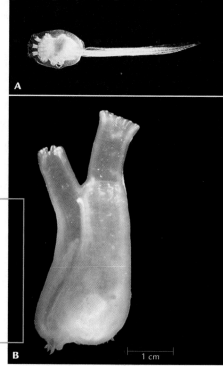

FIGURE 15.17 Tunicates
A Tunicate larva. It swims briefly, then glues its head to a surface and metamorphoses. Tissues of its tail, notochord, and much of the nervous system are remodeled into the adult body form **B**.

1 cm

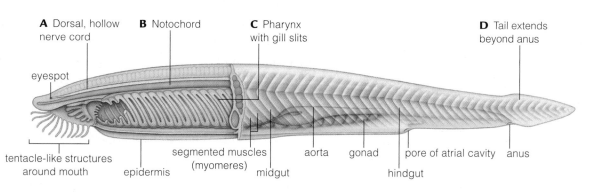

A Dorsal, hollow nerve cord **B** Notochord **C** Pharynx with gill slits **D** Tail extends beyond anus

eyespot

tentacle-like structures around mouth epidermis segmented muscles (myomeres) midgut aorta gonad pore of atrial cavity anus hindgut

FIGURE 15.18 Animated! Photo and body plan of a lancelet, a small filter-feeder. It retains the four characteristic chordate traits as an adult: **A** dorsal, hollow nerve cord; **B** supporting notochord; **C** pharynx with gill slits; **D** tail that extends past the anus.

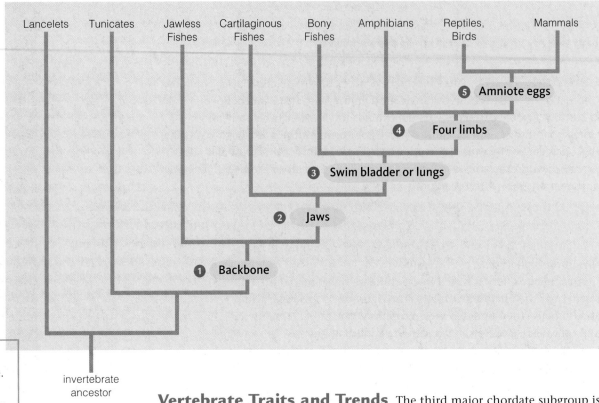

Lancelets   Tunicates   Jawless   Cartilaginous   Bony   Amphibians   Reptiles,   Mammals
                        Fishes   Fishes   Fishes   Birds

⑤ Amniote eggs

④ Four limbs

③ Swim bladder or lungs

② Jaws

① Backbone

invertebrate
ancestor

Vertebrate Traits and Trends The third major chordate subgroup is
the vertebrates. Figure 15.19 shows the chordate family tree and notes the innovations that define the various vertebrate groups.

The first major innovation was the backbone itself ❶. The notochord of a
vertebrate embryo develops into a **vertebral column**, or backbone. This flexible
but sturdy structure encloses and protects the spinal cord that develops from the
embryonic nerve cord.

The next major innovation was **jaws**, which are hinged skeletal elements
used in feeding ❷. Jaws evolved by expansion of bony parts that structurally supported the gill slits of early jawless fishes. Jaws opened up new feeding
opportunities, allowing fish to bite off chunks rather than simply sucking at
food. The vast majority of modern fishes have jaws.

One lineage of jawed fishes evolved a **swim bladder**, a gas-filled flotation
device that allows a fish to adjust its buoyancy ❸. In some descendant lineages,
the swim bladder would become modified into **lungs**, saclike organs inside
which blood exchanges gases with air.

One lineage of fish with lungs had bony fins that evolved into bony limbs.
The result was the first **tetrapods**, or four-legged walkers ❹.

A special type of waterproof egg, called an amniote egg ❺, evolved in one
tetrapod lineage. Modern members of that lineage, the amniotes (reptiles, birds,
and mammals), are the most successful tetrapods on land.

Take-Home Message **What trends shaped chordate evolution?**

- The earliest chordates were invertebrates, and two groups of invertebrate
  chordates (tunicates and lancelets) still exist. They are aquatic filter-feeders.

- Vertebrates evolved from an invertebrate chordate ancestor. All have a vertebral column, or backbone.

- Jaws, lungs, limbs, and waterproof eggs were key innovations that led to the
  adaptive radiation of vertebrates, first in the seas and then on the land.

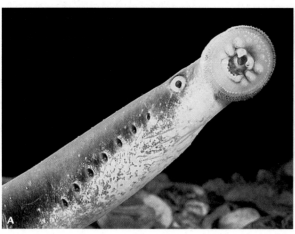

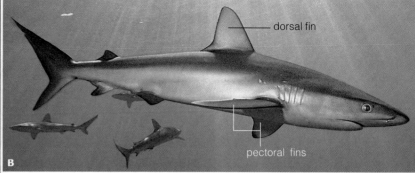

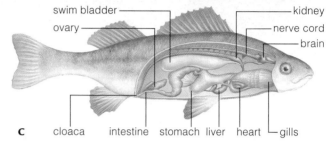

FIGURE 15.20 Fishes. **A** Parasitic lamprey, a jawless fish with no fins. **B** A shark, a cartilaginous fish. Its jaws, fins, and streamlined body make it a swift predator. **C** Body plan of a bony fish.

Fins to Limbs—Fishes and Amphibians

We begin our survey of vertebrate diversity with **fishes**, the first vertebrate lineage to evolve and the most fully aquatic (Figure 15.20). Before we discuss the fish subgroups, a brief overview of vertebrate anatomy is in order. All vertebrates have a brain, a closed circulatory system with one heart, and a urinary system with a pair of **kidneys**. The kidneys filter blood, adjust its volume and solute composition, and eliminate wastes. The digestive system is complete. In fishes and most other vertebrates, digestive and urinary wastes exit the body through a single opening, the **cloaca**, that also functions in reproduction.

The Jawless Lampreys The 50 or so species of lampreys are an evolutionarily ancient lineage of fishes. Lampreys do not have jaws or fins, but they do have a backbone of cartilage and thus are vertebrates.

As adults, some lampreys parasitize other fish. Figure 15.20**A** shows the distinctive mouth of an adult parasitic lamprey. Lacking jaws, the lamprey does not bite its prey. Instead, it uses an oral disk with horny teeth made of the protein keratin to attach to another fish. Once attached, the lamprey secretes enzymes and uses a tooth-covered tongue to scrape up bits of the host's tissues. The host fish often dies from blood loss or a resulting infection.

Jawed Fishes Most jawed fishes have paired fins and **scales**: hard, flattened structures that grow from and often cover the skin. Scales and an internal skeleton make a fish denser than water and prone to sinking. Highly active swimmers have fins with a shape that helps lift them, something like the way that wings help lift up an airplane. Water resists movements through it, so speedy swimmers typically have a streamlined body that reduces friction.

There are two groups of jawed fishes: cartilaginous fishes and bony fishes (Figure 15.20**B,C**). **Cartilaginous fishes** have a skeleton made of cartilage. They include 850 species, with sharks being the most well known. Some sharks are predators that swim in upper ocean waters. Others strain plankton from the water or are bottom feeders. Human surfers and swimmers resemble typical prey of some predatory sharks, and rare attacks by a few species have given all of the

cartilaginous fish Fish with a skeleton of cartilage, such as a shark.

cloaca Of most vertebrates, a body opening that functions in reproduction and elimination of urinary and digestive waste.

fish Aquatic vertebrate of the oldest and most diverse vertebrate group.

jaws Hinged skeletal elements used in feeding.

kidney Organ of the vertebrate urinary system that filters blood and adjusts its composition.

lung Of some vertebrates, saclike organ inside which blood exchanges gases with the air.

scales Hard, flattened elements that cover the skin of some vertebrates.

swim bladder Organ that adjusts buoyancy in bony fishes.

tetrapod Vertebrate with four limbs.

vertebral column Backbone.

gill cover

FIGURE 15.21 Bony fish innovations. **A** Protective gill cover. **B** Lobe-finned coelacanth with bones in its pectoral and pelvic fins. **C** A lungfish, with similar bony fins, also has lungs.

sharks an undeserved bad reputation. Worldwide, shark attacks kill about 25 people a year. For comparison, dogs kill about 30 people each year in the United States alone.

In **bony fishes**, an embryonic skeleton of cartilage becomes transformed to mostly bone in adults. Bony fishes also evolved a protective gill cover (Figure 15.21**A**). In contrast, look back at the lamprey and shark in Figure 15.20. Their gill slits are uncovered. The swim bladder is another bony fish innovation. By adjusting the volume of gas inside this organ, a bony fish can remain suspended in the water at any depth.

Ray-finned fishes and lobe-finned fishes are bony fish subgroups. **Ray-finned fishes** have flexible fins supported by thin rays derived from skin. With about 30,000 species, ray-finned fishes make up nearly half of all the vertebrates. They include most of the fish that end up on dinner plates, such as salmon, sardines, sea bass, swordfish, snapper, trout, tuna, halibut, carp, and cod.

The modern **lobe-finned fishes** include coelacanths and lungfishes (Figure 15.21**B**,**C**). Their pelvic and pectoral fins are fleshy body extensions with bony elements inside them. As the name suggests, lungfishes have gills *and* one or two lunglike sacs. The fish inflates these sacs by gulping air. Like the air in our lungs,

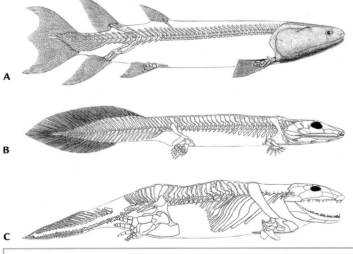

FIGURE 15.22 Transition to tetrapods. Skeleton of a Devonian lobe-finned fish **A**, and two early amphibians, *Acanthostega* **B**, and *Ichthyostega* **C**. The painting **D** shows what *Acanthostega* (foreground), and *Ichthyostega* (background) may have looked like.

air in the sacs exchanges gases with blood. Some lungfishes are totally dependent on this type of gas exchange and will drown if kept underwater.

Early Tetrapods All land vertebrates are descendants of an ancient lobe-finned fish. Fossils show how the skeleton became modified as fishes adapted to swimming evolved into four-legged walkers (Figure 15.22). The bones inside a lobe-finned fish's pelvic and pectoral fins are homologous with amphibian limb bones (Section 11.6). However, the transition to land was not only a matter of skeletal changes. Division of the heart into three chambers allowed blood to flow in two circuits, one to the body and one to the increasingly important lungs. Changes in the inner ear improved the ability to detect airborne sounds, and eyelids prevented the eyes from drying out.

What was the selective advantage to living on land? An ability to survive out of water would have been favored in seasonally dry places. Also, on land, individuals were safe from aquatic predators and had access to a new food resource: insects, which had recently evolved.

Modern Amphibians Amphibians are tetrapods that spend time on land, but require water to breed. Eggs and sperm are typically released into water, and the aquatic larvae have gills. Larvae feed and grow, then metamorphose into adults. Most species lose their gills and develop lungs during this transition. The adults are active predators with a three-chambered heart.

Of all amphibians, the 535 species of salamanders and newts are the most like early tetrapods in body form. Forelimbs and back limbs are of similar size and there is a long tail (Figure 15.23**A**). Frogs and toads belong to the most diverse amphibian lineage, with more than 5,000 species. Long, muscular, hindlimbs allow the tailless adults to swim, hop, and make leaps that can be spectacular, given their body size (Figure 15.23**B**). The forelimbs are much smaller and help absorb the impact of landings.

Larvae of salamanders look like small adults, except for the presence of gills. In contrast, frogs and toad larvae are markedly different from adults. The larvae, commonly called tadpoles, have gills and a tail but no limbs (Figure 15.23**C**).

Biologists are alarmed by ongoing declines in many amphibian populations. Shrinking or deteriorating habitats are part of the problem. People commonly fill low-lying areas where pooling of seasonal rains could provide breeding grounds for amphibians. Climate change causes additional harm by fostering the spread of pathogens and parasites, and increasing the amount of UV radiation that reaches Earth's surface. Amphibians are also highly sensitive to water pollution. Their thin skin, unprotected by scales, allows waste carbon dioxide to diffuse out of their body. Unfortunately, it also allows chemical pollutants to enter.

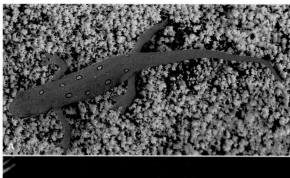

FIGURE 15.23 Amphibians.
A Salamander with four equally sized limbs. **B** A frog with long, muscular hindlimbs that allow leaping. **C** A larval frog, with a tail and gills.

Take-Home Message What are the traits of fishes and amphibians?

- Most modern fishes have jaws. Cartilaginous fishes such as sharks have a skeleton of cartilage. Bony fishes, with a skeleton of cartilage and bone, are the most diverse fishes.

- The lobe-finned fishes, a subgroup of the bony fishes, are the closest living relatives of the tetrapods.

- Amphibians are tetrapods with a three-chambered heart. They begin life in water as gilled larvae, then undergo metamorphosis. Adults typically have lungs and are carnivores on land.

amphibian Tetrapod with a three-chambered heart and scaleless skin that typically develops in water, then lives on land as a carnivore with lungs.
bony fish Jawed fish with a skeleton composed mainly of bone.
lobe-finned fish Bony fish with bony supports in its fins.
ray-finned fish Bony fish with fins supported by thin rays derived from skin.

Escape From Water—The Amniotes

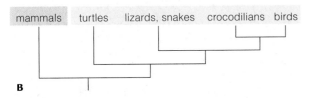

yolk sac   embryo   amnion   chorion   allantois

A   hardened shell   albumin ("egg white")

mammals   turtles   lizards, snakes   crocodilians   birds

B

FIGURE 15.24 Amniotes. **A** The egg of a bird, showing the membranes characteristic of amniotes. **B** Relationships among modern amniotes. All groups in the *blue* box belong to the reptile clade.

Amniote Innovations About 135 million years ago, in the early Carboniferous, the **amniotes** branched off from an amphibian ancestor. This new lineage became better adapted to life in dry places. Amniotes have **amniote eggs** with four membranes that allow embryos to develop away from water (Figure 15.24**A**). In addition, amniote skin is rich in the protein keratin, which makes it waterproof. A pair of well-developed kidneys help conserve water, and fertilization is usually internal.

An early branching of the amniote lineage separated ancestors of mammals from the common ancestor of the reptiles (Figure 15.24**B**). You probably do not think of birds as reptiles, but from a evolutionary perspective they are. To a biologist, **reptiles** include turtles, lizards, snakes, crocodilians, and birds, as well as the extinct dinosaurs.

Some early amniotes, including the ancestors of birds and mammals, evolved an ability to regulate their internal body temperature. Fish, amphibians, turtles, lizards, and snakes are **ectotherms**, which means "heated from outside." People commonly describe such animals as "cold-blooded." Ectotherms adjust their internal temperature by their behavior. They bask on a warm rock to heat up, or burrow into the soil to cool off. In contrast, **endotherms** such as birds and mammals produce their own heat through metabolic processes. Because endotherms use more energy keeping warm, they require more food than ectotherms. A bird or mammal requires far more calories than a lizard or snake of the same weight. However, because endotherms warm themselves, they can remain active at lower temperatures than endotherms.

Reptile Diversity Today, lizards are the most diverse reptile subgroup. The smallest can fit on a dime (Figure 15.25**A**). The typical lizard is a predator, although the iguanas are plant eaters. Most lizards lay eggs that develop outside the body, but some species give birth to live young. In the live-bearers, eggs develop inside the mother, but they are not nourished by her tissues.

Snakes evolved from ancestral lizards and some show evidence of this ancestry. As shown in Figure 11.3, they have bony remnants of ancestral hindlimbs.

A

B   **C**   **D**

FIGURE 15.25 Reptiles. **A** From the Dominican Republic, the smallest known lizard species. **B** Hognose snakes emerging from their leathery amniote eggs. **C** Galápagos tortoise. It can pull its head and limbs into its protective shell. **D** Spectacled caiman, a crocodilian, showing its peglike teeth.

FIGURE 15.26 An owl in flight. Birds fly by flapping their wings. The downstroke provides lift.

All snakes are carnivores. Many have flexible jaws that help them swallow prey whole. Rattlesnakes and other fanged species bite and subdue prey with venom made in modified salivary glands. Like lizards, most snakes lay eggs (Figure 15.24**B**), but some hold eggs inside the body and give birth to live young.

Turtles and tortoises have a bony, scale-covered shell that connects to the backbone. They do not have teeth. Instead a "beak" made of keratin covers their jaws. Tortoises are plant eaters adapted to a life spent on land (Figure 15.25**C**). Turtles spend most of their time in or near water. Sea turtles emerge from the ocean only to lay their eggs on beaches. All sea turtles are now endangered.

Crocodilians live in or near water and include crocodiles, alligators, and caimans. They are predators with powerful jaws, a long snout, and sharp teeth (Figure 15.25**D**). Crocodilians also have a highly efficient four-chambered heart, like mammals and birds. They are the closest living relatives of birds and like most birds they lay eggs, then protect and care for their young.

The Jurassic and Cretaceous periods (245–65 million years ago) are sometimes referred to as the "Age of Reptiles." During this time the dinosaurs underwent a great adaptive radiation and some reached enormous size. The largest stood as tall as a six-story building. Most members of the dinosaur lineage became extinct about 65 million years ago, during a mass extinction most likely caused by an asteroid impact (Section 11.1). However, as Section 15.1 explained, birds are the descendants of feathered dinosaurs.

Birds are the only modern amniotes with feathers. A bird's feathers are modified scales that help adapt it to flight. The bird forelimb is a wing that consists of feathers and bones attached to powerful muscles. Air cavities in the bones make them lightweight. The flight muscles attach to an enlarged breastbone and to the upper limb bones. When these muscles contract, they produce a powerful downstroke (Figure 15.26).

Flight uses up a lot of energy, and birds have the most efficient respiratory system of any vertebrate. This system provides the oxygen necessary for aerobic respiration. A four-chambered heart pumps oxygen-poor blood to the lungs and oxygen-rich blood to the rest of the body in separate circuits. Flight also requires a great deal of coordination, and much of the bird brain controls movement.

Modern birds differ in size, proportions, colors, and capacity for flight. The tiny bee hummingbird weighs 1.6 grams (0.6 ounce). The ostrich, the largest living bird, can weigh as much as 150 kilograms (330 pounds). It cannot fly, but instead sprints on powerful legs.

amniote Vertebrate that produces amniote eggs; a reptile, bird, or mammal.

amniote egg Egg with four membranes that allows an embryo to develop away from water.

bird Modern amniote with feathers.

ectotherm Animal that gains heat from the environment; commonly called "cold-blooded."

endotherm Animal that produces its own heat; commonly called "warm-blooded."

reptile Amniote subgroup that includes lizards, snakes, turtles, crocodilians, and birds.

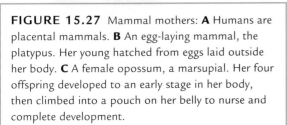

FIGURE 15.27 Mammal mothers: **A** Humans are placental mammals. **B** An egg-laying mammal, the platypus. Her young hatched from eggs laid outside her body. **C** A female opossum, a marsupial. Her four offspring developed to an early stage in her body, then climbed into a pouch on her belly to nurse and complete development.

Mammals **Mammals** are the only amniotes in which females nourish their offspring with milk secreted from mammary glands (Figure 15.27**A**). The group name is derived from the Latin *mamma*, meaning breast. Mammals are also the only animals that have hair or fur. Both are modifications of scales. Like birds, mammals are endotherms. A coat of fur or head of hair helps them maintain their core temperature. Mammals have a single lower jawbone and most have more than one type of tooth. By contrast, reptiles have a hinged two-part jaw, and all teeth are similarly shaped. Having a variety of different kinds of teeth allows mammals to eat more types of foods than most other vertebrates.

Mammals evolved early in the Jurassic, and early mouselike species coexisted with dinosaurs. By 130 million years ago, three lineages had evolved: **monotremes** (egg-laying mammals), **marsupials** (the pouched mammals), and **placental mammals** (mammals in which an organ called the placenta provides nutrients to developing offspring). Figure 15.27 shows examples of each group.

Only three species of monotremes survive, and most pouched mammals live in Australia or New Zealand. In contrast, placental mammals occur worldwide. What gives placental mammals their competitive edge? They have a higher metabolic rate, better body temperature control, and a more efficient way to nourish embryos. Compared to other mammals, placental mammals develop to a far more advanced stage inside their mother's body.

Rats and bats are the most diverse mammals. About half the 4,000 species of placental mammals are rodents and, of those, about half are rats. The next most diverse group is the bats, with about 375 species. Bats are the only flying mammals. Although some may look like flying mice, bats are more closely related to carnivores such as wolves and foxes than to rodents.

Take-Home Message What are amniotes?

- Amniotes are vertebrates that are adapted to life on land by their unique eggs, waterproof skin, and highly efficient kidneys.

- One amniote lineage includes snakes, lizards, turtles, and crocodiles as well as the birds. Birds are the only amniotes with feathers. Like mammals, birds are endotherms; metabolic heat maintains their internal temperature.

- Mammals are amniotes that nourish their young with milk. Most have hair or fur. The three lineages are egg-laying monotremes, pouched marsupials, and placental mammals. Placental mammals are the most widespread and diverse.

15.7

From Early Primates to Humans

Primate Groups and Trends **Primates**, the mammalian subgroup to which humans belong, have five-digit hands and feet capable of grasping (Figure 15.28). Some or all digits are tipped by flattened nails, rather than the sharp claws of most mammals.

Primates first appear in the fossil record about 55 million years ago in Europe, North America, and Asia. Prosimians ("before monkeys") are the most ancient of the surviving lineages. Tarsiers are an example. The anthropoid lineage, to which monkeys, apes, and humans belong (Table 15.1), branched from prosimians. A further branching produced the hominoids—apes and humans. Our closest living relatives are chimpanzees and bonobos (previously called pygmy chimpanzees). Humans and a variety of extinct humanlike species are referred to as **hominids**.

Five trends define the lineage that led to primates, hominids, and then to humans. *First*, the structure of the face changed. Ancestors of primates had a shrewlike snout and eyes at the side of the head. Later, primates evolved a flattened face with both eyes facing forward. Forward-facing eyes provided better depth perception, which was important when leaping from branch to branch. The earliest primates were tree dwellers. As the face flattened, the size of the snout and the importance of the sense of smell declined.

Second, skeletal changes allowed upright standing and walking, known as **bipedalism**. The backbone developed an S-shaped curve. The skull became modified so it sat atop the backbone, rather than being connected to it at the rear.

Third, hands became adapted for novel tasks. Changes to the arrangement of bone and muscle in the hands provided an ability to grasp objects in a fist, or to pick them up between fingertips and the thumb. These abilities allow us to manipulate objects in unique ways, and to use them as tools.

Fourth, teeth became less specialized. Modifications to the jaws and teeth accompanied a shift from eating insects to eating fruits and leaves, then to a mixed diet. Rectangular jaws and lengthy canines evolved in monkeys and apes. A bow-shaped jaw and smaller, more uniform teeth evolved in early hominids.

FIGURE 15.28 Examples of primates. **A** Tarsier, small enough to fit on a human hand, is a climber and leaper. **B** A gibbon has arms and legs adapted to swinging through the trees. **C** Male gorilla, a knuckle-walker. **D** A chimpanzee, showing off its grasping hands and feet.

Table 15.1	Primate Classification
Prosimians	Lemurs, tarsiers
Anthropoids	New World monkeys (e.g., spider monkeys)
	Old World monkeys (e.g., baboons, macaques)
	Hominoids:
	Apes (gibbons, orangutans, gorillas, chimpanzees, bonobos)
	Hominids (humans, extinct humanlike species)

bipedalism Standing and walking on two legs.

hominid Humans and extinct humanlike species.

mammal Vertebrate that nourishes its young with milk from mammary glands.

marsupial Mammals in which young complete development in a pouch on the mother's body.

monotreme Egg-laying mammal.

placental mammal Mammal in which young are nourished within the mother's body by way of a placenta.

primate Mammalian group that includes tarsiers, monkeys, apes, and humans.

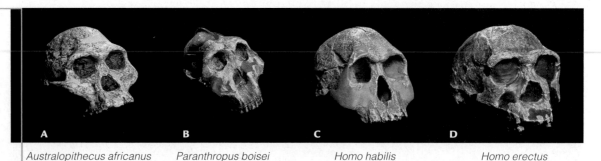

FIGURE 15.29 Hominid fossils. **A–D** A sampling of fossilized skulls from Africa, all to the same scale.

E Fossilized bones of Lucy, a female australopith (*Australopithecus afarensis*) that lived in Africa 3.2 million years ago.

F,G Footprints left by bipedal hominids 3.7 million years ago.

Australopithecus africanus   *Paranthropus boisei*   *Homo habilis*   *Homo erectus*

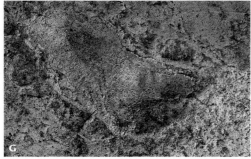

Fifth, the evolution of the brain, behavior, and culture became interlinked. Culture is the sum of all behavioral patterns of a social group, passed on through generations by learning and symbolic behavior, such as spoken and written language.

In brief, our uniquely human traits emerged through modification of traits that evolved earlier, in ancestral species.

Hominoids to Hominids The first hominoids, or apelike species, evolved perhaps as early as 23 million years ago and spread through Africa, Asia, and Europe. During that time, shifts in land masses and ocean circulation were causing a long-term change in climate. Africa became cooler and drier, with more pronounced seasonal changes. Tropical forests, with edible soft fruits, leaves, and abundant insects, gave way to open woodlands and later to grasslands. Food became drier, harder, and more difficult to find. Hominoids that had evolved in the lush forests had two options: Move into new adaptive zones or die out. Most died out, but the common ancestor of apes and humans survived.

Genetic comparisons indicate that hominids diverged from apelike ancestors about 6 to 8 million years ago. The period that followed was a "bushy" time for the hominid lineage in central, eastern, and southern Africa, with many new species branching off. *Ardipithecus ramidus* evolved by 4.4 million years ago. It is the earliest of the fossil hominids now informally known as **australopiths** (for "southern apes"). *Australopithecus* and *Paranthropus* are other australopiths (Figure 15.29**A,B** and Figure 15.30). We still do not know exactly how the australopiths are related to one another or to humans, although one or more species of *Australopithecus* are most likely among our ancestors.

Like apes, australopiths had a large face relative to their braincase, a relatively small brain, and a protruding jaw. But they differed in critical respects from earlier hominoids. For one thing, their thick-enameled molars could grind up harder food. For another, they walked upright. We know this from fossil bones (Figure 15.29**E**) and from footprints. About 3.7 million years ago, a group of australopiths walked across newly fallen volcanic ash. Transformation of the ash to stonelike hardness preserved their tracks (Figure 15.29**F,G**).

Emergence of Early Humans What do fossilized fragments of early hominids tell us about human origins? We do not have enough fossils to determine how all the diverse forms were related, let alone which were our ancestors. Besides, exactly which traits characterize **humans**—members of the genus *Homo*?

Our brain provides the basis of analytical and verbal skills, complex social behavior, and technological innovation—traits that set us apart from apes with a far smaller skull volume and brain size. Yet skull volume alone cannot tell us which hominids made the evolutionary leap to becoming human or when. The brain size of early humans was within the range of apes. They made simple tools, but so do chimps. We know little about early hominid social behavior.

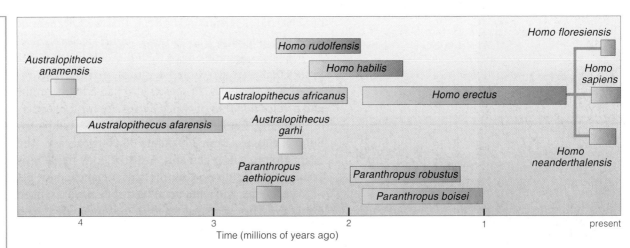

FIGURE 15.30 Estimated dates for the origin and extinction of three hominid genera. *Purple lines* show one view of how the human species relate to one another. Number of species, which fossils belong to each species, and how species relate remains a matter of debate.

Time (millions of years ago)

Without additional information about brain function or behavior, we can only speculate about how hominids used the traits for which we do have evidence: bipedalism, manual dexterity, larger brain volume, small face, and small, more thickly enameled teeth. These traits emerged in *Homo habilis* (Figure 15.29**C**). The name means "handy man."

All early forms of *Homo* are from Africa. Between 2.4 and 1.6 million years ago, australopiths and early forms of *Homo* lived together in woodlands of eastern and southern Africa (Figure 15.31). Fossilized teeth suggest that they ate hard-shelled nuts, dry seeds, soft fruits and leaves, and insects—all seasonal foods. Most likely, *H. habilis* had to think ahead, to plan when to gather and store foods for cool, dry seasons ahead.

H. habilis shared its habitat with carnivores such as saber-toothed cats. A big cat's teeth could impale prey and tear flesh but could not break open bones to get at the marrow. Marrow and bits of meat still clinging to bones may have provided an important protein source for early hominids. *H. habilis* and other forms may have enriched their diet by scavenging carcasses, perhaps breaking bones open with stones.

We have many stone tools dating to the time of *H. habilis*, although we cannot be sure if this species made them. Making even simple stone tools requires a fair amount of brainpower and dexterity. Toolmakers did not simply slam rocks together at random. They chose rocks that were suited for sharpening as their raw material, then shaped them with repeated angled blows from another rock.

Between 1.9 and 1.6 million years ago, strikingly new species evolved on the plains of Africa. Compared to *H. habilis*, they had bigger brains, stood taller, and were better adapted to walking upright for long distances. Some researchers place all of the more modern-looking hominids into one species, *H. erectus* (Figure 15.29**D**). Others distribute them into *H. erectus* and *H. ergaster*. All agree that the new hominids were more creative toolmakers, and that social organization and communication skills must have been well developed.

Some *H. erectus* left Africa and colonized other continents. By 1.7 million years ago, *H. erectus* populations had been established in places as far from Africa as the island of Java and eastern Europe. At the same time, African populations continued to thrive. Over thousands of generations, each of the geographically separated groups adapted to local environmental conditions. Some populations

FIGURE 15.31 Painting of a band of *Homo habilis* in an East African woodland. Two australopiths are shown in the distance at the left.

australopiths Collection of now-extinct hominid lineages, some of which may be ancestral to humans.
humans Members of the genus *Homo*.

Chapter 15 Animal Evolution **315**

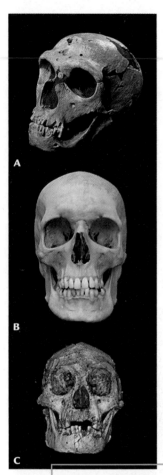

FIGURE 15.32 Recent *Homo* species. **A** *H. neanderthalensis*, **B** *H. sapiens* (modern human), and **C** the proposed species from Indonesia, *H. floresiensis*.

became so different from *H. erectus* that we call them new spieces: *H. neanderthalensis* (Neandertals), *H. floresiensis*, and *H. sapiens* (Figure 15.32).

We know from one fossil found in Ethiopia that *Homo sapiens* had evolved by 195,000 years ago. Compared to *H. erectus*, *H. sapiens* had smaller teeth, facial bones, and jawbones. *H. sapiens* also had a higher, rounder skull, a larger brain, and a voice box that provided a capacity for spoken language.

Similarly, fossils reveal that Neandertals lived in Africa, the Middle East, Europe, and Asia from 200,000 to 30,000 years ago. Their stocky body was an adaptation to colder climates. A stocky body has a lower ratio of surface area to volume than a thin one, so it retains metabolic heat better.

Neandertals had a big brain. Did they have a spoken language? We do not know. They vanished when *H. sapiens* entered the same regions. The new arrivals may have driven Neandertals to extinction through warfare or by outcompeting them for resources. Members of the two species may have occasionally mated, but comparisons between DNA from modern humans and DNA from Neandertal remains indicate that Neandertals did not contribute to the gene pool of modern *Homo sapiens*.

In 2003, human fossils about 18,000 years old were discovered on the Indonesian island of Flores. Like *H. erectus*, they had a heavy brow and a relatively small brain for their body size. Adults would have stood a meter tall. The scientists who found the fossils assigned them to a new species, *H. floresiensis*. Other scientists think the fossils are more likely *H. sapiens* individuals who had a disease or disorder of some sort.

Where Did Modern Humans Originate?

Neandertals evolved from *H. erectus* populations in Europe and western Asia. Where did *H. sapiens* originate? Two major models describe how *H. sapiens* evolved from *H. erectus* but differ over where the changes took place and how fast they occurred. Both attempt to explain the distribution of *H. erectus* and *H. sapiens* fossils, as well as genetic differences among modern humans who live in different regions.

By the **multiregional model**, populations of *H. erectus* in Africa and other regions evolved into populations of *H. sapiens* gradually, over more than a million years. Gene flow among populations maintained the species through the transition to fully modern humans (Figure 15.33**A**).

By this model, some of the genetic variation now seen among modern Africans, Asians, and Europeans began to accumulate soon after their ancestors branched from an ancestral *H. erectus* population. The model is based on interpretation of fossils. For example, faces of *H. erectus* fossils from China are said to look more like modern Asians than those of *H. erectus* that lived in Africa. This model emphasizes the differences between modern human groups and implies that they have a longstanding basis.

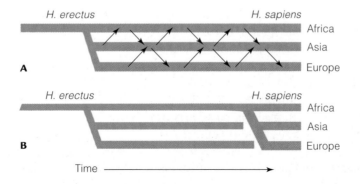

FIGURE 15.33 Two models for the origin of *H. sapiens*. **A** Multiregional model. *H. sapiens* slowly evolves from *H. erectus* in many regions. Arrows represent ongoing gene flow among populations. **B** Replacement model. *H. sapiens* rapidly evolves from one *H. erectus* population in Africa, then disperses and replaces *H. erectus* populations in all regions.

According to the more widely accepted "out-of-Africa model," or **replacement model**, *H. sapiens* arose from a single *H. erectus* population in sub-Saharan Africa within the past 200,000 years. Later, bands of *H. sapiens* entered regions already occupied by *H. erectus* populations, and drove them all to extinction (Figure 15.33**B**). If this model is correct, then the regional variations observed among modern *H. sapiens* populations arose relatively recently. This model emphasizes the enormous degree of genetic similarity among living humans.

Fossils support the replacement model. *H. sapiens* fossils date back to 195,000 years ago in East Africa and 100,000 years ago in the Middle East. In Australia, the oldest such fossils date to 60,000 years ago. In Europe, they date to 40,000 years ago. In addition, global comparisons of markers in mitochondrial DNA, and in the X and Y chromosomes, place the modern Africans closest to the root of the human family tree

Take-Home Message **How did humans evolve?**

- The primate lineage, to which humans belong, is characterized by grasping hands and feet, with five digits tipped by nails. Among modern primates, the apes (especially chimpanzees and bonobos) are our closest relatives.

- Australopiths are a group of upright-walking species that probably include human ancestors. They evolved from apes by 4.4 million years ago in Africa.

- The first humans, or members of the species *Homo*, arose in Africa and were tool makers. *Homo erectus* migrated out of Africa and into Europe and Asia.

- *Homo sapiens* is the only human species that survives. By the currently favored model, modern humans evolved in Africa, then migrated worldwide.

15.8 Impacts/Issues Revisited:
Windows on the Past

Interpreting vertebrate fossils can be a challenge. Scientists often disagree over whether a fossil represents a new species, a variant of a known species, or a collection of bones from more than one species. Such disputes are sorted out over time, as many scientists examine the fossils and publish their findings.

For example, consider what happened after one group of researchers uncovered hominid bones in Indonesia and declared their find a new species, *Homo floresiensis*. Some other scientists suggested that the small stature of the fossils could indicate that they were remains of *H. erectus* or *H. sapiens* with a genetic or nutritional disorder. A scientist who examined the fossil wrist disagreed, concluding that the bones are unlike those of *H. erectus* or *H. sapiens*, and so do represent a distinct species. Still other scientists noted that the fossil skull resembles that of some australopiths. They suggest that the fossils could be a new species that descended from australopiths who migrated out of Africa. Further study of existing fossils, a search for more fossils, and possibly DNA studies will help test the competing hypothesis.

How Would **YOU** ☑ **Vote?**

Collecting fossils is big business, but it encourages poaching on public lands and decreases access to fossils. Should the government regulate fossil sales? **See CengageNow for details, then vote online (cengagenow.com).**

multiregional model Model for origin of *H. sapiens*; human populations in different regions evolved from *H. erectus* in those regions.

replacement model Model for origin of *H. sapiens*; humans evolved in Africa, then migrated to different regions and replaced the other hominids that lived there.

Summary

Section 15.1 Fossils tell us about the relationships between lineages and how the traits that characterize each lineage evolved.

Section 15.2 Animals are multicelled heterotrophs that ingest food. Unwalled body cells are typically organized into tissues, organs, and organ systems. Animals reproduce sexually; many also reproduce asexually. They develop through embryonic stages. Most are motile during at least part of the life cycle. Animals probably evolved from colonial protists. Early ones may have resembled placozoans, the simplest modern animals.

Section 15.3 Sponges are simple filter-feeders with no body symmetry or tissues. The radially symmetrical cnidarians, including jellies, sea anemones, and corals, have tissues. Nematocysts help them capture prey.

Flatworms have the simplest organ systems. They have a saclike gut and no coelom. Planarians are free-living flatworms, and tapeworms and flukes are parasites.

Annelids are coelomate, highly segmented worms with a complete digestive system and a closed circulatory system. Earthworms are the best known.

Only mollusks form a mantle, which is an extension of the body mass that drapes back on itself like a skirt. Bivalves, gastropods, and cephalopods are examples.

Roundworms and arthropods have a cuticle that they molt as they grow. Roundworms are free-living or parasitic, have a complete gut, and have a false coelom. Arthropods have a coelom and a jointed exoskeleton. The body has modified segments and specialized appendages. Living arthropods include the arachnids, crustaceans, centipedes, and insects. Insects include the only winged invertebrates and are the most diverse animals.

Echinoderms such as sea stars have an exoskeleton of spines, spicules, or plates of calcium carbonate. A water–vascular system with tube feet allows adults to move about. Adults are radial, but bilateral ancestry is evident in their larval stages and other features.

CENGAGENOW Learn about the water flow through a sponge, action of a cnidarian nematocyst, flatworm organ systems, and the tapeworm life cycle with animation on CengageNow.

 **Section 15.4** Four features that develop in embryos help define the chordates: a notochord, a dorsal hollow nerve cord (which becomes a brain and spinal cord), a pharynx with gill slits, and a tail extending past the anus. Depending on the group, some or all of the features persist in adults.

Lancelets and tunicates are invertebrate filter-feeding chordates. Most chordates are vertebrates; they have a vertebral column (backbone) of cartilage or bone. Jaws, a swim bladder, and later, lungs, limbs, and waterproof eggs were key innovations that made the adaptive radiation of vertebrates possible.

CENGAGENOW Explore the body plan and chordate features of a lancelet with animation on CengageNow.

Section 15.5 Cartilaginous fishes and bony fishes are aquatic vertebrates. Bony fishes are the most diverse vertebrates, and one group gave rise to the tetrapods.

Amphibians, the first tetrapods on land, share traits with aquatic tetrapods and reptiles. Existing groups include salamanders, frogs, and toads. Many species are now threatened or endangered.

Section 15.6 Amniotes were the first vertebrates that did not require external water for reproduction. Their skin and kidneys conserve water, and they produce amniote eggs. Mammals and their ancestors are one amniote lineage. Reptiles, including birds (the only animals that have feathers), are another.

There are three lineages of mammals: egg-laying mammals, pouched mammals, and placental mammals. Placental mammals are the most diverse group and have displaced others in most habitats.

Section 15.7 The hominoid branch of the primate family tree evolved in Africa. It includes apes and hominids: humans and their most recent ancestors. Australopiths were early hominids, and they walked upright. The first known members of the genus *Homo* lived in Africa between 2.4 million and 1.6 million years ago. *H. erectus* was the first hominid to migrate out of Africa. *H. sapiens* evolved by 190,000 years ago.

Section 15.8 Fossils often inspire competing hypotheses that are tested by a community of researchers.

Self-Quiz   Answers in Appendix I

1. True or false? Animal cells do not have walls.

2. All animals _____ .
   a. are motile for at least some stage in the life cycle
   b. consist of tissues arranged as organs
   c. can reproduce asexually as well as sexually
   d. both a and b

3. A body cavity that is fully lined with tissue derived from mesoderm is a _____ .

4. Most animals have _____ symmetry.

5. Earthworms are most closely related to _____ .
   a. insects   c. leeches
   b. tapeworms   d. roundworms

6. The _____ have a cuticle and molt as they grow.
   a. roundworms   c. arthropods
   b. annelids   d. both a and c

7. List the four distinguishing chordate traits.

8. Which traits from Question 7 are retained by an adult tunicate?

9. True or false? A backbone is always made of bone.

Digging Into Data

Sustaining Ancient Horseshoe Crabs

Atlantic horseshoe crabs, *Limulus polyphemus*, are sometimes called living fossils. For more than a million years, their eggs have fed migratory shorebirds. More recently, people began to harvest horseshoe crabs for use as bait. More recently still, people started using horseshoe crab blood to test injectable drugs for potentially deadly bacterial toxins. To keep horseshoe crab populations stable, blood is extracted from captured animals, which are then returned to the wild. Concerns about the survival of animals after bleeding led researchers to do an experiment. They compared survival of animals captured and maintained in a tank with that of animals captured, bled, and kept in a similar tank. Figure 15.34 shows the results.

1. In which trial did the most control crabs die? In which did the most bled crabs die?

2. Looking at the overall results, how did the mortality of the two groups differ?

3. Based on these results, would you conclude that bleeding harms horseshoe crabs more than capture alone does?

	Control Animals		Bled Animals	
Trial	Number of crabs	Number that died	Number of crabs	Number that died
1	10	0	10	0
2	10	0	10	3
3	30	0	30	0
4	30	0	30	0
5	30	1	30	6
6	30	0	30	0
7	30	0	30	2
8	30	0	30	5
Total	200	1	200	16

FIGURE 15.34 Mortality of young male horseshoe crabs kept in tanks during the two weeks after their capture. Blood was taken from half the animals on the day of their capture. Control animals were handled, but not bled. This procedure was repeated eight times with different sets of horseshoe crabs.

10. All vertebrates are _____ but only some are _____ .
   a. tetrapods; mammals
   c. amniotes; hominids
   b. chordates; amniotes
   d. bipedal, australopiths

11. Amniote adaptations to land include _____ .
   a. waterproof skin
   d. specialized eggs
   b. internal fertilization
   e. a and c
   c. highly efficient kidneys
   f. all of the above

12. True or false. All hominids belong to the species *Homo*.

13. Birds and placental mammals _____ .
   a. are ectotherms
   c. have mammary glands
   b. lay eggs
   d. have a four-chambered heart

14. Match the organisms with the appropriate description.
   _____ sponges   a. most diverse vertebrates
   _____ cnidarians   b. no true tissues, no organs
   _____ flatworms   c. jointed exoskeleton
   _____ roundworms   d. mantle over body mass
   _____ annelids   e. segmented worms
   _____ arthropods   f. tube feet, spiny skin
   _____ mollusks   g. nematocyst producers
   _____ echinoderms   h. lay amniote eggs
   _____ amphibians   i. feed young secreted milk
   _____ fishes   j. complete gut, false coelom
   _____ reptiles   k. first terrestrial tetrapods
   _____ mammals   l. saclike gut, no coelom

15. Arrange the events in order from most ancient.
   _____ a. Cambrian explosion of diversity
   _____ b. origin of animals from colonial protist
   _____ c. tetrapods move onto the land
   _____ d. extinction of dinosaurs
   _____ e. first hominids evolve
   _____ f. first jawed vertebrates evolve

Additional questions are available on CENGAGENOW™

Critical Thinking

1. In the summer of 2000, only 10 percent of the lobster population in Long Island Sound survived after a massive die-off. Many lobstermen in New York and Connecticut lost small businesses that their families had owned for generations. Some believe the die-off followed heavier sprays of pesticides to control mosquitoes that carry the West Nile virus. Explain why a chemical substance that targets mosquitoes might also harm lobsters but not fish.

2. In 1798, a stuffed platypus specimen was delivered to the British Museum. At the time, many biologists were sure it was a fake, assembled by a clever taxidermist. Soft brown fur and beaverlike tail put the animal firmly in the mammalian camp. But a ducklike bill and webbed feet suggested an affinity with birds. Reports that the animal laid eggs only added to the confusion.

   We now know that platypuses burrow in riverbanks and forage for prey under water. Webbing on their feet can be retracted to reveal claws. The highly sensitive bill allows the animal to detect prey even with its eyes and ears tightly shut.

   To modern biologists, a platypus is clearly a mammal. Like other mammals, it has fur and the females produce milk. Young animals have more typical mammalian teeth that are replaced by hardened pads as the animal matures. Why do you think modern biologists can more easily accept the idea that a mammal can have some reptile-like traits, such as laying eggs? What do they know that gives them an advantage over scientists living in 1798?

3. Humans belong to the genus *Homo* and chimpanzees to the genus *Pan*. Yet studies of primate genes show that chimpanzees and humans are more closely related to one another than each is to any other ape. In light of this result, some researchers suggest that chimpanzees should be renamed as members of the genus *Homo*. What practical, scientific, and ethical issues might be raised by such a change in naming?

16

Population Ecology

Ecology is the study of interactions among organisms, and between organisms and their environment.

Impacts/Issues: A Honkin' Mess

Visit a grassy park or a golf course in many parts of the United States and you may need to watch where you step. Wide expanses of grass with a nearby body of water are extremely inviting to Canada geese, or *Branta canadensis* (Figure 16.1). These plant-eating birds produce an abundance of slimy, green feces that can soil shoes, stain clothes, and discourage picnics. Goose feces also wash into ponds and waterways. The added nutrients encourage bacteria and algae to grow, clouding the water and making swimming unappealing and possibly dangerous. Goose feces can contain microbes that sicken humans.

Canada geese are becoming increasingly common. For example, Michigan had about 9,000 in 1970, and has 300,000 today. Controlling their number is challenging, because several different Canada goose populations spend time in the United States. A **population** is a group of organisms of the same species who live in a specific location. Members of a population breed with one another more often than they breed with members of other populations. In the past, nearly all Canada geese seen in the United States were migratory. They nested in northern Canada, flew to the United States to spend the winter, then returned to Canada. The common name of the species reflects this tie to Canada.

Most Canada geese still migrate, but some populations are permanent residents of the United States. Canada geese breed where they were raised, and the nonmigratory birds are generally descendants of geese deliberately introduced to a park or hunting preserve. During the winter, migratory birds often mingle with nonmigratory ones. For example, a bird that breeds in Canada and flies to Virginia for the winter finds itself alongside geese that have never left Virginia.

Life is harder for migratory geese than for nonmigratory ones. Flying hundreds of miles to and from a northern breeding area takes lots of energy and is dangerous. A bird that does not migrate can devote more energy to producing young than a migratory one can. If the nonmigrant lives in a suburban or urban area, it also benefits from an unnatural abundance of food (grass) and an equally unnatural lack of predators. Not surprisingly, the biggest increases in Canada geese have been among nonmigratory birds living where humans are plentiful.

Migratory birds are protected under federal law and by international treaties. However, in 2006, increasing complaints about Canada geese led the U.S. Fish and Wildlife Service to exempt this species from some protections. The agency encouraged wildlife managers to look for ways to reduce nonmigratory Canada goose populations, without unduly harming migratory birds. To do so, these biologists need to know about the traits that characterize different goose populations, as well as how these populations interact with one another, with other species, and with their physical environment.

These sorts of questions are the focus of the science of **ecology**, the study of interactions among organisms, and between organisms and their physical environment. Ecology is not the same as environmentalism, which is advocacy for protection of the environment. However, environmentalists often cite the results of ecological studies when drawing attention to environmental concerns.

FIGURE 16.1 Canada geese in a California park during the summer.

Characteristics of Populations

When studying a population, ecologists collect information about the genes, reproductive traits, and behavior of its component individuals. They also look at **demographics**—vital statistics that describe the population, such as its size, density, and distribution.

Overview of the Demographics **Population size** refers to the number of individuals in a population. **Population density** is the number of individuals in some specified area or volume of a habitat, such as the number of frogs per acre of rain forest or the number of amoebas per liter of pond water. Density tells you how many individuals are in an area but not how they are dispersed through it.

Population distribution describes how individuals are spread through their habitat. It may be clumped, nearly uniform, or random (Figure 16.2**A**). In nature, individuals of most species tend to clump together for one or more reasons. First, each species is adapted to a limited set of ecological conditions that usually occur in patches. For example, geese tend to cluster in an area that has plenty of food and a body of water. Plants survive only where soil, water, and sunlight are adequate. Second, the offspring of many species do not disperse far from their parents. As the saying goes, the nut does not fall far from the tree. Many larvae and immature forms of animals are not very mobile. For example, sponges are usually clumped because larvae settle close to their parents. Asexual reproduction also results in clumping. Third, many animals spend their lives in social groups that offer protection and other advantages. For example, chimpanzees live in cooperative groups of about fifty individuals (Figure 16.2**B**).

When individuals are more evenly spaced than we would expect them to be by chance, they show a nearly uniform dispersal pattern. This pattern often arises as a result of territoriality or stiff competition for some essential resource. Some mature desert shrubs are uniformly distributed (Figure 16.2**C**). Their roots compete for resources, discouraging close growth.

We observe a random pattern of population dispersion only when habitat conditions are nearly uniform, resource availability is fairly constant, and members of a population neither attract nor avoid one another. This combination of factors is fairly unusual. Wolf spiders, which are solitary hunters on forest floors, are one example.

demographics Statistics that describe a population.
ecology The study of interactions among organisms, and between organisms and their environment.
population A group of organisms of the same species who live in a specific location and breed with one another more often than they breed with members of other populations.
population density The number of organisms of a particular population in a given area.
population distribution The way in which members of a population are spread out in their environment.
population size Number of individuals in a population.

clumped

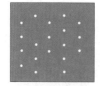

nearly uniform

A   random

FIGURE 16.2 Population distribution. **A** Three types of distribution patterns. **B** Chimpanzees, which live in social groups, have a clumped distribution. Among the advantages of social living is the ability to learn from others. These chimps are using sticks to "fish" for termites in their nest, a behavior learned by imitation. **C** Competition for resources among creosote bushes, a type of desert shrub, results in a nearly uniform distribution among mature plants.

The scale of the study area and timing of a study can influence the observed pattern of distribution. For example, seabirds often are spaced almost uniformly at a nesting site, but nesting sites are clustered along a shoreline. Also, these birds crowd together in the breeding season, but disperse at other times.

Using Demographic Data Wildlife managers use demographic information to decide how best to manage populations. For example, when the U.S. Fish and Wildlife Service wanted to set up a plan for management of nonmigratory Canada geese, its wildlife managers began by evaluating the size, density, and distribution of the nonmigratory populations. Based on this information, the Service decided to allow destruction of some eggs and nests, along with increased hunting opportunities during times when migratory Canada geese are least likely to be present. Scientists will continue to monitor the demographics of the populations to determine the effects of these measures.

Take-Home Message **How do we describe populations?**

- A population has characteristic demographics such as size, density, and distribution pattern.

- Environmental conditions and species interactions shape these characteristics, which may change over time.

16.3 Population Growth

Exponential Growth Birth rates, death rates, and movement of individuals affect population size. Population size increases as a result of births and immigration, and it falls as a result of deaths and emigration. **Immigration** is a permanent addition of individuals that previously belonged to another population. **Emigration** is a permanent loss of individuals to another population. **Migration** is a periodic movement between regions. Many species migrate daily or seasonally, but because individuals return to the starting point, we need not consider this effect in our initial study of population size.

We measure births and deaths in terms of rates per individual, or per capita. *Capita* means head, as in a head count. For example, if the birth rate in a population of 2,000 mice is 1,000 young per month, then the monthly per capita birth rate is 1,000/2,000, or 0.5. Subtract a population's per capita death rate from its per capita birth rate and you have its **per capita growth rate**. If the death rate in our population of 2,000 mice is 200 per month (0.1 per mouse per month), then the per capita growth rate is 0.5 − 0.1 = 0.4 per mouse per month.

A population with a per capita growth rate that is constant and greater than zero grows exponentially. With **exponential growth**, a population grows faster and faster, although the per capita growth rate stays the same. The process is like the compounding of interest in a bank account. The annual interest rate remains fixed, yet every year the amount of interest paid increases. Why? Annual interest paid into the account adds to the size of the balance, and the next interest payment will be calculated based on that larger balance.

In exponentially growing populations, the per capita growth rate is like the interest rate. In any interval the increase in size can be calculated as follows:

$$\begin{pmatrix} \text{population growth} \\ \text{per unit time} \end{pmatrix} = \begin{pmatrix} \text{per capita} \\ \text{growth rate} \end{pmatrix} \times \begin{pmatrix} \text{number of} \\ \text{individuals} \end{pmatrix}$$

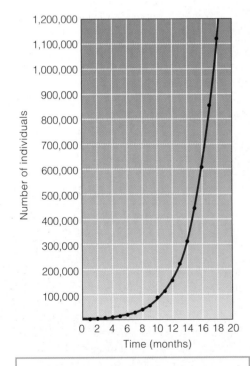

FIGURE 16.3 Animated!

Exponential growth. A plot of the increase in size over time of a population with a per capita growth rate of 0.4 individuals per month. Note that the curve is shaped like a J.

FIGURE 16.4 One example of a limiting factor. **A** Wood ducks nests only inside cavities of specific dimensions. With the clearing of old-growth forests, access to natural cavities of the correct size and position has become a limiting factor on wood duck population size. **B** Artificial nesting boxes are being placed in preserves to help ensure the survival of wood duck populations.

Population growth accelerates as population size increases. A population of 2,000 mice with a per capita growth rate of 0.4 individuals per month means 800 new mice are added in the first month (2,000 × 0.4). This brings the population size to 2,800. In the next month, 2,800 × 0.4, or 1,120 mice, are added, and so on. Plotting the size of this population against time produces the curve shown in Figure 16.3. Exponential growth always plots out as a J-shaped curve.

Biotic Potential and Limiting Factors
Species vary in their **biotic potential**: their maximum possible growth rate if resources were unlimited and there were no predators or pathogens. In nature, organisms seldom achieve their biotic potential or grow exponentially for more than a short time. As a population's size increases, more and more individuals must share a finite amount of resources. As the share of resources available to the average individual diminishes, fewer offspring are born, and deaths resulting from resource scarcity increase. The population's growth rate will slow until births are balanced (or outnumbered) by deaths.

Any essential resource that is in short supply is a **limiting factor** on population growth. Food, mineral ions, refuge from predators, and safe nesting sites are examples of limiting factors (Figure 16.4). Many factors can potentially limit population growth. Which specific factor is the first to be in short supply and thus limit growth varies from one environment to another.

To get a sense of the limits on growth, imagine a bacterial cell in a culture flask with all the glucose and other nutrients bacteria require for growth. Initially, growth will be exponential, but as the population grows, the cells will begin to use up the available nutrients. Lack of nutrients slows bacterial cell divisions, and then eventually stops them. When the nutrient supply becomes completely exhausted, the last cells will die of starvation.

Population growth would still eventually slow and then halt even if additional nutrients were continually added to the flask. Like other organisms,

biotic potential Maximum possible population growth rate under optimal conditions.

emigration Movement of individuals out of a population.

exponential growth A population grows by a fixed percentage in successive time intervals; the size of each increase is determined by the current population size.

immigration Movement of individuals into a population.

limiting factor A necessary resource, the depletion of which halts population growth.

migration Periodic movement of many individuals between one or more regions.

per capita growth rate The number of individuals added during some interval divided by the initial population size.

FIGURE 16.5 Animated! One example of logistic growth: what happens when a few deer are introduced to a new habitat with finite resources.

1 At first the population size increases slowly, because the number of breeding individuals is small.

2 The population grows exponentially, as long as resources are plentiful.

3 Then, as essential resources begin to run out, competition begins to affect the growth rate. Each individual gets a smaller and smaller share of the available food, so population growth slows, then levels off.

Population size plotted against time produces an S-shaped curve.

No population can grow exponentially forever. Remove one limiting factor and another one becomes limiting.

bacteria generate metabolic wastes. Over time, accumulated waste would poison the habitat and prevent further growth. No population can grow exponentially forever. Remove one limiting factor and another one becomes limiting. When nutrients are continually replenished, access to a waste-free environment becomes the limiting factor for the bacteria.

Carrying Capacity and Logistic Growth The quantity of resources in an environment affects how large a population it can sustain. **Carrying capacity** is the maximum number of individuals of a species that a given environment can sustain indefinitely.

A population in an environment with limited resources undergoes **logistic growth**. When there are few individuals relative to the amount of resources, the population grows quickly. As numbers rise and the population approaches carrying capacity, growth slows and then levels off. With logistic growth, a plot of population size over time produces a graph with an S-shaped curve (Figure 16.5).

In any environment, the carrying capacity for a specific species depends on physical and biological factors, both of which can change over time. A prolonged drought can lower the carrying capacity for a plant species and for any animals that depend on it. The carrying capacity for a species can also be affected by the presence of other species with similar resource needs. For example, a grassland can support only so many grazers. The presence of more than one grass-eating species lowers the carrying capacity for all of them.

Factors That Limit Population Growth Factors that control population growth fall into two categories.

Density-dependent factors decrease birth rates or increase death rates, and they come into play or worsen with crowding. Competition among members of a population for limited resources leads to density-dependent effects, as does infectious disease. Pathogens and parasites spread more easily when hosts are crowded than when host population density is low. The logistic growth pattern results from the effects of density-dependent factors on population size.

FIGURE 16.6 Demonstration of limiting factors. A reindeer herd introduced to a small island in 1944 increased in size exponentially, then crashed when a low food supply was coupled with an especially cold and snowy winter in 1963–64.

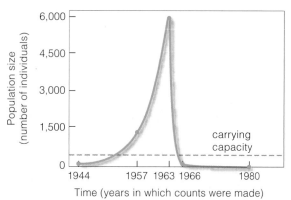

Density-independent factors also decrease births or increase deaths, but crowding does not influence the likelihood that these factors will occur, or the magnitude of their effects. Fires, snowstorms, earthquakes, and other natural disasters affect crowded and uncrowded populations alike. For example, in December 2004, a powerful tsunami (a giant wave caused by an earthquake) hit Indonesia. It killed about 250,000 people. The degree of crowding did not make the tsunami any more or less likely to happen, or to strike any particular island.

Density-dependent and density-independent factors can interact to determine the fate of a population. As an example, consider what happened after a herd of 29 reindeer was introduced to St. Matthew Island off the coast of Alaska in 1944 (Figure 16.6). Biologist David Klein visited the island in 1957 and found 1,350 well-fed reindeer munching on lichens. In 1963, Klein returned to the island and counted 6,000 reindeer. The population had soared far above the island's carrying capacity. Lichens had become sparser and the average body size of the reindeer had decreased. When Klein returned again in 1966, bleached-out reindeer bones littered the island and only 42 reindeer were alive. Only one was a male; it had abnormal antlers, which made it unlikely to reproduce. There were no fawns. Klein figured out that thousands of reindeer had starved to death during the winter of 1963–1964. The winter had been unusually harsh, in both temperature and amount of snow. The reindeer had already been in poor condition because of the increased competition for food, so most starved when deep snow covered their food source. By the 1980s, there were no reindeer on the island at all.

Take-Home Message How do populations increase in size?

- With exponential growth, population size increases by a fixed percentage of the whole in each interval, so the reproductive base gets larger and larger over time. A plot of population size against time produces a J-shaped curve.

- Populations typically do not grow exponentially for very long. A shortage of some essential resource generally slows growth.

- Carrying capacity is the maximum number of individuals of a population that can be sustained indefinitely by the resources in a given environment. The carrying capacity of an environment can change over time.

- With logistic growth, a low-density population increases in size slowly, goes through a rapid growth phase, then levels off once the carrying capacity for that species is reached.

- Density-dependent factors cause logistic growth, but the size of a population can also be altered by density-independent factors, such as harsh weather.

carrying capacity Maximum number of individuals of a species that an environment can sustain.

density-dependent factor Factor that limits population growth and has a greater effect in dense populations than less dense ones.

density-independent factor Factor that limits population growth and arises regardless of population density.

logistic growth A population grows slowly, then increases rapidly until it reaches carrying capacity and levels off.

FIGURE 16.7 Two individuals marked for population studies. **A** Florida Key deer with a neck ring and **B** a Costa Rican owl butterfly with numbers written on its wing.

16.4 Life History Patterns

So far, we have looked at populations as if all of their members are identical during any given interval. However, the individuals of a population are at many different stages in a life cycle. Individuals at each stage may interact in a different way with other organisms and with the environment. Those at different stages may also make use of different resources, as when caterpillars eat leaves and butterflies sip nectar. They may also be more or less vulnerable to predation or other threats. Species, and even populations within a species, vary in their **life history pattern**—a set of heritable traits such as rate of development, age at first reproduction, number of breeding events, and life span. In this section, we look at how such life history traits vary and the evolutionary basis for this variation.

Survivorship Patterns Each species has a characteristic life span, but only a few individuals survive to the maximum age possible. Death is more likely at some ages than others. To gather information about the age-specific risk of death, researchers focus on a **cohort**, a group of individuals that are all born at about the same time. Members of the cohort are tracked from birth until the last one dies. Marking the members of a cohort allows a researcher to keep track of them (Figure 16.7).

A **survivorship curve** is a plot showing how many members of a cohort remain alive over time. Three types of survivorship curves are common.

A type I curve indicates survivorship is high until late in life. Populations of large animals that bear one or, at most, a few offspring at a time and give these young extended parental care show this pattern (Figure 16.8**A**). For example, a female elephant has one calf at a time and cares for it for several years. Type I curves are typical of human populations with access to good health care.

A type II curve indicates that death rates do not vary much with age (Figure 16.8**B**). In lizards, small mammals, and big birds, old individuals are about as likely to die of disease or predation as young ones.

A type III curve indicates that the death rate for a population peaks early in life. It is typical of species that produce many small offspring and provide little or no parental care. Figure 16.8**C** shows how the curve plummets for sea urchins, which release great numbers of eggs. Sea urchin larvae are soft and tiny, so fish, snails, and sea slugs devour most of them before protective hard parts can develop. A type III curve is common for marine invertebrates, insects, fishes, fungi, and for annual plants such as dandelions.

Reproductive Strategies Natural selection influences the timing of reproduction and how much parental investment goes into each offspring. For any species, the evolutionarily favored reproductive strategy is the one that maximizes lifetime reproductive success.

Some species such as bamboo and Pacific salmon reproduce once, then die. Others such as oak trees, mice, and humans reproduce repeatedly. A one-shot strategy is evolutionarily favored when an individual is unlikely to have a second chance to reproduce. For Pacific salmon, reproduction requires a life-threatening journey from the sea to a stream. For bamboo, environmental conditions that favor reproduction occur only sporadically.

Some species produce lots of tiny young and do not care for them. Other species produce fewer, larger offspring. Population density influences whether an individual will do better by maximizing the quantity or the quality of its offspring. At a low population density, there will be little competition for resources, so individuals who turn resources into offspring fast are at a selective

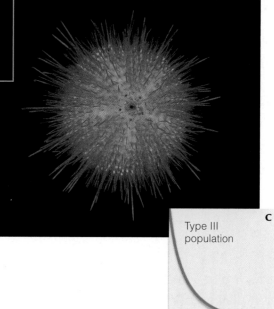

FIGURE 16.8 Three generalized types of survivorship curves and examples of species with each type.

A Elephants have type I survivorship, with low mortality until advanced age.

B Snowy egrets have a type II curve, with a fairly constant death rate throughout the life cycle.

C Sea urchins have a type III curve, with high early mortality. Larval sea urchins are tiny, soft-bodied, and highly vulnerable to predation. Adults, like this one, are well protected by a hard endoskeleton and spines. Some species of sea urchin can live for more than 100 years.

advantage. Such individuals maximize their reproductive success by reproducing while still young, producing many small offspring, and investing very little in each offspring. Selection that favors parents who produce many offspring as quickly as possible, is called **r-selection**. Species that frequently colonize new habitats or live in unpredictable habitats often are subject to r-selection. As a result, they tend to have small body size and a short generation time.

When population density nears the carrying capacity, individuals who can outcompete others for resources have the selective advantage. In species in which populations are often at or near carrying capacity, selection favors parents who maximize the competitive quality of a few offspring, rather than the number of offspring. Such selection is called **K-selection**. Species living in stable environments often are subject to K-selection. As a result, they tend to have a big body and long generation time.

Among plants, life history traits of dandelions and other opportunistic weeds reflect r-selection, whereas redwood trees have K-selected traits. Among mammals, mice have r-selected traits, and elephants K-selected ones.

Evolving Life History Traits
Often, different populations within a species live in slightly different environments and have history traits that reflect these differences. For example, in one long-term study, biologists David Reznick and John Endler documented the evolutionary effects of predation on the life history traits of one species of small fish, the guppy, *Poecilia reticulata*. Their study began with fieldwork in the mountains of Trinidad, an island in the

cohort Group of individuals born during the same interval.

life history pattern Set of traits related to growth, survival, and reproduction such as life span, age-specific mortality, age at first reproduction, and number of breeding events.

K-selection Selection favoring individuals that compete well in crowded conditions; selection for offspring quality.

r-selection Selection favoring individuals that produce many offspring fast; selection for offspring quantity.

survivorship curve Graph showing the decline in numbers of a cohort over time.

A Male guppy

B Killifish, preys mainly on small guppies

C Pike-cichlid, preys mainly on large guppies

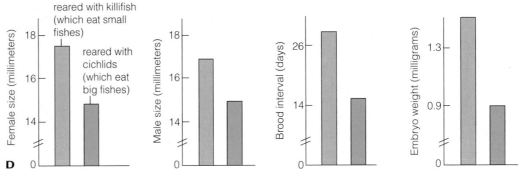

southern Caribbean Sea. Here, guppies live in shallow freshwater streams (Figure 16.9). Waterfalls in the streams act as barriers; they keep guppies from moving from one part of the stream to another, isolating them genetically. Waterfalls also keep predators from moving from one part of a stream to another. As a result, different guppy populations are preyed upon by different predators.

The two kinds of guppy predators in this habitat, killifishes and pike-cichlids, differ in their body size and feeding habits. A killifish is a relatively small fish. It preys on the small, immature guppies but ignores big adults. Pike-cichlids are large fish. They tend to pursue big, mature guppies, and ignore small ones. Many of the streams have one of these predators, but not the other.

Reznick and Endler discovered that the guppies living with pike-cichlids grow faster and have a smaller body size at maturity than guppies in streams containing killifish. Also, guppies in streams with pike-cichlids reproduce earlier, have more offspring at a time, and breed more often (Figure 16.9**D**).

To determine whether these life history differences have a genetic basis, guppies from the two habitats were reared for several generations in aquariums. The results support the hypothesis that the differences are inherited. Descendants of guppies from streams where there were pike-cichlids showed different life history traits than descendants of guppies from streams with killifishes, even when both groups were raised under identical conditions.

Based on their field and laboratory observations, Reznick and Endler hypothesized that predation pressure shapes guppy life history patterns. They set up a field experiment by introducing some guppies to a site upstream from one waterfall. Before the experiment, this waterfall had been an effective barrier to dispersal, preventing guppies and all predators except killifish from moving upstream. The guppies that were introduced to the experimental site were from a population that had evolved with pike-cichlids downstream from the waterfall. That part of the stream was designated the control site.

Eleven years later, researchers revisited the stream. They found that the guppy population at the upstream experimental site had evolved. Exposure to a novel predator had caused shifts in growth rate, age at first reproduction, and other life history traits. Guppies at the control site showed no such changes.

The evolution of life history traits in response to predation pressure is not merely interesting to researchers. It has great economic importance. Just as guppies evolved in response to predators, the North Atlantic codfish (*Gadus morhua*) evolved in response to human fishing. North Atlantic codfish can be quite large (Figure 16.10). From the mid-1980s to early 1990s, the number of fisherman pursuing codfish rose. Fishermen kept the largest fish, and threw smaller ones back. This human behavior put codfish that became sexually mature when they were young and small at an advantage, and such fish became increasingly common. As codfish numbers declined, fishermen kept smaller and smaller fish.

In 1992, Canada banned cod fishing in some areas. That ban, and later restrictions, came too late to stop the Atlantic cod population from plummeting as a result of overfishing. The population still has not recovered from this decline. Looking back, a rapid decline in age at first reproduction was a sign that the cod population was under great pressure. Had biologists recognized this change in a life history trait as a warning sign, they might have been able to save this fishery and protect the livelihood of thousands of workers. Monitoring the life history data for other economically important fishes may help prevent overfishing of other species in the future.

FIGURE 16.10 A fisherman with a large Atlantic codfish. A human preference for big codfish selected for fish that matured while still young and small.

Take-Home Message **What factors shape life history patterns?**

- Life history traits such as the age at which an organism first reproduces and the number of times it reproduces are subject to natural selection.

- When resources are plentiful and population size is low, individuals who produce lots of offspring have an advantage. When resources are scarce, the ability of offspring to compete becomes more important. In this case, producing fewer, higher quality offspring is favored.

- Predators influence the life history traits of their prey.

16.5 Human Populations

Population Size and Growth Rate More than 6.7 billion people now share the planet, and that number continues to grow by more than 200,000 every day. That adds up to more than 80 million additional people every year. Unless birth rates decline to match the still declining death rates, these annual

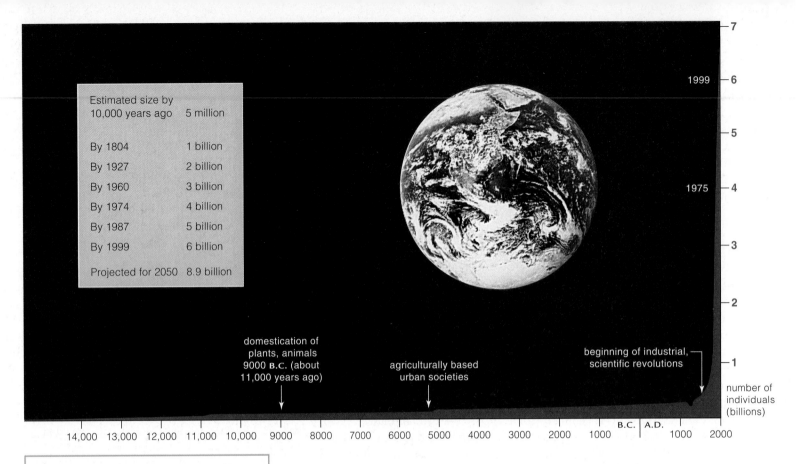

Estimated size by	
10,000 years ago	5 million
By 1804	1 billion
By 1927	2 billion
By 1960	3 billion
By 1974	4 billion
By 1987	5 billion
By 1999	6 billion
Projected for 2050	8.9 billion

domestication of plants, animals 9000 B.C. (about 11,000 years ago)

agriculturally based urban societies

beginning of industrial, scientific revolutions

number of individuals (billions)

FIGURE 16.11 Growth curve (*red*) for the world human population. The *orange* box lists how long it took for the human population to increase from 5 million to 6 billion.

additions will drive a larger absolute increase in the human population far into the foreseeable future

How did we get into this predicament? For most of its history, the human population grew very slowly. The growth rate began to increase about 10,000 years ago, and during the past two centuries, it soared (Figure 16.11). Three trends promoted the large increases. First, humans were able to migrate into new habitats and expand into new climate zones. Second, humans developed new technologies that increased the carrying capacity of existing habitats. Third, humans sidestepped some limiting factors that restrain growth of other species.

Early humans evolved in the dry woodlands of Africa, then moved into the savannas. We assume they subsisted mainly on plant foods, but they probably also scavenged bits of meat. Bands of hunter–gatherers moved out of Africa about 2 million years ago. By 44,000 years ago, their descendants were established in much of the world (Section 15.7).

Few species can expand into such a broad range of habitats, but the early humans had large brains that allowed them to develop the necessary skills. They learned how to start fires, build shelters, make clothing, manufacture tools, and cooperate in hunts. With the advent of language, knowledge of such skills did not die with the individual. Compared to most species, humans have a greater capacity to disperse fast over long distances and to become established in physically challenging new environments.

Beginning about 11,000 years ago, bands of hunter–gatherers were shifting to agriculture. Instead of counting on the migratory game herds, they were settling in fertile valleys and other regions that favored seasonal harvesting of fruits and grains. They developed a more dependable food supply. A pivotal factor was the domestication of wild grasses, including species ancestral to modern wheat and rice. Now people harvested, stored, and planted seeds all in one place. They domesticated animals as sources of food and to pull plows. They dug irrigation ditches and diverted water to croplands.

age structure Of a population, distribution of individuals among various age groups.
total fertility rate Average number of children born to a woman in a particular population.

Agricultural productivity became a basis for increases in population growth rates. Towns and cities formed. Later, food supplies increased yet again. Farmers started to use chemical fertilizers, herbicides, and pesticides to protect their crops. Transportation and food distribution improved. Even at its simplest, the management of food supplies through agricultural practices increased the carrying capacity for the human population.

Until about 300 years ago, malnutrition and infectious diseases kept the death rate high enough to more or less balance the birth rate. Infectious diseases are density-dependent factors. Plagues swept through crowded cities. In the mid-1300s, one-third of Europe's population was lost to a pandemic known as the Black Death. Waterborne diseases such as cholera that are associated with poor sanitation ran rampant. Then plumbing improved and vaccines and medications began to cut the death toll from disease. Births increasingly outpaced deaths, and exponential growth accelerated.

The industrial revolution took off in the middle of the eighteenth century. People had discovered how to harness the energy of fossil fuels, starting with coal. Within decades, cities of western Europe and North America became industrialized. World War I sparked the development of more technologies. After the war, factories that had produced wartime goods turned to mass production of cars, tractors, and other affordable goods. Advances in agricultural practices meant that fewer farmers were required to support a larger population.

In sum, by controlling disease agents and tapping into fossil fuels—a concentrated source of energy—the human population sidestepped many factors that had previously limited its rate of increase.

Where have our colonization of many habitats and ongoing advances in technology and infrastructure gotten us? It took more than 100,000 years for the human population size to reach 1 billion. It took just 123 years to reach 2 billion, 33 more to reach 3 billion, 14 more to reach 4 billion, and then 13 more to get to 5 billion. It took only 12 more years to arrive at 6 billion!

Fertility Rates and Future Growth Most governments now recognize that population growth, resource depletion, pollution, and quality of life are interconnected. Many offer family planning programs. The United Nations Population Division estimates that globally, more than 60 percent of married women use family planning methods.

The **total fertility rate** is the average number of children born to the women of a population during their reproductive years. In 1950, the worldwide total fertility rate averaged 6.5. By 2003, it had declined to 2.8 but was still above the replacement level, which is the number of children a couple must bear to replace themselves. At present, this replacement level is 2.1 for developed countries and as high as 2.5 in some developing countries. (It is higher in developing countries because more female children die before reaching reproductive age.)

World population is expected to reach 8.9 billion by 2050, and possibly to decline as the century ends. At present, China (with 1.3 billion people) and India (with 1.1 billion) dwarf other countries; together, they hold 38 percent of the world population. Next in line is the United States, with 305 million.

Compare the **age structure**, or distribution of individuals among age groups, for these three countries (Figure 16.12). Notice especially the size of the age groups that will be reproducing during the next fifteen years. The broader the base of an age structure diagram, the greater the proportion of young people, and the greater the expected growth. Government polices that favor couples who have only one child have helped China to narrow its pre-reproductive base.

Even if every couple now alive decides to bear no more than two children, world population growth will not slow for many years, because 1.9 billion people

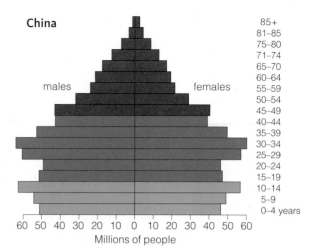

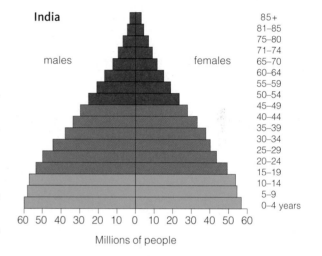

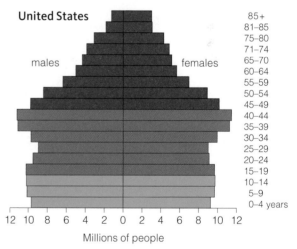

FIGURE 16.12 Animated! Age structure diagrams for the world's three most populous countries. The width of each bar represents the number of individuals in a 5-year age group. *Green* bars represent people in their pre-reproductive years. The *left* side of each chart indicates males; the *right* side, females.

Figure It Out: Which country has the largest number of people in the 45 to 49 age group?   Answer: China

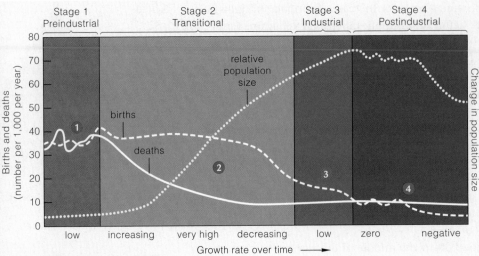

Stage 1 Preindustrial | Stage 2 Transitional | Stage 3 Industrial | Stage 4 Postindustrial

Births and deaths (number per 1,000 per year)

relative population size

births

deaths

Change in population size

low | increasing | very high | decreasing | low | zero | negative

Growth rate over time ⟶

FIGURE 16.13 Animated! Demographic transition model for changes in population growth rates and sizes, correlated with long-term changes in the economy.

Figure It Out: In which stage are death rates highest?

Answer: The preindustrial stage

demographic transition model Model describing the changes in human birth and death rates that occur as a region becomes industrialized.

are about to enter the reproductive age bracket. More than one-third of the world population is in the broad pre-reproductive base.

While some countries face overpopulation, others have a declining birth rate and an increasing average age. In some developed countries, the decreasing total fertility rate and increasing life expectancy have resulted in a high proportion of older adults. In Japan, people over 65 currently make up about 20 percent of the population. In the United States, the proportion of people over 65 is projected to reach this level by 2030. In 2050, there could be as many as 31 million Americans over age 85.

Economics and Population Growth We can correlate changes in population growth with changes that often unfold in four stages of economic development. These four stages are at the heart of the **demographic transition model** (Figure 16.13). By this model, living conditions are harshest during the preindustrial stage, before technological and medical advances become widespread. Birth and death rates are both high, so the growth rate is low ❶.

Next, in the transitional stage, industrialization begins. Food production and health care improve. The death rate drops fast, but the birth rate declines more slowly ❷. As a result, the population growth rate increases rapidly.

During the industrial stage, when industrialization is in full swing, the birth rate declines. People move from the country to cities, where birth control is available and couples tend to want smaller families. The birth rate moves closer to the death rate, and the population grows less rapidly ❸.

In the postindustrial stage, a population's growth rate becomes negative. The birth rate falls below the death rate, and population size slowly decreases ❹.

The United States, Canada, Australia, the bulk of western Europe, Japan, and much of the former Soviet Union have reached the industrial stage. Developing countries such as Mexico are now in the transitional stage, with people continuing to migrate from agricultural regions to cities.

The demographic transition model is based on analysis of what happened when western Europe and North America industrialized. Whether it can accurately predict changes in modern developing countries remains to be seen. Less developed countries now receive aid from highly developed countries, but must also compete against these countries in a global market. There are also regional differences in how the transition to an industrial stage is proceeding. In Asia, rising affluence is increasing life expectancy and reducing the birth rate, as predicted. However, in sub-Saharan Africa, the AIDS epidemic is keeping some countries from moving out of the lowest stage of economic development.

Moreover, a global shift toward industrialized life-styles is straining Earth's finite resources. The average person in an industrialized nation uses far more nonrenewable resources than a person in a less developed country. For example, the United States accounts for about 4.6 percent of the world's population, yet it uses about 25 percent of the world's minerals and energy supply. Billions of people living in India, China, and other less developed nations would like to own the same kinds of consumer goods that people in developed countries enjoy. Earth does not have the resources to make that possible. The World Resources Institute estimates that for everyone now alive to have an average American life-style would require the resources of four Earths. Finding ways to meet the wants and needs of expanding populations with limited resources will be a challenge.

For everyone now alive to live like an average American would require the resources of four Earths.

Take-Home Message What factors affect human population growth?

- Through expansion into new regions, invention of agriculture, and technological innovation, the human population has temporarily skirted environmental resistance to growth. Its size has been skyrocketing since the industrial revolution.

- Birth rates typically fall as nations become more industrialized, but per capita consumption of resources increases.

16.6
Impacts/Issues Revisited:
A Honkin' Mess

Like Canadian geese, white-tailed deer (*right*) are benefiting from human alteration of their habitat. States are struggling to control rising deer numbers. For example, Ohio had about 17,000 deer in 1977. It now has more than 700,000. The deer destroy landscaping and cause accidents when they try to cross roads. An overabundance of deer is also contributing to a rise in Lyme disease. The bacteria that cause this disease are transferred by ticks that feed on both humans and deer. More deer means more disease-carrying ticks are around and can bite humans.

Rising deer numbers adversely affect other native species. In West Virginia, deer are overbrowsing plants that grow on the forest floor. They eat the young seedlings that would otherwise replenish the forest. The deer are also decimating the wild ginseng population. Ginseng gathered from Appalachian forests is an important export crop. Biologist James McGraw argues that controlling deer is essential to saving West Virginia's forests, and will require either reintroducing big predators or increasing deer hunting.

How Would
YOU
☑ **Vote ?**

Encouraging hunting is the least expensive way to reduce the numbers of a wild species that has become a threat or nuisance because of its increased numbers. Do you favor hunting as a way to control the size of these populations? See CengageNow for details, then vote online (cengagenow.com).

Summary

Section 16.1 A population is a group of individuals that live in a given area and tend to mate with one another. By our actions, we humans have encouraged the growth of certain wild populations. The study of populations is one aspect of the field of biology known as ecology.

Section 16.2 Populations vary in their demographics, or characteristics such as density and distribution. Most often populations have a clumped distribution because of limited dispersal, a need for resources that are clumped, and/or the benefits of living in a social group.

CENGAGENOW™ Learn how scientists can measure population size with the animation on CengageNow.

Section 16.3 The growth rate for a population during a specified interval is determined by rates of birth, death, immigration, and emigration. In cases of exponential growth, the population increases by a fixed percentage of the whole with each successive interval. This trend plots out as a J-shaped curve. As long as a population's per capita birth rate is above its per capita death rate, its growth rate will be exponential.

Any essential resource that is in short supply becomes a limiting factor for population growth. The logistic growth model describes how population growth is affected by density-dependent factors, such as disease or competition for resources. The population slowly increases in size, goes through a rapid growth phase, then levels off in size once carrying capacity is reached. Carrying capacity is

the maximum number of individuals that can be sustained indefinitely by the resources available in their environment. Population size can also be reduced as a result of density-independent factors such as adverse weather. These factors can affect any population regardless of its size.

CENGAGENOW™ Learn about exponential and logistic population growth with the animation on CengageNow.

Section 16.4 The time to maturity, number of reproductive events, number of offspring per event, and life span are aspects of a life history pattern. Such patterns are often studied in a cohort, a group of individuals that were born in the same time interval.

Three types of survivorship curves are common: a high death rate late in life, a constant rate at all ages, or a high rate early in life. Life histories have a genetic basis and are subject to natural selection. Depending on the environments, individuals may maximize their reproductive success by reproducing once, or many times. At low population density, individuals who make many offspring fast have an advantage. At a higher population density, individuals who invest more in fewer, higher quality offspring are at an advantage.

Predation can also alter life history patterns. Predators (including humans) act as selective agents on prey populations.

Section 16.5 The human population has surpassed 6.7 billion. Expansion into new habitats and agriculture allowed early increases. Later, medical and technological innovations raised the carrying capacity and removed many limiting factors.

A population's total fertility rate is the average number of children born to women during their reproductive years. The global total fertility rate is declining and most countries have family planning programs of some sort. Even so, the pre-reproductive base of the world population is so large that population size will continue to increase for many years.

The demographic transition model predicts how human population growth rates will change with industrialization. Generally, the death rate and birth rate both fall with rising industrialization, but conditions in countries can vary in ways that affect this trend.

Developed nations have a much higher per capita consumption of resources than developing nations. Earth does not have enough resources to support the current population in the style of the developed nations.

CENGAGENOW™ Compare the age structures of populations in different countries and learn about the demographic transition model with the interaction on CengageNow.

Section 16.6 Increases in wildlife populations can have negative impacts on other species and on human health.

Self-Quiz   Answers in Appendix I

1. Most commonly, individuals of a population show a _____ distribution through their habitat.
   a. clumped   c. nearly uniform
   b. random   d. none of the above

2. The rate at which population size grows or declines depends on the rate of _____ .
   a. births   c. immigration   e. a and b
   b. deaths   d. emigration   f. all of the above

3. For a given species, the maximum rate of population increase under ideal conditions is the _____ .
   a. biotic potential   c. environmental resistance
   b. carrying capacity   d. density control

4. Resource competition, disease, and predation are _____ controls on population growth rates.
   a. density-independent   c. age-specific
   b. population-sustaining   d. density-dependent

5. An increase in the population of a prey species in an environment would most likely _____ the carrying capacity for that species' predators.
   a. increase   c. not affect
   b. decrease   d. stabilize

6. A life history pattern for a population is a set of traits such as the average _____ .
   a. longevity   c. age at first reproduction
   b. fertility   d. all of the above

7. True or false? Life history traits are shaped by natural selection.

8. The human population is now over 6.7 billion. It was about half that in _____ .
   a. 2004   b. 1960   c. 1802   d. 1350

9. Compared to the less developed countries, the highly developed ones have a higher _____ .
   a. death rate   c. total fertility rate
   b. birth rate   d. resource consumption rate

10. Match each term with its most suitable description.
    _____ carrying   a. change in birth and death rates
            capacity   with industrialization
    _____ logistic   b. group of individuals born
            growth   during the same period of time
    _____ exponential   c. population growth plots out
            growth   as an S-shaped curve
    _____ demographic   d. largest number of individuals
            transition   sustainable by the resources
    _____ limiting   in a given environment
            factor   e. population growth plots out
    _____ cohort   as a J-shaped curve
                           f. essential resource that restricts
                              population growth when scarce

Additional questions are available on CENGAGENOW™

Critical Thinking

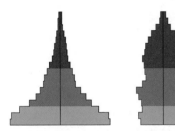

1. The age structure diagrams for two hypothetical human populations are shown at left. Describe the growth rate of each population and discuss the current and future social and economic problems that each is likely to face.

2. Each summer, the giant saguaro cactuses in deserts of the American Southwest produce tens of thousands of tiny black seeds apiece. Most die, but a few land in a sheltered spot and sprout the following spring. The saguaro is a CAM plant (Section 5.5) and it grows slowly. After 15 years, the saguaro may be only knee high, and it will not flower until it is about age 30. The cactus may survive to age 200. Saguaros share their desert habitat with annuals such as poppies, which sprout just after the seasonal rains, produce seeds, and die in just a few weeks. How would you describe these two life history patterns? How could such different life histories both be adaptive in this environment?

3. Most species consist of many populations. When determining whether a species need to be protected, the U.S. Fish and Wildlife Service holds hearings to determine whether or not specific populations are endangered. Imagine that you are a member of the panel at such a hearing. A biologist comes before you to discuss a population of whales that spend their summer in a nearby bay. What would you want to know about this whale population to decide whether it is threatened?

Digging Into Data

Monitoring Iguana Populations

In 1989, Martin Wikelski started a long-term study of marine iguana populations in the Galápagos Islands (Section 11.3). He marked the iguanas on two islands—Genovesa and Santa Fe—and collected data on how their body size, survival, and reproductive rates varied over time. The iguanas eat algae and have no predators, so deaths are usually the result of food shortages, disease, or old age. His studies showed that the iguana populations decline during El Niño events, when water surrounding the islands heats up.

In January 2001, an oil tanker ran aground and leaked a small amount of oil into the waters near Santa Fe. Figure 16.14 shows the number of marked iguanas that Wikelski and his team counted in their census of study populations just before the spill and about a year later.

1. Which island had more marked iguanas at the time of the first census?

2. How much did the population size on each island change between the first and second census?

3. Wikelski concluded that changes on Santa Fe were the result of the oil spill, rather than sea temperature or other climate factors common to both islands. How would the census numbers be different from those he observed if an adverse event had affected both islands?

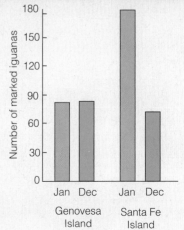

FIGURE 16.14 Shifting numbers of marked marine iguanas on two Galápagos islands. An oil spill occurred near Santa Fe just before the January 2001 census (*blue* bars). A second census was carried out in December 2001 (*green* bars).

17

Communities and Ecosystems

A species takes part in a web of relationships with other species, while also interacting with its nonliving environment.

Impacts/Issues: Fighting the Foreign Fire Ants

Accidentally step on a nest mound of red imported fire ants (RIFAs), and you will be sorry. When disturbed, these ants stream out from the ground and sting their attacker. Venom from the ant's stinger causes burning pain, and multiple stings can cause nausea, dizziness, and, in rare cases, lead to death.

RIFAs arrived in the United States from South America during the 1930s, as stowaways on a ship. The ants spread out from the Southeast and have been found as far west as California and as far north as Kansas and Delaware.

The species name of the most widespread RIFA is *Solenopsis invicta* (Figure 17.1**A**). Invicta means "invincible" in Latin, and *S. invicta* lives up to its name. So far, pesticides have not managed to slow the ants' spread. In fact, these chemicals might even be aiding the foreign ants by wiping out native ant populations.

S. invicta is an example of an **exotic species**, a species that evolved in one region but has been introduced to and become established in another. In its new home, an exotic species is often untroubled by competitors, predators, parasites, and diseases that kept it in check back home. Freed from these natural restrictions, the exotic species often outcompetes species native to its new habitat. For example, RIFAs outcompete and displace North America's native fire ants. As a result, the Texas horned lizard (a species that eats native ants but not RIFAs) is on the decline. A high population density of RIFAs also threatens some songbirds. The ants feed on defenseless nestlings, killing many of them.

Taking a cue from nature, biologists are now enlisting some of the ants' enemies. In South America, attacks of phorid flies (Figure 17.1**B**) help keep *S. invicta* populations in check. The flies kill the ants in a rather gruesome way. A female fly pierces the cuticle of an adult ant, then lays an egg in the ant's soft tissues. The egg hatches into a larva, which grows and eats its way through the body to the ant's head. After the larva gets big enough, it causes the ant's head to fall off (Figure 17.1**C**). The larva then develops into an adult within the detached head.

Phorid fly species have now been introduced in several southern states. The flies attack *S. invicta*, but ignore other ants. The flies are not expected to kill off all the *S. invicta* in affected areas, but they may reduce the density of colonies. When phorid flies are present, the exotic ants spend more time acting defensively, and less time foraging for food. The biologists hope that less food will lead to a reduction in *S. invicta* populations. Such a reduction may allow our native ants and the species that depend on them to recover.

A species takes part in a web of relationships with other species, and interacts with its nonliving environment. We refer to all the species that live and interact in a specific area as a **community**. A community together with its physical environment constitute an **ecosystem**. In this chapter, we consider interactions among species, as well as between species and the nonliving portion of their environment. In Chapter 18 we will take a closer look at the ways in which human activities alter these interactions, and the long-range implications of these changes.

FIGURE 17.1 Off with her head! **A** Group of red imported fire ants. **B** Phorid fly that can inject its egg into an ant's body. The fly larva that hatches feeds on the ant's tissues. When ready to become an adult, the larva moves into the ant's head and **C** causes the head to fall off.

Factors That Shape Communities

Communities vary in size and often are nested one inside another. For example, we find a community of microbial organisms inside the gut of a termite. That termite is part of a larger community of organisms that live on a fallen log. This community is in turn part of a still larger forest community.

Even communities that are similar in scale differ in their species diversity. There are two components to species diversity. The first, species richness, refers to the number of species that are present. The second is species evenness, or the relative abundance of each species. For example, a pond in which five fish species occur in nearly equal numbers has a higher species diversity than a pond in which one fish species is abundant and four others are rare.

Community structure is dynamic. The array of species and their relative abundances change over time. Communities change over a long time span as they form and then age. They also change suddenly as a result of natural or human-induced disturbances.

Nonbiological Factors The structure of a community is affected by geographic and climate variables. These variables include sunlight intensity, rainfall, and temperature along gradients in latitude, elevation, and—for aquatic habitats—depth. Species diversity varies with latitude. Tropical regions receive the most sunlight energy and have the most even temperature. For most plants and animal groups, the number of species is greatest in the tropical regions near the equator, and declines as you move toward the poles. For example, tropical rain forest communities are highly diverse (Figure 17.2), and forest communities in temperate regions are less so. Similarly, tropical reef communities are more diverse than comparable marine communities farther from the equator.

Biological Factors The evolutionary history and adaptations of the various species in a community affect community structure. Each species is adapted to a specific **habitat**, the type of place where it typically occurs. All species of a community share the same habitat, the same "address," but each also has a unique ecological role that sets it apart. This role is the species' **niche**, which we describe in terms of the conditions, resources, and interactions necessary for survival and reproduction. Aspects of an animal's niche include temperatures it can tolerate, the kinds of foods it can eat, and the types of places where it can breed or hide. A description of a plant's niche would include its soil, water, light, and pollinator requirements.

Species in a community interact in ways that affect community structure. Any increase in the numbers of one species often affects the abundance of other species. In some cases, the effect is indirect. For example, when songbirds eat caterpillars, the birds indirectly benefit the trees that the caterpillars feed on, while directly reducing the abundance of caterpillars. The types of direct species interactions among members of a community are the focus of the next section.

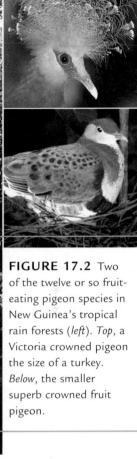

FIGURE 17.2 Two of the twelve or so fruit-eating pigeon species in New Guinea's tropical rain forests (*left*). *Top*, a Victoria crowned pigeon the size of a turkey. *Below*, the smaller superb crowned fruit pigeon.

Take-Home Message

What factors shape a biological community?

- Communities vary in their species diversity as a result of nonbiological factors such as differences in sunlight and nutrient availability.
- Species' resource requirements and interactions among species also influence community structure.

community All populations of all species in some area.
ecosystem A community plus its physical environment.
exotic species A species that evolved in one community and later became established in a different one.
habitat The type of place in which a species lives.
niche The role of a species in its community.

17.3 Species Interactions in Communities

We recognize five types of direct interactions among species in a community: commensalism, mutualism, competition, predation, and parasitism (Table 17.1). Three of these—parasitism, commensalism, and mutualism—can all be types of **symbiosis**, which means "living together." Symbiotic species, also known as symbionts, spend most or all of their life cycle in close association with each other. An **endosymbiont** is a species that lives inside its partner.

Regardless of whether one species helps or hurts another, two species that interact closely may coevolve over generations. In **coevolution**, each species is a selective agent that shifts the range of variation in the other (Section 12.6).

Commensalism and Mutualism

Commensalism benefits one species and does not affect the other. For example, some ferns live attached to the trunk or branches of a tree (Figure 17.3**A**). The fern benefits by getting a perch in the sun, and the tree is unaffected. As another example, many animals have commensal bacteria inside their gut. The bacteria get a warm, nutrient-rich place to live, and their presence neither helps nor harms their host.

Other gut bacteria help their host by aiding in digestion or synthesizing vitamins. Such an interaction, which benefits both species, is a **mutualism**. Flowering plants take part in mutualistic relationships with animals that pollinate them or disperse their seeds. Plants also have a mutualistic relationship with fungi that live on or in their roots. The fungi absorb mineral ions from the soil and share them with the plant. In return the fungi get sugars from the plant.

Mutualism is not a cozy cooperative venture, but rather a case of mutual exploitation. If there is a cost to participating, individuals who lower that cost will be at a selective advantage. For example, a plant that lures a pollinator with a small nectar reward has an advantage over one that provides more nectar.

Mutualisms can be essential to one or both partners. A pink anemone fish will be eaten by a predator unless it has a sea anemone to hide in (Figure 17.3**B**). The anemone's tentacles are covered by stinging cells that do not affect the anemone fish, but help keep predatory fish at bay. The anemone can survive on its own, but benefits by having a partner that chases away the few fish species that are able to feed on anemone tentacles.

Table 17.1	Interspecific Interactions	
Type of Interaction	Direct Effect on Species 1	Direct Effect on Species 2
Commensalism	Benefits	None
Mutualism	Benefits	Benefits
Competition	Harmed	Harmed
Predation	Benefits	Harmed
Parasitism	Benefits	Harmed

FIGURE 17.3 A Commensal fern growing on a tree trunk. **B** An anemone fish nestles among the tentacles of a sea anemone. In this mutualistic partnership, each species protects the other.

FIGURE 17.4 Interspecific competition among scavengers. **A** A golden eagle and a red fox face off over a moose carcass. **B** The eagle attacks the fox with its talons. After this attack, the fox retreated, leaving the eagle to exploit the carcass.

As a final example, mitochondria probably evolved from aerobic bacteria that were endosymbionts inside early eukaryotes (Section 13.3). The bacteria entered host cells or were taken in as food. Over generations, the bacteria lost the ability to live on their own and their hosts came to rely on ATP produced by the symbionts. Similarly, chloroplasts probably evolved from photosynthetic bacteria.

Competitive Interactions Individuals of the same species compete for resources. As Malthus and Darwin understood, these interactions can be fierce and they are important driving forces for natural selection (Section 11.3). In this chapter, we focus on **interspecific competition**, the competition between different species. Usually interspecific competition is not as intense as competition among members of the same species. Members of two species may be similar, but never as similar as members of the same species.

Sometimes, members of one species actively prevent members of another species from using a resource. For example, scavengers such as eagles and foxes fight over carcasses (Figure 17.4). Plants interfere with their competitors as well. For example, a sagebrush plant secretes chemicals that taint the soil around it, thus preventing potential competitors from taking root.

In other cases, competing species do not actively interfere with one another. Instead, all scramble for a share of the resource pie. For example, blue jays, deer, and squirrels eat acorns in oak forests. The animals do not fight over acorns, but they do compete. By eating or storing acorns, each species reduces the number of acorns available to the others.

When species compete for a resource, each gets less of that resource than it would if it lived alone. In this way, competition has a negative impact on all competitors. The more alike two species are, the more intensely they compete. If both species require the same limiting resource, the superior competitor will drive the lesser competitor to extinction in their shared habitat. We refer to this outcome of competition as **competitive exclusion**.

The more alike two species are, the more intensely they will compete.

coevolution The joint evolution of two closely interacting species; each species is a selective agent that shifts the range of variation in the other.

commensalism Species interaction that benefits one species and neither helps nor harms the other.

competitive exclusion When two species compete for a limiting resource, one drives the other to local extinction.

endosymbiont A species that lives in the body of another species in a commensal, mutualistic, or parasitic relationship.

interspecific competition Two species compete for a limited resource and both are harmed by the interaction.

mutualism Species interaction that benefits both species.

symbiosis One species lives in or on another in a commensal, mutualistic, or parasitic relationship.

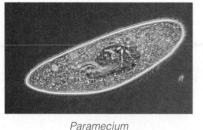

Paramecium

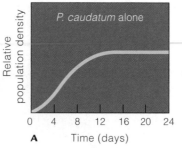

P. caudatum alone

Relative population density

0 4 8 12 16 20 24
A   Time (days)

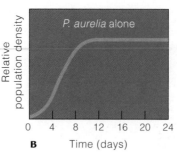

P. aurelia alone

Relative population density

0 4 8 12 16 20 24
B   Time (days)

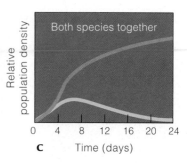

Both species together

Relative population density

0 4 8 12 16 20 24
C   Time (days)

FIGURE 17.5 Animated! Competitive exclusion. Two species of *Paramecium*, *P. caudatum* and *P. aurelia*, both feed on bacteria. When grown in separate test tubes, each does well **A**, **B**. When grown together **C** one species drives the other to extinction.

Figure It Out: Which species of *Paramecium* was the superior competitor?

Answer: P. aurelia

The concept of competitive exclusion grew out of experiments carried out by G. Gause in the 1930s. Gause grew two species of *Paramecium* separately and then in the same culture (Figure 17.5). Both of these protists ate bacteria and they competed intensely for food. When the two species were grown together, one species outgrew the other, driving it to extinction.

Competitors can coexist when their resource needs are not exactly the same. However, competition suppresses the population growth of all competing species. In each species, the individuals who differ most from competing species are favored. Over time, competing species may evolve in a way that lowers the intensity of competition between them. The result of this process is **resource partitioning**, a subdividing of an essential resource, which reduces competition among species that use it. For example, the two pigeons shown in Figure 17.2 are among the twelve or so species that feed on fruit in the forests of New Guinea. Although all of these species eat fruit, the birds differ in the size and type of fruit they eat and so can coexist.

Predator–Prey Interactions In **predation**, one free-living species (the **predator**) feeds on and typically kills another (the **prey**). Carnivores and omnivores are in this category. Herbivores are considered comparable to predators, although they typically do not kill a plant. Even so, herbivores and the plants they feed on coevolve the same way that carnivores and their prey do.

Predator and prey exert selective pressure on one another. Suppose a mutation that provides a more effective means of defense arises in a prey species. Over generations, directional selection will cause this gene to spread through the prey population. If some members of the predator species have a trait that makes them better at thwarting the improved defense, these predators and their

FIGURE 17.6 Prey defenses. Spiny structures protect the edible soft parts of **A** a cactus and **B** a porcupine. **C** The coloration of this yellow jacket wasp warns predators that it can deliver a painful sting. **D** This fly that cannot sting benefits by its resemblance to the wasp.

FIGURE 17.7 Camouflage. **A** Desert plants (*Lithops*) resemble stones and hide among them. **B** A pink praying mantis eats insects attracted to the flowers that it resembles.

descendants will be at an advantage. Thus, predators exert selection pressure that favors better defenses by prey, which in turn exerts selection pressure on predators, and so it goes over many generations.

You have already learned about some defensive adaptations. Many prey species have hard or sharp parts that make them difficult to eat (Figure 17.6**A**,**B**). Others contain chemicals that taste bad or sicken predators. Ricin-producing castor bean plants (Section 7.1) are a good example. Most defensive toxins in animals come from the plants that they eat. For example, monarch butterflies sicken birds with a chemical they took up from plants they ate as a caterpillar.

Another prey adaptation is **warning coloration**, a conspicuous pattern or color that predators learn to avoid. For example, many species of stinging wasps and bees have a pattern of black and yellow stripes (Figure 17.6**C**). Their similar appearance is also a type of **mimicry**, an evolutionary pattern in which one species comes to resemble another. In one type of mimicry, well-defended species benefit by looking alike. In another type of mimicry, prey masquerade as a species that has a defense that they lack. For example, some flies, which cannot sting, resemble stinging bees or wasps (Figure 17.6**D**). The fly benefits when predators avoid it after a run-in with the better-defended look-alike species.

Some prey have evolved a last-chance trick that can startle an attacking predator. Section 1.7 described how eye spots and a hissing sound protect some butterflies. Similarly, a lizard's tail may detach from the body and wiggle a bit as a distraction. Skunks squirt a foul-smelling, irritating repellent.

Many prey have **camouflage**—a form, patterning, color, and often behavior that allows them to blend into their surroundings and avoid detection. *Lithops*, a desert plant, looks like a rock. It flowers only during a brief rainy season, when there are plenty of other plants around to tempt herbivores (Figure 17.7**A**).

Predators have adaptive traits that help them detect, catch, and kill prey. Like prey, predators often benefit from camouflage (Figure 17.7**B**). Sharp teeth and claws help them break through protective hard parts. Herbivores have the ability to break down plant toxins. Sharp eyes help predators detect camouflaged prey. Speedy prey select for faster predators. The cheetah, the fastest land animal, can run as fast as 114 kilometers per hour (70 mph). What selective pressure caused the cheetah to become so fast? Its preferred prey, Thomson's gazelles, can run 80 kilometers per hour (50 mph).

camouflage A plant's or animal's appearance helps it blend into its surroundings.

mimicry Two or more species come to resemble one another.

predation One species feeds on another, usually killing it.

predator A species that eats another, usually killing it.

prey A species that is eaten by a predator.

resource partitioning Use of different portions of a limited resource; allows species with similar needs to coexist.

warning coloration Distinctive color or pattern that makes a well-defended prey species easy to recognize.

Parasites and Parasitoids **Parasites** obtain nutrients from a living host. You already know about some examples from earlier chapters. Some bacteria, protists, and fungi are parasites. Tapeworms and flukes are parasitic annelid worms. Some roundworms, insects, and crustaceans are parasites, as are all ticks (Figure 17.8**A**). Even a few plants are parasitic. Nonphotosynthetic plants, such as dodders, obtain sugars and nutrients from other plants (Figure 17.8**B**). Other parasitic plants, including mistletoes, can carry out photosynthesis but tap into nutrients and water in a host plant's tissues.

Many parasites are pathogens, organisms that cause disease in their hosts. Even when a parasite does not cause obvious symptoms, infection can weaken the host, making it more vulnerable to predation or less attractive to potential mates. Some parasitic infections cause sterility. Others shift the sex ratio of the offspring that their host produces.

Like predators and prey, parasites and their hosts exert selective pressure on one another. Some parasites have evolved the ability to alter their host's behavior to their own benefit. For example, *Plasmodium*, the protist that causes malaria, needs a mosquito to convey it from one human host to the next. When the protist is in its infectious stage, it "calls for a ride" by making its human host smell especially appetizing to mosquitoes.

Hosts' defenses against parasites include immune responses (a topic we consider in detail in Chapter 22) and behavioral responses. For example, many primates take turns removing ticks and other parasites from one another. Birds kill external parasites by preening their feathers, and their bill shape reflects this function, as well as its role in feeding. In pigeons, even a slight experimental modification of bill shape that has no effect on feeding can result in an increase in parasite numbers.

Parasitoids are free-living insects that lay eggs in other insects. Larvae hatch and devour their host from the inside, eventually killing the host. As many as 15 percent of all insects may be parasitoids. The phorid flies described in Section 17.1 are one example.

Parasitoids usually are very particular about their host. This makes them useful as biological controls. Introducing a parasitoid that attacks a pest allows for a more targeted approach than use of pesticides, which often affect nonpest species.

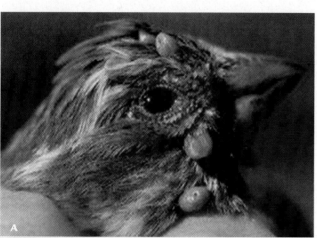

FIGURE 17.8 Two examples of parasites.
A Bloodsucking ticks on the head of a finch.
B Dodder (*Cuscuta*) is also known as strangle-weed or devil's hair. This parasitic flowering plant lacks chlorophyll. It winds around a host plant during growth. Modified roots penetrate the host's vascular tissues and draw water and sugars from them.

Take-Home Message How do species interactions affect a community?

- In commensalism, one species benefits and the other is unaffected.

- In mutualism, two species exploit one another in a way that benefits both.

- When two species compete for a limited resource, both are harmed. If both use the same resource in the same way, the stronger competitor drives the weaker one to local extinction, a process called competitive exclusion.

- Interspecific competition favors individuals of both species whose resource needs are most unlike those of the competing species. Over time, the competition acts as a selective pressure that alters traits related to resource use and leads to resource partitioning among species.

- Predators benefit at the expense of their prey, and parasites benefit at the expense of their hosts. As a result, predators and parasites select for defensive traits in prey and hosts. These defenses in turn select for new traits in the predators and parasites.

parasite A species that withdraws nutrients from another species (its host), usually without killing it.
parasitoid An insect that lays eggs in another insect, and whose young devour their host from the inside.

How Communities Change

Ecological Succession Community structure—the kinds of species and their relative abundance in a habitat—is in constant flux. In a process called **ecological succession**, the array of species gradually shifts over time as organisms alter their own habitat. One array of species is replaced by another, which is in turn replaced by another, and so on.

Primary succession is succession in habitats that lack soil and have few or no existing species (Figure 17.9). For example, a rocky area exposed by retreat of a glacier in Alaska undergoes primary succession. At first, no multicellular organisms are present, although there are prokaryotes and perhaps some single-celled algae ❶. The community begins to change as pioneer species gain a foothold ❷. **Pioneer species** are species that colonize new or vacated habitats. They often include lichens, mosses, and hardy annual plants with wind-dispersed seeds. As generations of pioneers live and die, they help build and improve the soil. In doing so, they set the stage for their own replacement. Seeds of shrubby species find shelter in mats of pioneers and take root ❸. Over time, organic wastes and remains build up and, by adding volume and nutrients to soil, this material allows tall trees to take hold ❹.

Secondary succession occurs after a natural or human disturbance removes the natural array of species, but not the soil. We observe this kind of succession after a fire destroys a forest or a plowed field is abandoned and wild species move in and take over.

The 1980 eruption of Mount Saint Helens, a volcano in Washington State, gave scientists an opportunity to observe primary succession in action (Figure 17.10). The eruption showered the area around the volcano with volcanic rock and ash, wiping out the existing plant life and covering the mature soil. Since then, plant life has colonized the area and succession is under way.

FIGURE 17.9 Artist's depiction of how primary succession in a previously glaciated area can lead to establishment of a forest community.

FIGURE 17.10 Example of ecological succession. **A** Mount Saint Helens erupted in 1980. Volcanic ash completely buried the community that previously existed at the base of this volcano. **B** In less than a decade, numerous pioneer species had become established. **C** Twelve years after the eruption, Douglas fir seedlings were taking hold in soils enriched with volcanic ash.

ecological succession A gradual change in a community in which one array of species replaces another.
pioneer species Species that can colonize a new habitat.
primary succession Ecological succession occurs in an area where there was previously no soil.
secondary succession Ecological succession occurs in an area where a community previously existed and soil remains.

FIGURE 17.11 Effect of fire on community structure. Some woody shrubs, such as this toyon, resprout from their roots after a fire. In the absence of occasional fire, they are outcompeted and displaced by faster-growing but less fire-resistant species.

Random Factors and Disturbances Communities are dynamic entities that are rarely, if ever, in a stable state. Disturbances such as windstorms, freezes, droughts, and fires alter community structure (Figure 17.11). It is often difficult to predict exactly how a community will change during succession or after a disturbance because of random factors. For example, the order in which the pioneer species arrive can have a dramatic effect on community structure.

The frequency and severity of disturbances influences the number of species in a community. Field studies and models indicate that species diversity is highest in communities where moderate disturbances occur with an intermediate frequency. Under these conditions, there are many opportunities for new species to enter the habitat after disturbances, but not enough time between disturbances for many species to become competitively excluded from it.

The Role of Keystone Species A **keystone species** has a disproportionately large effect on a community relative to its actual abundance. Robert Paine was the first to describe the importance of a keystone species on community structure. He studied interactions on the rocky shores of California's coast. Species in this rocky intertidal zone survive by clinging to rocks, and access to a space to hold onto is a limiting factor. Paine set up control plots with the sea star *Pisaster ochraceus* and its prey, which include snails, barnacles, and mussels. He removed all of the sea stars from his experimental plots.

With the sea stars gone, mussels took over in the experimental plots, crowding out seven other invertebrate species. The mussels are the preferred prey of the sea star, and the best competitors in its absence. Normally, predation by sea stars keeps the diversity of prey species high, because it restricts competitive exclusion by mussels. Remove all the sea stars, and the community shrinks from fifteen species down to eight. The mussels overgrow and crowd out many of their competitors.

Not all keystone species are predators. For example, beavers can be a keystone species. These large rodents cut down trees by gnawing through their trunks. The beaver then uses the felled trees to build a dam, thus creating a deep pool where a shallow stream would otherwise exist (Figure 17.12). By altering the physical conditions in a section of the stream, the beaver affects which types of fish and aquatic invertebrates can live there.

FIGURE 17.12 A beaver dam. The beaver is a keystone species that modifies its environment in a way that affects community composition. A beaver gnaws through trunks of small trees and uses the logs and branches to dam a stream. The resulting pond holds aquatic species that would not be present in a fast-moving stream.

Exotic Invaders As Section 17.1 explained, an exotic species is a resident of an established community that has dispersed from its home range and become established elsewhere. Unlike most imports, which never take hold outside their home range, an exotic species becomes a permanent part of its new community. Here, with no coevolved parasites, pathogens, or predators to keep the species in check, its numbers may soar.

More than 4,500 exotic species have established themselves in the United States. The red imported fire ants are one example. A look at three others will

give you an idea of some of their effects. One of the most notorious is the vine called kudzu (*Pueraria lobata*). Native to Asia, it was introduced to the American Southeast as a food for grazers and to control erosion, but it quickly became an invasive weed. Kudzu shoots overgrow trees, telephone poles, houses, and almost everything else in their path (Figure 17.13**A**).

Gypsy moths (*Lymantria dispar*) are an exotic species native to Europe and Asia. They entered the northeastern United States in the mid-1700s and have extended their range into the Southeast, Midwest, and Canada. Gypsy moth caterpillars (Figure 17.13**B**) prefer to feed on oaks. Loss of leaves to gypsy moths weakens the trees, making them more susceptible to parasites and less efficient competitors.

Nutrias (*Myocastor coypus*) are large semi-aquatic rodents that were imported from South America for their fur, and released into the wild in the 1940s. In the states near the Gulf of Mexico, they thrive in freshwater marshes, where they eat so much that they threaten the native vegetation (Figure 17.13**C**). In addition, their burrowing contributes to marsh erosion and damages levees, increasing the risk of flooding.

FIGURE 17.13 Three exotic species that are altering natural communities in the United States.

A Kudzu vines overgrowing trees in South Carolina.

B A gypsy moth caterpillar feeding on oak leaves.

C A nutria in a freshwater marsh in Texas.

To learn more about invasive species in the United States, visit the National Invasive Species Information Center online at www.invasivespeciesinfo.gov.

Take-Home Message What causes changes in community structure?

- In ecological succession, one array of species changes the habitat in a way that allows another array of species to take hold.
- Large and small disturbances shift community structure on an ongoing basis.
- A change in the presence or abundance of a keystone species has a big effect on other species in a habitat.
- An exotic species that leaves behind the predators, parasites, and competitors it evolved with can have dramatic effects on its adopted community.

keystone species A species that has a disproportionately large effect on community structure.

The Nature of Ecosystems

Overview of the Participants

The organisms of a community interact with their environment as an ecosystem. In any ecosystem, there is a one-way flow of energy and a cycling of essential materials (Figure 17.14).

An ecosystem's **producers** capture energy and use it to make their own food from inorganic materials in the environment. Usually the producer's energy source is sunlight and the producers are plants and photosynthetic prokaryotes and protists. An ecosystem's **consumers** get energy and carbon by feeding on tissues, wastes, and remains of producers and one another. Herbivores (plant eaters), predators, and parasites are consumers that feed on living organisms. **Detritivores** such as crabs and earthworms eat tiny bits of organic matter, or detritus. Finally, wastes and remains of organisms are broken down into inorganic building blocks by bacterial, protist, and fungal **decomposers**.

Light energy captured by producers is converted to bond energy in organic molecules, which is then released by metabolic reactions that give off heat. This is a one-way process because heat energy is not recycled—producers cannot convert heat into chemical bond energy. In contrast, nutrients are cycled within an ecosystem. The cycle begins when producers take up hydrogen, oxygen, and carbon from inorganic sources such as the air and water. They also take up dissolved nitrogen, phosphorus, and other necessary minerals. Nutrients move from producers into the consumers who eat them. Decomposition returns nutrients to the environment, where producers take them up again.

Food Chains and Webs

All organisms of an ecosystem take part in a hierarchy of feeding relationships referred to as **trophic levels** ("troph" means nourishment). When one organism eats another, energy and nutrients are transferred from the eaten to the eater. All organisms at the same trophic level are the same number of transfers away from the energy input into that system.

A **food chain** is the sequence of steps by which some energy captured by primary producers is transferred to higher trophic levels. For example, grasses and other plants are the main producers in a tallgrass prairie (Figure 17.15). They are at this ecosystem's first trophic level. In one food chain, energy flows from grasses to grasshoppers, to sparrows, and finally to hawks. Grasshoppers are primary consumers and are at the second trophic level. Sparrows that eat grasshoppers are second-level consumers and at the third trophic level. Hawks that eat sparrows are third-level consumers and at the fourth trophic level.

A number of food chains cross-connect with one another as a **food web**. Figure 17.16 shows just a small sample of the species that take part in an arctic food web. Nearly all food webs include two types of food chains. In a grazing

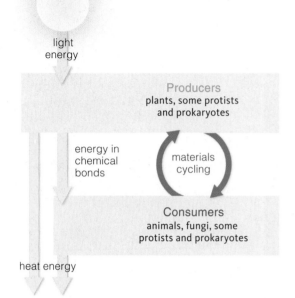

FIGURE 17.14 Generalized model for the one-way flow of energy (*yellow* arrows) and the cycling of materials (*blue* arrows) in an ecosystem. Producers convert sunlight energy to heat energy and energy in chemical bonds. Consumers use the chemical bond energy and also give off energy as heat. Eventually, all light energy that entered the ecosystem is converted to heat energy in the environment.

FIGURE 17.15 One food chain in a tallgrass prairie. Species at the first trophic level capture sunlight energy. Arrows represent the transfer of nutrients and energy from one trophic level to the next.

First Trophic Level	Second Trophic Level	Third Trophic Level	Fourth Trophic Level

   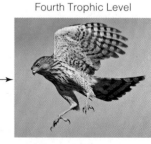

| Producer | Primary Consumer | Second-Level Consumer | Third-Level Consumer |
| Grass | Grasshopper | Sparrow | Hawk |

human (Inuk)

arctic wolf

arctic fox

HIGHER TROPHIC LEVELS

A sampling of carnivores that feed on herbivores and one another

gyrfalcon

snowy owl

ermine

SECOND TROPHIC LEVEL

Major parts of the buffet of primary consumers (herbivores)

vole

arctic hare

lemming

mosquito

flea

Parasitic consumers feed at more than one trophic level.

FIRST TROPHIC LEVEL

This is just part of the buffet of primary producers.

grasses, sedges

purple saxifrage

arctic willow

Detritivores and decomposers (nematodes, annelids, saprobic insects, protists, fungi, bacteria)

FIGURE 17.16 Animated! Arctic food web. Arrows point from eaten to eater.

food chain, the energy stored in producer tissues flows to herbivores, which tend to be relatively large animals. In a detrital food chain, the energy in producers flows to detritivores, which tend to be smaller animals, and to decomposers.

In most land ecosystems, the bulk of the energy that becomes stored in producer tissues moves through detrital food chains. For example, in an arctic ecosystem, grazers such as voles, lemmings, and hares eat some plant parts. However, far more plant matter becomes detritus that sustains soil-dwelling worms and insects, and decomposers such as soil bacteria and fungi.

Detrital food chains and grazing food chains interconnect to form the overall food web. For example, animals at higher trophic levels such as wolves eat both grazers and detritivores. Also, after grazers and their predators die, the energy in their tissues flows to detritivores and decomposers.

consumers Organisms that eat organisms or their remains.
decomposers Consumers that feed on remains and break them into their inorganic building blocks.
detritivores Consumers that feed on small bits of organic material (detritus).
food chain Linear description of who eats whom.
food web System of cross-connecting food chains.
producers Organisms that capture energy and make their own food from inorganic materials in the environment.
trophic level Position of an organism in a food chain.

When one animal eats another, it obtains nutrients and energy. However, it can also receive a dose of pesticides or other poisons in the prey's body. In a process known as **bioaccumulation**, or biological magnification, a chemical that can be stored in fat becomes increasingly concentrated in the tissues of animals as it moves up a food chain. For example, toxic mercury released into the air by coal-burning power plants and some industries falls to Earth in rain and gets into marine ecosystems. It travels up the food chain, accumulating in the tissues of each higher trophic level. The top predators, such as sharks and swordfish, can have very high mercury levels. Mercury interferes with development of the nervous system, so pregnant women, nursing women, and children should avoid eating these fish.

Primary Production and Inefficient Energy Transfers

The flow of energy through an ecosystem begins with **primary production**: the capture and storage of energy by producers. The rate of primary production varies among habitats and also varies seasonally in a given habitat. Scientists measure primary production in terms of the amount of carbon taken up per unit area. On average, primary production per unit area is higher on land than it is in the oceans. However, because the oceans cover about 70 percent of Earth's surface, they contribute about half of Earth's total primary production.

An **energy pyramid** is a graphic representation of how much of the energy captured by producers reaches higher trophic levels. Energy pyramids always have a large energy base, representing the producers, and taper up. Figure 17.17 shows an energy pyramid for a freshwater ecosystem in Silver Springs, Florida.

On average, only about 10 percent of the energy in tissues of organisms at one trophic level ends up in tissues of those at the next trophic level. Several factors limit the efficiency of transfers. All organisms lose energy as metabolic heat. This energy is not available to organisms at the next trophic level. Also, some energy becomes stored in molecules that most consumers cannot break down. For example, few herbivores can digest the cellulose in plant parts. Similarly, most carnivores cannot access the energy tied up in skeletal parts, scales, hair, feathers, or fur. The inefficiency of energy transfers explains why food chains typically do not extend for more than a few links.

When people promote a vegetarian diet by touting the ecological benefits of "eating lower on the food chain," they are referring to the inefficiency of energy transfers between plants, livestock, and humans. When a person eats a plant food, he or she gets all the available calories in that food. When the plant food is used to grow livestock, only a small percentage of the food's calories ends up in the meat that a person eats. Thus, feeding a population of meat-eaters requires far greater crop production than sustaining a population of vegetarians.

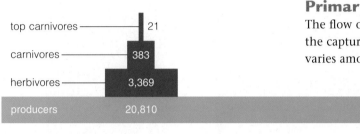

FIGURE 17.17 Energy pyramid for a freshwater ecosystem. Numbers are the energy in kilocalories per square meter per year.

Figure It Out: What percentage of the sunlight energy captured by the producers was transferred to the herbivores that ate them?

Answer: 3,369/20,810 X 100 = 16 percent

Take-Home Message

How do organisms and their environment interact in ecosystems?

- Materials cycle between organisms and their environment, but energy flow is one-way because high-quality energy is converted to unusable heat energy.

- Energy and raw materials are taken up by producers, then flow to consumers.

- Food webs and food chains describe routes by which energy and materials, including toxins, move from one trophic level to another.

- Energy transfers between trophic levels are inefficient because energy is lost as metabolic heat, and some energy becomes tied up in materials that are not easily digested by most consumers.

Nutrient Cycling in Ecosystems

In a **biogeochemical cycle**, ions or molecules of an essential substance flow among environmental reservoirs, and into and out of the world of life. Physical, chemical, and geological processes cause the slow movement of nutrients among their environmental reservoirs. An ecosystem's producers take up various nutrients from the air, soil, and water. Consumers take in nutrients when they drink water and when they eat producers or one another.

We focus on four biogeochemical cycles that move important elements: the water cycle, phosphorus cycle, nitrogen cycle, and carbon cycle.

The Water Cycle

The **water cycle** moves water from oceans to the atmosphere, onto land and into freshwater ecosystems, and back to the oceans (Figure 17.18). Solar energy drives evaporation of water from the oceans and from freshwater reservoirs. Water that enters the lower atmosphere spends some time aloft as vapor, clouds, and ice crystals. By the process of precipitation, water falls from the atmosphere mainly as rain and snow.

Precipitation that falls on land mostly seeps into soil or joins surface runoff toward streams. Some soil water is taken up by plants, which then release most of it by transpiration—the evaporation from their leaves.

Our planet has a lot of water, but 97 percent of it is saltwater. Of the 3 percent that is fresh water, most is tied up in glaciers. **Groundwater**, another freshwater reservoir, includes water in the soil, and water stored in porous rock layers called **aquifers**. About half of the population of the United States relies on aquifers for drinking water. Surface water—the water in streams, rivers, lakes, and freshwater marshes—is less than 1 percent of Earth's fresh water.

The Phosphorus Cycle

The water cycle helps move minerals through ecosystems. Water that falls on land and makes its way back to the sea carries with it silt and dissolved minerals, including phosphorus.

Most of the phosphorus on Earth is bonded to oxygen as phosphate (PO_4^{3-}), an ion that occurs in rocks and sediments. There is no commonly occurring gaseous form of phosphorus, so the atmosphere is not one of its main reservoirs.

aquifer Porous rock layer that holds some groundwater.

bioaccumulation A substance becomes more concentrated in organisms at increasingly higher levels of a food chain.

biogeochemical cycle A nutrient moves among environmental reservoirs and into and out of food webs.

energy pyramid Diagram that illustrates the energy flow in an ecosystem.

groundwater Water between soil particles and in aquifers.

primary production The energy captured by an ecosystem's producers.

water cycle Water moves from its main reservoir—the ocean—into the air, falls as rain and snow, and flows back to the ocean.

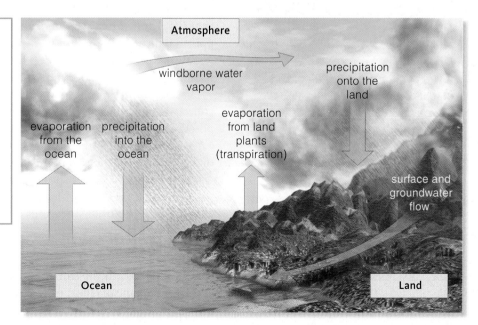

FIGURE 17.18 Animated! The water cycle. Water moves from the ocean into the atmosphere, onto land, and then back. The numbers below show the distribution of Earth's freshwater component among its environmental reservoirs.

Freshwater reservoir	Percent of fresh water
Polar ice, glaciers	68.7
Groundwater	30.1
Surface water	0.3
Other	0.9

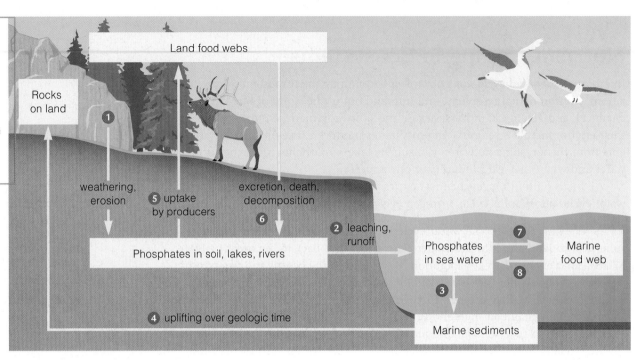

Land food webs

Rocks on land

weathering, erosion

❺ uptake by producers

excretion, death, decomposition

❷ leaching, runoff

Phosphates in soil, lakes, rivers

Phosphates in sea water

❼

❽

Marine food web

❸

❹ uplifting over geologic time

Marine sediments

nitrogen, carbon added

nitrogen, carbon, phosphorus added

FIGURE 17.20 Results of a nutrient enrichment experiment. Researchers put a plastic curtain across a channel between two basins of a lake. They added nitrogen, carbon, and phosphorus on one side of the curtain (here, the *lower* part of the lake) and added nitrogen and carbon on the other side. Within months, the phosphorus-rich basin had a dense layer of single-celled algae covering its surface.

All organisms need phosphorus. It is a component of the sugar–phosphate backbone of DNA, and the "P" in ATP stands for "phosphate." Your body's main store of phosphorus is your bones.

In the **phosphorus cycle**, phosphate moves among Earth's rocks, soil, and water, and into and out of food webs (Figure 17.19). Weathering and erosion move phosphate ions from rocks into soil, lakes, and rivers ❶. Leaching and runoff deliver phosphate ions to the ocean ❷, where most of the phosphorus comes out of solution and settles as deposits along the edges of continents ❸.

Over millions of years, movements of Earth's crust can uplift parts of the sea floor onto land ❹. Weathering releases phosphates from the rocks. Dissolved phosphates enter streams and rivers, and the environmental portion of the phosphorus cycle starts over again.

The biological portion of the phosphorus cycle begins when producers take up phosphorus. Roots of land plants take up dissolved phosphate from the soil water ❺. Land animals get phosphates by eating plants or one another. Phosphorus returns to the soil in wastes and remains ❻. In the seas, phosphorus enters food webs when producers take up dissolved phosphate from seawater ❼. As on land, wastes and remains replenish the supply ❽.

Vast quantities of phosphate-rich droppings from seabird or bat colonies are collected and used as fertilizer. Phosphate-rich rock is also mined for this purpose. Lack of phosphate commonly limits growth of aquatic producers. Figure 17.20 shows an experiment that demonstrated how adding phosphate to part of a lake basin can cause a population explosion of algae. Such algal blooms can pose a threat to other species that inhabit the lake. Humans cause the same problem when they allow phosphate-containing detergents, sewage, fertilizer runoff, and waste from stockyards to pollute aquatic environments.

The Nitrogen Cycle Gaseous nitrogen (N_2) makes up about 80 percent of Earth's atmosphere. In the **nitrogen cycle**, nitrogen moves among the atmosphere, reservoirs in the soil and water, and into and out of food webs (Figure 17.21). All organisms need nitrogen. Nucleotides such as those in DNA include nitrogen, as do the amino acids that make up proteins.

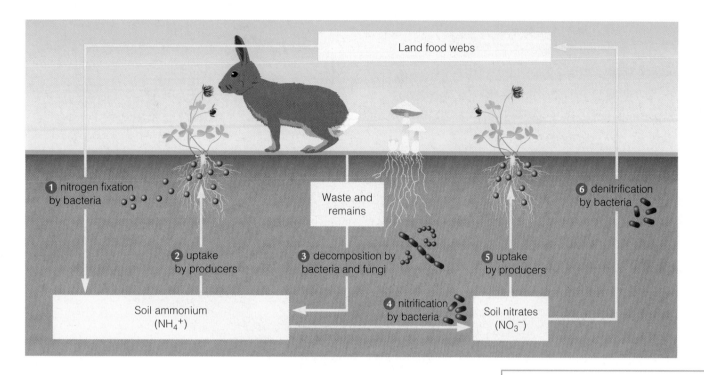

FIGURE 17.21 The nitrogen cycle. The main reservoir for nitrogen is the atmosphere. Activity of nitrogen-fixing bacteria makes gaseous nitrogen available to other living organisms.

Figure It Out: What are the two forms of nitrogen that can be taken up by plants?

Answer: Ammonium and nitrates

Plants cannot use gaseous nitrogen because they do not have an enzyme that breaks the triple covalent bond holding its two nitrogen atoms together. Some bacteria do have such an enzyme and can carry out **nitrogen fixation**. They break the bonds in N_2, then use the nitrogen atoms to form ammonia, which dissolves and forms ammonium ions (NH_4^+) ❶. Plant roots take up ammonium from the soil ❷ and use it in metabolic reactions. Consumers get nitrogen by eating plants or one another.

There is also another source of soil ammonium. Bacterial and fungal decomposers break down wastes and remains and form ammonium ❸.

Other soil bacteria (nitrifying bacteria) obtain energy by converting ammonium to nitrates (NO_3^-), a process called **nitrification** ❹. Like ammonium, nitrates can be taken up from the soil and used by producers ❺.

Ecosystems lose nitrogen by **denitrification**. Denitrifying bacteria use nitrate for energy and produce nitrogen gas that escapes into the atmosphere ❻.

Humans alter the nitrogen cycle by industrial fixation of nitrogen to make synthetic fertilizers. Scientists estimate that use of synthetic fertilizers has doubled the amount of nitrogen that enters land ecosystems. Burning fossil fuels further alters the nitrogen cycle by producing nitrogen oxides. These pollutants cause problems when rain deposits them in habitats where plants are adapted to low nitrogen levels.

The Carbon Cycle Carbon occurs abundantly in the atmosphere, combined with oxygen as carbon dioxide (CO_2). In the **carbon cycle**, carbon moves among rocks, water, and the atmosphere, and into and out of food webs. After water, carbon is the most abundant substance in living organisms. All organic molecules (carbohydrates, fats, lipids, and proteins) have a carbon backbone.

On land, plants take up and use carbon dioxide in photosynthesis. Plants and most other land organisms release carbon dioxide back to the atmosphere when they carry out aerobic respiration. Bicarbonate ions (HCO_3^-) form when carbon dioxide dissolves in water. Aquatic producers take up bicarbonate and convert it to carbon dioxide for use in photosynthesis. As on land, most aquatic organisms carry out aerobic respiration and release carbon dioxide.

carbon cycle Movement of carbon among rocks, water, the atmosphere, and living organisms.

denitrification Conversion of nitrates or nitrites to gaseous forms of nitrogen.

nitrogen cycle Movement of nitrogen among the atmosphere, soil, and water, and into and out of food webs.

nitrogen fixation Conversion of nitrogen gas to ammonia.

nitrification Conversion of ammonium to nitrates.

phosphorus cycle Movement of phosphorus among rocks, water, soil, and living organisms.

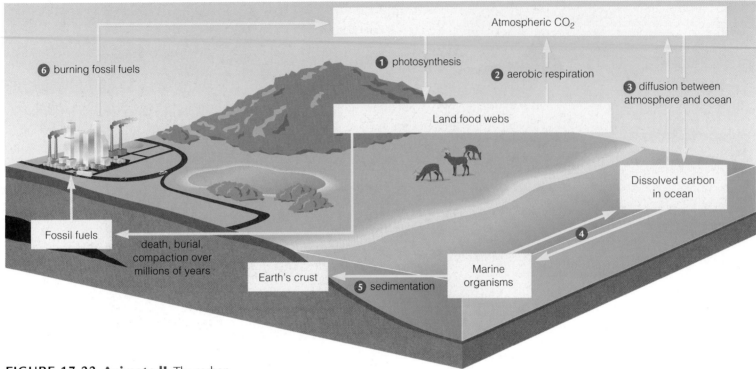

Atmospheric CO$_2$

6 burning fossil fuels

1 photosynthesis

2 aerobic respiration

3 diffusion between atmosphere and ocean

Land food webs

Dissolved carbon in ocean

Fossil fuels

death, burial, compaction over millions of years

Earth's crust

4

5 sedimentation

Marine organisms

FIGURE 17.22 Animated! The carbon cycle. Earth's crust is the largest reservoir.

1 Carbon enters land food webs when plants take up carbon dioxide from the air and carry out photosynthesis.

2 Carbon returns to the atmosphere as carbon dioxide when plants and other land organisms carry out aerobic respiration.

3 Carbon diffuses between the atmosphere and the ocean. Carbon dioxide becomes bicarbonate when it dissolves in ocean water.

4 Marine producers take up bicarbonate for use in photosynthesis, and marine organisms release carbon dioxide produced by aerobic respiration.

5 Many marine organisms incorporate carbon into their shells. After they die, these shells become part of the sediments. Over time, these sediments become carbon-rich rocks such as limestone and chalk in Earth's crust.

6 Burning of fossil fuels derived from the ancient remains of plants adds additional carbon dioxide into the atmosphere.

Earth's largest carbon reservoir consists of sedimentary rocks such as limestone. These rocks were formed over millions of years by the compaction of the carbon-rich shells of marine organisms. Plants do not take up dissolved carbon from the soil, so the carbon in these rocks is not readily accessible to organisms in land ecosystems. Fossil fuels are another reservoir of carbon that, under natural conditions, is largely unavailable to the world of life. Figure 17.22 shows the main carbon reservoirs and the processes that transfer carbon among them.

The Greenhouse Effect and Global Climate Change By burning fossil fuels and wood, humans are putting extra carbon dioxide and nitrogen oxides into the air. At the same time, we are cutting forests, thus decreasing the global uptake of carbon dioxide by plants. The result of these activities is a well-documented change in the composition of the atmosphere.

These atmospheric changes can affect Earth's climate because carbon dioxide and nitrogen oxides are among the greenhouse gases. **Greenhouse gases** are atmospheric gases that slow the movement of heat from Earth to space (Figure 17.23). The process by which heat emitted by Earth's atmosphere warms its surface is called the **greenhouse effect**.

Without the greenhouse effect, Earth's surface would be cold and lifeless. But there can be too much of a good thing. Scientists estimate that atmospheric carbon dioxide is at its highest level in 650,000 years. As the concentrations of greenhouse gases in the atmosphere have risen, so has the average global temperature. In the past 100 years, Earth's average temperature has risen by about 0.74°C (1.3°F), and the rate of warming is accelerating.

A rise of a degree or two in average temperature may not seem like a big deal, but it is enough to increase the rate of glacial melting, raise sea level, alter wind patterns, shift the distribution of rainfall and snowfall, and increase the frequency and severity of hurricanes. Scientists now refer to the many climate-related effects of the rise in greenhouse gases as **global climate change**. We discuss global climate change and its effects in detail in the next chapter.

FIGURE 17.23 Animated! The greenhouse effect.

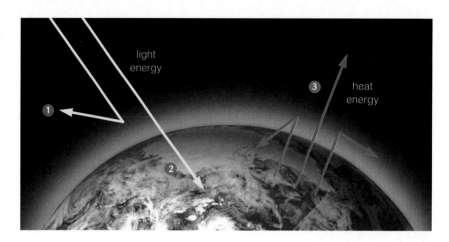

1 Some light energy from the sun is reflected by Earth's atmosphere or surface.

2 More light energy reaches Earth's surface, and warms the surface.

3 Earth's warmed surface emits heat energy. Some of this energy escapes through the atmosphere into space. But some is absorbed and then emitted in all directions by greenhouse gases. The emitted heat warms Earth's surface and lower atmosphere.

Take-Home Message How do major nutrients cycle between organisms and their environment?

■ Water moves on a global scale from the ocean (its main reservoir), through the atmosphere, onto land, then back to the ocean.

■ Phosphorus cycles between its main reservoir—rocks and sediments—and soils and water. Phosphorus enters food webs when producers take up dissolved phosphates. The atmosphere does not play a significant role in this cycle.

■ Nitrogen moves from its main reservoir—the atmosphere—into soils and water, and into and out of food webs. Bacteria play a pivotal role in the nitrogen cycle by producing forms of nitrogen that producers can take up and use.

■ Carbon's main reservoir is sediments, but most carbon enters food webs when producers take up dissolved carbon from water or carbon dioxide from the air.

■ Human activities cause nutrient imbalances in land and aquatic habitats. Our activities also add excess greenhouse gases such as carbon dioxide and nitrogen oxides to the air, which may affect global climate.

global climate change Wide-ranging changes in rainfall patterns, average temperature, and other climate factors that may result from rising concentrations of greenhouse gases.

greenhouse effect Warming of Earth's lower atmosphere and surface as a result of heat trapped by greenhouse gases.

greenhouse gas Atmospheric gas that helps keep heat from escaping into space and thus warms the Earth.

17.7 Impacts/Issues Revisited: Fighting the Foreign Fire Ants

Now that you have learned a bit about how communities and ecosystems work, let's take a second look at those red imported fire ants (RIFAs) we discussed at the beginning of this chapter. These ants are an introduced predator that is having a negative effect on competing ants and on prey species. RIFAs did not evolve in North America, so there are few predators, parasites, or pathogens to hold them in check. Freed from these natural restrictions, RIFAs have an easier time outcompeting native ants. As RIFA populations soar, populations of organisms that are their prey decline.

Global climate change is expected to help RIFAs extend their range. For example, RIFAs require mild winters and relatively high rainfall to do well. With winters becoming milder, the ants are expected to move farther north and ascend to higher altitudes. If current climate predictions are correct, the amount of habitat available to red imported fire ants in the United States is expected to increase by about 20 percent before the close of this century.

How Would YOU Vote? Currently, only a small fraction of the crates being imported to the United States are inspected for exotic species. Should the inspections increase? See CengageNow for details, then vote online (cengagenow.com).

Summary

Section 17.1 A species interacts with other species and its environment. All species in a defined area are a community. An exotic species evolved in one community, then became established in another. A community and its environment is an ecosystem.

Section 17.2 Each species occupies a certain habitat, or type of place. Every species in the community has its own niche—the conditions and resources it requires, and the interactions it takes part in. The types of species in a community and their relative abundance is determined by nonbiological factors such as climate, as well as biological ones such as how species interact.

Section 17.3 Species in an ecosystem interact with one another in a variety of ways. Commensalism benefits one species and neither helps nor harms the other. In mutualism, both participants exploit one another to their mutual benefit.

Interspecific competition harms both participants. When two (or more) species require identical resources, the superior competitor drives others to local extinction. Resource partitioning allows similar species to coexist by using different portions of shared resources.

Predators are free-living and typically kill their prey. Parasites live in or on their hosts and withdraw nutrients from them, usually without killing them. Parasitoids are free-living as adults, but their larvae develop inside and feed on a host, which they eventually kill. Predators, parasites, and parasitoids coevolve with their hosts.

CENGAGENOW™ Learn about competitive exclusion with the animation on CengageNow.

Section 17.4 Ecological succession is the sequential replacement of arrays of species within a community. Primary succession occurs in new habitats. Secondary succession takes place in disturbed ones. The first species in a community are pioneer species. Their presence may help other potential colonists.

Community structure is not easily predicted. It is affected by random events and disturbances such as storms, droughts, and fires, and it rarely stabilizes.

The presence of a keystone species has a large effect on community structure. Arrival of an exotic species can drastically alter a community.

Section 17.5 There is a one-way flow of energy through an ecosystem, and a cycling of materials among the organisms in it.

Sunlight is the initial energy source for almost all ecosystems. Producers convert sunlight energy to chemical bond energy and take up nutrients that they and the ecosystem's consumers require.

All organisms at the same trophic level in an ecosystem are the same number of steps away from the energy source for that ecosystem. Energy transfers are inefficient, so most ecosystems support no more than a few trophic levels.

The rate of primary production—the capture and storage of energy by producers—varies with climate, season, and other factors. Energy pyramids show how usable energy lessens as it flows through an ecosystem. Toxic chemicals also move up trophic levels and become increasingly concentrated in body tissues.

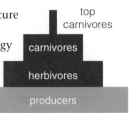

top carnivores

carnivores

herbivores

producers

CENGAGENOW™ Learn about trophic levels with animation on CengageNow.

Section 17.6 In a biogeochemical cycle, water or a nutrient moves through the environment, then through organisms, then back to an environmental reservoir.

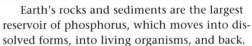

food webs

environmental reservoirs

In the water cycle, water moves from the ocean into the atmosphere, falls on land, and flows back to the ocean—its main reservoir.

Earth's rocks and sediments are the largest reservoir of phosphorus, which moves into dissolved forms, into living organisms, and back.

The atmosphere is the main reservoir for nitrogen, but plants cannot use gaseous forms of nitrogen. Bacteria and the fertilizer industry fix nitrogen (convert atmospheric nitrogen to forms that plants can take up).

The global carbon cycle moves carbon from its reservoirs in rocks and seawater, through its gaseous form (CO_2), and through living organisms.

Carbon dioxide and nitrogen oxides are among the greenhouse gases that trap heat in Earth's atmosphere and make life possible. An increase in their concentration in the atmosphere is causing global climate change.

CENGAGENOW™ Learn about biogeochemical cycles and the greenhouse effect with the animation and interactions on CengageNow.

Self-Quiz   Answers in Appendix I

1. A species' habitat is like its address, and its _____ is like its occupation.

2. True or false? A symbiosis always benefits both partners.

3. Match the species interaction with a suitable description.
   _____ mutualism
   _____ competition
   _____ predation
   _____ parasitism

   a. A snake eats a mouse.
   b. A bee pollinates a flower while sipping its nectar.
   c. An owl and a wood duck both need a tree cavity to nest.
   d. A mosquito sucks your blood.

4. Interspecific competition favors individuals of both species who are most _____ the competing species in their resource needs.
   a. similar to   b. different from

5. The establishment of a biological community on a newly formed volcanic island is an example of _____ .
   a. primary succession   c. competitive exclusion
   b. secondary succession   d. bioaccumulation

Digging Into Data

Biological Control of Fire Ants

Ant-decapitating phorid flies are just one of the biological control agents used to battle imported fire ants. Researchers have also enlisted the help of a fungal parasite that infects ants and slows production of offspring. As a result of infection, a colony dwindles in numbers and eventually dies out.

Are these biological controls useful against imported fire ants? To find out, USDA scientists treated infested areas with either traditional pesticides or pesticides plus biological controls (both flies and the parasite). The scientists left some plots untreated as controls. Figure 17.24 shows the results.

1. How did population size in the control plots change during the first four months of the study?

2. How did population size in the two types of treated plots change during this same interval?

3. If this study had ended after the first year, would you conclude that biological controls had a major effect?

4. How did the two types of treatment (pesticide alone versus pesticide plus biological controls) differ in their longer-term effects? Which is most effective?

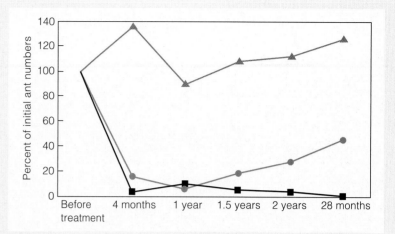

FIGURE 17.24 Effects of two methods of controlling red imported fire ants. The graph shows the numbers of red imported fire ants over a 28-month period. *Orange* triangles represent untreated control plots. *Green* circles are plots treated with pesticides alone. *Black* squares are plots treated with pesticide and biological control agents (phorid flies and a fungal parasite).

---

6. Match the terms with suitable descriptions.
   _____ producers   a. feed on plants
   _____ herbivores   b. feed on small bits of
   _____ decomposers   organic matter
   _____ detritivores   c. degrade organic wastes and
                    remains to inorganic forms
                 d. capture sunlight energy

7. A chemical that bioaccumulates in a prairie food web would be found at highest concentrations in the tissues of _____ .
   a. grasses   c. grasshopper-eating sparrows
   b. grasshoppers   d. sparrow-eating hawks

8. In an energy pyramid diagram of a prairie food web, which of these organisms would be in the lowest (and largest) tier?
   a. grasses   c. grasshopper-eating sparrows
   b. grasshoppers   d. sparrow-eating hawks

9. Match each substance with its largest environmental reservoir. One reservoir choice will be used more than once.
   _____ carbon   a. seawater
   _____ water   b. rocks and sediments
   _____ phosphorus   c. the atmosphere
   _____ nitrogen

10. Earth's largest reservoir of fresh water is _____ .
    a. lakes   c. glacial ice
    b. groundwater   d. water in living organisms

11. _____ convert nitrogen gas to a form producers can take up.
    a. Fungi   c. Carnivores
    b. Bacteria   d. Herbivores

12. Land plants take in _____ for photosynthesis from the air.
    a. carbon dioxide   c. ammonium ions
    b. phosphate ions   d. bicarbonate ions

13. Aquatic producers get carbon for photosynthesis from _____ .
    a. decaying organic matter   c. ammonium ions
    b. phosphate ions   d. bicarbonate ions

14. Greenhouse gases _____ .
    a. trap heat in the atmosphere
    b. are released by burning of fossil fuels
    c. may cause global climate change if they accumulate
    d. all of the above

15. True or false? Imbalances in both the carbon cycle and nitrogen cycle contribute to increases in greenhouse gases.

Additional questions are available on CENGAGENOW™

Critical Thinking

1. With antibiotic resistance rising (Section 12.3), researchers are looking for ways to reduce use of these drugs. Some cattle once fed antibiotic-laced food now get feed that includes helpful bacteria that can live in the animal's gut. The idea is that if a large population of beneficial bacteria is in place, then harmful bacteria with the same resource needs are less likely to thrive. Explain why this idea makes sense in terms of species interactions.

2. Figure 17.6 shows a harmless fly that mimics a stinging wasp. Studies have shown that in such mimicry systems, mimics benefit most when they are rare relative to the well-protected model. Can you explain why?

3. Ectotherms, or "cold-blooded" animals, such as invertebrates and fish, convert more of the energy they eat into body tissues than endotherms, or "warm-blooded" animals. What allows the ectotherms to make more efficient use of food energy?

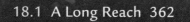

18 The Biosphere and Human Effects

18.1 Impacts/Issues: A Long Reach

We began this book with the story of biologists who ventured into a remote forest in New Guinea and found many previously unknown species. At the far end of the globe, a U.S. submarine surfaced in Arctic waters and found polar bears hunting on the ice-covered sea (Figure 18.1). The bears were 435 kilometers (270 miles) from the North Pole and 805 kilometers (500 miles) from the nearest land.

Even seemingly remote regions are no longer beyond the reach of human explorers and human influence. Increasing levels of greenhouse gases are raising the temperature of Earth's atmosphere and seas. In the Arctic, the warming is causing sea ice to thin and to break up earlier in the spring. This raises the risk that polar bears hunting far from land will become stranded, unable to return to solid ground before the ice thaws.

Polar bears are top predators and their tissues contain a high amount of mercury and organic pesticides. These pollutants enter the water and air far from the Arctic, in more temperate regions, and winds and ocean currents deliver them to polar realms. Migratory animals such as seabirds pick up pollutants in densely populated regions where they spend their winters, then carry them to the Arctic and release them in wastes and, if they die, remains.

In places less remote than the Arctic, human populations have a more direct effect. As we cover more and more of the world with our homes, factories, and farms, less habitat remains for other species. We also put species at risk by competing with them for resources, overharvesting them, and introducing nonnative competitors.

It would be presumptuous to think that humans alone have had a profound impact on the world of life. More than a billion years ago, the rise of oxygen-releasing photosynthetic cells changed the course of evolution by enriching the atmosphere with this gas (Section 5.6). Over life's long existence, the success of one species or group of organisms has often come at the expense of others. What is new is the increasing pace of change and the capacity of our own species to recognize and affect its role in this increase.

A century ago, Earth's physical and biological resources seemed inexhaustible. Now we know that many practices put into place when we were ignorant of how natural systems operate take a heavy toll on the biosphere. They threaten other species, ecosystem processes, and possibly our own long-term existence.

The **biosphere** is the sum of all places where we find life on Earth. With this chapter, we look at physical processes that affect the distribution of organisms through the biosphere. We also consider the ways that human activities are altering these processes and disrupting the world of life.

FIGURE 18.1 Three polar bears investigate an American submarine that surfaced in ice-covered Arctic waters.

FIGURE 18.2 Variation in intensity of solar radiation with latitude. For simplicity, we depict two equal parcels of incoming radiation on an equinox, a day when incoming rays are perpendicular to Earth's axis.

Rays that fall on high latitudes **A** pass through more atmosphere (*blue*) than those that fall near the equator **B**. Compare the length of the *green* lines. Atmosphere is not to scale.

Also, energy in the rays that fall at the high latitude is spread over a greater area than energy that falls on the equator. Compare the length of the *red* lines.

Factors Affecting Climate

Global Air Circulation Patterns **Climate** refers to average weather conditions, such as temperature, humidity, wind speed, cloud cover, and rainfall, over a long period. Many factors influence a region's climate, which in turn affects the types of species that can live there.

A region's latitude (how far north or south it is) determines how much light energy it receives. On any particular day, regions near the equator receive more sunlight than higher latitudes for two reasons (Figure 18.2). First, sunlight traveling to high latitudes passes through more atmosphere to reach Earth's surface than light traveling to the equator. Fine particles of dust, water vapor, and greenhouse gases absorb some solar radiation or reflect it back into space, so less light energy reaches the poles. Second, the energy in any incoming parcel of sunlight is spread out over a smaller surface area at the equator than at the higher latitudes. As a result of these factors, Earth's surface warms more at the equator than at the poles.

Latitudinal difference in surface warming, and the ways these differences affect air and water, cause global patterns of air circulation and rainfall (Figure 18.3). At the equator, intense sunlight warms the air and causes evaporation of water from the ocean. As the air heats up, it expands and rises ❶. The same effect causes a hot-air balloon to inflate and rise when the air inside it is heated. As the equatorial air mass rises, it flows north and south and begins to cool. Cool air can hold less moisture than warm air, so moisture leaves the air as rain that supports tropical rain forests.

By the time the air reaches about 30° north and south latitude, it has cooled and become dry. Being cool, the air sinks downward ❷. When it descends, it draws moisture from the soil. As a result, deserts often form at around 30° north and south latitude.

Air that continues flowing along Earth's surface toward the poles once again picks up heat and moisture. At a latitude of about 60° the air is warm and moist again and it rises, losing moisture as it does so ❸. In polar regions, cold air that holds little moisture descends ❹. Precipitation is sparse, and polar deserts form.

Landforms also affect rainfall. Deserts often form downwind of coastal mountains. Moisture-laden winds give up water as rain as they rise over mountains. The dry region beyond the mountains is referred to as a **rain shadow**.

Ocean Currents Like air, water expands when heated, and it is heated more at the equator than at higher latitudes. As a result, the sea level at the

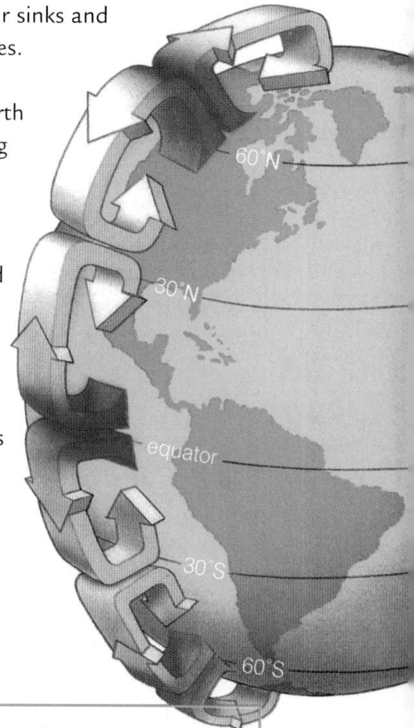

❹ At the poles, cold, dry air sinks and moves toward lower latitudes.

❸ Air rises again at 60° north and south, where air flowing poleward meets air coming from the poles.

❷ At around 30° north and south latitude, the air—now cooler and dry—sinks.

❶ Warm, moist air rises at the equator. As the air flows north and south, it cools and loses moisture as rain.

FIGURE 18.3 Animated! Air circulation patterns that result from latitudinal differences in the amount of solar radiation reaching Earth.

biosphere All portions of the Earth where life exists.

climate Average weather conditions in a region over a long time period.

rain shadow Dry region on the downwind side of a coastal mountain range.

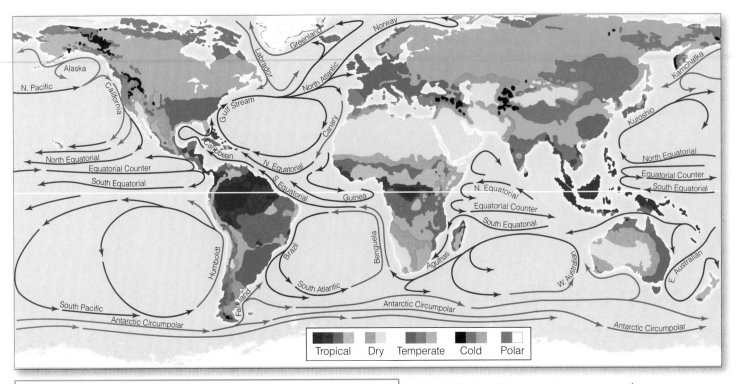

FIGURE 18.4 Animated! Major climate zones correlated with surface currents of the world ocean. Warm surface currents start moving from the equator toward the poles, but prevailing winds, Earth's rotation, gravity, the shape of ocean basins, and landforms influence the direction of flow. Water temperatures, which differ with latitude and depth, contribute to the regional differences in air temperature and rainfall.

↙ **warm surface current**   ↙ **cold surface current**

equator is about 8 centimeters (3 inches) higher than at either pole. The volume of water in this "slope" starts sea surface water moving "downhill," from the equator toward the poles.

Enormous volumes of water circulate as ocean currents. The force of major winds, Earth's rotation, and underwater mountains and trenches determine the directional movement of these currents. Surface currents circulate clockwise in the Northern Hemisphere and counterclockwise in the Southern Hemisphere (Figure 18.4).

Ocean currents affect climates. Coasts in the Pacific Northwest are cool and foggy in summer because the cold California current chills the air, so water condenses out as droplets. Boston and Baltimore are humid in summer because air masses pick up heat and moisture from the warm Gulf Stream, then deliver it to these cities.

Take-Home Message **What causes prevailing winds and ocean currents and how do they affect climate?**

- Longitudinal differences in the amount of solar radiation reaching Earth produce global air circulation patterns.

- Surface ocean currents, set in motion by latitudinal differences in solar radiation, are also affected by winds and by Earth's rotation.

- Collective effects of air masses, oceans, and landforms determine regional temperature and moisture levels.

Types of Land Ecosystems

Different climates support different types of plant life, which in turn support different kinds of animals. A **biome** is a large-scale ecosystem characterized by particular climate conditions and its main type of vegetation. Each type of biome can typically be found on more than one continent.

Deserts

Deserts cover about one-fifth of Earth's land. They receive little rainfall (less than 30 centimeters, or 11 inches, per year). The most common desert plants include cactuses or succulents that store water in their tissues, and shrubs with deep roots that tap water far underground (Figure 18.5**A**). Many desert animals have physiological adaptations that reduce their water needs.

As noted earlier, deserts tend to form at 30° north and south latitude, where dry air descends. In most biomes, moisture in the air has an insulating effect, slowing temperature increases during the day and heat loss at night. Deserts have low humidity, so they can heat up and cool off fast.

Grasslands and Chaparral

Grasslands form in the interior of continents between deserts and temperate forests. Shortgrass and tallgrass prairie are North America's main types of grasslands (Figure 18.5**B**). Summers are warm, and winters are cold. Annual rainfall of 25 to 100 centimeters (10–40 inches) keeps desert from forming, but is too little water to support forest. Constant trimming by grazing animals, along with periodic fires, also keeps trees and most shrubs from taking hold. Low-growing plants tolerate strong winds and intervals of drought.

We find dry shrublands or **chaparral** in South Africa, Mediterranean regions, and California, where winters are mild and rainy but summers are dry (Figure 18.5**C**). Chaparral shrubs are adapted to survive lightning-ignited fires. Seeds of some species germinate only after exposure to heat or smoke. This ensures that they sprout when seedlings face little competition from established plants.

biome Type of ecosystem that can be characterized by its climate and dominant vegetation.

chaparral Biome where mild, rainy winters and hot, dry summers support shrubs adapted to periodic fires.

desert Biome where little rain falls, humidity is low, and the main plants store water in their tissues or tap into water sources deep underground.

grasslands Biome in the interior of continents, where grasses and other low-growing plants are adapted to warm summers, cold winters, periodic fires, and grazing animals.

FIGURE 18.5 Examples of three low-moisture biomes in the United States. **A** The Sonoran Desert in Arizona. **B** Tallgrass prairie in Kansas. **C** Chaparral in California.

Deserts

Grassland, shrublands, and woodlands

FIGURE 18.6 Broadleaf forests.
A Tropical rain forest in Southeast Asia.
B Temperate deciduous forest in New England.

Broadleaf
forests

Tropical Rain Forests There are a variety of forest biomes. For example, **tropical rain forests** are found in parts of Southeast Asia, Africa, and South America where temperatures are high and rain is plentiful throughout the year (Figure 18.6**A**). The broadleaf (angiosperm) trees that dominate this biome remain leafy year-round. Tropical rain forests have a complex, layered structure with many vines and epiphytes (plants that grow on another plant, but do not take nutrients from it). Animal species are similarly diverse and abundant. By some estimates, about half of all animal species live in tropical rain forests.

Deciduous Broadleaf Forests Deciduous broadleaf forests occur in regions where trees can only grow during part of the year. Such habitats are dominated by **deciduous trees** that shed all their leaves at once and become dormant during a season when conditions are not favorable. For example, the main plants in New England's **temperate deciduous forests** are maples and other trees that shed their leaves in the fall (Figure 18.6**B**). The trees then remain dormant during the winter when cold temperatures would slow photosynthesis and frozen soil would limit water uptake.

Coniferous Forests Conifers (evergreen gymnosperms with seed-bearing cones) are the main plants in coniferous forests (Figure 18.7). Conifer leaves are typically needle-shaped, with a thick cuticle and stomata below the leaf surface. These adaptations help conifers conserve water during drought or times when soil water is frozen. As a group, conifers tolerate poorer soils and drier habitats than broadleaf trees.

The most extensive land biome is the coniferous forest that sweeps across northern Asia, Europe, and North America (Figure 18.7). It is known as boreal forest, or **taiga**, which means "swamp forest" in Russian. Rain falls in the summer, and winters are long, cold, and dry. Conifers also dominate temperate lowlands along the Pacific coast from Alaska into northern California. These forests hold the world's tallest trees, Sitka spruce to the north and coast redwoods to the south. Pine forest covers about one-third of the southeastern United States. The pines survive periodic fires that kill most hardwood species. If fires are suppressed, hardwoods will replace the pines.

Coniferous
forests

FIGURE 18.7 Coniferous forest.
Aerial view of Siberian taiga.

Tundra **Arctic tundra** forms between the polar ice cap and the belts of boreal forests in the Northern Hemisphere. Most is in northern Russia and Canada. It is Earth's youngest biome, having appeared about 10,000 years ago, when glaciers retreated at the end of the last ice age.

Arctic tundra is blanketed with snow for as long as nine months of the year. Annual snow and rain is usually less than 25 centimeters (10 inches). During a brief summer, plants grow fast under nearly continuous sunlight (Figure 18.8). Lichens and shallow-rooted, low-growing plants are a base for food webs. Many migratory birds nest here in summer, when the air is thick with insects.

Only the surface layer of tundra soil thaws during summer. Below that lies **permafrost**, a frozen layer that can be 500 meters (1,600 feet) thick in places. Permafrost acts as a barrier that prevents drainage, so the soil above it remains perpetually waterlogged. Cool, anaerobic conditions slow decay, so organic remains can build up. Organic matter in the permafrost makes the arctic tundra one of Earth's greatest stores of carbon.

Alpine tundra occurs at high altitudes around the world. At lofty elevations, where night temperatures usually fall below freezing, trees cannot grow. The soil is thin, nutrient-poor, and—unlike that of arctic tundra—fast draining, because there is no permafrost. The growing season is short and, even in summer, patches of snow persist. Strong winds whip the vegetation. Lichens, mosses, grasses, and small-leafed shrubs that form ground-hugging cushions and mats dominate the community.

Arctic tundra

FIGURE 18.8 Arctic tundra. Soil just below the surface remains frozen even during the summer.

Take-Home Message **What are the major biomes?**

- Low rainfall produces deserts at latitudes between 30° north and south. Latitudes with somewhat higher rainfall are covered by grasslands or chaparral.

- At the equator, high rainfall and temperature support tropical rain forests with broadleaf trees that remain green year-round. Deciduous broadleaf trees are adapted to regions that support trees, but cannot sustain year-round growth.

- Forests of hardy conifers dominate high latitudes in the Northern Hemisphere, as well as other regions where drought, poor soil, or periodic fires prevent broadleaf trees from taking hold.

- The most northerly and youngest biome is arctic tundra, where low plants are adapted to a short growing season and soil with a layer of permafrost. Alpine tundra is low-growing plants that live at high altitudes.

18.4

Types of Aquatic Ecosystems

We can distinguish between types of aquatic ecosystems, just as we do biomes on land. Temperature, salinity, rate of water movement, and depth influence the composition of aquatic communities.

Freshwater Ecosystems A lake is a body of standing fresh water. All but the shallowest lakes have zones that differ in their physical characteristics and species composition. Near shore, where sunlight penetrates all the way to the lake bottom, rooted aquatic plants and algae that attach to the bottom are primary producers. A lake's open waters include an upper well-lit zone and, if a lake is deep or cloudy, a zone where light does not penetrate. Producers in the well-lit water include photosynthetic protists and bacteria. In the deeper dark zone, consumers feed on organic debris that drifts down from above.

alpine tundra High-altitude biome dominated by low plants.

arctic tundra Northernmost biome dominated by low plants that grow over a layer of permafrost.

deciduous tree A tree that drops all its leaves annually just before a season that does not favor growth.

permafrost Layer of permanently frozen soil in the Arctic.

taiga Extensive northern biome dominated by conifers. Also called boreal forest.

temperate deciduous forest Biome dominated by trees that drop all their leaves and go dormant during a cold winter.

tropical rain forest Species-rich tropical biome in which continual warmth and rainfall allows the dominant broadleaf trees to grow all year.

Streams are flowing-water ecosystems. They typically originate from runoff or melting snow or ice. As they flow downslope, they grow and merge to form rivers. Properties of a stream or river vary along its length. The type of rocks a stream flows over can affect its solute concentration, as when limestone rocks dissolve and add calcium to the water. Shallow water that flows rapidly over rocks mixes with air and holds more oxygen than slower-moving, deeper water. Also, cold water holds more oxygen than warm water. As a result, different parts of a stream or river support species with different oxygen needs. For example, rainbow trout can only live in cool, well-oxygenated water.

Marine Ecosystems An **estuary** is a mostly enclosed coastal region where seawater mixes with nutrient-rich fresh water from rivers and streams. Water inflow continually replenishes nutrients, so estuaries have a high primary productivity. Estuaries serve as marine nurseries for many invertebrates and fishes. Migratory ducks and geese often use them as rest stops. Estuaries need a constant inflow of unpolluted, fresh water to remain healthy, so they are threatened by the diversion of fresh water for human uses and contamination by pollutants.

Along ocean shores, organisms are adapted to the mechanical force of the waves and to tidal changes. Many species are underwater during high tide, but are exposed to the air when the tide is low. Along rocky shores, where waves prevent detritus from piling up, algae that cling to rocks are the producers in grazing food chains. In contrast, waves continually rearrange loose sediments along sandy shores, and make it difficult for algae to take hold. Here, detrital food chains start with organic debris from land or offshore.

Warm, shallow, well-lit tropical seas hold **coral reefs**, formations made primarily of calcium carbonate secreted by generations of corals, which are invertebrate animals. Like tropical rain forests, tropical coral reefs are home to an extraordinary assortment of species (Figure 18.9**A**).

The main producers in a coral reef community are photosynthetic dinoflagellates that live inside the tissues of reef-building corals. These protists provide their coral hosts with sugars. If a coral becomes stressed by environmental change such as a shift in temperature, it expels its protistan symbionts in a response called **coral bleaching**. If conditions improve fast, the symbiont population in the coral can recover. But when adverse conditions persist, the symbionts will not be restored, and the coral will die, leaving only its bleached skeletons behind (Figure 18.9**B**).

The open waters of the ocean are the **pelagic province** (Figure 18.10**A**). These open waters include the water over continental shelves and the more extensive waters farther offshore. In the oceanic zone's upper, brightly lit waters, photosynthetic microorganisms such as algae and bacteria are the primary producers, and grazing food chains predominate. Depending on the region, some light may penetrate as far as 1,000 meters (more than a half mile) beneath the sea surface. Below that, organisms live in continual darkness, and organic material that drifts down from above is the basis of detrital food chains.

The **benthic province** is the ocean bottom, its rocks and sediments. Species richness is greatest on the edges of continents, or the continental shelves. The benthic province also includes some largely unexplored species-rich regions, including seamounts and at hydrothermal vents.

FIGURE 18.9 Coral reefs. **A** In Fiji, a healthy reef teems with life. **B** A "bleached" reef near Australia.

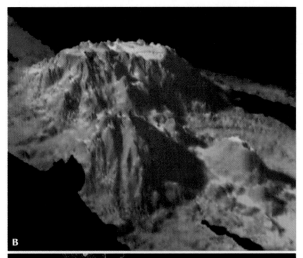

air at ocean surface | ←— water of the open ocean —→ | ←— water over continental shelf —→

Pelagic Province

continental shelf

sunlit water   0
"twilight" water   200
sunless water

1,000
2,000

4,000

Benthic Province

deep-sea trenches

11,000
depth (meters)

A

FIGURE 18.10 Animated! The deep ocean. **A** Oceanic zones. Zone dimensions are not drawn to scale. **B** Computer model of three seamounts on the sea floor off the coast of Alaska. Patton Seamount, at the rear, stands 3.6 kilometers (about 2 miles) tall, with its peak about 240 meters (800 feet) below the sea surface. **C** Tube worms (annelids) and crabs of a hydrothermal vent community on the sea floor.

Seamounts are underwater mountains that stand 1,000 meters or more tall, but are still below the sea surface (Figure 18.10**B**). They attract large numbers of fishes and are home to many marine invertebrates. Like islands, seamounts often are home to species that evolved there and live nowhere else.

Superheated water that contains dissolved minerals spews out from the ocean floor at **hydrothermal vents**. When this heated, mineral-rich water mixes with cold seawater, the minerals settle out as extensive deposits. Some prokaryotes can get energy from these deposits. The prokaryotes serve as primary producers for food webs that include diverse invertebrates such as tube worms and crabs (Figure 18.10**C**). As explained in Section 13.2, one hypothesis holds that life originated near hydrothermal vents.

Take-Home Message **What factors shape aquatic ecosystems?**

- Fast-flowing, cooler water holds more oxygen than warmer, slower-moving water, and the amount of available light decreases with the water's depth.

- In well-lit upper waters, producers such as aquatic plants, algae, and photosynthetic microbes are the base for grazing food chains. On coral reefs, the main producers are photosynthetic protists that live inside the coral's tissues.

- Detritus drifting down from above sustains most deep-water communities in lakes and oceans. However, hydrothermal vent communities on the ocean floor are sustained by energy that prokaryotes harvest from minerals.

benthic province The ocean's rocks and sediments.
coral bleaching Stress response in which a coral expels the photosynthetic protists in its tissues.
coral reef In tropical sunlit seas, a formation composed of secretions of coral polyps that serves as home to many other species.
estuary A highly productive ecosystem where nutrient-rich water from a river mixes with seawater.
hydrothermal vent Place where hot, mineral-rich water streams out from an underwater opening in Earth's crust.
pelagic province The ocean's open waters.
seamount An undersea mountain.

FIGURE 18.11 Living or extinct? Colorized photo of an ivory-billed woodpecker (*Campephilus principalis*). It is, or was, North America's largest woodpecker and a native of the southeastern states.

Human Effects on the Biosphere

The rising size of the human population and its increasing industrialization have far-reaching effects on the biosphere. We begin by discussing how human activities affect individual species, then turn to activities with wider impacts.

Increased Species Extinctions Extinction, like speciation, is a natural process. Species arise and become extinct on an ongoing basis. The rate of extinction picks up dramatically during a mass extinction, when many kinds of organisms in many different habitats all become extinct in a relatively short period. We are currently in the midst of such an event. Unlike historical mass extinctions, this one cannot be blamed on some natural catastrophe such as an asteroid impact. Rather, this mass extinction is the outcome of the success of a single species—humans—and their effect on Earth.

Consider, for example, the ivory-billed woodpecker. This spectacular bird is native to swamp forests of the American Southeast (Figure 18.11). The species declined when these forests were cut for lumber. Until recently, the bird was thought to have become extinct in the 1940s. A possible sighting in Arkansas in 2004 led to an extensive hunt for evidence of the bird. By the end of 2007, this search had produced some blurry photos, snippets of video, and a few recordings of what may or may not be ivory-bill calls and knocks. Definitive proof that the bird still lives remains elusive.

If the ivory-billed woodpecker is still around, it is an **endangered species**, a species that faces extinction in all or part of its range. A **threatened species** is one that is likely to become endangered in the near future. Keep in mind that not all rare species are threatened or endangered. Some species have always been uncommon. If their numbers remain stable, they are not at risk of extinction.

Nearly all currently endangered or threatened species owe their precarious position to human actions. When European settlers first arrived in North America, they found billions of passenger pigeons. In the 1800s, commercial hunting caused a steep decline in the bird's numbers. The last wild passenger pigeon was shot in 1900. The last captive bird died in 1914.

We continue to overharvest species. The crash of the Atlantic codfish population (described in Section 16.4), is a recent example. It is a sad commentary on human nature that the rarer a species becomes, the higher the price it fetches on the black market. Globalization means that species can be sold to high bidders anywhere in the world. For example, rhino horn from endangered animals in Africa ends up as a traditional medicine in Asia and as knife handles in Yemen.

Each species requires a specific type of habitat, and any loss, degradation, or fragmentation of that habitat reduces population numbers. An **endemic species**, one that is confined to the limited area in which it evolved, is more likely to go extinct than a species with a more widespread distribution. For example, giant pandas are endemic to China's bamboo forests. As China's human population soared, bamboo was cut for building materials and to make room for farms. As the bamboo forests disappeared, so did pandas. Their numbers, which may once have been as high as 100,000, have fallen to 1,000 or so animals in the wild.

In the United States, habitat loss affects nearly all of the more than 700 species of threatened or endangered flowering plants. For example, widespread destruction of prairies and meadows threaten prairie fringed orchids (Figure 18.12**A**). Humans also degrade habitats in less direct ways. For example, Edwards Aquifer in Texas consists of water-filled, underground limestone formations that supply drinking water to the city of San Antonio. Excessive withdrawals of water from this aquifer, along with pollution of the water that recharges it, endanger species that live in the aquifer's depths. The Texas blind salamander (Figure 18.12**B**) is one of these endangered species.

Species introductions also pose a threat, because exotic species often outcompete native ones. For example, rats that reached islands by stowing away on ships attack and endanger many island species. As another example, European brown trout and eastern brook trout were introduced into California's mountain streams for sport fishing. Both exotic species threaten native golden trout.

A species most often becomes endangered because of a number of simultaneous threats. Often, the decline or loss of one species endangers another. For example, running buffalo clover and the buffalo that grazed on it were once common in the Midwest. The plants thrived in the open woodlands that the buffalo favored. Here, the soil was enriched by buffalo droppings and periodically disturbed by the animals' hooves. Buffalo helped to disperse the clover's seeds, which survive passage through digestive systems. When buffalo were hunted to near extinction, buffalo clover populations declined. Now listed as endangered, the clover is further threatened by conversion of its habitat to housing developments, competition from introduced plants, and attacks by introduced insects.

Endangered species listings have historically focused on vertebrates. We have just begun to consider threatened invertebrates and plants. Our impact on protists and fungi is mostly unknown, and the World Conservation Union's IUCN Red List of Threatened Species (Table 18.1) does not address prokaryotes.

Table 18.1	Global List of Threatened Species (2007)*		
	Described Species	Evaluated for Threats	Found to Be Threatened
Vertebrates			
Mammals	5,416	4,863	1,094
Birds	9,956	9,956	1,217
Reptiles	8,240	1,385	422
Amphibians	6,199	5,915	1,808
Fishes	30,000	3,119	1,201
Invertebrates			
Insects	950,000	1,255	623
Mollusks	81,000	2,212	978
Crustaceans	40,000	553	460
Corals	2,175	13	5
Others	130,200	83	42
Land Plants			
Mosses	15,000	92	79
Ferns and allies	13,025	211	139
Gymnosperms	980	909	321
Angiosperms	258,650	10,771	7,899
Protists			
Green algae	3,715	2	0
Red algae	5,956	58	9
Brown algae	2,849	15	6
Fungi			
Lichens	10,000	2	2
Mushrooms	16,000	1	1

* IUCN–WCU Red List, available online at www.iucnredlist.org

FIGURE 18.12 Species under threat. **A** Habitat destruction threatens the eastern prairie fringed orchid. **B** Aquifer depletion and pollution endanger Texas blind salamanders. Generations of life in a dark aquifer, where there is no selection against mutations that impair eye development, have reduced this species' eyes to tiny black spots.

endangered species A species that faces extinction in all or part of its range.

endemic species A species that evolved in one place and is found nowhere else.

threatened species A species likely to become endangered in the near future.

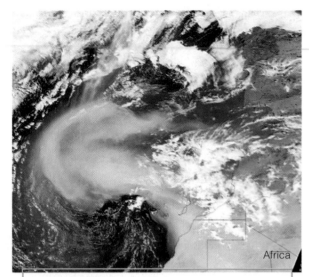

FIGURE 18.13 A symptom of desertification. Increasing amounts of dust blow from Africa's Sahara Desert into and across the Atlantic Ocean.

acid rain Rainfall contaminated by acidic pollutants.

deforestation Removal of all trees from a large tract of land.

desertification Conversion of grassland or woodlands to desertlike conditions.

pollutant A natural or man-made substance that is released into the environment in greater than natural amounts and that damages the health of organisms.

Desertification Some human activities can threaten entire ecosystems. For example, natural grasslands often are plowed under for farming or used to graze livestock. **Desertification**, a conversion of grasslands or woodlands to desertlike conditions, is one result. It begins when loss of plant cover allows wind to blow away topsoil, the nutrient-rich, upper soil layer.

Drought encourages desertification, which encourages drought. Normally, plant roots take up water, which then evaporates from leaves into the air. However, plants cannot thrive where topsoil has blown away. With fewer plants, less water enters the air and local humidity and rainfall decrease.

During the mid-1930s, large portions of prairie on the southern Great Plains were plowed under to plant crops. This plowing exposed the prairie topsoil to the force of the region's constant winds. Coupled with a drought, the result was an economic and ecological disaster. Winds lifted up more than a billion tons of topsoil as huge dust clouds, and the region came to be known as the Dust Bowl. A similar situation now exists in Africa, where the Sahara Desert is expanding south into the Sahel region. Overgrazing and poor farming practices in this region strip grasslands of their vegetation and allow winds to erode the soil. Winds carry the soil aloft and westward (Figure 18.13). African soil particles land as far away as the southern United States and the Caribbean.

Deforestation **Deforestation**, the removal of trees, threatens forest ecosystems. The amount of forested land is currently stable or increasing in North America, Europe, and China, but this increase is overshadowed by the staggering losses of tropical forests (Figure 18.14). Southeast Asia, Africa, and Latin America stretch across the tropical latitudes. Developing nations on these continents have the world's fastest-growing populations and high demands for food, fuel, and lumber. Of necessity, people turn to forests.

Cutting down forests puts forest species at risk of extinction. It also encourages flooding. When trees are cut down, more water flows into streams and rivers, raising the risk of downstream floods. In hilly areas, deforestation raises the risk of landslides because with no tree roots to keep the soil in place, waterlogged soil becomes more likely to slide.

Loss of tropical forests is not merely a regional concern. People in highly developed nations use most of the world's resources, including products from tropical forests. In addition, the disappearance of tropical forests could alter the global atmosphere. Trees take up and store carbon, and they release oxygen. Fewer forests means less carbon storage and less oxygen production. Also, burn-

FIGURE 18.14

Deforestation. Aerial view (*left*) and ground view (*right*) of a region in Mato Grosso, Brazil, where a tropical forest is being cleared for pasture and industrial-scale soybean, sugarcane, and corn plantations.

50 km

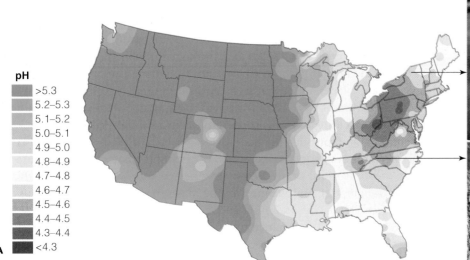

ing enormous tracts of tropical forest to make way for agriculture releases carbon dioxide. Ironically, concern about greenhouse gases released by the burning of fossil fuels may be encouraging some rain forest destruction. Areas of rain forest in the Amazon and Indonesia are being cleared to make way for plantations that grow soybeans or palms. Oils from these plants are exported, mainly to Europe, where they are used to produce biodiesel fuel.

Acid Rain **Pollutants** can harm forests and other ecosystems. A pollutant is a natural or man-made substance released into soil, air, or water in greater than natural amounts. It disrupts the physiological processes of organisms that evolved in its absence, or that are adapted to lower levels of it.

For example, coal-burning power plants, smelters, and factories emit sulfur dioxides into the air. Vehicles, power plants that burn gas and oil, and nitrogen-rich fertilizers emit nitrogen oxides. In dry weather, airborne sulfur and nitrogen oxides coat dust particles and fall as dry acid deposition. In moist air, they form nitric acid vapor, sulfuric acid droplets, and sulfate and nitrate salts. Winds typically disperse these pollutants far from their source. They fall to Earth in rain and snow. We call this a wet acid deposition, or **acid rain**.

The pH of typical rainwater is about 5 (Section 2.5). Acid rain can be 10 to 100 times more acidic. In much of eastern North America, rain is now thirty to forty times more acidic than it was even a few decades ago (Figure 18.15**A**). Most aquatic organisms cannot tolerate a dramatic change in pH. As one result, acid rain has caused fish to vanish from more than 200 lakes in New York's Adirondack Mountains (Figure 18.15**B**).

Acid rain contributes to the decline of forests by burning the leaves of trees. It also changes the composition of the soil by raising the level of nitrates and removing other nutrients. Plants weaken or die as a result (Figure 18.15**C**).

Other Sources of Water Pollution In addition to acid rain, aquatic ecosystems receive an onslaught of other pollutants. Industrial chemicals, oil, fertilizers, animal wastes, sewage, excreted prescription drugs, and sediments flow into lakes, streams, rivers, and the sea. Some of these pollutants come from a point source, an easily identifiable source, such as a factory or a sewage treatment facility that discharges wastes into a river. Pollutants that

> **FIGURE 18.15 Animated!** Acid rain.
>
> **A** Average precipitation acidities in the United States in 2006. The lower the pH, the more acidic the rain.
>
> **B** A biologist tests the pH of water in New York's Woods Lake. In 1979, the lake's pH was 4.8. Since then, experimental addition of calcite to soil around the lake has raised the pH to more than 6.
>
> **C** Dying trees in Great Smoky Mountains National Park, where acid rain harms leaves and soil.
>
> **Figure It Out:** Is rain more acidic on the East Coast or the West Coast of the United States?
>
> *Answer: The East Coast*

pH legend for map:
>5.3
5.2–5.3
5.1–5.2
5.0–5.1
4.9–5.0
4.8–4.9
4.7–4.8
4.6–4.7
4.5–4.6
4.4–4.5
4.3–4.4
<4.3

Two-thirds of plastic soft drink bottles and three-quarters of glass bottles are not recycled.

come from a point source are the easiest to control. Identify the source, and you can take action to halt the pollution.

Dealing with pollution from nonpoint sources is more challenging. Such pollution stems from widespread release of a pollutant. For example, oil from vehicles is a nonpoint source of water pollution. The oil drips from cars and trucks while they move along roads and are parked in parking lots. During rains, the accumulated oil washes from pavement into natural waterways or storm drains. In many places, water that flows into storm drains is discharged into a body of water without any sort of treatment. Halting nonpoint source pollution is difficult because it requires the cooperative action of many people.

Chemical pollutants such as oil, pesticides, or mercury cause harm by disrupting the metabolism of aquatic organisms. For example, the herbicide atrazine is widely used to control weeds. When atrazine gets into water, it disrupts the sexual development of amphibians, causing males to develop ovaries. We discuss the effects of atrazine and other endocrine-disrupting chemicals more in Chapter 25. Section 17.5 explained the process of **bioaccumulation**, in which mercury and some other toxins become increasingly concentrated as they move up through food chains.

Eutrophication, the nutrient enrichment of aquatic ecosystems, is another menace to aquatic life. Fertilizer runoff, sewage, and animal waste that enter water can encourage population explosions of photosynthetic protists. Such events are commonly called algal blooms. They are also known as red tides, because red dinoflagellates are often involved. Some protists involved in algal blooms make toxins that kill aquatic organisms. Even a bloom of nontoxic protists can deplete the water of oxygen, because after the protists die, aerobic bacteria that feed on their remains use up all the oxygen in the water.

Sediments in runoff from deforested areas, construction sites, and other places where soil is exposed can also damage aquatic organisms. Soil particles that wash into the water can clog gills and choke filter-feeders. In addition, suspended particles prevent light from penetrating the water, thus interfering with photosynthesis. Reef-building corals are especially vulnerable to damage from sediments. These corals filter food from the water and also rely on sugars produced by photosynthetic protists that live inside their tissues.

The Trouble With Trash Trash is a menace to water supplies and ocean life. Historically, humans buried unwanted material in the ground or dumped it out at sea. Trash was out of sight, and also out of mind. We now understand that chemicals seeping from buried trash can contaminate groundwater. Waste dumped into the sea disturbs marine ecosystems. For example, seabirds frequently eat floating bits of plastic or feed them to their chicks, with deadly results (Figure 18.16).

In 2006, the United States generated 251 million tons of garbage, which averages out to 2.1 kilograms (4.6 pounds) per person per day. By weight, about a third of that material was recycled, but there is plenty of room for improve-

FIGURE 18.16 A Recently deceased Laysan albatross chick, dissected to reveal the contents of its gut. **B** Scientists found more than 300 pieces of plastic inside the bird. One of the pieces had punctured its gut wall, resulting in the chick's death. The chick was fed this junk by its parents, who gathered floating plastic from the ocean surface, mistaking it for food.

ment. Two-thirds of plastic soft drink bottles and three-quarters of glass bottles are not recycled. Foam packaging from fast-food outlets, plastic shopping bags, and other material discarded as litter ends up in storm drains. From there it enters streams and rivers that can carry it to the sea. Seawater sampled near the mouth of the San Gabriel River in southern California had 128 times more plastic than plankton by weight. Once in the ocean, trash persists for a surprisingly long time. Components of a disposable diaper will last for more than 100 years, as will fishing line. A plastic bag will be around for more than 50 years, and a cigarette filter for more than 10.

Air Quality Just as sediments can clog the respiratory organs of aquatic organisms, airborne particles can harm the respiratory organs of air-breathers. Particles most dangerous to human health are smaller in diameter than a human hair. These particles are most likely to be inhaled all the way into the lungs.

Burning nearly anything produces particle pollution. Forest fires can cause a temporary rise in the amount of particles in the air. However, burning of fossil fuel and industrial processes sends a constant stream of particles skyward.

Release and burning of fossil fuels also causes an increase in the concentration of ozone (O_3) in the air near the ground. Ozone irritates the eyes and increases the effects of allergens on sensitive people. It also slows plant growth.

Ozone forms when nitrogen oxides and volatile organic compounds released by the burning or evaporation of fossil fuels are exposed to sunlight. Warm temperatures speed the reaction. Thus, ground-level ozone tends to vary daily (being higher during the daytime) and seasonally (being higher during the summer). You can help reduce ozone pollution by avoiding actions that put fossil fuels or their combustion products into the air at times that favor ozone production. On hot, sunny, still days, postpone filling your gas tank or using gasoline-powered appliances until the evening, when there is less sunlight to power the conversion of pollutants to ozone (Figure 18.17).

bioaccumulation A substance becomes more concentrated in organisms at increasingly higher levels of a food chain.

eutrophication Increase in nutrients in an aquatic ecosystem; may lead to oxygen depletion.

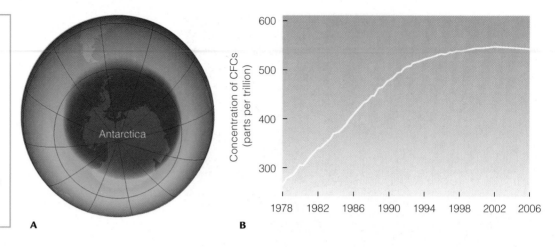

FIGURE 18.18 Animated!
Ozone and CFCs. **A** Ozone levels in September 2007, the Antarctic spring. *Purple* indicates the least ozone, with *blue*, *green*, and *yellow* indicating increasingly higher levels. Check current status of the ozone hole at NASA's web site: http://ozonewatch.gsfc.nasa.gov/.

B Concentration of CFCs high in the atmosphere. These pollutants encourage ozone destruction. A worldwide ban on CFCs successfully halted the rise.

A

B

The Ozone Hole Ozone is sometimes said to be "bad nearby, but good up high." You just learned why ozone is bad near the ground. The "good" ozone is high in Earth's atmosphere. Between 17 and 27 kilometers above sea level (10.5 and 17 miles), the ozone concentration is so great that scientists refer to this region as the **ozone layer**. The ozone layer benefits us by absorbing most of the ultraviolet (UV) radiation in incoming sunlight. UV radiation, remember, is a mutagen. It can damage DNA and cause mutations.

In the mid-1970s, scientists noticed that the ozone layer was thinning. Its thickness had always varied a bit with the season, but now there was steady decline from year to year. By the mid-1980s, the spring ozone thinning over Antarctica was so pronounced that people were referring to the low-ozone region as an "ozone hole" (Figure 18.18**A**).

Declining ozone quickly became an international concern. The ozone hole over the South Pole was a sign that the ozone layer was thinning worldwide. With a thinner ozone layer, people would be exposed to more UV radiation, the main cause of skin cancers. Higher UV levels also harm wildlife, which do not have the option of avoiding sunlight. In addition, exposure to higher-than-normal UV levels affects plants and other producers, slowing the rate of photosynthesis and the release of oxygen into the atmosphere.

Chlorofluorocarbons, or CFCs, are the main ozone destroyers. These odorless gases were once widely used as propellants in aerosol cans, as coolants, and in solvents and plastic foam. CFCs interact with ice crystals and UV light in the atmosphere. These reactions release chlorine radicals that degrade ozone. A single chlorine radical can break apart thousands of ozone molecules. Ozone thins the most at the poles because swirling winds concentrate CFCs in this region during dark, cold polar winters. In the spring, increasing daylight and the presence of ice clouds allow a surge in the formation of chlorine radicals from the highly concentrated CFCs.

In response to the potential threat posed by ozone thinning, countries worldwide agreed in 1987 to phase out the production of CFCs and other ozone-destroying chemicals. As a result of that agreement (the Montreal Protocol), the concentrations of CFCs in the atmosphere are no longer rising dramatically (Figure 18.18**B**). However, CFCs break down quite slowly, so scientists expect them to remain at a level that significantly impairs the ozone layer for several decades.

A Muir Glacier 1941

B Muir Glacier 2004

C Arctic perennial sea ice 1979

D Arctic perennial sea ice 2003

FIGURE 18.19 Dramatic evidence of a warming world. **A** Muir Glacier in south-eastern Alaska in 1941. **B** By 2004, the ice in the foreground of the earlier photo had melted. **C,D** The extent of perennial sea ice in the Arctic in 1979 and in 2003. Decreases in the amount of glacial ice and sea ice are expected to accelerate the warming trend. Ice reflects sunlight, whereas the water and land exposed by melting ice absorbs light and then radiates the absorbed energy as heat.

Global Climate Change Like ozone depletion, climate change affects ecosystems worldwide. How is the climate changing? Most notably, average temperatures are getting hotter. The effects of the temperature rise are illustrated by recent changes in the extent of glaciers and sea ice (Figure 18.19). However, global warming is just one aspect of **global climate change**. As you learned earlier in this chapter, the temperature of the land and seas affects evaporation, winds, and currents. As a result, many weather patterns are expected to change as temperature rises. For example, warmer temperatures are correlated with extremes in rainfall patterns, with periods of drought interrupted by unusually heavy rains. Also, warmer seas tend to increase the intensity of hurricanes.

global climate change Global warming and other changes in the current climate and weather patterns.

ozone layer Atmospheric layer with a high concentration of ozone that prevents much ultraviolet radiation from reaching Earth's surface.

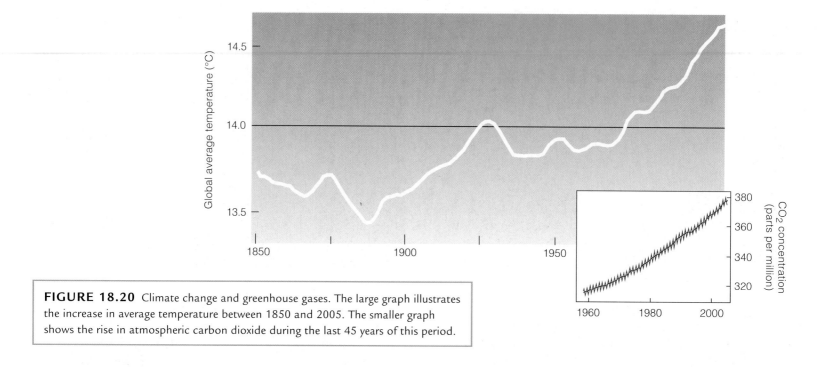

FIGURE 18.20 Climate change and greenhouse gases. The large graph illustrates the increase in average temperature between 1850 and 2005. The smaller graph shows the rise in atmospheric carbon dioxide during the last 45 years of this period.

Causes of Climate Change

Earth's climate has varied greatly over its long history. There have been ice ages, when much of the planet was covered by glaciers, and unusually hot periods, when tropical plants and coral reefs thrived at what are now cool latitudes. Scientists can correlate past large-scale temperature changes with shifts in Earth's orbit, which varies in a regular fashion over 100,000 years, and Earth's tilt, which varies over 40,000 years. Changes in solar output and volcanic eruptions also affect Earth's temperature. However, most scientists do not see evidence that these natural factors have a major role in the rise in temperature that we are currently experiencing.

In 2007, the Intergovernmental Panel on Climate Change reviewed the results of many scientific studies related to climate change. The panel, which includes hundreds of scientists from all over the world, concluded that the warming in recent decades is very likely due to a human-induced increase in atmospheric greenhouse gases.

Unlike other possible factors, the rise in greenhouse gases does correlate with the rise in temperature. Figure 18.20 shows the average global temperature between 1850 and 2005, and the rise in atmospheric carbon dioxide during the latter part of this interval. As Section 17.6 explained, greenhouse gases such as carbon dioxide act like insulation. They keep heat near Earth's surface from escaping into space. Burning of fossil fuels is the main source of the excess carbon dioxide entering the atmosphere (Figure 18.21).

Effects of Climate Change

As temperatures rise, so does the sea level. Meltwater from glaciers and sea ice runs into the ocean, increasing its volume. In addition, water expands as it warms. A rising sea level puts coastal communities at risk for increased erosion and flooding. It also allows salt water to seep into aquifers that supply drinking water to coastal regions. Global sea level rose 10 to 20 centimeters (4 to 8 inches) during the twentieth century. Scientists predict that if the current warming trend continues, sea level could rise as much as 0.6 meter (2 feet) over the next century. That may not sound like much of a change, but in the United States it would put about 13,000 square kilometers (5,000 square miles) of land underwater at high tide.

The warming climate is already having effects on biological systems. Temperature changes are important cues for many temperate-zone species, and warming has been more pronounced at temperate and polar latitudes than at the equator. Warmer-than-normal springs are causing deciduous trees to put leaves out earlier, and spring-blooming flowers to flower earlier. Animal migration times and breeding seasons are also shifting. Species arrays in biological communities are changing as warmer temperatures allow some species to move to higher latitudes or elevations. Of course, not all species can move, and warmer temperatures will certainly drive some of them to extinction. For example, warming of tropical waters is already stressing reef-building corals and increasing the frequency of coral bleaching events.

Reducing levels of greenhouse gases will be a challenge. Fossil fuels are in more widespread use than CFCs ever were. Still, efforts are under way to increase the efficiency of processes that require fossil fuels, shift to alternative energy sources such as solar and wind power, and develop innovative ways to store excess carbon dioxide.

FIGURE 18.21 Fueling global climate change. Burning of fossil fuels is the greatest source of excess carbon dioxide by far.

Take-Home Message **How do human activities affect the biosphere?**

- Humans are increasing the rate of species extinctions by degrading, destroying, and fragmenting natural habitats, overharvesting species, and introducing exotic species.

- Human activities turn grasslands into deserts, strip woodlands of trees, and generate pollutants and trash that kill animals and damage ecosystems.

- Some pollutants have global effects, as when CFCs cause thinning of the ozone layer, and rising levels of greenhouse gases bring about climate change.

18.6 Maintaining Biodiversity

The Value of Biodiversity Every nation enjoys several forms of wealth: material wealth, cultural wealth, and biological wealth, or biodiversity. We measure a region's **biodiversity** at three levels: the genetic diversity within its species, its species diversity, and its ecosystem diversity.

Why should we protect biodiversity? From a purely selfish standpoint, doing so is an investment in our future. Healthy ecosystems are essential to the survival of our own species. Other organisms produce the oxygen we breathe and the food we eat. They also remove waste carbon dioxide from the air and decompose and detoxify other wastes. Plants take up rain and hold soil in place, preventing erosion and reducing the risk of flooding.

Compounds discovered in wild species often serve as medicines. Two widely used chemotherapy drugs, vincristine and vinblastine, were extracted from the rosy periwinkle, a low-growing plant native to Madagascar's rain forests. Cone snails that live in tropical seas are the source of a highly potent pain reliever.

Wild relatives of crop plants are reservoirs of genetic diversity that plant breeders can draw on to protect and improve crops. Wild plants often have genes that make them more resistant to disease or adverse conditions than their domesticated relatives. Plant breeders can use traditional cross-breeding methods or biotechnology to introduce genes from wild species into domesticated ones, thus creating improved varieties.

biodiversity Of a region, the genetic diversity within its species, variety of species, and variety of ecosystems.

The extinction of a species removes its unique collection of traits from the world of life forever.

A decline in biodiversity warns us that the natural support system we depend on is in trouble. **Indicator species**, species that are particularly sensitive to environmental change, can be monitored as indicators of environmental health. For example, a decline in lichens tells us that air quality is deteriorating. The loss of mayflies from a stream tells us that the quality of water in a stream is declining.

In addition, there are ethical reasons to preserve biodiversity. As we have emphasized many times, all living species are the result of an ongoing evolutionary process that stretches back billions of years. Each species has a unique combination of traits. The extinction of a species removes its unique collection of traits from the world of life forever.

Conservation Biology Biodiversity is currently in decline at all three levels and in all regions. **Conservation biology** addresses these declines. The goals of this relatively new field of biology are to survey the range of biodiversity and find ways to maintain and use biodiversity in a manner that benefits human populations. The aim is to conserve as much biodiversity as possible by encouraging people to value it and use it in ways that do not destroy it.

With so many species and ecosystems at risk, conservation biologists are often forced to make difficult choices about which should be targeted for protection first. The biologists are working to identify the so-called **hot spots**, places that are richest in endemic species and under the greatest threat. Once identified, hot spots can take priority in worldwide conservation efforts. The goal of prioritizing is to save representative examples of all of Earth's existing biomes. By focusing on hot spots, rather than on individual species, scientists hope to maintain ecosystem processes that naturally sustain biological diversity.

Conservation scientists of the World Wildlife Fund have defined 867 distinctive land ecoregions that are considered the top priority for conservation efforts. Each has a large number of endemic species and is under threat. The Klamath-Siskiyou forest in southwestern Oregon and northwestern California is one hot spot. This forest is home to many rare conifers. Two endangered birds, the northern spotted owl and the marbled murrelet, nest in old-growth portions of the forest. Endangered coho salmon breed in streams that run through it. Logging is the main threat to this region, but a newly introduced pathogen that affects conifers is another concern.

FIGURE 18.22 Protecting a biodiversity hotspot. **A** Costa Rica's Monteverde Cloud Forest Reserve is a protected area that is home to many endangered species, including jaguars **B**.

Protecting biological diversity is often a tricky proposition. Even in developed countries, people often oppose environmental protections because they fear such measures will have adverse economic consequences. However, taking care of the environment can make good economic sense. With a bit of planning, people can both preserve and profit from their biological wealth.

Worldwide, many ecologically important regions are now protected and are providing benefits to local people. For example, during the 1970s, George Powell was studying birds in the Monteverde Cloud Forest in Costa Rica. This forest was rapidly being cleared and Powell got the idea to buy part of it as a nature sanctuary. His efforts inspired individuals and conservation groups to donate funds, and much of the forest is now protected as a private nature reserve (Figure 18.22). The reserve's plants and animals include more than 100 mammal species, 400 bird species, and 120 species of amphibians and reptiles. The reserve is one of the few habitats left for the jaguar, ocelot, puma, and their relatives. More than 50,000 tourists now visit the Monteverde Cloud Forest Reserve each year. A tourism industry centered on this reserve provides much-needed employment to local people.

Ecological Restoration Sometimes, an ecosystem is already damaged or there is so little of it left that conservation alone is not enough. **Ecological restoration** is work designed to bring about the renewal of a natural ecosystem that has been degraded or destroyed, fully or in part.

For example, more than 40 percent of the coastal wetlands in the United States are in Louisiana. These marshes, known locally as bayous, are an ecological and economic treasure. Millions of migratory birds overwinter in Louisiana's bayous and billions of dollars' worth of fish, shrimp, and shellfish are harvested from them. Despite their importance, the marshes are in trouble. Dams and levees built upstream of the marshes keep back sediments that would normally replenish sediments lost to the sea. Channels cut through the marshes for oil exploration and production have encouraged erosion, and the rising sea level threatens to flood the existing plants. Since the 1940s, Louisiana has lost an area of marshland the size of Rhode Island. Restoration efforts now under way aim to reverse some of those losses (Figure 18.23).

conservation biology Field of applied biology that surveys biodiversity and seeks ways to maintain and use it.

ecological restoration Actively altering an area in an effort to restore or create a functional ecosystem.

hot spots Threatened regions with great biodiversity that are considered a high priority for conservation efforts.

indicator species A species that is particularly sensitive to environmental changes and can be monitored to assess whether an ecosystem is threatened.

FIGURE 18.24 Resource extraction. Bingham copper mine near Salt Lake City, Utah. The mine is 4 kilometers (2.5 miles) wide and 1,200 meters (0.75 miles) deep.

Living Sustainably Ultimately, the health of our planet depends on our ability to recognize that the principles of energy flow and of resource limitation, which govern the survival of all systems of life, do not change. We must take note of these principles and find a way to live within our limits. The goal is **sustainable development**, meeting the needs of the present generation without reducing the ability of future generations to meet their own needs.

Promoting sustainability begins with recognizing the environmental consequences of our own life-style. People in industrial nations use enormous quantities of resources, and the extraction and delivery of these resources has negative effects on biodiversity. In the United States, the size of the average family has declined since the 1950s, while the size of the average home has doubled. All the materials used to build and furnish those larger homes come from the environment. For example, an average new home contains about 500 pounds of copper in its wiring and plumbing. Where does copper come from? Like most other mineral elements used in manufacturing, most of it is mined from the ground (Figure 18.24). Most mines generate pollution and are ecological dead zones.

Nonrenewable mineral resources are also used in electronic devices such as phones, computers, televisions, and MP3 players. Constantly trading up to the newest device may be good for the ego and the economy, but it is bad for the environment. Reducing consumption by fixing existing products promotes sustainability, as does recycling. Obtaining nonrenewable materials by recycling reduces the need for extraction of those resources from the environment and helps keep pollutants out of ecosystems.

Reducing your energy use is another way to promote sustainability. Fossil fuels such as petroleum, natural gas, and coal supply most of the energy used by developed countries. You already know that use of these nonrenewable fuels contributes to global warming and acid rain. In addition, extraction and transportation of these fuels have negative impacts. Oil harms many species when it leaks from pipelines or from ships. Renewable energy sources do not produce

sustainable development Using resources to meet current human needs in a way that takes into account the needs of future generations.

greenhouse gases, but they have their own drawbacks. For example, dams in rivers of the Pacific Northwest generate renewable hydroelectric power, but they also prevent endangered salmon from returning to streams above the dam to breed. Similarly, wind turbines can harm birds and bats. Panels used to collect solar energy are made using nonrenewable mineral resources, and manufacturing the panels generates pollutants.

In short, all commercially produced energy has some kind of negative environmental impact, so the best way to minimize that impact is to use less energy. Shop for energy-efficient appliances, use fluorescent bulbs instead of incandescent ones, and do not leave lights on in empty rooms. Walking, bicycling, and using public transportation are energy-efficient alternatives to driving. Shopping locally and purchasing locally produced goods also saves energy.

If you want to make a difference, learn about the threats to ecosystems in your own area. Are species threatened? If so, what are the threats? How can you help reduce those threats? Support efforts to preserve and restore local biodiversity. Many ecological restoration projects are supervised by trained biologists but carried out primarily through the efforts of volunteers (Figure 18.25).

Keep in mind that unthinking actions of billions of individuals are the greatest threat to biodiversity. Each of us may have little impact on our own, but our collective behavior, for good or for bad, will determine the future of the planet.

FIGURE 18.25 Volunteers restoring the Little Salmon River in Idaho so that salmon can migrate upstream to their breeding grounds.

Take-Home Message **What is biodiversity and how can we sustain it?**

- Biodiversity is the genetic diversity of individuals of a species, the variety of species, and the variety of ecosystems. Worldwide, biodiversity is declining at all of these levels.

- Conservation biologists are working to identify threatened regions with high biodiversity and prioritize which receive protection.

- Ecological restoration is the process of re-creating or renewing a diverse natural ecosystem that has been destroyed or degraded.

- Individuals can help maintain biodiversity by using resources in a sustainable fashion.

18.7

Impacts/Issues Revisited:
A Long Reach

The Arctic is not a separate continent like Antarctica, but rather a region that encompasses the northernmost parts of several continents. Eight countries, including the United States, Canada, and Russia, control parts of the Arctic and have rights to its extensive oil, gas, and mineral deposits. Until recently, ice sheets that covered the Arctic Ocean made it difficult for ships to move to and from the Arctic land mass, but those sheets are breaking up as a result of global climate change. At the same time, ice that covered the Arctic land mass is melting. These changes will make it easier for people to remove minerals and fossil fuels from the Arctic. With the world supply of fuel and minerals dwindling, pressure to exploit Arctic resources is rising. However, conservationists warn that extracting these resources will harm Arctic species such as the polar bear that are already threatened by global climate change.

How Would

YOU
☑ Vote ?

The Arctic has extensive deposits of minerals and fossil fuel, but tapping into these resources might pose a risk to species already threatened by global climate change. Should the United States exploit its share of the Arctic resources? **See CengageNow for details, then vote online (cengagenow.com).**

Summary

Section 18.1 Humans and the chemical pollutants that their activities produce have reached even remote parts of the biosphere.

Section 18.2 Climate refers to average weather conditions over time. Variations in climate depend largely upon differences in the amount of solar radiation reaching different parts of the Earth. The closer a region is to the equator, the more solar energy it receives. Warming of air at the equator is the start of global patterns of air circulation and ocean currents.

CENGAGENOW™ Learn how sunlight energy drives patterns of air circulation and observe major ocean currents with the animation on CengageNow.

Section 18.3 Biomes are categories of major ecosystems on land. They are maintained largely by regional variations in climate and are described mainly in terms of their dominant plant life.

Deserts occur at latitudes about 30° north and south, where dry air descends and annual rainfall is sparse. Grasslands form at midlatitudes in the interior of continents. Fire helps maintain grasslands. Chaparral, a biome dominated by shrubby plants, takes hold in places with cool, wet winters and hot, dry summers. Like grasses, chaparral plants are adapted to periodic fires.

Near the equator, high rainfall and mild temperatures support tropical rain forests dominated by trees that remain leafy and active all year. Deciduous broadleaf forests grow in temperate regions with warm summers and cold winters. The trees lose their leaves and become dormant during the winter.

Conifers tend to be more resistant to dryness and cold than broadleaf trees. Taiga, a biome dominated by conifers, extends across high latitudes where a cold, dry season alternates with a cool, rainy season. Coniferous forests also occur in other regions.

Tundra forms at high latitudes and high altitudes. Arctic tundra is the youngest biome and lies over permafrost.

CENGAGENOW™ Explore the distribution of Earth's biomes with the interaction on CengageNow.

Section 18.4 Aquatic ecosystems have gradients of sunlight penetration, water temperature, salinity, and dissolved gases. Lakes are standing bodies of water. Different communities of organisms live at different depths and distances from the shore.

Streams and rivers are flowing bodies of water. Physical characteristics of a stream or river that vary along its length influence the types of organisms that live in it. A semi-enclosed area where nutrient-rich water from a river mixes with seawater is an estuary, a highly productive habitat. Seashores may be rocky or sandy. Grazing food chains based on algae form on rocky shores. Detrital food chains dominate sandy shores.

Coral reefs are species-rich ecosystems that form in warm, well-lit tropical waters. The main producers on reefs are photosynthetic protists that live inside the coral's tissues.

The open ocean's upper waters hold photosynthetic organisms that form the basis for grazing food chains. Deeper water communities usually subsist on material that drifts down from above. However, prokaryotes that can strip energy from minerals serve as the producers at hydrothermal vent ecosystems.

CENGAGENOW™ Use the animation on CengageNow to learn about the oceanic zones.

Section 18.5 Humans are increasing the rate of species extinctions by overharvesting species, degrading and destroying habitat, and introducing exotic species.

Human activities can also threaten entire ecosystems, as when poor agricultural practices turn grasslands or woodlands into desert. Deforestation and pollutants such as those that cause acid rain threaten forest ecosystems. Acid rain also threatens freshwater ecosystems. Trash that gets into fresh water can make its way into the oceans, where it degrades marine ecosystems.

Ozone is a pollutant near the ground, but depletion of the ozone layer is a global threat caused by use of chemicals called CFCs. Global agreement to phase out CFC use lessens this threat. Climate change caused by rising concentrations of greenhouse gases is another global threat.

CENGAGENOW™ Learn how acid rain affects forests and how CFCs destroy ozone with the animation on CengageNow.

Section 18.6 Biodiversity includes the diversity of genes, species, and ecosystems. All three levels of biodiversity are threatened. A decrease in biodiversity can harm humans. We rely on ecosystems to produce oxygen and decompose waste. We also benefit from many compounds produced by wild species, and by tapping their genetic diversity to enhance our crops.

Conservation biologists document the extent of biodiversity and look for ways to preserve it, while benefiting humans. They give priority to hot spots, areas with high biodiversity that are most threatened.

Ecological restoration is the work of actively renewing an ecosystem that has been damaged or destroyed.

Ordinary individuals can help maintain biodiversity and ensure that resources remain for future generations by cutting their consumption, reusing and recycling materials, and reducing energy use.

Self-Quiz   Answers in Appendix I

1. The most solar radiation reaches the ground at _____ .
   a. the equator   c. midlatitudes
   b. the North Pole   d. the South Pole

2. When air is heated, it _____ and can hold _____ water.
   a. sinks, less   c. rises, less
   b. sinks, more   d. rises, more

3. In what direction do ocean currents circulate water in the Northern Hemisphere?
   a. clockwise   b. counterclockwise

4. Air circulation patterns cause dry air to descend and deserts to form at _____ degrees north and south latitude.

5. Plants in _____ are adapted to periodic fires.
   a. deserts
   b. taiga
   c. arctic tundra
   d. chaparral

6. Permafrost underlies _____ .
   a. grasslands
   b. arctic tundra
   c. temperate deciduous forests
   d. coniferous forests

7. The taiga is a broad northern biome dominated by _____ .
   a. conifers
   b. mosses and lichens
   c. grasses
   d. deciduous broadleaf trees

8. Prokaryotes that can obtain energy from minerals are the main producers at _____ .
   a. hydrothermal vents
   b. estuaries
   c. coral reefs
   d. sandy shores

9. Which would hold more oxygen?
   a. a fast-moving, cool stream   b. a warm pond

10. Match the terms with their most suitable description.
    _____ tundra
    _____ chaparral
    _____ desert
    _____ prairie
    _____ estuary
    _____ taiga
    _____ tropical rain forest

    a. freshwater and seawater mix
    b. low humidity, and little rainfall
    c. North American grassland
    d. fire-adapted shrubs
    e. low-growing plants at high latitudes or high elevations
    f. northern coniferous forest
    g. broadleaf trees are active all year

11. An _____ species has population levels so low it is at great risk of extinction in the near future.
    a. endemic
    b. endangered
    c. indicator
    d. exotic

12. An _____ species can be monitored to gauge the health of its environment.
    a. endemic
    b. endangered
    c. indicator
    d. exotic

13. The 1930s environmental disaster known as the Dust Bowl is an example of _____ .
    a. deforestation
    b. desertification
    c. ecological restoration
    d. species extinction

14. The main threat to the ozone layer is _____ .
    a. use of CFCs
    b. acid rain
    c. burning of fossil fuels
    d. deforestation

15. Acid rain _____ .
    a. harms aquatic ecosystems
    b. kills trees
    c. is a type of pollution
    d. all of the above

Additional questions are available on CENGAGENOW

Critical Thinking

1. Conifers have evergreen leaves and maintain a higher rate of photosynthesis than broadleaf trees at low temperatures. Conifers also are less likely to get flow-stopping bubbles in their tissues during spring thaws. How do these traits help them survive in taiga?

Digging Into Data

Arctic PCB Pollution

Winds carry chemical contaminants produced and released at temperate latitudes to the Arctic, where the chemicals enter food webs. By the process of bioaccumulation (Section 17.5), top carnivores in arctic food webs, including polar bears and people, end up with high concentrations of these chemicals in their body. Arctic people who eat a lot of local wildlife tend to have unusually high levels of polychlorinated biphenyls, or PCBs. The Arctic Monitoring and Assessment Programme studies the effects of these industrial chemicals on the health and reproduction of Arctic people. Figure 18.26 shows the average PCB levels for women in native populations in the Russian Arctic and the sex ratios at birth for each PCB level.

1. Which sex was more common in offspring of women with less than 1 microgram per milliliter of PCB in serum?

2. At what PCB concentrations were women more likely to have daughters?

3. In some villages in Greenland, nearly all recent newborns are female. Would you expect PCB levels in those villages to be above or under 4 micrograms per milliliter?

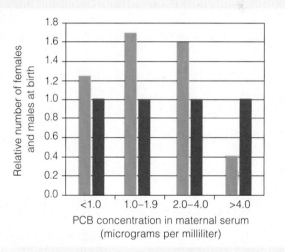

FIGURE 18.26 Effect of maternal PCB concentration on sex ratio of newborns in indigenous populations in the Russian Arctic. *Blue* bars indicate the relative number of males born per one female (*pink* bars).

2. In one seaside community in New Jersey, the U.S. Fish and Wildlife Service suggested trapping and removing feral cats (domestic cats that live in the wild). The goal was to protect some endangered wild birds (plovers) that nested on the town's beaches. Many residents were angered by the proposal, arguing that the cats have as much right to be there as the birds. Do you agree? Why or why not?

3. Burning fossil fuel puts excess carbon dioxide into the atmosphere, but deforestation and desertification also affect carbon dioxide concentration. Explain why a global decrease in the amount of vegetation is contributing to the rise in carbon dioxide.

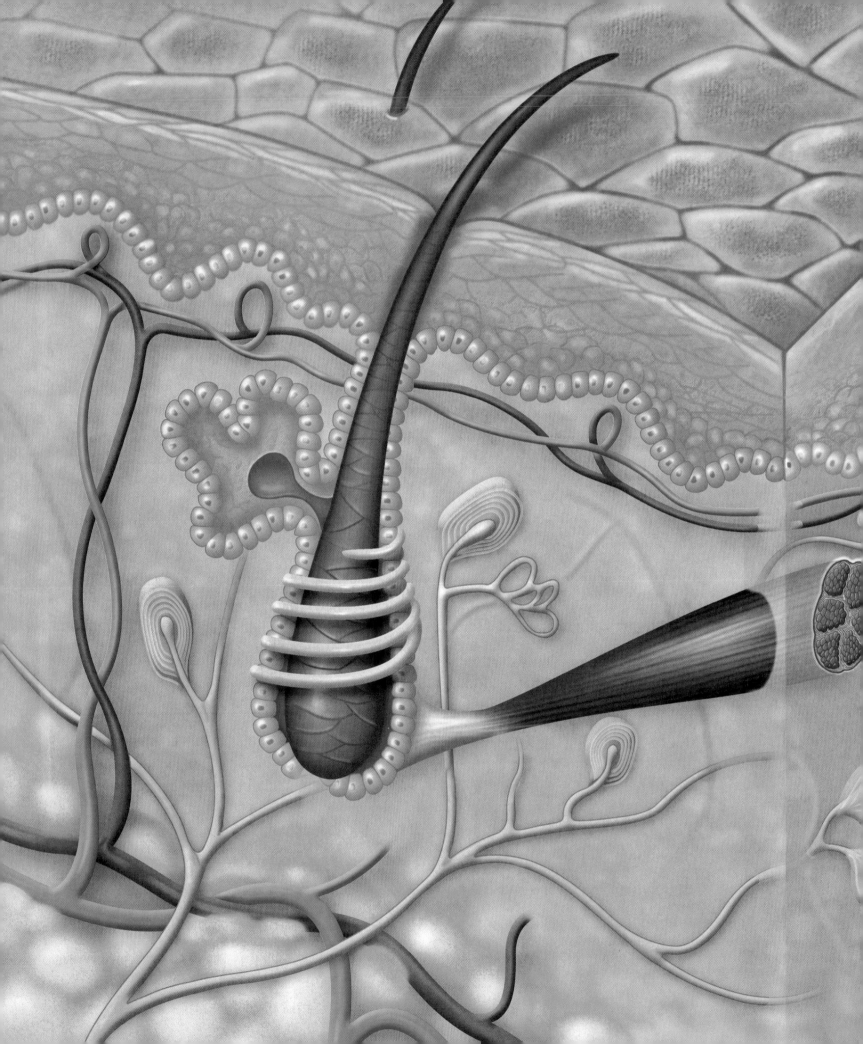

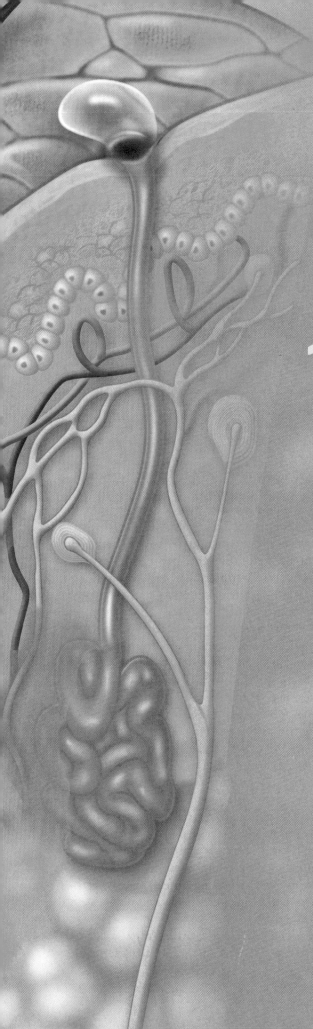

19 Animal Tissues and Organs

19.1 Impacts/Issues: It's All About Potential

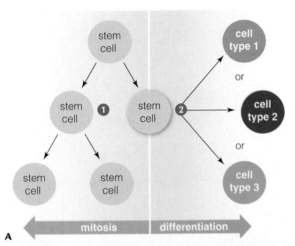

Imagine being able to grow new body parts to replace lost or diseased ones. This dream motivates researchers who study stem cells. As Figure 19.1**A** illustrates, **stem cells** are self-renewing cells that can divide and produce more stem cells ❶. In addition, stem cells can differentiate into specialized cell types that make up specific body parts ❷. In short, all cells in your body "stem" from stem cells.

Cell types that your body continually replaces, such as blood and skin, arise from adult stem cells. Adult stem cells are specialists that normally differentiate into a limited variety of cells. For example, stem cells in adult bone marrow can become blood cells, but not muscle cells or brain cells. Stem cells that can become nerve cells or muscle cells are rare in adults. Thus, unlike skin and blood cells, nerves and muscles are not replaced if they get damaged or die. This is why an injury to the nerves of the spinal cord can cause permanent paralysis. New nerves do not grow to replace the damaged ones.

Early embryos have stem cells that are more versatile. These cells are the source of all tissues in the new body. Embryonic stem cells form soon after fertilization, when cell division produces a pinhead-sized ball of cells. By birth, the stem cells have become somewhat differentiated and are less versatile.

In theory, embryonic stem cell treatments could provide new nerve cells for paralyzed people. They might also help treat other nerve and muscle disorders such as heart disease, muscular dystrophy, and Parkinson's disease. Despite the promise of embryonic stem cell research, some people oppose it. They are troubled by the original source of the cells: early human embryos. Some stem cells are from embryos that were produced in fertility clinics and never used. Others are from embryos created in the lab for this purpose.

So far, scientists have not found any adult stem cells that have the same potential as embryonic stem cells. However, they may be able to genetically engineer such cells. For example, James Thomson and Junying Yu (Figure 19.1**B**) used viruses to insert genes from embryonic cells into skin cells from a newborn. The genes were transcribed into proteins that caused the specialized cells to "dedifferentiate," to turn back the clock and behave like embryonic stem cells. Researchers in Japan developed a similar technique.

Such breakthroughs suggest that using embryonic stem cells may one day become unnecessary. However, there are still major obstacles. First, the viruses used to produce genetically engineered stem cells can cause cancer, so cells created by this method cannot safely be placed in a human body. Second, while the engineered cells seem to behave like embryonic stem cells in the lab, they might behave differently once implanted in a person. Further research will be necessary to see whether stem cells can be engineered in a safer way, and if they actually have the same potential as embryonic stem cells in a human body.

Stem cells, the source of all tissues and organs, are a fitting introduction to this unit, which deals with animal anatomy (how a body is put together) and physiology (how a body works).

FIGURE 19.1 Stem cells. **A** Stem cells can divide to form new stem cells or differentiate to form specialized cell types. **B** Junying Yu helped develop a method of genetically engineering skin cells so they behave like embryonic stem cells.

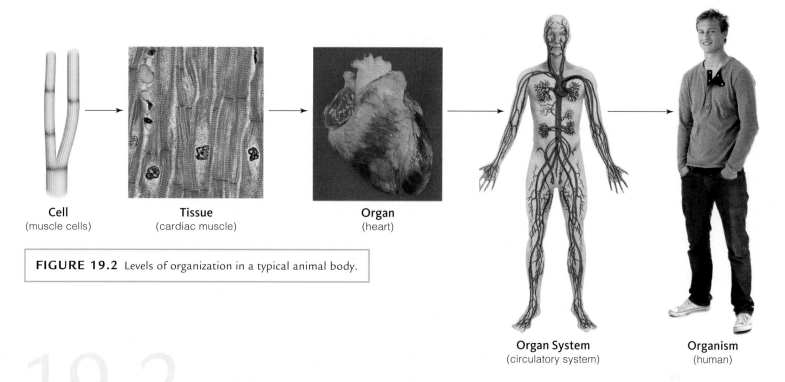

Cell
(muscle cells)

Tissue
(cardiac muscle)

Organ
(heart)

Organ System
(circulatory system)

Organism
(human)

FIGURE 19.2 Levels of organization in a typical animal body.

Animal Structure and Function

Organization and Integration All animals are multicelled, and nearly all have cells organized as tissues. A **tissue** consists of one or more cell types—and often an extracellular matrix—that collectively perform a specific task or tasks. Four types of tissue occur in all vertebrate bodies:

1. Epithelial tissues cover body surfaces and line internal cavities.
2. Connective tissues hold body parts together and provide structural support.
3. Muscle tissues move the body and its parts.
4. Nervous tissues detect stimuli and relay information.

Each tissue is characterized by the types of cells it includes and their proportions. For example, nervous tissue includes neurons, a type of cell not found in muscle tissue or epithelial tissue.

Typically, animal tissues are organized into organs. An **organ** is a structural unit of two or more tissues organized in a specific way and capable of carrying out specific tasks. For example, a human heart is an organ (Figure 19.2), and it includes all four tissue types. The wall of the heart is made up mostly of cardiac muscle tissue. A sheath of connective tissue encloses the heart, and the heart's internal chambers are lined with epithelial tissue. Nervous tissue delivers signals to and from the heart.

In **organ systems**, two or more organs and other components interact physically, chemically, or both in a common task. For example, the force generated by a beating heart moves blood through a system of blood vessels that extends throughout the body.

As an animal goes about its activities, it gains and loses solutes and water, and is exposed to temperature variations. Each living cell engages in metabolic activities that keep it alive. At the same time, organs and organ systems interact to keep solute concentrations and temperature of the internal environment within the range that cells can tolerate. Such interactions keep conditions in the internal environment within tolerable limits, a process we call **homeostasis**.

homeostasis The process of maintaining favorable conditions inside the body.

organ Structural unit that is composed of two or more tissues and adapted to carry out a particular task.

organ system Organs that interact closely in some task.

stem cell A cell that can divide and create more stem cells or differentiate to become a specialized cell type.

tissue A collection of one or more specific cell types that are organized in a way that suits them to a task.

Because evolution modifies existing structures, it often does not produce an entirely optimal body plan.

Evolution of Structure and Function Anatomical and physiological traits have a genetic basis and vary among individuals. In each generation, traits that best help individuals survive and reproduce in their environment are preferentially passed on. In this way, over countless generations, structural traits become optimized in ways that reflect their function in a specific environment. Thus, the structure of modern animals has evolved by natural selection.

For example, animals evolved in water and faced new challenges when they moved onto land (Section 15.5). Gases can only move into and out of an animal's body by moving across a moist surface. That is not a problem for aquatic organisms, but evaporation causes moist surfaces to dry out on land. The evolution of a new structural feature—lungs—allowed land animals to exchange gases with air across a moist surface deep inside their body.

Lungs are not modified fish gills. Rather, lungs evolved from outpouchings of the gut in fishes ancestral to land vertebrates. As this example illustrates, evolution typically does not produce entirely new tissues or organs. Instead, it modifies the structure and function of existing ones.

As anyone who has remodeled a home knows, modifying an existing structure requires compromises. Similarly, we detect evidence of evolutionary compromises in many animal structural traits. For example, as a legacy of the lungs' ancient connection to the gut, the human throat opens to both the digestive tract and respiratory tract. As a result, food sometimes goes where air should and a person chokes. It would be safer if food and air entered the body through separate passageways. However, because evolution modifies existing structures, it often does not produce an entirely optimal body plan.

Take-Home Message How does the structure of an animal body relate to its function?

- All animals are multicelled and most have cells organized as tissues, organs, and organ systems.

- All structural levels interact in homeostasis, the processes that keep conditions in the internal environment within the limits that cells tolerate.

- Structural traits evolve by natural selection. Existing structures are modified over generations in ways that better adapt their bearers to their environment.

- New structures evolve by modification of existing ones. This evolutionary remodeling often results in body plans that are less than optimal.

19.3

Types of Animal Tissues

Epithelial Tissues **Epithelial tissue**, or epithelium (plural, epithelia), is a sheetlike tissue with one free surface. The other surface typically attaches to connective tissue. Material secreted by epithelial cells forms a noncellular basement membrane that glues the epithelium to the underlying tissue (Figure 19.3). Cells in a sheet of epithelium have no extracellular matrix between them. They attach to one another by tight junctions (Section 3.7). Blood vessels do not run through epithelium, so nutrients reach cells by diffusing from vessels that run through the adjacent connective tissue.

FIGURE 19.3 General structure of an epithelium.

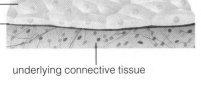

free surface of a simple epithelium

basement membrane (material secreted by epithelial cells)

underlying connective tissue

Most of what you see when you look in a mirror—your skin, hair, and nails—is epithelial tissue or structures derived from it. Epithelium even covers the outer surface of your eyeballs. On the inside, epithelium lines the body's many tubes and cavities, including blood vessels, airways, and the digestive, urinary, and reproductive tracts.

We describe an epithelial tissue in terms of the shape of the cells and number of cell layers. A simple epithelium is one cell thick. A stratified epithelium has multiple cell layers. Cells in squamous epithelium are flattened or platelike. Cells of cuboidal epithelium are short cylinders that look like cubes when viewed in cross-section. Cells in columnar epithelium are taller than they are wide. Figure 19.4 shows the three types of simple epithelium.

Different kinds of epithelia are adapted to different tasks. The thinnest type of epithelium, simple squamous epithelium, lines blood vessels and the inner surface of the lungs. It functions in exchange of materials. Gases and nutrients diffuse across it easily. In contrast, thicker stratified squamous epithelium has a protective function. This tissue makes up the outer layer of human skin.

Cells of cuboidal and columnar epithelium act in absorption and secretion. In some tissues, such as the lining of the kidneys and small intestine, finger-like projections called microvilli extend from the free surface of epithelial cells. These projections increase the surface area across which substances are absorbed. In other tissues, such as those of the upper airways and oviducts, the free epithelial surface is ciliated. The action of cilia in the airways moves mucus with inhaled particles away from the lungs. The action of cilia in the oviducts moves an egg toward the uterus (the womb).

Only epithelial tissue contains gland cells. Such cells secrete a substance that functions outside the cell. In most animals, secretory cells are clustered inside **glands**, organs that release substances onto the skin or into a body cavity or a body fluid. **Exocrine glands** have ducts or tubes that deliver their secretions onto an internal or external surface. Exocrine secretions include mucus, saliva, tears, milk, digestive enzymes, and earwax. **Endocrine glands** have no ducts. They release their products, signaling molecules called hormones, into some body fluid. Most commonly, hormones enter the bloodstream, which distributes them throughout the body.

Adults make few new muscle cells or nerve cells, but they do renew their epithelial cells. For example, each day you lose skin cells, and new ones replace them. An adult sheds about 0.7 kilogram (1.5 pounds) of skin each year. Similarly, the lining of your intestine is replaced every four to six days. All those cell divisions provide lots of opportunities for DNA replication errors that can lead to cancer. Epithelium is the tissue most likely to become cancerous.

Connective Tissues

Connective tissues are the most abundant and widely distributed vertebrate tissues. They range from soft connective tissues to the specialized types called bone tissue, cartilage, adipose tissue, and blood. Unlike epithelial cells, connective tissue cells are not tightly packed. Instead they are spread out in a secreted extracellular matrix.

The various types of soft connective tissues all include the same components but in different proportions. **Fibroblasts**, the main cell type in these tissues,

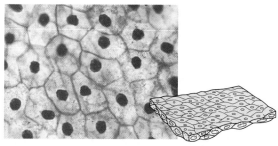

Simple squamous epithelium
• Lines blood vessels, the heart, and air sacs of lungs
• Allows substances to cross by diffusion

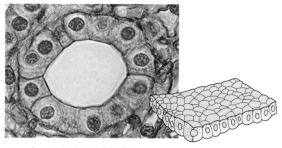

Simple cuboidal epithelium
• Lines kidney tubules, ducts of some glands, oviducts
• Absorbs, secretes, moves materials

mucus-secreting gland cell

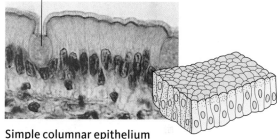

Simple columnar epithelium
• Lines some airways, parts of the gut
• Absorbs, secretes materials

FIGURE 19.4 Types of simple epithelia, with examples of their functions and locations.

connective tissue Animal tissue with an extensive extracellular matrix; provides structural and functional support.

endocrine gland Ductless gland that secretes hormones into a body fluid.

epithelial tissue Sheetlike animal tissue that covers outer body surfaces and lines internal tubes and cavities.

exocrine gland Gland that secretes milk, sweat, saliva, or some other substance through a duct.

fibroblast Main cell type in soft connective tissue; secretes collagen and other components of extracellular matrix.

gland Cluster of secretory epithelial cells.

make and secrete components of the extracellular matrix. These components include collagen, the body's most abundant protein, and polysaccharides. This noncellular material surrounds and supports the tissue's cells.

Loose connective tissue has fibroblasts and collagen fibers scattered widely in its matrix (Figure 19.5**A**). This tissue, the most common type in vertebrate bodies, holds organs and epithelia in place.

The two types of **dense connective tissue** both have a matrix packed full of fibroblasts and collagen fibers. In dense, irregular connective tissue, the fibers are oriented every which way, as in Figure 19.5**B**. This tissue makes up deep skin layers. It supports intestinal muscles and also forms capsules around organs that do not stretch, such as kidneys. Dense, regular connective tissue has fibroblasts in orderly rows between parallel, tightly packed bundles of fibers (Figure 19.5**C**). This organization helps keep the tissue from being torn apart when stretched. Tendons and ligaments are dense, regular connective tissue. The tendons connect skeletal muscle to bones. Ligaments attach one bone to another.

Specialized Connective Tissues All vertebrate skeletons include **cartilage**, which has a matrix of collagen fibers and rubbery glycoproteins. Cartilage cells secrete the matrix, which eventually imprisons them (Figure 19.5**D**). When you were an embryo, cartilage formed a model for your developing skeleton, then bone replaced most of it. Cartilage still supports the outer ears, nose, and throat. It cushions joints and acts as a shock absorber between vertebrae. Blood vessels do not extend through cartilage, so nutrients and oxygen must diffuse from vessels in nearby tissues. Also, unlike cells of other connective tissues, cartilage cells seldom divide in adults, so torn cartilage does not repair itself.

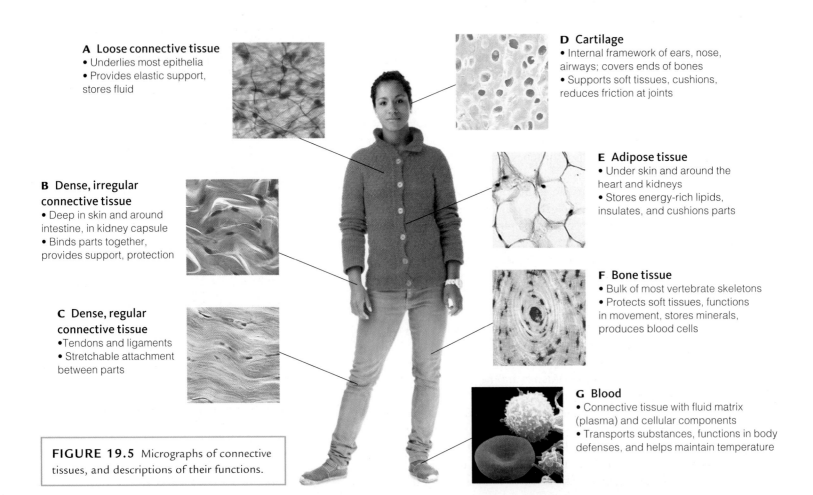

A Loose connective tissue
• Underlies most epithelia
• Provides elastic support, stores fluid

B Dense, irregular connective tissue
• Deep in skin and around intestine, in kidney capsule
• Binds parts together, provides support, protection

C Dense, regular connective tissue
• Tendons and ligaments
• Stretchable attachment between parts

D Cartilage
• Internal framework of ears, nose, airways; covers ends of bones
• Supports soft tissues, cushions, reduces friction at joints

E Adipose tissue
• Under skin and around the heart and kidneys
• Stores energy-rich lipids, insulates, and cushions parts

F Bone tissue
• Bulk of most vertebrate skeletons
• Protects soft tissues, functions in movement, stores minerals, produces blood cells

G Blood
• Connective tissue with fluid matrix (plasma) and cellular components
• Transports substances, functions in body defenses, and helps maintain temperature

FIGURE 19.5 Micrographs of connective tissues, and descriptions of their functions.

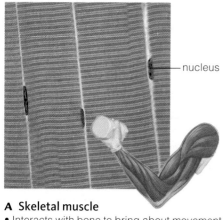

A Skeletal muscle
• Interacts with bone to bring about movement, maintain posture
• Reflex activated, but also voluntarily controlled

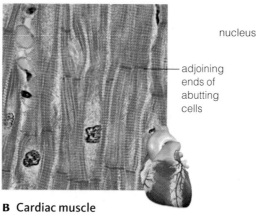

B Cardiac muscle
• Occurs only in the heart wall
• Contraction is not under voluntary control

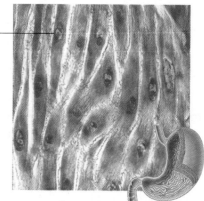

nucleus

C Smooth muscle
• Walls of digestive tract, arteries, reproductive tract, the bladder, other hollow organs
• Contraction is not under voluntary control

FIGURE 19.6 Micrographs of muscle tissues, and descriptions of their functions.

Many cells store some fat, but only the cells of **adipose tissue** can bulge with so much fat that cellular components, including the nucleus, get pushed to one side (Figure 19.5**E**). Small blood vessels that run through adipose tissue carry fats to and from cells. In addition to its energy-storage role, adipose tissue cushions and protects body parts, and a layer of this tissue under the skin serves as insulation.

Bone tissue is a connective tissue in which cells are surrounded by a calcium-hardened matrix (Figure 19.5**F**). Bone tissue is the main component of bones, which interact with skeletal muscles to move a body. Bones also support and protect internal organs, and some produce blood cells.

Blood is considered a connective tissue because its cells and platelets descend from cells in bone (Figure 19.5**G**). Red blood cells transport oxygen. White blood cells defend the body against pathogens. Platelets are cell fragments that help blood clot. Plasma, the fluid portion of blood, consists mostly of water. It transports gases, proteins, nutrients, hormones, and other substances.

Muscle Tissues Cells of muscle tissues contract—or forcefully shorten—in response to signals from nervous tissue. **Skeletal muscle tissue** attaches to bones and moves the body or its parts. It has parallel arrays of muscle fibers with multiple nuclei (Figure 19.6**A**). Skeletal muscle looks striated (striped) because it is made up of orderly arrays of contractile units. Skeletal muscles contract reflexively, but we also deliberately cause them to contract when we want to move a body part. That is why skeletal muscles are described as "voluntary" muscles.

Only the heart wall contains **cardiac muscle tissue** (Figure 19.6**B**). Its cells contract as a unit in response to signals that flow through gap junctions (Section 3.7) between their abutting plasma membranes. Each branching cardiac cell has a single nucleus. It has far more mitochondria than other types of muscle cells, because the heart's constant contractions require a steady supply of ATP. Cardiac muscle tissue also is striated, although less conspicuously so. Cardiac muscle and smooth muscle are both "involuntary" muscle; we cannot make them contract.

We find layers of **smooth muscle tissue** in the wall of many soft internal organs, including the stomach, bladder, and uterus (Figure 19.6**C**). Among other things, smooth muscle contractions propel material through the gut, shrink the diameter of blood vessels, and constrict the pupil of the eye. Smooth muscle cells are unbranched with tapered ends, and a single nucleus. Contractile units are not arranged in a repeating fashion, so smooth muscle is not striated. As in cardiac muscle, gap junctions relay signals between the cells.

adipose tissue Connective tissue with fat-storing cells.
blood Fluid connective tissue with cells that form inside bones.
bone tissue Connective tissue with cells surrounded by a mineral-hardened matrix of their own secretions.
cardiac muscle tissue Striated, involuntary muscle of the heart wall.
cartilage Connective tissue with cells surrounded by a rubbery matrix of their own secretions.
dense connective tissue Connective tissue with many fibroblasts and fibers in a random or a regular arrangement.
loose connective tissue Connective tissue with relatively few fibroblasts and fibers scattered in its matrix.
skeletal muscle tissue Striated, voluntary muscle that interacts with bone to move body parts.
smooth muscle tissue Involuntary muscle that lines blood vessels and hollow organs; not striated.

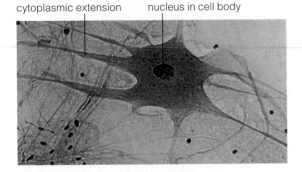

cytoplasmic extension   nucleus in cell body

FIGURE 19.7 Micrograph of a motor neuron. Its cytoplasmic extensions carry signals. The cell body contains organelles, including the nucleus.

Nervous Tissue **Nervous tissue** allows a body to detect and respond to internal and external changes. This tissue detects stimuli, integrates information, and controls the actions of muscles and glands. Figure 19.7 shows a **neuron**, one kind of excitable cell in nervous tissue. In this context, "excitable" means that electrical signals can move along the cell's plasma membrane. Neurons also communicate with one another and with muscles or glands by secreting chemical messages.

A neuron has a central cell body that holds the nucleus and other organelles. Long, thin cytoplasmic extensions project out from the nucleus. Some of these extensions are remarkably long. For example, cytoplasmic extensions that carry signals between a person's spinal cord and foot can be more than a meter in length. In a whale, the equivalent structure can be tens of meters long.

Neurons are not the only cells in nervous tissue. More than half of the volume of your nervous system consists of cells called neuroglia that support and protect the neurons.

Take-Home Message  **What are the features of the four types of tissues?**

- Sheetlike epithelial tissue covers the body surface and lines tubes and cavities. Cells in it are connected by junctions and have no matrix between them.

- Connective tissues consist of cells in a secreted extracellular matrix. Soft connective tissues hold body parts in place. Adipose tissue stores fat. Bone and cartilage support body parts and function in movement. Blood is a transport medium with cellular components that were made inside bones.

- Muscle tissue consists of cells that contract. Skeletal muscle moves bones, and you can make it contract at will. Cardiac muscle (found in the wall of the heart) and smooth muscle (in walls of hollow organs) are involuntary; you cannot control their contraction.

- Nervous tissue includes neurons. Neurons relay electrical signals along their plasma membrane and use chemicals to communicate with other cells.

19.4 Organs and Organ Systems

Tissues are organized into organs. The heart, stomach, and liver are organs, as are eyes, kidneys, and lungs. Most organs include all four types of tissues.

Organs in Body Cavities Many human organs are located inside body cavities. Like other vertebrates, humans are bilateral and have a lined body cavity known as a coelom (Section 15.2). A sheet of smooth muscle, the diaphragm, divides the human coelom into an upper thoracic cavity and a lower cavity that has abdominal and pelvic regions (Figure 19.8). The heart and lungs are in the thoracic cavity. Digestive organs such as the stomach, intestines, and liver lie in the abdominal cavity. The bladder and reproductive organs are in the pelvic cavity. A cranial cavity in the head holds the brain, and a spinal cavity in the back holds the spinal cord. These last two cavities are not derived from the coelom.

The Skin—Your Largest Organ Your largest organ is not inside your body. It is the skin that covers your surface (Figure 19.9). The outermost region of skin, the **epidermis**, is a stratified squamous epithelium. In other

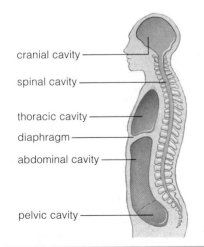

cranial cavity

spinal cavity

thoracic cavity

diaphragm

abdominal cavity

pelvic cavity

FIGURE 19.8 Body cavities that hold many human organs.

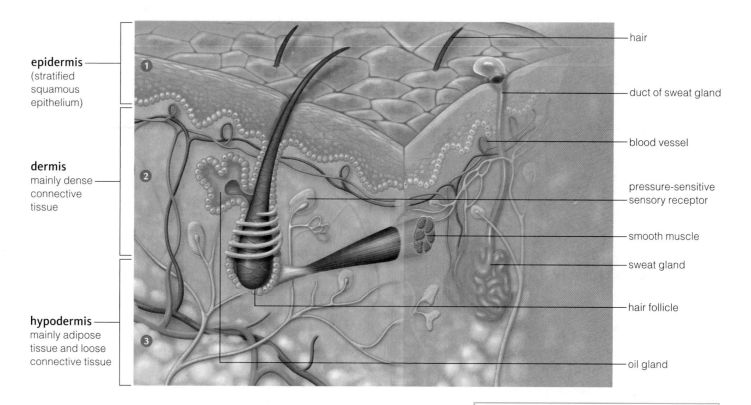

epidermis
(stratified squamous epithelium)

dermis
mainly dense connective tissue

hypodermis
mainly adipose tissue and loose connective tissue

hair

duct of sweat gland

blood vessel

pressure-sensitive sensory receptor

smooth muscle

sweat gland

hair follicle

oil gland

FIGURE 19.9 Animated! Structure of human skin, your largest organ.

Figure It Out: Is a sweat gland an endocrine gland or an exocrine gland?

Answer: An exocrine gland

words, it consists of many layers of squamous (flattened) cells ❶. Ongoing mitotic divisions of cells in the deepest epidermal layer continually push cells toward the sheet's free surface. Abrasion and the pressure exerted by the growing cell mass flatten and kill older cells as they move upwards. Dead cells at the skin surface get rubbed off or flake away.

The vast majority of epidermal cells are keratinocytes, cells that make filaments of the protein keratin. The presence of large amounts of keratin water-proofs the skin of reptiles, birds, and humans, and is an adaptation to life on land. The epidermis also contains melanocytes, cells that make the brownish protein melanin. Keratinocytes obtain melanin from melanocytes and store it in special organelles. Differences in the size and distribution of these organelles among human ethnic groups are the basis for differences in their skin color.

The **dermis** beneath the epidermis consists mostly of dense connective tissue ❷. Sensory receptors thread through the dermis, as do capillaries, small blood vessels that are tubes of epithelial tissue. Sweat glands in the dermis are made up of epidermal cells that migrated into the dermis during development. Remember, glands are epithelial tissue.

Epithelial tissue embedded in the dermis also forms hair follicles. The base of the hair follicle holds living hair cells, the fastest-dividing cells in the human body. As these cells divide, they push cells above them up, lengthening the hair. The portion of a hair that sticks out beyond the skin surface consists of the keratin-rich remains of dead cells. Attached to each hair is a smooth muscle. Hair stands upright when this muscle reflexively contracts in response to cold or fright. Hair is naturally kept soft and shiny by secretions from an oil gland next to each hair follicle. Like a sweat gland, an oil gland is epidermal tissue embedded in the dermis.

In most body regions, the skin sits atop a layer of loose connective tissue and adipose tissue called the hypodermis ❸. This layer contains larger blood vessels that connect to the small ones running through the dermis.

dermis Deep layer of skin that consists of connective tissue with nerves and blood vessels running through it.

epidermis Outermost, epithelial skin layer.

nervous tissue Animal tissue composed of neurons and supporting cells; detects stimuli and controls responses to them.

neuron Main type of cell in nervous tissue; transmits electrical signals along its plasma membrane and communicates with other cells through chemical messages.

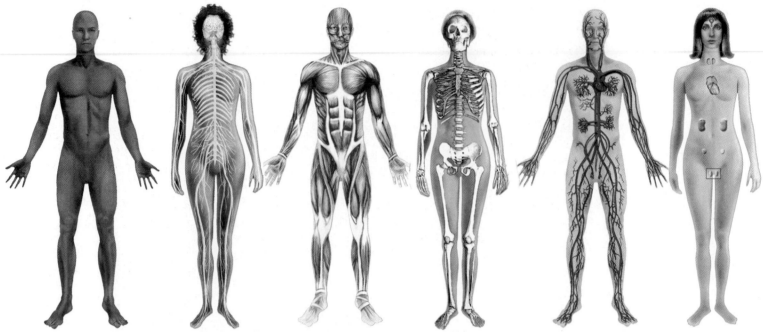

Integumentary System
Protects body from injury, dehydration, and some pathogens; controls its temperature; excretes certain wastes; receives some external stimuli.

Nervous System
Detects external and internal stimuli; controls and coordinates responses to stimuli; integrates all organ system activities.

Muscular System
Moves body and its internal parts; maintains posture; generates heat by increases in metabolic activity.

Skeletal System
Supports and protects body parts; provides muscle attachment sites; produces red blood cells; stores calcium, phosphorus.

Circulatory System
Rapidly transports many materials to and from cells; helps stabilize internal pH and temperature.

Endocrine System
Hormonally controls body functioning; with nervous system integrates short- and long-term activities. (Male testes added.)

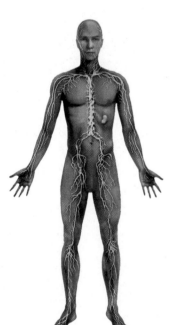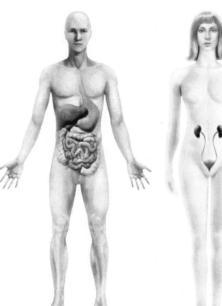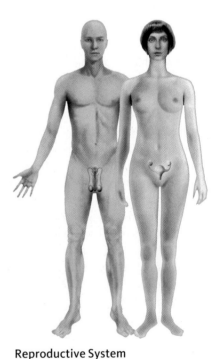

Lymphatic System
Collects and returns some tissue fluid to the bloodstream; defends against infection and tissue damage.

Respiratory System
Rapidly delivers oxygen to the tissue fluid that bathes all living cells; removes carbon dioxide wastes of cells; helps regulate pH.

Digestive System
Ingests food and water; mechanically, chemically breaks down food and absorbs small molecules into internal environment; eliminates food residues.

Urinary System
Maintains the volume and solute composition of internal fluids; excretes excess fluid, solutes, and dissolved wastes.

Reproductive System
Male: Produces and transfers sperm to the female. Hormones of both systems also influence other organ systems. Female: Produces eggs; after fertilization, affords a protected, nutritive environment for the development of a new individual.

FIGURE 19.10 Animated! Vertebrate organ systems.

Components of the skin work together to carry out a variety of tasks. The skin is a barrier that prevents pathogens from invading deeper tissues. Nerve endings in skin detect temperature, pressure, and touch and send signals to the spinal cord and brain. Skin also plays a role in temperature regulation.

Sun, Aging, and the Skin When skin is exposed to sunlight, melanin production increases, producing a "suntan." The melanin has a protective function. It absorbs ultraviolet (UV) radiation that might otherwise damage underlying skin layers. A bit of UV radiation is a good thing because it stimulates melanocytes to make vitamin D, which is essential to good health. However, excessive UV exposure damages collagen and causes elastin fibers to clump. Chronically tanned skin gets less resilient and leathery. UV also damages DNA in skin cells, increasing the risk of skin cancer.

As we age, epidermal cells divide less often. Skin thins and becomes less elastic as collagen and elastin fibers become sparse. Glandular secretions that kept the skin soft and moist dwindle. Wrinkles deepen. Many people needlessly accelerate the aging process by excessive tanning or by smoking, which shrinks the skin's blood supply.

Organ Systems Skin and structures derived from it such as hair, fur, hooves, claws, nails, and quills are a vertebrate's integumentary system, one of many vertebrate organ systems (Figure 19.10).

Organ systems work cooperatively to carry out reproduction and survival tasks. For example, several organ systems interact to provide cells with necessary substances and remove wastes (Figure 19.11). Food and water enter the body by way of the digestive system, which includes digestive organs such as the stomach and intestine, as well as organs that aid digestion such as the pancreas and gallbladder. The digestive system also eliminates undigested waste. Oxygen enters the body by way of the respiratory system, which includes lungs and airways that lead to them. The heart and blood vessels of the circulatory system deliver nutrients and oxygen to cells, and remove waste carbon dioxide and solutes from them. The circulatory system delivers the carbon dioxide to the respiratory system, which eliminates it. The circulatory system also moves excess water, salts, and soluble wastes to the urinary system. Organs of the urinary system include kidneys that filter wastes from the blood and make urine, and a bladder that stores the urine until it can be eliminated.

Figure 19.11 does not include the nervous, endocrine, muscular, and skeletal systems, but these too help you obtain essential substances and eliminate wastes. For example, the nervous system detects changes in internal levels of water, solutes, and nutrients. Signals from the nervous and endocrine systems cause conservation or elimination of water. Signals from the nervous system also result in movement, as when your muscles and bones interact to lift you off the couch and move you toward the refrigerator or the toilet.

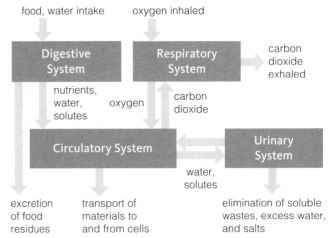

FIGURE 19.11 Flowchart of links between organ systems. This diagram shows some of the ways that organ systems interact to keep the body supplied with essential substances and eliminate unwanted wastes. Other organ systems that are not shown also take part in these tasks.

Take-Home Message   **What are the roles of organs and organ systems?**

- An organ has tissues organized in a way that allows it to carry out one or more essential tasks. For example, the skin—an organ with all four tissue types—protects the body, helps make vitamin D, detects events in the outside environment, and helps maintain body temperature.

- Organ systems are composed of interacting organs. All vertebrates have the same types of organ systems.

- Survival and reproduction depend on interactions among all organ systems.

Control of Body Temperature

Detecting and Responding to Change Homeostasis, again, is the process of keeping conditions in the body within limits. In vertebrates, homeostasis involves interactions among sensory receptors, the brain, and muscle and glands. A **sensory receptor** is a cell or cell component that detects a specific stimulus. Sensory receptors involved in homeostasis function like internal watchmen; they monitor the body for change. Information from sensory receptors throughout the body flows to the brain. The brain evaluates the incoming information, then signals muscles and glands to take any necessary actions to keep the body functioning properly.

Keeping Temperature Just Right Homeostasis often involves **negative feedback**, a process in which a change causes a response that reverses the change. A familiar nonbiological example is the way a furnace with a thermostat works. A person sets the thermostat to a desired temperature. When the temperature falls below this preset point, the furnace turns on and emits heat. When the temperature rises to the desired level, the thermostat turns off the furnace. Similarly, a negative feedback mechanism keeps your internal temperature near 37°C (98.6°F).

Think about what happens when you exercise on a hot day. Muscle activity generates heat, so the body's internal temperature rises. Sensory receptors in the skin detect the increase and signal the brain. The brain sends signals that bring about the body's response (Figure 19.12). Blood flow shifts, so more blood from the body's hot interior flows to the skin. The shift maximizes the amount of heat given off to the surrounding air. At the same time, sweat glands in the skin increase their output. Evaporation of sweat helps cool the body surface. You breathe faster and deeper, speeding the transfer of heat from the blood in your lungs to the air. Hormonal changes make you feel more sluggish. As your activity level slows and your rate of heat loss increases, your temperature falls.

Despite feedback mechanisms, hyperthermia—a dangerous rise in the body's temperature—kills about 175 Americans each year. Profuse sweating causes loss

negative feedback A change causes a response that reverses the change.

sensory receptor Cell or cell component that detects stimuli and signals the brain.

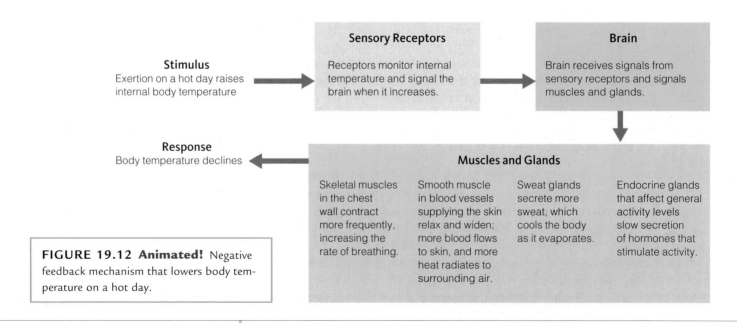

FIGURE 19.12 Animated! Negative feedback mechanism that lowers body temperature on a hot day.

of water and salts, changing the concentration of body fluids. Blood flow to the gut and liver decreases. Starved of nutrients and oxygen, these organs release toxins that interfere with the function of the nervous system, as well as other organ systems. As a result, a person may become incapable of recognizing and responding to seemingly obvious signs that something is wrong.

To stay safe outside on a hot day, drink plenty of water and avoid excessive exercise. If you must exert yourself, take breaks to monitor how you feel. Wear light-colored clothing that allows evaporation. Avoid direct sunlight, wear a hat and use sunscreen. Sunburn impairs the skin's ability to transfer heat to the air.

Receptors in the skin also notify the brain when conditions get chilly. The brain then diverts blood flow away from the skin, lessening movement of heat to the body surface, where it would be lost to the surrounding air. As another response, reflex contractions of smooth muscle in the skin cause hair to stand up, causing "goose bumps." Erect hair insulates better than hair that lies flat. With prolonged cold, the brain commands skeletal muscles to contract ten to twenty times a second. This shivering response increases heat production by muscles, but it has a high energy cost.

Continued exposure to cold can cause hypothermia in which the body's core temperature falls dangerously low. A decline to 35°C (95°F) alters brain functions. "Stumbles, mumbles, and fumbles" are the symptoms of early hypothermia. Severe hypothermia causes loss of consciousness, disrupts heart rhythm, and can be fatal.

Take-Home Message How does your body maintain its temperature?

- Sensory receptors detect a change in the temperature and notify the brain. Signals from the brain then cause muscles and glands to react to this change.

- Temperature regulation involves negative feedback in which a change triggers a response that reverses the change.

Impacts/Issues Revisited:
It's All About Potential

For stem cells to be of use in treatments, they have to be induced to become a specific cell type. The cells are like college freshmen who need to be directed to a particular major. Many researchers are working on methods to guide stem cells along a particular developmental pathway. For example, Su-Chun Zhang at the University of Wisconsin coaxed human embryonic stem cells to become the motor neurons shown at *right*. Green fibers are the neuron's cytoplasmic extensions. Other researchers have produced neurons from adult skin cells that were turned into stem cells using the viral technique described in Section 19.1.

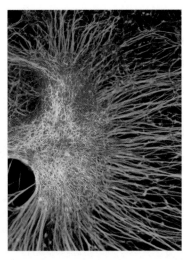

How Would YOU Vote? Should scientists be allowed to start new embryonic stem cell lines from early human cell clusters produced in fertility clinics but not used? See CengageNow for details, then vote online (cengagenow.com).

# Summary

Section 19.1 Stem cells can divide to form more stem cells or differentiate to become specialized cells. All body parts develop from

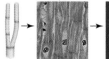

stem cells present in the embryo. Use of embryonic stem cells or of adult stem cells made to behave like embryonic ones could one day help treat diseases and disorders.

Section 19.2 Animal bodies are structurally and functionally organized on several levels. A tissue is a group of cells and intercellular substances that perform a common task. Tissues make up organs, which interact as organ systems. Activities at all levels contribute to homeostasis, the process of maintaining conditions inside the body within tolerable limits.

Animal structure and function have a genetic basis and are shaped by natural selection. However, because evolution modifies existing structures, body parts often are less than optimal for a task.

Section 19.3 Most animals are constructed of four types of tissue.

 Epithelial tissues cover external body surfaces and line internal cavities and tubes. Epithelium has one free surface exposed to some body fluid or the environment. Gland cells and glands are derived from epithelial tissue.

 Connective tissues bind, support, protect, and insulate other tissues. Soft connective tissues have different proportions and arrangements of protein fibers, fibroblasts, and other cells in an extracellular matrix. Specialized connective tissues include cartilage, bone tissue, adipose tissue, and blood.

 Muscle tissues contract (shorten) when stimulated. They help move the body or parts of it. The three types of muscle tissue are skeletal muscle, smooth muscle, and cardiac muscle.

 Neurons are communication lines of nervous tissue. They relay signals along their plasma membrane and send and receive chemical signals. Cells called neuroglia support neurons.

CENGAGENOW™ Use the interaction on CengageNow to investigate the different types of tissues.

Section 19.4 An organ is a structural unit of different tissues combined in definite proportions and patterns that allow them to perform a common task. Skin is an organ composed of all four tissue types. It functions in protection, temperature control, detection of shifts in the environment, vitamin D production, and defense. An organ system consists of two or more organs interacting in tasks that keep individual cells as well as the whole body functioning.

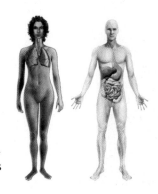

CENGAGENOW™ Explore the structure of skin and survey the organ systems of the vertebrate body with animation on CengageNow.

Section 19.5 Temperature regulation is an aspect of homeostasis. Human temperature regulation involves sensory receptors that detect changes, a brain that receives signals from receptors and coordinates responses, and muscles and glands that carry out responses. Temperature regulation involves negative feedback in which a change brings about a response that reverses the change.

CENGAGENOW™ Learn how a body regulates temperature with the animation on CengageNow.

Self-Quiz   Answers in Appendix I

1. _____ tissues are sheetlike with one free surface.
   a. Epithelial   c. Nervous
   b. Connective   d. Muscle

2. A _____ connects epithelium to underlying connective tissue.
   a. gap junction   c. keratinocyte
   b. basement membrane   d. plasma membrane

3. Most animals have glands derived from _____ tissue.
   a. epithelial   c. muscle
   b. connective   d. nervous

4. Only _____ cells have cilia or microvilli at their surface.
   a. epithelial   c. muscle
   b. connective   d. nervous

5. The most abundant protein in the human body is _____ , made by fibroblasts.
   a. collagen   c. melanin
   b. keratin   d. hemoglobin

6. _____ is mostly plasma.
   a. Irregular connective tissue   c. Cartilage
   b. Blood   d. Bone

7. Your body converts excess carbohydrates and proteins to fats, which accumulate in _____ .
   a. fibroblasts   c. adipose tissue cells
   b. neurons   d. melanocytes

8. _____ tissues are the body's most abundant and widely distributed tissue.
   a. Epithelial   c. Nervous
   b. Connective   d. Muscle

9. Cells of _____ can shorten (contract).
   a. epithelial tissue   c. muscle tissue
   b. connective tissue   d. nervous tissue

10. _____ muscle tissue has a striated appearance.
    a. Skeletal   c. Cardiac
    b. Smooth   d. a and c

11. _____ detects and integrates information about changes and controls responses to those changes.
    a. Epithelial tissue   c. Muscle tissue
    b. Connective tissue   d. Nervous tissue

Digging Into Data

Growing Skin To Heal Wounds

Diabetes is a disorder in which the blood sugar level is not properly controlled. Among other complications, this disorder reduces blood flow to the lower legs and feet. As a result, about 3 million diabetes patients have ulcers—open wounds that do not heal—on their feet. Each year, about 80,000 diabetics require amputations.

Fibroblasts and keratinocytes can be grown to produce a cultured skin substitute (Figure 19.13**A**). The cultured skin is placed over wounds to help them heal (Figure 19.13**B**). Figure 19.13**C** shows the results of a clinical experiment that tested the effect of the cultured skin product versus standard treatment for diabetic foot wounds. Patients were randomly assigned either to the experimental treatment group or to the control group. Their progress was monitored for 12 weeks.

1. What percentage of wounds had healed at 8 weeks when treated the standard way? When treated with cultured skin?

2. What percentage of wounds had healed at 12 weeks when treated the standard way? When treated with cultured skin?

3. How early was the healing difference between the control and treatment groups obvious?

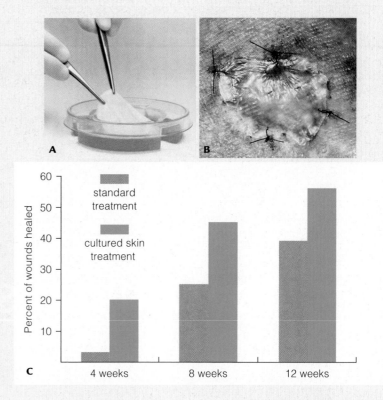

FIGURE 19.13 Cultured skin. **A** Cultured skin ready for use and **B** in place over a wound. **C** Results of a study comparing standard treatment for diabetic foot ulcers to use of a cultured cell product. Bars show the percentage of foot ulcers that had completely healed.

---

12. Thin cytoplasmic projections from cells called _____ carry signals between your spinal cord and your toes.

13. The functions of skin include _____ .
    a. defense against pathogens
    c. production of vitamin D
    b. helping to cool the body
    d. all of the above

14. When the level of sugar in your blood falls too low, your body senses this decline and converts glycogen to sugar, putting more sugar into your blood. This is an example of _____ .
    a. differentiation   c. homeostasis
    b. negative feedback   d. b and c

15. Match the terms with the most suitable description.
    _____ exocrine gland   a. outermost skin layer
    _____ endocrine gland   b. secretes through duct
    _____ epidermis   c. in heart only
    _____ cardiac muscle   d. support in ears and nose
    _____ cartilage   e. contracts, not striated
    _____ smooth muscle   f. darkens skin
    _____ blood   g. plasma, platelets, and cells
    _____ melanin   h. ductless hormone secretor

Additional questions are available on CENGAGENOW

Critical Thinking

1. Leukemia is a cancer in which the body makes too many white blood cells. Affected people are sometimes treated with a bone marrow transplant. Radiation or chemotherapy drugs are used to kill the existing bone marrow, then a bone marrow transplant replaces it. Why would a bone marrow transplant affect the production of white blood cells?

2. Radiation and chemotherapy drugs preferentially kill cells that divide frequently, most notably cancer cells. These cancer treatments also cause hair to fall out. Why?

3. Each level of biological organization has emergent properties that arise from the interaction of its component parts. For example, cells have a capacity for inheritance that the molecules making up the cell do not. Can you think of an emergent property of a tissue? Of an organ that contains that tissue?

4. The micrograph to the *right* shows cells from the lining of an airway leading to the lungs. The *gold* cells are ciliated and the *darker brown* ones secrete mucus. What type of tissue is this? How can you tell?

How Animals Move

Your body is like a machine that improves with use.
The more you use your muscles, the stronger you become.

Impacts/Issues:

Bulking Up Muscles

Like neurons, skeletal muscle cells do not divide after birth, so muscles bulk up by enlarging existing cells. Inside each muscle cell, protein filaments involved in muscle contraction are continually built and broken down. Exercise tilts this process in favor of synthesis, so muscle cells get bigger and the muscle they are in gets stronger. Your body is like a machine that improves with use. The more you use your muscles, the stronger you become.

Certain hormones also encourage increases in muscle mass. For example, the sex hormone testosterone causes muscle cells to build more proteins, among its other effects. Women make a little bit of testosterone, but men make much more, which is why men tend to be more muscular than women. Human growth hormone also stimulates synthesis of muscle proteins.

Synthetic versions of natural muscle-building, or "anabolic," hormones can also increase muscle mass and enhance athletic ability. However, use of these drugs is considered cheating in most sports.

Some people may have a natural genetic advantage when it comes to putting on muscle. In 2004, scientists in Germany reported on their study of an unusual child. Even at birth the boy had bulging biceps and thigh muscles. The boy does not make myostatin, a regulatory protein that normally slows production of muscle proteins. Genetic analysis revealed that he is homozygous for a mutation in the myostatin gene and so produces no myostatin. As you might expect, he is unusually strong. His mother is heterozygous for the myostatin gene and is a former professional athlete.

Additional evidence that myostatin level affects athletic ability comes from a study of whippets, a type of dog bred for racing. Some whippets are, like the German boy, homozygous for a mutation that prevents them from making myostatin. Such dogs, called bully whippets, are unusually heavily muscled compared to a normal whippet (Figure 20.1). Bully whippets do not match the breeders' ideal of how a whippet should look, so they are not usually raced or bred. However, dogs that are heterozygous for the mutated allele do race. Compared to normal whippets, they make less myostatin, are more muscular, and are more likely to win races.

Several drug companies are working to develop chemicals that inhibit myostatin production or activity. The goal is to come up with drugs for muscle-wasting disorders such as muscular dystrophy. Will some athletes use such drugs to push their bodies beyond normal limits? No doubt they will. Some are already buying nutritional supplements that purport to bind to myostatin and reduce its activity. In clinical tests, these supplements had no effect on strength, but hope for an extra edge keeps moving them off the shelf.

FIGURE 20.1 A bully whippet (*top*). The massive muscles of a bully whippet result from a myostatin deficiency. Compare a normal whippet (*bottom*).

20.2 The Skeletal System

Types of Skeletons Muscles bring about movement of a body or its parts by interacting with a skeleton. Many soft-bodied invertebrates have a **hydrostatic skeleton**, an internal, fluid-filled chamber or chambers that muscles exert force against. For example, earthworms have a coelom divided into many fluid-filled chambers, one per segment (Section 15.3). Muscles alter the shape of a body segment by squeezing its fluid-filled chamber. By analogy, think about how squeezing a water-filled balloon changes its shape. Coordinated changes in the shape of body segments move the worm.

In animals with an **exoskeleton**, the cuticle, shell, or other hard external body part receives the force of a muscle contraction. Muscles attach to and pull on an arthropod's hinged exoskeleton. For example, the muscles attached to a fly's thorax cause its wings to flap up and down when they contract.

An **endoskeleton** is an internal framework of hardened elements to which muscles attach. Echinoderms such as sea stars have an endoskeleton, as do all vertebrates, including humans.

The Human Skeleton

For a closer look at vertebrate skeletal features, think about a human skeleton, beginning with the skull (Figure 20.2). Flattened cranial bones fit together to form a braincase that surrounds and protects the brain. Facial bones include cheekbones and other bones around the eyes, the bone that forms the bridge of the nose, and bones of the jaw.

The **vertebral column**, or backbone, extends from the base of the skull to the pelvic girdle. It consists of 23 bones called **vertebrae**. Cartilage **intervertebral disks** separate adjoining vertebrae. The spinal cord runs through the backbone and connects with the brain at an opening at the base of the skull.

In humans, evolution of an ability to walk upright involved modification of the backbone. Viewed from the side, our backbone has an S shape that keeps our head and torso centered over our feet. Maintaining an upright posture requires that vertebrae and intervertebral disks stack one on top of the other, rather than being parallel to the ground, as in four-legged walkers. The stacking puts additional pressure on disks and, as people age, their disks often slip out of place or rupture (herniate), causing back pain.

The rib cage attaches to the vertebral column. Both males and females have twelve pairs of ribs. Ribs and the breastbone, or sternum, form a protective cage around the heart and lungs.

The scapula (shoulder blade) and clavicle (collarbone) are bones of the human pectoral girdle. When a person falls on an outstretched arm, excessive force transferred to the clavicle often causes the clavicle to break. The upper arm has one bone, the humerus. The forearm has two bones, the radius and ulna. The wrist and palm each have multiple small bones, as does each finger.

The pelvic girdle protects organs and supports the weight of the upper body when you stand upright. It consists of two sets of fused bones, one set on each side of the body.

The largest bone in the body is the femur, or thighbone. It attaches to the lower leg bones, the tibia and fibula, at the knee. The knee is protected by the patella (kneecap). The smaller bone of the lower leg, the fibula, is not necessary for normal function. When a person has cancer of the jaw, reconstructive surgeons sometimes remove the cancerous jawbone and replace it with bone from the person's fibula. The ankle consists of multiple bones, as do the sole of the foot and each toe.

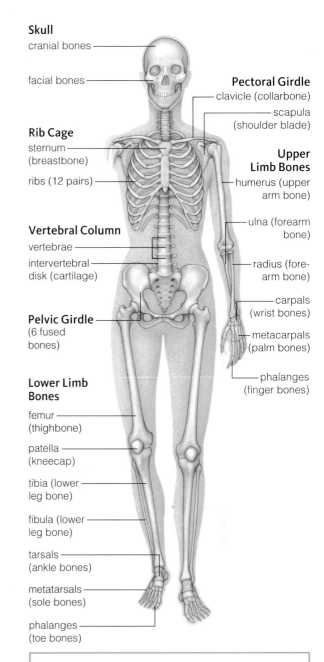

Skull
cranial bones
facial bones

Rib Cage
sternum (breastbone)
ribs (12 pairs)

Vertebral Column
vertebrae
intervertebral disk (cartilage)

Pelvic Girdle (6 fused bones)

Lower Limb Bones
femur (thighbone)
patella (kneecap)
tibia (lower leg bone)
fibula (lower leg bone)
tarsals (ankle bones)
metatarsals (sole bones)
phalanges (toe bones)

Pectoral Girdle
clavicle (collarbone)
scapula (shoulder blade)

Upper Limb Bones
humerus (upper arm bone)
ulna (forearm bone)
radius (forearm bone)
carpals (wrist bones)
metacarpals (palm bones)
phalanges (finger bones)

FIGURE 20.2 Animated! The human skeletal system. It consists of bone and cartilage.

endoskeleton Hard internal parts that muscles attach to and move.

exoskeleton Hard external parts that muscles attach to and move.

hydrostatic skeleton Fluid-filled chamber that muscles act on, redistributing the fluid.

intervertebral disk Cartilage disk between two vertebrae.

vertebrae Bones of the backbone.

vertebral column The backbone.

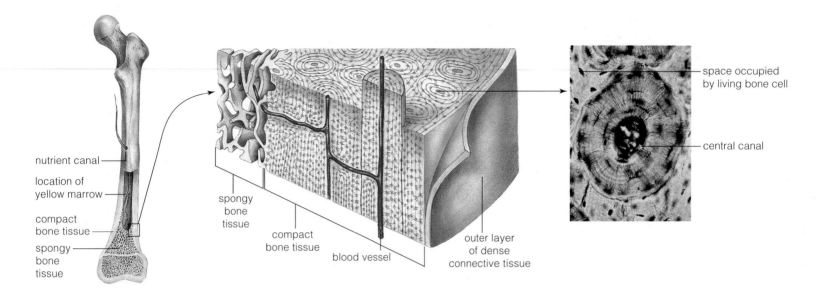

FIGURE 20.3 Animated!
Structure of a human thighbone (femur).

Bone Structure and Function Bones are organs with many functions. Together with skeletal muscles, bones maintain or change the orientation of the body or its parts. Bony chambers surround and protect soft internal organs. Some bones are sites for formation of blood cells, and all serve as reservoirs for calcium and phosphorus ions.

Each bone is wrapped in a dense connective tissue sheath that has nerves and blood vessels running through it. Bone tissue consists of bone cells in an extracellular matrix (Section 3.7). The matrix is mainly collagen (a protein) hardened with calcium and phosphorus.

There are two types of bone. **Compact bone** makes up the outer, weight-bearing part of a femur (Figure 20.3). Compact bone consists of many thin, concentric layers of matrix surrounding canals for nerves and blood vessels that service living bone cells. The shaft and bone ends also have **spongy bone**, a tissue that is lightweight and has many internal spaces. **Red marrow** fills spaces in the spongy bone of limb bones and some others, such as the breastbone. It is the main site for blood cell formation. The main cavity of adult limb bones is filled by **yellow marrow**, which is mostly fat cells.

Ongoing mineral deposits to bone and removals from it help maintain the normal blood concentrations of calcium and phosphorus. The level of calcium in the blood is one of the most tightly controlled aspects of metabolism. Calcium ions play a role in nerve cell function, muscle contraction, and other important processes. Bones and teeth store just about all of the body's calcium.

Until people are about twenty-four years old, they produce bone matrix faster than they break it down, so bone mass increases. As people age, matrix production falters and bone density declines. **Osteoporosis** is a disorder in which bone loss greatly outpaces bone formation. As a result, the bones become weaker and more likely to break. Osteoporosis is most common in postmenopausal woman because they no longer produce the sex hormones that encourage bone deposition. However, about 20 percent of osteoporosis cases occur in men.

To reduce your risk of osteoporosis, ensure that your diet provides plenty of vitamin D and calcium. A premenopausal woman needs 1,000 milligrams of calcium daily; a postmenopausal woman requires 1,500 milligrams a day. Avoid smoking and excessive alcohol intake, which slow bone deposition. Get regular exercise to encourage bone renewal and, if you are female, do not drink too many cola soft drinks. Studies have shown that such drinks affect bone metabolism, and women who have more than two a day have lowered bone density.

Where Bones Meet—Skeletal Joints A **joint** is an area where bones come together. Connective tissue holds bones securely in place at **fibrous joints** such as those that hold cranial bones together. Pads or disks of cartilage connect bones at **cartilaginous joints**. The flexible connection allows a bit of movement. Cartilaginous joints connect vertebrae to one another and connect some of the ribs to the sternum. Joints of the hip, shoulder, wrist, elbow, and knee are **synovial joints**, the most common type of joint. In a synovial joint the ends of bones are covered with cartilage and enclosed in a fluid-filled capsule. **Ligaments**, cords of dense connective tissue, hold the bones in place. Figure 20.4 shows some ligaments that hold bones together at the knee joint.

Different kinds of joints allow different kinds of movements. For example, ball-and-socket joints at the shoulders and hips allow rotational motion. At other joints, including some in the wrists and ankles, bones glide over one another. Joints at elbows and knees function like a hinged door, allowing the bones to move back and forth in one plane only.

A sprained ankle is the most common joint injury. A **sprain** occurs when one or more of the ligaments that hold bones together at a joint overstretches or tears. A sprained ankle is usually treated immediately with rest, application of ice, compression with an elastic bandage, and elevation of the affected area. After the ankle heals, exercises may help strengthen muscles that stabilize the joint and prevent future sprains.

Tearing cruciate ligaments in the knee joint may require surgery. *Cruciate* means cross, and these short ligaments cross one another in the center of the joint. They stabilize the knee and when they are torn completely, bones may shift so the knee gives out when a person tries to stand. A blow to the lower leg, as often happens in football, can injure a cruciate ligament, but so can a fall or misstep. Female athletes are at a higher risk for cruciate ligament tears than men who play the equivalent sport. For example, female soccer players tear these ligaments four times more often than male soccer players do.

A **dislocation** occurs when the bones of a joint move out of place. It is usually highly painful and requires immediate treatment. The bones must be placed back into proper position and immobilized for a time to allow healing.

Arthritis is chronic inflammation of a joint. As you will learn in Section 22.4, inflammation is a normal response to injury. However, with arthritis, inflammation and the associated pain and swelling do not go away. The most common type of arthritis, osteoarthritis, usually appears in old age, after cartilage wears down at a specific joint. For example, women who often wear high-heeled shoes increase their risk of osteoarthritis of the knees in their later years. Rheumatoid arthritis is a disorder in which the immune system mistakenly attacks all synovial joints. Rheumatoid arthritis can occur at any age, and women are two to three times more likely than men to be affected.

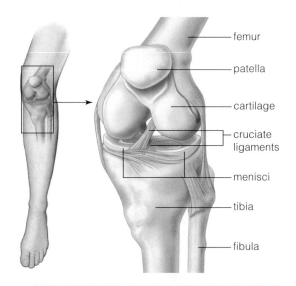

FIGURE 20.4 Anatomy of the knee, a hinge-type synovial joint. Bones are held in place by ligaments. The knee is also stabilized by wedges of cartilage called menisci (singular, meniscus). A torn ligament or damaged meniscus can impair knee function.

femur
patella
cartilage
cruciate ligaments
menisci
tibia
fibula

Take-Home Message   What are the features of animal skeletons?

- Animal skeletons consist of hard structural elements or fluid-filled chambers that muscle contractions act upon.
- Humans have a bony endoskeleton with some features such as an S-shaped backbone that are adaptations to upright posture and walking.
- Bones are wrapped in connective tissue and have marrow in their interior. They are continually remodeled; minerals are removed and added as needed.
- Bones meet at joints. At synovial joints, the most common type, bones are held in place by ligaments.

arthritis Chronic inflammation of a joint.
cartilaginous joint Joint where cartilage holds bones together and provides cushioning, as between vertebrae.
compact bone Dense bone with concentric layers of matrix.
dislocation Bones of a joint are out of place.
fibrous joint Joint where dense connective tissue holds bones firmly in place.
joint Region where bones come together.
ligament Dense connective tissue that holds bones together at a joint.
osteoporosis Disorder in which bones weaken.
red marrow Bone marrow that makes blood cells.
spongy bone Lightweight bone with many internal spaces.
sprain Ligaments of a joint are injured.
synovial joint Joint such as the knee that is lubricated by fluid and allows movement of bones around the joint.
yellow marrow Bone marrow that is mostly fat; fill cavity in most long bones.

How Bones and Muscles Interact

Muscles can only pull on bones; they cannot push.

Skeletal muscles are the functional partners of bones. A sheath of dense connective tissue encloses each muscle and extends beyond it as a cordlike or straplike **tendon**. Most often, tendons attach each end of the muscle to a bone. Muscles and bones act like a lever system in which a rigid rod is attached to a fixed point and moves about it. Muscles connect to bones (rigid rods) near a joint (fixed point). When a muscle contracts, it transmits force to the bones to which it is attached and makes them move.

Figure 20.5 shows two muscles of the upper arm, the biceps and the triceps. Two tendons attach the upper part of the biceps to the scapula (shoulder blade) ❶. At the opposite end of the muscle, a tendon attaches the biceps to the radius of the lower arm ❷. When the biceps contracts (shortens), the lower arm is pulled toward the shoulder. You can feel this happen if you extend your arm outward, place your other hand over the biceps, and slowly bend the elbow. Feel the biceps contract? Even when a biceps shortens only a bit, it causes a large motion of the bone connected to it.

Muscles can only pull on bones; they cannot push. Often two muscles work in opposition, with the action of one resisting or reversing action of another. For example, the triceps in the upper arm opposes the biceps. When you pull your arm toward your shoulder, the triceps muscle relaxes as the biceps contracts. Contraction of the triceps coupled with relaxation of the biceps reverses this movement, extending the arm.

Bear in mind that only skeletal muscle is the functional partner of bone. As Section 19.3 explained, smooth muscle is a component of some blood vessels and of the stomach, bladder, and other internal organs. Cardiac muscle is present only in the heart wall. We consider the functions of smooth muscle and cardiac muscle in later chapters.

The human body has close to 700 skeletal muscles, some near the surface, others deep inside the body wall. Collectively, skeletal muscles account for about 40 percent of the body weight of a young man of average fitness. Most skeletal muscles move bones, but some have other functions. Skeletal muscles that pull on facial skin cause changes in expression. Others attach to and move the eyeball, or open and close eyelids. The tongue is skeletal muscle, and sphincters of skeletal muscle allow voluntary control of defecation and urination.

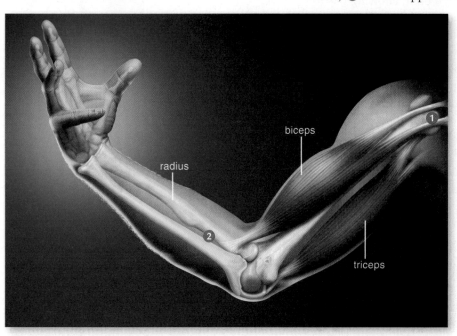

biceps

radius

triceps

FIGURE 20.5 Some bones and muscles of the human arm.

Take-Home Message

How do skeletal muscles work together with bones?

- Tendons of dense connective tissue attach skeletal muscles to bones.
- When skeletal muscle contracts, it pulls on the bone to which it is attached.
- Skeletal muscles often work in opposition, with the action of one muscle opposing or reversing the action of another.

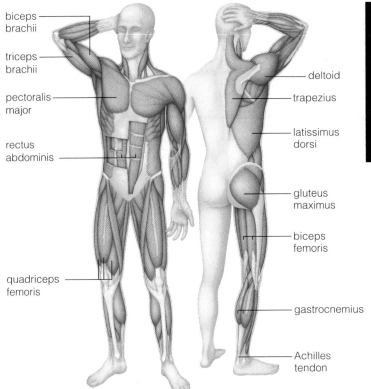

biceps brachii
triceps brachii
pectoralis major
rectus abdominis
quadriceps femoris

deltoid
trapezius
latissimus dorsi
gluteus maximus
biceps femoris
gastrocnemius
Achilles tendon

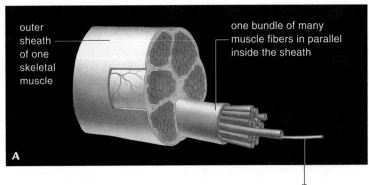

outer sheath of one skeletal muscle

one bundle of many muscle fibers in parallel inside the sheath

A

B one myofibril, made up of sarcomeres arranged end to end

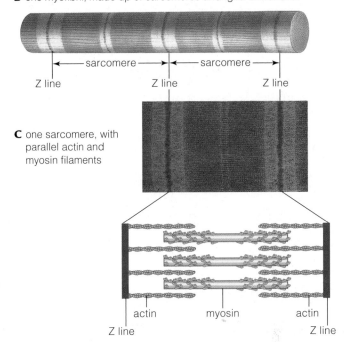

sarcomere — sarcomere

Z line   Z line   Z line

C one sarcomere, with parallel actin and myosin filaments

actin   myosin   actin

Z line   Z line

FIGURE 20.6 Animated! Skeletal muscle structure. *Above*, some of the major skeletal muscles and tendons. Tendons are shown in *light blue*.
A Each skeletal muscle consists of many muscle fibers bundled together inside a sheath of connective tissue.
B Each muscle fiber contains many myofibrils. Myofibrils consist of contractile units called sarcomeres arranged end to end.
C A sarcomere consists of thin actin filaments and thick myosin filaments in a parallel array. Dark regions called Z lines mark the end of the sarcomere.

Skeletal Muscle Structure and Function

Our many skeletal muscles are the voluntary muscles that allow us to type, to smile, and to dance. A skeletal muscle consists of many muscle fibers bundled together inside a sheath of dense connective tissue (Figure 20.6**A**). Each muscle fiber consists of many **myofibrils**, threadlike, cross-banded structures that parallel the muscle's long axis (Figure 20.6**B**). **Sarcomeres**, the basic units of contraction, are arranged end to end within the myofibril. Dark regions called Z lines mark the ends of each sarcomere. A sarcomere has thin and thick filaments arrayed parallel to one another and to the muscle's long axis (Figure 20.6**C**).

Within a sarcomere, two coiled strands of the globular protein **actin** make up each thin filament. **Myosin**, a motor protein with a club-shaped head, makes up thick filaments. The myosin heads of the thick filaments are just a few nanometers away from the thin actin filaments.

Muscle bundles, muscle fibers, myofibrils, and thick and thin filaments of a sarcomere are all in the same parallel orientation. As a result, they all pull in the same direction when a muscle contracts.

actin Globular protein; in thin filaments of muscle fibers.
myofibrils Threadlike, cross-banded skeletal muscle components that consist of sarcomeres arranged end to end.
myosin Motor protein with a club-shaped head; in thick filaments of muscle fibers.
sarcomere Contractile unit of skeletal and cardiac muscle.
tendon Strap of dense connective tissue that connects a skeletal muscle to bone.

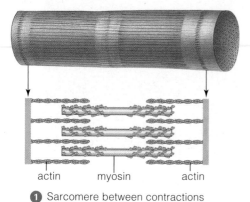

actin   myosin   actin

1 Sarcomere between contractions

FIGURE 20.7 Animated! The sliding-filament model, which explains skeletal muscle contraction.

1 A sarcomere in a muscle at rest. Actin and myosin filaments lie next to one another, but are not interacting.

2 Myosin heads in the thick filaments have been activated by a phosphate-group transfer from ATP. ADP and phosphate remain attached to the myosin.

3 Release of calcium from intracellular storage allows myosin to bind to actin.

4 The myosin head releases bound ADP and phosphate as it tilts toward the sarcomere center and slides the attached actin filaments along with it.

5 New ATP binds to the myosin heads, causing them to release their grip on actin and return to their original orientation, ready to act again.

6 Many myosin heads repeatedly bind to and pull on adjacent actin filaments. Their collective action makes the sarcomere shorten (contract).

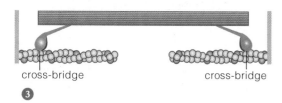

myosin head

2 one of many myosin-binding sites on actin

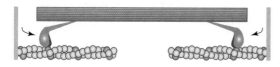

cross-bridge   cross-bridge

3

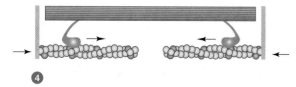

4

ATP   ATP

cross-bridge broken   cross-bridge broken

5

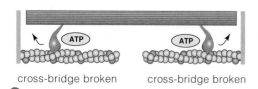

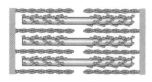

6   Same sarcomere, contracted

How a Muscle Contracts

How do sarcomeres shorten during muscle contraction? The **sliding-filament model** explains the process (Figure 20.7). According to this model, neither actin nor myosin filaments shorten during contraction. A myosin filament remains in place and heads at either end of it attach to adjacent actin filaments. After attaching, the myosin heads slide the actin toward the sarcomere's center. Because each actin filament is attached to the end of the sarcomere, pulling actin inward shortens the sarcomere.

Calcium and ATP are needed for muscle contraction. When a muscle is relaxed, the actin and myosin filaments of a sarcomere lie close to one another but do not interact **1**, **2**. In preparation for contraction, ATP binds to a myosin head, and is broken into ADP and phosphate, which remain attached.

All skeletal muscles contract in response to signals from motor neurons. The signals trigger the release of calcium ions from a special type of endoplasmic reticulum inside the muscle cell. The resulting increase in intracellular calcium allows myosin heads to bind to an adjacent actin filament **3**. Once actin and myosin bind, the myosin head tilts toward the center of the sarcomere, pulling the attached actin with it **4**. At the same time, the ADP and phosphate that were attached to the myosin head are released.

Binding of a new ATP frees the myosin head from actin and it resumes its original position **5**. The myosin head then attaches to another binding site on actin, tilts in another stroke, and so on. The sarcomere shortens as hundreds of myosin heads perform a series of strokes all along the actin filaments **6**.

Getting Energy for Contraction

Muscle contraction is work and so requires energy. Stored ATP is the first energy source a muscle uses, but cells usually store just enough for a few seconds of work. Once that ATP is used up, the muscle turns to its store of creatine phosphate (Figure 20.8, pathway 1). Phosphate transfers from creatine phosphate to ADP produce more ATP, and keep a muscle going for another 5 to 10 seconds. Some athletes take creatine supplements to increase the amount of creatine phosphate available in muscles. Supplements can improve strength, but excessive amounts can harm kidneys.

Most ATP used during prolonged, moderate activity is produced by aerobic respiration (Figure 20.8, pathway 2). Glucose derived from breakdown of glycogen fuels five to ten minutes of activity. Next, glucose and fatty acids delivered to muscle fibers by the blood are broken down. After a half hour of activity, fatty acids become the main source of fuel (Section 5.8).

Also, even in resting muscle, some pyruvate is converted to lactate by fermentation (Figure 20.8, pathway 3). Lactate production increases with exercise.

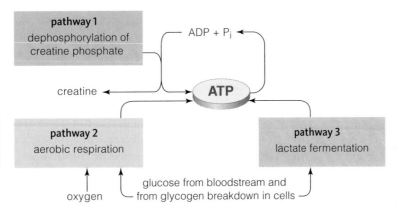

FIGURE 20.8
Animated! Three metabolic pathways by which muscles obtain the ATP molecules that fuel their contraction.

Figure It Out: Which pathway requires oxygen?

Answer: Aerobic respiration.

pathway 1
dephosphorylation of creatine phosphate

ADP + P$_i$

creatine

ATP

pathway 2
aerobic respiration

pathway 3
lactate fermentation

oxygen

glucose from bloodstream and from glycogen breakdown in cells

Although this pathway does not yield much ATP, it can operate even when a muscle is low on oxygen.

Take-Home Message How does skeletal muscle contract?

- A skeletal muscle shortens through combined decreases in the length of its numerous sarcomeres. Sarcomeres are the basic units of skeletal and cardiac muscle contraction.

- The parallel orientation of a skeletal muscle's components directs the force of contraction toward a bone.

- Interactions between myosin and actin filaments in the many sarcomeres of a muscle cell collectively bring about muscle contraction.

- Muscle contraction requires ATP, which can be provided by several different metabolic pathways.

motor unit One motor neuron and the muscle fibers it controls.
muscle twitch Brief muscle contraction.
sliding-filament model Explanation of how interactions among actin and myosin filaments shorten a sarcomere and bring about muscle contraction.

20.5 Properties of Whole Muscles

Muscle Tension A motor neuron has many endings that send signals to different fibers inside each muscle. Each motor neuron and the muscle fibers it controls constitute a **motor unit**. Brief stimulation from the motor neuron causes all fibers in the motor unit that it controls to contract for a few milliseconds. This brief contraction and relaxation is a **muscle twitch** (Figure 20.9).

When a new stimulus is applied before a response ends, the muscle twitches once again. When a motor unit is stimulated repeatedly during a short interval, the twitches all run together. The result is a sustained contraction that produces three or four times the force of a single twitch.

When the nervous system signals a motor unit to contract, all muscle fibers in that motor unit act together. The nervous system cannot make only some of the fibers in a motor unit contract. Because different muscles perform different tasks, the number of muscle fibers controlled by a single motor neuron varies. In motor units that control small, fine movements such as those that control eye muscles, one motor neuron signals only 5 or so muscle fibers. In the biceps of the arm, there are about 700 fibers per motor unit. Having many fibers contract together at once increases the force a muscle can generate.

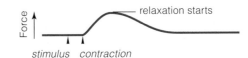

A A single, brief stimulus causes a twitch.

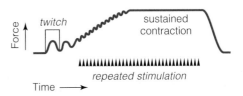

B Repeated stimulation results in a sustained contraction with several times the force of a twitch.

FIGURE 20.9 Animated! Recordings of force generated by a muscle fiber when the motor neuron controlling it is stimulated.

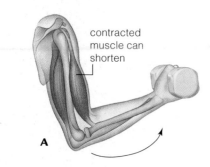

contracted muscle can shorten

A

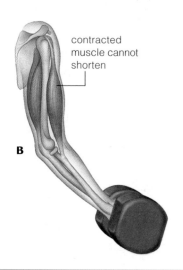

contracted muscle cannot shorten

B

FIGURE 20.10 Two types of muscle contraction. **A** Isotonic contraction. The load is less than a muscle's peak capacity to contract. The muscle can contract, shorten, and lift the load. **B** Isometric contraction. The load exceeds a muscle's peak capacity, so actin and myosin filaments interact but the muscle cannot shorten.

muscle fatigue Decrease in a muscle's ability to contract despite ongoing stimulation.
muscle tension Force exerted by a contracting muscle.

Muscle tension is the mechanical force exerted by a muscle on an object. It is affected by the number of fibers recruited into action. Opposing this force is a load: either an object's weight or gravity's pull on the muscle. Only when muscle tension exceeds an opposing force does a stimulated muscle shorten. Isotonically contracting muscles shorten and move a load (Figure 20.10**A**). Isometrically contracting muscles develop tension but cannot shorten, as when you attempt to lift something that is too heavy (Figure 20.10**B**).

Muscle fatigue is a decrease in a muscle's capacity to generate force; muscle tension declines despite ongoing stimulation. After a few minutes of rest, the fatigued muscle will contract again in response to stimulation.

Aerobic exercise—low intensity, but long duration—makes muscles more resistant to fatigue. It increases their blood supply and the number of mitochondria, the organelles that produce the bulk of ATP during aerobic respiration. By contrast, strength training (intense, short-duration exercise such as weight lifting) stimulates formation of more actin and myosin, as well more enzymes of glycolysis. Strong, bulging muscles develop, but these muscles do not have much endurance. They fatigue rapidly.

As people age, their muscles generally begin to shrink. The number of muscle fibers declines and the remaining fibers increase their diameter more slowly in response to exercise. Muscle injuries take longer to heal. Even so, exercise can be helpful at any age. Strength training can slow the loss of muscle tissue. Aerobic exercise can improve circulation. In addition, aerobic exercise by the middle-aged and elderly lifts major depression as well as many drugs can. Aerobic exercise may help improve memory and capacity to plan and organize complex tasks. No matter what your age, exercise is good for more than just muscles. It is also good for the brain.

Impaired Muscle Contraction Muscular dystrophies are genetic disorders in which skeletal muscles progressively weaken. With Duchenne muscular dystrophy, symptoms begin to appear in childhood. A mutation of a gene on the X chromosome causes this disorder. This affected gene encodes a protein (dystrophin) in the plasma membrane of muscle fibers. Having a defective form of this protein allows foreign material to enter a muscle fiber, causing the fiber to break down (Figure 20.11).

About 1 in 3,500 males are born with Duchenne muscular dystrophy. Like other X-linked disorders, it rarely causes symptoms in females, who nearly always have a normal version of the gene on their other X chromosome. Affected boys usually begin to weaken by the time they are three years old, and require a wheelchair in their teens. Most die in their twenties from respiratory failure that occurs when skeletal muscles involved in breathing stop functioning.

When motor neurons cannot signal muscles to contract, or signaling is impaired, skeletal muscles weaken or become paralyzed. For example, muscle paralysis often occurs after poliovirus infects and kills motor neurons. Poliovirus

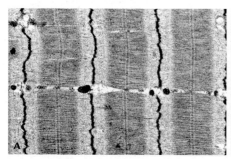

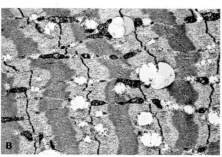

FIGURE 20.11 Electron micrographs of **A** normal skeletal muscle tissue and **B** muscle tissue of a person affected by muscular dystrophy.

most frequently infects children; those who survive an infection may be paralyzed or have a weakened voluntary muscle response as a result. Polio vaccines have been available since the 1950s, so the disease is on the decline. In the United States, it has been essentially eliminated. However, infections still occur in less developed countries. Also, some adults who had polio as children develop postpolio syndrome, the symptoms of which include fatigue and progressive muscle weakness.

Some bacterial toxins can interfere with muscle action by disrupting motor neuron function. For example, endospores of the toxin-producing bacteria *Clostridium tetani* are common in soil. Sometimes these spores get into a deep wound, resulting in a bacterial infection. The bacteria produce a toxin that causes unrelenting stimulation of motor neurons, resulting in the symptoms of the disease tetanus. Muscles stiffen and cannot be released from contraction. Tetanus is commonly called "lockjaw" because jaw muscles that become tightly clenched prevent the mouth from opening. The backbone becomes locked in an abnormally arching curve (Figure 20.12). Affected people die when the skeletal muscles involved in breathing become locked in contraction. Vaccines have eradicated tetanus in the United States, but the annual death toll remains over 200,000 worldwide. Most who die are newborns infected during an unsanitary delivery.

Amyotrophic lateral sclerosis (ALS) also kills motor neurons. It is sometimes called Lou Gehrig's disease, after a famous baseball player whose career was cut short by the disease in the late 1930s. ALS usually causes death by respiratory failure within three to five years of diagnosis. Some inherited mutations cause ALS, but most cases arise in people with no family history of the disorder.

FIGURE 20.12 Painting of a young victim of a contaminated battle wound dying of tetanus in a military hospital during the 1800s.

Take-Home Message What factors affect whole muscle function?

- Collectively, interacting filaments in sarcomeres exert a mechanical force, called muscle tension, in response to an external force acting on the muscle.

- Properties of muscles vary with age and activity. Exercise can be beneficial at any age. Diseases and disorders can interfere with contraction.

20.6
Impacts/Issues Revisited:
Bulking Up Muscles

Researchers have long been searching for ways to slow muscle loss resulting from muscular dystrophy, ALS, and even normal aging. Drugs that inhibit myostatin production or prevent myostatin from acting may help them reach this goal. One way to learn what sort of effects such drugs might have is to study mice in which the myostatin gene was knocked out. As one example, the larger mouse in the photo at the *right* is a knockout mutant. Its myostatin gene was disabled by genetic engineering, so it is larger and more muscular than the normal mouse beside it. The bad news is that such mice have unusually small, stiff, easily torn tendons. Thus, it is likely that increased tendon injuries will be a side effect of myostatin-inhibiting drugs.

How Would
YOU
☑ **Vote?**

Dietary supplements that claim to block myostatin are already for sale. Should dietary supplements be subject to testing for effectiveness and safety as drugs are? **See CengageNow for details, then vote online (cengagenow.com).**

Summary

Section 20.1 Exercise makes muscles bigger, not by adding cells but by adding proteins to existing cells. Hormones and other molecules regulate this process.

Section 20.2 Muscles interact with skeletons. Three categories of skeletal systems are common in animals— hydrostatic skeletons, exoskeletons, and endoskeletons. Different kinds apply contractile force against body fluids or structural elements, such as bones.

Humans, like other vertebrates, have an endoskeleton. The skeleton consists of skull bones, a vertebral column (backbone), a rib cage, a pelvic girdle, a pectoral girdle, and paired limbs. The vertebral column consists of individual segments called vertebrae, with intervertebral disks between them. The spinal cord runs through the vertebral column and connects with the brain through a hole in the base of the skull. The shape of the human backbone is an evolutionary adaptation to upright walking.

Bones are collagen-rich, mineralized organs. They function in mineral storage, movement, and protection and support of soft organs. Blood cells form in bones that contain red marrow. Ongoing mineral deposits and removals help maintain blood levels of calcium and phosphorus, and also adjust bone strength. Bone density declines with age.

Bones meet at joints. Fibrous joints hold bones tightly in place and cartilaginous joints let them move a bit. Synovial joints allow the most motion. Ligaments connect bones at synovial joints.

CENGAGENOW Use the interaction on CengageNow to learn about the bones of the human skeleton.

Section 20.3 Tendons of connective tissue attach skeletal muscles to bones. When skeletal muscles contract, they transmit force that makes the bones move. Muscles can only pull on bones; they cannot push them. Many muscles work as opposing pairs.

CENGAGENOW See how opposing muscle groups interact with the animation on CengageNow.

Section 20.4 Skeletal muscles contract in response to signals from the nervous system. The internal organization of a skeletal muscle promotes a strong, directional contraction. Many myofibrils make up a skeletal muscle fiber.

A myofibril consists of sarcomeres, units of muscle contraction lined up along its length. Each sarcomere has parallel arrays of actin and myosin filaments. The sliding-filament model describes how ATP-driven sliding of myosin filaments past actin filaments shortens the sarcomere.

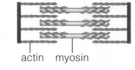

actin   myosin

Muscle fibers produce the ATP needed for contraction by way of three pathways: dephosphorylation of creatine phosphate, aerobic respiration, and lactate fermentation.

CENGAGENOW Explore the structure of a muscle, see how it contracts, and learn what fuels its contraction with the animation on CengageNow.

Section 20.5 Muscle tension is a mechanical force caused by muscle contraction. A muscle shortens only when muscle tension exceeds an opposing load.

A motor neuron and all muscle fibers that form junctions with its endings are a motor unit. All fibers of a motor unit contract at the same time. Repeated stimulation of a motor unit results in a strong, sustained contraction.

Genetic disorders that affect muscle structure impair muscle function. So do some diseases and toxins that affect motor neurons.

CENGAGENOW Use the interaction on CengageNow to see how repeated stimulation causes muscle contraction.

Self-Quiz   Answers in Appendix I

1. A hydrostatic skeleton consists of _____ .
   a. a fluid in an enclosed space
   b. hardened plates at the surface of a body
   c. internal hard parts
   d. none of the above

2. Bones are _____ .
   a. mineral reservoirs   c. sites where blood cells
   b. skeletal muscle's   form (some bones only)
      partners   d. all of the above

3. The human backbone _____ .
   a. consists of vertebrae and intervertebral cartilage disks
   b. has a shape adapted to upright walking
   c. encloses the spinal cord
   d. all of the above

4. Bones move when _____ muscles contract.
   a. cardiac   c. smooth
   b. skeletal   d. all of the above

5. A ligament connects _____ .
   a. bones at a joint   c. a muscle to a tendon
   b. a muscle to a bone   d. a tendon to bone

6. Binding of ATP to _____ activates it and prepares it to take part in muscle contraction.
   a. actin   c. collagen
   b. myosin   d. bone

7. A sarcomere shortens when _____ .
   a. thick filaments shorten
   b. thin filaments shorten
   c. both thick and thin filaments shorten
   d. none of the above

8. Release of _____ from intracellular storage allows actin and myosin to interact.
   a. sodium ions   c. calcium ions
   b. potassium ions   d. collagen

9. ATP for muscle contraction can be formed by _____ .
   a. aerobic respiration   c. creatine phosphate breakdown
   b. lactate fermentation   d. all of the above

Digging Into Data

Building Stronger Bones

Tiffany (*right*) was born with multiple fractures in her arms and legs. By age six, she had undergone surgery to correct more than 200 bone fractures. Her fragile, easily broken bones are symptoms of osteogenesis imperfecta (OI), a genetic disorder caused by a mutation in a gene for collagen. As bones develop, collagen forms a scaffold for deposition of mineralized bone tissue. The scaffold forms improperly in children with OI. Figure 20.13 shows the results of an experimental test of a new drug. Bones of treated children, all less than two years old, were compared to bones of a control group of similarly affected, same-aged children who did not receive the drug.

1. An increase in vertebral area during the 12-month period of the study indicates bone growth. How many of the treated children showed such an increase?

2. How many of the untreated children showed an increase in vertebral area?

3. How did the rate of fractures in the two groups compare?

4. Do the results shown support the hypothesis that this drug can increase bone growth and reduce fractures in young children who have OI?

Treated child	Vertebral area in cm^2 (Initial)	(Final)	Fractures per year
1	14.7	16.7	1
2	15.5	16.9	1
3	6.7	16.5	6
4	7.3	11.8	0
5	13.6	14.6	6
6	9.3	15.6	1
7	15.3	15.9	0
8	9.9	13.0	4
9	10.5	13.4	4
Mean	11.4	14.9	2.6

Control child	Vertebral area in cm^2 (Initial)	(Final)	Fractures per year
1	18.2	13.7	4
2	16.5	12.9	7
3	16.4	11.3	8
4	13.5	7.7	5
5	16.2	16.1	8
6	18.9	17.0	6
Mean	16.6	13.1	6.3

FIGURE 20.13 Results of a clinical trial of a drug treatment for osteogenesis imperfecta (OI), also known as brittle bone disease. The drug being tested is a compound that reduces the rate of bone breakdown. Nine children with OI received the drug. Six others were untreated controls. Surface area of specific vertebrae was measured before and after treatment. Fractures occurring during the 12 months of the trial were also recorded.

---

10. A motor unit is _____ .
    a. a muscle and the bone it moves
    b. two muscles that work in opposition
    c. the amount a muscle shortens during contraction
    d. a motor neuron and the muscle fibers it controls

11. A muscle can _____ bone.
    a. push against   b. pull on   c. both   d. neither

12. What protein makes up the thick filaments in a sarcomere?

13. The knee is a _____ joint.
    a. synovial   b. cartilaginous   c. fibrous

14. Match each bone with its description.
    _____ femur   a. part of skull
    _____ radius   b. thighbone
    _____ vertebrae   c. segments of backbone
    _____ sternum   d. breastbone
    _____ cranial bones   e. forearm bone

15. Match the words with their definitions.
    _____ muscle fatigue   a. genetic disorder of muscles
    _____ muscle twitch   b. makes ATP; requires oxygen
    _____ muscle tension   c. decline in bone density
    _____ muscular dystrophy   d. decline in muscle tension
    _____ osteoporosis   e. chronic joint inflammation
    _____ arthritis   f. brief contraction
    _____ sprain   g. force exerted by contraction
    _____ dislocation   h. makes ATP without oxygen
    _____ lactate fermentation i. misplaced bones
    _____ aerobic respiration   j. damage to ligaments

Additional questions are available on CENGAGENOW™

Critical Thinking

1. Humans can trace their ancestry back to reptile ancestors that walked on all fours. The graphic *below* shows what the skeleton of these reptiles was probably like. Compare this graphic with the human skeleton in Figure 20.2 and list three skeletal changes that occurred during the evolutionary transitions from early reptiles that walked on four legs to humans.

2. Compared to most people, long-distance runners have more cells in their skeletal muscles, and the cells have more mitochondria. By comparison, sprinters have skeletal muscle fibers that have more of the enzymes necessary for glycolysis, but fewer mitochondria. Think about the differences between these two forms of exercise. Explain why the two kinds of runners have such pronounced differences in their muscles.

3. Zachary's younger brother Noah had Duchenne muscular dystrophy and died at the age of 16. Zachary is now 26 years old, healthy, and planning to start a family of his own. However, he worries that his sons might be at high risk for muscular dystrophy. His wife's family has no history of this genetic disorder. Review Sections 9.6 and 20.5 and decide whether Zachary's concerns are well founded.

4. Athletes tend to have stronger bones than nonathletes, but different sports strengthen different bones. Volleyball, basketball, gymnastics, and soccer all thicken the femur, whereas swimming, skating, and bicycling have little effect on this bone. What does this tell you about how exercise promotes increases in bone density?

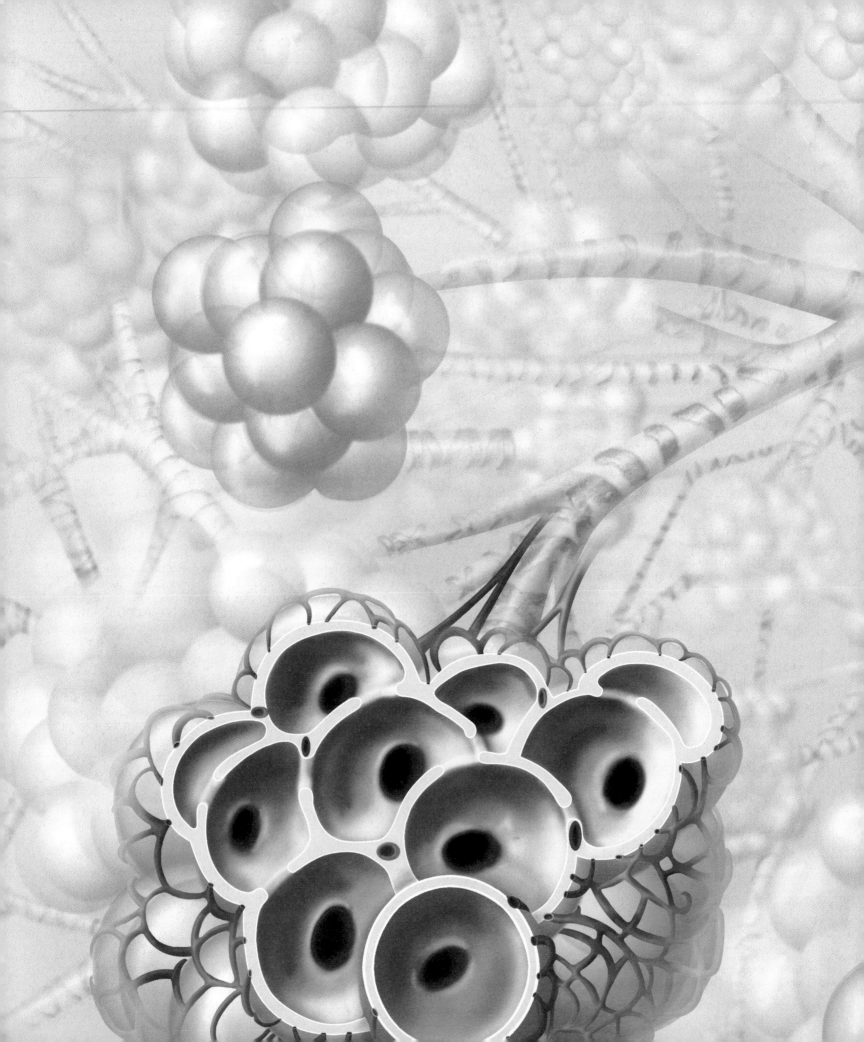

21 Circulation and Respiration

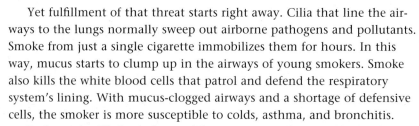

Impacts/Issues: Up in Smoke

21.1

Each day 3,000 or so teenagers join the ranks of habitual smokers in the United States. Most are not even fifteen years old. The first time they light up, they cough and choke on irritants in the smoke. They typically become dizzy and nauseated, and develop headaches.

Sound like fun? Hardly. So why do they ignore the signals of the threat to the body and work so hard to be a smoker? Mainly to fit in. To many adolescents, the perceived social benefits overwhelm the seemingly remote threat to their health.

Yet fulfillment of that threat starts right away. Cilia that line the airways to the lungs normally sweep out airborne pathogens and pollutants. Smoke from just a single cigarette immobilizes them for hours. In this way, mucus starts to clump up in the airways of young smokers. Smoke also kills the white blood cells that patrol and defend the respiratory system's lining. With mucus-clogged airways and a shortage of defensive cells, the smoker is more susceptible to colds, asthma, and bronchitis.

Each smoke-filled inhalation delivers nicotine that quickly reaches the circulatory system and brain. This highly addictive stimulant constricts blood vessels, which makes the heart beat harder and raises blood pressure. Smoking also raises the blood level of "bad" cholesterol (LDL) and lowers that of "good" cholesterol (HDL). It makes blood stickier and more sluggish, and it promotes the formation of clots. Clogged arteries, heart attacks, and strokes are among the physiological costs of a social pressure that leads to addiction.

Which smokers have not heard that they are inviting painful, deadly lung cancers? Maybe they do not know that carcinogens in cigarette smoke can induce cancers in organs throughout their body. For instance, we now know that females who start smoking when they are teenagers are about 70 percent more likely to develop breast cancer than those who never took up the habit.

The damaging effects of cigarette smoke are not confined to smokers. Their families, coworkers, and friends get an unfiltered dose of the toxins in tobacco smoke. Urine samples from nonsmokers who live with smokers show that their body contains high levels of carcinogens. The National Cancer Institute estimates that exposure to secondhand smoke causes about 3,000 lung cancer deaths and 46,000 heart disease deaths each year in the United States alone. Children exposed to secondhand smoke are more prone to develop respiratory ailments such as asthma, as well as middle ear infections.

This chapter focuses on respiration and circulation, and how they contribute to homeostasis in the internal environment. If you or someone you know is a smoker, you might use the chapter as a preview of what it does to internal operating conditions. For a more graphic preview, find out what goes on every day with smokers in emergency rooms and intensive care units. No glamor there. It is not cool, and it is not pretty.

Moving Substances Through a Body

Open and Closed Circulatory Systems All animals have to keep their cells supplied with nutrients and oxygen. Distributing these materials through the body is an important aspect of homeostasis, as is removing waste.

In some invertebrates, including flatworms and sea anemones, solutes and gases simply diffuse through the **interstitial fluid**, the fluid between cells. As explained in Section 4.5, diffusion is the movement of a substance from an area where it is concentrated to one where it is less concentrated. Animals that rely on diffusion alone to move materials within their body tend to be small, with all cells close to the body surface and digestive cavity. Substances do not have to diffuse very far to reach all of the cells in such bodies.

Animals with larger or more complex bodies typically have a circulatory system, which is an organ system that speeds the distribution of substances through the body. Circulatory systems include one or more **hearts**, muscular organs that pump a fluid through a system of tubular vessels.

There are two types of circulatory systems. In an **open circulatory system**, a heart or hearts pump blood into large vessels that empty into spaces around body tissues. The blood washes over body cells, then is drawn back up into the heart. A grasshopper has this type of system (Figure 21.1**A,B**).

All vertebrates and some invertebrates, including octopuses and earthworms, have a **closed circulatory system**. In such a system, a heart or hearts pump blood through a continuous series of vessels (Figure 21.1**C,D**). A closed circulatory system distributes substances faster than an open one. It is "closed" because blood does not flow out of blood vessels to bathe the tissues. Instead, most transfers between blood and the cells of other tissues take place by diffusion across the walls of **capillaries**, the smallest-diameter blood vessels. The network of capillaries supplying blood to an organ is called a capillary bed.

Evolution of Vertebrate Cardiovascular Systems All vertebrates have a closed circulatory system, with a single heart and a network of blood vessels. However, the structure of the heart and the circuits through which blood flows differs among vertebrate groups. In most fishes, the heart has

capillaries Smallest-diameter blood vessels; site of exchanges of gases and other materials with the tissues.
closed circulatory system Circulatory system in which blood flows through a continuous network of vessels.
heart Muscular organ that pumps fluid through a body.
interstitial fluid Fluid between cells of a multicelled body.
open circulatory system Circulatory system in which blood leaves vessels and flows among tissues.

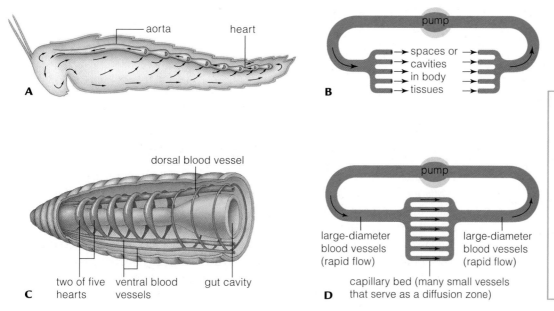

A

B

C two of five hearts · ventral blood vessels · gut cavity · dorsal blood vessel

D large-diameter blood vessels (rapid flow) · capillary bed (many small vessels that serve as a diffusion zone) · large-diameter blood vessels (rapid flow) · pump

FIGURE 21.1 Animated! Comparison of open and closed circulatory systems. **A,B** In a grasshopper's open system, the hearts pump blood through a vessel (an aorta). From there, blood moves into tissue spaces, mingles with fluid bathing cells, then reenters the hearts through openings in the heart wall. **C,D** Blood in the closed system of an earthworm stays inside pairs of muscular hearts near the head and many blood vessels.

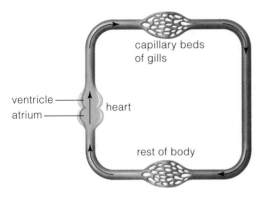

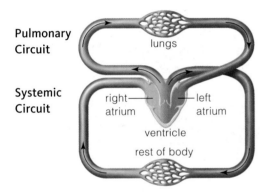

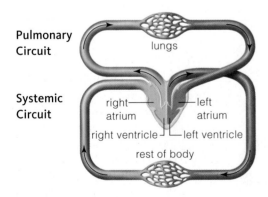

FIGURE 21.2 Animated! Vertebrate circulatory systems. *Red* represents oxygen-rich blood. *Blue* represents blood that has little oxygen.

A A fish heart has one atrium and one ventricle. Force of the ventricle's contraction propels blood through the single circuit.

B In amphibians and most reptiles, the heart has three chambers: two atria and one ventricle. Blood flows in two partially separated circuits. Oxygenated and oxygen-poor blood mix a bit in the ventricle.

C In crocodilians, birds, and mammals, the heart has four chambers: two atria and two ventricles. Oxygenated blood and oxygen-poor blood do not mix.

two main chambers, and blood flows in one circuit (Figure 21.2**A**). One chamber of the heart, an **atrium** (plural, atria), receives blood. From there, the blood enters a **ventricle**, a chamber that pumps blood out of the heart. The pressure exerted when the ventricle contracts drives the blood through a series of vessels, into capillary beds inside each gill, through capillary beds in body tissues and organs, and finally back to the heart. Capillaries dissipate the pressure from the ventricle's contraction, so the blood is not under much pressure when it leaves the gill capillaries, and even less as it travels back to the heart.

Adapting to life on land involved modifications of respiratory and circulatory systems. Amphibians and most reptiles have a three-chamber heart, with two atria emptying into one ventricle (Figure 21.2**B**). The heart of these animals moves blood in two circuits, increasing speed of blood flow. The force of one contraction propels blood through the **pulmonary circuit**, to the lungs and then back to the heart. (The Latin word *pulmo* means lung.) A second contraction sends the newly oxygenated blood through the **systemic circuit**, which runs through capillary beds in body tissues before returning to the heart.

In a three-chamber heart, oxygenated and oxygen-poor blood mix in the single ventricle. Oxygen delivery improved with evolution of a four-chamber heart, which has two atria and two ventricles (Figure 21.2**C**). Crocodilians, birds, and mammals have a four-chamber heart. With two fully separate circuits, only oxygen-rich blood flows to tissues, and blood pressure can be regulated independently in each circuit. Strong contraction of the heart's left ventricle keeps blood moving fast through the long systemic circuit. Less forceful contraction of the right ventricle protects delicate lung capillaries.

Take-Home Message   **What are features of animal circulatory systems?**

- In an open circulatory system, fluid leaves vessels and seeps around tissues.
- In a closed circulatory system, blood flows through a continuous network of blood vessels. Blood flows through a closed circulatory system faster than it flows through an open system. All vertebrates have a closed circulatory system.
- In fishes, blood flows in one circuit. In other vertebrates, it flow in two circuits. The pulmonary circuit carries blood to and from lungs. The systemic circuit carries blood to and from other organs.
- Crocodilians, birds, and mammals have the most efficient system. Their four-chambered heart keeps oxygen-rich blood separate from oxygen-poor blood.

21.3 Human Cardiovascular System

The human circulatory system is called a cardiovascular system. The Greek word *kardia* means heart, and the Latin word *vasculum* means vessel. Like other mammals, humans have a four-chambered heart that pumps blood through two circuits. Each circuit includes a network of blood vessels that carry blood from the heart to a capillary bed and back again (Figure 21.3**A**). In each circuit, the heart pumps blood out of a ventricle and into branching arteries. **Arteries** are wide-diameter blood vessels that carry blood away from the heart and to organs. Within an organ, arteries branch into smaller vessels called **arterioles**. Arterioles

FIGURE 21.3 Animated!

Blood flow in the human cardio-vascular system.

A Flowchart illustrating the components of each circuit.

B In the pulmonary circuit, blood flows from the heart's right ventricle into pulmonary arteries that carry it to the lungs. As it flows through pulmonary capillaries, it picks up oxygen. Pulmonary veins deliver the blood to the heart's left atrium.

C In the systemic circuit, blood flows from the heart's left ventricle into the aorta, the body's largest artery. Branches of the aorta carry the blood to arterioles, then to capillaries in various organs. Blood from these capillary beds collects in venules that feed into veins. The veins return the blood to the heart. They empty into the right atrium.

Figure It Out: Is blood in the pulmonary arteries oxygen-rich or oxygen-poor?

Answer: Oxygen-poor

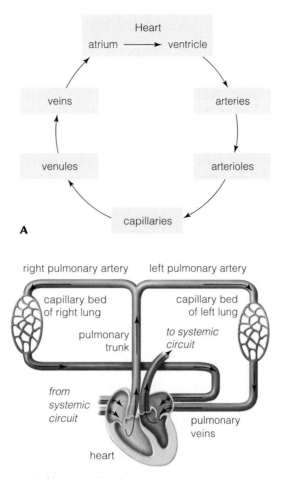

A

B Pulmonary Circuit

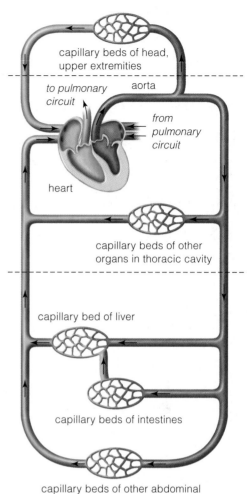

C Systemic Circuit

in turn branch into capillaries. As noted earlier, exchanges between the blood and interstitial fluid take place as blood flows through the capillaries. Several capillaries join up to form a **venule**, a vessel that carries blood to a vein. **Veins** are large-diameter vessels that return blood to the heart. Blood from veins empties into one of the two atria.

Now let's look at the circuits in detail. The shorter pulmonary circuit carries blood to and from lungs (Figure 21.3**B**). Oxygen-poor blood is pumped out of the heart's right ventricle into the pulmonary arteries. One pulmonary artery delivers blood to each lung. As blood flows through pulmonary capillaries, it picks up oxygen and gives up carbon dioxide. Oxygen-rich blood then returns to the heart by way of the pulmonary veins, which empty into the left atrium.

The oxygenated blood next travels through the longer systemic circuit. The heart's left ventricle pumps blood into the body's largest artery, the **aorta**. Arteries that branch from the aorta carry blood to various body organs. Each artery branches into arterioles and then capillaries. Blood gives up oxygen and picks up carbon dioxide as it flows through the capillaries. The now oxygen-poor blood returns to the heart's right atrium.

Most blood moving through the systemic circuit flows through only one capillary bed. However, blood that passes through the capillaries in the small intestine then flows through a vein to a capillary bed in the liver. This arrangement allows the blood to pick up glucose and other substances absorbed from the gut, and deliver them to the liver. The liver stores some of the absorbed glucose as glycogen. It also breaks down some absorbed toxins, including alcohol.

aorta Large artery that receives blood pumped out of the left ventricle.

arteriole Blood vessel that carries blood from an artery to a capillary bed.

artery Large-diameter blood vessel that carries blood from the heart to an organ.

atrium Heart chamber that pumps blood into a ventricle.

pulmonary circuit Circuit through which blood flows from the heart to the lungs and back.

systemic circuit Circuit through which blood flows from the heart to the body tissues and back.

vein Large-diameter vessel that returns blood to the heart.

ventricle Heart chamber that pumps blood out of the heart and into an artery.

venule Small-diameter blood vessel that carries blood from capillaries to a vein.

Contraction of ventricles is the driving force for blood circulation. Atrial contraction only helps fill the ventricles.

Take-Home Message What are the components of a human cardiovascular system and how do they interact?

- The human cardiovascular system has a four-chamber heart that pumps blood through a network of blood vessels in two circuits.

- Oxygen-poor blood pumped out of the heart's right ventricle travels through the pulmonary circuit. As it passes through the lung, the blood gives off carbon dioxide and picks up oxygen. It then returns to the heart at the left atrium.

- Oxygen-rich blood pumped out of the left ventricle flows through the systemic circuit. The blood gives up oxygen in capillary beds of the body and reenters the heart at the right atrium.

- Most blood flows through one capillary bed, but blood from capillaries in the gut travels through capillaries in the liver before returning to the heart.

21.4 The Human Heart

Heart Location and Structure The heart lies in the thoracic cavity, between the lungs. A double-layered sac of tough connective tissue (pericardium) protects and anchors the heart. Fluid between the layers of the sac provides lubrication for the heart's continual motions. The inner layer of the sac is the heart wall, which is mostly cardiac muscle. Endothelium, a special type of epithelium, lines the heart's inner wall, as well as the blood vessels.

As you already know, each side of the human heart has two chambers: an atrium that receives blood from veins, and a ventricle that pumps blood into arteries (Figure 21.4). The superior vena cava and inferior vena cava are large veins that deliver oxygen-poor blood from the body to the right atrium. The pulmonary veins deliver oxygen-rich blood to the left atrium.

To get from an atrium into a ventricle, or from a ventricle into an artery, blood has to pass through a heart valve. Heart valves act like one-way doors. High fluid pressure forces the valve open. When fluid pressure declines, the valve shuts and prevents blood from moving backwards. The "lup-dup" sound made by a beating heart arises from the closing first of the atrioventricular (AV) valves, then the simultaneous closing of aortic and pulmonary valves.

The Cardiac Cycle With each heartbeat, the heart's chambers go through a sequence of relaxation and contraction known as the **cardiac cycle** (Figure 21.5). As we begin our observation of the cardiac cycle, both the atria and ventricles are relaxed. Blood entering the atria has forced the AV valves open, and the ventricles are beginning to fill with blood ❶. More blood is forced into the relaxed ventricles as the atria contract ❷. Next, the ventricles begin contracting. The resulting rise in fluid pressure inside the ventricles causes the AV valves to shut. Their shutting makes the first heart sound. Pressure in the ventricles continues to rise until the aortic and pulmonary

FIGURE 21.4
Animated! Cutaway view of the human heart. Arrows indicate the path of oxygenated blood (*red*) and oxygen-poor blood (*blue*).

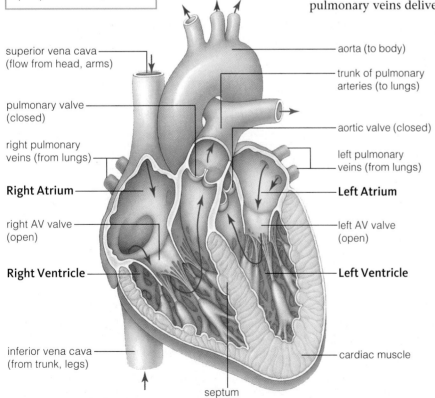

- superior vena cava (flow from head, arms)
- pulmonary valve (closed)
- right pulmonary veins (from lungs)
- **Right Atrium**
- right AV valve (open)
- **Right Ventricle**
- inferior vena cava (from trunk, legs)
- septum
- aorta (to body)
- trunk of pulmonary arteries (to lungs)
- aortic valve (closed)
- left pulmonary veins (from lungs)
- **Left Atrium**
- left AV valve (open)
- **Left Ventricle**
- cardiac muscle

valves open, and blood rushes out of the ventricles and into the aorta and pulmonary arteries ❸. As fluid pressure in the ventricles declines, aortic and pulmonary valves close, making the second heart sound. At this point, the atria are starting to fill again ❹.

Contraction of ventricles is the driving force for blood circulation. Atrial contraction only helps fill ventricles. The structure of the cardiac chambers reflects this difference in function. Atria need only to generate enough force to squeeze blood into the ventricles, so they have relatively thin walls. Ventricle walls are much thicker. Contraction of muscle in the ventricle walls has to be strong enough to create a pressure wave that propels blood through an entire circuit. The left ventricle, which pumps blood throughout the entire body, has thicker walls than the right ventricle, which pumps blood only to the lungs and back.

❶ Relaxed atria fill. Fluid pressure opens AV valves and blood flows into the relaxed ventricles.

❷ Atrial contraction squeezes more blood into the still-relaxed ventricles.

❹ As blood flows into the arteries, pressure in the ventricles declines and the aortic and pulmonary valves close.

❸ Ventricles start to contract and the rising pressure pushes the AV valves shut. A further rise in pressure causes the aortic and pulmonary valves to open.

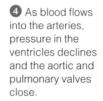

FIGURE 21.5 Animated! The cardiac cycle. **Figure It Out:** Which valves open in response to rising atrial pressure?

Answer: The AV valves

Setting the Pace of Contractions

Like skeletal muscle, cardiac muscle has orderly arrays of sarcomeres that contract by a sliding-filament mechanism (Section 20.4), but cardiac muscle cells are structurally unique. They are branching, short, and connected at their ends. Gap junctions (Section 3.7) connect the cytoplasm of adjacent cardiac muscle cells and allow signals that induce contraction to spread swiftly across the heart.

The sinoatrial (SA) node, a clump of specialized cardiac cells in the right atrium's wall, is the **cardiac pacemaker** (Figure 21.6). About seventy times a minute, it spontaneously generates excitatory signals that call for contraction. The signal from the SA node spreads through the atria, causing them to contract. Simultaneously, the signal travels along fibers to cells of the atrioventricular (AV) node. From the AV node, the signal travels along other fibers in the septum, between the heart's left and right halves. The fibers extend to the heart's lowest point and up the ventricle walls. Ventricles contract from the bottom up, with a twisting motion. The time it takes for a signal to travel from the atria to the ventricles allows ventricles to fill completely before they contract.

If the SA node malfunctions, the heart can stop beating. In people under age 35, such sudden cardiac arrest usually occurs because of an inborn heart defect. In older people, heart disease is usually the cause. Chances of surviving a cardiac arrest rise when **cardiopulmonary resuscitation** (CPR) is started immediately. With this technique, a person alternates mouth-to-mouth respiration with chest compressions that keep a victim's blood moving. However, CPR cannot restart a heart. That requires a **defibrillator**, a device that delivers an electric shock to the chest and resets the SA node. Automated electronic defibrillators (AEDs) are now available in many public places. They allow a trained layperson to give the required shock. A person who survives cardiac arrest may need an artificial pacemaker or an internal defibrillator installed to ensure normal function.

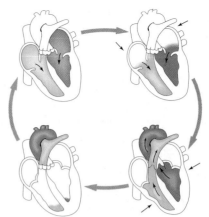

SA node (cardiac pacemaker)

AV node

fibers that relay signals

FIGURE 21.6 Animated! Components of the heart's signaling system. Signals from the SA node travel to the AV node and then on to the ventricles along junctional fibers.

Take-Home Message

How does the human heart function?

■ The four-chambered heart is a muscular pump. Contraction of the thick-walled ventricles drives blood circulation. Atrial contraction fills the ventricles.

■ The SA node is the cardiac pacemaker. Its spontaneous signals make cardiac muscle fibers of the heart wall contract in a coordinated fashion.

cardiac cycle Sequence of contraction and relaxation of heart chambers that occurs with each heartbeat.

cardiac pacemaker Group of heart cells (SA node) that emits rhythmic signals calling for muscle contraction.

cardiopulmonary resuscitation (CPR) Life-saving technique that keeps oxygen flowing to tissue when the heart stops beating; involves mouth-to-mouth respiration and chest compressions.

defibrillator Device that administers an electric shock to the chest wall to reset the SA node and restart the heart.

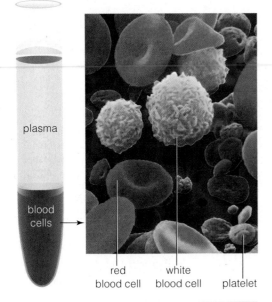

plasma

blood cells →

red blood cell   white blood cell   platelet

FIGURE 21.7 Components of blood.

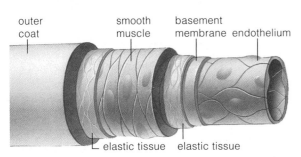

outer coat   smooth muscle   basement membrane   endothelium

elastic tissue   elastic tissue

A Artery

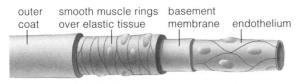

outer coat   smooth muscle rings over elastic tissue   basement membrane   endothelium

B Arteriole

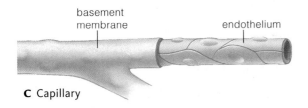

basement membrane   endothelium

C Capillary

FIGURE 21.8 Examples of blood vessel structure. Vessels are not shown to scale.

Blood and Blood Vessels

Components and Functions of Blood An average adult has about 4.5 liters (nearly 5 quarts) of blood. The blood consists of plasma, red blood cells, white blood cells, and platelets (Figure 21.7). All blood cells and platelets arise from stem cells in red bone marrow.

Plasma is mainly water and constitutes about 50 to 60 percent of the total blood volume. It is the transport medium for blood cells and platelets. Nutrients, wastes, signaling molecules, and hundreds of plasma proteins are dissolved in plasma. Some plasma proteins transport lipids and fat-soluble vitamins. Others help blood clot or act against pathogens. Gases such as oxygen and carbon dioxide travel in the plasma, as do the signaling molecules called hormones.

Disk-shaped erythrocytes, or **red blood cells**, are the most numerous cells in blood. As many as 30 trillion circulate in your body. Their red color comes from hemoglobin, an iron-containing pigment that can reversibly bind oxygen. As red blood cells develop in bone marrow, hemoglobin accumulates inside of them. Before they enter the circulation, they expel their nucleus and other organelles. Mature red blood cells circulate for about four months.

Your 10 billion leukocytes, or **white blood cells**, have a variety of roles in housekeeping and defense. Some types patrol tissues. Others accumulate in lymph nodes. White blood cells engulf and digest cellular debris such as aged red blood cells. They also defend the body against viruses, bacteria, and other pathogens. The next chapter discusses their role in defense in more detail.

Platelets are bits of cytoplasm wrapped in plasma membrane. A platelet lasts only five to nine days, but hundreds of thousands circulate in the blood. If an injury occurs, platelets release substances that initiate blood clotting. When small vessels are cut or torn, platelets clump together and temporarily fill the breach. They release substances that attract more platelets. At the same time, blood proteins called fibrinogens stick to the collagen fibers exposed by damage to the vessel wall. The fibrinogen molecules stick together and form long threads that trap blood cells and platelets, as shown in the micrograph at *right*. The entire mass is called a blood clot. Eventually, the clot retracts and forms a compact patch that seals the breach in the blood vessel.

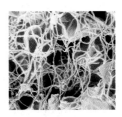

Rapid Transport in Arteries Blood pumped out of ventricles enters arteries that have a muscular wall reinforced with elastic tissue (Figure 21.8**A**). The arteries' structure helps keep blood flowing, even when ventricles are not contracting. With each ventricular contraction, the arteries bulge from pressure exerted when blood is forced into them. Then, while the ventricle relaxes, the artery wall springs back like a rubber band that has been stretched. As the artery wall recoils, it pushes blood inside the artery forward.

The bulging of an artery with each ventricular contraction is referred to as the **pulse**. You can feel a person's pulse if you put your finger on a pulse point, a place where an artery runs near the body surface. For example, to feel the pulse in your radial artery, put your fingers on your wrist near your thumb.

Blood pressure is the pressure exerted by blood against the walls of vessels that enclose it. Because the right ventricle contracts less forcefully than the left ventricle, blood entering the pulmonary circuit is under less pressure than blood entering the systemic circuit. In both circuits, blood pressure is highest in the arteries, and declines as blood flows through the circuit, being lowest in veins.

Blood pressure is usually measured in the brachial artery of the upper arm. Two pressures are recorded. **Systolic pressure**, the highest pressure of a cardiac cycle, occurs as the contracting ventricles are forcing blood into the arteries. **Diastolic pressure**, the lowest blood pressure of a cardiac cycle, occurs when the ventricles are fully relaxed. Blood pressure is typically written as systolic value/diastolic value. It is measured in millimeters of mercury (mm Hg), a standard unit for measuring pressure. Normal blood pressure is about 120/80 mm Hg. In conversation, you would say this is a pressure of "120 over 80."

Adjusting Resistance at Arterioles Depending on your need, your body alters the distribution of blood flow by adjusting the diameter of arterioles (Figure 21.8**B**). For example, after you eat, arterioles that supply blood to your gut dilate, increasing blood flow to this region. At the same time, arterioles supplying blood to skeletal muscles in your legs narrow. When you go for a run, more blood is directed to the legs and less goes to the gut.

Arterioles widen or narrow when smooth muscle that rings arteriole walls responds to the nervous and endocrine systems. Some signals make the smooth muscle relax, thus enlarging blood vessel diameter (vasodilation). Other signals make the smooth muscle contract, narrowing the arteriole (vasoconstriction).

Exchanges at Capillaries A capillary bed is a diffusion zone for exchanges between the blood and interstitial fluid. Between 10 billion and 40 billion capillaries service the human body. Collectively, they offer a tremendous surface area for gas exchanges. Every cell is near at least one capillary. Proximity to a capillary is essential because diffusion occurs too slowly to move substances effectively over long distances.

A capillary wall consists of a single layer of endothelial cells (Figure 21.8**C**). Materials move between capillaries and body cells in several ways. Gases such as oxygen and carbon dioxide diffuse across the plasma membranes of cells and through the interstitial fluid between them. In most tissues, the capillaries are leaky, with narrow spaces between their cells. At the arterial end of a capillary bed, pressure exerted by the beating heart forces fluid out through these spaces into the surrounding interstitial fluid (Figure 21.9**A**). Plasma fluid that leaks out of the capillaries carries oxygen, ions, and essential nutrients such as glucose into interstitial fluid. Most proteins are too big to exit through the spaces between cells and remain in the plasma.

As the blood flows through a capillary bed, blood pressure declines. Near the venous end of the capillary bed, water moves by osmosis from the interstitial fluid into the protein-rich plasma (Figure 21.9**B**). With all the leaking and reentry of fluid, there is a small *net* outward flow from a capillary bed. The extra fluid enters vessels of the lymphatic system and becomes lymph. Lymph vessels feed into lymphatic ducts that return fluid to veins near the heart.

Back to the Heart Blood from capillary beds enters venules, or "little veins," which merge into veins. Veins are large-diameter vessels with one-way valves that prevent blood from flowing backwards. The vein wall is flexible and can bulge greatly under pressure, so the veins serve as a blood reservoir. In a person standing at rest, veins can hold up to 70 percent of the total blood volume.

Like arteries and arterioles, veins have some smooth muscle in their wall. When blood needs to circulate faster, as during exercise, the muscle in vein walls contracts. The resulting increase in blood pressure drives more blood back to the heart. In addition, contractions of skeletal muscles help keep the blood moving. When these muscles contract, they bulge and press against nearby

FIGURE 21.9 Fluid movement at a capillary bed. **A** At a capillary's arteriole end, blood pressure forces plasma fluid out between cells of the capillary wall. Plasma proteins remain behind in the vessel, making the plasma more concentrated.

B Plasma, with its dissolved proteins, has a greater solute concentration than the interstitial fluid. Thus, at the opposite end of the capillary, where blood pressure is lower, water diffuses into the vessel by osmosis.

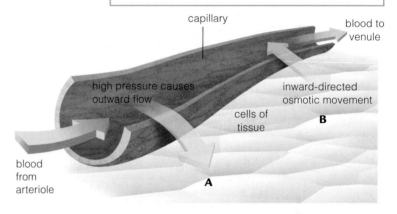

capillary

blood to venule

high pressure causes outward flow

inward-directed osmotic movement

cells of tissue

B

blood from arteriole

A

blood pressure Pressure exerted by blood against the walls of blood vessels.

diastolic pressure Blood pressure when the ventricles are relaxed.

plasma Fluid portion of blood.

platelet Cell fragment that helps blood clot.

pulse Brief stretching of artery walls that occurs when ventricles contract.

red blood cell Hemoglobin-filled blood cells that transport oxygen.

systolic pressure Blood pressure when the ventricles are contracting.

white blood cell Blood cell with a role in housekeeping and defense.

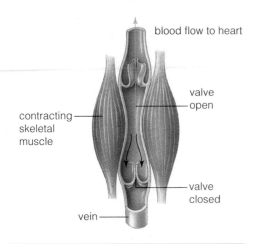

blood flow to heart

contracting skeletal muscle

valve open

valve closed

vein

FIGURE 21.10 Skeletal muscle's effect on a vein. When a skeletal muscle next to a vein contracts, the muscle's increased bulk pushes on the vein, forcing blood inside it through a one-way valve and in the direction of the heart.

veins. The additional pressure on the blood in the veins helps move it toward the heart (Figure 21.10).

If venous valves do not work properly, blood moves backward under the force of gravity and pools in lower body regions. Veins overfilled with blood become enlarged and bulge near the surface of skin. Valve failure commonly occurs in veins of the legs. They become bulging varicose veins. Failure of valves in veins around the anus causes hemorrhoids.

Take-Home Message How do blood and blood vessels function?

- Blood consists mainly of plasma, a protein-rich fluid with solutes and dissolved gases. Red blood cells distribute oxygen. White blood cells defend the body. Platelets are cell fragments that help blood clot after an injury.

- Blood flows into arteries under high pressure. Elastic arterial walls help propel blood forward when ventricles are relaxed. Adjusting arteriole diameter alters how much blood flows to different body regions. Exchanges between the blood and cells occur at capillary beds. Venules and veins return blood to the heart.

21.6 Animal Respiration

The Basis of Gas Exchange In most animals, the circulatory and respiratory systems work together to allow gas exchange. Gas exchange is necessary because animals make ATP mainly by aerobic respiration, which requires oxygen and produces carbon dioxide. While aerobic respiration is a cellular-level metabolic pathway, our focus for the remainder of this chapter is a physiological process called **respiration**. By this process, gases enter and leave an animal body by crossing a moist **respiratory surface**, usually a thin layer of epithelium. The surface must be moist because gases cannot diffuse across it unless they are dissolved in fluid.

The area of a respiratory surface affects the rate of exchange. The larger the surface, the more molecules cross it in any given interval. The need for a large respiratory surface relative to body area constrains body plans. As noted earlier (Section 3.2), as an object increases in size, its volume increases faster than its surface area does. As a result, for an animal to evolve a body plan in which some cells are far from the body surface, it must also evolve respiratory organs.

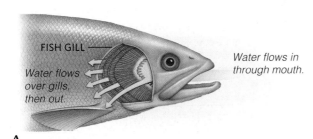

FISH GILL

Water flows over gills, then out.

Water flows in through mouth.

A

FIGURE 21.11 Animated! Fish respiration.

A Location of gills and direction of water flow.

B Bony gill arches support many gill filaments that collectively serve as the respiratory surface.

C A blood vessel carries oxygen-poor blood from the body to each gill filament. Another vessel carries oxygenated blood into the body. Blood flows from one vessel into the other, counter to the flow of water over the gill filament. The flow of blood and water in opposite directions maximizes the amount of oxygen that diffuses from the water into the blood.

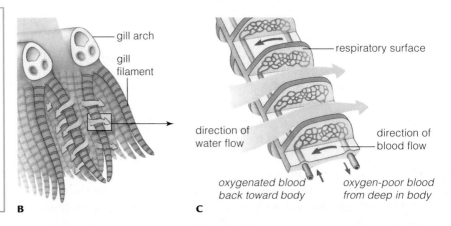

gill arch

gill filament

respiratory surface

direction of water flow

direction of blood flow

oxygenated blood back toward body

oxygen-poor blood from deep in body

B

C

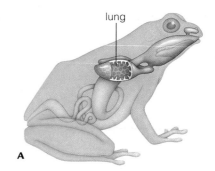

Invertebrate Respiration Some invertebrates that live in aquatic or continually damp land environments rely on **integumentary exchange**: the diffusion of gases across their entire body surface. Flatworms are an example. In other aquatic invertebrates, external **gills** (folds or extensions of the body) increase the surface area across which gases can diffuse.

The most successful air-breathing land invertebrates are insects. They have a hard exoskeleton that helps conserve water but also prevents gas exchange across the body surface. Insects have a **tracheal system** that consists of repeatedly branching, air-filled tubes. Tracheal tubes start at small openings across the body surface, and branch inside the body. Gas exchange occurs at the tips of the finest tracheal branches where gases dissolve in fluid and diffuse into cells.

Gills and Fish Respiration Most fishes have a pair of internal gills: rows of slits at the back of the mouth that extend to the body's surface (Figure 21.11**A**). Water flows into the mouth and pharynx, then over arrays of gill filaments (Figure 21.11**B**). Each filament has blood vessels that carry blood into and out of it (Figure 21.11**C**). Water flows first over the blood vessel that moves blood from the gill back to the rest of the body. Blood in this vessel has less oxygen than the water, so oxygen diffuses into the blood. Next, the water flows over a vessel moving oxygen-poor blood from the body into the gill. Although the water already gave up some oxygen, it still contains more than the blood in this vessel. So a second helping of oxygen diffuses into the blood.

Evolution of Paired Lungs Some fishes, nearly all amphibians, and all mammals and birds have **lungs**—internal organs with respiratory surfaces in the shape of a cavity or a sac. More than 450 million years ago, lungs started to form as outpouchings of the gut of some fishes. In oxygen-poor habitats, lungs provided a larger surface area for gas exchange. Later, an ability to breathe air allowed amphibians descended from these fish to live on land.

Frogs and toads have small lungs for oxygen uptake, but most carbon dioxide diffuses outward across their skin. Unlike us, frogs do not pull air into lungs. Instead, they raise the floor of the mouth and throat, and push the air inside this cavity into their lungs (Figure 21.12**A**). In reptiles, birds, and mammals, the skin is waterproof, so all gas exchange occurs in the lungs.

Birds have a unique system of air sacs connected to their lungs (Figure 21.12**B**). Gases are not exchanged in these sacs. Instead, the sacs increase the rate of gas exchange by causing air to move through lungs, rather than into and out of them. Air sacs are an adaptation to the high oxygen demands of flight. Because of their air sacs, birds have a constant flow of fresh air across their respiratory surface. In contrast, each mammalian inhalation mixes fresh air with some oxygen-depleted air that remained in the lungs after exhalation.

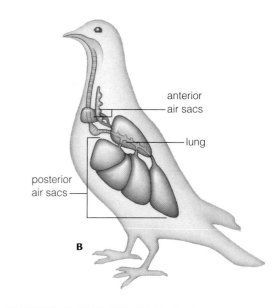

FIGURE 21.12 Animated! Examples of vertebrate lungs. **A** A frog has two small lungs. It fills its throat with air, then squeezes air into its lungs by raising the floor of the throat. **B** Large, stretchy air sacs attach to a bird's two small, inelastic lungs. Contraction and expansion of chest muscles force air into and out of this system. Air flows continually through the bird's lungs.

Take-Home
Message **What is respiration and how are animal bodies adapted to carry it out?**

- Respiration is a physiological process by which animals obtain oxygen and get rid of waste carbon dioxide by diffusion across a moist respiratory surface.

- In some invertebrates, the entire body surface is the respiratory surface. Other invertebrates have specialized external organs for gas exchange, or respiratory systems that bring gases into the body.

- In fish, gill filaments are the respiratory surface. Amphibians exchange gases across their skin and force air into and out of small lungs. In reptiles, birds, and mammals, the action of skeletal muscles draws air into lungs.

gills Folds or body extensions that increase the surface area for respiration.

integumentary exchange Gas exchange across the outer body surface.

lungs Internal saclike organs that serve as the respiratory surface in most land vertebrates and some fish.

respiration Physiological process by which gases enter and leave an animal body.

respiratory surface Moist surface across which gases are exchanged between animals cells and their environment.

tracheal system Tubes that deliver air from body surface to tissues of insects and some other land arthropods.

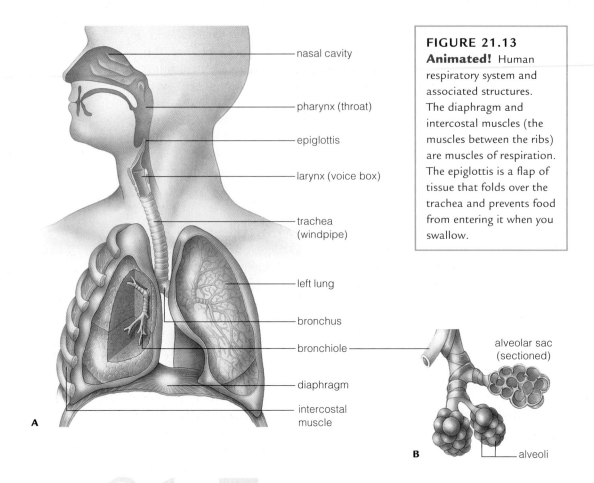

nasal cavity

pharynx (throat)

epiglottis

larynx (voice box)

trachea (windpipe)

left lung

bronchus

bronchiole

diaphragm

intercostal muscle

A

alveolar sac (sectioned)

alveoli

B

FIGURE 21.13 Animated! Human respiratory system and associated structures. The diaphragm and intercostal muscles (the muscles between the ribs) are muscles of respiration. The epiglottis is a flap of tissue that folds over the trachea and prevents food from entering it when you swallow.

glottis open

glottis closed

vocal cords

glottis (closed)

epiglottis

tongue's base

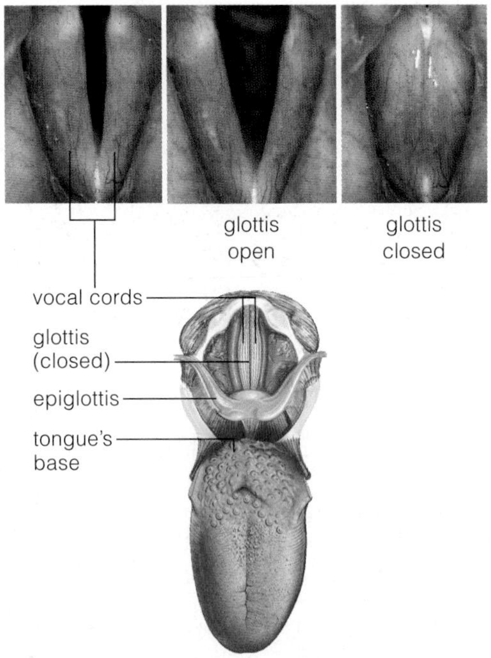

FIGURE 21.14 Animated! The glottis and vocal cords as viewed from above.

21.7 Human Respiratory Function

From Airways to Alveoli Take a deep breath. Now look at Figure 21.13 to get an idea of where the air traveled in your respiratory system. If you are healthy and sitting quietly, air probably entered through your nose, rather than your mouth. Air from the nostrils enters the nasal cavity, where it gets warmed and moistened. It flows next into the **pharynx**, or throat.

Air continues to the **larynx**, a short airway commonly known as the voice box because a pair of vocal cords projects into it (Figure 21.14). Each vocal cord is skeletal muscle with a cover of mucus-secreting epithelium. Contraction of the vocal cords changes the size of the **glottis**, the gap between them. Adjusting the size of the glottis, tension on vocal cords, and speed of air movement allows you to make different sounds. When you are breathing but not speaking, the glottis is wide open. At the entrance to the larynx is an **epiglottis**. When this tissue flap points up, air moves into the **trachea**, or windpipe. When you swallow, the epiglottis flops over, points down, and covers the larynx entrance, so food and fluids enter the esophagus. The esophagus connects the pharynx to the stomach.

The trachea branches into two airways, one to each lung. Each airway is a **bronchus** (plural, bronchi). Human lungs are cone-shaped organs in the thoracic cavity, one on each side of the heart. The rib cage encloses and protects the lungs. A two-layer-thick pleural membrane covers each lung's outer surface and lines the inner thoracic cavity wall.

Once inside a lung, air moves through finer and finer branchings of a "bronchial tree." The branches are called **bronchioles**. At the tips of the finest bronchioles are respiratory **alveoli** (singular, alveolus), little air sacs where gases

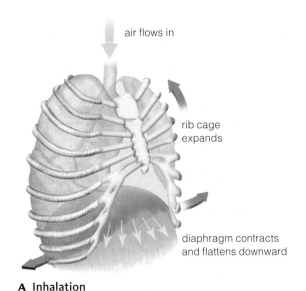

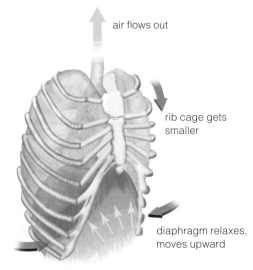

A Inhalation

B Exhalation

air flows in

rib cage expands

diaphragm contracts and flattens downward

air flows out

rib cage gets smaller

diaphragm relaxes, moves upward

FIGURE 21.15
Animated! How muscle actions alter the size of the thoracic cavity during one respiratory cycle.

are exchanged (Figure 21.13**B**). Air in alveoli exchanges gases with blood flowing through pulmonary capillaries. Blood oxygenated in these capillaries returns to the heart, which then pumps it to the body's tissues.

How You Breathe A broad sheet of smooth muscle beneath the lungs, the **diaphragm**, partitions the coelom into a thoracic cavity and an abdominal cavity. Of all smooth muscle, only the diaphragm can be controlled voluntarily. You can make it contract by deliberately inhaling. Actions of the diaphragm and **intercostal muscles**, the skeletal muscles between the ribs, allow you to breathe.

A **respiratory cycle** is one breath in (inhalation) and one breath out (exhalation). Inhalation is always active; muscle contractions drive it. Changes in the volume of the lungs and thoracic cavity cause the movement of air.

When you inhale, the diaphragm contracts and moves downward. At the same time, external intercostal muscles between the ribs contract, moving the rib cage upward and outward (Figure 21.15**A**). The thoracic cavity expands, and air is pulled into the lungs.

Exhalation is usually passive. When muscles that caused inhalation relax, the lungs passively recoil and lung volume decreases. This decrease in volume compresses alveolar sacs, and pushes air out of the lungs (Figure 21.15**B**).

Exhalation becomes active when you exercise vigorously or consciously attempt to expel more air. During active exhalation, internal intercostal muscles contract, pulling the thoracic wall inward and downward. At the same time, muscles of the abdominal wall contract. Abdominal pressure increases and pushes the diaphragm upward. As a result, the volume of the thoracic cavity decreases more than normal, and a bit more air is forced outward.

You do not have to think about breathing. Neurons in a part of your brain stem act as a pacemaker for respiration. When you rest, these neurons send out signals 10 to 14 times per minute. Nerves carry these signals to the diaphragm and intercostal muscles, causing the contractions that result in inhalation. When you are more active, muscle cells increase their rate of aerobic respiration and produce more CO_2. This CO_2 enters blood, causing changes that are detected by receptors in arteries and the brain. In response to these changes, the brain alters the breathing pattern, so you breathe faster and more deeply.

alveoli (singular, alveolus) Tiny, thin-walled air sacs that are the site of gas exchange in the lung.
bronchiole Small airway leading to alveoli.
bronchus (plural, bronchi) Airway connecting the trachea to a lung.
diaphragm Dome-shaped muscle at base of thoracic cavity that alters the size of this cavity during breathing.
epiglottis Tissue flap that folds down to prevent food from entering the airways when you swallow.
glottis Opening formed when the vocal cords relax.
intercostal muscles Muscles between the ribs; help alter the size of the thoracic cavity during breathing.
larynx Short airway containing the vocal cords (voice box).
pharynx Throat; opens to airways and digestive tract.
respiratory cycle One inhalation and one exhalation.
trachea Major airway leading to the lungs; windpipe.

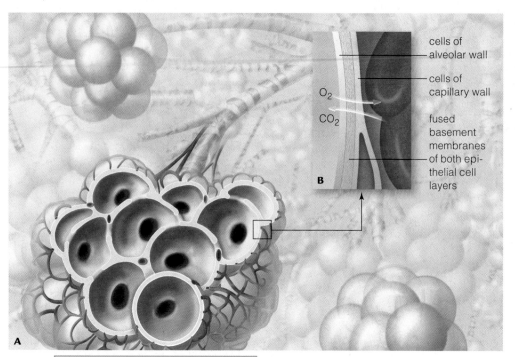

cells of alveolar wall

cells of capillary wall

O₂

CO₂

fused basement membranes of both epithelial cell layers

B

A

FIGURE 21.16 The site of gas exchange. **A** Cutaway view of a cluster of alveoli. Each alveolus is filled with air and surrounded by a network of pulmonary capillaries. **B** The respiratory surface consists of the wall of the alveolus, the wall of the capillary, and their fused basement membranes. Gases diffuse across this three-layered boundary.

Exchanges at the Respiratory Membrane Inhaled air exchanges gases with the blood at the alveoli. Alveolus means cavity in Latin, and each alveolus is a tiny sphere with a central air-filled cavity (Figure 21.16**A**). The wall of an alveolus is a sheet of simple squamous epithelium, a single layer of flattened cells. Pulmonary capillaries run across the outer surface of each alveolus. The basement membranes secreted by the alveolar and capillary cells fuse together as the respiratory membrane (Figure 21.16**B**). Gases diffuse quickly across this thin membrane between the air inside alveoli and the blood flowing through a capillary.

Oxygen and Carbon Dioxide Transport Some oxygen can dissolve in plasma, but not enough to meet the body's needs. In all vertebrates and many invertebrates, hemoglobin increases blood's oxygen-carrying capacity. Nearly all the oxygen you inhale becomes bound to hemoglobin in your red blood cells. Hemoglobin is a protein (globin) with an iron-containing cofactor called heme associated with it (Section 2.9). Oxygen binds reversibly to the iron in hemoglobin. Hemoglobin tends to let go of oxygen in regions where the oxygen concentration is low, the temperature is warm, the pH is low, and the concentration of carbon dioxide is high. Such conditions occur in contracting muscles and other metabolically active tissues. After giving up oxygen to tissues, red blood cells return to the pulmonary capillaries where they pick up more.

On their return trip from the tissues to the lungs, red blood cells carry a bit of carbon dioxide bound to hemoglobin. However, enzymes inside red blood cells convert most carbon dioxide to bicarbonate. Bicarbonate diffuses out of the cell, dissolves in the plasma, and is carried to the lungs. In the pulmonary capillaries, bicarbonate is converted back to carbon dioxide. The carbon dioxide diffuses into the air in alveoli and is exhaled.

Take-Home Message **How does the human respiratory system function?**

- Air flows to and from the lungs through a system of airways. Air enters through the nose or mouth, and flows through the pharynx, larynx (voice box), and trachea to the two bronchi that carry air into the lungs.

- Inhalation is always active. Contraction of the diaphragm and muscles of the rib cage increase the volume of the thoracic cavity and lungs, so air is sucked into the lungs.

- Exhalation is usually passive. As muscles of respiration relax, the volume of the thoracic cavity and lungs decrease, pushing air out of the lungs.

- Gas exchange takes place in alveoli. Oxygen diffuses from air inside alveoli into pulmonary capillaries. Carbon dioxide diffuses in the opposite direction.

- Blood carries gases between the lungs and body tissues. Most of the oxygen in blood is bound to hemoglobin in red blood cells. Most carbon dioxide is converted to bicarbonate, which dissolves in the plasma.

21.8 Cardiovascular and Respiratory Disorders

Too Many or Too Few Blood Cells In **anemias**, red blood cells are in short supply or abnormal. As a result, oxygen delivery to cells is less efficient than it should be. Shortness of breath, fatigue, and chills are symptoms of anemia. Sickle-cell anemia is a genetic disorder. Pathogens also cause anemia, as when the parasites that cause malaria kill red blood cells. Too little iron in the diet can also cause anemia, because iron is needed to make hemoglobin. Also, blood loss from heavy menstrual periods or an ulcer can deplete the body of iron and cause anemia.

Leukemias are cancers that arise when stem cells that give rise to white blood cells begin to divide uncontrollably. The excess of white blood cells impairs normal blood functions.

Good Clot, Bad Clot Too much or too little clotting can cause problems. Hemophilia is a genetic disorder in which clotting is impaired (Section 9.6). Other disorders cause clots to form spontaneously inside vessels. A clot that forms in a vessel and stays put is a **thrombus**. A clot that breaks loose and travels in blood is an **embolus**. Both clot types can block vessels and halt blood flow. For example, a **stroke** occurs when a vessel in the brain ruptures or gets blocked by a clot. Either way, blood flow to brain cells is disrupted. A person who survives a stroke often has impairments caused by death of brain cells.

Atherosclerosis With **atherosclerosis**, the space inside arteries becomes narrowed (Figure 21.17). Cholesterol plays a role in this process. Most cholesterol in the blood is bound to plasma proteins in clumps called low-density lipoproteins, or LDLs. Most cells take up LDLs. A smaller amount of cholesterol is bound to plasma proteins as high-density lipoproteins, or HDLs. The liver takes up HDLs and uses them to make bile, a substance that is secreted into the gut. With a high concentration of LDLs or a low concentration of HDLs, lipids build up on an artery's wall, rather than being eliminated by the digestive tract. The site of lipid accumulation becomes inflamed, and fibrous connective tissue is laid down. The resulting atherosclerotic plaque makes the arterial wall bulge inward, slowing flow through the vessel.

Narrowing of the small arteries that deliver blood to the heart tissue can lead to a **heart attack**, in which the impaired blood flow causes cardiac muscle cells to die. A person has about a 30 percent chance of dying with a first heart attack. Heart cells do not divide after birth, so the cells killed in a heart attack cannot be replaced. Thus, heart attack survivors have a weakened heart.

Several methods are used to treat clogged coronary arteries. With coronary bypass surgery, a bit of blood vessel from elsewhere in the body is stitched to the aorta and to the coronary artery below a clogged region (Figure 21.18**A**). In laser angioplasty, laser beams vaporize plaques. In balloon angioplasty, doctors inflate a small balloon in a blocked artery to flatten the plaques. A wire mesh tube called a stent is then inserted to help keep the vessel open (Figure 21.18**B**).

Hypertension—A Silent Killer **Hypertension** refers to chronically high blood pressure (above 140/90). Often the cause is unknown. Heredity is a factor, and African Americans have an elevated risk. Diet also plays a role. In some people, high salt intake causes water retention that raises blood pressure. Hypertension is sometimes described as a silent killer, because most affected

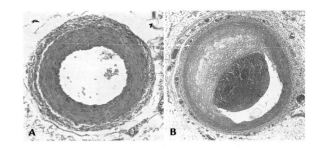

FIGURE 21.17 Atherosclerosis. **A** Normal artery. **B** Artery with its interior narrowed by an atherosclerotic plaque.

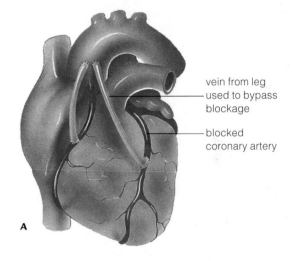

vein from leg used to bypass blockage

blocked coronary artery

A

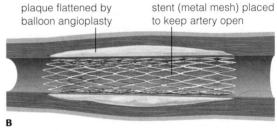

plaque flattened by balloon angioplasty

stent (metal mesh) placed to keep artery open

B

FIGURE 21.18 Two treatments for blocked coronary arteries. **A** Cardiac bypass. **B** Balloon angioplasty and placement of a stent.

anemia Fewer than normal or impaired red blood cells.

atherosclerosis Artery interior narrows because of lipid deposition and inflammation.

embolus Clot that forms in a blood vessel, then breaks loose.

heart attack Heart cells die because of impaired blood flow through cardiac arteries.

hypertension Chronically high blood pressure.

leukemia Cancer that increases white blood cell numbers.

stroke Brain cells die because a clot or vessel rupture disrupts blood flow within the brain.

thrombus Clot that forms in a vessel and remains there.

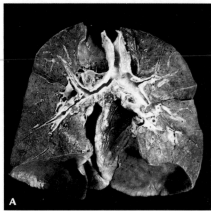

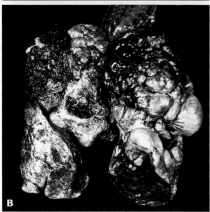

FIGURE 21.19 The effect of smoking on lungs. **A** Normal human lungs. **B** Lungs from a smoker with emphysema.

people feel fine and do not seek treatment. However, hypertension makes the heart work harder than normal, increasing the risk of heart attack. It also raises the risk of stroke and stresses the kidneys. An estimated 180,000 Americans die each year as a result of hypertension.

Respiratory Disorders Ciliated and mucus-secreting epithelial cells line the walls of your bronchioles and help protect you from respiratory infections. Cigarette smoke and other airborne pollutants harm this lining. Smoking may contribute to bronchitis. When epithelial cells lining the bronchioles are exposed to smoke they secrete excess mucus, which accumulates. Tiny particles stick to the mucus and bacteria begin to grow in it. Coughing helps brings up some of this mucus, but if smoking persists, so does the irritation, mucus accumulation, and coughing.

Initial attacks of bronchitis are usually treated with antibiotics, which help to keep bacteria in check. However, the bronchioles will stay inflamed unless smoking stops. Bacteria, chemical agents, or both eventually degrade the bronchiole walls, ciliated cells die off, and mucus-secreting cells multiply. Scar tissue forms and narrows or obstructs the airways. Thick mucus begins to clog the airways. Tissue-destroying bacterial enzymes go to work on the thin, stretchable walls of the alveoli. The walls crumble, inelastic fibrous tissue forms around them, and eventually they stop working. The lungs become distended and inelastic, which makes walking, running, and even exhaling difficult. These problems are among the symptoms of emphysema, a chronic ailment that affects over a million people in the United States. Once emphysema sets in, lung tissues cannot be repaired (Figure 21.19**B**).

A few people are genetically predisposed to develop emphysema. They do not have a functional gene for an enzyme that can inhibit bacterial attack on alveoli. Poor diet and persistent or recurring colds and other respiratory infections also raise the risk of emphysema later in life. However, smoking is the major cause of the disease.

Smoking's Impact Tobacco use kills 4 million people annually around the world. The number will probably rise to 10 million by 2030. It is estimated that each year in the United States, the direct medical costs of treating tobacco-induced respiratory disorders drain $22 billion from the economy. As G. H. Brundtland, the former director of the World Health Organization, once put it, tobacco remains the only legal consumer product that kills half of its regular users. If you are a smoker, you may wish to consider the information in Figure 21.20 carefully.

Increasingly, laws are being passed to prohibit any smoking in public buildings, workplaces, restaurants, airports, and similar enclosed spaces. Such regulations currently provide smoke–free areas to a little more than a third of the population of the United States. Worldwide, tobacco companies continue to expand their sales. Developing countries hold about 85 percent of the more than 1 billion current smokers. Women and children are increasingly among them.

Smoking marijuana (*Cannabis*) can also damage the respiratory system. Although marijuana contains fewer toxic particles, or "tar," than tobacco, it is usually smoked without a filter. Also, people smoking marijuana tend to inhale more deeply than tobacco smokers, to hold hot smoke in their lungs for longer periods, and to smoke their cigarettes down to stubs, where tar accumulates. As a result, long-term marijuana smokers, like tobacco smokers, have an increased risk of respiratory problems, and they tend to show lung damage earlier than cigarette smokers. On the other hand, unlike tobacco, marijuana has not been shown to increase the risk of lung cancer.

Risks Associated With Smoking	Reduction in Risks by Quitting
Shortened life expectancy Nonsmokers live about 8.3 years longer than those who smoke two packs a day from their midtwenties on.	Cumulative risk reduction; after 10–15 years, the life expectancy of ex-smokers approaches that of nonsmokers.
Chronic bronchitis, emphysema Smokers have 4–25 times higher risk of dying from these diseases than do nonsmokers.	Greater chance of improving lung function and slowing down rate of deterioration.
Cancer of lungs Cigarette smoking is the major cause.	After 10–15 years, risk approaches that of nonsmokers.
Cancer of mouth 3–10 times greater risk among smokers.	After 10–15 years, risk is reduced to that of nonsmokers.
Cancer of larynx 2.9–17.7 times more frequent among smokers.	After 10 years, risk is reduced to that of nonsmokers.
Cancer of esophagus 2–9 times greater risk of dying from this.	Risk proportional to amount smoked; quitting should reduce it.
Cancer of pancreas 2–5 times greater risk of dying from this.	Risk proportional to amount smoked; quitting should reduce it.
Cancer of bladder 7–10 times greater risk for smokers.	Risk decreases gradually over 7 years to that of nonsmokers.
Cardiovascular disease Cigarette smoking a major contributing factor in heart attacks, strokes, and atherosclerosis.	Risk for heart attack declines rapidly, for stroke declines more gradually, and for atherosclerosis it levels off.
Impact on offspring Women who smoke during pregnancy have more stillbirths, and the weight of liveborns is lower than the average (which makes babies more vulnerable to disease and death).	When smoking stops before fourth month of pregnancy, risk of stillbirth and lower birth weight eliminated.
Impaired immunity More allergic responses, destruction of white blood cells (macrophages) in respiratory tract.	Reduced to that of nonsmokers.
Bone healing Surgically cut or broken bones may take 30 percent longer to heal in smokers, perhaps because smoking depletes the body of vitamin C and reduces the amount of oxygen delivered to tissues. Reduced vitamin C and reduced oxygen interfere with formation of collagen fibers in bone (and many other tissues).	Reduced to that of nonsmokers.

FIGURE 21.20 From the American Cancer Society, a list of the major risks incurred by smoking and benefits of quitting. *Right*, a child in Mexico City already proficient at smoking cigarettes, a behavior that ultimately will endanger her capacity to breathe.

21.9

Impacts/Issues Revisited:
Up in Smoke

Advertisers sell tobacco products with images that suggest youth and sexuality. Yet use of these products causes premature aging and interferes with sexual function. It does so in part by disrupting the body's distribution of blood, causing arteries that supply some organs to constrict, while others widen.

For example, nicotine decreases blood flow to the skin. As a result, smoking a pack of cigarettes a week makes a person five times more likely to have wrinkles than a same-aged nonsmoker with the same amount of sun exposure. Each time smokers light up, they disrupt the flow of blood to their skin, depriving it of nutrients. Nicotine may also slow down synthesis of elastin, a protein that helps keep skin supple and tight.

As another effect, nicotine directs blood flow away from the genitals (external sex organs). A dose of nicotine interferes with a man's ability to achieve and sustain an erection. Smokers are twice as likely as nonsmokers to have erectile dysfunction. Nicotine also inhibits female sexual response.

How Would
YOU
☑ **Vote ?**

Should the United States encourage efforts to reduce tobacco use around the world, even if it means less profits for American tobacco companies? **See CengageNow for details, then vote online (cengagenow.com).**

Summary

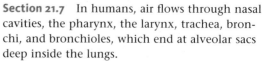

Section 21.1 Smoking, a habit that usually begins in the teens, impairs the health of smokers and the people around them.

Section 21.2 Most animals have a circulatory system, in which a heart pumps blood through blood vessels. In an open circulatory system, blood leaves vessels and flows among tissues. In a closed system, blood stays inside vessels. Some invertebrates and all vertebrates have a closed circulatory system.

CENGAGENOW™ Compare open and closed circulatory systems and investigate the closed systems of different vertebrate groups with the animations on CengageNow.

Sections 21.3, 21.4 The human heart pumps blood through two separate circuits. The pulmonary circuit carries oxygen-poor blood from the heart to the lungs, then returns oxygen-rich blood to the heart. The systemic circuit carries oxygen-rich blood from the heart to the tissues, then returns oxygen-poor blood to the heart.

Each half of the heart has two chambers: an upper atrium and a lower ventricle. In one cardiac cycle, signals from the cardiac pacemaker trigger contraction of the atria, then the ventricles. Ventricular contraction drives blood flow away from the heart. The sounds we associate with a heartbeat are caused by closing of heart valves.

CENGAGENOW™ Investigate the structure and function of the human cardiovascular system with the animation and interactions on CengageNow.

Section 21.5 Blood consists of plasma, blood cells, and platelets. Plasma is water with dissolved ions and molecules. Red blood cells are packed with hemoglobin and function in the transport of oxygen and, to a lesser extent, carbon dioxide. Many kinds of white blood cells function in housekeeping and defense. Platelets release substances that initiate blood clotting. All blood cells and platelets arise from stem cells in bone marrow.

Blood is pumped from ventricles into arteries, wide tubes for rapid transport. Blood pressure, the pressure exerted by blood on the walls of blood vessels, is usually measured in arteries. Arterioles adjust the distribution of blood through the body. Capillary beds are diffusion zones for exchanges between blood and interstitial fluid. Veins are a blood volume reservoir and move blood back to the heart. Valves keep blood from moving backward in veins.

CENGAGENOW™ See how blood pressure is measured with the animation on CengageNow.

Section 21.6 Aerobic respiration uses oxygen and produces carbon dioxide. Respiration is the physiological process by which these gases enter and leave an animal body by crossing a respiratory surface. In small, flattened animals, this may be the body surface. Most animals exchange gases across the respiratory membranes of respiratory organs, such as gills or lungs.

CENGAGENOW™ Compare the respiratory systems of fishes, frogs, and birds with animation on CengageNow.

Section 21.7 In humans, air flows through nasal cavities, the pharynx, the larynx, trachea, bronchi, and bronchioles, which end at alveolar sacs deep inside the lungs.

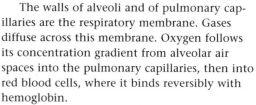

Each respiratory cycle consists of one inhalation and one exhalation. Inhalation requires energy. Contraction of the diaphragm and muscles between the ribs expand the chest cavity and pull air into the lungs. These events are reversed during exhalation, which is usually passive. A respiratory center in the brain controls the depth and rate of breathing.

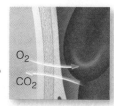

The walls of alveoli and of pulmonary capillaries are the respiratory membrane. Gases diffuse across this membrane. Oxygen follows its concentration gradient from alveolar air spaces into the pulmonary capillaries, then into red blood cells, where it binds reversibly with hemoglobin.

In capillary beds in the tissues, hemoglobin releases oxygen, which diffuses across interstitial fluid into cells. Carbon dioxide diffuses from cells into the interstitial fluid, then into the blood. In the lungs, carbon dioxide diffuses from the blood into air in alveoli, then is expelled.

CENGAGENOW™ Learn about components of the human respiratory system and what happens during breathing with the animation on CengageNow.

Section 21.8 Hypertension and atherosclerosis can lead to a heart attack or a stroke. Maintaining a moderate weight, eating a healthy diet, and getting regular exercise can reduce the risk of these cardiovascular disorders.

Bronchitis and emphysema are respiratory disorders that are caused or worsened by smoking. Worldwide, smoking is a leading cause of death.

Self-Quiz   Answers in Appendix I

1. All vertebrates have _____ .
   a. an open circulatory system   c. a four-chambered heart
   b. a closed circulatory system   d. both b and c

2. The SA node _____ .
   a. prevents backflow of blood   c. is located in the brain
   b. is the cardiac pacemaker   d. regulates breathing rate

3. Blood pressure is highest in the _____ and lowest in the _____ .
   a. arteries, veins   c. veins, arteries
   b. arterioles, venules   d. capillaries, arterioles

4. Contraction of _____ is the main force driving the flow of blood away from the heart.
   a. the atria   c. the ventricles
   b. arterioles   d. skeletal muscle

Digging Into Data

Risks of Radon

Radon is a colorless, odorless gas emitted by many rocks and soils. It is formed by the radioactive decay of uranium and is itself radioactive. There is some radon in the air almost everywhere, but routinely inhaling a lot of it raises the risk of lung cancer. Radon also seems to increase cancer risk far more in smokers than in nonsmokers. Figure 21.21 is an estimate of how radon in homes affects risk of lung cancer mortality. Note that this data shows only the death risk for radon-induced cancers. Smokers are also at risk from lung cancers that are caused by tobacco.

1. If 1,000 smokers were exposed to a radon level of 1.3 pCi/L over a lifetime (the average indoor radon level) how many would die of a radon-induced lung cancer?

2. How high would the radon level have to be to cause approximately the same number of cancers among 1,000 nonsmokers?

3. The risk of dying in a car crash is about 7 out of 1,000. Is a smoker in a home with an average radon level (1.3 pCi/L) more likely to die in a car crash or of radon-induced cancer?

	Risk of Cancer Death From Lifetime Radon Exposure	
Radon Level (pCi/L)	Never Smoked	Current Smokers
20	36 out of 1,000	260 out of 1,000
10	18 out of 1,000	150 out of 1,000
8	15 out of 1,000	120 out of 1,000
4	7 out of 1,000	62 out of 1,000
2	4 out of 1,000	32 out of 1,000
1.3	2 out of 1,000	20 out of 1,000
0.4	>1 out of 1,000	6 out of 1,000

FIGURE 21.21 Estimated risk of lung cancer death as a result of lifetime radon exposure. Radon levels are measured in picocuries per liter (pCi/L). The Environmental Protection Agency considers a radon level above 4 pCi/L to be unsafe. To learn about testing for radon and what to do if the radon level is high, visit the EPA's Radon Information Site at www.epa.gov/radon.

5. At rest, the largest volume of blood is in the _____ .
   a. arteries   c. veins
   b. capillaries   d. arterioles

6. In the blood, most oxygen is transported _____ .
   a. in red blood cells   c. bound to hemoglobin
   b. in white blood cells   d. both a and c

7. The _____ circuit carries blood from the heart to the lungs, then back to the heart.

8. The heart chamber with the thickest wall pumps blood into the _____ .
   a. aorta   c. pulmonary vein
   b. pulmonary artery   d. superior vena cava

9. In human lungs, gas exchange occurs at the _____ .
   a. bronchi   c. alveoli
   b. pericardium   d. epiglottis

10. When you breathe quietly, inhalation is _____ and exhalation is _____ .
    a. passive; passive   c. passive; active
    b. active; active   d. active; passive

11. During inhalation _____ .
    a. the thoracic cavity expands   c. the diaphragm relaxes
    b. the glottis closes   d. all of the above

12. The diaphragm is _____ muscle.
    a. smooth   b. skeletal   c. cardiac

13. Arrange these structure in the order through which air flows inward when inhaled.
    _____ larynx
    _____ pharynx
    _____ trachea
    _____ bronchus
    _____ bronchiole

14. Match the words with their descriptions.
    _____ plasma   a. receives blood from veins
    _____ alveolus   b. fluid component of blood
    _____ hemoglobin   c. cardiac pacemaker
    _____ veins   d. gap between vocal cords
    _____ SA node   e. site of gas exchange
    _____ trachea   f. drives blood flow from heart
    _____ glottis   g. windpipe
    _____ ventricle   h. blood volume reservoir
    _____ atrium   i. reversibly binds oxygen

Additional questions are available on CENGAGENOW

Critical Thinking

1. Sitting motionless for long periods, as on airline flights, allows blood to pool and clots to form in legs. Long-distance flights cause thrombus formation in about 1 percent of air travelers, and the risk is the same regardless of whether a person is in a first-class seat or an economy seat. Physicians recommend that air travelers drink plenty of fluids and periodically get up and walk around the cabin. Given what you know about blood flow in the veins, explain why these precautions can lower the risk of clot formation.

2. Blood entering systemic capillaries has the same oxygen and carbon dioxide concentration as the blood in pulmonary veins. Blood in systemic veins has the same oxygen and carbon dioxide concentration as the blood in pulmonary arteries. Explain the reason for each of these similarities.

3. High blood pressure can cause edema, a condition in which excess fluid pools in the tissues, causing them to appear swollen. Which type of blood vessel do you think the fluid escapes from, capillaries, arterioles, arteries, or veins? Explain your reasoning.

22 Immunity

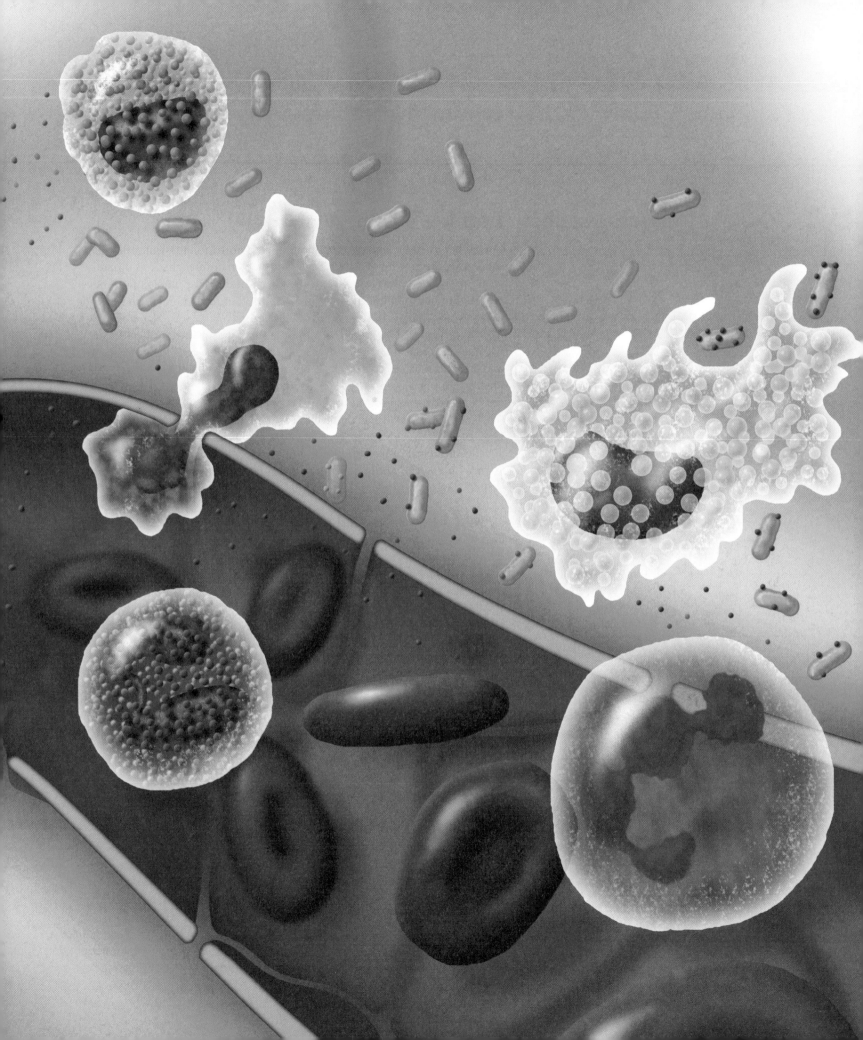

Impacts/Issues: Frankie's Last Wish

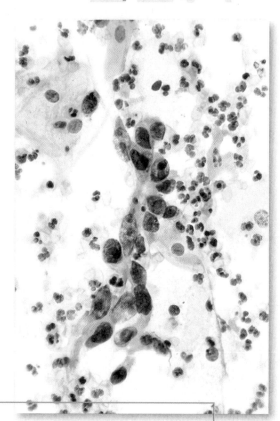

In October of 2000, Frankie McCullough had known for a few months that something was not quite right. She hadn't had an annual checkup in many years; after all, she was only 31 and had been healthy her whole life. It never occurred to her to doubt her own invincibility until the moment she saw the doctor's face change as he examined her cervix.

The cervix is the lowest part of the uterus, or womb. Cervical cells can become cancerous, but the process is usually slow. The cells pass through several precancerous stages that are detectable by routine Pap tests (Figure 22.1). Precancerous and even early-stage cancerous cells can be removed from the cervix before they spread to other parts of the body. However, plenty of women like Frankie do not take advantage of regular exams. Those who end up at the gynecologist's office with pain or bleeding may be experiencing symptoms of advanced cervical cancer, the treatment of which offers only about a 9 percent chance of survival. About 3,600 women die of cervical cancer each year in the United States. Many more than that die in places where routine gynecological testing is not common.

What causes cancer? At least in the case of cervical cancer, we know the answer to that question: Healthy cervical cells are transformed into cancerous ones by infection with human papillomavirus (HPV). HPV is a DNA virus that infects skin and mucous membranes. There are about 100 different types of HPV; a few cause warts on the hands or feet, or in the mouth. About 30 others that infect the genital area sometimes cause genital warts, but usually there are no symptoms of infection. Genital HPV is spread very easily by sexual contact. At least 80 percent of women have been infected with HPV by the age of 50.

A genital HPV infection usually goes away on its own, but not always. A persistent infection with one of about 10 strains is the main risk factor for cervical cancer. Types 16 and 18 are particularly dangerous: One of the two is found in more than 70 percent of all cervical cancers. In 2006, the FDA approved Gardasil, a vaccine against four types of genital HPV, including 16 and 18. The vaccine prevents cervical cancer caused by these HPV strains. It is most effective in girls who have not yet become sexually active, because they are least likely to have become infected with HPV.

The HPV vaccine came too late for Frankie McCullough. Despite radiation treatments and chemotherapy, her cervical cancer spread quickly. She died in 2001, leaving a wish for other young women: awareness. "If there is one thing I could tell a young woman to convince them to have a yearly exam, it would be not to assume that your youth will protect you. Cancer does not discriminate; it will attack at random, and early detection is the answer." She was right. Almost all women newly diagnosed with invasive cervical cancer have not had a Pap test in five years. Many have never had one.

Pap tests, HPV vaccines, and all other medical tests and treatments are direct benefits of our increasing understanding of the interplay of the human body with its pathogens, an interaction that we call immunity.

FIGURE 22.1 HPV and cervical cancer. *Above*, a Pap test reveals cancer cells (with enlarged, irregularly shaped nuclei) among normal epithelial cells of the cervix. Cells with multiple nuclei are characteristic of HPV infection. *Below*, Frankie McCullough (waving) died of cervical cancer in 2001.

Integrated Responses to Threats

Evolution of the Body's Defenses

Humans continually cross paths with a tremendous array of viruses, bacteria, fungi, parasitic worms, and other pathogens, but you need not lose sleep over this. Humans coevolved with these pathogens, so you have defenses that protect your body from them. **Immunity**, an organism's capacity to resist and combat infection, began long before multicelled eukaryotes evolved from free-living cells. Mutations introduced new patterns in membrane proteins, patterns that were unique in cells of a given type. As multicellularity evolved, so did mechanisms of identifying the patterns as self, or belonging to one's own body.

Table 22.1	Innate and Adaptive Immunity Compared	
	Innate Immunity	Adaptive Immunity
Response time	Immediate	About a week
How antigen is detected	Fixed set of receptors for molecular patterns typically found on pathogens	Billions of antigen receptors produced by random recombination of antigen genes
Specificity	None	Specific antigens targeted
Persistence	None	Long-term

By 1 billion years ago, nonself recognition had also evolved. The cells of all modern multicelled eukaryotes bear a set of receptors that collectively can recognize around 1,000 different molecular patterns that occur mainly on or in pathogens. When a cell's receptors bind to one of these patterns, they trigger a set of immediate, general defense responses. In mammals, for example, binding triggers activation of complement. **Complement** is a set of proteins that circulate in inactive form throughout the body. Activated complement binds to microorganisms, thereby destroying them or marking them for destruction by phagocytic cells (Section 4.6). Pattern receptors and the responses they initiate are part of **innate immunity**, a set of fast, general defenses against infection. All multicelled organisms start out life with these defenses, which do not change very much within the individual's lifetime.

Vertebrates have another set of defenses that is carried out by interacting cells, tissues, and proteins. This **adaptive immunity** tailors immune defenses to a vast array of specific pathogens that an individual may encounter during its lifetime. It is triggered by **antigen**: any molecule or particle recognized by the body as nonself.

Three Lines of Defense The mechanisms of adaptive immunity evolved within the context of innate immunity. The two systems were once thought to operate independently of each other, but we now know they function together. We describe both systems together in terms of three lines of defense. The first line includes the physical, chemical, and mechanical barriers that keep pathogens on the outside of the body. Innate immunity, the second line of defense, begins after tissue is damaged, or after an antigen is detected inside the body. Its general response mechanisms rid the body of many invaders before populations of them become established in the internal environment.

Activation of innate immunity triggers the third line of defense, adaptive immunity. White blood cells form huge populations that target a specific antigen and destroy anything bearing that antigen. Some of the white blood cells persist long after infection ends. If the same antigen returns, these memory cells mount a secondary response. Adaptive immunity can specifically target billions of different antigens. Table 22.1 compares innate and adaptive immunity.

The Defenders White blood cells carry out all immune responses. Many kinds circulate through the body in blood and lymph. Others populate the lymph nodes, spleen, and other tissues. Some white blood cells are phagocytic,

adaptive immunity Set of vertebrate immune defenses tailored to specific pathogens encountered by an organism during its lifetime.

antigen A molecule or particle that the immune system recognizes as nonself. Triggers an immune response.

complement A set of proteins that circulate in inactive form in blood. Part of innate immunity.

immunity The body's ability to resist and fight infections.

innate immunity Set of inborn, fixed general defenses against infection.

Neutrophil Fast-acting and most abundant phagocyte. Circulates in blood and migrates into damaged tissues.

Macrophage Antigen-presenting phagocyte (see Section 22.5). Secretes cytokines. Circulates in blood; enters damaged tissue.

Dendritic cell Antigen-presenting phagocyte (see Section 22.5). Works mainly in tissue fluids.

Eosinophil Granules contain enzymes that target parasitic worms. Circulates in blood; migrates into damaged tissues.

Basophil Granules contain histamine and other substances that cause inflammation. Circulates in blood.

Mast cell Anchored in tissues. Granules contain histamine and other substances that cause inflammation; part of allergies.

Lymphocytes (below) are central to adaptive immunity. After antigen recognition, clonal populations of these cells form and circulate in blood and tissue fluids.

B cell Recognizes antigens via membrane-bound antibodies (see Section 22.6). Only type of cell that produces antibodies.

T cell Helper T cells coordinate adaptive immune responses. Cytotoxic T cells kill infected, cancerous, or foreign cells.

Natural killer (NK) cell Kills antibody-tagged cells and stressed body cells.

FIGURE 22.2 White blood cells. Staining reveals details such as cytoplasmic granules that contain enzymes, toxins, and signaling molecules.

which means they move about and engulf other cells. All are secretory. The secretions include peptides and proteins called cytokines. Cytokines are signaling molecules that cells of the immune system use to communicate with one another. Communication allows the cells to coordinate their activities during immune responses.

Figure 22.2 shows common white blood cells. The different types are specialized for specific tasks, such as phagocytosis. **Neutrophils** are the most abundant of the circulating phagocytic cells. Phagocytic **macrophages** that patrol tissue fluids are mature monocytes, which patrol the blood. **Dendritic cells** are phagocytes that alert the adaptive immune system to the presence of antigen in tissues.

Some white blood cells have granules. These secretory vesicles contain cytokines, enzymes, or pathogen-busting toxins that the cell releases in response to a trigger such as binding to antigen. **Eosinophils** target parasites too big for phagocytosis. **Basophils** circulating in blood and **mast cells** anchored in tissues secrete substances contained by their granules in response to injury or antigen. Mast cells also respond to chemical signals from the nervous system.

Lymphocytes are a special category of white blood cells that are central to adaptive immunity. **B** and **T lymphocytes** (B and T cells) have the capacity to collectively recognize billions of specific antigens. There are several kinds of T cells, including some called cytotoxic T cells that target infected or cancerous body cells. **Natural killer cells** (NK cells) can destroy infected or cancerous body cells that are undetectable by cytotoxic T cells.

Take-Home Message **What is immunity?**

- The innate immune system is a set of general defenses against a fixed number of antigens. It acts immediately to prevent infection.

- Vertebrate adaptive immunity is a system of defenses that can specifically target billions of different antigens.

- White blood cells are central to both systems; signaling molecules such as cytokines integrate their activities.

22.3 Surface Barriers

Your skin is in constant contact with the external environment, so it picks up many microorganisms. It normally teems with about 200 different kinds of yeast, protozoa, and bacteria (Figure 22.3**A**). If you showered today, there are probably thousands of microorganisms on every square inch of your external surfaces. If you did not, there may be billions. Microorganisms tend to flourish in warm, moist areas, such as between the toes. Huge populations inhabit cavities and tubes that open out on the body's surface, including the eyes, nose, mouth, and anal and genital openings.

Microorganisms that typically live on human surfaces, including the interior tubes and cavities of the digestive and respiratory tracts, are called **normal flora**. Our surfaces provide them with a stable environment and nutrients. In return, their populations deter more aggressive species from colonizing (and penetrating) body surfaces; help us digest food; and make nutrients that we depend on, including a vitamin (B_{12}) made only by bacteria.

skin surface
dead skin cells
epidermis
dividing cells

2 µm

0.1 mm

FIGURE 22.3 Examples of normal flora and surface defenses against them.

A *Staphylococcus epidermidis*, the most common colonizer of human skin.

B *Staphylococcus aureus* cells (*yellow*) stuck in mucus secreted by human nasal epithelial cells. The coordinated beating of cilia on the surface of the cells sweeps trapped bacteria to the throat for disposal.

C The thick layer of dead cells in human epidermis keeps normal skin flora on the outside of the body.

Normal flora are helpful only on the outside of body tissues. If they invade the internal environment, normal flora can cause or worsen serious illnesses such as pneumonia; ulcers; colitis; whooping cough; meningitis; abscesses of the lung and brain; and colon, stomach, and intestinal cancers. The bacterial agent of tetanus, *Clostridium tetani*, passes through our intestines so often that it is considered a normal inhabitant. The bacteria responsible for diphtheria, *Corynebacterium diphtheriae*, was normal skin flora before widespread use of the vaccine eradicated the disease. *Staphylococcus aureus*, a resident of human skin and linings of the mouth, nose, throat, and intestines (Figure 22.3**B**), is also a leading cause of human bacterial disease. Antibiotic-resistant strains of *S. aureus* are now widespread. A particularly dangerous kind, MRSA (methicillin-resistant *S. aureus*), is resistant to a wide range of antibiotics. MRSA is now a permanent resident of most hospitals around the world.

In contrast to body surfaces, the blood and tissue fluids of healthy people are typically sterile (free of microorganisms). Surface barriers usually prevent normal flora from entering the body's internal environment. Epidermis, the tough outer layer of vertebrate skin (Section 19.4 and Figure 22.3**C**), is an example. Microorganisms flourish on skin's waterproof, oily surface, but they rarely penetrate the thick epidermis.

The thinner epithelial tissues that line the body's interior tubes and cavities also have surface barriers. Figure 22.3**B** shows how sticky mucus secreted by cells of these linings can trap microorganisms. The mucus contains **lysozyme**, an enzyme that kills bacteria. In the sinuses and respiratory tract, the coordinated beating of cilia sweeps trapped microorganisms away before they have a chance to breach the delicate walls of these structures.

Your mouth is a particularly inviting habitat for microorganisms because it offers plenty of nutrients, warmth, moisture, and surfaces for colonization. Accordingly, it harbors huge populations of normal flora. Microorganisms that normally inhabit the mouth resist lysozyme in saliva. Those that are swallowed are typically killed by gastric fluid, a potent brew of protein-digesting enzymes and acid in the stomach. Those that survive to reach the small intestine are usually killed by bile salts. The hardy ones that make it to the large intestine must compete with about 500 resident species that are specialized to live there and have already established large populations. Any that displace normal flora are typically flushed out by diarrhea.

B lymphocyte (B cell) Type of white blood cell that makes antibodies.

basophil Circulating white blood cell; role in inflammation.

dendritic cell Phagocytic white blood cell that patrols tissue fluids.

eosinophil White blood cell that targets internal parasites.

lysozyme Antibacterial enzyme that occurs in body secretions such as mucus.

macrophage Phagocytic white blood cell that patrols tissue fluids.

mast cell White blood cell that is anchored in many tissues; factor in inflammation.

natural killer cell (NK cell) White blood cell that kills infected or cancerous cells.

neutrophil Circulating phagocytic white blood cell.

normal flora Microorganisms that typically live on human surfaces, including the interior tubes and cavities of the digestive and respiratory tracts.

T lymphocyte (T cell) Circulating white blood cell central to adaptive immunity; some kinds target sick body cells.

Lactic acid produced by *Lactobacillus* helps keep the vaginal pH outside the range of tolerance of most fungi and other bacteria. Urination's flushing action usually stops pathogens from colonizing the urinary tract.

Take-Home Message What prevents ever-present microorganisms from entering the body's internal environment?

- Surface barriers keep microorganisms that contact or inhabit vertebrate surfaces from invading the internal environment.

- Skin's tough epidermis is one barrier. Mucus, lysozyme, and often the sweeping action of cilia protect the linings of internal tubes and cavities.

- Populations of normal flora deter more dangerous types from colonizing the body's internal and external surfaces.

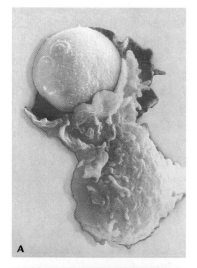

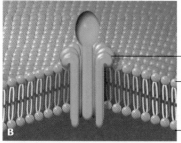

one membrane attack complex (cutaway view)

lipid bilayer of a bacterium

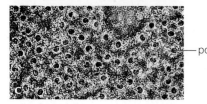

pore

FIGURE 22.4 Mechanisms of adaptive immunity. **A** Macrophage caught in the act of phagocytosis. **B** Activated complement assembles into membrane attack complexes that insert themselves into the lipid bilayer of bacteria. The resulting pores **C** bring about lysis of the cell.

22.4 The Innate Immune Response

What happens if a pathogen slips by surface barriers and enters the body's internal environment? Animals are normally born with a set of fast-acting, off-the-shelf immune defenses that can keep an invading pathogen from becoming established in the body's internal environment. These innate defenses include phagocyte and complement action, inflammation, and fever—all general mechanisms that normally do not change much over an individual's lifetime.

Phagocytes and Complement Macrophages are large phagocytes that engulf and digest essentially everything except undamaged body cells (Figure 22.4**A**). They patrol interstitial fluid, so they are often the first white blood cells to encounter an invading pathogen. When receptors on a macrophage bind to antigen, the cell begins to secrete cytokines. These signaling molecules attract more macrophages, neutrophils, and dendritic cells to the site of invasion.

Antigen also triggers complement activation. In vertebrates, about 30 different types of complement protein circulate in inactive form in blood and interstitial fluid. Some become activated when they encounter antigen, or an antibody bound to antigen (we will return to antibodies in Section 22.6). Activated complement proteins activate other complement proteins, which activate other complement proteins, and so on. These cascading reactions quickly produce huge concentrations of activated complement at the site of invasion.

Activated complement attracts phagocytes. Like bloodhounds, these cells follow complement gradients back to an affected tissue. Some complement proteins attach directly to pathogens. Phagocytes have complement receptors, so a pathogen coated with complement is recognized and engulfed faster than an uncoated pathogen. Other activated complement proteins assemble into complexes that puncture bacterial cell walls or plasma membranes (Figure 22.4**B**).

Inflammation Activated complement and cytokines trigger **inflammation**, a local response to tissue damage or infection (Figure 22.5 ❶). Inflammation occurs when pattern receptors on basophils, mast cells, or neutrophils bind to antigen, or when mast cells directly bind to activated complement. In response to the binding, the cells release signaling molecules such as histamine and prostaglandins into the affected tissue ❷. The signaling molecules have two local effects. First, they cause nearby arterioles to widen. As a result, blood flow to

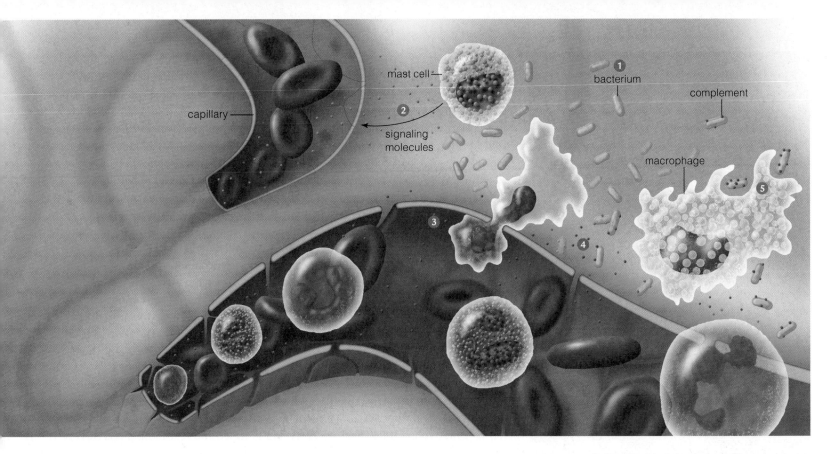

the tissue increases. The increased flow speeds the arrival of more phagocytes, attracted to the cytokines. Second, the signaling molecules cause spaces between cells in capillary walls to widen, so capillaries in an affected tissue become "leakier." Phagocytes squeeze between the cells, out of the blood vessel ❸. By now, invading bacteria have triggered complement activation and are coated with complement ❹, which makes them easy targets for the phagocytes ❺.

The symptoms of inflammation include redness and warmth that are outward indications of the area's increased blood flow. Swelling and pain occur because plasma proteins that escape from the leaky capillaries make interstitial fluid hypertonic with respect to blood. More water diffuses into tissue, which becomes painful as it swells with fluid.

Fever Fever is a temporary rise in body temperature above the normal 37°C (98.6°F) that often occurs in response to infection. Some cytokines stimulate brain cells to make and release prostaglandins, which act on the hypothalamus to raise the body's internal temperature set point. As long as the temperature of the body is below the new set point, the hypothalamus sends out signals that cause blood vessels in the skin to constrict, which reduces heat loss from the skin. The signals also trigger reflexive movements called shivering, or "chills," that increase the metabolic heat output of muscles. Both responses raise the body's internal temperature.

Fever enhances immune defenses by increasing the rate of enzyme activity, thus speeding up metabolism, tissue repair, and the formation and activity of phagocytes. Some pathogens multiply more slowly at the higher temperature, so white blood cells can get a head start in the proliferation race against them. A fever is a sign that the body is fighting something, so it should never be ignored. However, a fever of 40.6°C (105°F) or less does not necessarily require treatment in an otherwise healthy adult. Body temperature usually will not rise above that value, but if it does, immediate hospitalization is recommended because a fever of 42°C (107.6°F) can result in brain damage or death.

FIGURE 22.5 Animated! An innate immune response to a bacterial infection.

❶ Bacteria invade a tissue.

❷ Pattern receptors on mast cells in the tissue recognize and bind to bacterial antigen. The mast cells release signaling molecules (*blue* dots) that cause arterioles to widen. The resulting increase in blood flow reddens and warms the tissue.

❸ The signaling molecules also increase capillary permeability, which allows phagocytes to squeeze through the vessel walls into the tissue. Plasma proteins leak out of the capillaries, and the tissue swells with fluid.

❹ Bacterial antigens activate complement (*purple* dots). Activated complement binds to the bacteria.

❺ Phagocytes in the tissue quickly recognize and engulf the complement-coated bacteria.

fever An internally induced rise in core body temperature above the normal set point as a response to infection.
inflammation A local response to tissue damage or infection; characterized by redness, warmth, swelling, and pain.

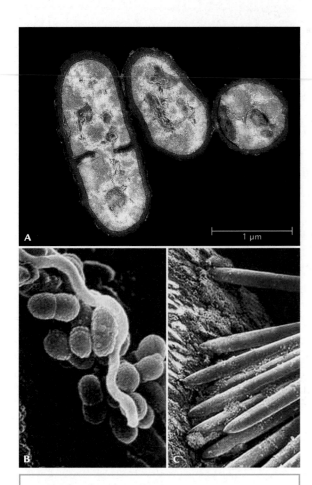

FIGURE 22.6 Normal flora that cause health problems. **A** *Propionibacterium acnes*, the bacterial cause of acne. **B** The main contributor to dental plaque, *Streptococcus mutans*. **C** Micrograph of toothbrush bristles scrubbing plaque on a tooth surface.

The adaptive immune system "adapts" to different antigens an individual encounters during its lifetime.

Examples of Innate Responses Acne is one effect of an innate response. This skin condition is caused in part by a major constituent of normal flora, *Propionibacterium acnes* (Figure 22.6**A**). This bacterial species feeds on sebum, a greasy mixture of fats and waxes that lubricates hair and skin. Glands in the skin secrete sebum into hair follicles. During puberty, higher levels of steroid hormones trigger an increase in sebum production. Excess sebum combines with dead, shed skin cells and blocks the openings of hair follicles.

P. acnes can survive on the surface of the skin, but far prefer anaerobic habitats such as the interior of blocked hair follicles. There, they multiply to tremendous numbers. Secretions of flourishing *P. acnes* populations leak into internal tissues, attracting neutrophils that initiate inflammation in the tissue around the follicles. The resulting pustules are called acne.

Another example of a common innate response occurs in the mouth, which is normally inhabited by about 400 species of microorganisms (Figure 22.6**B**). A few of these normal flora cause dental **plaque**, a thick biofilm of various bacteria and the occasional archaean, their extracellular products, and saliva proteins. Plaque sticks tenaciously to teeth (Figure 22.6**C**). Some bacteria that live in it are fermenters. They break down bits of carbohydrate that stick to teeth and then secrete acids. The acids etch away tooth enamel and make cavities.

In young, healthy people, tight junctions (Section 3.7) seal gum epithelium to teeth, forming a barrier that prevents oral microorganisms from entering gum tissue. As we age, the connective tissue beneath the epithelium thins, so the barrier becomes vulnerable. Deep pockets form between the teeth and gums. A very nasty collection of anaerobic bacteria and archaea accumulates in the pockets. Their noxious secretions, including destructive enzymes and acids, cause inflammation of the surrounding gum tissues—a condition called periodontitis.

Porphyromonas gingivalis is one of those anaerobic species. Along with every other species of oral bacteria associated with periodontitis, *P. gingivalis* is also found in atherosclerotic plaque (Section 21.8). Periodontal wounds are an open door to the circulatory system and its arteries.

Atherosclerosis is now known to be a disease of inflammation. What role the oral microorganisms play in this scenario is not yet clear, but one thing is certain—they contribute to the inflammation that fuels coronary artery disease.

Take-Home Message **What is innate immunity?**

- Innate immunity is the body's built-in set of general immune defenses.
- Complement, phagocytes, inflammation, and fever quickly eliminate most invaders from the body before their populations become established.

22.5 Overview of Adaptive Immunity

If innate immune mechanisms do not quickly rid the body of an invading pathogen, populations of pathogenic cells may become established in body tissues. By that time, long-lasting mechanisms of adaptive immunity have already begun to target the invaders specifically, by way of their antigens.

Life is so diverse that the number of different antigens is essentially unlimited. No system can recognize all of them, but adaptive immunity comes close. Unlike innate immunity, the adaptive immune system changes: It "adapts" to

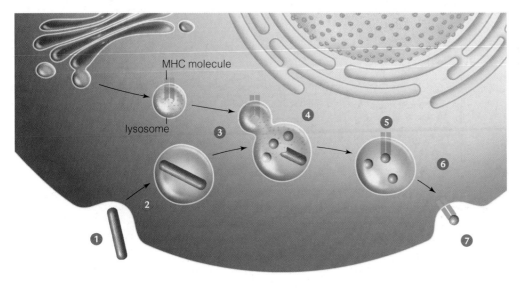

FIGURE 22.7 Antigen processing. From encounter to display, what happens when a B cell, macrophage, or dendritic cell engulfs an antigenic particle—in this case, a bacterium.

❶ A phagocytic cell engulfs a bacterium.

❷ An endocytic vesicle forms around the bacterium.

❸ The vesicle fuses with a lysosome, which contains enzymes and MHC molecules.

❹ Lysosomal enzymes digest the bacterium to molecular bits.

❺ Inside the vesicle, antigenic bits of bacterium bind to MHC molecules.

❻ The vesicle fuses with the cell membrane by exocytosis. When it does, the antigen–MHC complex becomes part of the plasma membrane.

❼ The antigen–MHC complex becomes displayed on the surface of the white blood cell. The combination is a signal that provokes an adaptive immune response targeting the antigen.

Figure It Out: From what organelles do the lysosomes bud?

Answer: Golgi bodies

different antigens an individual encounters during its lifetime. Lymphocytes and phagocytes interact to effect the four defining characteristics of adaptive immunity, which are self/nonself recognition, specificity, diversity, and memory:

Self/nonself recognition starts with the molecular patterns that give each kind of cell a unique identity. The plasma membrane of your cells bears **MHC markers**, which are recognition proteins named after the genes that encode them. Your T cells also bear antigen receptors called **T cell receptors**, or **TCRs**. Part of a TCR recognizes MHC markers as self, and another part recognizes an antigen as nonself.

Specificity means the adaptive immune response can be tailored to combat specific antigens.

Diversity refers to the antigen receptors on a body's collection of B and T cells. There are potentially billions of different antigen receptors, so an individual has the potential to counter billions of different threats.

Memory refers to the capacity of the adaptive immune system to "remember" an antigen. It takes a few days for B and T cells to respond in force the first time they encounter an antigen. If the same antigen shows up again, they make a faster, stronger response.

First Step—The Antigen Alert
Recognizing a specific antigen is the first step of the adaptive immune response. A new B or T cell is naive, which means that no antigen has bound to its receptors yet. Once the cell binds to an antigen, it begins to divide by mitosis, and tremendous populations form.

T cell receptors do not recognize antigen unless it is presented by an antigen-presenting cell. Macrophages, B cells, and dendritic cells do the presenting. First, they engulf something bearing antigens (Figure 22.7 ❶). A vesicle that contains the antigen-bearing particle forms in the cells' cytoplasm ❷ and fuses with a lysosome ❸. Lysosomal enzymes then digest the particle into molecular bits ❹ (Section 3.6). The lysosomes also contain MHC markers that bind to some of the antigen bits ❺. The resulting antigen–MHC complexes become displayed at the cell's surface when the vesicles fuse with (and become part of) the plasma membrane ❻. The display of MHC markers paired with antigen fragments serves as a call to arms ❼.

MHC markers Self-recognition protein on the surface of body cells. Triggers adaptive immune response when bound to antigen fragments.

plaque On teeth, a thick biofilm composed of bacteria, their extracellular products, and saliva proteins.

T cell receptor (TCR) Antigen-binding receptor on the surface of T cells; also recognizes MHC markers.

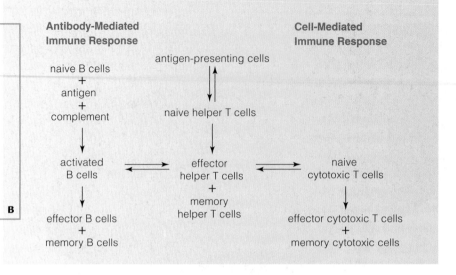

FIGURE 22.8 How adaptive immunity works. **A** Primary and secondary immune responses. A first exposure to an antigen causes a primary immune response in which effector cells fight the infection. Memory cells also form in a primary response but are set aside, sometimes for decades. If the antigen returns, the memory cells initiate a faster, stronger secondary response. **B** Overview of key interactions between antibody-mediated and cell-mediated responses—the two arms of adaptive immunity.

A
B

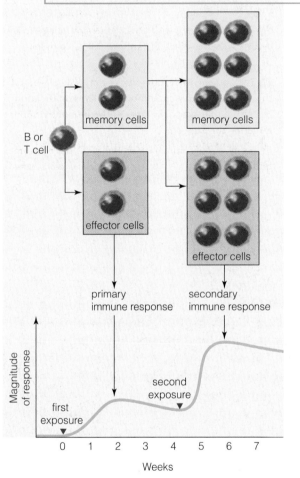

Any T cell that bears a receptor for this antigen will bind the antigen–MHC complex displayed on the surface of the white blood cell. Once bound, the T cell starts secreting cytokines, which signal all other B or T cells with the same antigen receptor to divide again and again. Huge populations of B and T cells form after a few days; all of the cells recognize the same antigen. Most are **effector cells**, differentiated lymphocytes that act at once. Some are **memory cells**, long-lived B and T cells reserved for future encounters with the antigen. Memory cells can persist for decades after the initial infection ends. If the same antigen enters the body at a later time, these memory cells will initiate a secondary response (Figure 22.8**A**). In a secondary immune response, larger populations of effector cell clones form much more quickly than they did in the primary response.

Two Arms of Adaptive Immunity Like a boxer's one-two punch, adaptive immunity has two separate arms: the antibody-mediated and the cell-mediated immune responses (Figure 22.8**B**). These two responses work together to eliminate diverse threats. Why two arms? Not all threats present themselves in the same way. For example, bacteria, fungi, or toxins can circulate in blood or interstitial fluid. These cells are intercepted quickly by B cells and other phagocytes that interact in an **antibody-mediated immune response**. In this response, B cells produce antibodies, which are proteins that can bind to specific antigen-bearing particles (we return to antibodies in the next section).

Some kinds of threats are not targeted by B cells. For example, B cells cannot detect body cells altered by cancer. As another example, some viruses, bacteria, fungi, and protists can hide and reproduce inside body cells; B cells can detect them only briefly, when they slip out of one cell to infect others. Such intracellular pathogens are targeted primarily by the **cell-mediated immune response**, which does not involve antibodies. In this response, cytotoxic T cells and NK cells detect and destroy altered or infected body cells.

Intercepting Antigen After engulfing an antigen-bearing particle, a dendritic cell or macrophage migrates to a lymph node, where it presents antigen to many T cells that filter through the node (Figure 22.9). Every day, about 25 billion T cells pass through each node. T cells that recognize and bind to antigen presented by a phagocyte initiate an adaptive response.

Antigen-bearing particles in interstitial fluid flow through lymph vessels to a lymph node, where they meet up with resident B cells, dendritic cells, and macrophages. These phagocytes engulf, process, and present antigen to T cells that are passing through the node. During an infection, the lymph nodes swell

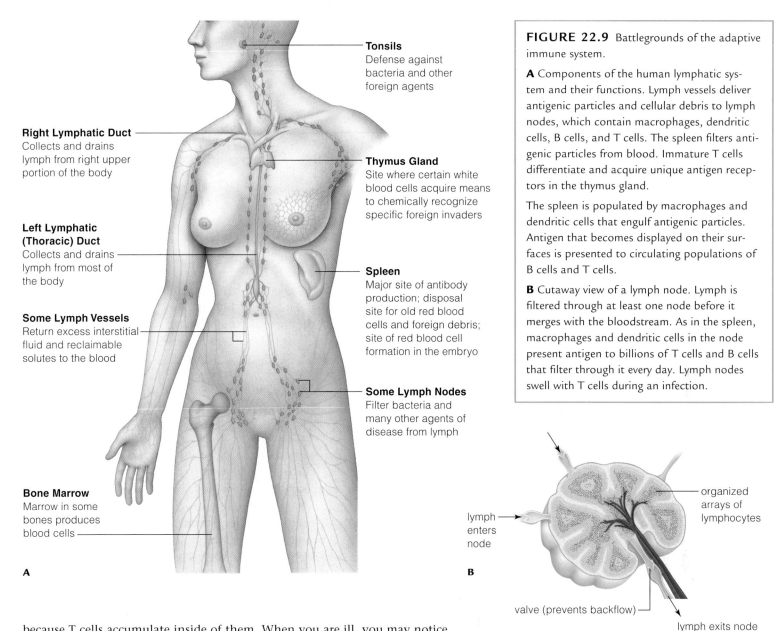

Tonsils
Defense against bacteria and other foreign agents

Right Lymphatic Duct
Collects and drains lymph from right upper portion of the body

Left Lymphatic (Thoracic) Duct
Collects and drains lymph from most of the body

Some Lymph Vessels
Return excess interstitial fluid and reclaimable solutes to the blood

Bone Marrow
Marrow in some bones produces blood cells

Thymus Gland
Site where certain white blood cells acquire means to chemically recognize specific foreign invaders

Spleen
Major site of antibody production; disposal site for old red blood cells and foreign debris; site of red blood cell formation in the embryo

Some Lymph Nodes
Filter bacteria and many other agents of disease from lymph

A

FIGURE 22.9 Battlegrounds of the adaptive immune system.

A Components of the human lymphatic system and their functions. Lymph vessels deliver antigenic particles and cellular debris to lymph nodes, which contain macrophages, dendritic cells, B cells, and T cells. The spleen filters antigenic particles from blood. Immature T cells differentiate and acquire unique antigen receptors in the thymus gland.

The spleen is populated by macrophages and dendritic cells that engulf antigenic particles. Antigen that becomes displayed on their surfaces is presented to circulating populations of B cells and T cells.

B Cutaway view of a lymph node. Lymph is filtered through at least one node before it merges with the bloodstream. As in the spleen, macrophages and dendritic cells in the node present antigen to billions of T cells and B cells that filter through it every day. Lymph nodes swell with T cells during an infection.

lymph enters node

organized arrays of lymphocytes

valve (prevents backflow)

lymph exits node

B

because T cells accumulate inside of them. When you are ill, you may notice your swollen lymph nodes as tender lumps under the jaw or elsewhere.

The tide of battle turns when the effector cells and their secretions destroy most antigen-bearing agents. With less antigen present, fewer immune fighters are recruited. Complement proteins assist in the cleanup by binding antibody–antigen complexes, forming large clumps that can be quickly cleared from the blood by the liver and spleen. Immune responses subside after the antigen-bearing particles are cleared from the body.

Take-Home Message **What is the adaptive immune system?**

- Phagocytes and lymphocytes interact to bring about vertebrate adaptive immunity, which has four defining characteristics: self/nonself recognition, specificity, diversity, and memory.

- The two arms of adaptive immunity work together. Antibody-mediated responses target antigen in blood or interstitial fluid; cell-mediated responses target altered body cells.

antibody-mediated immune response Immune response in which antibodies are produced in response to an antigen.
cell-mediated immune response Immune response involving cytotoxic T cells and NK cells that destroy infected or cancerous body cells.
effector cell Antigen-sensitized B cell or T cell that forms in an immune response and acts immediately.
memory cell Antigen-sensitized B or T cell that forms in a primary immune response but does not act immediately.

Antibody-Mediated Immunity

Antigen Receptors Most humans can make billions of unique antigen receptors. The diversity arises by random splicing of antigen receptor genes in newly forming B and T cells. Before a new B cell leaves bone marrow, it already is making its unique antigen receptor proteins. In time the B cell's membrane will bristle with more than 100,000 of these B cell receptors. T cells also form in bone marrow, but they mature only after they take a tour in the thymus gland. There, they encounter hormones that stimulate them to produce MHC receptors and their own unique T cell receptors.

So what, exactly, is an antigen receptor? T cell receptors are one kind. Antibodies are another. **Antibodies** are Y-shaped proteins made only by B cells (Figure 22.10). Each antibody can bind to the antigen that prompted its synthesis. B cell receptors are membrane-bound antibodies. Other antibodies circulate in blood and enter interstitial fluid during inflammation, but they do not kill pathogens directly. Instead, they activate complement, promote phagocytosis, prevent pathogens from attaching to body cells, and neutralize some toxins.

An Antibody-Mediated Response If we liken B cells to assassins, then each one has a genetic assignment to liquidate one particular target: an antigen-bearing extracellular pathogen or toxin. Antibodies are their molecular bullets, as the following example illustrates (Figure 22.11).

Suppose that you accidentally nick your finger. Being opportunists, some *Staphylococcus aureus* cells on your skin invade your internal environment. Complement in interstitial fluid quickly attaches to carbohydrates in the bacterial cell walls, and cascading complement activation reactions begin. Within an hour, complement-coated bacteria tumbling along in lymph vessels reach a lymph node in your elbow. There, they filter past an army of naive B cells.

One of the naive B cells residing in that lymph node makes antigen receptors that recognize a polysaccharide in *S. aureus* cell walls ❶. The B cell binds to the polysaccharide on one of the bacteria. The complement coating stimulates the B cell to engulf the bacterium. The B cell is now activated.

Meanwhile, more *S. aureus* cells have been secreting metabolic products into interstitial fluid around your cut. The secretions attract phagocytes. One of the phagocytes, a dendritic cell, engulfs several bacteria, then migrates to the lymph

---

If we liken B cells to assassins, then each has a genetic assignment to liquidate a particular antigen.

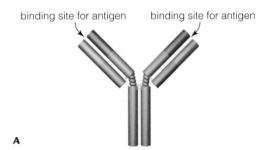

binding site for antigen   binding site for antigen

A

FIGURE 22.10 Antibody structure. **A** An antibody molecule consists of four polypeptide chains joined in a Y-shaped configuration. Each has two binding sites for antigen. **B** The antigen-binding sites of each antibody are unique. They only bind to an antigen that has complementary bumps, grooves, and charge distribution.

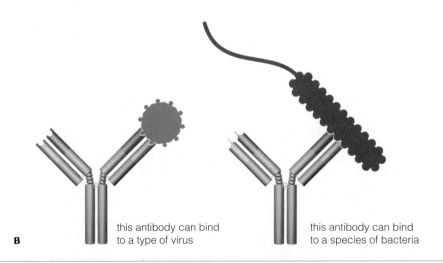

this antibody can bind to a type of virus

this antibody can bind to a species of bacteria

B

FIGURE 22.11 Animated! Example of an antibody-mediated immune response.

1 The B cell receptors on a naive B cell bind to an antigen on the surface of a bacterium. The bacterium's complement coating triggers the B cell to engulf it. Fragments of the bacterium bind MHC markers, and the complexes become displayed at the surface of the now-activated B cell.

2 A dendritic cell engulfs the same kind of bacterium that the B cell encountered. Digested fragments of the bacterium bind to MHC markers, and the complexes become displayed at the dendritic cell's surface. The dendritic cell is now an antigen-presenting cell.

3 The antigen–MHC complexes on the antigen-presenting cell are recognized by antigen receptors on a naive T cell. Binding causes the T cell to divide and differentiate into effector and memory helper T cells.

4 Antigen receptors of one of the effector helper T cells bind antigen–MHC complexes on the B cell. Binding makes the T cell secrete cytokines.

5 The cytokines induce the B cell to divide, giving rise to many identical B cells. The cells differentiate into effector B cells and memory B cells.

6 The effector B cells begin making and secreting huge numbers of antibodies, all of which recognize the same antigen as the original B cell receptor. The new antibodies circulate throughout the body and bind to any remaining bacteria.

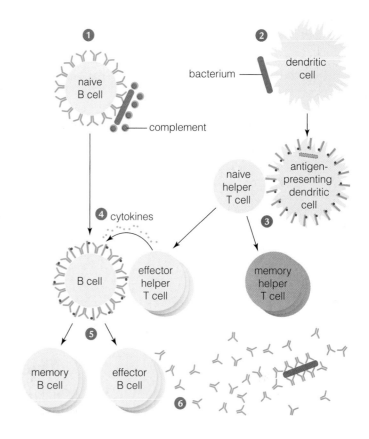

node in your elbow. By the time it gets there, it has digested the bacteria and is displaying their fragments as antigens bound to MHC markers on its surface **2**.

Every hour, about 500 different naive T cells travel through the lymph node, inspecting resident dendritic cells. In this case, one of the T cells has TCRs that recognize and bind the *S. aureus* antigen–MHC complexes displayed by the dendritic cell **3**. The T cell is called a helper T cell because it helps other lymphocytes produce antibodies and kill pathogens. The helper T cell and the dendritic cell interact for about 24 hours. When the two cells disengage, the helper T cell returns to the circulatory system and begins to divide. A gigantic population of identical helper T cells forms. These clones differentiate into effector and memory cells, each of which has receptors that recognize the same *S. aureus* antigen.

Let's go back to that B cell in the lymph node. By now, it has digested the bacterium, and it is displaying bits of *S. aureus* bound to MHC molecules on its plasma membrane. One of the new helper T cells recognizes the antigen–MHC complexes displayed by the B cell. Like long-lost friends, the B cell and the helper T cell stay together for a while and communicate **4**.

One of the messages that is communicated consists of cytokines secreted by the helper T cell. The cytokines stimulate the B cell to begin mitosis after the two cells disengage. The B cell divides again and again to form a huge population of genetically identical cells, all with receptors that can bind to the same *S. aureus* antigen **5**.

The B cell clones differentiate into effector and memory cells. The effector B cells start working immediately. They begin to produce and secrete antibodies instead of making membrane-bound B cell receptors **6**. The new antibodies

antibody Y-shaped antigen receptor protein made only by B cells.

recognize the same *S. aureus* antigen as the original B cell receptor. Antibodies now circulate throughout the body and attach themselves to any remaining bacterial cells. An antibody coating prevents bacteria from attaching to body cells, and flags them for phagocytosis.

Take-Home Message

What happens during an antibody-mediated immune response?

■ Antigen-presenting cells, T cells, and B cells interact in an antibody-mediated immune response targeting a specific antigen.

■ The adaptive immune system has the potential to recognize billions of different antigens via receptors on B cells and T cells.

■ Populations of B cells form; these make and secrete antibodies that recognize and bind the antigen. Antibodies are antigen receptors made only by B cells.

22.7 The Cell-Mediated Immune Response

If B cells are like assassins, then cytotoxic T cells are specialists in cell-to-cell combat.

If B cells are like assassins, then cytotoxic T cells are specialists in cell-to-cell combat. Antibody-mediated immune responses target pathogens that circulate in blood and interstitial fluid, but they are not as effective against pathogens inside cells. As part of a cell-mediated immune response, cytotoxic T cells kill ailing body cells that may be missed by an antibody-mediated response. Such cells typically display certain antigens. For example, cancer cells display altered body proteins, and body cells infected with intracellular pathogens display polypeptides of the infecting agent. Both types of cell are detected and killed by cytotoxic T cells.

A typical cell-mediated response begins in interstitial fluid during inflammation when a dendritic cell recognizes, engulfs, and digests a sick body cell or the remains of one (Figure 22.12). The dendritic cell begins to display antigen that was part of the sick cell, and migrates to the spleen or a lymph node ❶. There, the dendritic cell presents its antigen–MHC complexes to huge populations of naive helper T cells and naive cytotoxic T cells. Some of the naive cells have TCRs that recognize the complexes on the dendritic cell. Helper T cells ❷ and cytotoxic T cells ❸ that bind the antigen–MHC complexes displayed by the dendritic cell become activated.

The activated helper T cells divide and differentiate into populations of effector and memory helper T cells. The effector cells immediately begin to secrete cytokines ❹. Activated cytotoxic T cells recognize the cytokines as signals to divide and differentiate, and tremendous populations of effector and memory cytotoxic T cells form. All of them recognize and bind the same antigen—the one displayed by that first ailing cell. As in an antibody-mediated response, memory cells form in a primary cell-mediated response. These long-lasting cells do not act immediately. If the antigen returns at a later time, the memory cells will mount a secondary response.

The effector cytotoxic T cells start working immediately. They circulate throughout blood and interstitial fluid, and bind to any other body cell displaying the original antigen together with MHC markers ❺. After it is bound to an ailing cell, a cytotoxic T cell releases perforin and proteases (Figure 22.13). These toxins poke holes in the sick cell and cause it to die.

FIGURE 22.12 Animated! A cell-mediated immune response.

1 A dendritic cell engulfs a virus-infected cell. Digested fragments of the virus bind to MHC markers, and the complexes are displayed at the dendritic cell's surface. The dendritic cell, now an antigen-presenting cell, migrates to a lymph node.

2 Receptors on a naive helper T cell bind to antigen–MHC complexes on the dendritic cell. The interaction activates the helper T cell, which then begins to divide.

A large population of descendant cells forms. Each cell has T cell receptors that recognize the same antigen. The cells differentiate into effector and memory cells.

3 Receptors on a naive cytotoxic T cell bind to the antigen–MHC complexes on the surface of the dendritic cell. The interaction activates the cytotoxic T cell.

4 The activated cytotoxic T cell recognizes cytokines secreted by the effector helper T cells as signals to divide. A large population of descendant cells forms. Each cell bears T cell receptors that recognize the same antigen. The cells differentiate into effector and memory cells.

5 The new cytotoxic T cells circulate through the body. They recognize and touch-kill any body cell that displays the viral antigen–MHC complexes on its surface.

Figure It Out: What do the large red spots represent?

Answer: Viruses

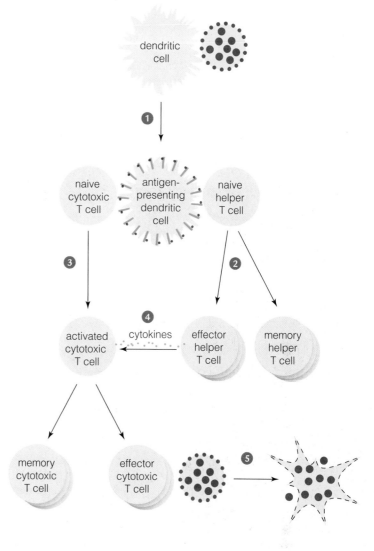

In order to kill a body cell, cytotoxic T cells must recognize the MHC molecules on the surface of the cell. However, some infections or cancer can alter a body cell so that it is missing part or all of its MHC markers. NK ("natural killer") cells are crucial for fighting such cells. Unlike cytotoxic T cells, NK cells can kill body cells that lack MHC markers. Cytokines secreted by helper T cells **4** also stimulate NK cell division. The resulting populations of effector NK cells attack body cells tagged by antibodies for destruction. They also recognize certain proteins displayed by body cells that are under stress. Stressed body cells with normal MHC markers are not killed; only those with altered or missing MHC markers are destroyed.

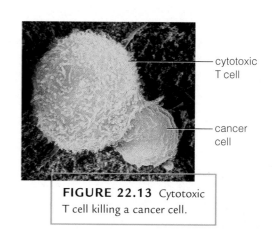

FIGURE 22.13 Cytotoxic T cell killing a cancer cell.

Take-Home Message
What happens during a cell-mediated immune response?

- Antigen-presenting cells, T cells, and NK cells interact in a cell-mediated immune response targeting infected body cells or those that have been altered by cancer.

Allergies In millions of people, exposure to normally harmless substances stimulates an immune response. Any substance that is ordinarily harmless yet provokes such responses is an **allergen**. Sensitivity to an allergen is called an **allergy**. Drugs, foods, dust mites, fungal spores, poison ivy, insect venom, and pollen (Figure 22.14**A**) are among the most common allergens.

Some people are genetically predisposed to having allergies. Infections, emotional stress, and changes in air temperature can trigger reactions. A first exposure to an allergen stimulates the immune system to make antibody that becomes anchored to mast cells and basophils. With later exposures, antigen binds to the antibody. Binding triggers the anchoring cell to secrete histamine and cytokines that initiate inflammation. If this reaction occurs at the lining of the respiratory tract, a copious amount of mucus is secreted and the airways constrict; sneezing, stuffed-up sinuses, and a drippy nose result (Figure 22.14**B**). Contact with an allergen that penetrates the skin's outer layers causes the skin to redden, swell, and become itchy.

Antihistamines relieve allergy symptoms by dampening the effects of histamines. These drugs act on histamine receptors, and also inhibit the release of cytokines and histamines from basophils and mast cells.

In some people who are hypersensitive to drugs, insect stings, foods, or vaccines, a second exposure to the allergen can result in a severe, whole-body allergic reaction called anaphylactic shock. Huge amounts of cytokines and histamines released in all parts of the body provoke an immediate, systemic reaction. Fluid leaking from blood into tissues causes the blood pressure to drop too much (shock), and tissues to swell. Swelling tissue constricts airways and may block them. Anaphylactic shock is rare but life-threatening and requires immediate treatment (Figure 22.14**C**). It may occur at any time, upon exposure to even a tiny amount of allergen. Risk factors include a prior allergic reaction of any kind.

FIGURE 22.14 Allergies. **A** Ragweed pollen, a common allergen. **B** A mild allergy may cause upper respiratory symptoms. **C** Anaphylactic shock, a severe allergic reaction that requires immediate hospitalization.

Autoimmune Disorders Sometimes lymphocytes and antibody molecules fail to discriminate between self and nonself. When that happens, they mount an **autoimmune response**, or an immune response that targets one's own tissues. For example, autoimmunity occurs in rheumatoid arthritis, a disease in which self antibodies form and bind to the soft tissue in joints. The resulting inflammation leads to eventual disintegration of bone and cartilage in the joints (Section 20.2).

Antibodies to self proteins may bind to hormone receptors, as in Graves' disease. Self antibodies that bind stimulatory receptors on the thyroid gland cause it to produce excess thyroid hormone, which quickens the body's overall metabolic rate. Antibodies are not part of the feedback loops that normally regulate thyroid hormone production. So, antibody binding continues unchecked, the thyroid continues to release too much hormone, and the metabolic rate spins out of control. Symptoms of Graves' disease include uncontrollable weight loss; rapid, irregular heartbeat; sleeplessness; pronounced mood swings; and bulging eyes.

Immune responses tend to be stronger in women than in men, and autoimmunity is far more frequent in women. We know that estrogen receptors are part of gene expression controls throughout the body. T cells have receptors for estrogens, so these hormones may enhance T cell activation in autoimmune diseases. Women's bodies have more estrogen, so interactions between their B cells and T cells may be amplified.

Table 22.2	Global HIV and AIDS Cases	
Region	AIDS Cases	New HIV Cases
Sub-Saharan Africa	22,000,000	1,900,000
South/Southeast Asia	4,200,000	330,000
Latin America	1,700,000	140,000
Central Asia/East Europe	1,500,000	110,000
North America	1,200,000	54,000
East Asia	740,000	52,000
Western/Central Europe	730,000	27,000
Middle East/North Africa	380,000	40,000
Caribbean Islands	230,000	20,000
Australia/New Zealand	74,000	13,000
Approx. worldwide total	33,000,000	2,700,000

Source: Joint United Nations Programme HIV/AIDS, 2008 report

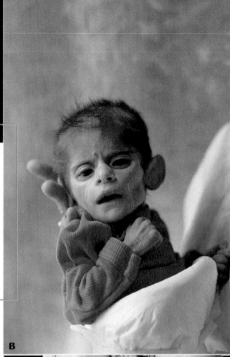

FIGURE 22.15 AIDS. **A** A human T cell (*blue*), infected with HIV (*red*). **B** This Romanian baby contracted AIDS from his mother's breast milk. He did not live long enough to develop lesions of Kaposi's sarcoma **C**, a cancer that is a common symptom of HIV infection in older AIDS patients.

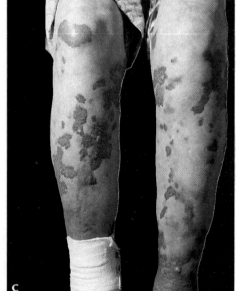

Immunodeficiency and AIDS Impaired immune function is dangerous and sometimes lethal. Immune deficiencies render individuals vulnerable to infections by opportunistic agents that are typically harmless to those in good health. Primary immune deficiencies, which are present at birth, are the outcome of mutations. Severe combined immunodeficiencies (SCIDs) are examples. Secondary immune deficiency is the loss of immune function after exposure to an outside agent, such as a virus. **AIDS** (acquired immune deficiency syndrome) is the most common secondary immune deficiency. AIDS is a constellation of disorders that occur as a consequence of infection with HIV, the human immunodeficiency virus (Section 13.6 and Figure 22.15**A**). This virus cripples the immune system, so it makes the body very susceptible to infections and rare forms of cancer. Worldwide, approximately 33 million individuals currently have AIDS (Table 22.2 and Figure 22.15**B**).

At first, an infected person appears to be in good health, perhaps fighting "a bout of flu." But symptoms eventually emerge that foreshadow AIDS: fever, many enlarged lymph nodes, chronic fatigue and weight loss, and drenching night sweats. Then, infections caused by normally harmless microorganisms strike. Yeast infections of the mouth, esophagus, and vagina often occur, as well as a form of pneumonia caused by the fungus *Pneumocystis carinii*. Colored lesions erupt. These lesions are evidence of Kaposi's sarcoma, a type of cancer that is common among AIDS patients (Figure 22.15**C**).

HIV mainly infects macrophages, dendritic cells, and helper T cells. When virus particles enter the body, dendritic cells engulf them. The dendritic cells then migrate to lymph nodes, where they present processed HIV antigen to naive T cells. An army of HIV-neutralizing antibodies and HIV-specific cytotoxic T cells forms. We have just described a typical adaptive immune response. It rids the body of most—but not all—of the virus. In this first response, HIV infects a few helper T cells in a few lymph nodes. For years or even decades, the antibodies keep the level of HIV in the blood low, and the cytotoxic T cells kill HIV-infected cells.

Patients are contagious during this stage, although they might show no symptoms of AIDS. HIV viruses persist in a few of their helper T cells, in a few lymph nodes. Eventually, the level of virus-neutralizing antibody in the blood plummets, and T cell production slows. Why the antibody level decreases is still

AIDS Acquired immune deficiency syndrome. A collection of diseases that develops after a virus (HIV) weakens the immune system.

allergen A normally harmless substance that provokes an immune response in some people.

allergy Sensitivity to an allergen.

autoimmune response Immune response that targets one's own tissues.

a major topic of research, but its effect is certain: The adaptive immune system becomes less and less effective at fighting the virus. The number of virus particles rises; up to 1 billion HIV viruses are built each day. Up to 2 billion helper T cells become infected. Half of the viruses are destroyed and half of the helper T cells are replaced every two days. Lymph nodes begin to swell with infected T cells.

Eventually, the body's capacity for adaptive immunity is destroyed as it makes fewer and fewer replacement helper T cells. Other types of viruses make more particles in a day, but the immune system eventually wins. HIV demolishes the immune system. Secondary infections and tumors kill the patient.

HIV is transmitted most frequently by having unprotected sex with an infected partner. HIV also travels in tiny amounts of infected blood in the syringes shared by intravenous drug abusers, or by patients in hospitals of poor countries. Infected mothers can transmit HIV to a child during pregnancy, labor, delivery, or breast-feeding. There is no way to rid the body of the HIV virus, no cure for those already infected. At present, our best option for halting the spread of HIV is prevention, by teaching people how to avoid being infected. The best protection against AIDS is to avoid unsafe behaviors. In most circumstances, HIV infection is the consequence of a choice: either to have unprotected sex, or to use a shared needle for intravenous drugs.

Take-Home Message What happens when the immune system does not function as well as it should?

- Misdirected or compromised immunity, which sometimes occurs as a result of mutation or environmental factors, can have severe or lethal outcomes.

- AIDS occurs as a result of infection by HIV, a virus that infects lymphocytes and so cripples the human immune system.

22.9 Vaccines

Immunization refers to processes designed to induce immunity. In active immunization, a preparation that contains antigen—a **vaccine**—is administered orally or injected. The first immunization elicits a primary immune response, just as an infection would. A second immunization, or booster, elicits a secondary immune response for enhanced immunity.

In passive immunization, a person receives antibodies purified from the blood of another individual. The treatment offers immediate benefit for someone who has been exposed to a potentially lethal agent, such as tetanus, rabies, Ebola virus, or a venom or toxin. Because the antibodies were not made by the recipient's lymphocytes, effector and memory cells do not form, so benefits last only as long as the injected antibodies do.

The first vaccine was the result of desperate attempts to survive smallpox epidemics that swept repeatedly through cities all over the world. Smallpox is a severe disease that kills up to one-third of the people it infects (Figure 22.16). Before 1880, no one knew what caused infectious diseases or how to protect anyone from getting them, but there were clues. In the case of smallpox, survivors seldom contracted the disease a second time. They were immune, or protected from infection.

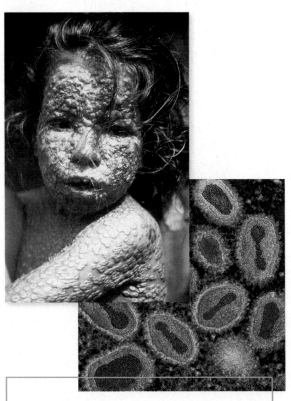

FIGURE 22.16 Young survivor and the cause of her disease, smallpox viruses. Worldwide use of the vaccine eradicated naturally occurring cases of smallpox; vaccinations for it ended in 1972.

The idea of acquiring immunity to smallpox was so appealing that people had been risking their lives on it for more than two thousand years. Many people poked into their skin bits of smallpox scabs or threads soaked in pus from smallpox sores. Some survived the crude practices and became immune to smallpox, but many others did not.

By the late 1700s, it was widely known that dairymaids did not get smallpox if they had already recovered from cowpox (a mild disease that affects cattle as well as humans). In 1796, Edward Jenner, an English physician, injected liquid from a cowpox sore into the arm of a healthy boy. Six weeks later, Jenner injected the boy with liquid from a smallpox sore. Luckily, the boy did not get smallpox. Jenner's experiment showed directly that the agent of cowpox elicits immunity to smallpox. Jenner named his procedure "vaccination," after the Latin word for cowpox (*vaccinia*). The use of Jenner's vaccine spread quickly through Europe, then to the rest of the world. The last known case of naturally occurring smallpox was in 1977, in Somalia. The vaccine had eradicated the disease.

We now know that the cowpox virus is an effective vaccine for smallpox because the antibodies it elicits also recognize smallpox virus antigens. Our knowledge of how the immune system works has allowed us to develop many other vaccines that save millions of lives yearly. These vaccines are an important part of worldwide public health programs (Table 22.3).

We still do not have an effective vaccine against AIDS despite decades of high-priority, top-notch research all over the world. At this writing, 42 different HIV vaccines are being tested. Most of them consist of proteins or peptides isolated from the HIV virus, and many deliver these antigens in viral vectors. Live, weakened HIV virus is an effective vaccine in chimpanzees, but the risk of HIV infection from the vaccines themselves outweighs their potential benefits in humans. Other types of HIV vaccines are notoriously ineffective. Antibody exerts selective pressure on the virus, which has a very high mutation rate because it replicates so fast. The human immune system just cannot produce antibodies fast enough to keep up with the mutations.

Take-Home Message **How does immunization work?**

- Immunization is the administration of an antigen-bearing vaccine designed to elicit immunity to a specific disease.

22.10

Impacts/Issues Revisited:

Frankie's Last Wish

The Gardasil HPV vaccine consists of viral capsid proteins that self-assemble into virus-like particles (VLPs). These proteins are produced by a recombinant yeast, *Saccharomyces cerevisiae*. The yeast carries genes for one surface protein from each of four strains of HPV, so the VLPs carry no viral DNA. Thus, the VLPs are not infectious, but the antigenic proteins they consist of elicit an immune response at least as strong as infection with HPV virus (*right*).

Table 22.3	Recommended Immunization Schedule
Vaccine	**Age of Vaccination**
Hepatitis B	Birth to 2 months
Hepatitis B boosters	1–4 months and 6–18 months
Rotavirus	2, 4, and 6 months
DTP: diphtheria, tetanus, and pertussis (whooping cough)	2, 4, and 6 months
DTP boosters	15–18 months, 4–6 years, and 11–12 years
HiB (*Haemophilus influenzae*)	2, 4, and 6 months
HiB booster	12–15 months
Pneumococcal	2, 4, and 6 months
Pneumococcal booster	12–15 months
Inactivated poliovirus	2 and 4 months
Inactivated poliovirus boosters	6–18 months and 4–6 years
Influenza	Yearly, 6 months–18 years
MMR (measles, mumps, rubella)	12–15 months
MMR booster	4–6 years
Varicella (chicken pox)	12–15 months
Varicella booster	4–6 years
Hepatitis A series	1–2 years
HPV series	11–12 years
Meningococcal	11–12 years

Source: Centers for Disease Control and Prevention (CDC), 2008

How Would YOU ☑ Vote?

Clinical trials of some vaccines take place in underdeveloped countries that have fewer regulations governing human testing than the United States. Should clinical trials be held to the same ethical standards no matter where they take place? **See CengageNow for details, then vote online (cengagenow.com).**

immunization A procedure designed to promote immunity to a disease.

vaccine A preparation introduced into the body in order to elicit immunity to an antigen.

Summary

Section 22.1, 22.2   Three lines of immune defense protect vertebrates from infection. An antigen-bearing pathogen that breaches surface barriers triggers innate immunity, a set of general defenses that usually prevents populations of pathogens from becoming established in the internal environment. Adaptive immunity, which can specifically target billions of different antigens, follows. Complement and signaling molecules coordinate the activities of white blood cells that carry out both types of responses.

 Section 22.3   Skin and linings of the body's tubes and cavities offer barriers to infection by microorganisms. Normal flora that colonize external surfaces can be beneficial, but only on the outside of the body. Many cause disease when they breach these surface barriers.

Section 22.4   An innate immune response includes fast, general responses that can eliminate invaders before an infection can become established: complement activation, the action of phagocytes, inflammation, and fever.

CENGAGENOW™   View the animation of inflammation and complement action on CengageNow.

Section 22.5   Four main characteristics of adaptive immunity are self/nonself recognition, sensitivity to specific threats, potential to intercept a tremendous number of diverse pathogens, and memory. B and T lymphocytes (B cells and T cells) are central to adaptive immunity. Antibody-mediated and cell-mediated immune responses work together to rid the body of a specific pathogen. Many effector cells form and target the antigen-bearing particles in a primary response. Memory cells that also form are reserved for a later encounter with an antigen, in which case they trigger a faster, stronger secondary response.

Section 22.6   B cells, assisted by T cells and signaling molecules, carry out antibody-mediated immune responses. B cells make antibodies. Antigen receptors—T cell receptors and antibodies—recognize specific antigens. These receptors are the basis of the immune system's capacity to recognize billions of different antigens.

CENGAGENOW™   See how antigen receptor diversity arises and follow an antibody-mediated immune response with the animations on CengageNow.

 Section 22.7   Antigen-presenting cells, T cells, and NK cells interact in cell-mediated responses. They target and kill body cells altered by infection or cancer.

CENGAGENOW™   Observe a cell-mediated immune response with the animation on CengageNow.

 Section 22.8   The consequences of malfunctioning immune mechanisms range from annoying to deadly. Allergens are normally harmless substances that induce an immune response. Immune deficiency is a reduced capacity to mount an immune response. In an autoimmune response, a body's own cells are inappropriately recognized as foreign and attacked. AIDS is caused by HIV, a virus that destroys the immune system mainly by infecting helper T cells.

Section 22.9   Immunization with vaccines designed to elicit immunity to specific diseases saves millions of lives each year as part of worldwide health programs.

Self-Quiz   Answers in Appendix I

1. Which of the following is not among the first line of immune defenses against infection?
   a. skin, mucous membranes   d. resident bacterial populations
   b. tears, saliva, gastric fluid   e. complement activation
   c. urine flow   f. lysozyme

2. Which of the following is not part of an innate immune response?
   a. phagocytic cells   e. inflammation
   b. fever   f. complement activation
   c. histamines   g. antigen-presenting cells
   d. cytokines   h. none of the above

3. Which of the following is not part of an adaptive immune response?
   a. phagocytic cells   e. antigen receptors
   b. antigen-presenting cells   f. complement activation
   c. histamines   g. antibodies
   d. cytokines   h. none of the above

4. Activated complement proteins _____ .
   a. form pore complexes   c. attract phagocytes
   b. promote inflammation   d. all of the above

5. _____ trigger immune responses.
   a. Cytokines   d. Antigens
   b. Lysozymes   e. Histamines
   c. Antibodies   f. all of the above

6. Name one defining characteristic of innate immunity.

7. Name one defining characteristic of adaptive immunity.

8. Antibodies are _____ .
   a. antigen receptors   c. proteins
   b. made only by B cells   d. all of the above

9. Antibody-mediated responses work against _____ .
   a. intracellular pathogens   d. both a and b
   b. extracellular pathogens   e. both b and c
   c. cancerous cells   f. a, b, and c

10. Cell-mediated responses work against _____ .
   a. intracellular pathogens   d. both a and b
   b. extracellular pathogens   e. both a and c
   c. cancerous cells   f. a, b, and c

11. _____ are targets of cytotoxic T cells.
   a. Extracellular virus particles in blood
   b. Virus-infected body cells or tumor cells
   c. Parasitic flukes in the liver
   d. Bacterial cells in pus
   e. Pollen grains in nasal mucus

12. Allergies occur when the body responds to _____ .
   a. pathogens   c. normally harmless substances
   b. toxins   d. all of the above

13. Match the immune cell with a description.

    _____ mast cell   a. kills virus-infected cells

    _____ B cell   b. targets parasitic worms

    _____ helper T cell   c. activates cytotoxic T cells

    _____ NK cell   d. kills cancer cells

    _____ cytotoxic T cell   e. factor in allergic reactions

    _____ dendritic cell   f. antigen-presenter

    _____ eosinophil   g. makes antibodies

14. Match the immunity concepts.

    _____ anaphylactic shock   a. neutrophil

    _____ antibody secretion   b. effector B cell

    _____ phagocyte   c. general defense

    _____ immune memory   d. immune response

    _____ autoimmunity   against one's own body

    _____ antigen receptor   e. secondary response

    _____ inflammation   f. B cell receptor

                                  g. hypersensitivity to an allergen

Additional questions are available on CENGAGENOW™

Critical Thinking

1. As described in Section 22.9, Edward Jenner was lucky. He performed a potentially harmful experiment on a boy who managed to survive the procedure. What would happen if a would-be Jenner tried to do the same experiment today in the United States?

2. Transplant rejection occurs when a transplanted organ or tissue fails to be accepted by the recipient's body. The recipient's immune system attacks transplanted tissues or organs even if they are from a human donor. Which type of lymphocyte is mainly responsible for transplant rejection? Explain your answer.

3. Before each flu season, you get a flu shot, an influenza vaccination. This year, you get "the flu" anyway. What happened? There are at least three explanations.

4. Elena developed chicken pox when she was in first grade. Later in life, when her children developed chicken pox, she remained healthy even though she was exposed to countless virus particles daily. Explain why.

5. Monoclonal antibodies are produced by immunizing a mouse with a particular antigen, then removing its spleen. Individual B cells producing mouse antibodies specific for the antigen are isolated from the mouse's spleen and fused with cancerous B cells from a myeloma cell line.

    The resulting hybrid myeloma cells—hybridoma cells—are cloned, or grown in tissue culture as separate cell lines. Each line produces and secretes antibodies that recognize the antigen to which the mouse was immunized. These monoclonal antibodies can be purified and used for research or other purposes.

    Monoclonal antibodies are sometimes used in passive immunization. They are effective, but only in the immediate term. Antibodies produced by one's own immune system can last up to about six months in the bloodstream, but monoclonals delivered in passive immunization often last for less than a week. Why the difference?

Digging Into Data

Cervical Cancer Incidence in HPV-Positive Women

In 2003, Michelle Khan and her coworkers published their findings on a 10-year study in which they followed cervical cancer incidence and HPV status in 20,514 women. All women who participated in the study were free of cervical cancer when the test began. Pap tests were taken at regular intervals, and the researchers used a DNA probe hybridization test to detect the presence of specific types of HPV in the women's cervical cells.

    The results are shown in Figure 22.17 as a graph of the incidence rate of cervical cancer by HPV type. Women who are HPV positive are often infected by more than one type, so the data were sorted into groups based on the women's HPV status ranked by type: either positive for HPV16; or negative for HPV16 and positive for HPV18; or negative for HPV16 and 18 and positive for any other cancer-causing HPV; or negative for all cancer-causing HPV.

1. At 110 months into the study, what percentage of women who were not infected with any type of cancer-causing HPV had cervical cancer? What percentage of women who were infected with HPV16 also had cervical cancer?

2. In which group would women infected with both HPV16 and HPV18 fall?

3. Is it possible to estimate from this graph the overall risk of cervical cancer that is associated with infection of cancer-causing HPV of any type?

4. Do these data support the conclusion that being infected with HPV16 or HPV18 raises the risk of cervical cancer?

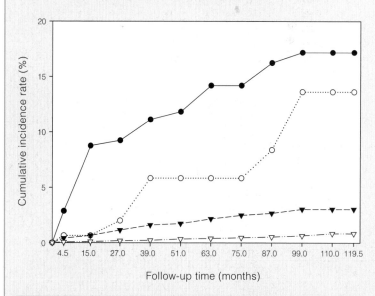

FIGURE 22.17 Cumulative incidence rate of cervical cancer correlated with HPV status in 20,514 women aged 16 years and older. The data were grouped as follows: HPV16 positive (closed circles); HPV16 negative and HPV18 positive (open circles); and all other cancer-causing HPV types combined (closed triangles). Open triangles: no cancer-causing HPV type was detected.

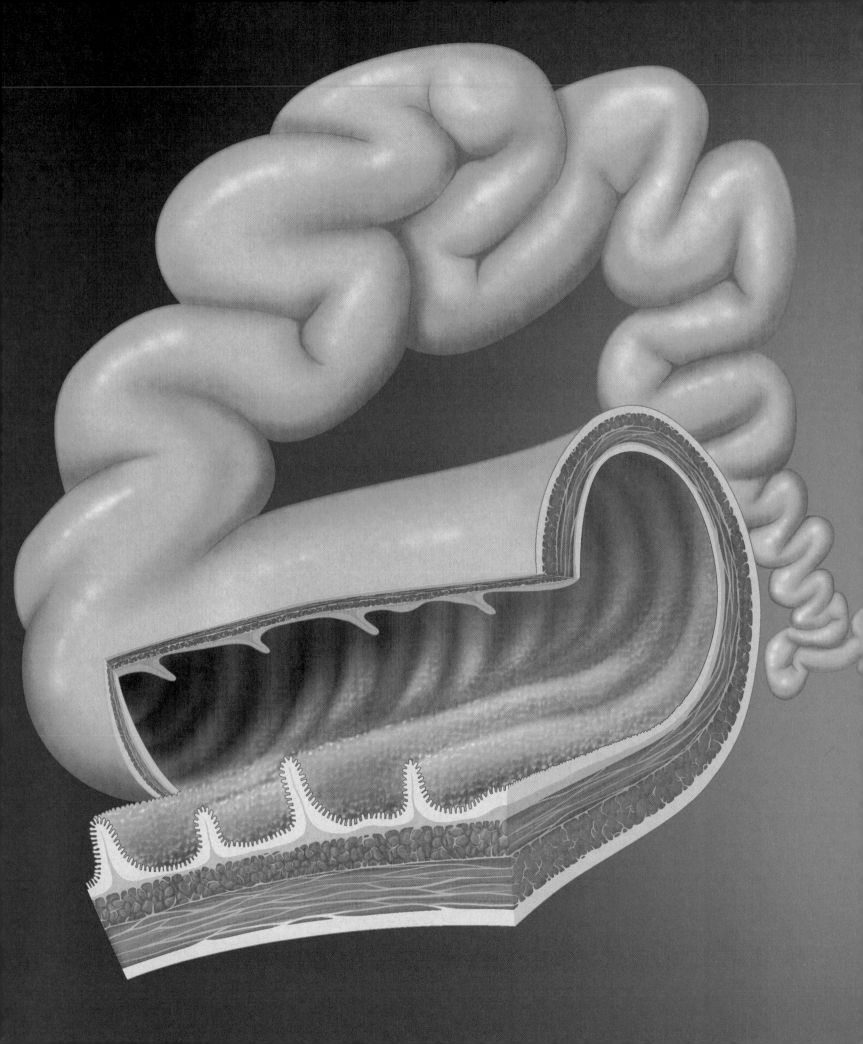

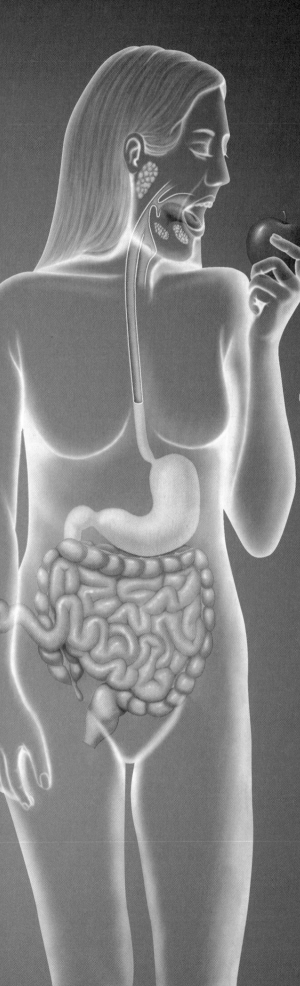

23

Digestion and Excretion

The number of adipose cells you have cannot increase, so when you put on weight more fat gets stored in each cell.

23.1 Impacts/Issues: Regulating Appetite

Like other mammals, humans have adipose tissue that consists of fat-storing cells. This energy warehouse served our ancestors well. As foragers, they could seldom be certain of their next meal. Filling adipose cells with fat when food was abundant helped them survive when food became scarce.

Food scarcity is not a problem for most Americans. In fact, 60 percent of us are overweight or obese. "Obesity" means there is so much fat in adipose tissue that it impairs health. Researchers have long known that obesity raises the risks of diabetes, cardiovascular disease, and some cancers. They have recently started to understand why too much fat contributes to these health problems.

The number of adipose cells in your body cannot increase, so when you put on weight more fat gets stored in each cell. Excess fat distends adipose cells and impairs their function. Overstuffed adipose cells send out distress signals. The signals cause a chronic inflammatory reaction that harms organs throughout the body (Section 22.4).

Adipose cells also make **leptin**, a signaling molecule that acts in the brain and decreases appetite. Mice that cannot make leptin just eat and eat, ballooning to enormous size (Figure 23.1). When researchers injected the obese mice with leptin, the mice ate less and began to slim down. Can leptin injections help humans? Not likely; lack of leptin is extremely rare in humans. Obese people, having more fat, make more leptin than slim ones, but their bodies do not heed leptin's call to stop eating.

Ghrelin is a signaling molecule that increases appetite. Some of the cells in the stomach lining and the brain secrete ghrelin when the stomach is empty. The secretion slows after a big meal. In one study of ghrelin's effects, a group of obese volunteers stayed on a low-fat, low-calorie diet for six months. They lost weight, but the concentration of ghrelin in their blood climbed dramatically, so they felt hungrier than ever!

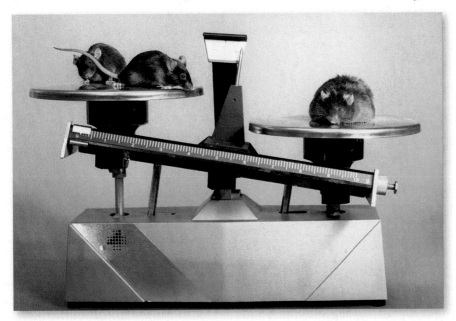

FIGURE 23.1 Two normal mice (*left*) and a mouse that cannot make leptin (*right*).

Some extremely obese people undergo gastric bypass surgery, which effectively reduces the size of the stomach and small intestine. The surgery makes people feel full faster. It also reduces the amount of nutrients that they absorb from food. The weight loss can be dramatic; however, the surgery raises risk for vitamin and mineral deficiencies.

Gastric bypass is more effective than standard weight loss methods. After the surgery, patients are far less likely to regain pounds. One reason may be that they secrete less ghrelin after the bypass surgery, so they feel less hungry.

Questions about food intake and body weight lead us into the world of nutrition. As you will see in this chapter, many organ systems interact to maintain the internal environment regardless of whether nutrients are scarce or abundant.

23.2 Animal Digestive Systems

All animals obtain energy and nutrients from their food. Most do so through the actions of a digestive system, an organ system that carries out the following tasks.

1. **Ingestion**: Taking in food.
2. **Digestion**: Breaking food down into small nutrient molecules.
3. **Absorption**: Moving nutrient molecules into the internal environment.
4. **Elimination**: Expelling leftover material that was not digested and absorbed.

Different animals accomplish the tasks listed above in different ways. Flatworms are among the simplest animals with organ systems. Their branching gut has a single opening that takes in food and expels wastes (Figure 23.2**A**). Nutrients released by digestion diffuse across the wall of the gut cavity and through interstitial fluid to reach body cells.

Most invertebrates and all vertebrates have a tubular gut, with a mouth at one end and an anus at the other. Material moves through the gut, from the mouth to the anus. Along the way, it passes through regions specialized for digestion, nutrient absorption, and waste concentration and elimination.

Figure 23.2**B** shows a frog's digestive system, with gut regions such as a stomach, small intestine, and large intestine. A liver, gallbladder, and pancreas are accessory organs that assist digestion by secreting enzymes and other products into the small intestine.

Variations in the structure of the vertebrate digestive system are adaptations that suit animals to a specific diet and way of feeding. For example, a pigeon has a big crop, a saclike food-storing region above the stomach (Figure 23.2**C**). This is an adaptation to a diet of seeds. A pigeon quickly fills its crop with seeds, then flies off and digests them later. An eat-and-run strategy reduces the amount of time a bird spends on the ground, where it is most vulnerable to predators.

Birds do not have teeth. They grind up food inside a gizzard, a muscular stomach chamber lined with hardened protein. A pigeon has a relatively large gizzard for its size, because it takes lots of grinding to break up seeds. In contrast, a meat-eating bird such as a hawk has a relatively small crop and gizzard.

Like seeds, grasses take a lot of processing to break down. Most animals cannot digest the cellulose in plant cell walls. Cattle and other hoofed grazers accomplish this task with a stomach divided into four chambers. Microbes that live inside the first two stomach chambers carry out fermentation reactions that break down cellulose in plant cell walls. Solids accumulate in the second chamber, forming "cud" that is regurgitated—moved back into the mouth for a second round of chewing. Nutrient-rich fluid moves from the second chamber to the third and fourth chambers, and finally to the intestine. Repeated chewing and processing of food allows cattle to maximize the amount of nutrients they extract from plant material.

Take-Home Message What are the functions of a digestive system?

- Digestive systems take in food, break it into small molecules, absorb nutrient molecules, and eliminate undigested wastes.

- Most animals have a tubular gut with a mouth at one end and an anus at the other. Differences in digestive anatomy adapt animals to their specific diet.

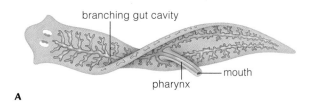

A

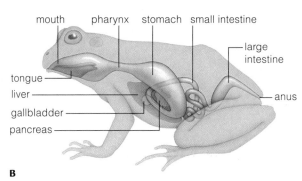

B

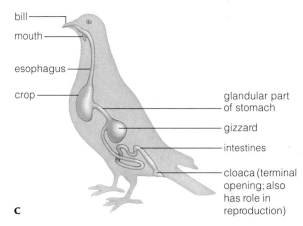

C

FIGURE 23.2 Animated!
Examples of animal digestive systems.

absorption Movement of nutrient molecules from the gut into the body's internal environment.

digestion Breakdown of food into smaller bits or molecules.

elimination Expulsion of unabsorbed material from the digestive tract.

ghrelin Molecule made in the stomach and brain that increases appetite.

ingestion Taking food into the digestive system.

leptin Molecule made by adipose cells that suppresses appetite.

23.3 The Human Digestive System

Figure 23.3 shows components of the human digestive system. This system consists of the tubular organs through which food travels, and the organs that secrete substances into that tube.

The Teeth and Mechanical Digestion Mechanical digestion physically breaks food into smaller pieces. This process begins in the mouth, or oral cavity, where teeth rip food and crush it into pieces. Each tooth consists mostly of bonelike material called dentin (Figure 23.4**A**). Living cells (the tooth's pulp) lie deep inside the tooth. **Enamel**, the hardest material in the body, covers the tooth's exposed surface (its crown) and reduces wear. An adult typically has thirty-two teeth of four types (Figure 23.4**B**).

Tooth decay occurs when bacteria convert food stuck to a tooth into acidic plaque. The plaque dissolves a hole in the tooth, creating what we commonly call a cavity. If an unfilled cavity expands into a tooth's pulp, bacteria can enter, irritating nerves and causing pain. Once an infection or injury damages the pulp, a root canal procedure may be necessary. With this procedure, living material inside a tooth is removed and replaced with inert material.

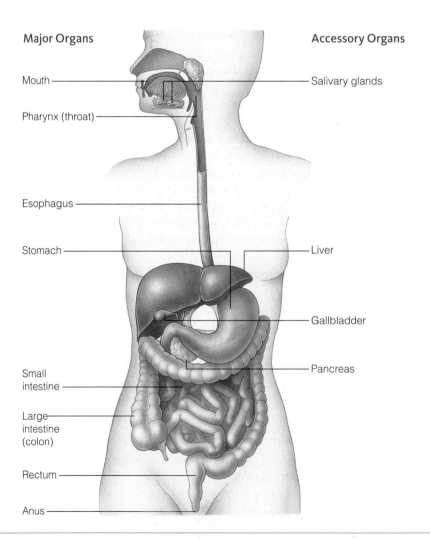

Major Organs

Mouth

Pharynx (throat)

Esophagus

Stomach

Small intestine

Large intestine (colon)

Rectum

Anus

Accessory Organs

Salivary glands

Liver

Gallbladder

Pancreas

FIGURE 23.3 Animated! Major components and accessory organs of the human digestive system.

FIGURE 23.4 Human teeth. **A** Anatomy of a tooth. **B** Arrangement of adult teeth. Chisel-shaped incisors shear off bits of food. Cone-shaped canines tear up meats. Premolars and molars have broad bumpy crowns that are platforms for grinding and crushing food.

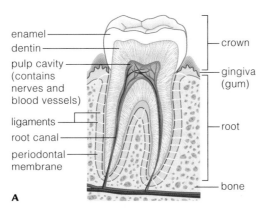

enamel
dentin
pulp cavity (contains nerves and blood vessels)
ligaments
root canal
periodontal membrane

crown
gingiva (gum)
root
bone

A

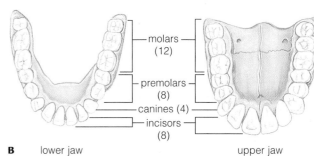

molars (12)
premolars (8)
canines (4)
incisors (8)

B   lower jaw   upper jaw

FIGURE 23.5 Animated! How to perform the Heimlich maneuver on an adult who is choking.

1. Determine that the person is actually choking; a person who has an object lodged in the trachea cannot cough or speak.

2. Stand behind the person and place one fist below the rib cage, just above the navel, with your thumb facing inward as in **A**.

3. Cover the fist with your other hand and thrust inward and upward with both fists as in **B**. Repeat until the object is expelled.

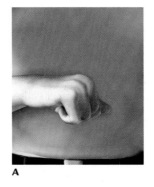

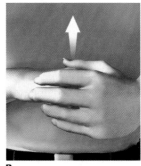

A   **B**

Saliva and Chemical Digestion

Chemical digestion is the enzymatic breakdown of large molecules into smaller ones that can be absorbed. Like mechanical digestion, chemical digestion begins in the mouth, where movements of the tongue help mix food with saliva from salivary glands. An enzyme in saliva (**salivary amylase**) breaks starch molecules into disaccharides. Saliva also contains proteins that combine with water and form mucus. The mucus makes small food bits stick together in easy-to-swallow clumps.

Swallowing—Down the Tube

A human pharynx, or throat, is the entrance to the digestive and respiratory tracts (Section 21.8). The tongue pushes food to the back of the throat, triggering a swallowing reflex. As you swallow, the flaplike epiglottis flops down and the vocal cords constrict, so the route between the pharynx and larynx is blocked. This reflex directs food into the **esophagus**, the muscular tube that connects the pharynx to the stomach.

If food or drink accidentally enters the larynx, a reflexive cough generally expels it. However, sometimes food gets through the larynx and obstructs the trachea. When this happens, a person chokes. A choking person cannot cough, speak, or breathe. A procedure called the Heimlich maneuver is designed to force food out of the trachea and save a person's life. Figure 23.5 illustrates how to carry out this procedure.

When swallowing proceeds normally and food enters the esophagus, waves of smooth muscle contraction, called peristalsis, move food to the stomach. At the lower end of the esophagus, where it joins with the stomach, is a **sphincter**. A sphincter is a ring of muscle that controls passage of material through a tube or a body opening. This particular sphincter opens to allow food into the stomach, then closes to prevent stomach acid from splashing into the esophagus. In some people, this sphincter does not shut properly, and acid reflux occurs. The acidic stomach fluids irritate the esophagus, causing "heartburn."

chemical digestion Breakdown of food molecules into smaller subunits by enzymes.

enamel Hard material covering exposed surface of teeth.

esophagus Muscular tube between the throat and stomach.

mechanical digestion Breaking of food into smaller pieces by mechanical processes such as chewing.

salivary amylase Enzyme in saliva that begins carbohydrate digestion by breaking starch into disaccharides.

sphincter Ring of muscle that controls passage through a tubular organ or body opening.

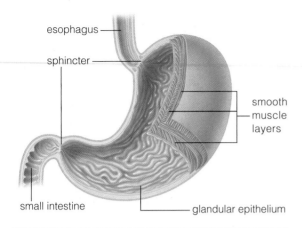

esophagus

sphincter

smooth muscle layers

small intestine

glandular epithelium

FIGURE 23.6 Stomach structure and function. The innermost layer is glandular epithelium that secretes gastric fluid. Three smooth muscle layers differ in their orientation, and their contractions cause a wringing action that mixes and moves food. Sphincters at either end of the stomach can close to keep acidic chyme in the stomach.

bile Mix of salts, pigments, and cholesterol produced by the liver; aids in fat digestion.

chyme Mix of food and gastric fluid.

feces Unabsorbed food material and cellular waste that is produced by digestion.

gallbladder Organ that receives bile from the liver and expels it into the small intestine.

gastric fluid Fluid secreted by the stomach lining; contains enzymes, acid, and mucus.

large intestine Organ that receives digestive waste from the small intestine and concentrates it as feces.

liver Organ that produces bile, stores glycogen, and detoxifies many substances.

microvilli Thin projections that increase the surface area of brush border cells.

pancreas Organ that secretes digestive enzymes into the small intestine and hormones into the blood.

small intestine Longest portion of the digestive tract, and the site of most digestion and absorption.

stomach Muscular organ that mixes food with gastric fluid that it secretes.

villi Multicelled projections from the lining of the small intestine.

The Stomach The **stomach** is a muscular, stretchable sac with three functions. First, it mixes and stores ingested food. Second, it secretes substances that begin to break down food, especially proteins. Third, it helps control the passage of food into the small intestine.

The stomach has three layers of smooth muscle and a lining of glandular epithelium (Figure 23.6). When the stomach is empty, this lining has numerous folds. As food begins to fill the stomach, these folds flatten. Arrival of food in the stomach also stimulates smooth muscle contractions and causes cells in the stomach lining to secrete **gastric fluid**. This fluid includes a strong acid, enzymes, and mucus. Stomach contractions mix food and gastric fluid together and form a semiliquid mass called **chyme**.

Protein digestion begins in the stomach. The acidity of gastric fluid causes proteins to unfold. Then an enzyme in gastric fluid (pepsin) cuts the proteins into polypeptides. Why doesn't this process break down proteins in the stomach wall? Mucus secreted by the stomach lining protects the underlying tissue.

When something disrupts the protective mucus layer, gastric fluid and enzymes can erode the stomach lining, causing an ulcer. Most ulcers occur after certain bacteria (*Helicobacter pylori*) make their way through gastric mucus and infect cells of the stomach lining. Antibiotics can halt the bacterial infection and allow the ulcer to heal. Another cause of ulcers is continual use of aspirin, ibuprofen, and other nonsteroidal anti-inflammatory drugs (NSAIDs). These drugs interfere with chemical signals that maintain the health of the stomach lining.

Digestion in the Small Intestine From the stomach, chyme is forced though a sphincter into the **small intestine**. The small intestine is "small" only in its diameter—about 2.5 cm (1 inch). It is the longest segment of the gut. Uncoiled, it would extend for about 5 to 7 meters (16 to 23 feet). Most digestion and absorption takes place in the small intestine.

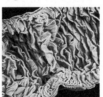

The lining of the small intestine is highly folded (*left* and Figure 23.7 ❶). The surface of each fold has many tiny multicelled projections, each about a millimeter long, called **villi** (singular, villus) ❷. The millions of villi that project from the intestinal lining give it a velvety appearance. Blood vessels and lymph vessels run through each villus.

The sides and tip of a villus are covered mostly with brush border cells ❸. These cells have thin projections called **microvilli** (singular, microvillus) at their membrane surface ❹. Each cell has 1,700 or so microvilli that make its outer edge look somewhat like a brush. Collectively, the many folds and projections of the small intestinal lining increase its surface area by hundreds of times, providing a huge area for digestion and absorption.

Enzymes at the surface of brush border cells and enzymes secreted into the small intestine by the adjacent **pancreas** carry out digestion. Together, these enzymes break carbohydrates, proteins, fats, and nucleic acids into smaller subunits. The pancreas also secretes a buffer into the intestine. The buffer decreases the acidity of chyme, protecting the intestinal lining and ensuring that intestinal enzymes function properly.

In addition to enzymes, fat digestion requires **bile**, a mix of salts, pigments, and cholesterol produced in the **liver**. In between meals, the main bile duct from the liver to the small intestine is closed and bile enters the **gallbladder**, which stores it. Eating a fatty meal stimulates the gallbladder to contract and expel bile into the small intestine.

Bile enhances fat digestion by helping to keep fat droplets from clumping together. Compared to bigger globules, tiny droplets present a greater surface area to fat-digesting enzymes. Fats are insoluble in water and tend to cluster as

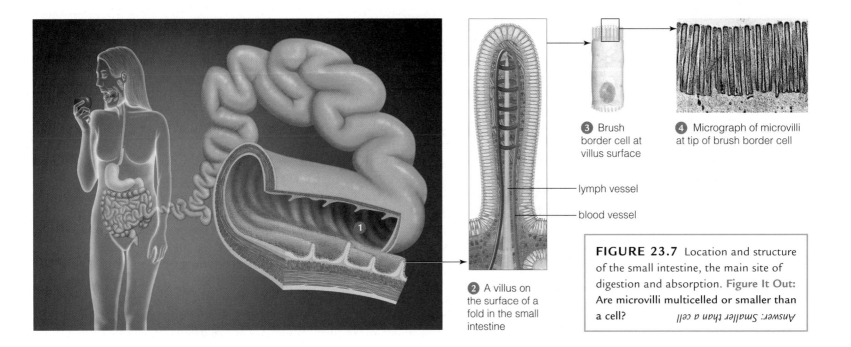

3 Brush border cell at villus surface

4 Micrograph of microvilli at tip of brush border cell

lymph vessel

blood vessel

2 A villus on the surface of a fold in the small intestine

FIGURE 23.7 Location and structure of the small intestine, the main site of digestion and absorption. **Figure It Out:** Are microvilli multicelled or smaller than a cell? *Answer: Smaller than a cell*

large globules in chyme. As the muscle layers in the intestinal wall move and shake the chyme, fat globules break into smaller droplets that become coated with bile salts. Bile salts carry negative charges, so the coated fat droplets repel each other and stay separated.

Gallstones are hard pellets of cholesterol or pigment that sometimes form in the gallbladder. Most are harmless, but some can block a bile duct and cause pain, especially after a fatty meal. When gallstones cause problems, the gallbladder can be removed surgically. After surgery, bile from the liver flows directly from the liver into the small intestine.

Absorption in the Small Intestine Absorption is the passage of nutrients, water, salts, and vitamins into the internal environment (Figure 23.8). The large absorptive surface of the intestinal wall aids the process. So does the action of rings of smooth muscle in this wall. Their repeated contraction and relaxation sloshes chyme back and forth, forcing it against the absorptive lining.

Products of carbohydrate digestion (simple sugars) and products of protein digestion (amino acids) move through brush border cells, diffuse through the interstitial fluid inside the villus, then enter a capillary. Products of fat digestion (triglycerides) also cross the brush border cell and interstitial fluid, but they enter lymph vessels, which eventually carry them to the blood.

Concentrating and Eliminating Wastes Not everything that enters the small intestine can be or should be absorbed. Undigestible material moves from the small intestine into the **large intestine** (colon) along with cells shed from the gut lining and some other wastes. This collection of waste material is called **feces**.

Most of the water that enters the gut is absorbed in the small intestine, but a final round of waste concentration takes place in the large intestine. Wastes become concentrated as water leaves the feces. Ions and vitamins made by bacteria that live in the large intestine are also absorbed into the blood.

Following a meal, nervous signals stimulate large regions of the large intestine to contract. The contractions move material already in the tube farther along making room for incoming material.

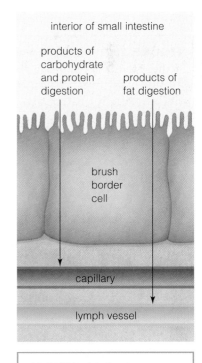

interior of small intestine

products of carbohydrate and protein digestion

products of fat digestion

brush border cell

capillary

lymph vessel

FIGURE 23.8 Absorption. Nutrients move across brush border cells into a villus, then enter the blood or lymph.

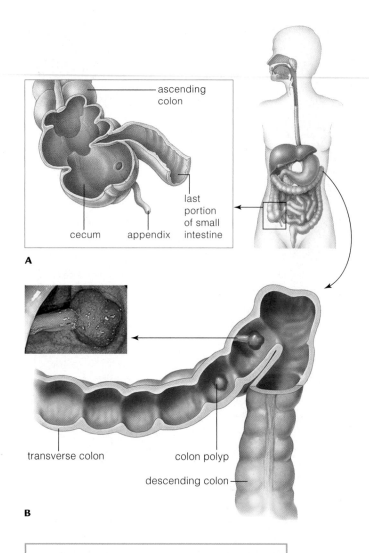

ascending colon

last portion of small intestine

cecum   appendix

A

transverse colon   colon polyp

descending colon

B

FIGURE 23.9 The large intestine. **A** Location of the appendix. **B** Sketch and photo of polyps.

The final length of the large intestine is the **rectum**. The rectum stores feces until enough accumulates to stretch rectal walls and trigger a defecation reflex. The brain controls the timing of defecation. It sends out signals that cause contraction or relaxation of skeletal muscle in the anal sphincter.

The **appendix** is a wormlike projection that extends from the cecum, a pouch at the start of the large intestine (Figure 23.9**A**). The appendix has no known digestive function, but is full of white blood cells that help defend the body. In the United States, about 7 percent of people will have appendicitis (an inflamed appendix) at some time in their life. Usually this occurs when bits of waste clog the opening to the appendix, obstructing blood flow and allowing fluids to build up inside it. Tissue begins to die and infection sets in. Unless the inflamed appendix is removed by surgery, it can rupture. If it does, bacteria can enter the abdominal cavity and cause a life-threatening infection.

A diet high in fiber (undigestible plant material) can help prevent appendicitis. Fiber also helps speed movement of material through the colon. Healthy adults typically defecate about once a day. Emotional stress, a diet low in fiber, minimal exercise, dehydration, and some medications can lead to constipation. With constipation, defecation occurs fewer than three times a week, is difficult, and yields small, hardened, dry feces. Occasional constipation usually goes away on its own. A chronic problem should be discussed with a doctor. Diarrhea, frequent passing of watery feces, can result from bacterial infection or problems with nervous controls. If prolonged, it can cause dehydration and disrupt solute levels in the blood.

Some people are genetically predisposed to develop colon polyps, small growths that project from the wall of the large intestine (Figure 23.9**B**). Polyps are benign growths, but some can become cancerous. If colon cancer is detected early, it is highly curable. Even so, it kills about 50,000 people in the United States every year. Blood in the stool and changes in bowel habits can be symptoms and should be discussed with a physician. Anyone over the age of 50 should have a colonoscopy, a procedure in which the wall of the large intestine is examined using a camera at the end of a flexible tube. Polyps detected by this procedure can be removed immediately, eliminating the possibility that they will later become cancerous.

Take-Home Message

How do humans digest their food, absorb nutrients, and eliminate digestive waste?

- Mechanical and chemical breakdown of food begins in the mouth. Saliva contains enzymes that start carbohydrate digestion.

- Gastric fluid contains acid and digestive enzymes that begin the process of protein digestion in the stomach.

- In the small intestine, intestinal and pancreatic enzymes complete digestion of proteins and carbohydrates, and bile from the gallbladder aids in fat digestion.

- With its richly folded lining, millions of villi, and hundreds of millions of microvilli, the small intestine has a vast surface area for absorbing nutrients.

- The large intestine, or colon, absorbs water and mineral ions from the gut contents and compacts undigested residues. The resulting feces are stored in the rectum before being expelled through the anus.

Human Nutrition

23.4

What happens to the nutrients we absorb? They are burned as fuel in energy-releasing pathways, stored, and used as building blocks. Section 5.8 described how carbohydrates, fats, and proteins are converted to intermediates that enter the ATP-producing pathway of aerobic respiration. Figure 23.10 summarizes the major routes by which the body uses organic compounds from food.

Energy-Rich Carbohydrates Fresh fruits, whole grains, and vegetables—especially legumes such as peas and beans—contain abundant complex carbohydrates (Section 2.7). The body breaks the starch in these foods into glucose, your primary source of energy. Glucose absorbed from the gut enters the blood, from which it is taken up by cells and used for metabolism. When glucose absorption exceeds the body's immediate needs, the excess is stored. Liver and muscle cells store excess glucose as glycogen and adipose tissue uses glucose to make fats. Between meals, when glucose is not being absorbed from the gut, the body turns to glycogen and fat stores to meet its energy needs. Liver cells break down glycogen and adipose tissue cells dismantle fats.

Carbohydrates are also the body's source of dietary fiber. There are two types of fiber. Soluble fiber consists of polysaccharides that form a gel when mixed with water. Eating foods high in soluble fiber helps lower cholesterol level and may reduce the risk of heart disease. Insoluble fiber such as cellulose does not dissolve in water and passes through the human digestive tract more or less intact. A diet with plenty of insoluble fiber helps prevent constipation. Processed carbohydrates such as white flour, refined sugar, and corn syrup are sometimes said to provide only "empty calories" because these foods provide little in the way of vitamins or fiber.

Good Fat, Bad Fat Fats are used as an energy source and as building blocks of cell membranes. They also help you take up fat-soluble vitamins. Your body can make most fats you need from carbohydrates. Fats the body cannot make, such as linoleic acid, are **essential fatty acids**. Vegetable oils are a good source of these fatty acids.

appendix Wormlike projection from the first part of the large intestine.

essential fatty acids Fatty acids that the body cannot make and must obtain from the diet.

rectum Portion of the large intestine that stores feces until they are expelled.

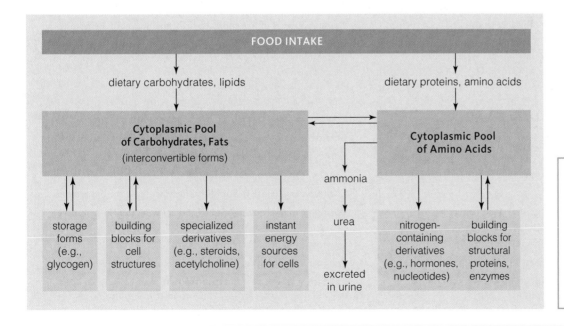

FIGURE 23.10 Fate of nutrients absorbed into the body from the gut.

Figure It Out: The liver converts ammonia, a toxic product of digestion, to urea that is excreted in the urine. What type of nutrient (carbohydrate, lipid, or protein) is the source of the ammonia?

Answer: Protein

Butter, cheese, and fatty meats are rich in saturated fats and cholesterol. Overconsumption of these foods increases the risk for heart disease and stroke, as well as for cancers. *Trans* fatty acids, or "*trans* fats," are also bad for the heart. All food labels are now required to show the amounts of *trans* fats, saturated fats, and cholesterol per serving.

Body-Building Proteins Your body uses amino acid components of dietary proteins for its own protein-building programs. Of the twenty common types, eight are **essential amino acids**. Your cells cannot synthesize them; you must get them from food. They are methionine (or cysteine, its metabolic equivalent), isoleucine, leucine, lysine, phenylalanine (or tyrosine), threonine, tryptophan, and valine.

Most proteins in animal products are *complete*; their amino acid ratios can satisfy human nutritional needs. Nearly all plant proteins are *incomplete*—they lack one or more essential amino acids. You can meet all of your amino acid requirements with plant-based foods, but doing so requires a bit more planning. The amino acids missing from one plant food must be provided by others. As an example, rice and beans together provide all necessary amino acids, but rice alone or beans alone do not. You do not have to eat the two complementary foods at the same meal, but both should be consumed within a 24-hour period.

Vitamins and Minerals **Vitamins** are organic substances that are required in small amounts for normal metabolism (Table 23.1).

Vitamins A, E, D, and K are fat-soluble vitamins. Such vitamins are not harmed by heating, and are abundant in both cooked and fresh foods. Fat-soluble vitamins get stored in the body's own fat. Because the body stores these vitamins, they do not need to be replenished daily and deficiencies rarely occur. In fact, taking megadoses of fat-soluble vitamins can cause problems. For exam-

> **Water-soluble vitamins must be replenished each day to prevent a deficiency.**

Table 23.1	Sources and Functions of Major Vitamins	
Vitamin	Common Sources	Main Functions
Fat-soluble vitamins		
A	Yellow/orange fruits, green vegetables, milk, egg yolk, fish, liver	Used in synthesis of visual pigments, bone, teeth; maintains the skin
D	Fish liver oils, egg yolk, fortified milk; also synthesized in skin	Promotes bone health; enhances calcium absorption
E	Whole grains, dark green vegetables, vegetable oils	Counters effects of free radicals; helps maintain cell membranes; blocks breakdown of vitamins A, C in gut
K	Made by bacteria in large intestine; also in green leafy vegetables, cabbage	Acts in blood clotting; ATP formation via electron transfer chains
Water-soluble vitamins		
B_1 (thiamin)	Legumes, whole grains, green leafy vegetables, meats, eggs	Connective tissue formation; folate utilization; coenzyme action
B_2 (riboflavin)	Whole grains, poultry, fish, egg white, milk	Coenzyme action in carbohydrate and amino acid metabolism
B_3 (niacin)	Green leafy vegetables, potatoes, peanuts, poultry, fish, meats	Coenzyme action in carbohydrate metabolism
B_6	Spinach, tomatoes, potatoes, meats	Coenzyme in amino acid metabolism
Pantothenic acid	Meats, yeast, egg yolk	Coenzyme in glucose metabolism, fatty acid and steroid synthesis
Folate (folic acid)	Dark green vegetables, whole grains, yeast, lean meats; made by bacteria in large intestine	Coenzyme in nucleic acid and amino acid metabolism; required for normal nervous system development of embryo
B_{12}	Poultry, fish, red meat, dairy foods (not butter)	Coenzyme in nucleic acid metabolism
Biotin	Legumes, egg yolk; made by bacteria in large intestine	Coenzyme in fat, glycogen formation, and amino acid metabolism
C (ascorbic acid)	Fruits and vegetables, especially citrus, berries, cantaloupe, green pepper, cabbage, broccoli	Collagen synthesis; used in carbohydrate metabolism; structural role in bone, teeth, cartilage; may counter free radicals

ple, taking an excessive amount of vitamin A during pregnancy can cause birth defects. Some studies have found that taking large amounts of vitamin E can actually increase the risk of death.

Water-soluble vitamins such as vitamin C and the B vitamins are not stored by the body. It is difficult to accumulate too much of these vitamins, because any excess is eliminated in the urine. Water-soluble vitamins must be replenished each day to prevent a deficiency. These vitamins are destroyed by heat, so fresh foods are the best source. Many of the B vitamins are coenzymes. For example, the coenzyme NADH, which plays an essential role in aerobic respiration (Section 5.6), is derived from vitamin B_3 (niacin).

Minerals are inorganic substances that are required in small amounts for normal growth and metabolism. For example, all cells use iron in electron transfer chains, and red blood cells use it to make hemoglobin. You can obtain iron from meat, and many cereals are now fortified with iron. As another example, calcium is necessary for normal function of nerves and muscles, and to build strong bones. Dairy products and tinned fish such as sardines are good sources of calcium. Some cereal products are also fortified with calcium.

People who are in good health get all the vitamins and minerals they need from a balanced diet. Vitamin and mineral supplements are necessary only for strict vegetarians, the elderly, and people who have a condition that interferes with how the body absorbs or processes nutrients.

USDA Dietary Recommendations

Every five years, the U.S. Department of Agriculture issues a set of dietary guidelines based on nutrition research and input from food industry representatives. The guidelines are designed to help prevent obesity, diabetes, heart disease, osteoporosis, and other chronic health problems. In 2005, the USDA replaced its previous one-size-fits-all food pyramid with a new Internet-based program that generates recommendations specific for a person's age, sex, height, weight, and activity level (Figure 23.11). To generate your own healthy eating plan, visit the USDA web site: www.mypyramid.gov.

The new guidelines recommend that Americans reduce their intake of refined grains, saturated fats, *trans* fatty acids, added sugar or caloric sweeteners, and salt (no more than a teaspoon per day). They also recommend eating more vegetables and fruits with a high potassium and fiber content, fat-free or low-fat milk products, and whole grains. About 55 percent of daily caloric intake should come from carbohydrates.

Food Group	Amount Recommended
USDA Nutrition Guidelines	
Vegetables	2.5 cups/day
Dark green vegetables	3 cups/week
Orange vegetables	2 cups/week
Legumes	3 cups/week
Starchy vegetables	3 cups/week
Other vegetables	6.5 cups/week
Fruits	2 cups/day
Milk products	3 cups/day
Grains	6 ounces/day
Whole grains	3 ounces/day
Other grains	3 ounces/day
Fish, poultry, lean meat	5.5 ounces/day
Oils	24 grams/day

FIGURE 23.11 Example of nutritional guidelines from the United States Department of Agriculture (USDA). These recommendations are for females between ages ten and thirty who get less than 30 minutes of vigorous exercise daily. Portions add up to a 2,000-kilocalorie daily intake.

Take-Home Message How do we use the nutrients in our food?

- Carbohydrates are broken down to glucose, the body's main energy source. Foods rich in complex carbohydrates also supply fiber and vitamins.

- Fats are burned for energy and used as building materials. Essential fatty acids can be obtained from plant oils. Excessive consumption of saturated fats and *trans* fats raises the risk of heart disease.

- Proteins are the source of amino acids that the body uses to build its own proteins. Meat provides all essential amino acids. Most plant foods lack one or more amino acids, but when combined correctly these foods can meet all human amino acid needs.

- Vitamins are organic substances required in small amounts for metabolism. Fat-soluble vitamins can be stored in the body. Water-soluble vitamins must be replenished daily. Minerals are inorganic substances required in small amounts for essential metabolic tasks.

essential amino acid Amino acid that the body cannot make and must obtain from food.

mineral Inorganic substance that is required in small amounts for normal metabolism.

vitamin Organic substance required in small amounts for normal metabolism.

Balancing Calories In and Out

When you think about your ideal weight, you probably focus on how you would like to look. As a culture, we are obsessed with being thin, but are also fond of fast and fatty foods. Billions of dollars in sales of diet books, plans, and products reflect our fat phobia. Steadily increasing waistlines underline the failure of most current approaches to weight control or loss.

Looks aside, excess weight is a pressing public health issue that continues to worsen. Since the 1970s, the percentage of overweight children and adolescents has doubled. One recent study concluded that if current trends continue, weight-related deaths could decrease the life expectancy in the United States by as much as 5 years over the next decade.

Is your current weight healthy? One way to find out is to calculate your **body mass index** (BMI), using this formula:

$$\text{BMI} = \frac{\text{weight (pounds)} \times 700}{\text{height (inches)}^2}$$

A BMI of 25 to 29.9 is defined as overweight, and a BMI of 30 or more is defined as obese. Having a BMI higher than 25 carries an increased risk of premature death due to weight-related illness. Fat distribution also affects risk. Belly fat poses more of a health risk than fat on the legs or hips. Males with a waist size of more than 47 inches (120 cm) and females with a waist size that exceeds 35 inches (88 cm) are especially likely to have health problems.

It is difficult to lower weight-related risks with diet alone. Eat less, and the body slows its metabolic rate to conserve energy. To maintain your weight, you must balance energy intake with energy used for metabolic activities.

The energy stored in food is measured as kilocalories, or Calories (with a big C). To figure out how many kilocalories you should take in daily to maintain a preferred weight, multiply that weight (in pounds) by 10 if you are not active physically, 15 if moderately active, and 20 if highly active. Next, subtract one of the following amounts from the multiplication value:

	Age	Subtract
	25–34	0
	35–44	100
	45–54	200
	55–64	300
	Over 65	230

Age is a factor because resting metabolic rate drops as a person gets older, so less food is required. So, if you are 25, are highly active, and want to maintain a weight of 120 pounds, you need to eat $120 \times 20 = 2,400$ kilocalories. Such calculations are only a rough estimate of caloric needs. Other factors, including height, must also be considered. An active person who is 5 feet, 2 inches tall does not require as many calories as an active 6-footer of the same weight.

Take-Home Message How can you maintain a healthy weight?

- To maintain your weight, your energy input from food must equal the energy you burn in metabolism and physical activity.

- The amount of calories needed to maintain a specific weight depends on height, age, and activity level.

body mass index Formula used to determine a healthy weight based on height.

kidney Organ that filters blood and forms urine.

ureter Tube that carries urine from a kidney to the bladder.

urethra Tube through which urine from the bladder flows out of the body.

urinary bladder Hollow, muscular organ that stores urine.

urinary system Organ system that filters blood, and forms, stores, and expels urine.

urine Mix of water and soluble wastes formed and excreted by the urinary system.

23.6 The Urinary System

Maintaining Body Fluids By weight, all organisms consist mostly of water with salts and other solutes. Fluid outside cells—the extracellular fluid (ECF)—functions as the body's internal environment. In humans and other vertebrates, extracellular fluid consists mostly of interstitial fluid (the fluid in spaces between cells) and plasma (the fluid portion of the blood).

Keeping the solute composition and volume of the extracellular fluid within the range that living cells can tolerate is a major aspect of homeostasis. Water and solute gains need to be balanced by water and solute losses. Eating and drinking adds water and nutrients to the ECF. Metabolism adds wastes such as urea, a nitrogen-containing by-product of protein digestion. In all vertebrates, a urinary system rids the body of unwanted solutes and excess water.

The Human Urinary System The human **urinary system** consists of two kidneys, two ureters, a bladder, and a urethra (Figure 23.12). The **kidneys** are a pair of bean-shaped organs, each about as large as an adult's fist. Each kidney is enclosed in a protective outer capsule of connective tissue.

The kidneys filter water, mineral ions, nitrogen-rich wastes, and other substances from blood. Then they adjust the composition of the filtrate and return nearly all water and nonwaste solutes to blood. **Urine**, the fluid that remains, consists of urea and other wastes, water, and excess solutes.

Urine flows from each kidney into a **ureter**, a tube that empties into the **urinary bladder**. This hollow muscular organ stores urine until a sphincter at its lower end opens and urine flows into the **urethra**. Urine leaves the body through the urethra.

An urge to urinate arises when the stretch receptors in the bladder are stimulated. Smooth muscle in the bladder wall contracts in a reflex response. After the age of 2 or 3 years, urination is voluntary. It occurs when a sphincter between the bladder and urethra is relaxed voluntarily so that the force of bladder contraction can propel urine out of the body.

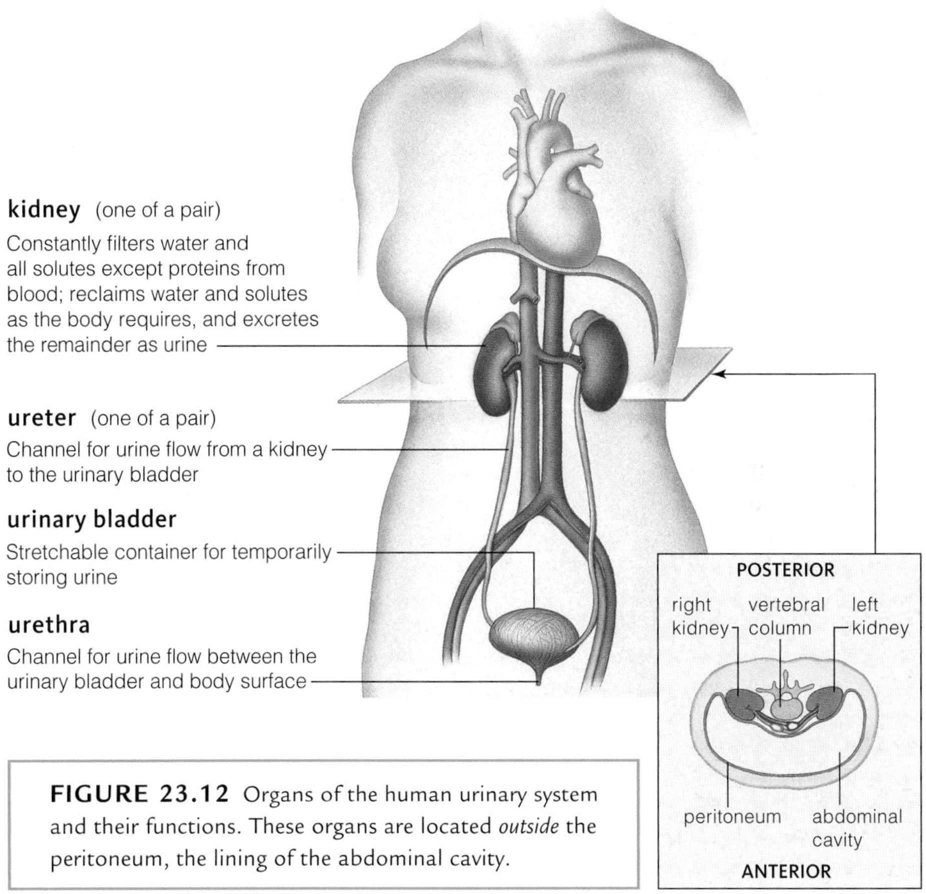

kidney (one of a pair)
Constantly filters water and all solutes except proteins from blood; reclaims water and solutes as the body requires, and excretes the remainder as urine

ureter (one of a pair)
Channel for urine flow from a kidney to the urinary bladder

urinary bladder
Stretchable container for temporarily storing urine

urethra
Channel for urine flow between the urinary bladder and body surface

POSTERIOR

right kidney · vertebral column · left kidney

peritoneum · abdominal cavity

ANTERIOR

FIGURE 23.12 Organs of the human urinary system and their functions. These organs are located *outside* the peritoneum, the lining of the abdominal cavity.

Take-Home Message

What is the function of a urinary system?

- A urinary system regulates the composition and volume of extracellular fluid. Kidneys filter the blood and form urine.
- Urine from kidneys flows through ureters to a bladder, where it is stored until it flows out of the body through the urethra.

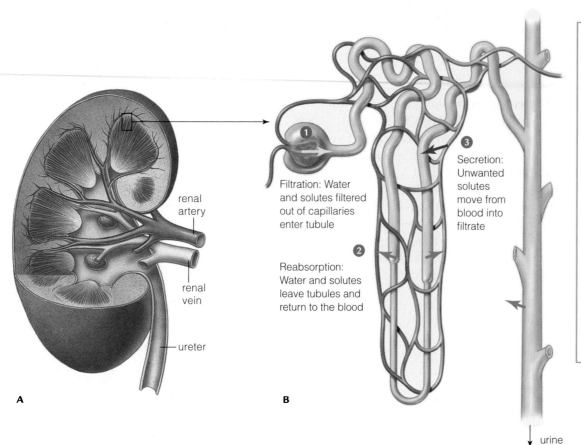

FIGURE 23.13 Animated!
Urine formation in the kidney.

A Structure of a human kidney.

B One nephron, the functional unit of a kidney. Nephron tubules interact with capillaries to form urine by three processes.

1 Filtration (*yellow* arrow): Blood pressure forces fluid out of capillaries and into the entrance to a kidney tubule, forming the filtrate.

2 Reabsorption (*green* arrows): Water and solutes leave kidney tubules and return to blood of adjacent capillaries.

3 Secretion (*purple* arrow): Solutes move from the blood in capillaries into adjacent tubules.

renal artery

renal vein

ureter

A

Filtration: Water and solutes filtered out of capillaries enter tubule

Reabsorption: Water and solutes leave tubules and return to the blood

Secretion: Unwanted solutes move from blood into filtrate

B

urine

23.7

Kidney Function

How Urine Forms A kidney is a blood-cleansing organ. A renal artery carries blood to be cleansed to each kidney, and a renal vein transports cleansed blood away from it (Figure 23.13**A**). The Latin word *renal* means "relating to kidneys." Inside a kidney, blood flows into arterioles. The arterioles branch into capillaries that associate with kidney tubules as a **nephron**.

Nephrons cleanse the blood and form urine by three processes: filtration, reabsorption, and secretion (Figure 23.13**B**). Each nephron starts in the outer region of the kidney, where the wall of a kidney tubule cups around a cluster of capillaries. These capillaries are structurally specialized for filtering; they are much leakier than capillaries in most parts of the body. By the process of **filtration**, blood pressure forces water and solutes out through spaces between cells in the walls of the capillaries **1**. Proteins are too big to get out through these spaces, so they remain in the blood.

The solute-rich filtrate forced out of the capillaries enters the first part of the kidney tubule, where **reabsorption** begins. By this process, water and essential solutes that left the blood return to it as the filtrate flows through tubules **2**. About 99 percent of the water returns to the blood, along with all the glucose and amino acids. Wastes such as urea and the breakdown product of hemoglobin remain in the filtrate. Degraded hemoglobin is what gives your urine its yellow color.

As the filtrate flows through the tubule, unwanted solutes from the blood are moved into it, a process called **tubular secretion** **3**. Secretion is essential to regulating the pH of the internal environment. Enzymes can function only within

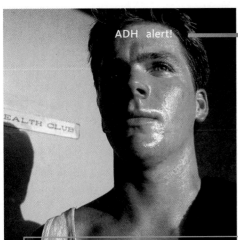
ADH alert!

Stimulus

1 Water loss lowers the volume of blood and makes it more salty. The hypothalamus in the brain senses these changes and signals the adjacent pituitary gland to release ADH.

2 ADH travels through the blood to the kidney, where it affects kidney tubules. The ADH makes the tubules more permeable to water.

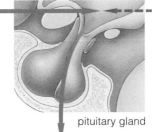
hypothalamus

Response

3 The hypothalamus senses the change in blood volume and concentration and stops calling for ADH secretion.

pituitary gland

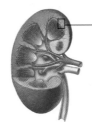

4 More water is reabsorbed and less is lost in urine, so blood volume rises and blood becomes more dilute.

FIGURE 23.14 Negative feedback control of ADH secretion. A negative feedback loop from kidneys to the brain helps adjust the volume of extracellular fluid. Nephrons in the kidneys reabsorb more water when we do not take in enough water or lose too much, as by profuse sweating.

a limited pH range. An excess of hydrogen ions (H$^+$) can disrupt many essential metabolic processes. Kidneys help maintain acid–base balance by regulating how much H$^+$ is secreted into the urine.

Feedback Control of Urine Formation The concentration of urine can vary. Sip soda all day and your urine will be dilute and light in color. Spend 8 hours or so without drinking, as when you sleep, and your urine becomes concentrated and darker in color. Your urine also becomes more concentrated when you lose a lot of water by sweating.

Antidiuretic hormone (ADH) is a signaling molecule that increases water reabsorption, thus making the urine more concentrated (Figure 23.14). When you lose water, the resulting change in the volume and concentration of body fluid is detected by the hypothalamus, a brain region that is a control center for many homeostatic responses **1**. The hypothalamus stimulates the pituitary gland to secrete ADH. The ADH travels through the blood to cells of the kidney tubules, where it causes the tubules to become more permeable to water **2**. More water is reabsorbed, so less departs in the urine. Increased water reabsorption returns the body fluid's salt concentration and volume back to normal **3**. The hypothalamus senses the change and stops calling for ADH secretion **4**.

Alcohol inhibits ADH secretion. As a result, more fluid remains in the filtrate and the drinker needs to urinate more frequently. The excessive urination can lead to dehydration, which contributes to the symptoms of a hangover.

Impaired Kidney Function Sometimes the substances dissolved in urine come out of solution and form hard kidney stones inside the kidney. About one million Americans receive treatment for kidney stones each year, with men four times as likely as women to be affected.

Most kidney stones do not cause symptoms, but some enter a ureter and cause extreme pain. When the ureter contracts around the stone, the ureter wall can be damaged, making the urine bloody. Usually the fluid pressure exerted on the stone by the urine pent up behind it forces the stone through the urinary tract and out of the body. If the kidney stone does not pass after a reasonable

antidiuretic hormone Hormone produced by the pituitary gland; increases water reabsorption by the kidney.

filtration Blood pressure forces water and small solutes, but not blood cells or proteins, out across the walls of capillaries.

nephron Kidney tubule and associated capillaries; filters blood and forms urine.

reabsorption Water and solutes enter capillaries.

tubular secretion Substances are moved out of capillaries and into kidney tubules.

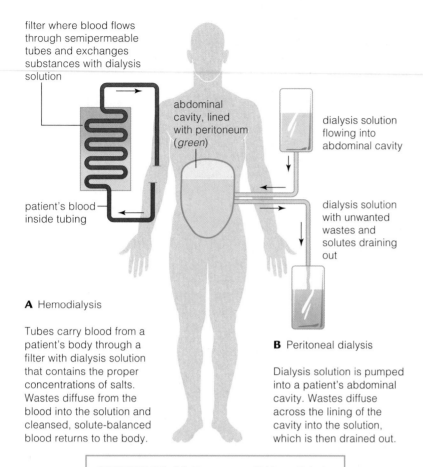

filter where blood flows through semipermeable tubes and exchanges substances with dialysis solution

patient's blood inside tubing

abdominal cavity, lined with peritoneum (*green*)

dialysis solution flowing into abdominal cavity

dialysis solution with unwanted wastes and solutes draining out

A Hemodialysis

Tubes carry blood from a patient's body through a filter with dialysis solution that contains the proper concentrations of salts. Wastes diffuse from the blood into the solution and cleansed, solute-balanced blood returns to the body.

B Peritoneal dialysis

Dialysis solution is pumped into a patient's abdominal cavity. Wastes diffuse across the lining of the cavity into the solution, which is then drained out.

FIGURE 23.15 Two types of kidney dialysis.

anorexia A disorder in which a person does not eat enough to maintain a healthy weight, despite having access to food.
kidney dialysis Procedure used to cleanse blood and restore proper solute concentrations in a person with impaired kidney function.

time, causes constant pain, or blocks flow of urine, it can be removed surgically or broken up by shock waves administered from outside the body.

The best way to prevent kidney stones is to drink lots of fluids. High-protein, low-carbohydrate diets can encourage kidney stone formation, even when the dieter drinks plenty of water. These diets also stress the kidney because they cause ketone formation. Ketones are acidic by-products of fat and protein digestion that must be filtered from the blood by the kidney. No one who has impaired kidney function or a history of kidney stones should begin a high-protein diet without discussing it with a doctor.

Kidney function is measured as the rate of filtration. When the filtration rate falls by half, a person is said to be in kidney failure. The vast majority of kidney failure occurs as complications of diabetes mellitus or high blood pressure. These disorders damage small blood vessels, including capillaries that filter blood into nephrons.

Kidney failure can be fatal because wastes build up in the blood and interstitial fluid. The pH of body fluids rise, and changes in concentrations of other ions interfere with metabolism. **Kidney dialysis** is used to restore proper solute balance in a person with kidney failure. "Dialysis" refers to the exchange of solutes between two solutions that are separated by a semipermeable membrane.

With hemodialysis, a dialysis machine is connected to a patient's blood vessel (Figure 23.15**A**). The machine pumps a patient's blood through semipermeable tubes submerged in a warm solution of salts, glucose, and other substances. As the blood flows through the tubes, wastes dissolved in the blood diffuse out and solute concentrations return to normal levels. Cleansed, solute-balanced blood is returned to the patient's body. Typically a person has hemodialysis three times a week at a dialysis center. Each treatment takes several hours.

Peritoneal dialysis can be done at home. Each night, dialysis solution is pumped into a patient's abdominal cavity (Figure 23.15**B**). Wastes diffuse across the peritoneal lining into the fluid, which is drained out the next morning. With this method of dialysis, the peritoneal lining functions as the semipermeable membrane.

Kidney dialysis can keep a person alive through an episode of temporary kidney failure. When kidney failure is permanent, dialysis must be continued for the rest of a person's life, or until a donor kidney becomes available for transplant. The National Kidney Foundation estimates that every day, 17 people die of kidney failure while waiting for a transplant. Most kidneys used as transplants come from people who had arranged to be organ donors after their death. However, an increasing number of kidneys are removed from a living donor, most often a relative. A kidney transplant from a living donor has a better chance of success than one from a deceased person. One kidney is adequate to maintain good health, so the risks to a living donor are mainly related to the surgery—unless a donor's remaining kidney fails.

The benefits of organs from living donors, a lack of donated organs, and high dialysis costs have led some to suggest that people should be allowed to sell a kidney. Critics argue that it is unethical to tempt people to risk their health for money. Section 10.4 describes another potential alternative. Some day, genetically modified pigs may provide organs for transplant.

Take-Home Message *How does urine form?*

- Filtration forces water and small solutes into the kidney tubules. The driving force for filtration is the pressure generated by the beating heart.

- Most of the filtered water and solutes are reabsorbed into capillaries around the tubules.

- Wastes that are not absorbed end up in the urine, along with any excess solutes secreted into the filtrate.

- The hormone ADH is secreted by the brain and acts in the kidney to promote water reabsorption, thus concentrating the urine.

23.8

Impacts/Issues Revisited:
Regulating Appetite

Kidneys normally break down leptin, the signaling molecule that is made by fat cells and decreases appetite. In people with chronic kidney disease, leptin breakdown is impaired, so there is excess leptin in the blood. The brain interprets the high leptin level to mean there is plenty of fat on the body, so people with kidney disease do not feel hungry when they should. The result is **anorexia**, a general term for a disorder in which a person who has access to food eats too little to maintain a healthy weight.

Scientists are also investigating the role of leptin in the psychological disorder anorexia nervosa. People with this type of anorexia have a distorted body image. They see themselves as fat, even when they are dangerously thin. Because they have little fat, they produce little leptin and feel hungry, but psychological factors prevent them from acting to relieve their hunger. The resulting malnutrition has damaging effects throughout the body and can be fatal.

Leptin levels can affect the ability to recover from anorexia nervosa. As a recovering anorexic begins to regain weight and more fat gets stored in adipose cells, blood leptin level begins to increase. Studies have shown that recovered anorexics who have a high leptin level are more likely to suffer a relapse than those with lower leptin levels.

Bulimia nervosa is another eating disorder. Like anorexia nervosa it is most common among young women. Affected people "binge and purge." They go on an eating binge, then try to get rid of the food by inducing vomiting. Many also abuse laxatives. Bulimics tend to be near normal weight, but perceive themselves as overweight. Induced vomiting bathes the esophagus and teeth in acidic gastric fluid. The acid harms the enamel of the teeth, causing them to become pitted and brittle. Repeated exposure to acid increases the risk of cancer of the esophagus. Frequent vomiting can also cause life-threatening changes in the body's ion balance. Loss of gastric fluid depletes the body of hydrogen ions, making body fluids more basic than normal. The shift in pH can cause apnea (a halt in breathing), disrupt normal heart rhythm, and cause convulsions.

How Would **YOU** ☑ Vote ?

Obesity may soon replace smoking as the number one cause of preventable deaths in the United States. Fast food is contributing to the problem. Should fast food items be required to carry health warnings like alcohol and tobacco products do? **See CengageNow for details, then vote online (cengagenow.com).**

Summary

Section 23.1 The body's fat cells produce leptin, a molecule that suppresses appetite. Leptin deficiency causes obesity in mice, but is rare in humans. Another molecule, ghrelin, increases appetite.

Section 23.2 A digestive system takes in food and breaks it down to molecules small enough to be absorbed into the internal environment, then eliminates unabsorbed residues.

Most animals have a complete digestive system: a tube with two openings (mouth and anus), and specialized areas between them.

CENGAGENOW™ Use the animation on CengageNow to compare animal digestive systems.

Section 23.3 Like other vertebrates, humans have a complete digestive system. In the mouth, chewing mixes food with amylase-containing saliva that begins the process of starch digestion. A swallowing reflex moves food from the throat into the esophagus, which opens into the stomach. Protein digestion starts in the stomach, a muscular sac with a lining that secretes gastric fluid.

The small intestine receives chyme (food mixed with gastric fluid) from the stomach, enzymes from the pancreas, and bile from the gallbladder. Bile, which assists in fat digestion, is made by the liver and is stored in the gallbladder. Digestion of all nutrients is completed in the small intestine.

Most nutrients and fluid are absorbed into the internal environment across the wall of the small intestine. Intestinal folds are covered with multicelled projections (villi). Brush border cells at the villus surface have membrane extensions called microvilli. All of the folding and extensions greatly increase the surface area for absorption.

The large intestine absorbs minerals and water, and concentrates undigested residues as feces, which are expelled through the anus.

CENGAGENOW™ Investigate the structure and function of the human digestive system with the animation on CengageNow.

Sections 23.4, 23.5 The body burns carbohydrates as fuel and whole grains provide fiber. Fats are another source of energy. Most fatty acids can be made from carbohydrates, but some must be obtained from the diet. Proteins are the source of amino acids used to build the body's own proteins. Vitamins and minerals are necessary in small amounts for normal metabolism. To maintain body weight, energy intake must balance energy output. Obesity raises the risk of health problems and shortens life expectancy.

CENGAGENOW™ Use the interactions on CengageNow to determine your body mass index and the calories needed to maintain your weight.

Sections 23.6, 23.7 The body must maintain the volume and composition of its fluid components within a narrow range. The human urinary system interacts with other organ systems to balance the intake and output of solutes and water. The urinary system consists of a pair of kidneys, a pair of ureters, a urinary bladder, and a urethra. Kidneys contain an enormous number of nephrons, or blood-filtering units. A nephron has a cup-shaped entrance around one set of blood capillaries and tubular regions that associate with another set of blood capillaries. Urine forms in nephrons by three processes: filtration, reabsorption, and secretion.

Filtration. Blood pressure drives water and most solutes except proteins out of the capillaries at the start of a nephron. These enter the nephron's tubular parts.

Reabsorption. Water and solutes to be conserved move out of the tubular parts, then into the capillaries that thread around the nephron. A small volume of water and solutes remains in the nephron.

ADH (antidiuretic hormone) is a hormone that adjusts urine concentration by promoting water reabsorption. As a result, more water returns to the blood rather than leaving in the urine.

Secretion. H^+ and some other substances move from capillaries into the nephron for excretion.

Good kidney function is essential to life. Dialysis or a transplant are the only options for those with permanent kidney damage.

CENGAGENOW™ Study the human urinary system, view the structure of a nephron, and see how urine is formed with the animation on CengageNow.

Self-Quiz   Answers in Appendix I

1. A digestive system functions in _____ .
   a. secreting enzymes
   b. absorbing nutrients
   c. eliminating wastes
   d. all of the above

2. Protein digestion begins in the _____ .
   a. mouth
   b. stomach
   c. small intestine
   d. colon

3. Digestion is completed and most nutrients are absorbed in the _____ .
   a. mouth
   b. stomach
   c. small intestine
   d. colon

4. Bile has roles in _____ digestion and absorption.
   a. carbohydrate
   b. fat
   c. protein
   d. amino acid

5. Most water that enters the gut is absorbed across the lining of the _____ .
   a. stomach
   b. small intestine
   c. large intestine
   d. esophagus

6. Match each organ with a digestive function.
   _____ gallbladder   a. makes bile
   _____ large intestine   b. compacts undigested residues
   _____ liver   c. secretes enzymes, bicarbonate
   _____ small intestine   d. absorbs most nutrients
   _____ stomach   e. secretes gastric fluid
   _____ pancreas   f. stores, secretes bile

7. Essential fatty acids are _____ .
   a. *trans* fats
   b. saturated fats
   c. vitamins
   d. not made by the body

8. Can vitamin C be stored in body fat?

9. Iron is an example of a _____ .
   a. vitamin
   b. mineral
   c. essential fatty acid
   d. essential amino acid

10. Urea forms as a breakdown product of _____ .
    a. nucleic acids
    b. simple sugars
    c. saturated fats
    d. proteins
    e. complex carbohydrates
    f. a through d

11. Filtration moves _____ into kidney tubules.
    a. water
    b. fiber
    c. proteins
    d. all of the above

12. Water loss triggers a(n) _____ in ADH secretion.
    a. increase
    b. decrease

13. Kidneys return water and small solutes to the blood by the process of _____ .
    a. filtration
    b. reabsorption
    c. secretion
    d. both a and b

14. Kidneys adjust the blood acidity by increasing or decreasing the _____ of H^+.
    a. filtration
    b. reabsorption
    c. secretion
    d. both a and b

15. Match each structure with a function.
    _____ ureter
    _____ bladder
    _____ urethra
    _____ nephron
    _____ pituitary gland

    a. stores urine
    b. delivers urine to body surface
    c. carries urine from kidney to bladder
    d. secretes ADH
    e. many inside a kidney

Additional questions are available on CENGAGENOW™

Critical Thinking

1. Starch and sugar have the same number of calories per gram. However, not all vegetables are equally calorie dense. For example, a serving of boiled sweet potato provides about 1.2 calories per gram, while a serving of kale yields only 0.3 calories per gram. What could account for the difference in the calories your body obtains from these two foods?

2. The reported incidence of anorexia nervosa has soared during the past 20 years. The overwhelming majority of cases occur in young women. Anorexia nervosa has complex causes, including some recently discovered genetic factors. Is it likely that a sudden rise in the frequency of alleles that put people at risk for anorexia is responsible for the rise in reported cases?

3. Oats and beans have lots of soluble fiber. A diet high in soluble fiber has been found to lower cholesterol, but it also can increase flatulence (gas). The gas is produced when bacteria living in the digestive tract break down the soluble fiber. In what part of the digestive tract do you think the most gas is produced?

Digging Into Data

Pesticides and Organic Food

Products labeled as "organic" fill an increasing amount of space on supermarket shelves. What does this label mean? A food that carries the USDA's organic label must be produced without pesticides such as malathion and chlorpyrifos, which conventional farmers typically use on fruits, vegetables, and many grains.

Does eating organic food significantly affect the level of pesticide residues in a child's body? Chensheng Lu of Emory University used urine testing to find out (Figure 23.16). For fifteen days, the urine of twenty-three children (aged 3 to 11) was monitored for breakdown products of pesticides. During the first five days, children ate their standard, nonorganic diet. For the next five days, they ate organic versions of the same types of foods and drinks. Then, for the final five days, the children returned to their nonorganic diet.

1. During which phase of the experiment did the children's urine contain the lowest level of the malathion metabolite?

2. During which phase of the experiment was the maximum level of the chlorpyrifos metabolite detected?

3. Did switching to an organic diet lower the amount of pesticide residues excreted by the children?

4. Even in the nonorganic phases of this experiment, the highest pesticide metabolite levels detected were far below those known to be harmful. Given these data, would you spend more to buy organic foods?

Study Phase	No. of Samples	Malathion Metabolite		Chlorpyrifos Metabolite	
		Mean (µg/liter)	Maximum (µg/liter)	Mean (µg/liter)	Maximum (µg/liter)
1. Nonorganic	87	2.9	96.5	7.2	31.1
2. Organic	116	0.3	7.4	1.7	17.1
3. Nonorganic	156	4.4	263.1	5.8	25.3

FIGURE 23.16 *Above*, levels of metabolites (breakdown products) of malathion and chlorpyrifos detected in the urine of children taking part in a study of the effects of an organic diet. The difference in the mean level of metabolites in the organic and inorganic phases of the study was statistically significant. *Right*, the USDA organic food label.

4. Diabetes insipidus is a medical disorder in which a person produces an unusually large amount of highly dilute urine. Some cases are caused by a gene mutation, but in most people diabetes insipidus arises after a head injury. How might an injury to the head affect kidney function?

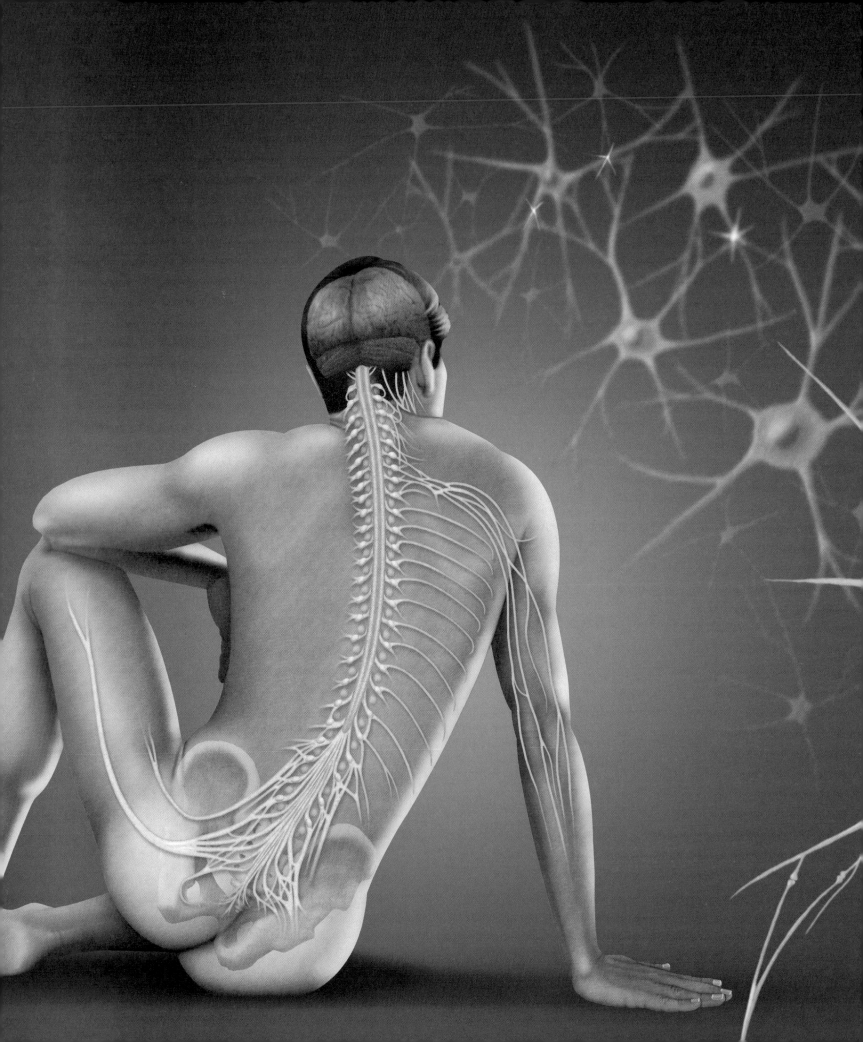

24 Neural Control and the Senses

24.1 Impacts/Issues: In Pursuit of Ecstasy

Ecstasy, an illegal drug, can make you feel socially accepted, less anxious, and more aware of your surroundings and of sensory stimuli. It also can leave you dying in a hospital, bleeding from all body openings as your temperature skyrockets. It can send your family and friends spiraling into horror and disbelief as they watch you stop breathing. Lorna Spinks ended life that way at age 19. Her anguished parents released the photographs in Figure 24.1 because they wanted others to know what their daughter did not: Ecstasy can kill.

Ecstasy is a psychoactive drug; it alters brain function. The active ingredient, MDMA (3,4-methylenedioxymethamphetamine), is a type of amphetamine, or "speed." As one effect, it makes neurons release an excess of the signaling molecule serotonin. The serotonin saturates receptors on target cells and cannot be cleared away, so cells cannot be released from overstimulation.

The abundance of serotonin promotes feelings of energy, empathy, and euphoria. But the unrelenting stimulation also causes rapid breathing, dilated eyes, restricted urine formation, and a racing heart. Blood pressure soars, and the body's internal temperature can rise out of control. Spinks became dizzy, flushed, and incoherent after taking just two Ecstasy tablets. She died because her increased temperature caused her organ systems to shut down.

Few Ecstasy overdoses end in death. Panic attacks and fleeting psychosis are more common short-term effects. We do not know much about the drug's long-term effects; users are unwitting guinea pigs for unscripted experiments.

We know that Ecstasy use depletes the brain's store of serotonin and that this shortage can last for some time. In animals, multiple doses of MDMA alter the structure and number of serotonin-secreting neurons. This is a matter of concern because low serotonin levels in humans are associated with inability to concentrate, memory loss, and depression.

Ecstasy users have impaired memory, and the more often they use the drug, the worse their memory becomes. Fortunately, at least over the short term, a person's memory seems to be restored when Ecstasy use stops. However, undoing the neural imbalances often takes many months.

The human nervous system is a legacy of millions of years of evolution in cellular signaling pathways, first in invertebrates, then through vertebrate lineages that led to humans. Among living species, we alone can make a conscious choice to harm the nervous system that gave us our remarkable capacities.

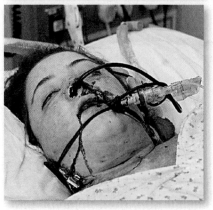

FIGURE 24.1 Photos of Lorna Spinks alive (*top*), and just after her death (*bottom*). She died after taking two Ecstasy tablets. If you suspect someone is having a bad reaction to Ecstasy or any other drug, get medical help fast and be honest about the drug use. Immediate, informed medical action may save a life.

Neurons—The Great Communicators

Neurons and Supporting Cells Of all multicelled organisms, animals respond fastest to external stimuli. Communication lines made up of neurons are key to these quick responses. A **neuron** is a cell that relays electrical signals along its plasma membrane and communicates with other cells by way of chemical messages. Three classes of neurons interact in most nervous systems (Figure 24.2). **Sensory neurons** detect stimuli. When activated by a stimulus such as light or touch, they send signals to interneurons. **Interneurons** integrate signals from sensory neurons and other interneurons, and send signals of their own to motor neurons. Nearly all of a vertebrate's interneurons are in the brain and spinal cord. **Motor neurons** control muscles and glands.

Figure 24.3 shows the structure of a motor neuron. A neuron's nucleus and organelles are inside its cell body. Special cytoplasmic extensions receive and send messages. **Dendrites** are branched extensions that receive information from other cells, and convey that information to the cell body. An **axon** conducts signals away from the cell body, and is typically a longer extension. Axon endings release chemical signals that affect other cells.

All neurons are metabolically assisted, protected, insulated, and held in place by diverse cells collectively known as **neuroglia**.

Organization of Nervous Tissue All cells have an electric gradient across their plasma membrane. Their cytoplasmic fluid contains more negatively charged ions and proteins than the interstitial fluid outside the cell does. As in a battery, these separated charges have potential energy. We call the potential energy across a membrane the **membrane potential**. The membrane

FIGURE 24.2 How the three classes of neurons interact in a vertebrate nervous system.

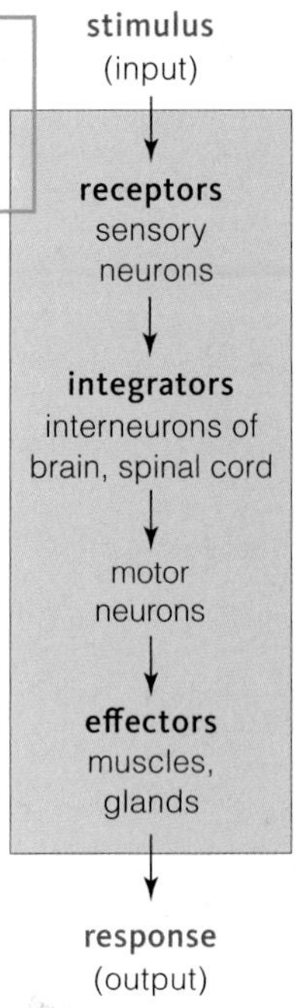

stimulus (input)

↓

receptors
sensory neurons

↓

integrators
interneurons of brain, spinal cord

↓

motor neurons

↓

effectors
muscles, glands

↓

response (output)

axon Cytoplasmic extension of a neuron; transmits electrical signals along its length and chemical signals at its endings.

dendrite Neuron's signal-receiving cytoplasmic extension.

interneurons Neurons that receive signals from other neurons and integrate information.

membrane potential Potential energy across the membrane of a resting neuron.

motor neurons Neurons that control muscles and glands.

neuroglia Cells that support neurons.

neurons Cells that make up the communication lines of nervous systems.

sensory neurons Neurons that detect specific stimuli.

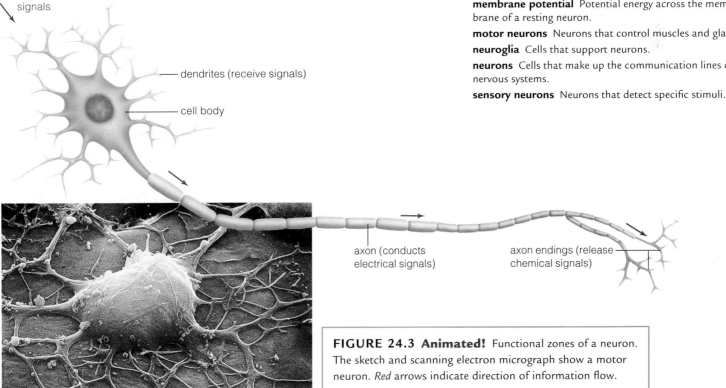

signals

dendrites (receive signals)

cell body

axon (conducts electrical signals)

axon endings (release chemical signals)

FIGURE 24.3 Animated! Functional zones of a neuron. The sketch and scanning electron micrograph show a motor neuron. *Red* arrows indicate direction of information flow.

10 µm

potential of a neuron that is not being stimulated is its **resting potential**. When a neuron is at rest, fluid inside and outside the cell differ not only in charge, but also in their concentration of sodium ions (Na⁺) and potassium ions (K⁺). There are more sodium ions in the fluid just outside a cell than there are inside it. The reverse is true for potassium ions. We can illustrate these ion concentration gradients as follows (the larger text represents higher concentrations):

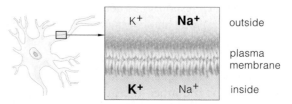

The concentration gradients for sodium and potassium are set up by sodium–potassium pumps in the plasma membrane. These active transport proteins pump sodium out of the neuron and pump potassium into it (Section 4.6).

The Action Potential Neurons and muscle cells are said to be "excitable" cells because, when properly stimulated, they undergo an action potential. An **action potential** is a brief reversal of the electric gradient across the plasma membrane. It is caused by the opening and closing of channel proteins with gates that open at a particular voltage, or membrane potential (Figure 24.4).

An action potential begins in a trigger region, the part of the axon adjacent to the cell body. When a neuron is at rest, this region's gated channels for sodium and potassium are closed, and there are more negatively charged ions inside the axon than outside it ❶.

A stimulus such as a signal from another neuron shifts the membrane potential. If the stimulus is large enough, the trigger zone reaches **threshold potential**. This is the membrane potential at which gated sodium channels in the trigger zone open and an action potential begins.

Opening of the gated sodium channels allows sodium to follow its concentration gradient, and diffuse from the interstitial fluid into the neuron ❷. Inward flow of sodium reverses the potential across the membrane. At the height of an action potential, the fluid outside the membrane is more negatively charged than the fluid inside.

In an example of positive feedback, gated sodium channels open in an accelerating way after threshold potential is reached. As sodium flows into the axon's cytoplasm, the cytoplasm becomes more positively charged, so more sodium channels open:

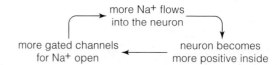

An action potential is an all-or-nothing event. Once threshold is reached, an action potential always occurs, and all action potentials are the same size. The action potential lasts only a millisecond because the charge reversal causes gated sodium channels to close and gated potassium channels to open ❸. As potassium diffuses outward, the axon cytoplasm again becomes more negatively charged than the fluid outside the axon.

Even as an action potential ends in one region, some of the sodium that flowed in is diffusing inside the axon to adjacent regions. When some of this sodium arrives in an adjoining region a bit farther along the axon, it drives that region to threshold and opens sodium gates ❹. As sodium gates swing open in

action potential Brief reversal of the charge difference across a neuron membrane.
resting potential Membrane potential of a neuron at rest.
threshold potential Membrane potential at which voltage-gated sodium channels in a neuron axon open, causing an action potential.

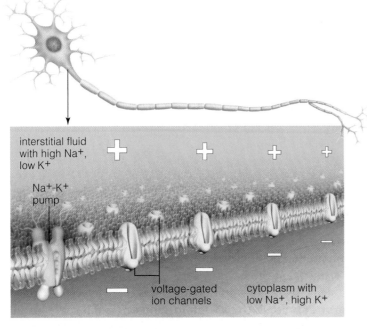

interstitial fluid
with high Na+,
low K+

Na+-K+
pump

voltage-gated
ion channels

cytoplasm with
low Na+, high K+

1 Close-up of the trigger zone of a neuron. One sodium–potassium pump and some of the voltage-gated ion channels are shown. At this point, the membrane is at rest and the voltage-gated channels are closed. The cytoplasm's charge is negative relative to interstitial fluid.

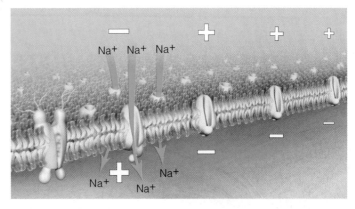

Na+ Na+ Na+

Na+

Na+ Na+

2 Arrival of a sufficiently large signal in the trigger zone raises the membrane potential to threshold level. Gated sodium channels open and sodium (Na+) flows down its concentration gradient into the cytoplasm. Sodium inflow reverses the voltage across the membrane.

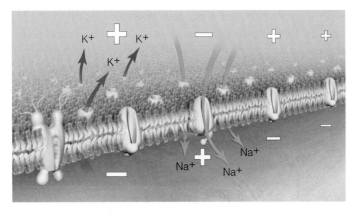

K+ K+

K+

Na+
Na+ Na+

3 The charge reversal makes gated Na+ channels shut and gated K+ channels open. The K+ outflow restores the voltage difference across the membrane. The action potential is propagated along the axon as positive charges spreading from one region push the next region to threshold.

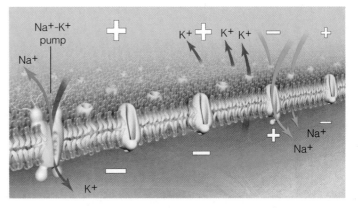

Na+-K+
pump

Na+

K+ K+ K+

Na+
Na+

K+

4 After an action potential, gated Na+ channels are briefly inactivated, so the action potential moves one way only, toward axon terminals. Na+ and K+ gradients changed by action potentials are restored by diffusion of ions that were put into place by activity of sodium–potassium pumps.

FIGURE 24.4 Animated! Propagation of an action potential along part of a motor neuron's axon. *Blue* arrows depict the flow of sodium (Na+) ions. *Red* arrows show the flow of potassium (K+) ions.

Figure It Out: Which type of protein in the axon membrane allows positive ions to flow into the neuron during an action potential?

Answer: Gated sodium channels allow positively charged sodium ions to diffuse into the neuron.

one region after the next, the action potential moves toward the axon endings without weakening.

Action potentials move in one direction only, away from the cell body and toward axon endings. This one-way transmission occurs because after sodium gates close, they are briefly unable to open. Incoming sodium diffuses in both directions, but it only affects sodium gates that are ready to open—the gates that are closer to the axon endings.

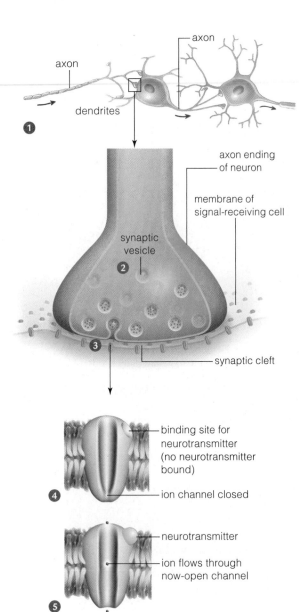

FIGURE 24.5 **Animated!** Chemical synapse.

1 Axon endings of one neuron send signals to another neuron at a chemical synapse.

2 Neurotransmitter (*green*) is stored in vesicles inside an axon ending.

3 Arrival of an action potential at the axon ending causes release of neurotransmitter.

4 The membrane of the signal-receiving cell has receptors for neurotransmitter.

5 Neurotransmitter binding opens a channel through the receptor. The opening allows ions to flow into the post-synaptic cell.

How Messages Flow From Cell to Cell Action potentials can travel along an axon to its endings, but they cannot jump from cell to cell. Chemical signals carry the message between cells (Figure 24.5). A **chemical synapse** is the communication point between two neurons, or between a neuron and a muscle or gland **1**. At the synapse, a narrow synaptic cleft separates the axon ending of a signal-sending neuron from a signal-receiving cell. Chemical signals called **neurotransmitters** convey information across this space. Axon endings store neurotransmitter in synaptic vesicles in their cytoplasm **2**. Arrival of an action potential at an axon's endings causes synaptic vesicles to move to the plasma membrane and fuse with it. As fusion occurs, neurotransmitter enters the synaptic cleft **3**.

The signal-receiving cell has plasma membrane proteins that are receptors for neurotransmitters **4**. Neurotransmitter molecules diffuse across the cleft and bind to these receptors. Different receptors respond to neurotransmitters in different ways. Some receptors are ion channels. Binding of a neurotransmitter opens a tunnel through their interior. Ions diffuse into or out of the signal-receiving cell through the tunnel **5**.

Some neurotransmitters excite a signal-receiving cell, bringing it closer to its threshold potential. Other neurotransmitters have an inhibitory effect. They make the signal-receiving cell less likely to have an action potential.

A Sampling of Signals Different types of neurons make different neurotransmitters. Table 24.1 lists some of the major neurotransmitters and their effects. For example, acetylcholine (ACh) acts on skeletal muscle, smooth muscle, the heart, many glands, and the brain. When motor neurons release acetylcholine into synapses with skeletal muscle cells, the muscle undergoes an action potential. The muscle then contracts. ACh has a different effect at synapses with heart muscle. Here it inhibits contraction. How can the same neurotransmitter have different effects? Different kinds of receptors respond differently to each neurotransmitter. Skeletal muscle cells and heart cells have different types of ACh receptors.

After neurotransmitter molecules have acted, they must be quickly removed from synaptic clefts so that new signals can be sent. Some neurotransmitter molecules simply diffuse away. Membrane transport proteins actively pump others back into the neuron or into neighboring neuroglial cells. Enzymes secreted into the cleft break down still others. For example, the enzyme acetylcholinesterase breaks down ACh. Nerve gases such as sarin exert their deadly effects by inactivating this enzyme. After being inhaled, they bind to acetylcholinesterase and inhibit ACh breakdown. The resulting ACh accumulation causes confusion, headaches, skeletal muscle paralysis, and, if the dosage is high enough, death.

Disrupted Signaling—Disorders and Drugs Some people have neurological disorders caused by disrupted signaling at synapses. Others deliberately disrupt synaptic function by using psychoactive drugs.

Table 24.1	Major Neurotransmitters and Their Effects
Neurotransmitter	Examples of Effects
Acetylcholine (ACh)	Induces skeletal muscle contraction, slows cardiac muscle contraction rate, affects mood and memory
Epinephrine and norepinephrine	Speed heart rate; dilate the pupils and the airways to lungs; slow gut contractions; increase anxiety
Dopamine	Dampens excitatory effects of other neurotransmitters; has roles in memory, learning, fine motor control
Serotonin	Elevates mood; role in memory
GABA	Inhibits release of other neurotransmitters

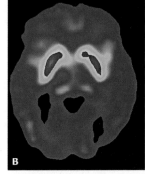
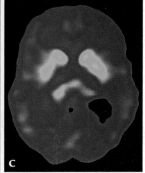

Alzheimer's disease and Parkinson's disease involve disrupted signaling. A low ACh level in the brain contributes to memory loss in Alzheimer's disease. Affected people often can recall long-known facts, such as a childhood address, but have trouble remembering recent events. Parkinson's disease occurs when dopamine-secreting neurons in a brain region governing motor control die or are impaired (Figure 24.6). Head injuries raise the risk for Parkinson's. Hand tremors are often the earliest symptom. Later, the sense of balance may be affected, movement can be difficult, and speech becomes slurred.

Drugs prescribed to treat mood disorders act at synapses in the brain. The drug fluoxetine (Prozac) can sometimes lift depression by raising serotonin levels. Diazepam (Valium) and alprazolam (Xanax) lower anxiety by increasing the output of GABA. This neurotransmitter inhibits the release of other neurotransmitters in the brain.

All addictive drugs stimulate the release of dopamine, a neurotransmitter with a role in reward-based learning. In nearly all animals, dopamine release provides pleasurable feedback when the animal engages in behavior that enhances survival or reproduction. This response is adaptive. It helps animals learn to repeat the behaviors that benefit them. However, when drugs cause dopamine release, they tap into this ancient learning pathway. Drug users inadvertently teach themselves that taking the drug is essential to their well-being.

Stimulants make users feel alert but also anxious, and they can interfere with fine motor control. Nicotine is a stimulant that, among other effects, blocks brain receptors for ACh. Cocaine and amphetamines are also stimulants. Ecstasy is a type of amphetamine, as is methamphetamine.

Depressants such as alcohol and barbiturates slow motor responses by inhibiting ACh output. Alcohol stimulates the release of endorphins and GABA, so users typically experience a brief euphoria followed by depression. Combining alcohol with barbiturates can be deadly.

Narcotic analgesics such as morphine, codeine, heroin, fentanyl, and oxycodone mimic the body's natural painkillers. However, they can cause a rush of euphoria and are highly addictive. Ketamine and PCP (phencyclidine) belong to a different class of analgesics. They give users an out-of-body experience and numb the extremities by slowing removal of neurotransmitter from synapses.

Hallucinogens distort sensory perception and bring on a dreamlike state. LSD (lysergic acid diethylamide) resembles serotonin and binds to receptors for it. LSD is not addictive. However, LSD users can be injured or killed because they do not perceive and respond to hazards such as oncoming cars. Flashbacks, which are brief distortions of perception, may occur years after the last intake of LSD. Two related drugs, mescaline and psilocybin, have weaker effects.

Marijuana comes from *Cannabis* plants. Smoking a lot of marijuana can cause hallucinations. More often, users become relaxed, sleepy, uncoordinated,

FIGURE 24.6 Battling Parkinson's disease. **A** This neurological disorder affects former heavyweight champion Muhammad Ali, actor Michael J. Fox, and about half a million other people in the United States. **B** A normal PET scan and **C** one from an affected person. *Red* and *yellow* indicate high metabolic activity in dopamine-secreting neurons.

chemical synapse Region where a neuron's axon endings transmit signals to another cell.

neurotransmitter Chemical signal released by a neuron's axon endings.

and inattentive. The active ingredient, THC (delta-9-tetrahydrocannabinol), alters brain levels of dopamine, serotonin, norepinephrine, and GABA. Chronic use can impair short-term memory and decision-making ability.

Take-Home Message **What are neurons and how do they transmit signals?**

■ Neurons are the cells that make up the communication lines of animal nervous systems.

■ In response to a stimulus, electrical signals called action potentials move along a neuron's axon to the axon endings.

■ The arrival of an action potential triggers axon endings to release chemical signals called neurotransmitters. The neurotransmitter molecules diffuse across a narrow space and bind to receptors on an adjacent cell. The binding may have excitatory or inhibitory effects on the signal-receiving cell.

■ Alzheimer's disease and Parkinson's disease result from impaired neuron signaling. Psychoactive drugs exert their effects by altering neuron signaling.

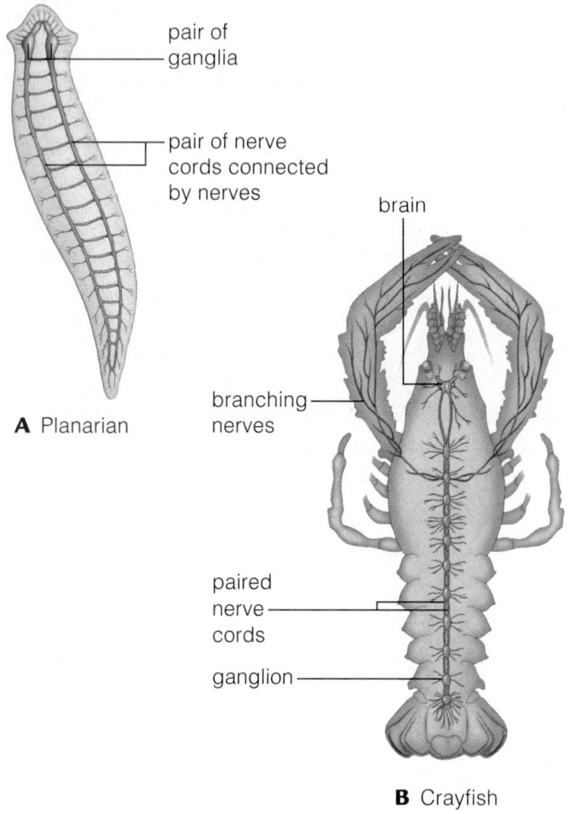

A Planarian

pair of ganglia

pair of nerve cords connected by nerves

brain

branching nerves

paired nerve cords

ganglion

B Crayfish

FIGURE 24.7 Invertebrate nervous systems. **A** Planarian with paired ganglia (clusters of nerve cell bodies) in the head, and nerve cords extending out from them. **B** A crayfish has a brain, and paired nerve cords with ganglia in each segment.

central nervous system Brain and spinal cord.

ganglion Cluster of nerve cell bodies.

myelin Insulating material around most vertebrate axons that increases speed of signal transmission.

nerve net Mesh of interacting neurons with no central control, as in the nervous system of jellies.

nerve Axons bundled in connective tissues.

peripheral nervous system Nerves that extend through the body and carry signals to and from the central nervous system.

24.3 Animal Nervous Systems

Invertebrate Nervous Systems Cnidarians such as sea anemones and jellies are among the simplest animals that have neurons. These radial, aquatic animals have a **nerve net**, which is a mesh of interconnected neurons. Information can flow in any direction among cells of the nerve net, and there is no centralized, controlling organ that functions like a brain. By causing cells in the body wall to contract, the nerve net can alter the size of the animal's mouth, change the shape of the body, or shift the position of tentacles.

Most animals have a bilaterally symmetrical body (Section 15.2). Evolution of bilateral body plans was accompanied by a concentration of sensory neurons and interneurons at the body's anterior, or head, end.

For example, a planarian's head end has a pair of ganglia (Figure 24.7**A**). A **ganglion** (plural, ganglia), is a cluster of neuron cell bodies that functions as an integrating center. A planarian's ganglia receive signals from eye spots and chemical-detecting cells on its head. The ganglia also connect to a pair of nerve cords that run the length of the body. Nerves cross the body between the cords, giving the nervous system a ladderlike appearance. A **nerve** is a bundle of axons wrapped in connective tissue.

Crayfish and other arthropods have paired nerve cords that connect to a simple brain (Figure 24.7**B**). In addition, a pair of ganglia in each body segment provides local control over that segment's muscles.

The Vertebrate Nervous System The nervous system of vertebrates has two divisions. Nearly all interneurons are located in the **central nervous system**—the brain and spinal cord. Nerves that extend through the rest of the body make up the **peripheral nervous system**.

Figure 24.8 illustrates the location of your brain ❶, spinal cord ❷, and some peripheral nerves. Each peripheral nerve has both sensory and motor components. Sensory fibers convey signals from sensory receptors to the central nervous system. Motor fibers carry signals away from the central regions to muscles. For example, a sciatic nerve extends from the spinal cord down each leg ❸. This nerve is composed of the bundled extensions of hundreds of neurons. Some

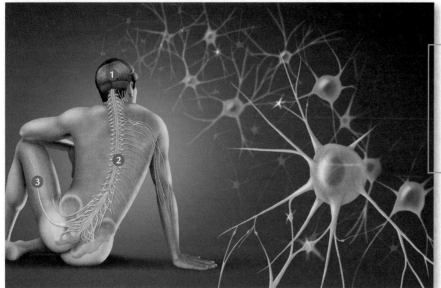

FIGURE 24.8 Components of vertebrate nervous systems. The brain **1** and spinal cord **2** make up the central nervous system. The peripheral nervous system includes spinal nerves, cranial nerves, and their branches, which extend through the rest of the body. Peripheral nerves such as the sciatic nerve **3** carry signals to and from the central nervous system.

axons in this bundle carry signals to skeletal muscles that move the leg, or to smooth muscle that adjusts the blood flow through the leg's arterioles. Other axons in the bundle convey messages from sensory neurons in the leg's muscles, joints, and skin.

Figure 24.9 shows the structure of a vertebrate nerve. Each nerve consists of bundles of axons enclosed in connective tissue. When a nerve is cut, it can be repaired by microsurgery. Using a microscope, the surgeon identifies corresponding bundles of axons within each part of the severed nerve and sews their connective tissue coverings back together. The part of the axon beyond the cut will die. But the axon can grow back through the repaired tunnel of connective tissue to the correct place.

Axons of most vertebrate nerves are wrapped in a fatty substance called **myelin**. Neuroglial cells produce this myelin sheath. The myelin functions like insulation on an electrical wire, increasing the speed of action potential transmission at least tenfold.

The importance of myelin becomes apparent when it is damaged. Multiple sclerosis (MS) is a neurological disorder in which white blood cells attack and destroy myelin sheaths of axons in the brain and spinal cord. The resulting disruption in information flow can lead to muscle weakness, severe fatigue, numbness, tremors, dizziness, and visual problems. An estimated 500,000 people in the United States have MS. Genes have a role in

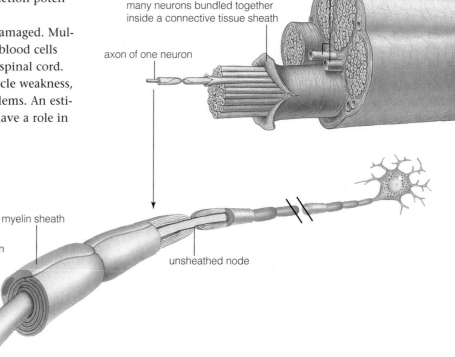

outer connective tissue of one nerve
blood vessels

many neurons bundled together
inside a connective tissue sheath

axon of one neuron

myelin sheath

axon

unsheathed node

FIGURE 24.9 Animated! Structure of a peripheral nerve. The myelin sheath around axons in the nerve increases the speed at which action potentials travel. Ions cannot enter or leave the axon in the sheathed region, so they diffuse more quickly inside the axon. At each unsheathed node, arrival of sodium that diffused from the previous node triggers an action potential. Thus an action potential in a myelinated axon "jumps" from node to node.

susceptibility to the disease, but viral infection or other environmental factors may set it in motion. There is currently no cure for MS, but drug treatments can prevent or slow the otherwise progressive disability.

Take-Home Message How are animal nervous systems organized?

- The simplest nervous system is a nerve net, a mesh of interconnecting neurons with no central integrating region. Jellies have this type of system.

- Most animals have a bilaterally symmetrical body with an integrating center at their head end. Nerves that run through the body carry signals to and from this center.

- Vertebrates have a central nervous system (brain and spinal cord) and a peripheral nervous system (nerves that extend through the body and carry signals to and from the central nervous system).

- Myelin sheaths around axons in vertebrate nerves speed information flow.

24.4 The Peripheral Nervous System

You can feel a pinch or wiggle your toes because of signals that travel along somatic nerves.

The vertebrate peripheral nervous system has two types of nerves (Figure 24.10). Nerves of the **somatic nervous system** relay commands to skeletal muscles. This is the only part of the nervous system that is normally under voluntary control. The somatic nervous system also relays information from sensory receptors in the skin and joints to the central nervous system. You can feel a pinch or wiggle your toes because of signals that travel along somatic nerves.

Nerves of the **autonomic nervous system** relay signals to smooth and cardiac muscle, and to glands. They also relay signals about internal conditions to the central nervous system. Signals traveling along autonomic nerves keep you breathing, adjust your heart rate, and inform your brain of your blood pressure.

There are two divisions of the autonomic system: sympathetic and parasympathetic. Both divisions service most organs and work antagonistically, meaning that the signals from one division oppose signals from the other (Figure 24.11).

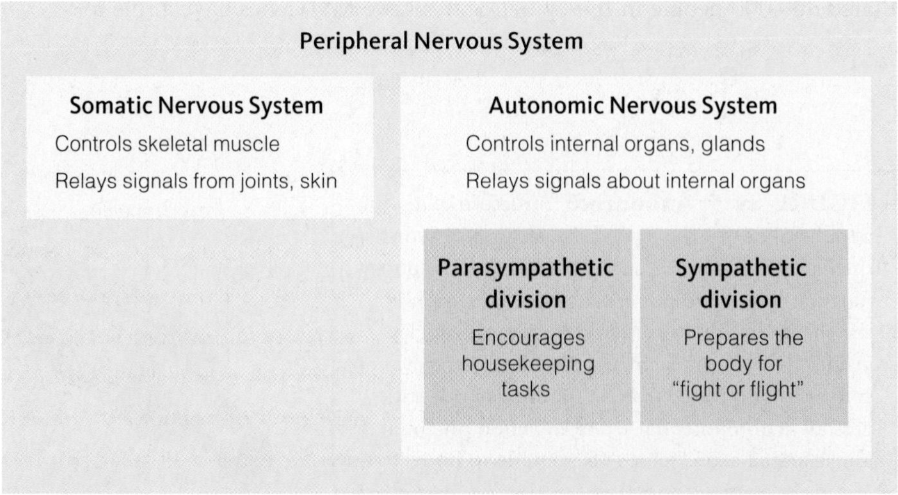

Peripheral Nervous System		
Somatic Nervous System Controls skeletal muscle Relays signals from joints, skin	**Autonomic Nervous System** Controls internal organs, glands Relays signals about internal organs	
	Parasympathetic division Encourages housekeeping tasks	**Sympathetic division** Prepares the body for "fight or flight"

FIGURE 24.10 Functional divisions of the peripheral nervous system. This system consists of spinal nerves, cranial nerves, and their branches, which extend through the body.

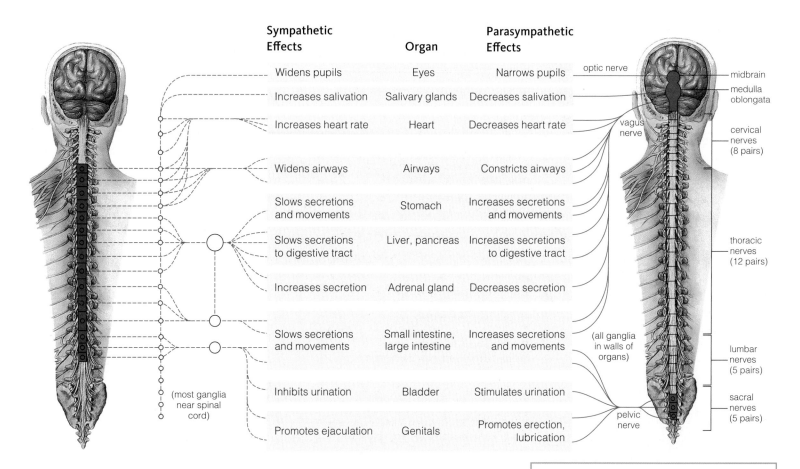

Sympathetic Effects	Organ	Parasympathetic Effects	
Widens pupils	Eyes	Narrows pupils	optic nerve
Increases salivation	Salivary glands	Decreases salivation	midbrain / medulla oblongata
Increases heart rate	Heart	Decreases heart rate	vagus nerve / cervical nerves (8 pairs)
Widens airways	Airways	Constricts airways	
Slows secretions and movements	Stomach	Increases secretions and movements	
Slows secretions to digestive tract	Liver, pancreas	Increases secretions to digestive tract	thoracic nerves (12 pairs)
Increases secretion	Adrenal gland	Decreases secretion	
Slows secretions and movements	Small intestine, large intestine	Increases secretions and movements	(all ganglia in walls of organs) / lumbar nerves (5 pairs)
Inhibits urination	Bladder	Stimulates urination	sacral nerves (5 pairs)
Promotes ejaculation	Genitals	Promotes erection, lubrication	pelvic nerve

(most ganglia near spinal cord)

FIGURE 24.11 Animated! Autonomic nerves and their effects. Autonomic signals travel to organs by a two-neuron path. The first neuron has its cell body in the brain or spinal region (indicated in *red*). This neuron synapses on a second neuron at a ganglion. Sympathetic ganglia are close to the spinal cord. Parasympathetic ganglia are in or near the organs they affect. Axons of these second neurons then synapse with the organ.

Figure It Out: What effect does stimulation of the vagus nerve (a parasympathetic nerve) have on heartbeat?

Answer: It slows the heartbeat.

Sympathetic neurons are most active in times of excitement or danger. Their axon terminals release norepinephrine. **Parasympathetic neurons** are most active in times of relaxation. Release of ACh from their axon terminals promotes housekeeping tasks such as digestion and urine formation.

When something startles or scares you, parasympathetic neurons signal less and sympathetic ones increase their output. The sympathetic signals raise your heart rate and blood pressure, make you sweat more and breathe faster, and induce adrenal glands to secrete epinephrine (also known as adrenaline). The signals put you in a state of intense arousal, so you are ready to fight or make a fast getaway. This reaction is commonly described as the "fight or flight" response.

Take-Home Message How does the peripheral nervous system function?

- The peripheral nervous system consists of nerves that extend through the body and connect with the central nervous system.
- Neurons of the somatic part of this system control skeletal muscle and convey information about the external environment to the central nervous system.
- The autonomic system carries information about internal conditions to the brain and sends signals to smooth muscle, cardiac muscle, and glands. Most organs receive both sympathetic and parasympathetic stimulation.
- Sympathetic stimulation prepares a body for "fight or flight." Parasympathetic stimulation encourages housekeeping tasks.

autonomic nervous system Set of nerves that relay signals to and from internal organs and to glands.

parasympathetic neurons Neurons of the autonomic system that encourage housekeeping tasks.

somatic nervous system Set of nerves that control skeletal muscle and relay signals from joints and skin.

sympathetic neurons Neurons of the autonomic system that prepare the body for danger or excitement.

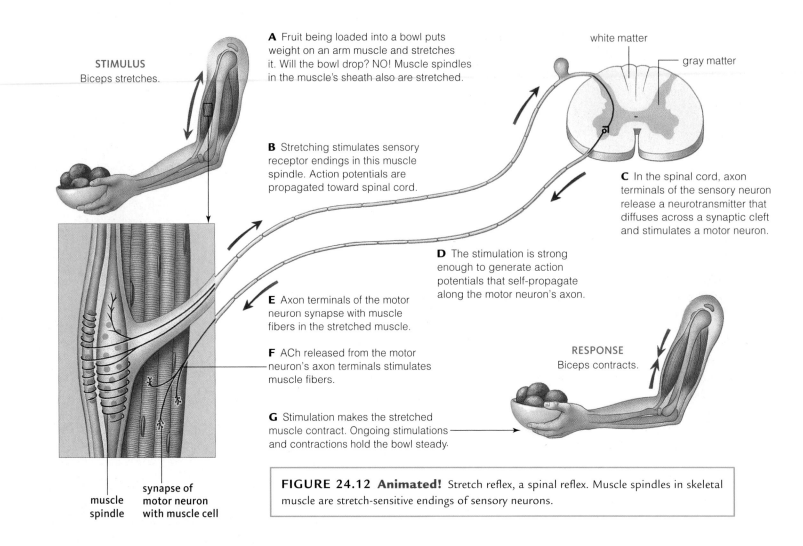

STIMULUS
Biceps stretches.

A Fruit being loaded into a bowl puts weight on an arm muscle and stretches it. Will the bowl drop? NO! Muscle spindles in the muscle's sheath also are stretched.

white matter

gray matter

B Stretching stimulates sensory receptor endings in this muscle spindle. Action potentials are propagated toward spinal cord.

C In the spinal cord, axon terminals of the sensory neuron release a neurotransmitter that diffuses across a synaptic cleft and stimulates a motor neuron.

D The stimulation is strong enough to generate action potentials that self-propagate along the motor neuron's axon.

E Axon terminals of the motor neuron synapse with muscle fibers in the stretched muscle.

F ACh released from the motor neuron's axon terminals stimulates muscle fibers.

RESPONSE
Biceps contracts.

G Stimulation makes the stretched muscle contract. Ongoing stimulations and contractions hold the bowl steady.

muscle spindle

synapse of motor neuron with muscle cell

FIGURE 24.12 Animated! Stretch reflex, a spinal reflex. Muscle spindles in skeletal muscle are stretch-sensitive endings of sensory neurons.

24.5 The Central Nervous System

The brain and spinal cord are the organs of the central nervous system (CNS). Three protective membranes, the **meninges**, enclose and protect these organs, which are bathed in colorless **cerebrospinal fluid**. This fluid forms when water and solutes seep out of capillaries in the brain. The composition of the cerebrospinal fluid is controlled by a **blood–brain barrier** that prevents most unwanted substances from entering the fluid. The barrier consists of the wall of blood capillaries and some supporting neuroglial cells. Sometimes bacteria or viruses breach the blood-brain barrier and enter the meninges and cerebrospinal fluid. The resulting infection, called meningitis, can be life-threatening.

Two visibly different types of tissue occur in both the brain and spinal cord. **White matter** consists of bundles of myelin-sheathed axons. In the CNS, such bundles are called tracts, rather than nerves. The tracts carry information from one part of the central nervous system to another. **Gray matter** consists of cell bodies, dendrites, and supporting neuroglial cells.

The Spinal Cord Your **spinal cord** is about as thick as your thumb. It runs through the vertebral column and connects peripheral nerves with the brain. An injury that disrupts the signal flow through the spinal cord can cause a loss of sensation and paralysis. Symptoms depend on where the cord is dam-

blood–brain barrier Protective mechanism that prevents unwanted substances from entering cerebrospinal fluid.
cerebellum Hindbrain region that coordinates voluntary movements.
cerebrospinal fluid Fluid around brain and spinal cord.
cerebrum Forebrain region that controls higher functions.
gray matter Tissue in brain and spinal cord consisting of cell bodies, dendrites, and neuroglial cells.
hypothalamus Homeostatic control center in forebrain.
medulla oblongata Hindbrain region that controls breathing rhythm and reflexes such as coughing and vomiting.
meninges Membranes that surround and protect the brain and spinal cord.
pons Hindbrain region between medulla oblongata and midbrain; has a role in control of breathing.
reflex Automatic response to a stimulus.
spinal cord Portion of central nervous system that connects peripheral nerves with the brain.
white matter Tissue of brain and spinal cord consisting of myelinated axons.

aged. Nerves carrying signals to and from the upper body originate higher in the cord than nerves that govern the lower body. An injury to the lower region of the cord often paralyzes the legs. An injury to the highest cord regions can paralyze all limbs, as well as muscles used in breathing. More than 240,000 Americans now live with a spinal cord injury.

In addition to transmitting signals to and from the brain, the spinal cord functions in some reflex pathways. A **reflex** is an automatic response to a stimulus—it is a movement or other action that does not require thought. The stretch reflex, illustrated in Figure 24.12, is an example. This reflex causes a muscle to contract after gravity or some other force stretches it. For example, stretching of the biceps causes action potentials in sensory receptors called muscle spindles. The action potentials travel along an axon to the spinal cord, where they synapse with motor neurons that control the biceps. Signals from the sensory neurons cause action potentials in the motor neurons. In response, the biceps contracts and steadies the arm.

Regions of the Human Brain

The average human brain weighs 1,240 grams, or 3 pounds. It contains about 100 billion interneurons, and neuroglia make up more than half of its volume. During development, the brain becomes organized as three functional regions: forebrain, midbrain, and hindbrain (Figure 24.13).

The hindbrain sits atop the spinal cord. The portion just above the cord, the **medulla oblongata**, influences the strength of heartbeats and the rhythm of breathing. It also controls reflexes such as swallowing, coughing, vomiting, and sneezing. Above the medulla oblongata lies the **pons**, which also affects breathing. Pons means "bridge," and tracts extend through the pons to the midbrain.

The **cerebellum** lies at the back of the brain and is about the size of a plum. It is densely packed with neurons, having more than all other brain regions combined. The cerebellum controls posture and coordinates voluntary movements. Excessive alcohol consumption disrupts coordination by impairing neurons in the cerebellum. That is why police often ask a person suspected of driving while drunk to walk a straight line.

In humans, the midbrain is the smallest of the three brain regions. It plays an important role in reward-based learning. The pons, medulla, and midbrain are collectively referred to as the brain stem.

The forebrain contains the **cerebrum**, the largest part of the human brain. A fissure divides the cerebrum into right and left hemispheres. A thick band of tissue, the corpus callosum, connects the hemispheres. Each hemisphere has an outer layer of gray matter called the cerebral cortex. Our large cerebral cortex is the part of the brain responsible for our unique capacities such as language and abstract thought.

Most sensory signals destined for the cerebrum pass through the adjacent thalamus, which sorts them and sends them to the proper region of the cerebral cortex.

The **hypothalamus** ("under the thalamus") is the center for homeostatic control of the internal environment. It receives signals about the state of the body and regulates thirst, appetite, and body temperature. It also controls sex drive and is an endocrine gland that interacts with the adjacent pituitary gland.

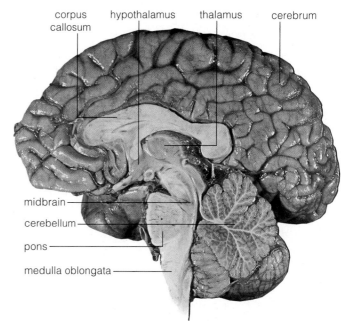

Forebrain	
Cerebrum	Localizes, processes sensory inputs; initiates, controls skeletal muscle activity; governs memory, emotions, abstract thought
Thalamus	Relays sensory signals to and from cerebral cortex; has a role in memory
Hypothalamus	With pituitary gland, functions in homeostatic control. Adjusts volume, composition, temperature of internal environment; governs behaviors that ensure homeostasis (e.g., thirst, hunger)

Midbrain	Relays sensory input to the forebrain

Hindbrain	
Pons	Bridges cerebrum and cerebellum, also connects spinal cord with forebrain. With the medulla oblongata, controls rate and depth of respiration
Cerebellum	Coordinates motor activity for moving limbs and maintaining posture, and for spatial orientation
Medulla oblongata	Relays signals between spinal cord and pons; functions in reflexes that affect heart rate, blood vessel diameter, and respiratory rate. Also involved in vomiting, coughing, other reflexive functions

FIGURE 24.13 Animated! Human brain. *Top*, Right half of a brain, with major structures labeled. *Bottom*, three main brain regions, major structures in each region, and their functions.

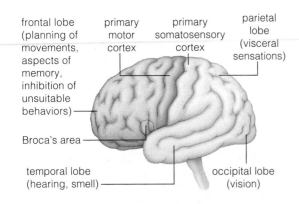

frontal lobe (planning of movements, aspects of memory, inhibition of unsuitable behaviors)

primary motor cortex

primary somatosensory cortex

parietal lobe (visceral sensations)

Broca's area

temporal lobe (hearing, smell)

occipital lobe (vision)

FIGURE 24.14 Lobes of the cerebrum and the location of some association and sensory areas of the cerebral cortex.

A Closer Look at the Cerebral Cortex The cerebral cortex is a 2-millimeter-thick, highly folded outer layer of gray matter. Prominent folds in the cortex are used as landmarks to define the cerebrum's frontal, parietal, temporal, and occipital lobes (Figure 24.14).

Much of each frontal lobe is devoted to association areas that integrate information and bring about conscious actions. Planning ahead and interacting socially involves the frontal lobe. The importance of the frontal lobe becomes obvious when it is damaged. During the 1950s, more than 20,000 people had their frontal lobes deliberately damaged when they underwent frontal lobotomy. This surgical procedure was intended to treat mental illness, personality disorders, and even severe headaches. One physician carried it out by inserting an ice pick through the bone at the back of the eye and wiggling it to destroy tissue of the frontal lobe. Frontal lobotomies sometimes made patients calmer, but the procedure also blunted emotions and impaired their ability to plan, to concentrate, and to behave appropriately in social situations.

Near the rear of each frontal lobe is the primary motor cortex, the region that controls skeletal muscles. Each hemisphere controls and receives signals from the opposite side of the body. For example, signals to move your right arm originate in the motor cortex of your left hemisphere.

The two hemispheres differ somewhat in their function. For example, in the 90 percent of people that are right-handed, the left hemisphere is more active in controlling movement and in language. Broca's area, which is the part of the frontal lobe that helps us translate thoughts into speech, is usually in the left hemisphere. It controls the tongue, throat, and lip muscles and gives humans our capacity to speak complex sentences. Damage to Broca's area often prevents normal speech, although an affected person can still understand language.

The abilities of each hemisphere are flexible. When a stroke or injury damages tissue on one side of the brain, the other hemisphere often can take on new tasks. People can even function with a single hemisphere.

Sensory areas of the cerebral cortex allow us to perceive sensations. The primary somatosensory cortex of the parietal lobe is the receiving area for sensory input from the skin and joints. When someone taps you on your left shoulder, signals that arrive in the primary somatosensory cortex of your right parietal lobe alert you to the tap. Another sensory area in the parietal lobe receives signals about taste. In the occipital lobe, the primary visual cortex integrates incoming signals from both eyes. The perception of sounds and odors arises in the primary sensory areas of the temporal lobe.

Take-Home Message **What are the functions of the spinal cord and brain?**

- The spinal cord relays information between peripheral nerves and the brain. It also has a role in some reflexes.

- The brain stem (hindbrain and midbrain) regulates breathing and adjusts heart rate. It also controls reflexes such as swallowing and coughing. The brain stem's cerebellum coordinates voluntary movements.

- The hypothalamus deep inside the forebrain has important roles in thirst, temperature regulation, and other responses related to homeostasis.

- The cerebrum makes up the bulk of the forebrain. Its outer layer, the cerebral cortex, controls voluntary activity, sensory perception, abstract thought, and language and speech.

- The two cerebral hemispheres differ somewhat in their function. In most people, the left hemisphere dominates.

24.6

The Senses

The Sensory Receptors Sensory receptors are the sentries of a nervous system. Each is a peripheral ending of a certain type of sensory neuron, and each detects a particular stimulus.

We classify receptors by the types of stimuli they detect. **Thermoreceptors** are sensitive to heat or to cold. **Mechanoreceptors** detect changes in pressure, position, or acceleration. Pain receptors (nociceptors) detect injury. **Chemoreceptors** detect substances dissolved in the fluid that bathes them. **Photoreceptors** include light-sensitive pigments that respond to light energy.

Animals monitor the environment in different ways depending on the kinds and numbers of sensory receptors they have. Many animals have evolved sensory capacities we lack. For example, pythons have thermoreceptors that allow them to detect warm-blooded prey (Figure 24.15). Bats have mechanoreceptors that detect ultrasound (high-pitched sound) beyond our range of hearing.

All sensory receptors convert stimulus energy into action potentials. Given that action potentials are "all-or-nothing" events that do not vary in size, how does the brain know how strong a stimulus is? Three mechanisms provide this information. First, an animal's brain is prewired to interpret action potentials of certain nerves in certain ways. For example, the brain always interprets signals from the optic nerve as "light."

Second, as stimulus strength increases, so does the rate at which a sensory receptor has action potentials. The same receptor detects the sound of a polite whisper or an exuberant whoop. The brain interprets the difference from variations in the rate of action potentials.

Third, a stronger stimulus can recruit more sensory receptors than a weaker stimulus. Gently touch a small patch of skin on one of your arms and you activate a few receptors. Press harder and you activate more. The brain interprets the heightened activity as an increase in stimulus intensity.

In some cases, the rate of action potentials in a sensory neuron falls or stops even though a stimulus continues at constant strength. A reduced response to an ongoing stimulus is called sensory adaptation. Some mechanoreceptors in skin adapt quickly to sustained stimulation. For example, soon after you put on clothing, receptors adapt and stop notifying you of its presence.

Somatic and Visceral Sensations Sensory neurons responsible for somatic sensations are located in skin, muscle, tendons, and joints. **Somatic sensations** are localized to a specific body part. **Visceral sensations** arise from neurons in the walls of soft internal organs, and are often difficult to pinpoint. It is easy to determine exactly where someone is touching you, but less easy to say exactly where you feel a stomachache.

You and other mammals detect touch, pressure, cold, warmth, and pain near the body surface. Regions with the most sensory receptors, such as the fingertips, are the most sensitive. Less sensitive regions, such as the back of the hand, have far fewer receptors.

Mechanoreceptors near skeletal muscle, joints, and ligaments detect limb motions. Muscle spindles that trigger the stretch reflex described in Figure 24.12 are an example. The response of these receptors depends on how much and how fast a muscle is stretched.

Pain is the perception of a tissue injury. Signals from pain receptors in skin, skeletal muscles, joints, and tendons give rise to somatic pain. Visceral pain is associated with organs inside body cavities. It occurs as a response to a smooth

FIGURE 24.15 Examples of sensory receptors. **A** Thermoreceptors in pits above and below a python's mouth allow it to detect body heat, or infrared energy, of nearby prey. **B** Mechanoreceptors inside a bat's ear allow the animal to detect high-pitched, or ultrasonic, pressure waves.

chemoreceptor Sensory receptor that responds to the binding of a particular chemical.

mechanoreceptor Sensory receptor that responds to pressure, position, or acceleration.

pain Perception of tissue injury.

photoreceptor Light-sensitive sensory receptor.

somatic sensations Sensations that arise when sensory neurons in skin and near joints and muscle are activated.

thermoreceptor Temperature-sensitive sensory receptor.

visceral sensations Sensations that arise when sensory neurons in soft internal organs are activated.

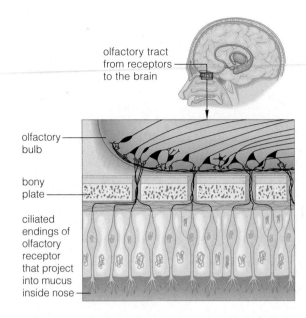

olfactory tract from receptors to the brain

olfactory bulb

bony plate

ciliated endings of olfactory receptor that project into mucus inside nose

FIGURE 24.16 Pathway from sensory endings of olfactory receptors in the human nose to the cerebral cortex. Receptor axons pass through holes in a bony plate between the lining of the nasal cavities and the brain.

muscle spasm, inadequate blood flow to an internal organ, over-stretching of a hollow organ, and other abnormal conditions.

Injured or distressed body cells release chemicals that stimulate nearby pain receptors. Signals from the pain receptors then travel along the axons of sensory neurons to the spinal cord. Here, the sensory axons synapse with the spinal interneurons that relay signals about pain to the brain.

Pain is an adaptive mechanism that alerts an animal to tissue damage. The animal's response to pain gives the damaged part time to heal. However, incapacitating pain can prevent an animal from responding to threats. Therefore, when a vertebrate is injured, its brain releases some natural pain relievers called **endorphins**. Endorphins dampen the flow of pain-related signals to the brain and promote a feeling of well-being. Synthetic opiates such as morphine mimic the effect of endorphins. Aspirin reduces pain in a different way. It slows production of prostaglandins, local signaling molecules that increase the sensitivity of pain receptors. Cells release prostaglandins in response to tissue damage.

The Chemical Senses—Smell and Taste Smell (olfaction) and taste are chemical senses. Stimuli are detected by chemoreceptors that send signals when a specific chemical dissolved in the surrounding fluid binds to them. Binding of a chemical to an olfactory receptor causes action potentials to flow along axons to one of the brain's two olfactory bulbs. In these small brain structures, axons synapse with groups of cells that sort out components of a scent. From there, information flows along the olfactory tract to the cerebrum, where it is further processed (Figure 24.16).

Many animals use olfactory cues to find food and avoid predators. Many also communicate by use of pheromones. **Pheromones** are signaling molecules that are secreted by one individual and affect another member of the same species. For example, olfactory receptors on antennae of a male silk moth allow him to locate a pheromone-secreting female more than a kilometer upwind.

In reptiles and most mammals, a cluster of sensory cells forms a vomeronasal organ that detects pheromones. In primates, including humans, a reduced version of this organ is located in the nasal septum separating the two nostrils. Whether the human vomeronasal organ is functional, and what role, if any, it plays in human behavior, is still unclear.

Depending on the animal, the chemoreceptors involved in taste can be on antennae, legs or tentacles, or inside the mouth. On the surface of your mouth, throat, and especially the upper part of the tongue, chemoreceptors are positioned in about 10,000 sensory organs called taste buds (Figure 24.17).

You perceive many different tastes, but all arise from a combination of five primary sensations: sweet (elicited by simple sugars), sour (acids), salty (NaCl or other salts), bitter (alkaloids), and umami (amino acids such as glutamate, which provides the savory taste typical of aged cheese and meat). MSG (monosodium glutamate), a commonly used artificial flavor enhancer, stimulates the receptors that are responsible for the sensation of umami.

Detecting Light Vision involves detection of light by photoreceptors, and processing of information from those receptors to form a mental image of objects in the environment. Some invertebrates such as earthworms have some photoreceptors at their body

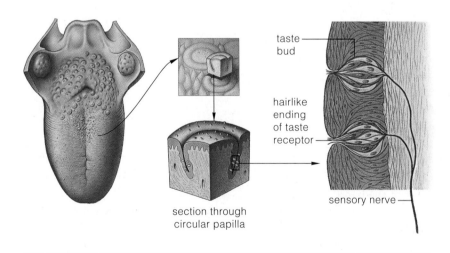

taste bud

hairlike ending of taste receptor

sensory nerve

section through circular papilla

FIGURE 24.17 Taste receptors in the human tongue. The structures called circular papillae enclose epithelial tissue that contains taste buds. A human tongue has approximately 10,000 of these sensory organs, each of which has as many as 150 chemoreceptors.

surface. These animals use light as a cue to orient their body or adjust biological clocks. However, they do not form a mental image of their surrounding as animals with eyes do.

Eyes are sensory organs with a dense array of photoreceptors. The most effective eyes have a **lens**, a transparent structure that bends light rays passing through it so the light falls on photoreceptors. Insects have compound eyes with many separate light-detecting units, each with its own lens. Compound eyes cannot detect visual details, but they are highly sensitive to movement. Cephalopod mollusks such as squids and octopuses have the most complex eyes of any invertebrate. Their camera eyes have an adjustable opening that allows light to enter a dark chamber. Each eye's single lens focuses incoming light onto a **retina**, a tissue densely packed with photoreceptors. The retina of a camera eye is analogous to the light-sensitive film used in a traditional film camera.

Vertebrates also have camera eyes, and because they are not closely related to cephalopod mollusks, camera eyes presumably evolved independently in the two lineages. This is an example of morphological convergence (Section 11.6).

The Human Eye

A human eyeball sits inside a protective, cuplike, bony cavity called the orbit. Skeletal muscles that run from the rear of the eye to bones of the orbit move the eyeball.

Eyelids, eyelashes, and tears protect the delicate eye tissues. Periodic blinking is a reflex that spreads a film of tears over the eyeball's exposed surface. A protective mucous membrane, the **conjunctiva**, lines the inner surface of the eyelids and folds back to cover most of the eye's outer surface. Conjunctivitis, commonly called pinkeye, is an inflammation of this membrane caused by a viral or bacterial infection.

The eyeball is spherical and has a three-layered structure (Figure 24.18). The front portion of each eye is covered by a **cornea** made of transparent crystallin protein. A dense, white, fibrous **sclera** covers the rest of the eye's outer surface.

The eye's middle layer includes the choroid, iris, and ciliary body. The blood vessel–rich **choroid** is darkened by the brownish pigment melanin. This dark layer prevents light reflection within the eyeball. Attached to the choroid and suspended behind the cornea is a muscular, doughnut-shaped **iris**. It too has melanin. Whether your eyes are blue, brown, or green depends on the amount of melanin in your iris.

Light enters the eye's interior through the **pupil**, an opening at the center of the iris. Muscles of the iris adjust pupil diameter in response to light conditions. In bright light the pupil contracts, so less light gets in. In low light the pupil widens, so more light enters the eye. Sympathetic stimulation also widens the pupil, presumably allowing a better look at the source of danger or excitement.

A ciliary body of muscle, fibers, and secretory cells attaches to the choroid and holds the lens in place behind the pupil. The stretchable, transparent lens is about 1 centimeter (1/2 inch) across and shaped like a bulging disk.

The eye has two internal chambers. The ciliary body produces the fluid that fills the anterior chamber. Called aqueous humor, this fluid bathes the iris and lens. A jellylike vitreous body fills the larger chamber behind the lens. The innermost layer of the eye, the retina, is at the back of this chamber. The retina contains the light-detecting photoreceptors.

The cornea and lens bend light rays coming from different points so they all converge at the back of the eye, on the retina. The image formed on the retina is an upside-down mirror image of the real world. However, the brain interprets this image so you perceive the world in its correct orientation.

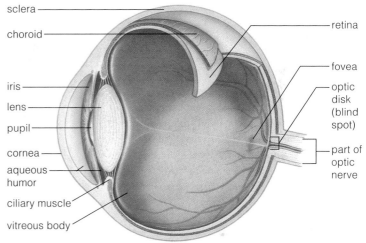

FIGURE 24.18 Animated! The human eye.

choroid Pigmented middle layer of the wall of the eye.
conjunctiva Mucous membrane of eyelids and around eye.
cornea Outermost layer at the front of the eye; bends light.
endorphins Natural pain-reliever molecules.
iris Ring of smooth muscle with pupil at its center; adjusts how much light enters the eye.
lens Structure that focuses light on an eye's photoreceptors.
pheromones Signaling molecules that affect another member of the same species.
pupil Opening through which light enters the eye.
retina Layer of eye that contains photoreceptors.
sclera Protective covering at sides and back of eyeball.

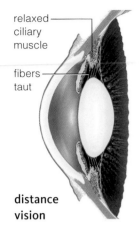
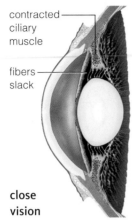

relaxed ciliary muscle

fibers taut

contracted ciliary muscle

fibers slack

distance vision

close vision

A Relaxation of ciliary muscle pulls fibers taut so the lens is stretched into a flat shape that focuses light from a distant object on the retina.

B Contraction of ciliary muscle allows fibers to slacken so the lens rounds up and focuses light from a close object on the retina.

FIGURE 24.19 Animated! How the eye focuses. The lens is encircled by ciliary muscle. Elastic fibers attach the muscle to the lens. The shape of the lens is adjusted by contracting or relaxing the ciliary muscle, increasing or decreasing the tension on the fibers, and changing the shape of the lens.

When you see an object, you are perceiving light rays that are reflected from that object. The properties of light reflected from near and distant objects hit the eye at different angles, but adjustments of the lens ensure that all light rays become focused on the retina. Lens adjustments are carried out by a ciliary muscle that encircles the lens and attaches to it. When you focus on something close up, the ciliary muscle in each eye contracts, causing the lens to bulge outward. As a result, rays of light from the nearby object are bent so they are focused onto the retina (Figure 24.19**A**). When an object is farther away, light rays do not have to be bent as much to be focused on the retina. Ciliary muscles relax a bit, allowing the lens to flatten (Figure 24.19**B**).

Continuous viewing of a close object such as a computer screen or book keeps the ciliary muscles contracted. To reduce eyestrain, take breaks to focus on more distant objects. This will allow your ciliary muscles to relax.

About 150 million Americans have disorders that impair their eyes' ability to focus. With astigmatism, an unevenly curved cornea does not properly focus incoming light on the lens. Nearsightedness occurs when the distance from the front to the back of the eye is longer than normal or when ciliary muscles react too strongly. In either case, distance vision is poor because rays from distant objects converge in front of the retina instead of on it. Close vision is normal. In farsightedness, the distance from front to back of the eye is unusually short, ciliary muscles are weak, or the lens is not flexible enough to stretch easily. As a result, light rays from nearby objects get focused behind the retina. The lens typically loses its flexibility as a person ages. That is why most people who are over age forty have impaired close vision and require reading glasses.

Glasses, contact lenses, or surgery can correct some focusing problems. About 1.5 million Americans undergo laser surgery (LASIK) annually. This procedure reshapes the cornea. Typically, LASIK eliminates the need for glasses during most activities, although older adults usually continue to require reading glasses.

At the Retina As explained in the previous section, the cornea and lens bend light rays so they fall on the retina. Figure 24.20 shows what a physician sees when she uses a lighted magnifying instrument to examine the retina inside the eyeball. The **fovea**, the area of the retina that is richest in photoreceptors, appears as a reddish spot in an area relatively free of blood vessels. With normal vision, most light rays are focused on the fovea.

FIGURE 24.20 Viewing the retina. **A** Doctor using a lighted magnifying instrument to look in through the pupil and see the retina. **B** What the doctor sees. The fovea is the area with the greatest concentration of photoreceptors. With normal vision, light rays are bent so they fall on the fovea. Also visible is the start of the optic nerve.

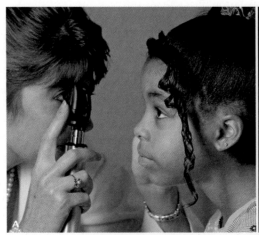
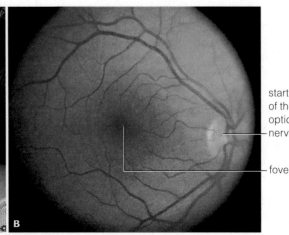

start of the optic nerve

fovea

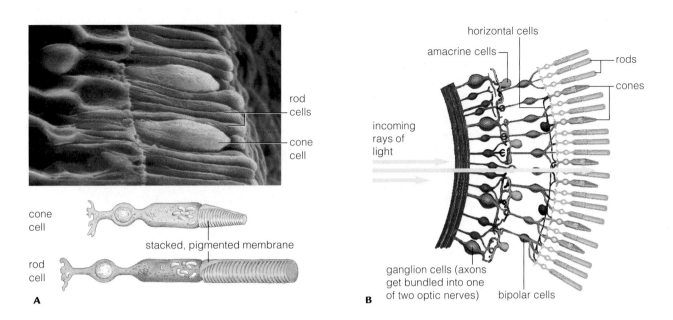

cone cell

rod cell

cone cell

stacked, pigmented membrane

rod cell

A

horizontal cells

amacrine cells

rods

cones

incoming rays of light

ganglion cells (axons get bundled into one of two optic nerves)

bipolar cells

B

FIGURE 24.21 Animated! Structure of the retina. **A** Rod cells and cone cells are photoreceptors that contain light-sensitive pigment. **B** Location of rods, cones, and signal processing cells in the retina.

Figure It Out: Which is closer to the eye's pupil, a rod cell or a ganglion cell?   *Answer: A ganglion cell*

The retina consists of multiple cell layers. Rod cells and cone cells, the photoreceptors, lie beneath several layers of interneurons that process visual signals (Figure 24.21). **Rod cells** detect very dim light and respond to changes in light intensity across the visual field. The sense of color and of acute daytime vision starts with light absorption by **cone cells**. There are three types, each having a different pigment. One cone cell pigment absorbs mainly red light, another absorbs mainly blue, and a third absorbs green. Normal human color vision requires all three kinds of cones. Individuals who lack one or more cone types are color blind (Section 9.6). They often have trouble telling red from green in dim light, and some cannot distinguish between the two even in bright light.

Signal integration and processing start in the retina. Cells that lie above the rods and cones accept information and process signals from these photoreceptors. The processed signals then flow to ganglion cells.

Bundled axons of ganglion cells leave the retina as the beginning of the optic nerve. The part of the retina where the optic nerve exits lacks photoreceptors. It cannot respond to light and thus is a "blind spot." We all have a blind spot in each eye, but usually do not notice it because information about the region that is missed by one eye is provided to the brain by the other.

Hearing Sounds are waves of compressed air—one form of mechanical energy. When you clap your hands or shout, you create pressure waves that move through the air. The amplitude of those waves determines the loudness of the sound, which we measure in decibels. The human ear can detect a one-decibel difference between sounds. An increase of 10 on this scale is an increase of tenfold in loudness. Normal conversation is about 60 decibels, a food blender at high speed is about 90 decibels, and a chain saw is about 100 decibels. Music at a rock concert is about 120 decibels, as is the sound heard through the earbuds of an iPod or similar device cranked up to its maximum volume.

cone cell Photoreceptor that provides sharp vision and allows detection of color.

fovea Region of vertebrate eye with densest concentration of photoreceptors.

rod cell Photoreceptor active in dim light; provides coarse perception of image and detects motion.

FIGURE 24.22
Animated!
Anatomy of the
human ear and
how we hear.

inner ear
vestibular
apparatus,
cochlea

outer ear
pinna,
auditory
canal

middle ear
eardrum,
ear bones

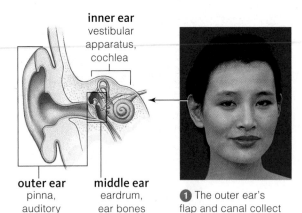

1 The outer ear's flap and canal collect sound waves.

oval window
(behind stirrup)

middle ear bones:
stirrup
anvil
hammer

auditory nerve

auditory
canal

eardrum

round
window

cochlea

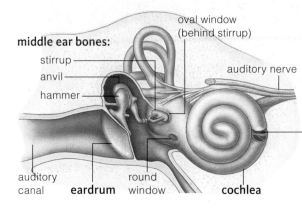

2 The eardrum and middle ear bones amplify sound.

Figure 24.22 shows the three regions of ear. The outer ear gathers sounds. Its pinna, a skin-covered, sound-collecting, folded flap of cartilage, projects outward from the side of the head **1**. The outer ear also includes the auditory canal that carries sound to the middle ear.

The middle ear amplifies and transmits air waves to the inner ear **2**. Pressure waves funneled into the auditory canal cause the **eardrum**, a thin membrane, to vibrate. Behind the eardrum is an air-filled cavity with a set of small bones, known as the hammer, anvil, and stirrup. These bones transmit the force of sound waves from the eardrum to the smaller surface of the oval widow. This flexible membrane is the boundary between the middle ear and inner ear.

The inner ear contains the vestibular apparatus, which functions in balance (discussed below), and the cochlea. The pea-sized, fluid-filled **cochlea** resembles a coiled snail shell (the Greek word *koklias* means snail). Internal membranes divide the cochlea into three fluid-filled ducts **3**.

Pressure from the stirrup on the oval window produces pressure waves in the fluid in these ducts. As these waves travel through the cochlear fluid, they cause the membranous walls of the ducts to vibrate.

The **organ of Corti**, the organ responsible for hearing, sits on a membrane (the basilar membrane) in one of the ducts **4**. Inside the organ are arrays of mechanoreceptors called hair cells. Specialized cilia extend from the hair cells into an overlying membrane. When pressure waves move the membranes, the hair cells bend and undergo action potentials. These signals then travel along an auditory nerve into the brain.

The brain determines the volume and pitch of a sound by assessing how many action potentials flow along the optic nerve, and which portion of the cochlea those potentials come from. The greater the volume of the sound, the more the hair cells bend. The pitch of a sound determines where along the cochlear coil hair cells bend the most. The basilar membrane is narrow and stiff at the entrance to the coil, then it broadens and becomes more flexible deeper in the coil. As a result, different parts of the coil vibrate in response to sounds of different pitches. Higher-pitched sounds set up vibrations early in the coil, whereas lower-pitched sounds cause vibrations deeper in the coil.

Sense of Balance In humans, the sense of balance depends on the vestibular apparatus, an organ with three fluid-filled semicircular canals (Figure 24.23). When the head rotates, the fluid shifts and mechanoreceptors are stimulated. Other mechanoreceptors at the bases of the canals respond to starts and stops. The brain compares the signals from organs of equilibrium in both

cochlea Coiled structure in the inner ear that holds the sound-detecting organ of Corti.
eardrum Membrane of middle ear that vibrates in response to sound waves.
organ of Corti Sound-detecting organ in the cochlea.

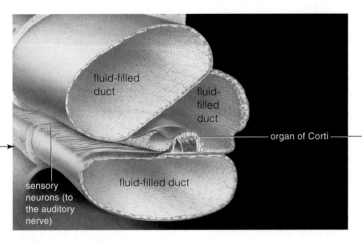

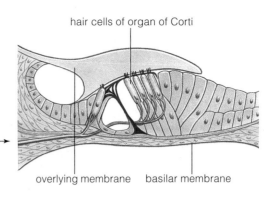
hair cells of organ of Corti

organ of Corti

overlying membrane   basilar membrane

③ One coil of the cochlea in cross-section. The organ of Corti detects pressure waves in fluid-filled ducts inside the cochlea.

fluid-filled duct

fluid-filled duct

fluid-filled duct

sensory neurons (to the auditory nerve)

④ Pressure waves cause the basilar membrane beneath the organ of Corti to move upward. The movement pushes hair cells against an overlying membrane. The resulting action potentials travel along the auditory nerve to the brain.

ears. At the same time, it receives input from receptors in the skin, joints, and tendons. The collective information allows your brain to control eye muscles and keep the visual field in focus, even as you shift position or rotate your head. Integration also helps maintain awareness of the body's position and motion.

Take-Home Message   **How do sensory organs function?**

- Different animals have different types of receptors, and so have different sensory experiences.
- Somatic and visceral sensations are detected by receptors located throughout the body.
- Smell and taste are chemical senses. They involve chemoreceptors that are activated by binding of a particular chemical.
- Vision occurs when the various components of the eye focus light onto photoreceptors on the retina at the back of the eye.
- Hearing is the detection of pressure waves. Components of the ear collect, amplify, and sort out these waves, which are detected by mechanoreceptors. Mechanoreceptors in the inner ear also have a role in sense of balance.

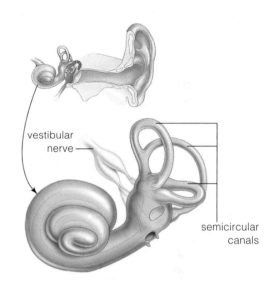
vestibular nerve

semicircular canals

FIGURE 24.23 Animated! The vestibular apparatus and the adjacent cochlea.

24.7
Impacts/Issues Revisited:
In Pursuit of Ecstasy

Now that you know a bit about how the brain functions, take a moment to reconsider the effects of MDMA, the active ingredient in Ecstasy. MDMA harms and possibly kills brain interneurons that produce the neurotransmitter serotonin. Neurons do not divide, so damaged ones are not replaced. Another concern is possible damage to the blood–brain barrier. In one study of rats, the blood–brain barrier remained impaired for as long as 10 weeks after MDMA use. Damage to this protective mechanism allows harmful molecules to slip into the cerebrospinal fluid.

How Would
YOU
☑ **Vote ?**

Some argue that addiction is a disorder and that it is better to send drug addicts to be treated in rehabilitation programs than to jail. Do you agree? See CengageNow for details, then vote online (cengagenow.com).

Summary

Section 24.1 Ecstasy is an illegal drug that enhances mood and energizes a person through its effects on the brain. However, it can also cause memory impairment, panic attacks, and even death.

Section 24.2 Neurons are electrically excitable cells that signal other cells by means of chemical messages. Sensory neurons detect stimuli. Interneurons relay signals between neurons. Motor neurons signal effectors (muscles and glands). Neuroglia support neurons.

Signals reaching neuron input zones (the cell body or dendrites) disturb the ion distribution across the plasma membrane. Strong stimulation may drive the disturbance to the threshold of an action potential that travels down to the neuron's axon endings.

An action potential is an abrupt, brief reversal in the voltage across a neuron's plasma membrane. The reversal triggers an action potential at an adjacent membrane patch, then the next, and on to axon endings. Action potentials are "all or nothing"; they occur only if a cell reaches threshold level and are always the same size.

Neurons interact at chemical synapses. Arrival of an action potential triggers the axon endings of a neuron to release neurotransmitters. These signaling molecules bind to receptors on a signal-receiving cell. The response of that cell to any given signal depends in part on what other signals are arriving at the same time. Psychoactive drugs act by altering signaling at chemical synapses.

CENGAGENOW™ Learn about neuron structure, action potentials, and what happens at a synapse with the animation on CengageNow.

Section 24.3 The simplest nervous system is a nerve net, a mesh of neurons with no central command. Most animals have a bilateral nervous system with a cluster of ganglia or a brain at the head end.

The vertebrate nervous system is functionally divided into the central nervous system (brain and spinal cord) and the peripheral nervous system (nerves that connect the spinal cord and brain with the body).

Nerves are bundles of axons inside connective tissue sheaths. Most vertebrate axons are enclosed in insulating myelin that speeds signal transmission.

CENGAGENOW™ Learn about nerve structure and the effect of myelin with the animation on CengageNow.

Section 24.4 Nerves of the peripheral nervous system extend from cell bodies in the brain or spinal cord out through the body. They relay signals in both directions. Somatic nerves of the peripheral system connect to skeletal muscles and to receptors in joints and the skin. The autonomic nerves (sympathetic and parasympathetic) connect to internal organs and often work in opposition. In a fight–flight situation, sympathetic signals predominate. In a relaxed state, parasympathetic signals predominate.

CENGAGENOW™ Explore the divisions of the peripheral nervous system and compare the effects of sympathetic and parasympathetic stimulation using the animation on CengageNow.

Section 24.5 The organs of the central nervous system are enclosed in protective membranes (meninges) and surrounded by fluid filtered from the blood (cerebrospinal fluid). The blood–brain barrier keeps unwanted substances from entering this fluid.

The spinal cord relays signals from the body to the brain. It also is the integrating center for reflexes that do not involve the brain. A reflex is an automatic response to a stimulus.

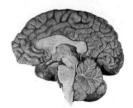

The brain's hindbrain and midbrain constitute the brain stem, which regulates breathing and reflexes such as swallowing and coughing. The brain stem also includes the cerebellum, which coordinates voluntary movements. The forebrain contains the hypothalamus, which regulates functions related to homeostasis. The bulk of the forebrain is the cerebrum. Its thin surface layer of gray matter, the cerebral cortex, governs complex functions such as language and abstract thought. It has specific areas that receive different types of sensory input or control voluntary movements.

CENGAGENOW™ Learn about the structure and function of the brain and spinal cord with the animation on CengageNow.

Section 24.6 Sensory receptors respond to a specific stimulus, such as light or pressure, by undergoing action potentials. The brain evaluates information from sensory receptors based on which nerve delivers them, the rate of action potentials, and the number of axons firing in a given interval.

The somatic sensations such as touch arise from sensory receptors located in skin, or near muscles or joints. Visceral sensations arise from receptors near organs in body cavities. Receptors for taste, smell, hearing, balance, and vision are in specific sensory organs.

The senses of taste and smell (olfaction) involve chemoreceptors that detect specific chemicals. In humans, taste receptors are concentrated in taste buds on the tongue and walls of the mouth. Olfactory receptors line human nasal passages.

Human vision requires focusing of light onto a dense array of photoreceptors (rods and cones) of the retina. Other cells of the retina integrate and process signals, which travel along optic nerves to the brain for final processing and interpretation.

In hearing, the outer ear collects sound waves, the middle ear amplifies them, and the inner ear sorts them and converts them to action potentials. The vestibular apparatus in the inner ear detects movement and changes in the body's position.

CENGAGENOW™ Investigate the mechanisms of sight and sound with the animation on CengageNow.

Self-Quiz   Answers in Appendix I

1. Neurons send signals at _____ .
   a. axon endings   c. dendrites
   b. the cell body   d. myelin sheaths

2. _____ occur mainly in the brain and spinal cord.
   a. Sensory neurons   c. Interneurons
   b. Motor neurons   d. b and c

3. An action potential occurs when _____ .
   a. a neuron reaches threshold potential
   b. gated sodium channels close
   c. sodium–potassium pumps stop working
   d. gated potassium channels open

4. Neurotransmitters are released by _____ .
   a. dendrites   c. myelin sheaths
   b. axon endings   d. both a and b

5. When you sit quietly on the couch reading, output from the _____ system prevails.
   a. sympathetic   c. both
   b. parasympathetic d. neither

6. Skeletal muscles are controlled by _____ nerves.
   a. sympathetic   c. somatic
   b. parasympathetic d. both a and b

7. A _____ is an automatic response that does not require thought.

8. The two halves of the cerebrum _____ .
   a. have identical functions
   b. are part of the autonomic nervous system
   c. are connected by the corpus callosum
   d. consist mainly of motor neurons

9. The blood–brain barrier controls what enters _____ .
   a. blood   c. peripheral nerves
   b. cerebrospinal fluid d. both a and b

10. Which is a somatic sensation?
    a. hearing   c. touch   e. both a and c
    b. smell   d. taste   f. all of the above

11. _____ is a reduced response to an ongoing stimulus.
    a. Propagation   c. Sensory adaptation
    b. Perception   d. Synaptic integration

12. Chemoreceptors play a role in the sense of _____ .
    a. hearing   b. smell   c. vision   d. pain

13. In a vertebrate eye, photoreceptors are in the _____ .
    a. retina   c. lens
    b. cornea   d. conjunctiva

14. Color blindness arises when _____ are missing or defective.
    a. hair cells   c. cone cells
    b. rod cells   d. neuroglial cells

15. Match each structure with its function.
    _____ rod cell   a. detects pheromones
    _____ cochlea   b. connects to spinal cord
    _____ cerebellum   c. sorts out pressure waves
    _____ brain stem   d. protects brain and spinal
    _____ cerebral cortex   cord from some toxins
    _____ taste bud   e. speeds signal transmission
    _____ myelin   f. contains chemoreceptors
    _____ neurotransmitter   g. secreted at synapse
    _____ blood–brain   h. governs higher thought
          barrier   i. coordinates voluntary moves
    _____ vomeronasal organ   j. detects light

Additional questions are available on CENGAGENOW™

Digging Into Data

Prenatal Effects of Ecstasy

Animal studies are often used to assess effects of prenatal exposure to illicit drugs. For example, Jack Lipton used rats to study the behavioral effect of prenatal exposure to MDMA, the active ingredient in Ecstasy. He injected female rats with either MDMA or saline solution when they were 14 to 20 days pregnant. This is the period when their offspring's brains were forming. When those offspring were 21 days old, Lipton tested their response to a new environment. He placed each young rat in a new cage and used a photobeam system to record how much each rat moved around before settling down. Figure 24.24 shows his results.

1. Which rats moved around most (caused the most photobeam breaks) during the first 5 minutes in a new cage, those prenatally exposed to MDMA or the controls?

2. How many photobeam breaks did the MDMA-exposed rats make during their second 5 minutes in the new cage?

3. Which rats moved around the most during the last 5 minutes of the study?

4. Does this study support the hypothesis that MDMA affects a developing rat's brain?

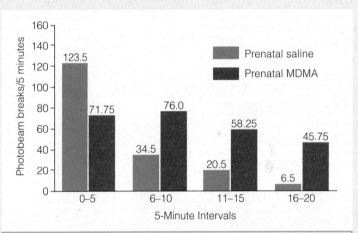

FIGURE 24.24 Effect of prenatal exposure to MDMA on activity levels of 21-day-old rats placed in a new cage. Movements were detected when the rat interrupted a photobeam. Rats were monitored at 5-minute intervals for a total of 20 minutes. *Blue* bars are results for rats whose mothers received saline, *red* bars are rats whose mothers received MDMA.

Critical Thinking

1. The blood–brain barrier in human newborns is not yet fully developed. Explain why this makes careful monitoring of diet and environmental chemical exposure particular important.

2. The axons of some human motor neurons extend from the base of the spinal cord to the big toe, a distance of more than a meter. In a giraffe, the longest axons are several meters long. What are some of the functional challenges involved in the development and maintenance of such lengthy cellular extensions?

25 Endocrine Control

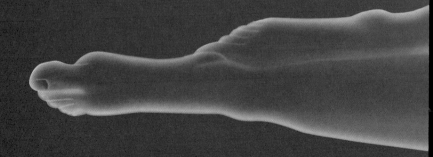

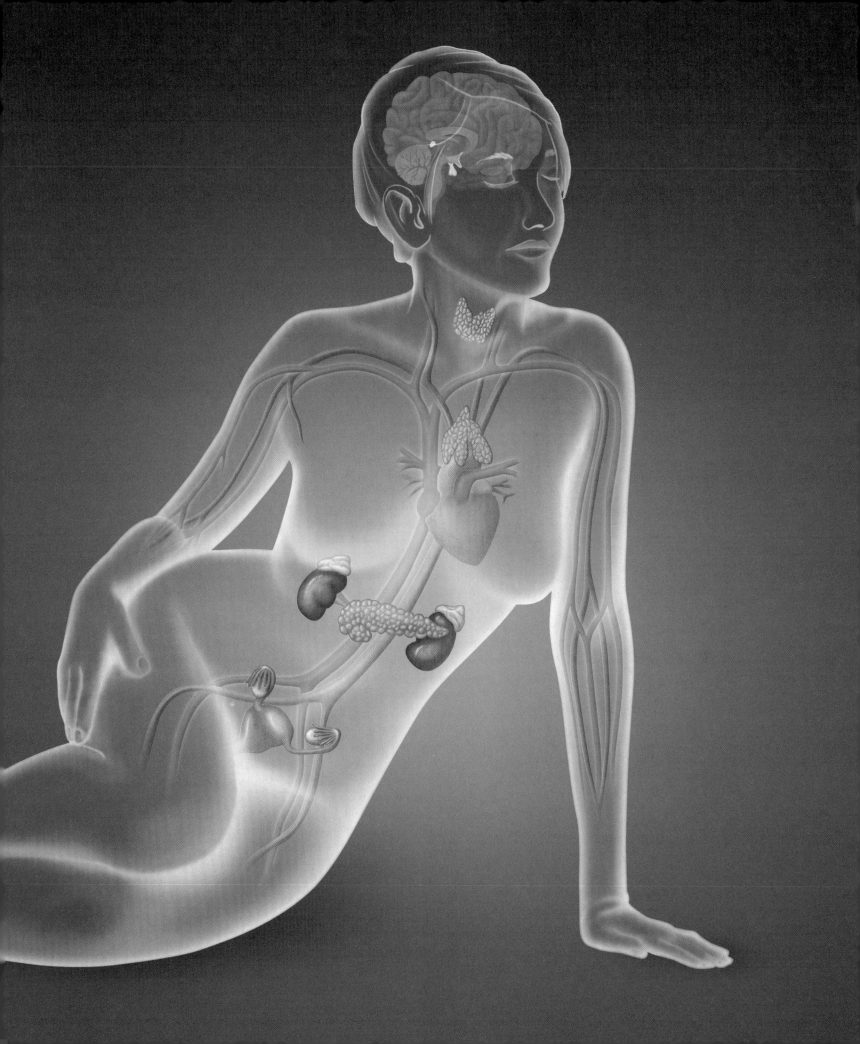

25.1

Impacts/Issues: ## Hormones in the Balance

We live in a world awash in synthetic chemicals. We drink from plastic bottles, wear clothing made of synthetic fabrics, slather ourselves with synthetic skin products, and dose our food with synthetic pesticides. Enormous numbers of man-made compounds are used to manufacture our computers and other electronic gadgets. What do we know about the safety of all these substances?

We have learned by sad experience that some synthetic chemicals harm the environment and threaten human health. For example, DDT (a pesticide) and PCBs (used to make electronic products, caulking, and solvents) are endocrine disruptors. Endocrine disruptors are molecules that interfere with action of hormones, the signaling molecules secreted by endocrine glands. As a result of this action, they can impair animal growth, development, and reproduction. DDT was banned in 1972 and PCBs in 1979. However, because both were widely used for years and are highly stable, they still persist in the environment.

Some chemicals still in use are also suspected endocrine disrupters. Atrazine has been used widely as an herbicide for more than forty years. Each year in the United States, about 76 million pounds of it are sprayed, mostly to kill weeds in cornfields (Figure 25.1**A**). From fields, atrazine-contaminated runoff seeps into soil and water. Atrazine breaks down within a year, but until then it accumulates in ponds, streams, wells, and other sources of drinking water.

Biologist Tyrone Hayes (Figure 25.1**B**) discovered that atrazine can cause frogs that are genetically male to develop both male and female sex organs. To find out if atrazine has this effect in the wild, Hayes collected frogs from ponds and ditches across the Midwest. In every atrazine-contaminated pond he sampled, some males had abnormal sex organs. In the pond with the highest atrazine concentration, 92 percent of the male frogs had feminized sex organs. Hayes hypothesizes that atrazine feminizes males by increasing the body's production of aromatase, an enzyme that converts a male sex hormone (testosterone) into a female sex hormone (estrogen).

Since Hayes sounded the alarm, scientific scrutiny of atrazine has increased. Recently, a team headed by Holly Ingraham reported that atrazine has feminizing effects on zebrafish. When fish embryos were exposed to atrazine levels comparable to levels in runoff from atrazine-treated fields, the number of embryos that developed as females increased dramatically.

It is possible that atrazine's effect on sexual development could help push some fish and frogs to extinction. If this seems a remote concern, consider this: The results of Hayes's and Ingraham's studies of aquatic vertebrates suggest that atrazine could be disrupting human hormones. All vertebrates have similar hormone-secreting glands and endocrine systems. Keep this point in mind when you think about the endocrine disruptors and their effects. What you learn in this chapter will help you evaluate the costs and benefits of synthetic chemicals that affect hormone action.

FIGURE 25.1 Benefits and costs of herbicide applications. **A** Atrazine can keep cornfields nearly weed-free; no need for constant tilling that causes soil erosion. **B** Tyrone Hayes suspects that the chemical scrambles amphibian hormonal signals.

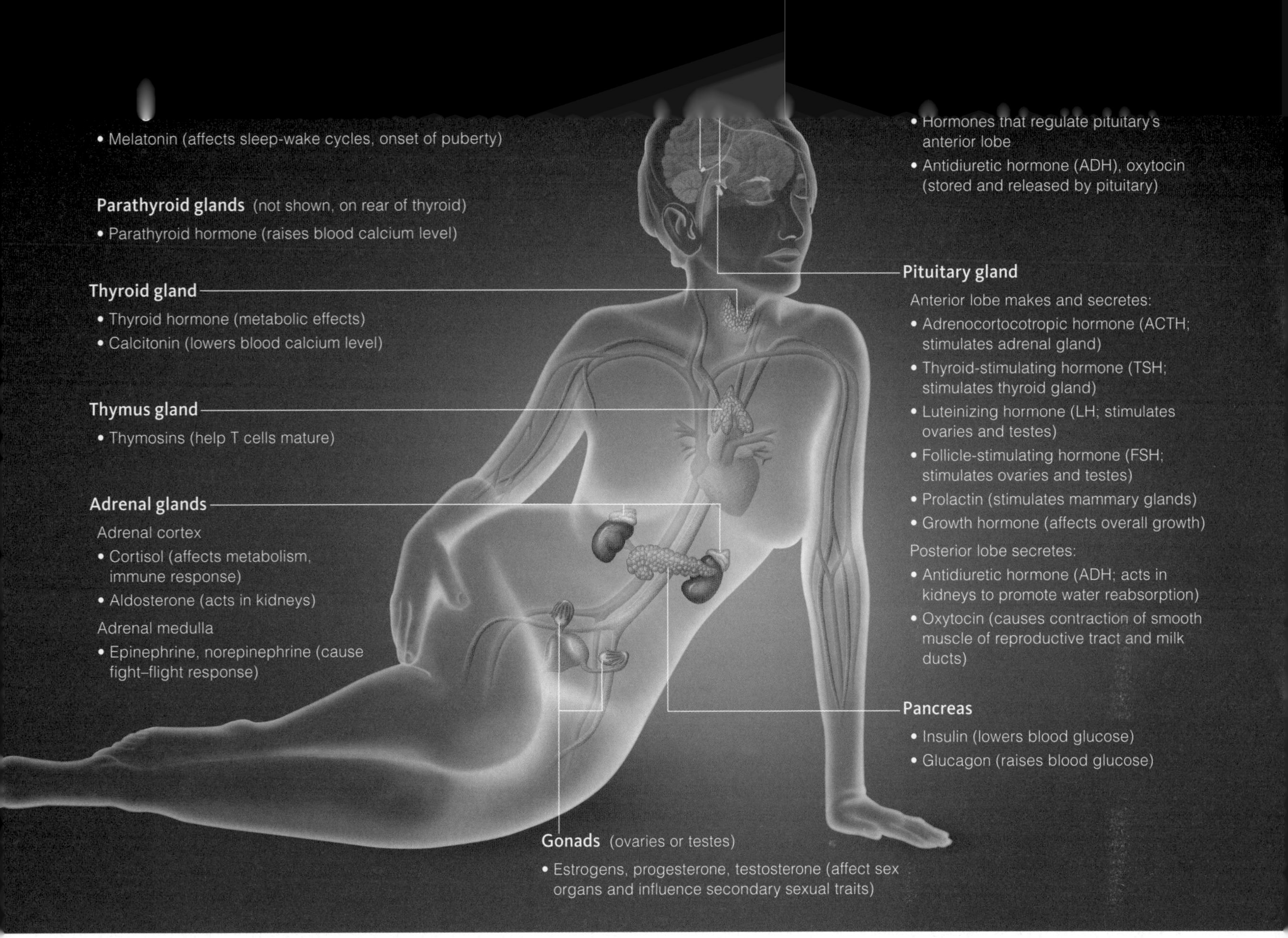

- Melatonin (affects sleep-wake cycles, onset of puberty)

Parathyroid glands (not shown, on rear of thyroid)

- Parathyroid hormone (raises blood calcium level)

Thyroid gland

- Thyroid hormone (metabolic effects)
- Calcitonin (lowers blood calcium level)

Thymus gland

- Thymosins (help T cells mature)

Adrenal glands

Adrenal cortex
- Cortisol (affects metabolism, immune response)
- Aldosterone (acts in kidneys)

Adrenal medulla
- Epinephrine, norepinephrine (cause fight–flight response)

- Hormones that regulate pituitary's anterior lobe
- Antidiuretic hormone (ADH), oxytocin (stored and released by pituitary)

Pituitary gland

Anterior lobe makes and secretes:
- Adrenocortocotropic hormone (ACTH; stimulates adrenal gland)
- Thyroid-stimulating hormone (TSH; stimulates thyroid gland)
- Luteinizing hormone (LH; stimulates ovaries and testes)
- Follicle-stimulating hormone (FSH; stimulates ovaries and testes)
- Prolactin (stimulates mammary glands)
- Growth hormone (affects overall growth)

Posterior lobe secretes:
- Antidiuretic hormone (ADH; acts in kidneys to promote water reabsorption)
- Oxytocin (causes contraction of smooth muscle of reproductive tract and milk ducts)

Pancreas

- Insulin (lowers blood glucose)
- Glucagon (raises blood glucose)

Gonads (ovaries or testes)

- Estrogens, progesterone, testosterone (affect sex organs and influence secondary sexual traits)

FIGURE 25.2 Overview of major glands of the human endocrine system and primary effects of their main hormones. The endocrine system also includes endocrine cells in many organs, such as the heart, kidneys, stomach, liver, small intestine, and skin. For example, Section 23.1 described the effects of leptin and other hormones that affect body weight.

Hormones and the Endocrine System

Animal hormones are signaling molecules that are secreted into the blood by endocrine glands and cells. Blood carries hormones to target cells in the body. Like neurotransmitters, hormones affect cells by binding to receptor proteins. We define a hormone's target as a cell that has receptors for that hormone.

In humans and other vertebrates, the many sources of hormones interact as an **endocrine system**. Figure 25.2 provides an overview of the main components of the human endocrine system. Other vertebrates have a similar system.

Invertebrates also produce hormones, although they do not have the same endocrine organs as vertebrates. For example, an insect must molt its cuticle periodically as it grows, and the molting process is under hormonal control. Some insecticides work by disrupting the hormonal control of molting.

animal hormone Signaling molecule secreted by an endocrine gland or cell.
endocrine system Hormone-producing glands and secretory cells of a vertebrate body.

Table 25.1	Types and Examples of Hormones
Steroids	Testosterone, estrogens, progesterone, aldosterone, cortisol
Amines	Melatonin, epinephrine, thyroid hormone
Peptides	Glucagon, oxytocin, antidiuretic hormone, calcitonin, parathyroid hormone
Proteins	Growth hormone, insulin, prolactin, follicle-stimulating hormone, luteinizing hormone

Types of Hormones There are two categories of hormones: steroid hormones, and hormones derived from amino acids. **Steroid hormones** are lipids derived from cholesterol. The sex hormones testosterone and estrogen are examples (Table 25.1). Being lipids, these hormones can diffuse through a cell's plasma and nuclear membranes (Figure 25.3**A**). The hormones bind to receptors inside the cell, forming a hormone–receptor complex. Often, this complex binds to a promoter in DNA and influences gene expression.

Some steroid hormones also produce effects by binding to receptors at the cell membrane. For example, a cell in breast tissue has estrogen receptors both on its surface and inside the cell. In some types of breast cancer, binding of estrogen to surface receptors helps the cancer grow. The drug tamoxifen binds to estrogen receptors at the surface of breast cancer cells, prevents estrogen from binding to these receptors, and so slows the growth of the cancer.

Hormones derived from amino acids include amines (modified amino acids), peptides (short amino acid chains), and proteins. Peptide and protein hormones cannot diffuse across a lipid bilayer. Instead, they bind to receptors at a target cell's surface. Often the binding starts a cascade of enzyme reactions. For example, the pancreas secretes the peptide hormone glucagon. Receptors for glucagon span the plasma membrane of liver cells (Figure 25.3**B**). The binding of glucagon to these receptors activates an enzyme that converts ATP to another compound called cAMP. The cAMP serves as a **second messenger**. It relays signals from outside the cell to target molecules inside the cell. Inside liver cells, cAMP activates enzymes that activate other enzymes, and so on. The series of reactions eventually activates enzymes that break glycogen into its glucose subunits. The glucose enters the blood and blood sugar rises.

The Importance of Receptors Blood flow distributes a hormone throughout the body, but the hormone only affects cells that have receptors for it. All hormone receptors are proteins, so mutations can interfere with their function. For example, development of male genitals in an XY embryo requires the action of the hormone testosterone. XY individuals who have total androgen insensitivity syndrome secrete testosterone, but a mutation alters receptors for it. Without functional receptors, it is as if testosterone is not present. As a result, the embryo forms testes, but they do not descend into the scrotum, and genitals appear female. Such individuals are often not diagnosed until their teens. Because they are genetically male, they do not menstruate.

Variations in receptors affect how tissues respond to a hormone. The receptor for a hormone in one tissue may cause a different cellular response than a receptor for the same hormone in a different tissue. Also, most tissues have receptors for many hormones. The response called up by one hormone may oppose or reinforce that of another. For example, every skeletal muscle fiber has receptors for glucagon, insulin, cortisol, epinephrine, estrogen, testosterone, growth hormone, somatostatin, and thyroid hormone, as well as others. The blood level of all of these hormones influences what happens in the muscle.

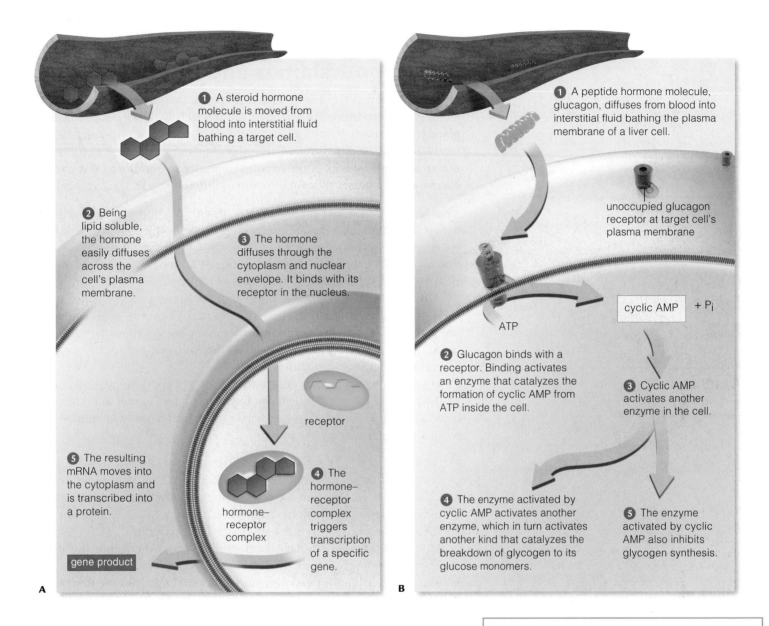

A

① A steroid hormone molecule is moved from blood into interstitial fluid bathing a target cell.

② Being lipid soluble, the hormone easily diffuses across the cell's plasma membrane.

③ The hormone diffuses through the cytoplasm and nuclear envelope. It binds with its receptor in the nucleus.

receptor

⑤ The resulting mRNA moves into the cytoplasm and is transcribed into a protein.

hormone–receptor complex

④ The hormone–receptor complex triggers transcription of a specific gene.

gene product

B

① A peptide hormone molecule, glucagon, diffuses from blood into interstitial fluid bathing the plasma membrane of a liver cell.

unoccupied glucagon receptor at target cell's plasma membrane

ATP

cyclic AMP + P_i

② Glucagon binds with a receptor. Binding activates an enzyme that catalyzes the formation of cyclic AMP from ATP inside the cell.

③ Cyclic AMP activates another enzyme in the cell.

④ The enzyme activated by cyclic AMP activates another enzyme, which in turn activates another kind that catalyzes the breakdown of glycogen to its glucose monomers.

⑤ The enzyme activated by cyclic AMP also inhibits glycogen synthesis.

FIGURE 25.3 Animated! A Example of steroid hormone action inside a target cell. **B** Example of peptide hormone action. Cyclic AMP, a second messenger, relays the signal from a plasma membrane receptor into the cell.

Take-Home Message How do hormones influence cells?

■ Components of the endocrine system produce and secrete hormones. Blood distributes a hormone through the body, but only cells with a receptor protein for that hormone are affected by it. Hormone molecules reversibly bind with a target cell's receptors.

■ Steroid hormones are derived from cholesterol. They can diffuse into a cell and bind to receptors inside it. Many receptor–hormone complexes bind to DNA, and influence protein synthesis. Some steroid hormones also can bind to receptors at a cell's surface.

■ Peptide hormones and protein hormones consist of amino acids. These hormones bind to receptors in the plasma membrane. Binding often sets in motion a chain of enzyme activations inside the cell.

■ Different types of cells have different receptors for the same hormone. The effect of a hormone varies with the properties of the receptor it binds to.

■ Most cells have receptors for, and are influenced by, many different hormones.

second messenger Molecule that forms inside a cell when a hormone binds at the cell surface; sets in motion reactions that alter enzyme activity inside the cell.

steroid hormone Hormone derived from cholesterol.

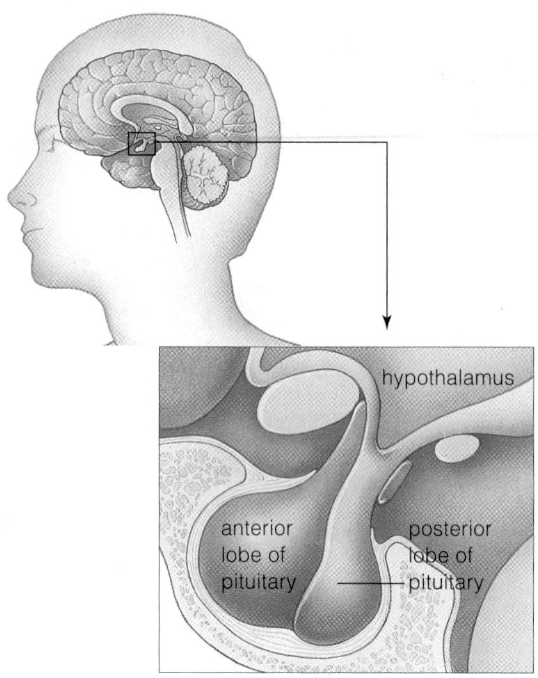

FIGURE 25.4 Location of the hypothalamus and pituitary gland.

The Hypothalamus and Pituitary Gland

The **hypothalamus** is the body's main center for control of the internal environment. It lies deep inside the forebrain and interacts with the pea-sized **pituitary gland** that connects to it by a slender stalk (Figure 25.4). The pituitary gland has two lobes that differ somewhat in their function. Its posterior lobe secretes hormones made by cells in the hypothalamus. Its anterior lobe makes its own hormones but releases them in response to hormones from the hypothalamus.

Posterior Pituitary Function Figure 25.5 illustrates the relationship between the hypothalamus and the posterior lobe of the pituitary gland. The cell bodies of specialized neurons in the hypothalamus produce hormones ❶. You learned in Section 23.7 how one of these hormones—antidiuretic hormone (ADH)—affects kidneys and reduces urine output.

ADH produced by cell bodies in the hypothalamus gets transported through axons to axon terminals in the posterior pituitary ❷. Arrival of an action potential at these axon terminals causes the posterior pituitary to release ADH into the blood. When ADH reaches the kidney, it binds to target cells in kidney tubules and causes them to reabsorb more water ❸.

The hormone oxytocin is also produced in the hypothalamus and released by the posterior pituitary. In females, oxytocin triggers muscle contractions that bring about childbirth. It also makes milk move into ducts of mammary glands when a female is nursing her young.

Anterior Pituitary Function The anterior pituitary makes hormones of its own, but hormones from the hypothalamus control their secretion. Most hypothalamic hormones that act on the anterior pituitary are **releasers**, which encourage secretion of hormones by target cells. Hypothalamic **inhibitors** of the anterior pituitary call for a reduction in target cell secretions.

Most of the anterior pituitary hormones influence other endocrine glands. Adrenocorticotropic hormone (ACTH) stimulates the release of hormones by adrenal glands. Thyroid-stimulating hormone (TSH) regulates the thyroid gland. Follicle-stimulating hormone (FSH) and luteinizing hormone (LH) affect sex hormone secretion and production of gametes by gonads—a male's testes or a female's ovaries.

The anterior pituitary hormone prolactin targets mammary glands, the exocrine glands that secrete milk. Prolactin stimulates and sustains milk production after a woman has given birth.

Growth hormone (GH) is an anterior pituitary hormone that has widespread effects throughout the body. It encourages the growth of bone and soft tissues in the young, and it influences metabolism in adults.

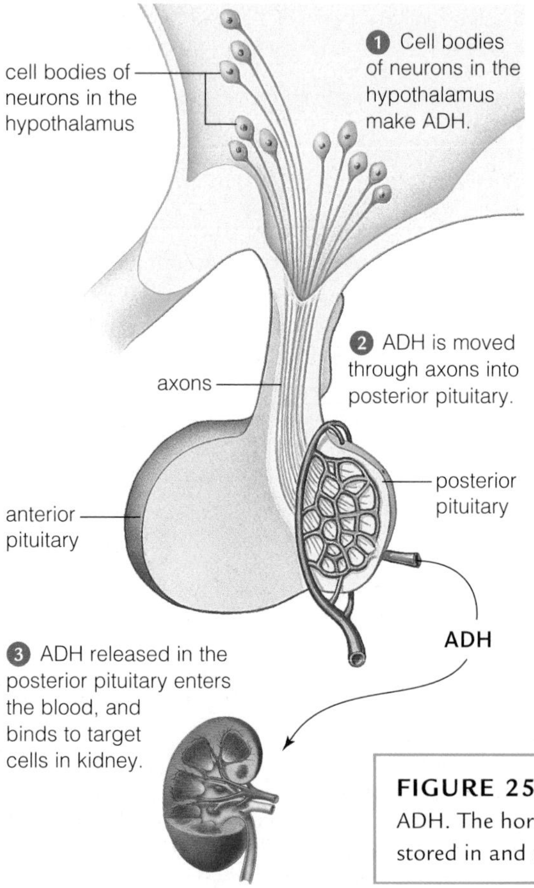

cell bodies of neurons in the hypothalamus

❶ Cell bodies of neurons in the hypothalamus make ADH.

❷ ADH is moved through axons into posterior pituitary.

axons

posterior pituitary

anterior pituitary

ADH

❸ ADH released in the posterior pituitary enters the blood, and binds to target cells in kidney.

FIGURE 25.5 **Animated!** Production and secretion of ADH. The hormone is produced in the hypothalamus, but stored in and released from the posterior pituitary gland.

FIGURE 25.6 Effects of disrupted growth hormone function.

A Bao Xishun, one of the world's tallest men, is 2.36 meters (7 feet 9 inches) tall.

B A woman before and after she became affected by acromegaly. Symptoms usually appear in middle age. Excess growth hormone secretion thickens fingers, enlarges ears, lips, and nose, and makes the brow and chin protrude.

C Dr. Hiralal Maheshwari, *right*, with two men who have an inherited form of dwarfism. Dr. Maheshwari found that the men make little GH because their anterior pituitary does not respond normally to signals from the hypothalamus.

Growth Disorders Normally, a surge of GH production during teenage years causes a growth spurt. GH production then declines with age. Excessive secretion of GH during childhood causes gigantism. Affected people have a normally proportioned body, but are unusually large (Figure 25.6**A**). Excess production of GH during adulthood causes acromegaly, a condition in which the bones thicken. Bones of the hands and feet, and facial bones are the most obviously affected (Figure 25.6**B**). The Greek word *acro* means extremities, and *megas* means large. Both gigantism and acromegaly usually arise as the result of a benign (noncancerous) pituitary tumor.

Pituitary dwarfism occurs when the body produces too little GH or receptors do not respond to it properly during childhood. Affected individuals are short but normally proportioned (Figure 25.6**C**). Pituitary dwarfism can be inherited, or it can result from a pituitary tumor or injury. Injections of human growth hormone can increase a child's growth rate. However, this treatment is controversial. Many people object to the idea that short stature is a defect to be "cured."

Take-Home Message How do the hypothalamus and pituitary function?

- The hypothalamus produces two hormones, antidiuretic hormone (ADH) and oxytocin, that are secreted by the pituitary gland's posterior lobe. ADH promotes water reabsorption by kidneys. Oxytocin causes contraction of muscle during childbirth and nursing.

- The hypothalamus also makes releasers and inhibitors that control secretions by the pituitary's anterior lobe. Hormones secreted by the anterior pituitary govern secretions of the thyroid gland, adrenal glands, gonads, and mammary glands. The anterior pituitary also secretes growth hormone, which has effects on growth and development of tissues throughout the body.

growth hormone (GH) Anterior pituitary hormone that promotes growth and development throughout the body.

hypothalamus Forebrain region that controls processes related to homeostasis and has endocrine functions.

inhibitor Hypothalamic hormone that acts in the anterior pituitary to slow hormone secretion.

pituitary gland Pea-sized endocrine gland in the forebrain that interacts closely with the adjacent hypothalamus.

releaser Hormone that is secreted by one endocrine gland and stimulates secretion by another.

The Thyroid, Parathyroids, and Thymus

The thyroid gland, parathyroid glands, and thymus are endocrine glands of the neck and upper chest.

The Thyroid Gland The **thyroid gland** lies at the base of the neck, in front of the trachea. In children and adolescents, the gland secretes calcitonin, which encourages bone to take up calcium. Normal adults produce little of this hormone. The thyroid gland also secretes two molecules (triiodothyronine and thyroxine) that we refer to collectively as thyroid hormone. Thyroid hormone binds to target cells throughout the body and increases the metabolic rate.

The anterior pituitary gland and hypothalamus regulate thyroid hormone secretion by way of a negative feedback loop (Figure 25.7**A**). When the level of thyroid hormone in the blood declines, the hypothalamus secretes thyroid-releasing hormone (TRH) that acts in the anterior lobe of the pituitary. The releaser causes the pituitary to secrete thyroid-stimulating hormone (TSH). TSH in turn induces the thyroid gland to release thyroid hormone. As a result, the blood level of thyroid hormone rises back to its normal level. Once that point is reached, secretion of TRH, TSH, and thyroid hormone declines.

Thyroid hormone contains the mineral iodine, so a lack of iodine in the diet can cause thyroid hormone deficiency, or hypothyroidism. When iodine is in short supply, signals calling for thyroid hormone secretion cannot be turned off by the normal feedback mechanism. Instead, ongoing stimulation causes thyroid enlargement, or goiter (Figure 25.7**B**). In the United States, use of iodized salt ensures that most people get plenty of iodine. However, goiter remains common in some impoverished parts of the world where such salt is not available.

Hypothyroidism can also cause developmental problems. If a mother lacks iodine during her pregnancy, her child's nervous system may not form properly.

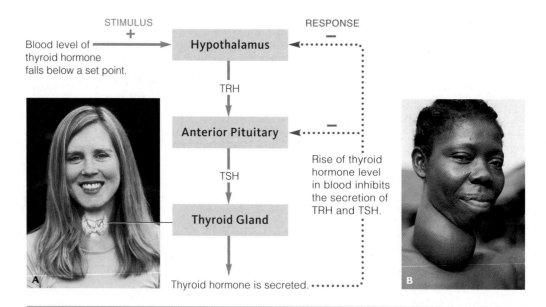

STIMULUS

RESPONSE

Blood level of thyroid hormone falls below a set point.

Hypothalamus

TRH

Anterior Pituitary

TSH

Thyroid Gland

Rise of thyroid hormone level in blood inhibits the secretion of TRH and TSH.

Thyroid hormone is secreted.

FIGURE 25.7 Thyroid function. **A** Negative feedback loop that governs thyroid hormone secretion. **B** Enlarged thyroid gland (goiter). This one was caused by an iodine-deficient diet.

Figure It Out: What effect does a high level of thyroid hormone have on the hypothalamus?

Answer: It inhibits secretion of thyroid-releasing hormone (TRH).

A low level of thyroid hormone during infancy or early childhood also stunts growth and impairs mental ability. The resulting syndrome is know as cretinism.

In the United States, thyroid disorders usually arise when the body's immune system mistakenly attacks the thyroid. Depending on the type of cells attacked, the amount of thyroid hormone can increase or decrease. In either case, the thyroid may become enlarged. Because thyroid hormone increases the body's metabolic rate, a deficiency typically causes fatigue, confusion, increased sensitivity to cold temperature, and weight gain. By contrast, an excess of thyroid hormone causes nervousness and irritability, a chronic fever, and weight loss. Often, altered metabolism causes tissues behind the eyeball to swell, causing eyes to bulge in their sockets. As with other autoimmune disorders, autoimmune thyroid disorders are much more common in women than in men.

Parathyroid Glands **Parathyroid glands** are the main regulators of the blood's calcium level. There are four of these glands, each about the size of a grain of rice, on the thyroid's rear surface. When the calcium level in the blood declines, the glands release parathyroid hormone (PTH). The main targets of PTH are cells in bones and the kidneys. PTH increases the breakdown of bone, thus putting calcium ions into the bloodstream. In the kidneys, PTH increases calcium reabsorption, so less is calcium is lost in the urine. PTH also increases production of an enzyme that activates vitamin D. This vitamin helps the intestine take up calcium from food.

Children who do not get enough vitamin D absorb too little calcium to build healthy new bone. Their low blood calcium level also causes oversecretion of PTH, which encourages breakdown of existing bone. The resulting disorder is called rickets. Bowed legs and pelvic deformities are symptoms (Figure 25.8).

In adults, a benign parathyroid tumor sometimes causes excessive PTH secretion that leads to osteoporosis. The excess calcium released from the bone into the blood also raises the risk of kidney stones.

The Thymus The **thymus** is an endocrine gland beneath the breastbone. It secretes thymosins, hormones required for differentiation of naive T cells into their active forms (Section 22.6). The thymus grows until it reaches the size of an orange at puberty, then begins to shrink.

The adult thymus is largely inactive. For most of us, this is not a problem. Divisions of memory T cells made when our thymus was highly active safeguard our immunity. However, AIDS kills all types of T cells. An affected person loses memory T cells and cannot activate naive T cells to replace them. Thus, researchers are now looking for ways to restore the immune function of AIDS patients by reactivating the thymus.

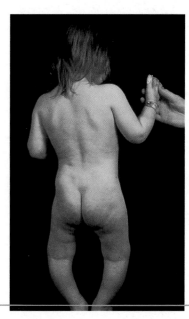

FIGURE 25.8 Characteristic bowed legs of a child with rickets. Lack of vitamin D caused excessive PTH secretion that softened leg bones, allowing them to bend.

Take-Home Message **What are the functions of the thyroid, parathyroids, and thymus?**

- The thyroid gland regulates the metabolic rate of cells throughout the body. Normal thyroid function requires a diet with adequate iodine.

- The parathyroid glands control the level of calcium in the blood. Parathyroid hormone increases breakdown of bone and lessens calcium output in urine, thus increasing the blood calcium level.

- The thymus is most active in childhood. The thymosins it secretes encourage the maturation of T cells, a type of white blood cell.

parathyroid glands Four small endocrine glands whose hormone product increases the level of calcium in blood.

thymus Endocrine gland beneath breastbone that, during childhood, produces hormones that help T cells mature.

thyroid gland Endocrine gland at the base of the neck; produces thyroid hormone, which increases metabolism.

Opposing effects of two pancreatic hormones, insulin and glucagon, keep the blood glucose level within its ideal range.

The Pancreas

The pancreas sits in the abdominal cavity, just behind the stomach (Figure 25.9). We saw in Section 23.3 that the pancreas secretes digestive enzymes into the small intestine, but it also secretes hormones into the bloodstream. The bulk of the organ produces the digestive enzymes. The hormone producing cells are grouped in clusters called pancreatic islets.

Controlling Blood Glucose Food intake and intracellular demands for glucose vary and have the potential to increase or decrease the amount of glucose in the blood beyond its ideal range. However, opposing effects of two pancreatic hormones, glucagon and insulin, keep the blood glucose level within its ideal range. **Glucagon** is a hormone secreted by some cells of the pancreatic islets. It targets cells in the liver, where it causes the breakdown of glycogen into glucose. Thus, glucagon raises the level of glucose in blood. **Insulin** is a hormone secreted by other cells of the pancreatic islets. Its main targets are liver, fat, and skeletal muscle cells. Thus, insulin lowers the level of glucose in the blood.

When blood glucose level rises above a set point, the pancreas secretes less glucagon and more insulin (Figure 25.9). As glucose is taken up and stored inside cells, blood glucose declines.

In contrast, a decline in blood glucose below the set point increases glucagon secretion and slows insulin secretion. The resulting release of glucose from the liver causes the blood glucose level to rise.

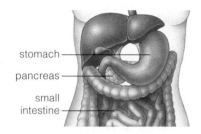

stomach
pancreas
small intestine

FIGURE 25.9 Animated! Maintaining blood glucose level. *Above*, location of the pancreas. *Right*, how cells that secrete insulin and glucagon respond to a change in the level of glucose circulating in blood. These two hormones work antagonistically to maintain the glucose level in its normal range.

❶ *After* a meal, glucose enters blood faster than cells can take it up, so its concentration in blood increases.

❷ In the pancreas, the increase stops cells from secreting glucagon.

❸ It also stimulates other cells to secrete insulin.

❹ In response to insulin, adipose and muscle cells take up and store glucose; cells in the liver and muscle make more glycogen.

❺ As a result, insulin *lowers* the blood level of glucose.

❻ *Between* meals, the glucose level in blood declines.

❼ The decrease causes glucagon secretion.

❽ It also slows insulin secretion.

❾ In the liver, glucagon causes cells to break glycogen down into glucose, which enters the blood.

❿ As a result, glucagon raises the blood glucose level.

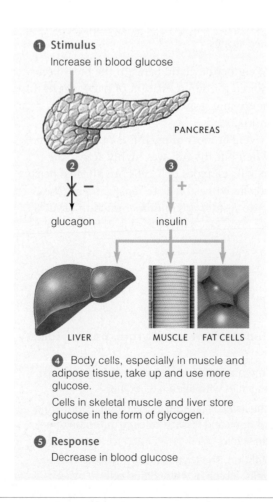

❶ **Stimulus**
Increase in blood glucose

PANCREAS

❷ ❸

glucagon   insulin

LIVER   MUSCLE   FAT CELLS

❹ Body cells, especially in muscle and adipose tissue, take up and use more glucose.

Cells in skeletal muscle and liver store glucose in the form of glycogen.

❺ **Response**
Decrease in blood glucose

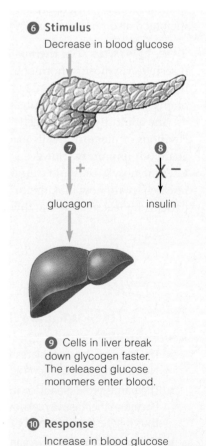

❻ **Stimulus**
Decrease in blood glucose

❼ ❽

glucagon   insulin

❾ Cells in liver break down glycogen faster. The released glucose monomers enter blood.

❿ **Response**
Increase in blood glucose

Table 25.2	Some Complications of Diabetes
Eyes	Changes in lens shape and vision; damage to blood vessels in retina; blindness
Skin	Increased susceptibility to bacterial and fungal infections; patches of discoloration; thickening of skin on the back of hands
Digestive system	Gum disease; delayed stomach emptying that causes heartburn, nausea, vomiting
Kidneys	Increased risk of kidney disease and kidney failure
Heart and blood vessels	Increased risk of heart attack, stroke, high blood pressure, and atherosclerosis
Hands and feet	Impaired sensations of pain; formation of calluses, foot ulcers; poor circulation in feet especially sometimes leads to tissue death that can only be treated by amputation

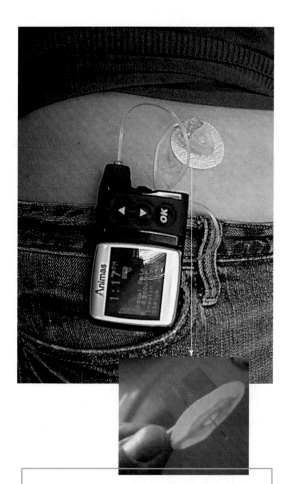

FIGURE 25.10 A diabetic with an insulin pump. The device is programmed to deliver insulin through a hollow tube that projects through the skin and into the body. The pump helps smooth out fluctuations in blood sugar, thus lowering risk of complications that arise from excessively low or high blood sugar.

Diabetes Mellitus Diabetes mellitus is a disorder whose name loosely translates as "passing honey-sweet water." Diabetics produce sweet urine because their liver, fat, and muscle cells do not take up and store glucose as they should. The result is high blood sugar, or hyperglycemia, which disrupts normal metabolism. The cells that do not take up glucose as they should have to break down proteins and fats for energy (Section 5.8). Breakdown of the substances produces harmful waste products. At the same time, high blood sugar causes some cells to overdose on glucose and produce other harmful substances. Accumulation of harmful molecules causes the many complications associated with diabetes (Table 25.2).

There are two main types of diabetes. Type 1 diabetes occurs when an immune reaction, genetic defect, tumor, or toxin destroys insulin-producing cells. Loss of these cells means that less or no insulin is produced. Type 1 diabetes accounts for only 5 to 10 percent of diabetes cases. It is also known as juvenile-onset diabetes because symptoms usually appear in childhood and adolescence. Affected people must monitor the amount of sugar in their blood and supply themselves with insulin. New devices, called insulin pumps, monitor blood sugar and provide a continual supply of insulin (Figure 25.10).

The more common form of diabetes mellitus is type 2 diabetes, in which insulin level is normal, but target cells stop responding to it. As a result, blood sugar levels remain high. Symptoms typically appear in middle age, as insulin production declines. Genetics is a factor, but obesity increases the risk. Diet, exercise, and oral medications can control most cases of type 2 diabetes. Some type 2 diabetics also eventually require insulin injections.

Take-Home Message How do pancreatic hormones regulate blood glucose?

- The pancreas secretes two hormones that work in opposition to control the level of glucose in the blood.
- Insulin is secreted in response to high blood glucose and it increases glucose uptake and storage by cells.
- Glucagon is secreted in response to low blood glucose and it increases breakdown of glycogen to glucose.

glucagon Pancreatic hormone that causes liver cells to break down glycogen and release glucose, thus raising the blood glucose level.

insulin Pancreatic hormone that causes cells to take up glucose from the blood, thus lowering blood glucose level.

The Adrenal Glands

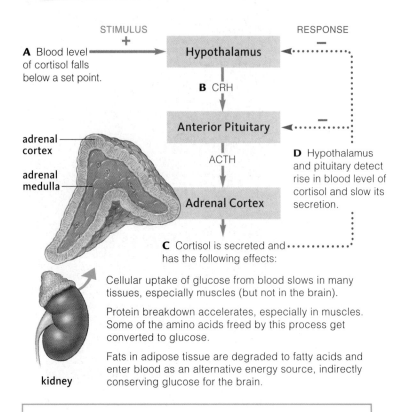

A Blood level of cortisol falls below a set point.

Hypothalamus

B CRH

Anterior Pituitary

ACTH

Adrenal Cortex

D Hypothalamus and pituitary detect rise in blood level of cortisol and slow its secretion.

C Cortisol is secreted and has the following effects:

Cellular uptake of glucose from blood slows in many tissues, especially muscles (but not in the brain).

Protein breakdown accelerates, especially in muscles. Some of the amino acids freed by this process get converted to glucose.

Fats in adipose tissue are degraded to fatty acids and enter blood as an alternative energy source, indirectly conserving glucose for the brain.

kidney

FIGURE 25.11 Animated! Structure and function of the adrenal gland. A negative feedback loop governs cortisol secretion. However, the nervous system can override the feedback controls during times of stress or danger.

adrenal cortex Outer portion of adrenal gland; secretes aldosterone and cortisol.

adrenal gland Endocrine gland located above kidney; secretes hormones with roles in urine formation and stress responses.

adrenal medulla Inner portion of the adrenal gland; secretes epinephrine and norepinephrine.

cortisol Adrenal cortex hormone that influences metabolism and immunity; secretions rise with stress.

estrogens Sex hormones produced by a female's ovaries.

gonads Ovaries and testes; organs that produce gametes and secrete sex hormones.

progesterone Sex hormone produced by a female's ovaries.

puberty Period when reproductive organs mature and begin to function.

testosterone Sex hormone produced by a male's testes.

On top of each kidney is a grape-sized **adrenal gland**. (In Latin *ad–* means near, and *renal* refers to the kidney.) Each adrenal gland has two regions: an outer **adrenal cortex**, and an inner **adrenal medulla**. The two regions of the gland are controlled by different mechanisms, and secrete different hormones.

The adrenal cortex secretes aldosterone and cortisol. Aldosterone promotes reabsorption of sodium and water in the kidney. Like ADH, aldosterone concentrates the urine.

The hormone **cortisol** has wide-ranging effects on metabolism and immunity. Usually, negative feedback loops with the anterior pituitary and hypothalamus maintain blood cortisol levels. Figure 25.11 shows what happens when the level of cortisol in blood decreases below its set point. This decline triggers secretion of corticotropin-releasing hormone (CRH) by the hypothalamus. CRH causes the anterior pituitary to secrete ACTH (adrenocorticotropin). ACTH in turn causes the adrenal cortex to secrete cortisol.

Cortisol induces liver cells to break down their stored glycogen, and suppresses uptake of glucose by other cells. It prompts adipose cells to degrade fats, and skeletal muscles to degrade proteins. The breakdown products of fats and proteins serve as alternative energy sources (Section 5.8).

Cortisol secretion rises dramatically with injury, illness, anxiety, or starvation. Under these circumstances, the nervous system overrides the feedback loop that stabilizes the cortisol level, and the cortisol level in the blood soars.

At the same time, the adrenal medulla increases its output. The adrenal medulla contains specialized neurons of the sympathetic division (Section 24.4). Like other sympathetic neurons, those in the adrenal medulla release norepinephrine and epinephrine. However, in this case, the norepinephrine and epinephrine do not act as neurotransmitters at a synapse. Instead, they enter the blood and function as hormones. Epinephrine and norepinephrine released into the blood travel to organs, where they have the same effect as direct stimulation by a sympathetic nerve. Remember that sympathetic stimulation plays a role in the fight–flight response. Epinephrine and norepinephrine dilate the pupils, increase breathing, and make the heart beat faster. They prepare the body to deal with an exciting or dangerous situation.

The increased secretion of cortisol, epinephrine, and norepinephrine during stressful times helps the body deal with an immediate threat by diverting resources from maintenance tasks. In the short term, a fear or stress response can be highly adaptive. When you are threatened with starvation or being pursued by a predator, short-term survival is of paramount importance. However, a chronically a high cortisol level can cause health problems.

Very high cortisol levels cause Cushing syndrome. Affected people have a puffy, rounded "moon face" and tend to put on fat around their torso. Blood pressure and blood glucose become unusually high. Impaired immunity makes infections common. The skin thins, bone density declines, and muscles shrink. Wounds may be slow to heal. Women's menstrual cycles are erratic or nonexistent, and men may be impotent. Cushing syndrome is usually caused by use of synthetic steroids such as cortisone shots and prednisone. These drugs are given to relieve pain, inflammation, and other ailments. The body converts them to cortisol. A tumor of the pituitary or adrenal gland can also cause oversecretion of cortisol and Cushing syndrome.

Because cortisol levels rise with starvation or improper nutrition, anorexics, bulimics, and people on fad diets tend to have an elevated cortisol level. So do people who are facing anxiety-provoking situations. As a group, poor people have more health problems than those who are well off—even controlling for differences, such as variations in diet and access to health care. By one hypothesis, stress-related high cortisol levels may be an important link between poverty and poor health.

An abnormally low cortisol level, called Addison's disease, is also unhealthy. It can be caused by some infections such as tuberculosis. However, in developed countries, it usually results from autoimmune attacks on the adrenal glands. President John F. Kennedy had the autoimmune form of Addison's disease (Figure 25.12). Symptoms often include fatigue, weakness, depression, weight loss, and darkening of the skin. If cortisol levels get too low, blood sugar and blood pressure can fall to life-threatening levels. People with Addison's disease are treated with a synthetic form of cortisol.

FIGURE 25.12 President John F. Kennedy. He had Addison's disease and received daily shots of cortisone.

Take-Home Message **What are the functions of the adrenal glands?**

- The adrenal cortex secretes aldosterone and cortisol. Aldosterone makes the urine more concentrated. Cortisol affects metabolism and the stress response.

- The adrenal medulla releases epinephrine and norepinephrine, which prepare the body for excitement or danger.

25.7

Hormones and Reproductive Behavior

The Gonads The **gonads** are primary reproductive organs: a male's testes (singular, testis) and a female's ovaries. The testes and ovaries synthesize the same steroid sex hormones—estrogens, progesterone, and testosterone—in different proportions. Ovaries produce mostly estrogen and progesterone. Testes produce mainly testosterone.

Synthesis of sex hormones steps up during **puberty**, the period during which the body's reproductive organs and structures mature and begin to function. In females, estrogen secretion results in development of female secondary sexual traits such as enlarged breasts and wider hips. **Estrogens** and **progesterone** also regulate maturation of eggs and prepare the body for pregnancy.

In males, **testosterone** production rises at puberty. The hormone causes development of male secondary sexual traits such as facial hair and regulates production of sperm. The small amount of estrogen produced by the testes may play an essential role in sperm development.

The hypothalamus and anterior pituitary control the secretion of sex hormones (Figure 25.13). In both males and females, the hypothalamus produces GnRH (gonadotropin-releasing hormone). This releaser causes the anterior pituitary to secrete follicle-stimulating hormone (FSH) and luteinizing hormone (LH). FSH and LH cause the gonads to secrete sex hormones. We will return to the subject of sex hormones and their role in reproduction in Chapter 26.

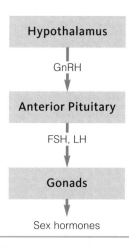

Hypothalamus

GnRH

Anterior Pituitary

FSH, LH

Gonads

Sex hormones

FIGURE 25.13 Control of sex hormone secretion.

Routinely working night shifts or having poor sleep habits disrupts melatonin secretion and raises the risk of cancer.

The Pineal Gland The **pineal gland** lies deep inside the brain. This small, pine cone–shaped gland secretes **melatonin**. Melatonin secretion declines when the retina detects light and signals flow along the optic nerve to the brain. Light varies with time of day and daylength varies seasonally, so melatonin secretion rises and falls in daily and seasonal cycles.

Fluctuations in melatonin level influence seasonal behavior in many species. For example, in some male songbirds long winter nights lead to an increase in melatonin production. The rise in melatonin indirectly prevents singing and other courtship behaviors by slowing testosterone secretion. As long as the levels of testosterone stay low, the bird will not sing. In the spring, there are fewer hours of darkness and the melatonin levels in the blood decline. With the inhibitory effects of melatonin lifted, the bird's testes secrete additional testosterone. The result is the chorus of birdsong we enjoy in the spring.

Melatonin also affects sex hormone production in humans. A drop in the blood level of melatonin is correlated with the onset of puberty and may trigger it. Also, certain endocrine disorders that increase the production of melatonin delay puberty. Others that lower melatonin production accelerate it.

Melatonin regulates human sleep–wake cycles, in part by affecting body temperature. Just after sunrise, melatonin secretion decreases, body temperature rises, and we awaken. At night, melatonin secretion causes a decline in body temperature and we become sleepy. Because cycles of melatonin secretion that affect sleep–wake rhythms are set by exposure to light, travelers who fly across many time zones are advised to spend some time in the sun. The exposure to light can help them reset their internal clock and minimize the effects of jet lag.

Similarly, some people in latitudes where there are big seasonal shifts in daylength have seasonal affective disorder (SAD), commonly called the "winter blues." During winter months, people with SAD tend to feel lethargic and depressed. Researchers have found that affected people secreted more melatonin in winter than in summer, whereas most people secrete the same amount year round. Exposure to bright artificial light in the early morning is used to treat SAD.

Melatonin has a protective effect against some cancers. Routinely working night shifts or having poor sleep habits disrupts melatonin secretion and raises the risk of cancer. For example, one study showed that female night-shift nurses have lower melatonin levels and a higher risk of breast cancer than their day-shift counterparts. Other studies suggest that a woman's risk of breast cancer decreases with the length of her average night's sleep. In men, night-shift work increases the risk of prostate cancer.

How does melatonin protect against cancer? There may be multiple mechanisms. Melatonin directly inhibits division of cancer cells in animals. It also suppresses production of sex hormones, and high levels of those hormones can encourage the growth of certain cancers.

Take-Home Message

What are the endocrine roles of the gonads and the pineal gland?

- A female's ovaries and a male's testes are gonads that make sex hormones as well as gametes. Males make mostly testosterone. Females make mostly estrogen and progesterone.

- In both sexes, sex hormone secretion is regulated by hormones secreted by the pituitary gland, in response to a releaser from the hypothalamus.

- The pineal gland deep inside the brain produces melatonin, which influences sleep–wake cycles and onset of puberty. Melatonin also has a protective effect against some cancers.

melatonin Hormone produced by the pineal gland; affects onset of puberty, sleep–wake cycles.
pineal gland Endocrine gland in the forebrain that secretes melatonin; secretion declines when the eye is exposed to light.

Impacts/Issues Revisited:
Hormones in the Balance

In 1996, the United States passed a bill charging the Environmental Protection Agency with the task of screening pesticides and other chemicals currently in use for endocrine-disrupting effects. The EPA was expected to begin testing chemicals in 1999. However, it did not come up with its first list of chemicals to be tested and a set of testing procedures until 2007. All the chemicals on the initial list are components of pesticides. They include atrazine (the herbicide discussed in Section 25.1), glyphosate (the herbicide sold as Roundup), and captan (a widely used fungicide).

Meanwhile, evidence that chemicals widely used in consumer products are endocrine disruptors continues to mount. For example, there have long been concerns about phthalates (pronounced THAL-aytes). Phthalates are used in soft plastics, cleaning products, and personal care products such as shampoos, lotions, nail polishes, and cosmetics. In 1999, the U.S. Consumer Product Safety Commission asked manufacturers to stop using phthalates in rattles and toys designed for teething infants. The next year, Osovalado Rosarios of the University of Puerto Rico reported that a high phthalate level was correlated with premature breast development in Puerto Rican girls. Most affected girls he studied showed some breast development by age 2 (Figure 25.14). This effect most likely arises because estrogens stimulate breast development and phthalates mimic estrogens.

Phthalates can also suppress the testosterone secretion that is necessary for normal male development. In 2005, Shanna Swan of the University of Rochester Medical School reported that the level of phthalates in the body is correlated with a woman's chances of giving birth to a son with somewhat feminized genitals, smaller genitals, or an undescended testicle. That same year, the U.S. National Toxicology Program, a board of government-appointed scientists reviewed Swan's study and others. They concluded that there was insufficient evidence that exposure to phthalates harms the developing male reproductive tract, but encouraged further studies.

Swan's study garnered widespread publicity, and consumer groups began pushing to have phthalates removed from children's toys. In 2008, the United States passed a toy safety bill that imposed a temporary ban on the use of phthalates in products for children under age 12. An industry group, the American Chemical Council, objected to the ban, saying it was based on politics rather than good science. However, a spokesperson for the group also noted that the ban would have little negative effect on phthalate producers, because less than 3 percent of phthalates are used in manufacturing toys.

It is certainly true that banning phthalates in toys will do little to lower prenatal phthalate exposure. Pregnant women, children, and everyone else will continue to be exposed to phthalates in other products. For example, most commonly used artificial fragrances contain phthalates. Pediatrician Sheela Sathyanarayan found that the more scented powders, shampoos, and lotions a mother used on her infant, the higher the level of phthalates in the child's urine. Her findings led the American Academy of Pediatrics to recommend that parents choose unscented products for use on infants and young children. Similarly, many obstetricians recommend that pregnant women use personal care products that are labeled as organic or free of phthalates. Phthalates do not have to be listed as ingredients if they are a component of a fragrance.

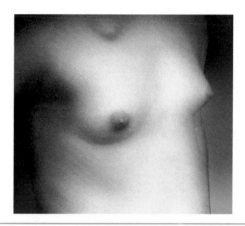

FIGURE 25.14 Breast development in a 23-month old girl. Exposure to phthalates may increase risk of early breast development.

How Would
YOU
☑ **Vote?**

Farmers currently use agricultural chemicals to maximize crop yields. Some of these chemicals may disrupt hormone function in animals and possibly humans. Should chemicals that are under suspicion remain in use while researchers investigate them? **See CengageNow for details, then vote online (cengagenow.com).**

Summary

Section 25.1 Synthetic chemicals such as the weed-killer atrazine get into the environment, where they can have endocrine-disrupting effects on aquatic animals. Endocrine disruptors may also pose a threat to human health.

Section 25.2 All vertebrates have an organ system consisting of endocrine glands and cells that secrete hormones. In most cases, hormones travel through the bloodstream to target cells elsewhere in the body. A cell is a target of a hormone only if it has receptors that can bind that hormone.

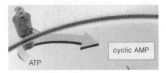

Steroid hormones are lipid soluble; they can enter a target cell and interact directly with DNA. Peptide and protein hormones bind to membrane receptors at the cell surface. Binding commonly leads to formation of a second messenger molecule. The second messenger causes a series of enzyme activations in the cytoplasm.

CENGAGENOW™ Compare the mechanisms of steroid and protein hormones by viewing the animation on CengageNow.

Section 25.3 The hypothalamus, a forebrain region, is structurally and functionally linked with the pituitary gland as a major center for homeostatic control.

The posterior pituitary releases antidiuretic hormone (ADH) and oxytocin made by cells in the hypothalamus. Oxytocin causes contraction of smooth muscle in milk ducts and the uterus. ADH acts on kidney tubules and promotes concentration of the urine.

The anterior pituitary makes and secretes its own hormones: Adrenocorticotropin tells the adrenal cortex to secrete cortisol; thyroid-stimulating hormone calls for thyroid hormone secretion; and follicle-stimulating hormone and luteinizing hormone stimulate hormone production by male and female gonads and have roles in gamete formation. Prolactin stimulates milk production by the mammary glands. Growth hormone has growth-promoting effects on cells throughout the body.

CENGAGENOW™ Use the animation on CengageNow to study how the hypothalamus and pituitary interact.

Section 25.4 A feedback loop to the anterior pituitary and hypothalamus governs the thyroid gland in the base of the neck. Iodine is required for synthesis of thyroid hormone, which increases the metabolic rate and plays an important role in development.

The four parathyroid glands located at the rear of the thyroid gland are the main regulators of calcium levels in the blood. They release parathyroid hormone in response to low calcium levels. Parathyroid hormone acts on cells in bone, the kidneys, and the intestine to raise calcium levels in the blood.

The thymus beneath the breastbone produces hormones that help some white blood cells mature. It increases in size until puberty, then becomes small and largely inactive in adults.

Section 25.5 Insulin and glucagon are pancreatic hormones secreted in response to shifts in the blood level of glucose. Insulin stimulates glucose uptake by muscle and liver cells and thus lowers the blood glucose level. Glucagon stimulates the release of glucose, which increases blood levels. Diabetes mellitus is a metabolic disorder in which blood glucose is chronically high.

CENGAGENOW™ Use the animation on CengageNow to see how the actions of insulin and glucagon regulate blood sugar.

Section 25.6 There is an adrenal gland atop each kidney. The adrenal cortex secretes aldosterone (which targets the kidney) and cortisol (the stress hormone). Cortisol secretion is governed by a negative feedback loop to the anterior pituitary gland and hypothalamus. In times of stress, the nervous system overrides feedback controls over cortisol secretion so cortisol level rises.

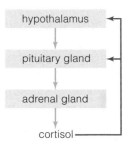

Norepinephrine and epinephrine released by neurons of the adrenal medulla influence organs the same way that sympathetic stimulation does. They cause a fight–flight response.

CENGAGENOW™ Watch the animation on CengageNow to see how cortisol levels are maintained by negative feedback.

Section 25.7 Gonads (ovaries and testes) secrete estrogens, progesterone, and testosterone. These sex hormones are steroids that act in reproduction and in development of secondary sexual traits.

Melatonin secretion by the vertebrate pineal gland is part of a biological clock, a type of internal timing mechanism. In humans, it affects the daily sleep–wake cycle and the onset of puberty. Exposure to light suppresses melatonin production.

Self-Quiz   Answers in Appendix I

1. _____ , signaling molecules released by endocrine cells and glands, enter the blood and are distributed through the body.
   a. Hormones   c. Endocrine disruptors
   b. Neurotransmitters   d. all of the above

2. Antidiuretic hormone and oxytocin are hormones produced in the hypothalamus but released from the _____ .
   a. anterior lobe of pituitary   c. pancreas
   b. posterior lobe of pituitary   d. pineal gland

3. Protein hormones typically bind to receptors _____ .
   a. in the DNA   c. at the plasma membrane
   b. in the cytoplasm   d. both a and c

4. Match each pituitary hormone with its target.
   _____ antidiuretic hormone   a. gonads (ovaries, testes)
   _____ oxytocin   b. mammary glands, uterus
   _____ luteinizing hormone   c. kidneys
   _____ growth hormone   d. most body cells

5. Overproduction of _____ causes acromegaly and pituitary gigantism.

6. The _____ regulate(s) calcium levels in the blood.
   a. hypothalamus
   b. pancreas
   c. pineal gland
   d. parathyroid glands

7. _____ lowers blood sugar levels; _____ raises it.
   a. Glucagon; insulin
   b. Insulin; glucagon
   c. Melatonin; insulin
   d. Cortisol; glucagon

8. A rise in hormone concentration in the blood slows production of that hormone in a _____ feedback loop.
   a. positive
   b. negative

9. The _____ produces digestive enzymes and hormones.
   a. hypothalamus
   b. pancreas
   c. pineal gland
   d. parathyroid gland

10. A diet lacking in iodine can cause _____ .
    a. rickets
    b. a goiter
    c. diabetes
    d. gigantism

11. True or false? Only women make follicle-stimulating hormone and luteinizing hormone.

12. A person with an overly active thyroid gland is more likely to be unusually _____ .
    a. heavy
    b. anxious
    c. cold
    d. both a and b

13. During stressful situations, the adrenal glands increase their output of _____ .
    a. cortisol
    b. epinephrine
    c. norepinephrine
    d. all of the above

14. The male sex hormone testosterone is secreted in response to secretion of hormones by the _____ .
    a. testes
    b. ovaries
    c. pituitary gland
    d. pancreas

15. Match the hormone source listed at left with the most suitable description at right.
    _____ adrenal cortex
    _____ thyroid gland
    _____ thymus gland
    _____ parathyroid glands
    _____ pancreatic islets
    _____ pineal gland
    _____ hypothalamus
    _____ testes

    a. make gametes and hormones
    b. major control center
    c. blood calcium effect
    d. allows T cell maturation
    e. stress increases secretions
    f. light inhibits secretion
    g. hormones require iodine
    h. insulin, glucagon source

Additional questions are available on CENGAGENOW™

Critical Thinking

1. Suppose a woman affected by type 1 (insulin-dependent) diabetes miscalculates and injects herself with too much insulin. She begins to feel confused and sluggish. She calls for medical assistance and injects herself with glucagon her doctor prescribed for such an emergency. An ambulance arrives and she is given dextrose (a sugar) intravenously. What caused her symptoms? How did the glucagon injection help?

Digging Into Data

Sperm Counts Down on the Farm

Contamination of water by agricultural chemicals affects the reproductive function of some animals. Are there effects on humans? Epidemiologist Shanna Swan and her colleagues studied sperm collected from men in four cities in the United States (Figure 25.15). The men were partners of women who had become pregnant and were visiting a prenatal clinic, so all were fertile. Of the four cities, Columbia, Missouri, is located in the county with the most farmlands. New York City in New York is in an area with no agriculture.

1. In which cities did researchers record the highest and lowest sperm counts?

2. In which cities did samples show the highest and lowest sperm motility (ability to move)?

3. Aging, smoking, and sexually transmitted diseases adversely affect sperm. Could differences in any of these variables explain the regional differences in sperm count?

4. Do these data support the hypothesis that living near farmlands can adversely affect male reproductive function?

	Location of Clinic			
	Columbia, Missouri	Los Angeles, California	Minneapolis, Minnesota	New York, New York
Average age	30.7	29.8	32.2	36.1
Percent nonsmokers	79.5	70.5	85.8	81.6
Percent with history of STD	11.4	12.9	13.6	15.8
Sperm count (million/ml)	58.7	80.8	98.6	102.9
Percent motile sperm	48.2	54.5	52.1	56.4

FIGURE 25.15 Data from a study of sperm collected from men who were partners of pregnant women that visited prenatal health clinics in one of four cities. STD stands for sexually transmitted disease.

2. A tumor in the pituitary gland can cause a woman to produce milk even when she is not pregnant. In which lobe of the pituitary would such a tumor be located, and what hormone would it affect?

3. Women who are completely blind tend to undergo puberty at an earlier age than sighted women. They also are less likely to have breast cancer. By one hypothesis, blindness encourages early puberty and discourages breast cancer by its effect on melatonin secretion. Based on this information, would you expect the average melatonin level in blind women to be higher or lower than that of sighted women? Explain your answer.

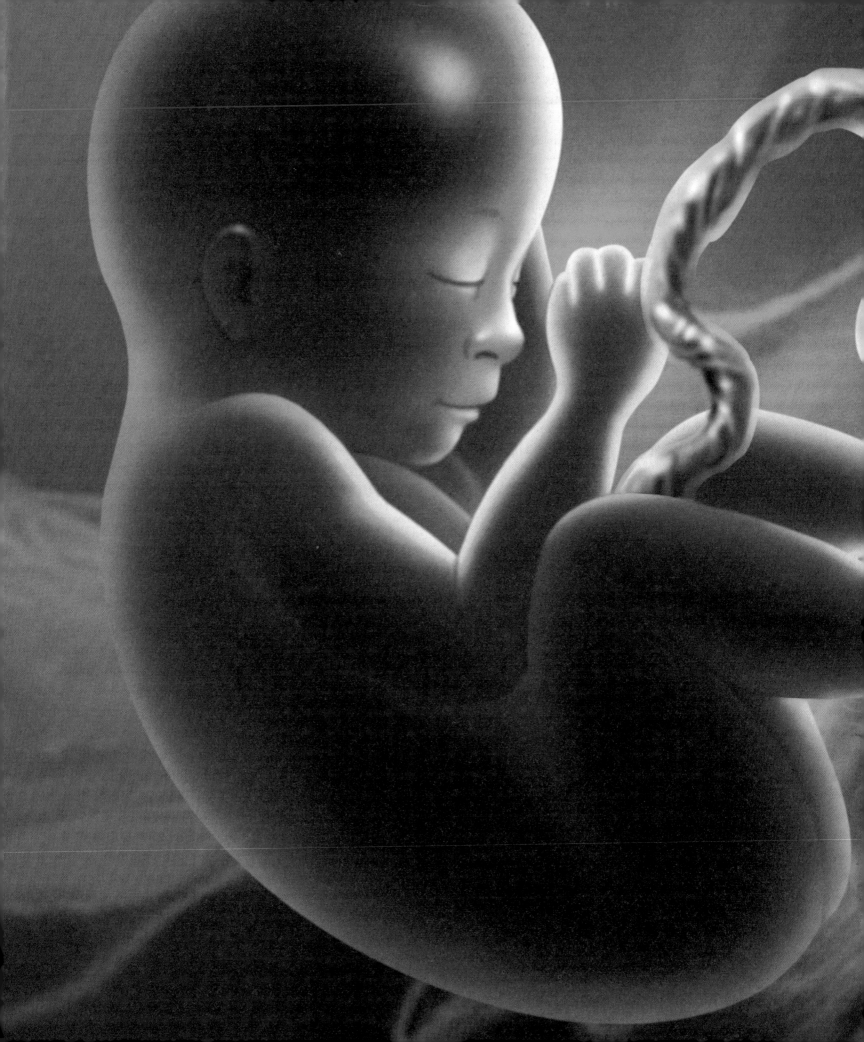

26 Reproduction and Development

26.1 Impacts/Issues: Mind-Boggling Births

In December of 1998, Nkem Chukwu of Houston, Texas, gave birth to six girls and two boys. They were the first set of human octuplets to be born alive. The births were premature. In total, all eight newborns weighed a bit more than 4.5 kilograms (10 pounds). Odera, the smallest, weighed about 300 grams (less than 1 pound). She died six days after birth, when her heart and lungs gave out. Two others required surgery. The seven survivors had to spend months in the hospital before going home (Figure 26.1). However, all are now in good health.

FIGURE 26.1 Outcome of fertility-enhancing drugs, seven survivors of a set of octuplets.

Why did octuplets form in the first place? Chukwu had trouble getting pregnant. Her doctors gave her hormone injections, which caused many of her eggs to mature and be released at the same time. When the doctors realized that she was carrying a large number of embryos, they suggested reducing the number. Chukwu chose instead to try to carry all of them to term. Her first child was thirteen weeks premature. The others were surgically delivered two weeks later.

Over the past two decades, the number of multiple births has increased by almost 60 percent. There have been four times as many higher-order multiple births (triplets or more). Increased use of fertility drugs is one cause for these increases. A woman's fertility peaks in her midtwenties. By thirty-nine, her chance of conceiving naturally has declined by about half. Yet the number of first-time mothers more than forty years old doubled in the past decade. Many of these women required fertility drugs to become pregnant, and such drugs raise the likelihood of multiple births.

Carrying more than one embryo increases the risk of miscarriage, premature delivery, or stillbirth. Multiple-birth newborns weigh less than normal and are more likely to have birth defects. For these reasons, many doctors consider the rise in multiple births a troubling trend.

In this chapter, we turn to one of life's most amazing dramas, animal reproduction and development. How does a single fertilized egg of a human—or frog or bird or any other animal—give rise to so many specialized kinds of cells? How does development yield an adult with all the complex tissues and organs discussed throughout this unit?

Answers to these questions will emerge as we consider the developmental processes common to all animals. We will also delve into the story of human reproduction, learning how humans develop from a single cell to an adult body with trillions of specialized cells.

How Animals Reproduce and Develop

Asexual Reproduction With **asexual reproduction**, a single individual makes offspring that are genetically identical to the parent and one another. As a result, the parent has all its genes represented in each offspring. Asexual reproduction can be advantageous in a stable environment. Gene combinations that make the parent successful can be expected to do the same for offspring. However, being locked into a particular gene combination may be a disadvantage if the environment is not stable.

Many invertebrates reproduce asexually. Some do this by fragmentation—a piece breaks off and grows into a new individual. In others, new individuals bud from existing ones (Figure 26.2**A**). Offspring of insects, fish, amphibians, and lizards develop from unfertilized eggs by a process called **parthenogenesis**. No mammals reproduce asexually.

Sexual Reproduction With **sexual reproduction**, two parents produce gametes that combine at fertilization. Each resulting offspring has a unique combination of paternal and maternal genes (Section 8.8).

Sexual reproducers have higher genetic and energetic costs than asexual reproducers. On average, only half of a sexually reproducing parent's genes end up in each offspring. However, most animals live where resources and threats change over time. In such environments, production of offspring that differ from both parents and from one another can be advantageous. By reproducing sexually, a parent increases the likelihood that some offspring will inherit a combination of genetic traits that suits them to a newly changed environment.

Some species switch between sexual and asexual reproduction. Plant-sucking insects called aphids offer one example. During the summer, a female aphid settles on a plant and gives birth to many wingless daughters that have developed inside the mother's body from unfertilized eggs. An aphid's daughters live on the same plant as their mother, or on an adjacent one, so they all experience more or less similar conditions. In autumn, males are produced and sexual reproduction yields genetically variable females. These females spend the winter in a resting state. In the spring, they seek out new plants on which to raise the new generation of genetically identical daughters.

Variations on Sexual Reproduction Sexually reproducing animals that produce both eggs and sperm are **hermaphrodites**. Tapeworms and some roundworms are simultaneous hermaphrodites. They produce eggs and sperm at the same time, and can fertilize themselves. Earthworms and slugs are simultaneous hermaphrodites too, but they swap gametes with a partner. So do barred hamlets, a type of marine fish (Figure 26.2**B**). During a bout of mating, hamlet partners take turns in the "male" and "female" roles. Other fishes are sequential hermaphrodites. They switch from one sex to another during the course of a lifetime. More typically, vertebrates have separate sexes that remain fixed for life. Each individual is either male or female.

Most aquatic invertebrates, fishes, and amphibians release gametes into the water, where they combine during external fertilization. As Section 15.6 explained, the amniotes have internal fertilization, meaning sperm and egg meet inside the female's body. In most reptiles and mammals, a specialized organ (a penis) delivers sperm into a female's reproductive tract (Figure 26.2**C**). Internal fertilization reduces the number of gametes that must be produced and, in some groups, allows offspring to develop inside the mother's body.

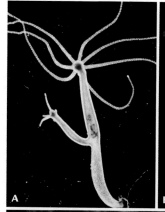

FIGURE 26.2 Animal reproduction. **A** Hydra reproducing asexually by budding. **B** Mating barred hamlets. The fish are hermaphrodites that fertilize eggs externally. During mating, each fish alternates between laying eggs and fertilizing its partner's eggs. **C** Male elephant inserting his penis into his female partner. Eggs will be fertilized inside the mother's body, and offspring will develop there.

asexual reproduction Reproductive mode by which offspring arise from one parent and inherit that parent's genes only.

hermaphrodite Individual animal that produces both eggs and sperm.

parthenogenesis Process by which individuals develop from unfertilized eggs.

sexual reproduction Reproductive mode by which offspring arise from two parents and inherit genes from both.

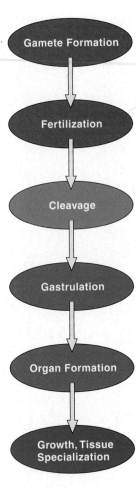

A Eggs form and mature in female reproductive organs. Sperm form and mature in male reproductive organs.

B A sperm penetrates an egg. Their nuclei fuse. A zygote has formed.

C Mitotic cell divisions form a ball of cells, a blastula. Each cell gets regionally different parts of the egg cytoplasm.

D A gastrula, an early embryo that has primary tissue layers, forms by cell divisions, cell migrations, and rearrangements.

E Details of the body plan fill in as different cell types interact and form tissues and organs in predictable patterns.

F Organs grow in size, take on mature form, and gradually assume specialized functions.

FIGURE 26.3 Overview of reproductive and developmental processes that occur in animals with tissues and organs.

blastula Hollow ball of cells that forms early in animal development.

cleavage Mitotic division of an animal cell.

differentiation Process by which cells become specialized; occurs as different cell lineages express different subsets of their genes.

ectoderm Outermost tissue layer of an animal embryo.

endoderm Innermost tissue layer of an animal embryo.

fertilization Egg and sperm combine and form a zygote.

gastrula Three-layered developmental stage formed by gastrulation in an animal.

gastrulation Animal developmental process by which cell movements produce a three-layered gastrula.

mesoderm Middle tissue layer of a three-layered animal embryo.

yolk Nutritious material in many animal eggs.

zygote Cell formed by fusion of gametes; first cell of a new individual.

Stages in Reproduction and Development

Animals develop through a series of stages. Figure 26.3 is an overview of the six stages of reproduction and development that occur in animals that have tissues and organs.

Reproduction begins with gamete formation. Gametes (eggs or sperm) form by meiosis of germ cells in primary reproductive organs (gonads). A sperm consists of paternal DNA and a bit of cellular equipment that helps it reach and penetrate an egg. There is little cytoplasm. An egg is much larger than a sperm and more complex internally. As it matures, enzymes, mRNA transcripts, and other materials are stored in its cytoplasm. Most animal eggs also include **yolk**, a mixture of lipids and proteins that nourishes a developing individual.

Fertilization begins when a sperm penetrates an egg. It ends when sperm and egg nuclei fuse, forming a **zygote**, the first cell of the new individual. A zygote's cytoplasm is almost entirely derived from the egg.

Developmental processes transform a single-celled zygote to an adult animal. Figure 26.4 show the development of one vertebrate, a frog. Development begins with **cleavage**, during which mitotic cell divisions increase the number of cells without adding to a zygote's original volume ❶. Because materials are not randomly distributed in the egg cytoplasm, cleavage distributes different mRNAs to different cells. Which of these "maternal messages" are inherited helps determine the fate of a cell lineage. For example, cells of one lineage alone might receive maternal mRNA for a regulatory protein that turns on the embryo's gene for a particular hormone.

Cleavage ends with the formation of the **blastula**, a hollow ball of cells around a fluid-filled center ❷. After the blastula forms, cell divisions slow and the blastula undergoes **gastrulation**. This is a process of structural organization, during which some cells on the blastula's surface move inward. These cellular movements produce a **gastrula**, an embryonic form with three primary tissue layers (also called germ layers) ❸. Cellular descendants of these primary tissues give rise to all tissues and organs of the adult.

Ectoderm, the outermost primary tissue, emerges first. It is the source of the nervous system and outer portions of the body coverings. Inner **endoderm** is the source of the gut's inner lining and organs derived from it. In most animals, **mesoderm** forms between the outer and inner primary tissue layers. It gives rise to muscles, most of the skeleton, the circulatory, reproductive, and excretory systems, and connective tissues of the gut and skin.

Because all cells of an embryo are descended from the same zygote, they all have the same number and kinds of genes. How then, do different kinds of cells arise? From gastrulation onward, different cell lineages express some groups of genes and not others. As Section 7.7 explained, selective gene expression begins the process of **differentiation**. By this process, cell lineages become specialized. For example, all cells use genes for enzymes that carry out glycolysis. However, only cells that will become the lens of the eye turn on a gene for crystallin, a transparent protein. Thus, the end of gastrulation and the beginning of cell differentiation marks the onset of organ formation ❹. Growth and tissue specialization is the sixth and final stage of animal development ❺. This stage usually continues into early adulthood.

Tissues and organs develop in a genetically programmed, orderly sequence. The body takes on its characteristic form as cells of various lineages divide, migrate, alter their size and shape, and die at specific times and in specific locations. For example, a frog larva (a tadpole) has a tail. During metamorphosis to the adult form, the death of cells in the tail makes the tail disappear. A similar process shapes a human hand. The hand starts out as a paddlelike structure with digits connected by webs of flesh. Death of cells in the webbing sculpts a five-fingered hand.

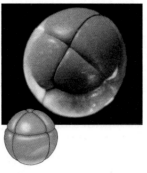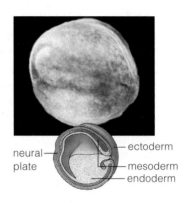

gray
crescent

blastocoel

blastula

1 Here we show the first three divisions of cleavage, a process that carves up the zygote's cytoplasm. In this species, cleavage results in a blastula, a ball of cells with a fluid-filled cavity.

2 Cleavage is over when the blastula forms.

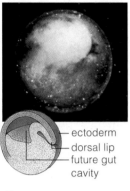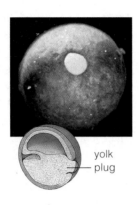

ectoderm
dorsal lip
future gut
cavity

yolk
plug

neural
plate

ectoderm
mesoderm
endoderm

neural
tube

notochord
gut cavity

3 The blastula becomes a three-layered gastrula by a process called gastrulation. At the dorsal lip, a fold of ectoderm above the first opening that appears in the blastula, cells migrate inward and start rearranging themselves.

4 Organs begin to form as a primitive gut cavity opens up. A neural tube, then a notochord and other organs form from the primary tissue layers.

Tadpole, a swimming larva with segmented muscles and a notochord extending into a tail.

Limbs grow and the tail is absorbed during metamorphosis to the adult form.

Sexually mature, four-legged adult leopard frog.

5 The frog's body form changes as it grows and its tissues specialize. The embryo becomes a tadpole, which metamorphoses into an adult.

> **FIGURE 26.4 Animated!**
> Stages in the development of a vertebrate, the leopard frog.

Take-Home Message **How do animals reproduce and develop?**

■ Some animals reproduce asexually, producing offspring identical to one another and to their parents. Most animals reproduce sexually, and so produce genetically variable offspring.

■ In sexually reproducing animals, gametes combine and form a zygote. Cleavage (mitotic divisions) produces a blastula. This hollow ball of cells undergoes gastrulation, a series of cellular rearrangement that yields a three-layered gastrula. Cells of the gastrula differentiate to form specialized tissues.

■ Cell divisions, migrations, shape changes, and deaths shape organs and give rise to the final body form.

The Human Reproductive System

We turn now to details of reproduction and development in our own species. In humans, as in other sexually reproducing animals, reproduction begins with gamete formation. Gametes form by meiosis inside a special pair of organs called gonads. Sperm form inside male gonads, the **testes** (singular, testis). Eggs form inside female gonads, the **ovaries**. The male reproductive tract also has components that store sperm and move them from the male's body into the female reproductive tract. A female's reproductive system has components that receive sperm and sustain developing offspring.

Male Reproductive System In both males and females, the gonads form deep inside the body. Before a male is born, the testes typically descend into the **scrotum**, a pouch suspended below the pelvic girdle (Figure 26.5). Smooth muscle inside this pouch encloses the testes. Reflexive contraction and relaxation of this muscle adjusts the position of the testes. When a man is cold or frightened, muscle contractions draw the testes closer to his body. When he feels warm, relaxation of muscle in the scrotum allows his testes to hang lower, so the sperm-making cells do not overheat. These cells function best just a bit below normal body temperature.

A male enters puberty (the stage of development when reproductive organs mature) sometime between the ages of 11 and 16 years. Testes enlarge and sperm production begins. Secretion of testosterone increases and leads to development of secondary sexual traits: thickened vocal cords that deepen the voice; increased growth of hair on the face, chest, armpits, and pubic region; and an altered distribution of fat and muscle.

Immature sperm formed in a testis are moved into the epididymis (plural, epididymides), a coiled duct perched on a testis. The Greek *epi–* means upon and *didymos* means twins. In this context, the "twins" refers to the two testes. Secretions from the epididymis wall nourish the sperm and help them mature.

The last region of each epididymis stores mature sperm and is continuous with the first portion of a **vas deferens** (plural, vasa deferentia). In Latin, *vas* means vessel, and *deferens*, to carry away. A vas deferens is a duct that carries sperm away from an epididymis, and to a short ejaculatory duct. Ejaculatory ducts deliver sperm to the urethra, the duct that extends through a male's penis to an opening at the body surface.

Beneath the skin of the penis, connective tissue encloses three elongated cylinders of spongy tissue. When a male becomes sexually excited, signals from the nervous system cause blood to flow into the spongy tissue faster than it flows out. As fluid pressure rises, the normally limp penis becomes erect.

Sperm stored in the epididymides typically continue their journey toward the body surface only when a male reaches the peak of sexual excitement and ejaculates. During **ejaculation**, smooth muscle in the walls of the epididymides and vasa deferentia undergoes rhythmic contractions that propel sperm and accessory gland secretions out of the body as a thick, white fluid called semen.

Semen is a complex mixture of sperm, proteins, nutrients, ions, and signaling molecules. Sperm constitute less than 5 percent of semen's volume; accessory glands secretions make up the other 95 percent. **Seminal vesicles**, exocrine glands near the base of the bladder, secrete fructose-rich fluid into the vasa deferentia. Sperm use fructose (a sugar) as their energy source. The **prostate gland**, an exocrine gland that encircles the urethra and secretes fluid into it, is the other major contributor to semen volume. Its secretions help raise the pH of

ejaculation Smooth muscle contractions that propel semen through the penis and out of the body.

ovary Organ in which eggs form.

prostate gland Exocrine gland that encircles a male's urethra; its secretions contribute to semen.

scrotum Pouch that holds the testes.

semen Sperm mixed with fluid secreted by exocrine glands.

seminal vesicles Exocrine glands that secrete fluid into the vasa deferentia; main source of semen volume.

testes In animals, organs that produce sperm.

vas deferens Main duct that conveys mature sperm toward the urethra.

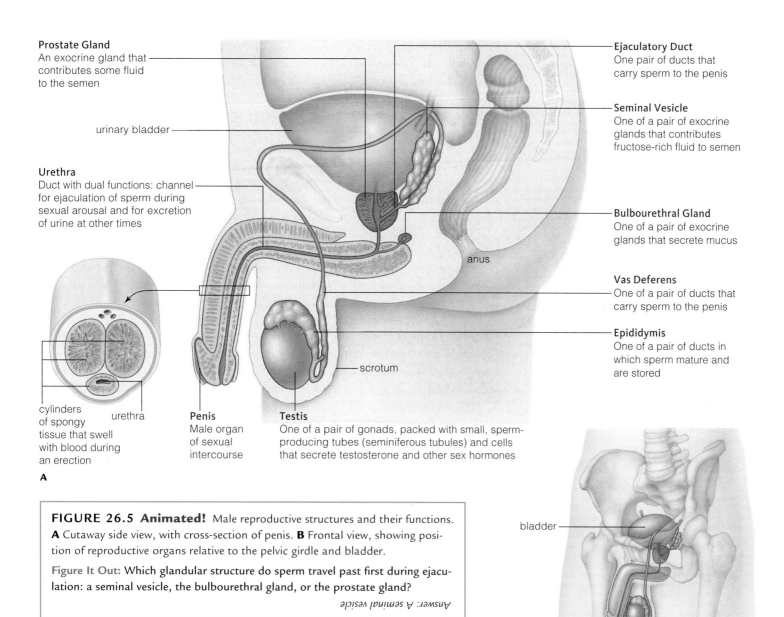

Prostate Gland
An exocrine gland that contributes some fluid to the semen

urinary bladder

Urethra
Duct with dual functions: channel for ejaculation of sperm during sexual arousal and for excretion of urine at other times

cylinders of spongy tissue that swell with blood during an erection

urethra

Penis
Male organ of sexual intercourse

Testis
One of a pair of gonads, packed with small, sperm-producing tubes (seminiferous tubules) and cells that secrete testosterone and other sex hormones

scrotum

anus

Ejaculatory Duct
One pair of ducts that carry sperm to the penis

Seminal Vesicle
One of a pair of exocrine glands that contributes fructose-rich fluid to semen

Bulbourethral Gland
One of a pair of exocrine glands that secrete mucus

Vas Deferens
One of a pair of ducts that carry sperm to the penis

Epididymis
One of a pair of ducts in which sperm mature and are stored

A

bladder

B

FIGURE 26.5 Animated! Male reproductive structures and their functions. **A** Cutaway side view, with cross-section of penis. **B** Frontal view, showing position of reproductive organs relative to the pelvic girdle and bladder.

Figure It Out: Which glandular structure do sperm travel past first during ejaculation: a seminal vesicle, the bulbourethral gland, or the prostate gland?

Answer: A seminal vesicle

the female reproductive tract, so sperm can swim more efficiently. Two pea-sized bulbourethral glands secrete mucus into the urethra. The mucus helps clear the urethra of urine prior to ejaculation.

In a young, healthy man, the prostate gland is about the size of a walnut. However, the prostate often enlarges with age. Because the urethra runs through the prostate gland, prostate enlargement can squeeze this tube and cause difficulty urinating. Medication, laser treatments, and surgery are used to shrink the prostate or create a tunnel through it and make room for the urethra.

Prostate enlargement can be a symptom of prostate cancer, the second most common fatal cancer in men (surpassed only by lung cancer). Many prostate cancers grow relatively slowly, but some grow fast and spread to other parts of the body. Risk factors for prostate cancer include advancing age, a diet rich in animal fats, smoking, and obesity. Genes also play a role. If a man has an affected father or brother, his own risk of prostate cancer doubles.

Doctors can diagnose prostate cancer by blood tests that detect increases in prostate-specific antigen (PSA) and by physical examination. Surgery and radiation therapy can cure cancers that are detected early.

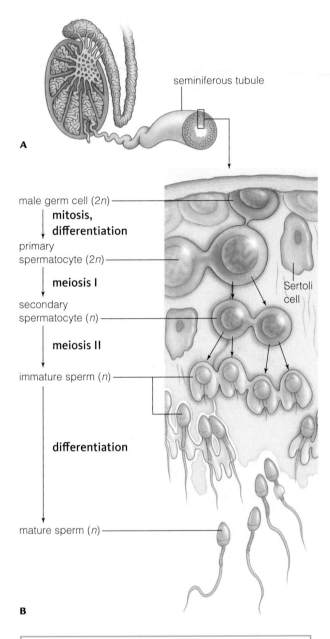

male germ cell (2n) ——

  **mitosis,
  differentiation**

primary
spermatocyte (2n) ——

  meiosis I

secondary
spermatocyte (n) ——

  meiosis II

immature sperm (n) ——

  differentiation

mature sperm (n) ——

A

seminiferous tubule

Sertoli
cell

B

FIGURE 26.6 Animated! Sperm formation.
A Location of seminiferous tubules in a testis.
B Steps in sperm formation inside a tubule.
C *Below*, a mature sperm.

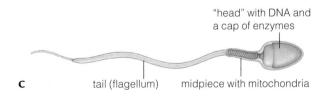

"head" with DNA and
a cap of enzymes

C   tail (flagellum)   midpiece with mitochondria

How Sperm Form

Although smaller than a golf ball, a testis holds coiled seminiferous tubules that would extend for 125 meters (longer than a football field) if stretched out (Figure 26.6**A**). Cells between the tubules secrete the hormone testosterone. Diploid male germ cells line the inner wall of the tubule (Figure 26.6**B**). These cells divide repeatedly by mitosis, and their offspring differentiate to become primary spermatocytes. Primary spermatocytes undergo meiosis I, forming secondary spermatocytes. Completion of meiosis yields immature sperm, which differentiate to form mature sperm. Sertoli cells, also in the tubules, metabolically support the developing sperm.

A mature sperm is a haploid, flagellated cell (Figure 26.6**C**). It uses the flagellum, or "tail," to swim toward an egg. Mitochondria in an adjacent midpiece supply the energy required for movement. A sperm's "head" is packed full of DNA and has an enzyme-containing cap. The enzymes in the cap help the sperm penetrate an egg during fertilization. Sperm formation takes about 100 days, from start to finish. An adult male makes sperm continually, so on any given day he has millions of them in various stages of development.

Hormones control gamete formation. Gonadotropin-releasing hormone (GnRH) is one of the hypothalamic hormones that targets the pituitary gland (Section 25.3). GnRH stimulates anterior pituitary cells to secrete luteinizing hormone (LH) and follicle-stimulating hormone (FSH). These two hormones have a role in both male and female reproductive function.

In males, LH and FSH target cells inside the testes. LH stimulates cells that lie between the seminiferous tubules to secrete testosterone. FSH targets Sertoli cells inside seminiferous tubules. In combination with testosterone, FSH causes Sertoli cells to produce substances that encourage sperm formation, development, and maturation.

Female Reproductive System

A female's ovaries lie deep inside her body (Figure 26.7). Each is about the size of an almond. A ciliated **oviduct** connects each ovary to the **uterus**, a hollow, pear-shaped organ above the urinary bladder. The uterus is commonly called the womb. If fertilization occurs, an embryo forms and develops in the uterus. A thick layer of smooth muscle makes up most of the uterine wall. The uterine lining consists of glandular epithelium, connective tissues, and blood vessels. The lowest portion of the uterus narrows, forming the **cervix**, which connects to the vagina.

The **vagina** is a muscular tube that extends from the cervix to the body's surface. The vagina functions as the female organ of intercourse and also as the birth canal in childbirth. Two pairs of skin folds enclose the surface openings of the vagina and urethra. Fatty tissue fills the pair of outer folds (the labia majora). Thin inner folds (the labia minora) have a rich blood supply and swell during sexual arousal.

The tip of the clitoris, a highly sensitive sex organ, is positioned between the two inner folds, just in front of the urethra. The clitoris and penis develop from the same embryonic tissue. Both have an abundance of highly sensitive touch receptors, and both swell with blood and become erect during sexual arousal.

Egg Formation and the Ovarian Cycle

Unlike male germ cells, female germ cells do not divide after birth. A girl is born with about 2 million primary oocytes in her ovaries. An **oocyte** is a general term for an immature egg. A primary oocyte is an immature egg that has entered meiosis but stopped in prophase I. When a female reaches puberty, her primary oocytes begin to mature, one at a time, in an approximately 28-day ovarian cycle.

Figure 26.8 shows how an oocyte matures over the course of this cycle. A primary oocyte and the cells around it constitute an **ovarian follicle** ❶. In the

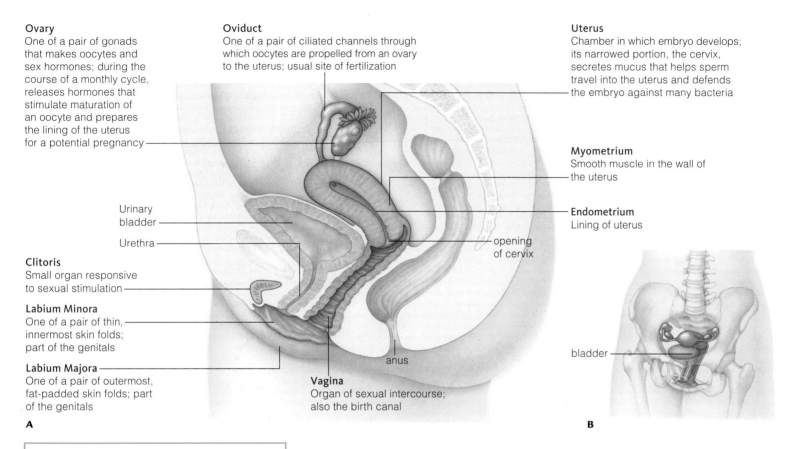

Ovary
One of a pair of gonads that makes oocytes and sex hormones; during the course of a monthly cycle, releases hormones that stimulate maturation of an oocyte and prepares the lining of the uterus for a potential pregnancy

Oviduct
One of a pair of ciliated channels through which oocytes are propelled from an ovary to the uterus; usual site of fertilization

Uterus
Chamber in which embryo develops; its narrowed portion, the cervix, secretes mucus that helps sperm travel into the uterus and defends the embryo against many bacteria

Myometrium
Smooth muscle in the wall of the uterus

Urinary bladder

Urethra

Endometrium
Lining of uterus

opening of cervix

Clitoris
Small organ responsive to sexual stimulation

Labium Minora
One of a pair of thin, innermost skin folds; part of the genitals

Labium Majora
One of a pair of outermost, fat-padded skin folds; part of the genitals

anus

Vagina
Organ of sexual intercourse; also the birth canal

bladder

A

B

FIGURE 26.7 Animated! Components of the human female reproductive system and their functions. **A** Cutaway side view. **B** Frontal view, showing position of reproductive organs relative to the pelvic girdle and bladder.

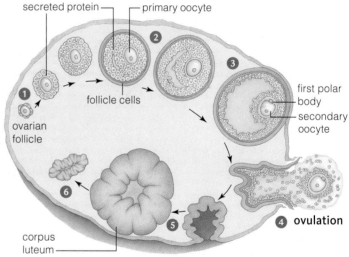

secreted protein   primary oocyte

follicle cells

ovarian follicle

first polar body

secondary oocyte

ovulation

corpus luteum

FIGURE 26.8 Animated! Changes in a follicle during one ovarian cycle.

first part of the ovarian cycle, cells around the oocyte divide repeatedly as the oocyte enlarges and secretes a layer of proteins ❷. As the follicle matures, a fluid-filled cavity opens around the oocyte. Often more than one follicle begins to develop, but typically only one becomes fully mature. In that follicle, the primary oocyte completes meiosis I and undergoes unequal cytoplasmic division. The result is a large secondary oocyte and a tiny polar body ❸. The secondary oocyte starts meiosis II, then halts in metaphase II. The polar body has no reproductive function and will later disintegrate.

About two weeks after the cycle began, the follicle wall ruptures and **ovulation** occurs: The secondary oocyte and polar body are ejected into the adjacent oviduct ❹. Once in the oviduct, an oocyte must meet up with sperm within 12 to 24 hours for fertilization to occur. The secondary oocyte will not complete meiosis until it has been penetrated by a sperm.

Meanwhile, back in the ovary, the cells of the ruptured follicle develop into a hormone-secreting structure called the **corpus luteum** ❺. (The name means "yellow body" in Latin and refers to its yellowish color.) If pregnancy does not occur, the corpus luteum breaks down ❻. Once the corpus luteum is gone, a new follicle can begin to mature.

cervix Narrowed region of uterus; connects to vagina.
corpus luteum Hormone-secreting structure that forms from follicle cells after ovulation.
oocyte Immature egg.
ovarian follicle Immature egg and surrounding cells.
oviduct Ciliated tube connecting an ovary to the uterus.
ovulation Release of a secondary oocyte from an ovary.
uterus Muscular chamber where offspring develop; womb.
vagina Female organ of intercourse and birth canal.

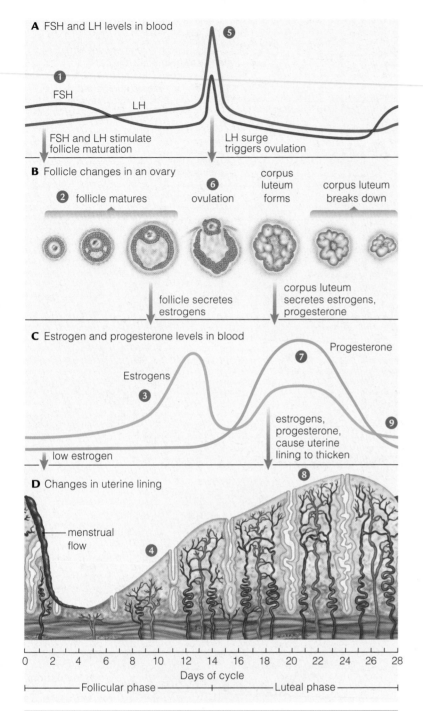

A FSH and LH levels in blood

FSH

LH

FSH and LH stimulate follicle maturation

LH surge triggers ovulation

B Follicle changes in an ovary

❷ follicle matures   ovulation   corpus luteum forms   corpus luteum breaks down

follicle secretes estrogens

corpus luteum secretes estrogens, progesterone

C Estrogen and progesterone levels in blood

Progesterone

Estrogens

❸

low estrogen

estrogens, progesterone, cause uterine lining to thicken

D Changes in uterine lining

menstrual flow

0 2 4 6 8 10 12 14 16 18 20 22 24 26 28
Days of cycle

Follicular phase —————— Luteal phase

FIGURE 26.9 Animated! Changes in a human ovary and uterus correlated with changing hormone levels. Onset of menstrual flow is considered Day 1 of the approximately 28-day menstrual cycle.

A,B Prompted by GnRH from the hypothalamus, the anterior pituitary secretes FSH and LH, which stimulate a follicle to grow and an oocyte to mature in an ovary. A midcycle surge of LH triggers ovulation and the formation of a corpus luteum. A decline in FSH after ovulation stops more follicles from maturing.

C,D Early on, estrogen from a maturing follicle encourages repair and rebuilding of uterine lining. After ovulation, the corpus luteum secretes some estrogen and more progesterone that primes the uterus for pregnancy. If pregnancy occurs, the corpus luteum will persist, and its secretions will stimulate the maintenance of the uterine lining.

Hormones and the Menstrual Cycle

Cyclic events in the ovaries are coordinated with cyclic changes in the uterus. We refer to the approximately monthly changes in the uterus as the **menstrual cycle**. The first day of the menstrual cycle is marked by the onset of **menstruation**: the flow of bits of uterine lining and a small amount of blood from the uterus, through the cervix, and out of the vagina.

Figure 26.9 shows how hormone levels and the thickness of the uterine lining change during the menstrual cycle. It also shows how those events are tied to the ovarian cycle.

Like testes, ovaries are under the control of GnRH. As the menstrual cycle begins, GnRH from the hypothalamus is making cells in the anterior lobe of the pituitary increase their secretion of FSH and LH ❶. As its name implies, follicle-stimulating hormone stimulates an ovarian follicle to begin maturing ❷. As a follicle matures, cells surrounding the oocyte secrete estrogens (a type of female sex hormone). Rising estrogens ❸ encourage the uterine lining to thicken ❹.

The pituitary detects the rising level of estrogens in the blood and responds with an outpouring of LH ❺. The LH surge encourages the primary oocyte to complete meiosis I and undergo cytoplasmic division. The LH surge also causes the follicle to swell and burst. Thus, a midcycle surge of LH is the trigger for ovulation ❻.

Immediately after ovulation, the estrogen level declines until the corpus luteum forms. The corpus luteum secretes some estrogen and a lot of progesterone (another type of female sex hormone) ❼. Estrogens and progesterone cause the uterine lining to thicken and encourage blood vessels to grow through it. The uterus is now ready for a pregnancy ❽.

If a pregnancy does not occur, the corpus luteum persists for about twelve days. The estrogen and progesterone it secretes keep the hypothalamus from secreting FSH, so no other follicles start to mature. When the corpus luteum begins to break down, estrogen and progesterone levels fall ❾. The hypothalamus senses this decline, and stimulates the pituitary to begin secreting FSH and LH once again. In the uterus, the decline in estrogens and progesterone causes thickened lining to break down, and menstruation begins.

Menstruation-Related Disorders

Many women occasionally experience some discomfort a week or so before they menstruate. Body tissues swell because premenstrual changes influence aldosterone secretion. This adrenal gland hormone stimulates reabsorption of sodium and, indirectly, water (Section 25.6). Breasts may become tender because hormonal changes cause the milk ducts inside them to enlarge. Cycle-induced hormonal changes can also cause depression, irritability, or anxiety. Headaches and sleeping problems are also common.

Regular recurrence of these symptoms is known as premenstrual syndrome (PMS). A balanced diet and exercise make PMS less likely and less severe. Taking oral contraceptives minimizes hormone swings and usually prevents PMS.

During menstruation, local signaling molecules (prostaglandins) stimulate contractions of smooth muscle in the uterine wall. Many women barely notice these contractions, but others experience a dull ache or sharp pain, commonly called menstrual cramps.

Menstruation usually lasts for five to seven days. Unusually heavy or persistent menstrual flow can be caused by fibroids. About one-third of women over age thirty have these benign tumors of the uterus. Most fibroids cause no symptoms, but some result in pain, long menstrual periods, and excessive bleeding. A woman who needs to change pads or tampons on an hourly basis should discuss this condition with her doctor. Fibroids can be surgically removed, usually without impairing fertility.

Excessive exercise or malnutrition can halt menstrual cycling. Usually, the cycle starts up again when physical activity lessens or diet improves. Menstrual cycles halt permanently and fertility ends when a woman completes **menopause**, usually at about age 50. Hormonal changes related to menopause cause many women to have hot flashes. A woman gets abruptly and uncomfortably hot, flushed, and sweaty as blood surges to her skin. Hormone replacement therapy (taking estrogen and in some cases progesterone) can relieve symptoms, but raises some health risks.

Take-Home Message **What are the roles of human reproductive organs?**

- A male's testes produce sperm and secrete the sex hormone testosterone. Sperm and secretions from accessory glands form semen. Semen is propelled through a series of ducts and leaves the body through an opening in the penis.

- A woman's ovaries make eggs and secrete sex hormones (estrogens and progesterone). Oviducts connect the ovaries to the uterus, where offspring develop. The vagina is the organ of intercourse and also the birth canal.

- In men, germ cells divide continually to produce sperm. By contrast, a woman is born with all the immature eggs she will ever have.

- From puberty until menopause, a woman's eggs mature—one at a time—in an approximately monthly cycle. During each cycle, the uterine lining thickens in preparation for pregnancy. If pregnancy does not occur, the woman sheds the uterine lining (menstruates), and the cycle begins again.

26.4 How Pregnancy Happens

Sexual Intercourse For males, intercourse requires an **erection**. Cylinders of spongy tissue make up the bulk of the penis. When a male is not sexually aroused, his penis remains limp, because the large blood vessels that transport blood to the spongy tissue are constricted. When a male becomes aroused, signals from the parasympathetic nervous system induce arteries supplying the spongy tissue to widen. Inward flow of blood now exceeds outward flow, and the increase in fluid pressure expands the internal chambers. As a result, the penis enlarges and stiffens, so it can be inserted into a female's vagina.

The ability to achieve and sustain an erection peaks during the late teens. As a man ages, he may have episodes of erectile dysfunction. With this disorder, the penis cannot stiffen enough for intercourse. Men who have circulatory problems are most often affected. Smoking also increases risk. The National Institutes of Health estimates that 30 million men are affected in the United States alone.

erection Penis becomes erect as a result of increased fluid pressure inside its spongy tissue.

menopause Of a human female, the end of fertility and of menstrual cycles.

menstrual cycle Approximately monthly cycle in which the uterine lining thickens in preparation for pregnancy, then is shed if pregnancy does not occur.

menstruation Flow of blood and bits of shed uterine lining out through the vagina.

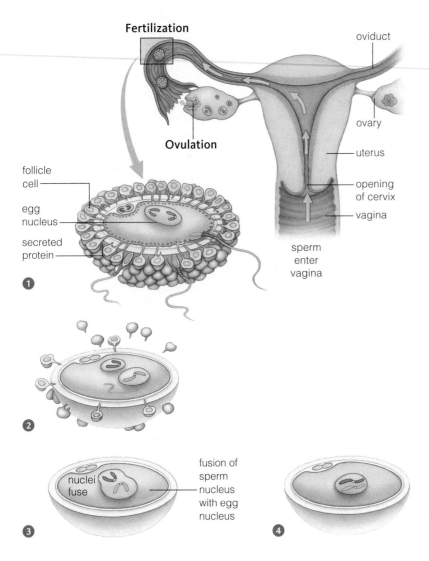

FIGURE 26.10 Animated! Fertilization.

❶ Sperm surround a secondary oocyte and release enzymes that digest their way through the protein layer around the oocyte.

❷ When a sperm penetrates the oocyte, granules in the egg cortex move to the surface and follicle cells are shed. Penetration of a sperm into the oocyte stimulates meiosis II in the oocyte's nucleus.

❸ The sperm tail degenerates; its nucleus enlarges and fuses with the oocyte nucleus.

❹ At nuclear fusion, a zygote forms and fertilization is completed.

Viagra and similar drugs prescribed for erectile dysfunction cause blood vessels that carry blood into the penis to widen and deliver more blood. Such drugs can cause headaches and (rarely) sudden hearing loss. They also interact with other drugs, and should never be taken without a prescription.

In a sexually excited female, the vaginal wall, labia, and clitoris swell with blood. Her cervix and glands near the entrance to the vagina increase their secretion of mucus.

During intercourse, mechanical stimulation of the male's penis and the female's clitoris and vaginal wall may lead to **orgasm**, or sexual climax. During orgasm, neurotransmitters flood the brain, causing sensations of release and pleasure. In females, muscles of the vaginal wall and pelvis contract rhythmically. In males, similar reflexive contractions force sperm and contents of seminal vesicles and the prostate gland into the urethra. These substances mix, forming semen, which is ejaculated in the vagina.

You may have heard that a female cannot become pregnant unless she reaches orgasm. Don't believe it!

Fertilization On average, an ejaculation will put 150 million to 350 million sperm into the vagina. Within thirty minutes some have made their way into the female's oviducts. Just a few hundred will survive the long journey to the upper region of the oviduct, where eggs usually are fertilized (Figure 26.10).

If sperm and oocyte meet, receptors in the plasma membrane of the sperm's head bind to proteins in the secreted layer that encloses the oocyte. The sperm's entry causes granules in the oocyte cortex (the region just beneath the plasma membrane) to move to the cell's surface. The granules contain enzymes that digest the outer portion of the protein coat. With this outer coat gone, other sperm have nothing to attach to and are unable to enter the oocyte.

Once inside the oocyte, a sperm breaks down, but its nucleus remains intact. Sperm entry prompts the secondary oocyte to complete meiosis II, forming a mature egg, or **ovum** (plural, ova), and a second polar body. The egg nucleus and sperm nucleus fuse, restoring the diploid number for the zygote, the first cell of the new individual.

Sometimes two eggs mature and are released at the same time, and each is fertilized by a different sperm. The result is fraternal twins. Such twins are no more similar than any other siblings.

Take-Home Message **How does a pregnancy occur?**

■ In preparation for intercourse, a male's penis becomes inflated by the inflow of blood and a female's reproductive tract becomes lubricated with mucus.

■ During orgasm, smooth muscles of the reproductive tract undergo rhythmic contractions. In males, ejaculation forces sperm out of the penis.

■ Fertilization usually occurs in the oviduct. Usually an oocyte changes after one sperm penetrates it, preventing other sperm from binding. The result of fertilization is a zygote, the first cell of the new individual.

26.5

Fertility and Reproductive Health

Contraception Methods used to prevent pregnancy are termed **contraception**. Table 26.1 lists common options and compares their effectiveness. The most effective option is abstinence—no sex— which is 100 percent effective. However, it can take great self-discipline. Rhythm methods are forms of abstinence; a woman simply avoids sex during her fertile period. She calculates when she is ovulating by recording how long her menstrual cycles last, checking her temperature each morning, monitoring the thickness of her cervical mucus, or by checking some combination of these indicators.

Withdrawal, or removing the penis from the vagina before ejaculation, requires great willpower. It also may be ineffective because sperm can be in pre-ejaculation fluids. Similarly, rinsing out the vagina (douching) immediately after intercourse is unreliable. Typically, some sperm swim through the cervix within seconds of ejaculation.

Surgical methods are highly effective, but are meant to make a person permanently sterile. Men may opt for a vasectomy. A doctor makes a small incision into the scrotum, then cuts and ties off each vas deferens. A tubal ligation blocks or cuts a woman's oviducts.

Other fertility control methods use physical and chemical barriers to stop sperm from reaching an egg. Spermicidal foams and jellies poison sperm. They are not always reliable, but their use with a condom or diaphragm reduces the chance of pregnancy. A diaphragm is a flexible, dome-shaped device that is positioned inside the vagina so it covers the cervix. A diaphragm is relatively effective if it is first fitted by a doctor and used correctly with a spermicide. A cervical cap is a similar but smaller device. Condoms are thin, tight-fitting sheaths worn over the penis during intercourse. Latex ones have the advantage of also protecting against sexually transmitted diseases (STDs). However, all condoms can tear or leak.

An intrauterine device, or IUD, is inserted into the uterus by a physician. Some IUDs thicken cervical mucus so sperm cannot swim through it. Others shed copper, which keeps an early embryo from implanting in the uterus.

The birth control pill is the most common fertility control method in developed countries. "The Pill" is a mix of synthetic estrogens and progesterone-like hormones that prevents both maturation of oocytes and ovulation. When taken consistently, it is highly effective. Its use lowers risk of ovarian and uterine cancer but raises risk of breast, cervical, and liver cancer. A birth control patch is a small, flat adhesive patch applied to skin. It delivers the same mixture of hormones as an oral contraceptive, and it blocks ovulation the same way.

Both birth control pills and the birth control patch raise the risk of stroke and other serious cardiovascular disorders. The risk of these complications is greatest in older women and women who smoke.

Table 26.1	Common Methods of Contraception	
Method	Description	Pregnancy Rate*
Abstinence	No intercourse	0% per year
Rhythm method	Avoid intercourse in female's fertile period	25% per year
Withdrawal	End intercourse before male ejaculates	27% per year
Douche	Wash semen from vagina after intercourse	60% per year
Vasectomy	Cut or close off male's vasa deferentia	>1% per year
Tubal ligation	Cut or close off female's oviducts	>1% per year
Condom	Enclose penis, block sperm entry to vagina	15% per year
Diaphragm, cervical cap	Cover cervix, block sperm entry to uterus	16% per year
Spermicides	Kill sperm	29% per year
Intrauterine device	Prevent sperm entry to uterus or prevent implantation of embryo	>1% per year
Oral contraceptives	Prevent ovulation	>1% per year
Hormone patches, implants, or injections	Prevent ovulation	>1% per year
Emergency contraception pill	Prevent ovulation	15–25% per use**

* Percent of users who become pregnant with consistent, correct usage.

** Not meant for regular use. 15–25 percent of users become pregnant despite use of pill.

Typically, some sperm swim through the cervix within seconds of ejaculation.

contraception Methods used to prevent pregnancy.
orgasm Sexual climax during which neurotransmitter release causes smooth muscle contractions accompanied by a feeling of release and pleasure.
ovum Mature egg.

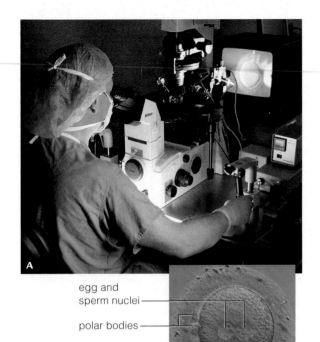

egg and sperm nuclei

polar bodies

FIGURE 26.11

Assisted reproduction. **A** Doctor using a micromanipulator to inject a human sperm into an egg. The monitor shows what he sees in the microscope. **B** Fertilized human egg produced by *in vitro* fertilization.

Hormone injections or implants prevent ovulation. Each injection acts for several months, whereas the implant lasts for three years. Both methods are quite effective, but may cause sporadic, heavy bleeding.

Some women turn to emergency contraception after a condom tears, or after unprotected consensual sex or rape. "Morning-after pills" are now available without a prescription to women over age 18. They prevent ovulation and work best if taken immediately after intercourse. However, they can be effective up to three days later. The pills are not meant to be used on a regular basis. Nausea, vomiting, abdominal pain, headache, and dizziness are side effects. Insertion of a copper IUD can also be used as a form of emergency contraception. Like any other use of an IUD, this requires the assistance of a physician.

Abortion About 10 percent of women who learn that they are pregnant suffer a spontaneous abortion, or miscarriage. The risk is higher for women older than age 35. Many more pregnancies end without ever being detected. It is estimated that about 50 percent of fertilized eggs are spontaneously aborted, most often because they have genetic defects.

An induced abortion is the deliberate dislodging and removal of an embryo or a fetus from the uterus. In the United States, about half of unplanned pregnancies end in induced abortion. From the clinical standpoint, abortion usually is a brief, safe, and relatively painless procedure, especially in the first trimester.

The drug mifepristone (RU-486) together with a prostaglandin can be used to induce abortion during the first nine weeks of pregnancy. The two chemicals interfere with progesterone receptors in the uterus. The uterine lining cannot be maintained, and neither can pregnancy.

Assisted Reproduction About 15 percent of couples in the United States are infertile; the woman either does not become pregnant or repeatedly miscarries. When a couple make normal sperm and oocytes but cannot conceive naturally they may turn to ***in vitro* fertilization**. By this process, an egg and sperm are combined outside the body. The process was first carried out in 1978, and has since produced about 45,000 pregnancies in the United States alone.

In preparation for *in vitro* fertilization, a woman receives injections of a drug that increases the number of oocytes that mature during a cycle. The physician then harvests mature oocytes from the woman's body. Often sperm are injected into the harvested oocytes (Figure 26.11). This overcomes any problems with sperm motility. The zygotes that result from fertilization are allowed to divide, forming small cell clusters. One or more of these clusters are then transferred to a woman's uterus to undergo development.

Assisted reproduction attempts are costly and most fail. In 30-year-old women, about one-third of *in vitro* attempts result in a birth. In 40-year-olds, only one in six attempts succeeds.

Overview of Sexually Transmitted Disease Each year, pathogens that cause **sexually transmitted diseases**, or STDs, infect about 15 million Americans (Table 26.2). Two-thirds of those infected are under age twenty-five; one-quarter are teenagers. More than 65 million Americans have a viral STD that cannot be cured. The social consequences of STDs are enormous. Women are more easily infected than men, and they develop more complications. For example, pelvic inflammatory disease (PID), a secondary outcome of bacterial STDs, scars the female reproductive tract and can cause infertility, chronic pain, and tubal pregnancies (Figure 26.12**A**). Infected women commonly pass chlamydia on to their newborn (Figure 26.12**B**). Besides causing skin sores, type 2 herpes virus kills half of infected fetuses and causes neural defects in one-fourth of the rest.

Table 26.2	New STD Cases Annually	
STD	In the U.S.	Worldwide
HPV infection	6,200,000	400,000,000
Trichomoniasis	7,400,000	174,000,000
Chlamydia	3,000,000	92,000,000
Genital herpes	1,000,000	24,000,000
Gonorrhea	650,000	62,000,000
Syphilis	70,000	12,000,000
HIV infection	56,000	2,500,000

Common STDs Infection by human papillomaviruses (HPV) is one of the most widespread and fastest spreading STDs in the United States. Of about 100 HPV strains, a few cause genital warts. In women, the bumplike growths form on the vagina, cervix, and genitals. In men, they form on the penis and scrotum. A few strains of HPV are the main cause of cervical cancer. Sexually active females should have an annual Pap smear, in which cells from the cervix are examined using a microscope. A recently approved vaccine can prevent HPV infection if given before viral exposure (Section 22.1).

Trichomoniasis is caused by the flagellated protozoan *Trichomonas vaginalis*. Symptoms in women include a yellowish discharge, soreness, and itching of the vagina. Men are usually symptom-free. In both sexes, untreated infections cause infertility. A single dose of an antiprotozoal drug can quickly cure the infection. Both sexual partners must receive treatment to prevent reinfection.

Chlamydial infection is primarily a young person's disease, because 40 percent of those infected are still in their teens. A bacterium, *Chlamydia trachomatis*, causes the disease, and antibiotics quickly kill it. Most infected women have no symptoms and are not even diagnosed. Half of infected men have discharge from the penis and painful urination. Untreated men risk inflammation of the epididymides and infertility.

About 45 million Americans have genital herpes, which is caused by type 2 herpes simplex virus. An initial infection commonly causes small sores at the site of infection. The sores heal, but once a person has been infected, he or she harbors the virus for life. In most people, the infection periodically becomes reactivated, causing tingling or itching, which may or may not be accompanied by visible sores. Antiviral drugs can promote healing of sores caused by an active infection and lessen the likelihood of viral reactivation.

The bacterium *Neisseria gonorrhoeae* is the cause of gonorrhea. Less than one week after a male is infected, yellow pus oozes from the penis. Urination becomes frequent and painful. An infected woman usually has few symptoms at first, but the pathogen can infect oviducts, causing cramps, scarring, and sterility. Oral contraceptives promote infection by altering vaginal pH. Altered pH causes the number of resident bacteria in the vagina to decline, allowing *N. gonorrhoeae* to move in. Even though antibiotic treatment quickly cures gonorrhea, the disease remains common. One reason may be a tendency of women to overlook the slight symptoms that arise in the early stages of the disease.

Syphilis is caused by a spiral-shaped bacteria (*Treponema pallidum*). Painless chancres (skin ulcers) appear in the primary stage. They heal, but the bacteria multiply inside the spinal cord, brain, eyes, bones, joints, and mucous membranes. More chancres form in the infectious secondary stage (Figure 26.12**C**). Continued infection can damage the liver, bones, and other internal organs. Chronic immune reactions to the bacteria may severely damage the brain and spinal cord, leading to blindness and paralysis. The incidence of syphilis declined in the 1990s, but began rising in 2000 and continues to increase.

Infection by HIV (or human immune deficiency virus) can cause AIDS (acquired immunodeficiency syndrome), in which the immune system weakens and infectious organisms take hold. Section 13.6 described the HIV virus and Section 22.8 explained how its effects on the immune system lead to AIDS. With regard to sexual transmission, oral sex is least likely to transmit an infection. Unprotected anal sex is 5 times more dangerous than unprotected vaginal sex and 50 times more dangerous than oral sex. To reduce the risk of HIV transmission during vaginal or anal sex, doctors recommend use of a latex condom and a lubricant. The lubricant helps prevent small abrasions that could allow the virus to enter the body. If you think you could have been exposed to HIV, get tested as soon as possible. Early treatment may prevent development of AIDS.

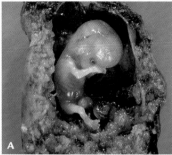

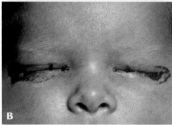

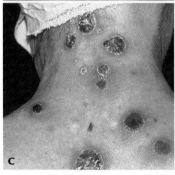

FIGURE 26.12 A few consequence of STDs.

A Increased risk of tubal pregnancy. Scarring caused by STDs can cause an embryo to implant itself in an oviduct. Untreated tubal pregnancies can rupture an oviduct and cause bleeding, infection, and death.

B An infant with chlamydia-inflamed eyes. The bacterial pathogen passed from mother to child during the birth process.

C Chancres (open sores) caused by syphilis.

in vitro **fertilization** Infertility treatment in which gametes are combined outside the body.

sexually transmitted disease Infectious disease caused by a pathogen that can live in the reproductive tract.

Human Prenatal Development

Development that occurs before birth is referred to as prenatal development. In humans, prenatal development typically lasts for 38 weeks after fertilization. During that time, cell divisions transform a single-celled zygote into a newborn with about a trillion cells of many different types.

Cleavage and Implantation Fertilization of a human egg typically occurs in one of the oviducts (Figure 26.13). Cleavage starts within a day or two of fertilization, as a zygote travels through the oviduct toward the uterus ❶, ❷. After four divisions, there is a cluster of sixteen cells called a morula ❸. (Morula is the Latin word for mulberry, which the cluster resembles.)

A **blastocyst**, the mammalian version of a blastula, forms by the fifth day ❹. It consists of a few hundred cells arranged as an outer cell layer, a cavity filled with their secretions (a blastocoel), and an inner cell mass. The embryo develops from the inner cell mass. The outer cells will form the membranes that surround the embryo.

By about six days after fertilization, the blastocyst is in the uterus ❺, where it expands by cell divisions and uptake of fluid. Implantation begins when the blastocyst attaches to the lining of the uterus (the endometrium) and burrows into it. During implantation, the inner cell mass develops into two flattened layers of cells called the embryonic disk ❻.

amnion Extraembryonic membrane that encloses an amniote embryo and the amniotic fluid.

blastocyst Mammalian version of the blastula.

chorion Outermost extraembryonic membrane of amniotes; major component of the placenta in placental mammals.

human chorionic gonadotropin (HCG) Hormone produced by the chorion and later by the placenta; encourages maintenance of the uterus, thus sustaining a pregnancy.

placenta Of placental mammals, organ composed of maternal and embryonic tissues that allows the exchange of substances between a developing individual and its mother.

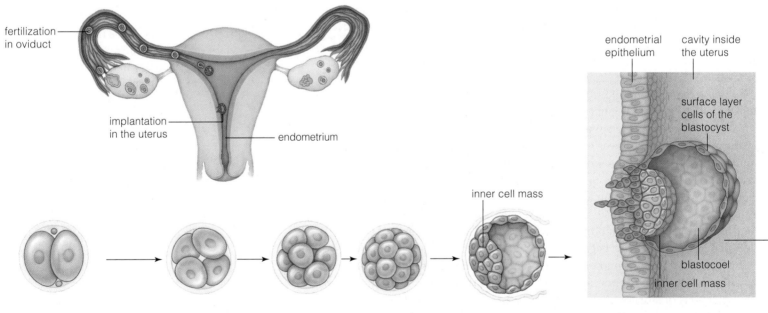

fertilization in oviduct

implantation in the uterus

endometrium

endometrial epithelium

cavity inside the uterus

surface layer cells of the blastocyst

inner cell mass

blastocoel

inner cell mass

❶ DAYS 1–2. The first cleavage furrow extends between the two polar bodies. Later cuts are angled, so cells become asymmetrically arranged. Until the eight-cell stage forms, they are loosely organized, with space between them.

❷ DAY 3. After the third cleavage, cells abruptly huddle into a compacted ball, which tight junctions among the outer cells stabilize. Gap junctions formed along the interior cells enhance intercellular communication.

❸ DAY 4. By 96 hours there is a ball of sixteen to thirty-two cells shaped like a mulberry. It is a morula (after *morum*, Latin for mulberry). Cells of the surface layer will function in implantation and will give rise to a membrane, the chorion.

❹ DAY 5. A blastocoel (fluid-filled cavity) forms in the morula as a result of surface cell secretions. By the thirty-two-cell stage, differentiation is occurring in an inner cell mass that will give rise to the embryo proper. This embryonic stage is the blastocyst.

❺ DAYS 6–7. Some of the blastocyst's surface cells attach themselves to the endometrium and start to burrow into it. Implantation has started.

actual size

Humans are amniotes. Thus human development involves formation of the same four extraembryonic membranes described in Section 15.6—the amnion, allantois, yolk sac, and chorion.

As a human blastocyst begins to implant, a fluid-filled amniotic cavity appears between the embryonic disk and the blastocyst surface. Many cells migrate around the wall of the cavity and form the **amnion**, a membrane that will enclose the embryo. Fluid in the amniotic cavity acts as a buoyant cradle in which an embryo grows and moves freely, protected from temperature changes and mechanical impacts.

As the amnion is forming, other cells move around the inner wall of the blastocyst, forming the lining of a yolk sac. In animals with shelled eggs, this sac holds yolk. A human yolk sac does not hold yolk and does not support the embryo nutritionally. Some cells that form in a human yolk sac give rise to the embryo's first blood cells. Others migrate to the gonads and become germ cells.

After a human blastocyst has implanted, an outpouching of the yolk sac will become the allantois. Part of the allantois functions as the embryo's urinary bladder and part will be a component of the placenta. The **placenta**, an organ composed of embryonic and maternal tissues, allows the exchange of substances between a mother and her developing offspring. The outermost membrane, the **chorion**, has tiny fingerlike projections (chorionic villi) that extend into the lining of the uterus **7**, **8**. They anchor a blastocyst and form part of the placenta.

Implantation of a blastocyst in the uterus prevents menstruation because the chorion secretes **human chorionic gonadotropin (HCG)**. This is the first hormone that a new human produces and it prevents degeneration of the corpus luteum. By the beginning of the third week, HCG can be detected in a mother's blood or urine. At-home pregnancy tests have a "dip stick" that changes color when exposed to urine that contains the embryonic hormone HCG.

At-home pregnancy tests have a "dip stick" that changes color when exposed to urine containing HCG, a hormone produced by the early embryo.

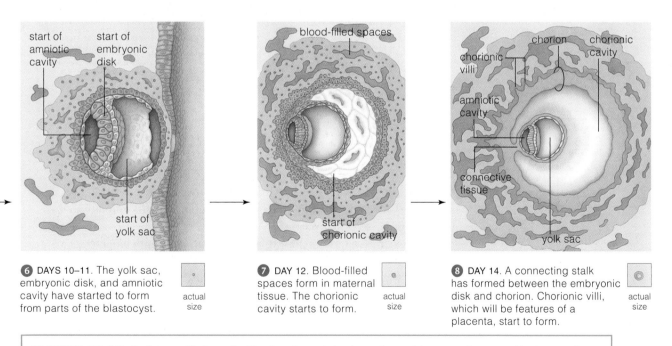

6 DAYS 10–11. The yolk sac, embryonic disk, and amniotic cavity have started to form from parts of the blastocyst.
actual size

7 DAY 12. Blood-filled spaces form in maternal tissue. The chorionic cavity starts to form.
actual size

8 DAY 14. A connecting stalk has formed between the embryonic disk and chorion. Chorionic villi, which will be features of a placenta, start to form.
actual size

FIGURE 26.13 Animated! From fertilization through implantation. A blastocyst forms, and its inner cell mass becomes an embryonic disk two cells thick. It will later become the embryo. Three extraembryonic membranes (the amnion, chorion, and yolk sac) start forming. A fourth membrane (allantois) forms after implantation is over.

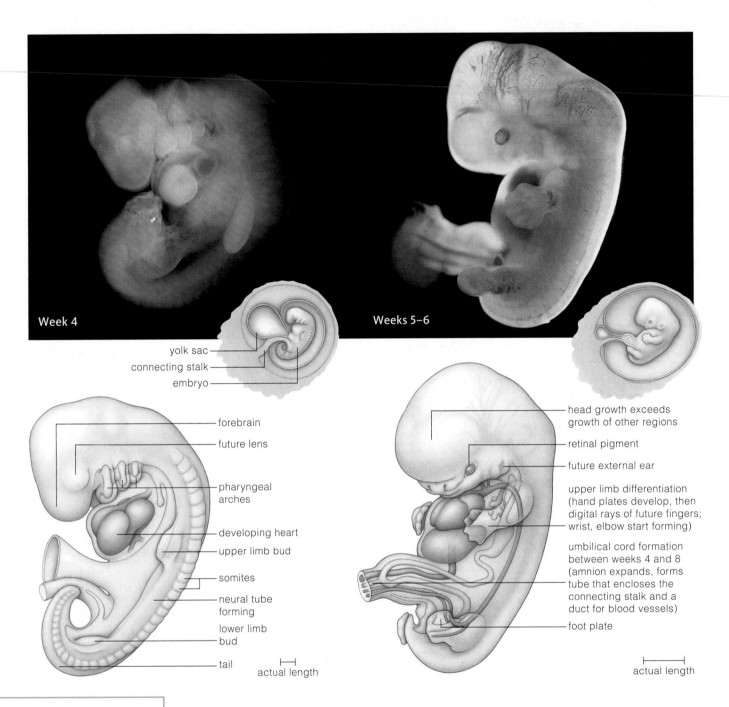

Week 4

Weeks 5–6

yolk sac
connecting stalk
embryo

forebrain

future lens

pharyngeal
arches

developing heart

upper limb bud

somites

neural tube
forming

lower limb
bud

tail

actual length

head growth exceeds
growth of other regions

retinal pigment

future external ear

upper limb differentiation
(hand plates develop, then
digital rays of future fingers;
wrist, elbow start forming)

umbilical cord formation
between weeks 4 and 8
(amnion expands, forms
tube that encloses the
connecting stalk and a
duct for blood vessels)

foot plate

actual length

FIGURE 26.14 Human embryonic and fetal development.

Figure It Out: Is a human embryo ever as big as your fist?

Answer: No. By the time the developing individual is as big as a fist, it is a fetus.

Embryonic Development Gastrulation occurs about two weeks after fertilization. Cells from the outer layer of the embryonic disk migrate inward and form a three-layered embryo.

By the fourth week, the neural tube and notochord that characterize all vertebrate embryos have formed (Figure 26.14). Multiple paired bumps of mesoderm called somites become visible. These are the embryonic sources of most bones, of skeletal muscles of the head and trunk, and of the dermis overlying these regions. Pharyngeal arches appear at an embryo's head end. They look like the arches that give rise to fish gills, but they do not ever support gills. They develop into bone, cartilage, and muscles of the face, ears, larynx, and pharynx.

When the fourth week ends, the embryo is 500 times its starting size. Now the pace slows as details of the embryo's organs begin to fill in. Limbs form as toes and fingers are sculpted from paddles.

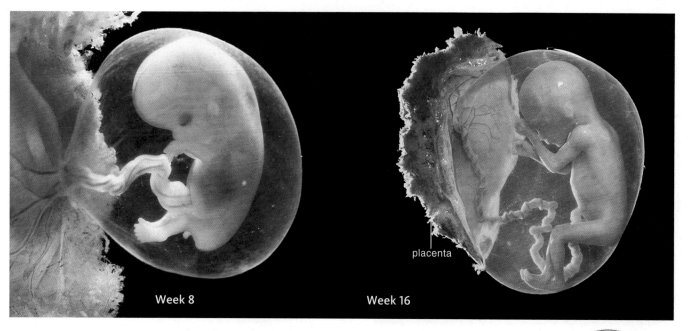

placenta

Week 8   Week 16

final week of embryonic
period; embryo looks
distinctly human
compared to other
vertebrate embryos

upper and lower limbs well
formed; fingers and then
toes have separated

primordial tissues of
all internal, external
structures now developed

tail has become stubby

actual length

Length: 16 centimeters
          (6.4 inches)
Weight: 200 grams
          (7 ounces)

WEEK 29
Length: 27.5 centimeters
          (11 inches)
Weight: 1,300 grams
          (46 ounces)

WEEK 38 (full term) ——————→
Length: 50 centimeters
          (20 inches)
Weight: 3,400 grams
          (7.5 pounds)

During fetal period, length
measurement extends
from crown to heel (for
embryos, it is the longest
measurable dimension, as
from crown to rump).

Fetal Development At the end of the eighth week, all organ systems
have formed and the developing individual is now considered a **fetus**. A fetal
heartbeat can be detected at about five months, and a mother usually begins to
feel reflexive fetal movements about five to six months after fertilization. Soft,
fuzzy hair, the lanugo, covers the fetus. The fetal skin is wrinkled, reddish, and
protected from abrasion by a thickened, cheeselike coating. Eyelids form and the
eyes open in the seventh month.

The optimal birthing time is about 38 weeks after the estimated time of
fertilization. A fetus born prematurely (before 22 weeks) cannot survive. Births
before 28 weeks (7 months) are risky, mainly because the lungs have not fully
developed. The risks start to drop after this. By 36 weeks, the survival rate is 95
percent. However, a fetus born between 36 and 38 weeks still has some trouble
breathing and maintaining a normal body temperature.

fetus Developing human between 9 weeks of age and birth.

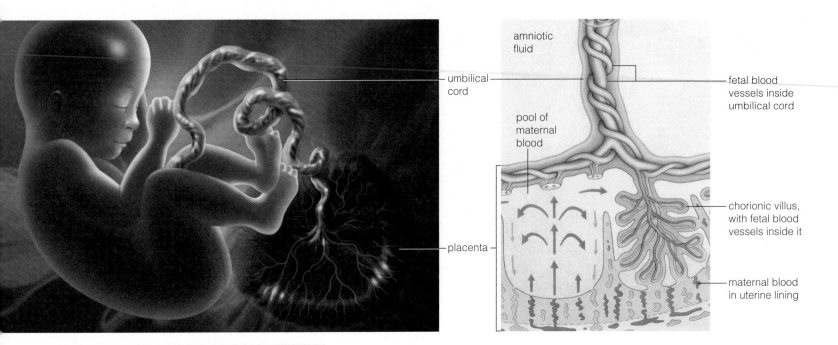

amniotic fluid

umbilical cord

fetal blood vessels inside umbilical cord

pool of maternal blood

placenta

chorionic villus, with fetal blood vessels inside it

maternal blood in uterine lining

FIGURE 26.15 Animated! The life-support system of a developing human. *Left*, artist's depiction of the view inside the uterus, showing a fetus connected by an umbilical cord to the pancake-shaped placenta. *Right*, the structure of the placenta. It consists of maternal and fetal tissue. Fetal blood flowing in vessels of chorionic villi exchanges substances by diffusion with maternal blood around the villi. However, the bloodstreams do not mix.

Functions of the Placenta All exchange of materials between an embryo and its mother takes place by way of the placenta, a pancake-shaped, blood-engorged organ made of uterine lining and extraembryonic membranes (Figure 26.15).

The placenta begins forming early in pregnancy. By the third week, maternal blood has begun to pool in spaces in the lining of the uterus tissue. Chorionic villi—the tiny fingerlike projections from the chorion—grow into the pools of maternal blood. Embryonic blood vessels extend through the umbilical cord to the placenta, and into the chorionic villi, where they are surrounded by the pooled maternal blood. The maternal and embryonic bloodstreams never mix. Instead, substances move between maternal and embryonic blood by diffusing across the walls of the embryonic vessels in the chorionic villi. Oxygen and nutrients diffuse from pooled maternal blood into embryonic vessels in the villi. Wastes diffuse the other way, and the mother's body disposes of them.

The placenta also has a hormonal role. From the third month on, it produces large amounts of HCG, progesterone, and estrogens. These hormones encourage the ongoing maintenance of the uterine lining.

Maternal Effects on Prenatal Development An embryo or fetus depends on its mother to supply nutrients. It is also exposed to pathogens and toxins that enter the mother's body.

A pregnant woman must eat enough to gain 20 to 25 pounds, on average. If she does not, her newborn may be seriously underweight. A low birth weight is correlated with a higher risk of postdelivery complications and impaired brain function. Supplements of some minerals and vitamins can help ensure that development proceeds normally. For example, taking B-complex vitamins during early pregnancy helps reduce the risk of severe neural tube defects in the embryo. Folate (folic acid) is especially important in this regard.

Some pathogens can cross the placenta. Diseases are especially dangerous in the first six weeks after fertilization, when organ systems are still forming. For example, if a pregnant woman gets rubella (German measles) in this critical period, there is a 50 percent chance that some of her child's organs will not form properly. For instance, if she is infected while embryonic ears are forming, her newborn may be deaf. If she is infected from the fourth month of pregnancy onward, the disease will have no notable effect. A woman can avoid risks associates with rubella by getting vaccinated against the virus before pregnancy.

Another pathogen that can affect development is a protist that sometimes lurks in garden soil, cat feces, and undercooked meat. It causes toxoplasmosis. The disease often does not cause symptoms, so a pregnant woman may become infected and not realize it. If the parasite crosses the placenta, it can infect her child and lead to developmental problems, a miscarriage, or stillbirth. To minimize the risk, pregnant women should eat well-cooked meat and avoid cat feces. Let someone else change that litter box.

Alcohol passes across the placenta, so when a pregnant woman drinks alcohol, her embryo or fetus feels the effects. Alcohol exposure can cause fetal alcohol syndrome, or FAS. A small head and brain, facial abnormalities, slow growth, mental impairment, heart problems, and poor coordination characterize affected infants (Figure 26.16). The damage is permanent. Children affected by FAS never catch up, physically or mentally. As a result, most doctors now advise women who are pregnant or attempting to become pregnant to avoid alcohol.

Caffeine interferes with nervous system development in animals, and may also harm human embryos. A recent study showed that women who took in 200 milligrams a day of caffeine (the amount in one and a half cups of coffee) had twice as many miscarriages as those who avoided caffeine. The study's authors advise pregnant women to chose decaffeinated or caffeine-free beverages.

Smoking or exposure to secondhand smoke increases the risk of miscarriage and adversely affects fetal growth and development. Carbon monoxide in the smoke can outcompete oxygen for the binding sites on hemoglobin, so the embryo or fetus of a smoker gets less oxygen than that of a nonsmoker. In addition, levels of the addictive stimulant nicotine in amniotic fluid can be even higher than those in the mother's blood.

Some medications cause birth defects. Isotretinoin (Accutane), a highly effective treatment for severe acne, is often prescribed for young women. If taken early in a pregnancy, it can cause heart problems or facial and cranial deformities in the embryo. Some antidepressants also increase the risk of birth defects. Paroxetine (Paxil) and related drugs inhibit the reuptake of serotonin. Use of these drugs during early pregnancy increases the likelihood of heart malformations. Taking them later in pregnancy increases risk that an infant will have fatal heart and lung disorders.

Highly addictive drugs such as methamphetamine and cocaine also cross the placenta. They increase the risk of a prenatal stroke and decrease birth weight.

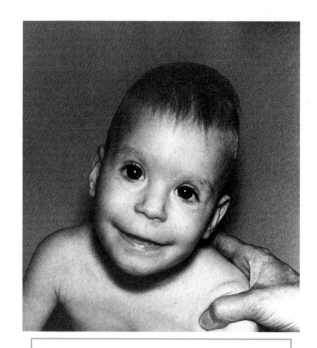

FIGURE 26.16 Child with fetal alcohol syndrome. Obvious symptoms are low and prominently positioned ears, improperly formed cheekbones, and an abnormally wide, smooth upper lip. Growth-related complications, heart problems, and nervous system abnormalities are also common.

Take-Home Message **How does a human develop?**

- Cleavage produces a blastocyst, which implants in the uterus. Projections from its surface invade maternal tissue and form a placenta. The placenta permits exchanges between a mother and embryo without intermingling bloodstreams. Pathogens and toxins, as well as nutrients, can cross the placenta.

- Gastrulation occurs about two weeks after fertilization, and produces a three-layered embryo. Chordate features such as a neural tube, notochord, and pharyngeal arches form later in the embryonic period.

- The fetal period begins at the end of the eighth week. Heartbeats are detected at about five months, and limb movements are felt at five to six months.

- Lungs are not fully formed until about 28 weeks (7 months), so births before this point are highly risky. A fetus born before 22 weeks will not survive.

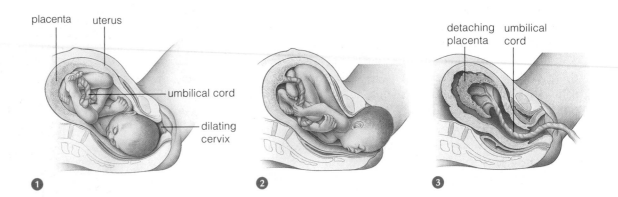

FIGURE 26.17 Childbirth.

1 The fetal head puts pressure on the cervix, stimulating hormone (oxytocin) release that causes muscle contractions.

2 Muscle contractions propel the fetus out of the uterus and through the vagina.

3 The placenta detaches and is also expelled.

placenta   uterus

umbilical cord

dilating cervix

detaching   umbilical
placenta   cord

26.7

From Birth Onward

The Process of Birth Properties of the cervix are altered as a fetus nears full term. Up to this point, the cervix has remained firm and has helped keep the fetus from slipping out of the uterus prematurely. In the final weeks of pregnancy, the connective tissue of the cervix weakens. Collagen fibers become less tightly linked and the cervix gets thinner, softer, and more flexible. This prepares it to stretch enough to allow the passage of the fetus.

The birth process is known as **labor**, or parturition. Often a women finds she is about to begin labor when "her water breaks." This common phrase describes the rupturing of the amnion, which causes a large amount of amniotic fluid to drain out of the vagina. During labor, the cervix dilates and the fetus moves through it into the vagina, and then into the outside world (Figure 26.17).

The hormone oxytocin stimulates smooth muscle contractions during labor. As the fetus nears full term, it usually "drops," or shifts down, so the head touches the cervix. Receptors in the cervix sense mechanical pressure and signal the hypothalamus, which calls for oxytocin secretion by the posterior pituitary.

The binding of oxytocin to smooth muscle of the uterus increases the strength of contractions, which causes more mechanical pressure. This leads to more oxytocin secretion, and so on. The result is a positive feedback cycle—stretching of the cervix triggers more oxytocin secretion, which results in more stretching. This continues until the fetus has been expelled and there is no longer any mechanical pressure on the cervix. Synthetic oxytocin is sometimes given intravenously to induce or increase contractions.

Strong contractions help detach the placenta from the uterus and expel it as the afterbirth. Contractions also help to stop the bleeding where the placenta was attached to the wall of the uterus, by causing blood vessels at this ruptured attachment site to constrict. The umbilical cord is cut and tied off. A few days after shriveling up, the cord's stump has become the navel.

Nourishing the Newborn Once the lifeline to the mother is severed, a newborn enters the extended time of dependency and learning that is typical of all primates. Early survival requires an ongoing supply of milk or a nutritional equivalent. Lactation, or milk production, occurs in mammary glands inside a mother's breasts (Figure 26.18). Before pregnancy, breast tissue is largely adipose tissue and a system of undeveloped ducts. During pregnancy, estrogens and progesterone stimulate the development of a glandular system for milk production. Prolactin, a hormone secreted by the anterior pituitary gland, stimulates the growth of cells that will produce milk proteins.

labor Expulsion of a placental mammal from its mother's uterus by muscle contractions.

After birth, the mechanical stimulus of suckling by a newborn increases maternal secretion of prolactin and oxytocin. The prolactin stimulates the production of milk proteins. For the first few days after birth, mammary glands produce a clear fluid rich in proteins and lactose (milk sugar). Then, milk production begins. Oxytocin causes smooth muscle contractions that force fluid into milk ducts. It also causes smooth muscle contractions in the uterus, which helps this organ shrink to its pre-pregnancy size.

In addition to its nutrients, human breast milk has antibodies that enhance resistance to infection. Other components of breast milk stimulate the growth of symbiotic bacteria in an infant's gut. Alcohol, drugs, mercury, and other toxins in a mother's body can also be secreted in milk. HIV and some other viruses are carried by milk as well.

Postnatal Development Like many animals, humans change in size and proportion until reaching sexual maturity. Postnatal growth is most rapid between years thirteen and nineteen. Sex hormone secretions increase and bring about development of secondary sexual traits as well as sexual maturity. Not until adulthood are the bones fully mature. Body tissues are usually well maintained into early adulthood. As more years pass, they slowly deteriorate through processes collectively described as aging.

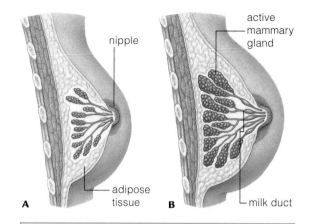

FIGURE 26.18 A Breast of a woman who is not lactating. **B** Breast of a lactating woman. Her mammary glands have been activated by exposure to prolactin and are producing milk. Milk ducts are lined with smooth muscle, which contracts to force milk out of the breast through ducts in the nipple.

Take-Home Message **What happens during and after birth?**

- Hormonal changes prepare a woman's body for labor and nursing.
- During labor, hormone-stimulated contraction of uterine smooth muscle forces the fetus out of the body.
- Milk secreted by mammary glands provides nutrition and also has antibodies that protect the newborn from infection. Some pathogens and toxins can be transmitted in breast milk.
- Growth and development continues after birth. Bones are not fully mature until early adulthood.

26.8 Impacts/Issues Revisited:
Mind-Boggling Births

A high level of FSH, the hormone that stimulates egg maturation, increases the likelihood of fraternal twins and other multiple births. FSH level and the prevalence of fraternal twinning varies among families and among ethnic groups. A woman who is herself a fraternal twin has double the average chance of giving birth to fraternal twins. If a woman has one set of fraternal twins, her odds of having a second set triples. Fraternal twins are most common among women of African descent, less common among Caucasians, and rare among Asians.

Age also has an effect. A woman's FSH levels rise from puberty through her midthirties, causing her likelihood of having fraternal twins to rise. Thus, a trend toward later childbearing is contributing to a rise in fraternal twinning.

FSH level does not influence formation of identical twins. Such twins arise when a zygote or early embryo splits, and two genetically identical individuals develop. A split is a chance event; a tendency to produce identical twins does not run in families and such twins occur with equal frequency among women of all ethnic groups and ages.

How Would YOU Vote? Should use of fertility drugs, which increase the risk of multiple births, be discouraged?

See CengageNow for details, then vote online (cengagenow.com).

Summary

Section 26.1 Multiple births are on the increase, largely as a result of fertility treatments. The increase is a matter of concern because multiple births are more likely than single births to have complications.

Section 26.2 Asexual reproduction produces genetic copies of the parent. Sexual reproduction is energetically more costly, and a parent does not have as many of its genes represented among the offspring. However, sexual reproduction produces variable offspring, which may be advantageous in environments where conditions change from one generation to the next.

Most animals reproduce sexually and have separate sexes, but some are hermaphrodites that produce both eggs and sperm. With external fertilization, gametes are released into water. Most animals on land have internal fertilization; gametes meet in a female's body.

Sexual reproduction begins with the formation of gametes—sperm and eggs—that combine at fertilization to form a zygote. During cleavage, mitosis increases the number of cells but not the original volume of the zygote. Cleavage yields a blastula, a hollow, fluid-filled cluster of cells (*right*).

During gastrulation, cells become organized into primary tissue layers. Organs start to form as cells in these layers differentiate. The development of body form involves cell migrations, cell shape changes, and cell death. In the final stage, growth and tissue specialization, organs enlarge and develop specialized properties.

CENGAGENOW™ Watch a frog develop and see how the neural tube is formed with the animation on CengageNow.

Section 26.3 Testes are primary male reproductive organs (Table 26.3). They produce sperm and the sex hormone testosterone. Accessory glands produce the other components of semen. These components are added as the sperm pass through a series of ducts that convey them out of the body. Ovaries are the female's primary reproductive organs (Table 26.3). They produce eggs and sex hormones: estrogens and progesterone. Pituitary hormones (LH, FSH) control production of both sperm and eggs.

Males produce sperm continually. A female is born with immature eggs. After puberty, the eggs mature one at a time in an approximately monthly cycle. In this cycle, FSH encourages follicle maturation, then a surge of LH near the midpoint of the cycle triggers ovulation. After ovulation, the remains of a follicle become a hormone-secreting corpus luteum.

Hormones secreted by the maturing follicle and the corpus luteum encourage thickening of the uterine lining. If pregnancy does not occur, the corpus luteum breaks down, and the uterine lining is shed during menstruation.

CENGAGENOW™ Use the animation and interaction on CengageNow to learn about human reproductive structures, sperm formation, egg formation, and the cyclic changes in the ovary and uterus.

Section 26.4 Hormones and nerves govern physiological changes that occur during arousal and intercourse. Millions of sperm are ejaculated, but usually only one penetrates the secondary oocyte. Fertilization usually occurs in an oviduct.

CENGAGENOW™ Observe fertilization with the animation on CengageNow.

Section 26.5 Pregnancy can be avoided by behavior, or through diverse methods that prevent ovulation, fertilization, or implantation. *In vitro* fertilization brings sperm and eggs together for fertilization outside the body. An abortion ends a pregnancy.

Sexually transmitted diseases (STDs) are caused by protozoal, bacterial, and viral pathogens and are spread by unsafe sex. Viral STDs are incurable. Bacterial and protozoal STDs can be cured. If untreated, they can cause sterility and harm the health of parents and their offspring.

Section 26.6 After fertilization, a blastocyst forms by cleavage and burrows into the wall of the uterus. Membranes form around the blastocyst. Some combine with the maternal tissue to form a placenta that passes oxygen and nutrients between mother and embryo. The placenta can also deliver toxins.

By the end of the eighth week, when the embryo becomes a fetus, all organ systems have formed. After 7 months it can survive if born prematurely.

CENGAGENOW™ Observe fertilization and implantation and learn how the placenta forms and functions with the animation on CengageNow.

Section 26.7 Hormones prepare a woman's body for pregnancy,

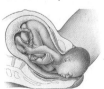

labor, and nursing. During labor, uterine contractions expel the fetus and placenta. Milk secreted by mammary glands supplies nutrients and also contains antibodies that help a newborn resist infection. Development and growth continue until adulthood.

Table 26.3	Organs of the Human Reproductive Tracts
Male Reproductive Tract	
Testes	Sperm production; sex hormone production
Epididymides	Sperm maturation and storage
Vasa deferentia	Convey sperm from epididymides to ejaculatory ducts
Ejaculatory ducts	Convey sperm from vasa deferentia to the urethra
Urethra	Conveys sperm through penis and out of the body
Penis	Organ of sexual intercourse
Female Reproductive Tract	
Ovaries	Egg production, maturation; sex hormone production
Oviducts	Convey oocyte from ovary to uterus
Uterus	Chamber in which new individual develops
Cervix	Entrance into uterus; mucus secretion
Vagina	Organ of sexual intercourse; birth canal

Self-Quiz   Answers in Appendix I

1. _____ reproduction produces genetic copies of a parent.

2. A(n) _____ produces both eggs and sperm.
   a. hermaphrodite   c. ovary
   b. asexual organism   d. testis

3. A _____ is a hollow ball of cells that becomes a _____ by the process of gastrulation.

4. Meiotic divisions of germ cells in the _____ give rise to sperm.
   a. seminiferous tubule   c. penis
   b. prostate gland   d. epididymis

5. True or false? A female's germ cells do not divide after birth.

6. During a menstrual cycle, a midcycle surge of _____ secreted by the pituitary triggers ovulation.
   a. estrogen   b. progesterone   c. LH   d. FSH

7. After ovulation, the corpus luteum secretes _____ .
   a. LH   c. progesterone
   b. FSH   d. prolactin

8. A _____ implants in the lining of the uterus.
   a. zygote   c. blastocyst
   b. gastrula   d. fetus

9. Number these events in human development in the correct order.
   _____ gastrulation occurs
   _____ blastocyst forms
   _____ zygote forms
   _____ neural tube forms
   _____ heart starts beating

10. Which of the following STDs are caused by bacteria?
    a. chlamydia   d. trichomoniasis
    b. gonorrhea   e. a and b
    c. genital warts   f. all of the above

11. The placenta includes _____ .
    a. maternal tissue   c. mesoderm
    b. chorion tissue   d. a and b

12. Secretion of oxytocin stimulates _____ .
    a. contraction of uterine smooth muscle
    b. expulsion of milk from mammary glands
    c. ovulation   e. a and b
    d. blastocyst implantation   f. all of the above

13. Match each human reproductive structure with the most suitable description.
    _____ testis   a. maternal and fetal tissues
    _____ cervix   b. adds fluid to sperm in semen
    _____ placenta   c. produces testosterone
    _____ vagina   d. produces estrogen and
    _____ ovary   progesterone
    _____ oviduct   e. usual site of fertilization
    _____ prostate gland   f. secretes milk
    _____ mammary gland   g. birth canal
                          h. entrance to uterus

Additional questions are available on CENGAGENOW™

Digging Into Data

Multiple Births and Birth Defects

People considering fertility treatments should be aware that such treatments raise the risk of multiple births, and that multiple pregnancies are associated with an increased risk of some birth defects.

Figure 26.19 shows the results of Yiwei Tang's study of birth defects reported in Florida from 1996 to 2000. Tang compared the incidence of various defects among single and multiple births. She calculated the relative risk for each type of defect based on type of birth, and corrected for other differences that might increase risk such as maternal age, income, race, and medical care during pregnancy. A relative risk of less than 1 means a defect occurs less often with multiple births than single births. A relative risk greater than 1 means that multiples are more likely to have a defect.

1. What was the most common type of birth defect in the single-birth group?

2. Was that type of defect more or less common among multiple-birth newborns than among single births?

3. Tang found that multiples have more than twice the risk of single newborns for one type of defect. Which type?

4. Does a multiple pregnancy increase the relative risk of chromosomal defects in offspring?

	Prevalence of Defect		Relative Risk
	Multiples	Singles	
Total birth defects	358.50	250.54	1.46
Central nervous system defects	40.75	18.89	2.23
Chromosomal defects	15.51	14.20	0.93
Gastrointestinal defects	28.13	23.44	1.27
Genital/urinary defects	72.85	58.16	1.31
Heart defects	189.71	113.89	1.65
Musculoskeletal defects	20.92	25.87	0.92
Fetal alcohol syndrome	4.33	3.63	1.03
Oral defects	19.84	15.48	1.29

FIGURE 26.19 Prevalence, per 10,000 live births, of various types of birth defects among multiple and single births. Relative risk for each defect is given after researchers adjusted for mother's age, race, previous adverse pregnancy experience, education, Medicaid participation during pregnancy, and the infant's sex and number of siblings.

Critical Thinking

1. Signals from parasympathetic nerves are necessary for a male to have an erection. Explain why fear or anxiety can make it difficult for a man to achieve an erection.

2. Of all birds, the flightless kiwi has the highest proportion of yolk in its eggs and the longest incubation time. Explain why yolk increases with the length of time a bird spends inside an egg.

27 Plant Form and Function

27.1 Impacts/Issues: Leafy Cleanup Crews

From World War I until the 1970s, the United States Army tested and disposed of weapons at J-Field, Aberdeen Proving Ground in Maryland (Figure 27.1**A**). Obsolete chemical weapons and explosives were burned in open pits, together with plastics, solvents, and other wastes. Lead, arsenic, mercury, and other metals heavily contaminated the soil and groundwater. So did highly toxic organic compounds, including trichloroethylene (TCE). TCE damages the nervous system, lungs, and liver, and exposure to large amounts can be fatal. Today, the toxic groundwater is seeping toward nearby marshes and the Chesapeake Bay.

To protect the bay and clean up the soil, the Army and the Environmental Protection Agency turned to phytoremediation: the use of plants to take up and concentrate or degrade environmental contaminants. They planted poplar trees that cleanse groundwater by taking up TCE and other organic pollutants from it (Figure 27.1**B**).

Like other vascular plants, poplar trees take up soil water through their roots. Along with the water come nutrients and chemical contaminants, including TCE. Although TCE is toxic to animals, it does not harm the trees. The poplars break down some of it, but they release most of the toxin into the atmosphere. Airborne TCE is the lesser of two evils: It breaks down much more quickly in air than it does in groundwater.

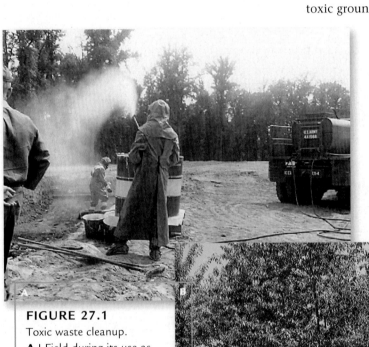

FIGURE 27.1
Toxic waste cleanup.
A J-Field during its use as a testing site for weapons.
B J-Field today. Poplar trees are helping to remove toxic waste left behind in the soil.

Phytotoremediation uses the ability of plants to take up toxic chemicals dissolved in soil water. Plants have this capacity not for our benefit, but rather to meet their own needs for water and nutrients. Scientists who know about plant physiology are finding ways to put plants to work as leafy cleanup crews. Compared to other methods of cleaning up toxic waste sites, phytoremediation is usually less expensive—and it is more appealing to neighbors.

In this chapter and the next, we look at the structure and function of plants. We focus on seed-bearing vascular plants, especially the flowering types that are integral to our lives. These plants are adapted to life on land, where they must obtain water from soil, hold their bodies upright, and keep themselves from drying out. The anatomy of stems, leaves, flowers, and roots are adaptations to these challenges. So are the physiological processes that plants use to obtain nutrients, distribute materials through their body, and reproduce.

Organization of the Plant Body

With 260,000 species, flowering plants dominate the plant kingdom. Its major groups are the magnoliids, eudicots (true dicots), and monocots (Section 14.7). We focus on eudicots and monocots, many of which have a body plan similar to the one in Figure 27.2. Aboveground shoots consist of plant stems together with their leaves. Stems provide a structural framework for plant growth, and pipelines inside of them conduct water and nutrients between leaves and roots. Roots are structures that absorb water and dissolved minerals as they grow down and outward through the soil. They often anchor the plant. Root cells store food for their own use, and some types also store it for the rest of the plant body.

Three Plant Tissue Systems

All plant parts consist of the same three types of tissues (Figure 27.3). The **ground tissue system**, which makes up the bulk of the plant, has many essential functions such as photosynthesis and food storage. Pipelines of the **vascular tissue system** thread through ground tissue. They distribute water and nutrients to all parts of the plant body. The **dermal tissue system** covers and protects the plant's exposed surfaces.

All three plant tissue systems consist of cells that are organized into simple and complex tissues. Simple tissues consist primarily of one type of cell; examples include parenchyma, collenchyma, and sclerenchyma. Complex tissues have two or more cell types. Xylem, phloem, and epidermis are examples.

Simple Tissues

Parenchyma tissue makes up most of the soft primary growth of roots, stems, leaves, and flowers, and it also has storage and secretion functions. **Parenchyma** is a simple tissue that consists mainly of parenchyma cells, which are typically thin-walled, flexible, and many-sided. These cells are alive in mature tissue, and they can continue

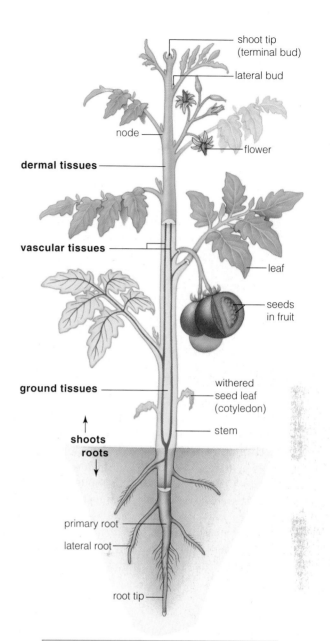

shoot tip (terminal bud)

lateral bud

node

flower

dermal tissues

vascular tissues

leaf

seeds in fruit

ground tissues

withered seed leaf (cotyledon)

stem

shoots
roots

primary root

lateral root

root tip

FIGURE 27.3 Three plant tissue types in a cross-section of a buttercup stem.

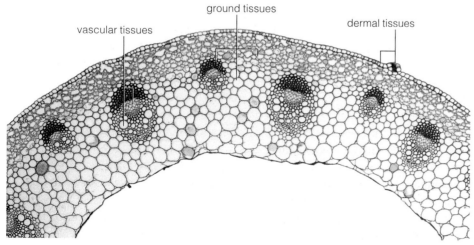

vascular tissues

ground tissues

dermal tissues

FIGURE 27.2 Animated! Body plan of a tomato plant. Vascular tissues (*purple*) conduct water and solutes. They thread through ground tissues that make up most of the plant body. Epidermis, a type of dermal tissue, covers root and shoot surfaces.

dermal tissue system Tissue system that covers and protects the plant body.
ground tissue system Tissue system that makes up the bulk of the plant body; includes most photosynthetic cells.
parenchyma Simple plant tissue made up of living cells; the main component of the ground tissue system.
vascular tissue system Tissue system that distributes water and nutrients through a plant body.

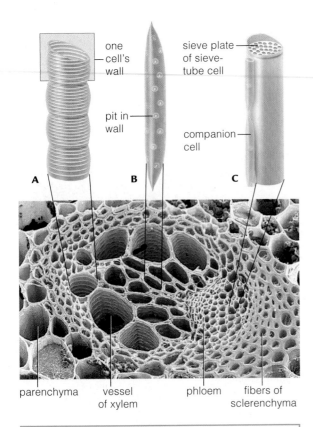

one cell's wall

pit in wall

sieve plate of sieve-tube cell

companion cell

A   B   C

parenchyma   vessel of xylem   phloem   fibers of sclerenchyma

FIGURE 27.4 Simple and complex tissues in a stem. In xylem, a part of a column of vessel members **A**, and a tracheid **B**. **C** One of the living cells that interconnect as sieve tubes in phloem. Companion cells closely associate with sieve tubes. Fibers of sclerenchyma and parenchyma cells are also visible in the micrograph.

Table 27.1	Overview of Flowering Plant Tissues	
Tissue Type	Main Components	Main Functions
Simple Tissues		
Parenchyma	Parenchyma cells	Photosynthesis, storage, secretion, tissue repair, other tasks
Collenchyma	Collenchyma cells	Pliable structural support
Sclerenchyma	Fibers or sclereids	Structural support
Complex Tissues		
Vascular		
Xylem	Tracheids, vessel members; parenchyma cells; sclerenchyma cells	Water-conducting tubes and reinforcing components
Phloem	Sieve-tube members, parenchyma cells; sclerenchyma cells	Tubes of living cells that distribute sugars; supporting cells
Dermal		
Epidermis	Undifferentiated cells and specialized cells (e.g., guard cells)	Secretion of cuticle; protection; control of gas exchange and water loss
Periderm	Cork cambium; cork cells; parenchyma	Forms protective cover on older stems, roots

to divide. Plant wounds are repaired by dividing parenchyma cells. Mesophyll, the only photosynthetic ground tissue, is a type of parenchyma.

Collenchyma is a simple tissue that consists mainly of collenchyma cells, which are elongated and alive in mature tissue. This stretchable tissue supports rapidly growing plant parts, including young stems and leaf stalks. A polysaccharide called pectin imparts flexibility to a collenchyma cell's primary wall, which is thickened where three or more of the cells abut.

Cells of **sclerenchyma** are variably shaped and dead at maturity, but the lignin-containing walls that remain help this tissue resist compression. Lignin, an organic compound, protects and structurally supports upright plants, and helped them evolve on land (Section 14.2). Fibers and sclereids are typical sclerenchyma cells. Fibers are long, tapered cells that structurally support the vascular tissues in some stems and leaves. They flex and twist, but resist stretching. We use certain fibers as materials for cloth, rope, paper, and other commercial products. The stubbier and often branched sclereids strengthen hard seed coats, such as peach pits, and make pear flesh gritty.

Vascular Tissues Xylem and phloem are vascular tissues. Both consist of elongated conducting tubes that are often sheathed in sclerenchyma fibers and parenchyma. **Xylem**, the vascular tissue that conducts water and mineral ions, consists of two types of cells: tracheids and vessel members (Figure 27.4A,B). Both cell types are dead at maturity, but their walls, which are stiffened and waterproofed with lignin, remain. These walls interconnect to form conducting tubes, and they lend structural support to the plant. The perforations in adjoining cell walls align, so water moves laterally between the tubes as well as upward through them.

Phloem conducts sugars and other organic solutes. Its main cells, sieve-tube members, are alive in mature tissue. They connect end to end at sieve plates, forming sieve tubes that transport sugars from photosynthetic cells to all parts of the plant (Figure 27.4C). Phloem's **companion cells** are parenchyma cells that load sugars into the tubes.

collenchyma Simple plant tissue composed of living cells with unevenly thickened walls; provides flexible support.
companion cell In phloem, parenchyma cell that loads sugars into sieve tubes.
cotyledon Seed leaf; part of a flowering plant embryo.
epidermis Outer tissue layer.
phloem Complex vascular tissue that distributes sugars through a plant body.
sclerenchyma Simple plant tissue; dead at maturity, its lignin-reinforced cell walls structurally support plant parts.
vascular bundle Multistranded, sheathed cord of primary xylem and phloem in a stem or leaf.
xylem Complex vascular tissue; distributes water and solutes through tubes that consist of interconnected walls of dead cells.

Dermal Tissues The first dermal tissue to form on a plant is **epidermis**, which is usually a single layer of cells. Epidermal cells secrete substances such as cutin, a polymer of fatty acids, on their outward-facing cell walls. The waxy deposits form a cuticle that helps the plant conserve water and repel pathogens. The epidermis of leaves and young stems also has specialized cells. For example, pairs of specialized epidermal cells form stomata (small gaps across the epidermis) when the cells swell with water. Plants control the diffusion of water vapor, oxygen, and carbon dioxide gases across their epidermis by opening and closing stomata.

Eudicots and Monocots Table 27.1 lists the main types of flowering plant tissues, their components, and their functions. The same tissues form in all flowering plants, but they do so in different patterns. Monocots and eudicots are named after their **cotyledons**, which are leaf-like structures that contain food for a plant embryo. These "seed leaves" wither after the seed germinates and the developing plant begins to make its own food by photosynthesis. Cotyledons consist of the same types of tissues in all plants that have them, but the seeds of eudicots have two cotyledons and those of monocots have only one.

Figure 27.5 shows some other differences between monocots and eudicots. Most shrubs and trees, such as rose bushes and maple trees, are eudicots. Lilies, orchids, and corn are examples of typical monocots.

Take-Home Message **What is the basic structure of a flowering plant?**

- Plants typically have aboveground shoots and stems, and belowground roots, all of which consist of ground, vascular, and dermal tissues.

- Ground tissues make up most of a plant. Cells of parenchyma have diverse roles, including photosynthesis. Collenchyma and sclerenchyma support and strengthen plant parts.

- Vascular tissues that thread through ground tissue distribute water and solutes. In xylem, water and ions flow through tubes of dead tracheid and vessel member cells. In phloem, sieve tubes that consist of living cells distribute sugars.

- Dermal tissues cover and protect plant surfaces. Epidermis covers young plant surfaces.

- The patterns in which plant tissues are organized differ between eudicots and monocots.

27.3 Primary Shoots and Roots

Inside a Stem Inside stems and leaves, the ground, vascular, and dermal tissue systems are organized in predictable patterns. In most flowering plants, the cells of primary xylem and phloem are bundled together as long, multistranded cords called **vascular bundles**. Vascular bundles thread lengthwise through the ground tissue system of all stems, leaves, and roots, but the arrangement of the bundles in these structures differs between eudicots and monocots. The vascular bundles of most eudicot

A Eudicots

In seeds, two cotyledons (seed leaves of embryo)

Flower parts in fours or fives (or multiples of four or five)

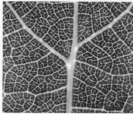

Leaf veins usually forming a netlike array

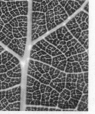

Pollen grains with three pores or furrows

Vascular bundles organized in a ring in ground tissue of stem

B Monocots

In seeds, one cotyledon (seed leaf of embryo)

Flower parts in threes (or multiples of three)

Leaf veins usually running parallel with one another

Pollen grains with one pore or furrow

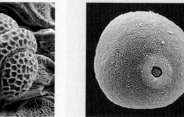

Vascular bundles throughout ground tissue of stem

FIGURE 27.5 Animated! Comparing some of the characteristics of eudicots **A** and monocots **B**.

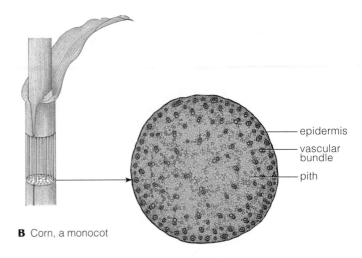

B Corn, a monocot

epidermis
vascular bundle
pith

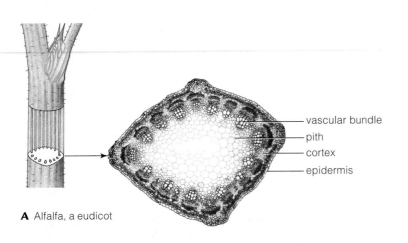

A Alfalfa, a eudicot

vascular bundle
pith
cortex
epidermis

FIGURE 27.6 Animated!
Stem structure.

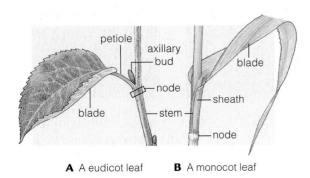

petiole
axillary
bud
blade
node
sheath
blade
stem
node

A A eudicot leaf **B** A monocot leaf

FIGURE 27.7 Typical leaf forms of eudicots and monocots.

apical meristem In shoot and root tips, mass of undifferentiated cells, the division of which lengthens plant parts.

guard cell One of a pair of cells that define a stoma across the epidermis of a leaf or stem.

meristem Zone of undifferentiated plant cells that can divide rapidly.

primary growth Plant growth from apical meristems in root and shoot tips.

stems form a cylinder that runs parallel with the long axis of the shoot. Figure 27.6**A** shows how the cylinder divides the parenchyma of ground tissue into cortex (parenchyma between the vascular bundles and the epidermis) and pith (parenchyma inside the cylinder of vascular bundles). Stems of monocots have a different arrangement. Their vascular bundles are distributed all throughout the ground tissue (Figure 27.6**B**).

Leaf Structure A leaf is a sugar factory that has many photosynthetic cells. A typical leaf has a flat blade and, in eudicots, a petiole (stalk) attached to the stem (Figure 27.7**A**). The leaves of most monocots are flat blades, the base of which forms a sheath around the stem (Figure 27.7**B**). Grasses are examples.

Leaf shapes and orientations are adaptations that help a plant intercept sunlight and exchange gases. Most leaves are thin, with a high surface-to-volume ratio; many reorient themselves during the day so that they stay perpendicular to the sun's rays. Typically, adjacent leaves project from a stem in a pattern that allows sunlight to reach them all. However, the leaves of plants native to arid regions may stay parallel to the sun's rays, reducing heat absorption and thus conserving water. Thick or needlelike leaves of some plants also conserve water.

Epidermis covers every leaf surface exposed to the air (Figure 27.8). This surface tissue may be smooth, sticky, or slimy, with hairs, scales, spikes, hooks, and other specializations. A translucent, waxy secreted cuticle slows water loss from the sheetlike array of epidermal cells ❶.

The bulk of the leaf consists of mesophyll, a photosynthetic parenchyma with air spaces between cells ❷. Leaves oriented perpendicular to the sun have two layers of mesophyll. Palisade mesophyll is attached to the upper epidermis. The elongated parenchyma cells of this tissue have more chloroplasts than cells of the spongy mesophyll layer below. Blades of grass and other monocot leaves that grow vertically can intercept light from all directions. The mesophyll in such leaves is not divided into two layers.

Leaf veins are vascular bundles typically strengthened with fibers ❸. Inside the bundles, continuous strands of xylem rapidly transport water and dissolved ions to mesophyll. Continuous strands of phloem rapidly transport the products of photosynthesis (sugars) away from mesophyll. In most eudicots, large veins branch into a network of minor veins embedded in mesophyll. In most monocots, all veins are similar in length and run parallel with the leaf's long axis.

Most leaves have stomata mainly on their lower surface ❹. The **guard cells** on either side of each stoma are the only photosynthetic cells in the leaf epider-

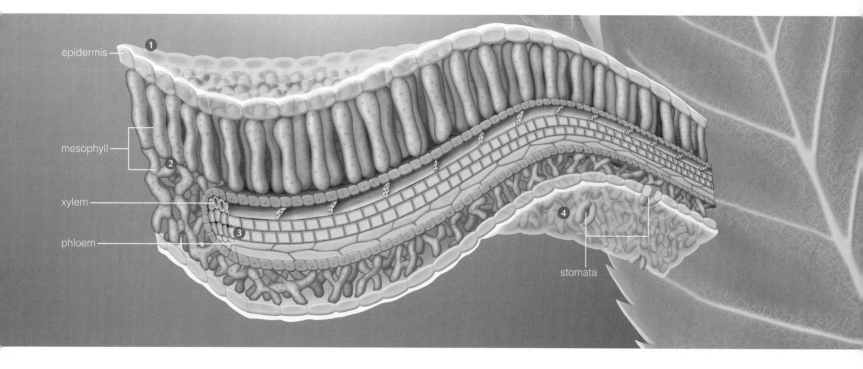

mesophyll ②

xylem

phloem ③

④

stomata

FIGURE 27.8 Animated!
Anatomy of a eudicot leaf.

❶ The upper leaf surface is epidermis with a secreted layer of cuticle.

❷ The bulk of the leaf is mesophyll, a type of photosynthetic parenchyma.

❸ Vascular bundles of xylem (*gold*) and phloem (*pink*) form veins that thread through the leaf.

❹ The lower leaf surface is also covered with cuticle. Gas exchanges between air inside and outside of the leaf occur at stomata in the lower leaf epidermis.

mis. As you will see in Section 27.6, shape changes of the guard cells close the stomata to prevent water loss, or open the stomata to allow gases to cross the epidermis. Carbon dioxide needed for photosynthesis enters the leaf through stomata, then through air in the spaces between mesophyll cells. Oxygen released by photosynthesis diffuses in the opposite direction.

Primary Growth of a Shoot
In animals, growth occurs as cells throughout the body divide. In plants, growth occurs only in localized regions called **meristems**, where cells continually divide. **Primary growth** is the lengthening of roots or shoots. It occurs at **apical meristems**—masses of undifferentiated cells that can divide rapidly in the tips of roots and shoots. Terminal buds at shoot tips are the main zone of primary growth (Figure 27.9). Just beneath a terminal bud's surface, cells of apical meristem divide continually during the growing season. Populations of cells that form here are the immature forerunners of dermal tissues, primary vascular tissues, and ground tissue.

Buds may be naked or encased in modified leaves called bud scales. Small regions of tissue that bulge out near the sides of a bud's apical meristem will become new leaves. As the stem lengthens, the leaves form and mature in orderly tiers, one after the next. A region of stem where one or more leaves form is called a node; the region between two successive nodes is called an internode.

A lateral bud is a dormant shoot of mostly meristematic tissue. Each bud forms inside a leaf axil, the point at which the leaf is attached to the stem. Division of cells in an axillary bud can give rise to a branch, leaf, or flower.

Structure and Development of Primary Roots
The underground portion of a plant, the root system, is just as extensive and essential as the shoot system. Roots grow through the soil and take up water and mineral nutrients. Root primary growth results in one of two kinds of root systems. The taproot system of eudicots consists of a primary root and its lateral branchings. Carrots, oak trees, and poppies are among the plants with a taproot system

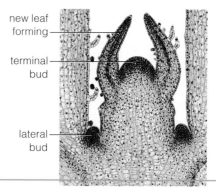

new leaf forming

terminal bud

lateral bud

FIGURE 27.9 Cross-section through a shoot tip. Division of meristem cells in the terminal bud gives rise to primary growth.

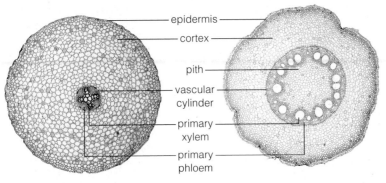

epidermis
cortex

pith
vascular
cylinder
primary
xylem
primary
phloem

B Eudicot root cross-section

C Monocot root cross-section

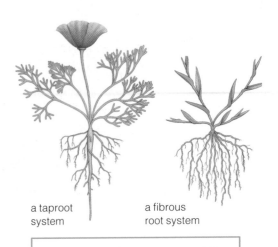

a taproot
system

a fibrous
root system

FIGURE 27.10 Comparing different types of root systems.

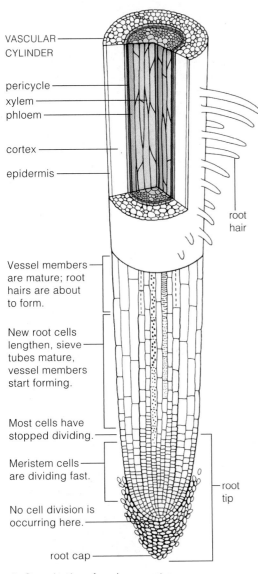

VASCULAR
CYLINDER

pericycle
xylem
phloem

cortex

epidermis

root
hair

Vessel members
are mature; root
hairs are about
to form.

New root cells
lengthen, sieve
tubes mature,
vessel members
start forming.

Most cells have
stopped dividing.

Meristem cells
are dividing fast.

No cell division is
occurring here.

root
tip

root cap

A Organization of a primary root

FIGURE 27.11 Animated! Tissue organization in typical roots. **Figure It Out:** Where does most cell division occur in a primary root? *Answer: At the apical meristem.*

(Figure 27.10). By comparison, the primary root of most monocots is replaced by adventitious roots that grow outward from the stem. Lateral roots that are similar in diameter and length branch from adventitious roots. Together, adventitious and lateral roots of such plants form a fibrous root system.

Look at the root tip in Figure 27.11**A**. Some descendants of root apical meristem give rise to a root cap, a dome-shaped mass of cells that protects the soft, young root as it grows through soil. Other descendants give rise to lineages of cells that lengthen, widen, or flatten when they differentiate as part of the dermal, ground, and vascular tissue systems.

Ongoing divisions push cells away from the active root apical meristem. Some of their descendants form epidermis. The root epidermis is the plant's absorptive interface with soil. Many of its specialized cells send out fine extensions called **root hairs**, which collectively increase the surface area available for taking up soil water along with dissolved oxygen and mineral ions.

Descendants of meristem cells also form the root's **vascular cylinder**, a central column of conductive tissue. The root vascular cylinder of typical eudicots is mainly primary xylem and phloem (Figure 27.11**B**). In a typical monocot, the vascular cylinder divides the ground tissue into two zones, cortex and pith (Figure 27.11**C**). The vascular cylinder is sheathed by a pericycle, an array of parenchyma cells one or more layers thick. These cells divide repeatedly in a direction perpendicular to the axis of the root. Masses of cells erupt through the cortex and epidermis as the start of new lateral roots.

Take-Home Message **How are primary roots and shoots organized?**

- The arrangement of vascular bundles, which are multistranded cords of vascular tissue, differs between eudicot and monocot stems and roots.

- Leaves are structurally adapted to intercept sunlight and distribute water and nutrients. Components include mesophyll (photosynthetic cells), veins (bundles of vascular tissue), and cuticle-secreting epidermis.

- A shoot lengthens as cells of apical meristem tissue in a terminal bud divide and differentiate. New leaves or branches develop from lateral buds.

- Roots provide a plant with a tremendous surface area for absorbing water and solutes. Inside each root is a vascular cylinder, with long strands of primary xylem and phloem.

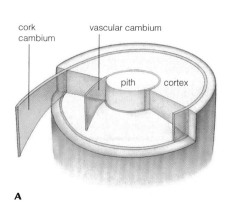

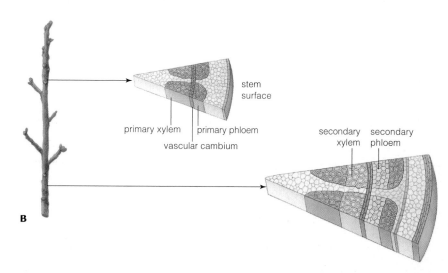

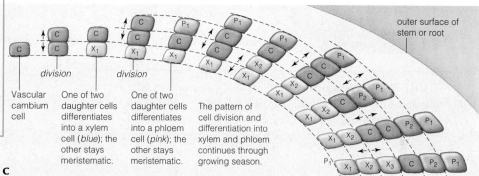

FIGURE 27.12 **Animated!** Secondary growth. **A** Secondary growth occurs at two lateral meristems: Vascular cambium gives rise to secondary vascular tissues, and cork cambium gives rise to periderm. **B** Pattern of growth at vascular cambium. **C** Divisions of meristem cells in vascular cambium expand the inner core of xylem, displacing the vascular cambium (*orange*) toward the surface.

Secondary Growth

Roots and shoots of many plants thicken and become woody over time. Such thickening is called **secondary growth**. Cell divisions in lateral meristems give rise to secondary growth. **Lateral meristems** are cylindrical layers of meristem that run lengthwise through stems and roots (Figure 27.12**A**). Woody plants have two types of lateral meristems: vascular cambium and cork cambium (*cambium* is the Latin word for change).

The **vascular cambium** is the lateral meristem that produces secondary vascular tissue, a few cells thick, inside older stems and roots. Divisions of vascular cambium cells produce secondary xylem on the cylinder's inner surface, and secondary phloem on its outer surface (Figure 27.12**B**). As the core of xylem thickens, it also displaces the vascular cambium toward the surface of the stem or root. The displaced cells of the vascular cambium divide in a widening circle (Figure 27.12**C**).

A core of secondary xylem, or **wood**, contributes up to 90 percent of the weight of some plants. As seasons pass, the expanding inner core of xylem continues to put pressure on the stem or root surface. In time, the pressure ruptures the cortex and the outer secondary phloem. Then, another lateral meristem, the **cork cambium**, forms and gives rise to periderm. **Periderm** is a dermal tissue that consists of parenchyma and cork, as well as the cork cambium that produces them. What we call **bark** is secondary phloem and periderm. Bark consists of all of the living and dead tissues outside of the vascular cambium. The **cork** component of bark has densely packed rows of dead cells, the walls of which are thickened with a waxy substance. Cork protects, insulates, and waterproofs the stem or root surface. Cork also forms over wounded tissues. When leaves drop from the plant, cork forms at the places where petioles had attached to stems.

bark In woody plants, secondary phloem and periderm.
cork Component of bark; waterproofs, insulates, and protects the surfaces of woody stems and roots.
cork cambium In plants, a lateral meristem that gives rise to periderm.
lateral meristem Sheetlike cylinder of meristem inside older stems and roots; vascular cambium or cork cambium.
periderm Plant dermal tissue that replaces epidermis on older stems and roots.
root hairs Hairlike, absorptive extensions of a young cell of root epidermis.
secondary growth A thickening of older stems and roots at lateral meristems.
vascular cambium Ring of meristematic tissue that produces secondary xylem and phloem.
vascular cylinder Sheathed, cylindrical array of primary xylem and phloem in a root.
wood Accumulated secondary xylem.

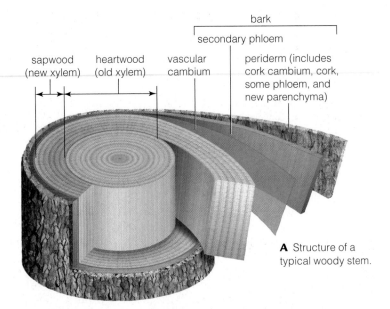

sapwood (new xylem)

heartwood (old xylem)

vascular cambium

secondary phloem

bark

periderm (includes cork cambium, cork, some phloem, and new parenchyma)

A Structure of a typical woody stem.

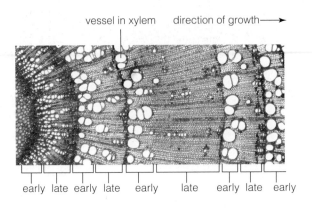

vessel in xylem   direction of growth⟶

early late early late   early   late   early late early

B Early and late wood in an ash tree. Early wood forms during wet springs. Late wood indicates that a tree did not waste energy making large-diameter xylem cells for water uptake during a dry summer or drought.

FIGURE 27.13 Animated! Structure of a woody stem. The rings visible in the heartwood and sapwood are regions of early and late wood. In most temperate zone trees, one ring forms each year.

Wood's appearance and function change as a stem or root ages. Metabolic wastes, such as resins, tannins, gums, and oils, clog and fill the oldest xylem so much that it no longer is able to transport water and solutes. These substances darken and strengthen the wood, which is called heartwood.

Sapwood is moist, still-functional xylem between heartwood and vascular cambium. In trees of temperate zones, dissolved sugars travel from roots to buds through sapwood's secondary xylem in spring. The sugar-rich fluid is sap. Each spring, New Englanders collect maple tree sap to make maple syrup.

Vascular cambium is inactive during cool winters or long dry spells. When the weather warms or moisture returns, the vascular cambium gives rise to early wood, with large-diameter, thin-walled cells. Late wood, with small-diameter, thick-walled xylem cells, forms in dry summers. A cross-section through an older trunk reveals alternating bands of early and late wood (Figure 27.13). Each band is a growth ring, or "tree ring."

Trees in regions where growth slows or is suspended during the winter tend to add one growth ring each year. In tropical regions where rainfall is continuous, trees grow at the same rate year-round and do not have growth rings.

Take-Home Message What is secondary growth in plants?

- Secondary growth thickens the stems and roots of older plants.
- Divisions of cells in vascular cambium and cork cambium give rise to secondary growth. These lateral meristems form in cylinders that run horizontally through older roots and shoots.
- Secondary vascular tissues form at vascular cambium. Cork cambium gives rise to periderm, which is part of a protective covering of bark.
- Wood is mainly accumulated secondary xylem.

humus Decaying organic matter in soil.

27.5

Plant Nutrition

Plant Nutrients and Properties of Soil Plant growth requires the sixteen elements listed in Table 27.2. Nine elements are macronutrients, which means they are required in amounts above 0.5 percent of the plant's dry weight (its weight after all the water has been removed from the plant body). Seven other elements are micronutrients, which make up traces—typically a few parts per million—of dry weight. Carbon, oxygen, and hydrogen are abundant in air and water. Plants obtain the other nutrients when their roots take up minerals dissolved in soil water.

Some soils are better for plant growth than others. Most plants do best in soil with some decaying organic matter, or **humus**. Humus provides nutrients and also retains water. The amounts and proportions of mineral particles also matters. Sand, silt, and clay—mineral particles that form by the weathering of rocks—differ in size and chemical properties. The biggest sand grains are big enough to see with the naked eye, about one millimeter in diameter. Silt particles are hundreds or thousands of times smaller than sand grains, so they are too small to see. Clay particles are even smaller. A clay particle consists of thin, stacked layers of negatively charged crystals with sheets of water alternating between the layers. Because clay is negatively charged, it attracts positively charged mineral ions dissolved in the soil water. Thus, clay retains dissolved nutrients that would otherwise trickle past roots too quickly to be absorbed.

Even though they do not bind mineral ions as well as clay, sand and silt are also necessary for growing plants. Water and air occupy spaces between the particles and organic bits in soil. Without enough sand and silt to intervene between tiny particles of clay, soil packs so tightly that it excludes air, and without air, root cells cannot secure enough oxygen for aerobic respiration.

Table 27.2	Plant Nutrients and Symptoms of Deficiencies				
Macronutrient	Functions	Symptoms of Deficiency	Micronutrient	Functions	Symptoms of Deficiency
Carbon Hydrogen Oxygen	Raw materials for second stage of photosynthesis	None; all are abundantly available in water and carbon dioxide	Chlorine	Role in root and shoot growth and photolysis	Wilting; chlorosis; some leaves die
Nitrogen	Component of proteins, coenzymes, chlorophyll, nucleic acids	Stunted growth; light-green older leaves; older leaves yellow and die (these are symptoms of a condition called chlorosis)	Iron	Roles in chlorophyll synthesis and in electron transport	Chlorosis; yellow and green striping in leaves of grass species
Potassium	Activation of enzymes; contributes to water-solute balances that influence osmosis*	Reduced growth; curled, mottled, or spotted older leaves; burned leaf edges; weakened plant	Boron	Roles in germination, flowering, fruiting, cell division, nitrogen metabolism	Terminal buds, lateral branches die; leaves thicken, curl, become brittle
Calcium	Component in control of many cell functions; cementing cell walls	Terminal buds wither, die; deformed leaves; poor root growth	Manganese	Chlorophyll synthesis; coenzyme action	Dark veins, but leaves whiten and fall off
Magnesium	Chlorophyll component; activation of enzymes	Chlorosis; drooped leaves	Zinc	Role in forming auxin, chloroplasts, starch; enzyme component	Chlorosis; mottled or bronzed leaves; root abnormalities
Phosphorus	Component of nucleic acids, ATP, phospholipids	Purplish veins; stunted growth; fewer seeds, fruits	Copper	Component of several enzymes	Chlorosis; dead spots in leaves; stunted growth
Sulfur	Component of most proteins, two vitamins	Light-green or yellowed leaves; reduced growth	Molybdenum	Part of enzyme used in nitrogen metabolism	Pale green, rolled or cupped leaves

* All mineral elements contribute to water–solute balances.

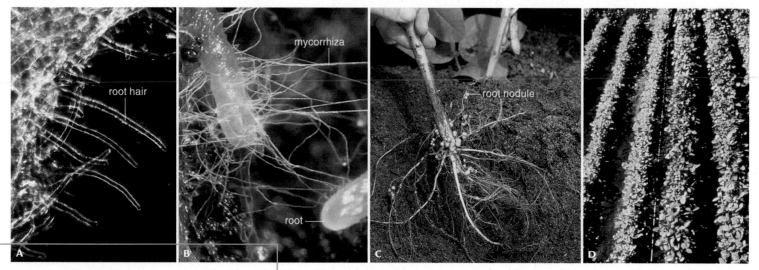

FIGURE 27.14 Examples of root specializations that help plants take up nutrients.

A The hairs on this root of a white clover plant are about 0.2 mm long. **B** Mycorrhizae (*white* hairs) extending from the tip of these roots (*tan*) greatly enhance their surface area for absorbing scarce minerals from the soil.

C Root nodules on this soybean plant fix nitrogen from air, and share it with the plant. **D** Soybean plants growing in nitrogen-poor soil show how root nodules affect growth. Only the plants in the rows at *right* were infected with nitrogen-fixing bacteria.

Root Adaptations for Nutrient Uptake Certain root specializations help plants take up water and nutrients from soil and air. For example, as a plant adds primary growth, its root system may develop billions of tiny root hairs (Figure 27.14**A**). Collectively, these thin extensions of root epidermal cells enormously increase the surface area available for absorbing water and nutrients. Root hairs are fragile and no more than a few millimeters long. They do not develop into new roots; each lasts only a few days.

You already know something about "fungus–roots," or **mycorrhizae** (singular, mycorrhiza). A mycorrhiza is a symbiotic interaction between a young root and a fungus (Section 14.9). Hyphae—filaments of the fungus—form a velvety cover around roots or penetrate the root cells. Collectively, the hyphae have a large surface area, so they absorb mineral ions from a larger volume of soil than roots alone (Figure 27.14**B**). The fungus absorbs some sugars and nitrogen-rich compounds from root cells. The root cells get some scarce minerals that the fungus is better able to absorb.

Some types of anaerobic bacteria in soil are mutualists with clover, peas, and other legumes. Like all other plants, legumes require nitrogen for growth. Nitrogen gas (N≡N, or N_2) is abundant in the air, but plants do not have an enzyme that can break it apart. The bacteria do. The enzyme uses ATP to convert nitrogen gas to ammonia (NH_3). The metabolic conversion of gaseous nitrogen to ammonia is called **nitrogen fixation**. Other soil bacteria convert ammonia to nitrate (NO_3^-), a form of nitrogen that plants readily absorb. Nitrogen-fixing bacteria infect roots and thus become symbionts in localized swellings called **root nodules** (Figure 27.14**C**,**D**). The bacteria get an oxygen-free environment and sugars from the plant. The plant gets some fixed nitrogen from the bacteria.

Take-Home Message How do plants get the nutrients they need?

- Nutrients are essential elements for plant growth. Plants require nine macronutrients and seven micronutrients. All are available in air, water, and soil.

- Humus (decaying organic matter) is rich in nutrients. Clay in soil attracts water and nutrients, but can pack tightly, excluding air. Most plants do best in soil with a mixture of particle sizes and a moderate amount of humus.

- Root hairs, root nodules that contain bacteria, and mycorrhizae greatly enhance the uptake of water and dissolved nutrients.

cohesion–tension theory Explanation of how transpiration creates a tension that pulls a cohesive column of water through xylem, from roots to shoots.
mycorrhiza Fungus–plant root partnership.
nitrogen fixation Conversion of nitrogen gas to ammonia.
root nodules Swellings of some plant roots that contain nitrogen-fixing bacteria.

27.6 Water and Solute Movement in Plants

Water that enters a root moves into xylem of the root vascular cylinder. From there, it travels through pipelines of the xylem to the rest of the plant body. Tracheids and vessel members of xylem are dead at maturity; only their lignin-impregnated walls remain. Being dead, these cells cannot expend any energy to pump water upward against gravity. So how does water move from roots to leaves that may be more than 100 meters (330 feet) above the soil? Figure 27.15 illustrates the **cohesion–tension theory**. By this theory, water in xylem is pulled upward by air's drying power, which creates a continuous negative pressure called tension. The tension extends all the way from leaves to roots.

FIGURE 27.15 Animated! Cohesion–tension theory of water transport in vascular plants.

① Water evaporates from aboveground plant parts.

② The evaporation exerts tension (pulls) on the narrow columns of water that fill xylem tubes. The tension extends from leaves to roots. Hydrogen bonds hold water molecules together inside xylem tubes (cohesion).

③ As long as evaporation continues, the tension it creates drives the uptake of water molecules from soil.

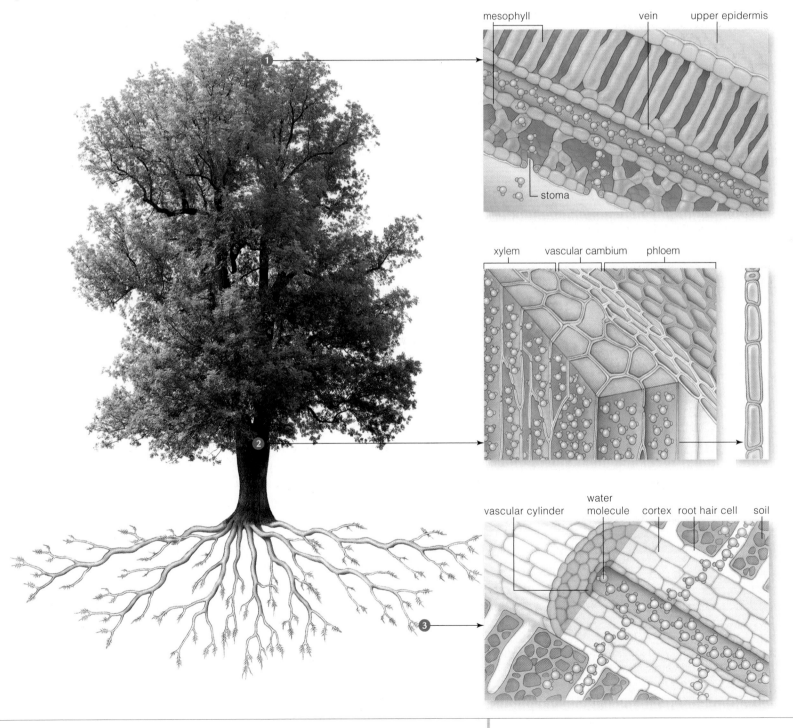

mesophyll   vein   upper epidermis

stoma

xylem   vascular cambium   phloem

vascular cylinder   water molecule   cortex   root hair cell   soil

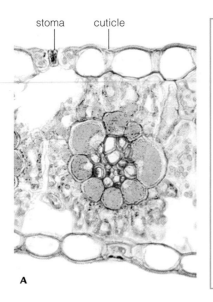

stoma   cuticle

A

FIGURE 27.16 Water-conserving structures in plants. **A** Stomata and cuticle in a cross-section of basswood leaf.

B Open stoma. When guard cells swell with water, they bend so that a gap opens between them. The gap allows the plant to exchange gases with air. Guard cells are the only type of epidermal cell with chloroplasts (round structures inside the cells).

C Closed stoma. The guard cells, which are not plump with water, are collapsed against each other so there is no gap between them. A closed stoma limits water loss, but it also limits gas exchange.

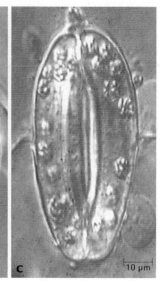

guard cell   guard cell

B   **C**   10 µm

The evaporation of water from aboveground plant parts, particularly at stomata, is called **transpiration**. Its effect on water inside a plant is a bit like what happens when you suck a drink through a straw. Transpiration puts negative pressure (pulls) on continuous columns of water that fill the narrow conductive tubes of xylem. The water resists breaking into droplets because its molecules are all connected to one another by hydrogen bonds (Section 2.4). A pull on one also pulls on the others. Thus, the negative pressure created by transpiration (tension) pulls on the entire column of water that fills a xylem tube. Because of water's cohesion, the tension extends from leaves, down through stems, into young roots where water is being absorbed from the soil.

The movement of water through plants is driven mainly by transpiration. However, evaporation is only one of many processes in plants that involve the loss of water molecules. Photosynthesis and many other processes contribute to the negative pressure that results in water movement.

Controlling Water Loss Even mildly water-stressed plants would wilt and die without a cuticle. This water-impermeable layer coats the walls of all plant cells exposed to air (Figure 27.16**A**). It consists of the waxy, waterproof secretions of epidermal cells. The cuticle is translucent, so it does not prevent light from reaching photosynthetic tissues.

Only about 2 percent of the water that enters a root is used in metabolism. Despite the water-conserving cuticle, most of the rest evaporates right out of the plant, through stomata. A pair of guard cells defines each stoma. When these two cells swell with water, they bend slightly so a gap forms between them (Figure 27.16**B**). The gap is the stoma. When the cells lose water, they collapse against each other, so the gap closes (Figure 27.16**C**).

Environmental cues such as water availability, the level of carbon dioxide inside the leaf, and light intensity affect whether stomata open or close. These cues trigger osmotic pressure changes in the cytoplasm of guard cells. For example, when the sun comes up, the light causes guard cells to begin pumping potassium ions into their cytoplasm. The resulting buildup of potassium ions causes water to enter the cells by osmosis. The guard cells plump up, so the gap between them opens. Carbon dioxide from the air diffuses into the plant's tissues, and photosynthesis begins.

Distribution of Sugars Unlike conducting tubes of xylem, sieve tubes in phloem consist of living cells. Sieve-tube cells are positioned side by side and

pressure flow theory Explanation of how flow of fluid through phloem is driven by differences in pressure and sugar concentration between a source and a sink.
translocation Process that moves organic compounds through phloem.
transpiration Evaporation of water from plant parts.

end to end, and their abutting end walls (sieve plates) are porous. Dissolved organic compounds flow through the tubes.

Companion cells that are pressed against the sieve tubes actively transport the organic products of photosynthesis into them. Some of the molecules are used in the cells that make them, but the rest travel through the sieve tubes to all other parts of the plant.

Translocation is the formal name for the process that moves organic molecules through phloem. The molecules flow from a source to a sink. A source is any region of the plant where companion cells are loading organic compounds into the sieve tubes. Photosynthetic tissues in leaves are usually a plant's main source region. A sink is any region where molecules are being used or stored. Roots are sink regions, as are growing fruits.

Why do organic compounds flow from a source to a sink? According to the **pressure flow theory**, internal pressure builds up at the source end of the sieve tube. The pressure pushes the solute-rich fluid toward a sink (Figure 27.17). By energy-requiring reactions, companion cells in leaf veins load sugars into sieve-tube members ❶. The sugars increase the solute concentration of the fluid inside the tubes, so water also moves into the tubes, by osmosis ❷. The water increases the fluid volume, which in turn increases the internal pressure (turgor) inside the sieve tubes ❸. Pressure in phloem can be five times higher than the pressure in an automobile tire. That high pressure pushes the sugar-laden fluid toward sink regions, which have a lower internal pressure ❹. The sugars are unloaded at sink regions, and water diffuses out of phloem there, again by osmosis ❺.

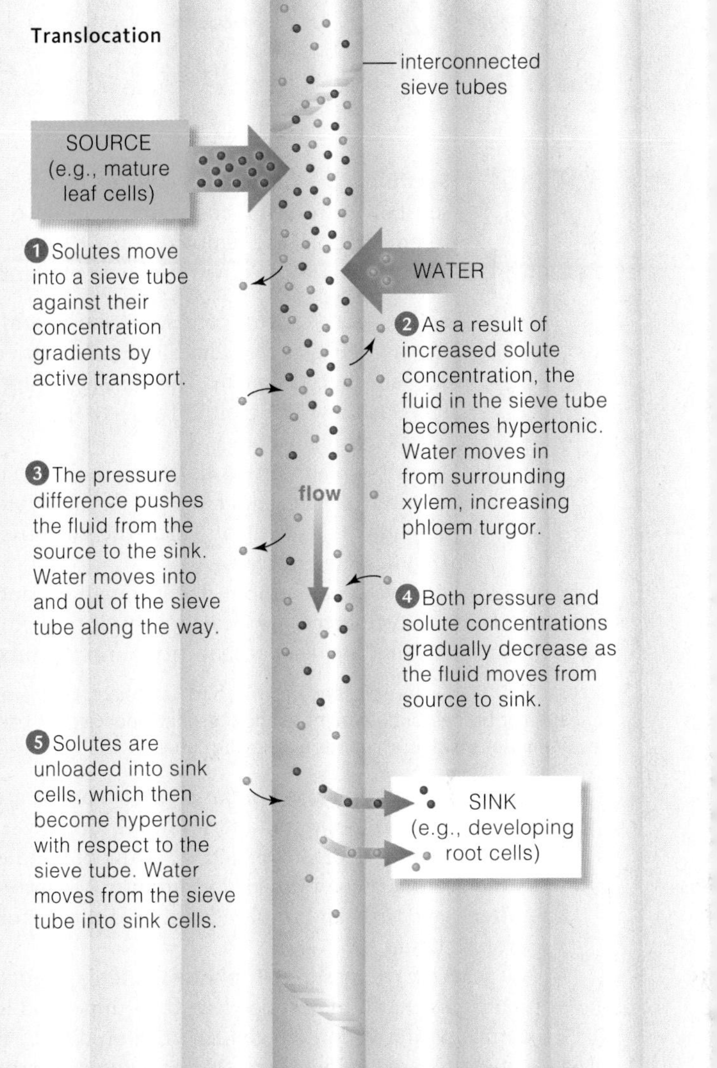

Take-Home Message How do substances move through a plant body?

- Open stomata allow gas exchange between the air and the plant, but they also allow water loss by transpiration.

- Transpiration puts water in xylem into a continuous state of negative tension from roots to leaves to roots. The tension pulls columns of water in xylem upward through the plant.

- Concentration and pressure gradients move organic compounds inside sieve tubes of phloem, from sources to sinks.

Translocation

interconnected sieve tubes

SOURCE (e.g., mature leaf cells)

WATER

❶ Solutes move into a sieve tube against their concentration gradients by active transport.

❷ As a result of increased solute concentration, the fluid in the sieve tube becomes hypertonic. Water moves in from surrounding xylem, increasing phloem turgor.

❸ The pressure difference pushes the fluid from the source to the sink. Water moves into and out of the sieve tube along the way.

flow

❹ Both pressure and solute concentrations gradually decrease as the fluid moves from source to sink.

❺ Solutes are unloaded into sink cells, which then become hypertonic with respect to the sieve tube. Water moves from the sieve tube into sink cells.

SINK (e.g., developing root cells)

FIGURE 27.17 Animated! Translocation.

27.7

Impacts/Issues Revisited:
Leafy Cleanup Crews

With elemental pollutants such as lead or mercury, the best phytoremediation strategies use plants that take up toxins and store them in aboveground tissues. The toxin-laden plant parts can then be harvested for safe disposal. Researchers have genetically modified some plants to enhance their absorptive and storage capacity. Dr. Kuang-Yu Chen, pictured at *left*, is analyzing zinc and cadmium levels in plants that can take up these elements. In the case of organic toxins such as TCE, the best phytoremediation strategies use plants with biochemical pathways that break down the compounds to less toxic molecules. Researchers are beefing up these pathways in many plants. Some are transferring genes from bacteria or animals into plants; others are enhancing expression of genes that encode molecular participants in the plants' own detoxification pathways.

How Would

YOU

 **Vote?**

Plants can be genetically engineered to take up toxins more effectively. Do you support the use of such plants to help clean up toxic waste sites? **See CengageNow for details, then vote online (cengagenow.com).**

Summary

Section 27.1 The ability of plants to take up substances from soil and water is the basis for phytoremediation (the removal of toxic substances from a contaminated area with the help of plants).

Section 27.2 A plant consists of shoots and roots. The vascular tissue system distributes nutrients and water through the plant. The dermal system protects plant surfaces. The ground tissue system makes up the bulk of the plant body.

Simple plant tissues have only one type of cell. Complex plant tissues have two or more. Monocots and eudicots (true dicots) have the same tissues organized in different ways.

Water and dissolved mineral ions flow through conducting tubes of xylem. The interconnected, perforated walls of tracheids and vessel members (cells that are dead at maturity) form the tubes.

CENGAGENOW™ Explore the body plan of a tomato plant, compare monocot and eudicot tissues, and learn more about ground and vascular tissues with the animations and interactions on CengageNow.

Section 27.3 Stems support upright growth and conduct water and solutes through vascular bundles. Most eudicot stems have a ring of bundles dividing the ground tissue into cortex and pith. Monocot stems often have vascular bundles distributed throughout their ground tissue.

Leaves have mesophyll (photosynthetic parenchyma) and veins (vascular bundles) between their upper and lower epidermis. Water vapor, oxygen, and carbon dioxide cross the cuticle-covered epidermis at stomata. Roots absorb water and dissolved mineral ions, which become distributed to aboveground parts. Eudicots typically have a taproot system, and many monocots have a fibrous root system.

Lengthening of roots and shoots (primary growth) occurs when cells of apical meristems at root tips or shoot tips divide.

CENGAGENOW™ Look inside plant stems and examine the structure of a leaf and a root with the animations on CengageNow.

Section 27.4 Woody plants thicken (add secondary growth) by cell divisions at lateral meristems. Divisions at vascular cambium add secondary xylem (wood) and secondary phloem. Divisions at cork cambium add periderm. Bark is periderm and secondary phloem of a woody stem.

CENGAGENOW™ Learn about secondary growth and the structure of a woody stem with the animations on CengageNow.

Section 27.5 Plant nutrition requires water, mineral ions, and carbon dioxide. Plants obtain nine macronutrients (such as carbon, oxygen, hydrogen, nitrogen, and phosphorus) and seven micronutrients (such as iron) from air, water, and soil. The properties of soil affect accessibility of water, oxygen, and nutrients to plants. Root hairs increase the surface area for absorption of water and nutrients. Mycorrhizae and bacterial symbionts in root nodules help many plants take up nutrients.

Section 27.6 Transpiration is the evaporation of water from plant parts, mainly at stomata, into air. By the cohesion–tension theory, transpiration pulls water upward by creating a continuous negative pressure (or tension) inside xylem from leaves to roots. Hydrogen bonds among water molecules keep the columns of fluid continuous inside the narrow vessels.

A cuticle and stomata balance a plant's loss of water with its needs for gas exchange. Stomata are gaps across the cuticle-covered epidermis of leaves and other plant parts. Each is defined by a pair of guard cells. Closed stomata limit the loss of water, but also prevent the gas exchange required for photosynthesis and aerobic respiration. Water moving into guard cells plumps them, which opens the stoma between them. Water diffusing out of the cells causes them to collapse against each other, so the gap closes.

The movement of sugars and other organic compounds through sieve tubes of phloem is driven by pressure and concentration gradients between source and sink regions. Companion cells load sugars from photosynthetic cells into sieve tubes. The sugars are unloaded at actively growing regions or in areas used for storage.

CENGAGENOW™ See how stomata work, and how vascular plants transport water and organic compounds with the animations on CengageNow.

Self-Quiz   Answers in Appendix I

1. Roots and shoots lengthen through activity at _____ .
   a. apical meristems   c. vascular cambium
   b. lateral meristems   d. cork cambium

2. _____ conducts mainly water and mineral ions, and _____ conducts mainly sugars.
   a. Phloem; xylem   b. Xylem; phloem

3. The bulk of a leaf is made up of mesophyll, a type of _____ .
   a. sclerenchyma   c. collenchyma
   b. parenchyma   d. dermal tissue

4. Which of the following cell types are alive in mature tissue? Choose all that apply.
   a. collenchyma cells   c. vessels   e. companion cells
   b. sieve tubes   d. tracheids   f. sclerenchyma cells

5. Division of cells of the vascular cambium produces _____ .
   a. epidermis   c. cork
   b. secondary phloem   d. all of the above

6. Water evaporation from plant parts is called _____ .
   a. translocation   c. transpiration
   b. cohesion   d. tension

7. When guard cells swell, _____ .
   a. transpiration ceases   c. a stoma opens
   b. sugars enter phloem   d. root cells die

8. A waxy cuticle is secreted by _____ .
   a. ground tissue   c. a stoma
   b. epidermal cells   d. root hairs

9. Plants obtain carbon, hydrogen, and oxygen from _____ .
   a. soil   c. air
   b. water   d. b and c

10. Decomposing matter in soil is called _____ .

11. In many plant species, older roots and stems thicken by activity at _____ .
   a. apical meristems   c. vascular cambium
   b. lateral meristems   d. both b and c

12. Water transport from roots to leaves occurs by _____ .
   a. pressure flow
   b. differences in source and sink solute concentrations
   c. the pumping force of xylem vessels
   d. transpiration, tension, and cohesion of water

13. Sugar transport from leaves to roots occurs by _____ .
   a. pressure flow
   b. differences in source and sink solute concentrations
   c. the pumping force of xylem vessels
   d. transpiration, tension, and cohesion of water

14. Match the plant parts with the suitable description.
   _____ apical meristem   a. mass of secondary xylem
   _____ lateral meristem   b. source of primary growth
   _____ xylem   c. distribution of sugars
   _____ phloem   d. source of secondary growth
   _____ vascular cylinder   e. distribution of water
   _____ wood   f. central column in roots

Additional questions are available on CENGAGENOW™

Critical Thinking

1. Suppose you hammer a nail into an oak tree trunk five feet above the ground. If you return 15 years later, after the tree has grown 20 feet taller, how high will the nail be?

2. Iron and manganese are components of chlorophyll. How does a plant obtain these minerals? What effect would you expect a deficiency in one of these minerals to have on plant growth?

3. Farmers sometimes plant clovers or other legumes in fields during the season when they are not growing crops. When crop-planting season approaches, the "cover crop" is plowed under. Explain how this procedure enhances soil fertility.

4. Is the plant with the yellow flower *below* a eudicot or a monocot? What about the plant with the purple flower?

Digging Into Data

Transgenic Plants for Phytoremediation

Plants used for phytoremediation take up organic pollutants from the soil or air, then transport the chemicals to plant tissues, where they are stored or broken down. Researchers are now designing transgenic plants that have enhanced ability to take up or break down toxins.

In 2007, Sharon Doty and her colleagues published the results of their efforts to design plants useful for phytoremediation of soil and air containing organic solvents. The researchers used *Agrobacterium tumefaciens* (Section 10.4) to deliver a mammalian gene into poplar plants. The gene encodes cytochrome P450, an enzyme involved in the metabolism of a range of organic molecules, including solvents such as TCE. The results of one of the researchers' tests on these transgenic plants are shown in Figure 27.18.

1. How many transgenic plants did the researchers test?

2. In which group did the researchers see the slowest rate of TCE uptake? The fastest?

3. On day 6, what was the difference between the TCE content of air around transgenic plants and that around vector control plants?

4. Assuming no other experiments were done, what two explanations are there for the results of this experiment? What other control might the researchers have used?

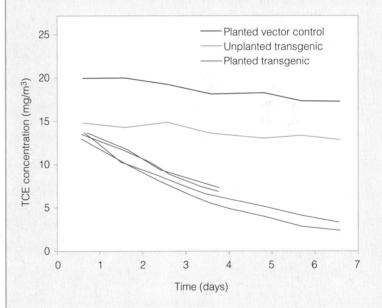

FIGURE 27.18 Results of tests on transgenic poplar trees. Planted trees were incubated in sealed containers with an initial 15,000 micrograms of TCE (trichloroethylene) per cubic meter of air. Samples of the air in the containers were taken daily and measured for TCE content. Controls included a tree transgenic for a Ti plasmid with no cytochrome P450 in it (vector control), and a bare-root transgenic tree (one that was not planted in soil).

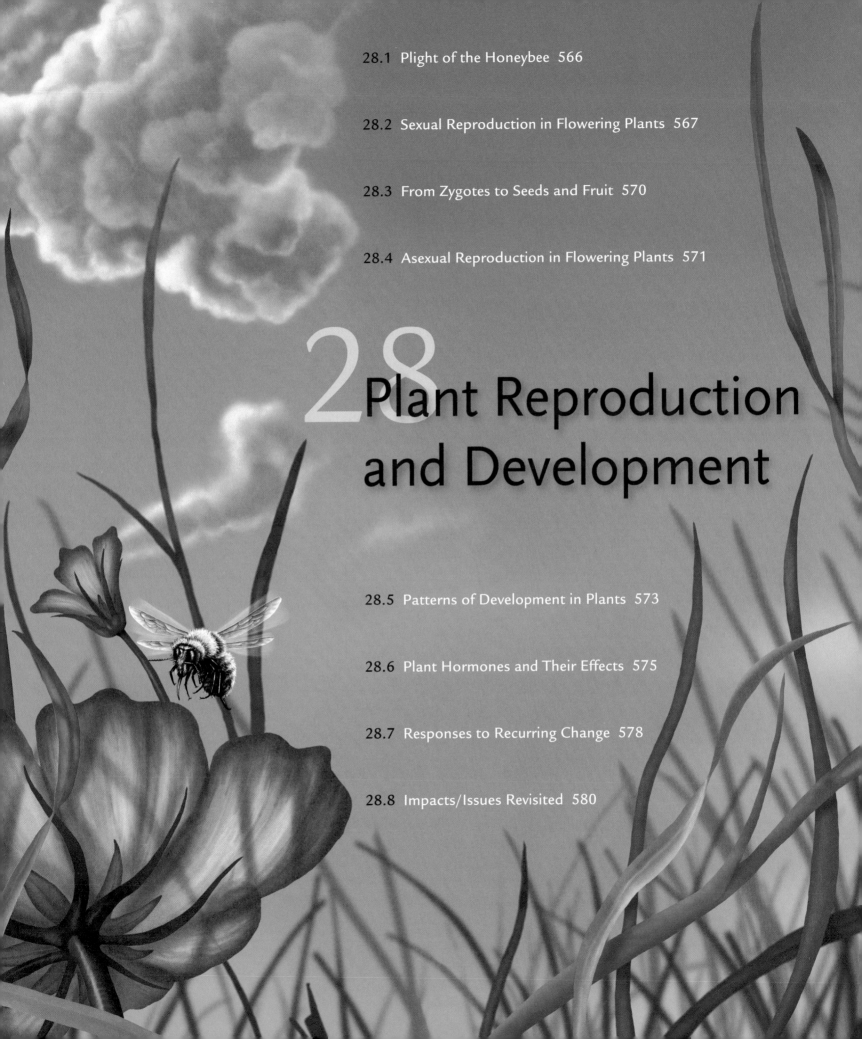

28 Plant Reproduction and Development

Impacts/Issues: Plight of the Honeybee

In the fall of 2006, commercial beekeepers in Europe, India, and North America began to notice something was amiss in their honeybee hives. The bees were dying off in unusually high numbers. Many colonies did not survive through the following winter. By spring, the phenomenon had a name: colony collapse disorder. Farmers and biologists began to worry about what would happen if the honeybee populations continued to decline. Honey production would suffer, but many commercial crops would fail too.

Nearly all of our crops are flowering plants. These plants make pollen grains that consist of a few cells, one of which produces sperm. Honeybees are **pollinators**; they carry pollen from one plant to another, pollinating flowers as they do. Typically, a flower will not develop into a fruit unless it receives pollen from another flower. Even plants with flowers that can self-pollinate tend to make bigger fruits and more of them when they are cross-pollinated (Figure 28.1).

Many types of insects pollinate plants, but honeybees are especially efficient pollinators of a variety of plant species. They are also the only ones that tolerate living in man-made hives that can be loaded onto trucks and carted wherever crops require pollination. Loss of their portable pollination service is a huge threat to our agricultural economy.

We still do not know what causes colony collapse disorder. Honeybees can be infected by a variety of parasites and diseases that may be part of the problem. Parasitic mites and Israeli acute paralysis virus have been detected in many affected hives, and pesticides may also be taking a toll. In the past few years, neonicotinoids have become the most widely used insecticides in the United States. These chemicals are systemic insecticides, which means they are taken up by all plant tissues, including the nectar and pollen that honeybees collect. Neonicotinoids are highly toxic to honeybees.

Colony collapse disorder is currently in the spotlight because it affects our food supply. However, other pollinator populations are also dwindling. Habitat loss is probably the main factor, but pesticides that harm honeybees also harm other pollinators.

Flowering plants rose to dominance in part because they coevolved with animal pollinators. Most flowers are specialized to attract and be pollinated by a specific type of pollinator or even a specific species. Those adaptations put the plants at risk of extinction if coevolved pollinator populations decline. Wild animal species that depend on the plants for fruits and seeds will also be affected. Recognizing the prevalence and importance of these interactions is our first step toward finding workable ways to protect them.

FIGURE 28.1 Importance of insect pollinators. **A** Honeybees are efficient pollinators of a variety of flowers, including berries. **B** Raspberry flowers can pollinate themselves, but the fruit that forms from a self-fertilized flower is of lower quality than that of a cross-pollinated flower. The two berries on the *left* formed from self-pollinated flowers. The one on the *right* formed from an insect-pollinated flower.

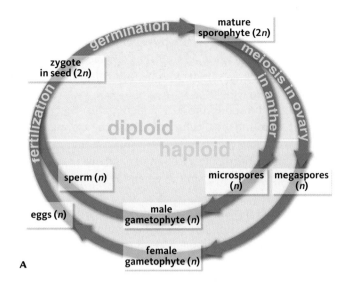

A

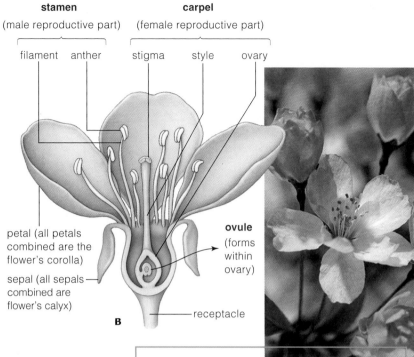

B

FIGURE 28.2 Animated! Sexual reproduction in flowering plants. **A** Life cycle of a typical angiosperm. **B** Like many flowers, a cherry blossom has several stamens and one carpel. The male reproductive parts are stamens, which consist of pollen-bearing anthers atop slender filaments. The female reproductive part is the carpel, which consists of stigma, style, and ovary.

Sexual Reproduction in Flowering Plants

Regarding Flowers A sporophyte (Section 14.2) is a diploid plant body that grows by mitotic divisions of a fertilized egg. Flowers are the specialized reproductive shoots of sporophytes. Haploid reproductive bodies (spores) form by meiosis in flowers. Spores develop into haploid gametophytes, which make eggs or sperm (Figure 28.2**A**).

A flower forms when a lateral bud along a stem develops into a short, modified branch called a receptacle. The petals and other parts of a typical flower are modified leaves that form in four spirals or rings (whorls) at the end of the floral shoot. The outermost whorl develops into a calyx, which is a ring of leaflike sepals (Figure 28.2**B**). The sepals of most flowers are photosynthetic and inconspicuous; they protect the flower's reproductive parts. Just inside the calyx, petals form in a whorl called the corolla (from the Latin *corona*, or crown). Petals are usually the largest and most brightly colored parts of a flower. They function mainly to attract pollinators.

A whorl of stamens forms inside the ring of petals. **Stamens** are the male parts of a flower. In most flowers, they consist of a thin filament with an anther at the tip. Inside a typical anther are two pairs of elongated pouches called pollen sacs. Meiosis of diploid cells in each sac produces haploid, walled spores. The spores differentiate into pollen grains, which are immature male gametophytes.

The innermost whorl of modified leaves are folded and fused into **carpels**, the female parts of a flower. Carpels are sometimes called pistils. Many flowers have one carpel; others have several carpels, or several groups of carpels, that may be fused. The upper region of a carpel is a sticky or hairy stigma that is specialized to trap pollen grains. Typically, the stigma sits on top of a slender stalk called a style. The lower, swollen region of a carpel is the **ovary**, which contains one or more ovules. An **ovule** is a tiny bulge of tissue inside the ovary. A cell in the ovule undergoes meiosis and develops into the haploid female gametophyte.

At fertilization, a diploid zygote forms when male and female gametes meet inside an ovary. The ovule then matures into a seed. The life cycle is completed when the seed sprouts, and a new sporophyte forms and matures.

carpel Female reproductive structure of a flower; a sticky or hairlike stigma above a chamber (ovary) that contains one or more ovules.

ovary In flowering plants, the enlarged base of a carpel, inside which one or more ovules form and eggs are fertilized.

ovule In a seed-bearing plant, structure in which a haploid, egg-producing female gametophyte forms; after fertilization, matures into a seed.

pollinator An organism that moves pollen grains from one plant to another.

stamen The male reproductive part of a flowering plant; consists of a pollen-producing anther on a filament.

From Spores to Zygotes Figure 28.3 zooms in on a flowering plant life cycle. On the male side, masses of diploid, spore-producing cells form by mitosis in the anthers. Typically, walls develop around the cell masses to form four pollen sacs ❶. Each cell inside the sacs undergoes meiosis, forming four haploid **microspores** ❷.

Mitosis and differentiation of microspores produce pollen grains. Each pollen grain consists of a coat that surrounds two cells, one inside the cytoplasm of the other ❸. A pollen grain's durable coat is a bit like a suitcase that carries and protects the cells inside on their journey to meet an egg. After a period of arrested growth, or **dormancy**, the pollen sacs split open and release pollen from the anther ❹.

Pollination refers to the arrival of a pollen grain on a receptive stigma. Interactions between the two structures stimulate the pollen grain to resume metabolic activity. One of the two cells in the pollen grain then develops into a tubular outgrowth called a pollen tube. The other cell undergoes mitosis and cytoplasmic division, producing two sperm cells (the male gametes) within the pollen tube. A pollen tube together with its contents of male gametes constitutes the mature male gametophyte ❺.

On the female side, a mass of tissue—the ovule—starts growing on the inner wall of an ovary ❻. One cell in the middle of the mass undergoes meiosis and cytoplasmic division, forming four haploid **megaspores** ❼.

Three of the four megaspores typically disintegrate. The remaining megaspore undergoes three rounds of mitosis without cytoplasmic division. The outcome is a single cell with eight haploid nuclei ❽. The cytoplasm of this cell divides unevenly, and the result is a seven-celled embryo sac that constitutes the female gametophyte ❾. The gametophyte is enclosed and protected by cell layers, called integuments, that developed from the outer layers of the ovule. One of the cells in the gametophyte, the **endosperm mother cell**, has two nuclei ($n + n$). Another cell is the egg.

The pollen tube grows from its tip down through the carpel and ovary toward the ovule, carrying with it the two sperm cells. Chemical signals secreted by the female gametophyte guide the tube's growth to the embryo sac within the ovule. Many pollen tubes may grow down into a carpel, but only one typically penetrates an embryo sac. The sperm cells are then released into the sac ❿. Flowering plants undergo **double fertilization**: One of the sperm cells from the pollen tube fuses with (fertilizes) the egg and forms a diploid zygote. The other fuses with the endosperm mother cell, forming a triploid ($3n$) cell. This cell gives rise to triploid **endosperm**, a nutritious tissue that forms only in seeds of flowering plants. When a seed sprouts, endosperm sustains rapid growth of the sporophyte seedling until true leaves form and photosynthesis begins.

Take-Home Message **How does sexual reproduction work in angiosperms?**

- The male parts of flowers are stamens, which typically consist of a filament with an anther at the tip. Pollen forms inside anthers. Pollen gives rise to the male gametophytes.

- The female parts of flowers are carpels, which typically consist of stigma, style, and ovary. Haploid, egg-producing female gametophytes form in an ovule inside the ovary. Pollination occurs when pollen arrives on a receptive stigma.

- A pollen grain germinates on a stigma as a pollen tube containing male gametes. The pollen tube grows into the carpel and enters an ovule. Double fertilization occurs when one of the male gametes fuses with the egg, the other with the endosperm mother cell.

dormancy Period of arrested growth.
double fertilization Mode of fertilization in flowering plants in which one sperm cell fuses with the egg, and a second sperm cell fuses with the endosperm mother cell.
endosperm Nutritive tissue in the seeds of flowering plants.
endosperm mother cell A cell with two nuclei ($n + n$) that is part of the female gametophyte of a flowering plant.
megaspore Haploid spore that forms in ovary of seed plants; gives rise to a female gametophyte with an egg cell.
microspore Walled haploid spore of seed plants; gives rise to a pollen grain.
pollination The arrival of pollen on a receptive stigma.

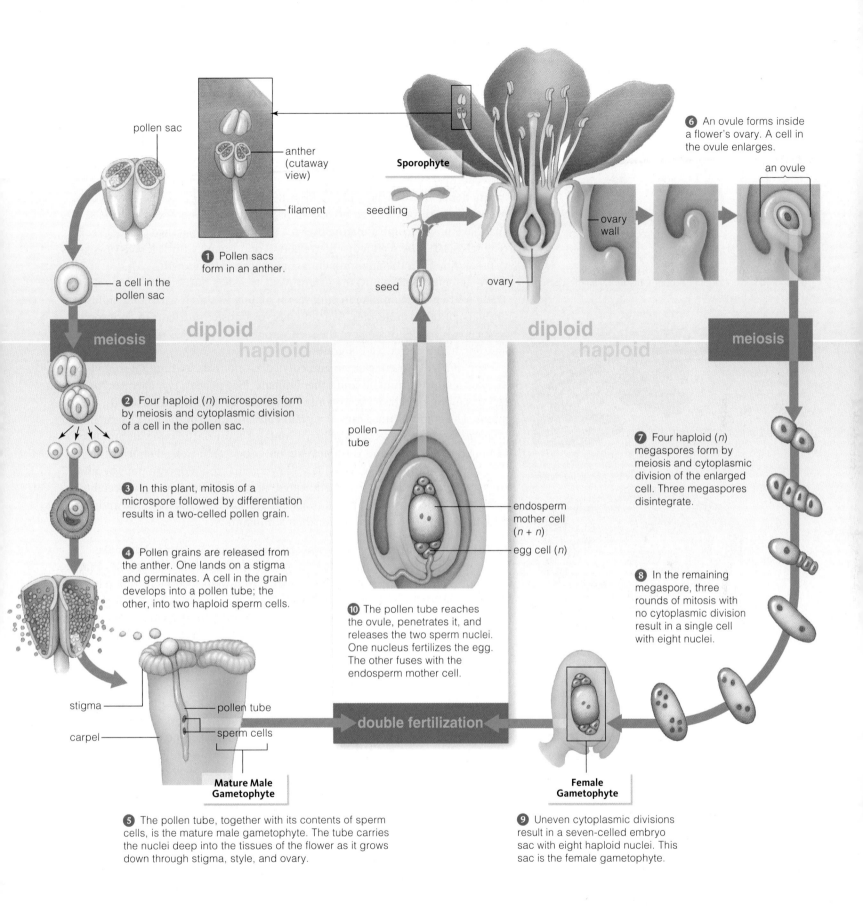

pollen sac

anther
(cutaway
view)

filament

Sporophyte

6 An ovule forms inside
a flower's ovary. A cell in
the ovule enlarges.

an ovule

seedling

ovary
wall

1 Pollen sacs
form in an anther.

a cell in the
pollen sac

seed

ovary

diploid

meiosis

haploid

diploid

haploid

meiosis

2 Four haploid (*n*) microspores form
by meiosis and cytoplasmic division
of a cell in the pollen sac.

pollen
tube

7 Four haploid (*n*)
megaspores form by
meiosis and cytoplasmic
division of the enlarged
cell. Three megaspores
disintegrate.

3 In this plant, mitosis of a
microspore followed by differentiation
results in a two-celled pollen grain.

4 Pollen grains are released from
the anther. One lands on a stigma
and germinates. A cell in the grain
develops into a pollen tube; the
other, into two haploid sperm cells.

endosperm
mother cell
(*n* + *n*)

egg cell (*n*)

8 In the remaining
megaspore, three
rounds of mitosis with
no cytoplasmic division
result in a single cell
with eight nuclei.

stigma

pollen tube

carpel

sperm cells

**Mature Male
Gametophyte**

10 The pollen tube reaches
the ovule, penetrates it, and
releases the two sperm nuclei.
One nucleus fertilizes the egg.
The other fuses with the
endosperm mother cell.

double fertilization

**Female
Gametophyte**

5 The pollen tube, together with its contents of sperm
cells, is the mature male gametophyte. The tube carries
the nuclei deep into the tissues of the flower as it grows
down through stigma, style, and ovary.

9 Uneven cytoplasmic divisions
result in a seven-celled embryo
sac with eight haploid nuclei. This
sac is the female gametophyte.

FIGURE 28.3 Animated! Life cycle of cherry tree, a eudicot. **Figure It Out:**
What structure gives rise to a pollen grain by mitosis?   *Answer: A microspore*

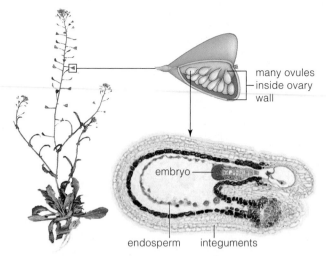

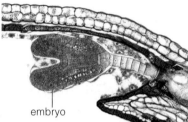

A After fertilization, the flower's ovary develops into a fruit. An embryo surrounded by integuments forms inside each of the ovary's many ovules.

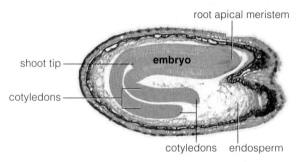

B The embryo is heart-shaped when cotyledons start forming. Endosperm tissue expands as the parent plant transfers nutrients into it.

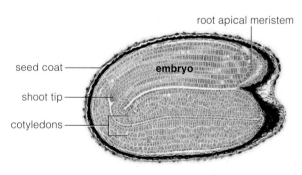

C The developing embryo becomes torpedo-shaped when the enlarging cotyledons bend inside the ovule.

D A layered seed coat that formed from the layers of integuments surrounds the mature embryo sporophyte. In eudicots like shepherd's purse, nutrients have been transferred from endosperm into two cotyledons.

FIGURE 28.4 Animated! Embryonic development of shepherd's purse, a eudicot.

From Zygotes to Seeds and Fruit

The Embryo Sporophyte Forms In flowering plants, double fertilization produces a zygote and a triploid (3*n*) cell. By mitotic cell divisions, the zygote develops into an embryo sporophyte, and the triploid cell develops into endosperm (Figure 28.4**A**–**C**). When the embryo approaches maturity, the integuments of the ovule separate from the ovary wall and become layers of a protective seed coat. The embryo sporophyte, its reserves of food, and the seed coat have now become a mature ovule, a self-contained package called a **seed** (Figure 28.4**D**). The seed may enter a period of dormancy until it receives signals that conditions in the environment are appropriate for growth.

As an embryo develops, the parent plant transfers nutrients to the ovule. These nutrients accumulate in endosperm mainly as starch with some lipids, proteins, or other molecules. Eudicot embryos transfer nutrients in endosperm to their two cotyledons before the seed sprouts. The embryos of monocots tap endosperm only after seeds sprout.

The nutrients in endosperm and cotyledons nourish seedling sporophytes. They also nourish humans and other animals. Rice, wheat, rye, oats, and barley are among the grasses commonly cultivated for their nutritious seeds, or grains. The embryo (the germ) of a grain contains most of the seed's protein and vitamins, and the seed coat (the bran) contains most of the minerals and fiber. Milling removes bran and germ, leaving only the starch-packed endosperm.

Maize, or corn, is the most widely grown grain crop. Popcorn pops because the moist endosperm steams when heated; pressure builds inside the seed until it bursts. Cotyledons of bean and pea seeds are valued for their starch and protein; those of coffee and cacao, for their stimulants.

Fruits Only flowering plants form seeds in ovaries, and only they make fruits. A **fruit** is a seed-containing mature ovary, often with fleshy tissues that develop from the ovary wall. In some plants, fruit tissues develop from parts of the flower other than the ovary wall (such as petals, sepals, stamens, or receptacles). Simple fruits, such as pea pods, acorns, and shepherd's purse, are derived from one ovary. Strawberries and other aggregate fruits form from separate ovaries of one flower; they mature as a cluster of fruits. Multiple fruits form from fused ovaries of separate flowers. The pineapple is a multiple fruit that forms from fused ovary tissues of many flowers.

An embryo or seedling can use the nutrients stored in endosperm or cotyledons, but not in fruit. The function of fruit is to protect and disperse seeds. Dispersal increases reproductive success by minimizing competition for resources among parent and offspring, and by expanding the area colonized by the species. Thus, fruits have adaptations that facilitate their dispersal by environmental factors such as water or wind, or by mobile organisms such as birds or insects.

Water-dispersed fruits have water-repellent outer layers. The fruits of sedges native to American marshlands have seeds encased in a bladderlike envelope that floats (Figure 28.5**A**). Buoyant fruits of the coconut palm have thick, tough husks that can float for thousands of miles in seawater.

Many plant species use wind as a dispersal agent. Part of a maple fruit is a dry outgrowth of the ovary wall that extends like a pair of thin, lightweight wings (Figure 28.5**B**). The fruit breaks in half when it drops from the tree; as the halves drop to the ground, wind currents that catch the wings spin the attached seeds away. Tufted fruits of thistle, cattail, dandelion, and milkweed may be blown as far as 10 kilometers (6 miles) from the parent plant (Figure 28.5**C**).

The fruits of cocklebur, bur clover, and many other plants have hooks or spines that stick to the feathers, feet, fur, or clothing of more mobile species (Figure 28.5**D**). The dry, podlike fruit of plants such as California poppy propel their seeds through the air when they pop open explosively (Figure 28.5**E**).

Colorful, fleshy, fragrant fruits attract insects, birds, and mammals that disperse seeds (Figure 28.5**F**). The animal may eat the fruit and discard the seeds, or eat the seeds along with the fruit. Abrasion of the seed coat by digestive enzymes in an animal's gut can help the seed sprout after it departs in feces.

Take-Home Message How do seeds and fruits develop?

- After fertilization, the zygote develops into an embryo, the endosperm becomes enriched with nutrients, and the ovule's integuments develop into a seed coat.

- A seed is a mature ovule. It contains an embryo sporophyte.

- A mature ovary, with or without accessory tissues that develop from other parts of a flower, is a fruit.

- Seeds and fruits are adapted for dispersal by specific environmental factors.

FIGURE 28.5 Examples of adaptations that aid fruit dispersal. **A** Air-filled bladders that encase the seeds of certain sedges allow the fruits to float in their marshy habitats.

B Wind lifts the "wings" of maple (*Acer*) fruits, which spin the seeds away from the parent tree.

C Wind that catches the hairy modified sepals of a dandelion fruit (*Taraxacum*) lifts the seed away from the parent plant.

D Curved spines make cocklebur (*Xanthium*) fruits stick to the fur of animals (and clothing of humans) that brush past it.

E The fruits of the California poppy (*Eschscholzia californica*) are long, dry pods that split open suddenly. The movement jettisons the seeds.

F The red, fleshy fruit of crabapples attracts birds such as cedar waxwings.

28.4 Asexual Reproduction in Flowering Plants

Unlike most animals, most flowering plants can reproduce asexually. New roots and shoots can grow from extensions or pieces of a parent plant, a process called **vegetative reproduction**. Each new plant is a clone, a genetic replica of its parent. As long as conditions in the environment favor growth, such clones are as close as any organism gets to being immortal. The oldest known plant is a clone: the one and only population of King's holly, which consists of several hundred stems growing along 1.2 kilometers (0.7 miles) of a river gully in Tasmania. Radiometric dating of the plant's fossilized leaf litter show that the clone is at least 43,600 years old—predating the last ice age!

The ancient King's holly is triploid. With three sets of chromosomes, it is sterile—it can only reproduce asexually. Why? During meiosis, an odd number of chromosome sets cannot be divided equally between the two spindle poles. If meiosis does not fail entirely, unequal segregation of chromosomes during meiosis results in aneuploid offspring, which rarely survive.

fruit Mature ovary, often with accessory parts, from a flowering plant.

seed The mature ovule of a seed plant; contains an embryo sporophyte.

vegetative reproduction Growth of new roots and shoots from extensions or fragments of a parent plant; form of asexual reproduction in plants.

FIGURE 28.6 Examples of asexual reproduction in plants.

A Strawberry plants can propagate themselves by sending out stolons. New plants develop at nodes in the stolons.

B Potatoes are tubers that grow on stolons. New plants sprout from their nodes, or "eyes."

C New plants form at nodes on the leaflike cladodes of many succulents such as this devil's backbone plant.

King's holly probably arose by a chance genetic event. Many other types of plants reproduce asexually when new roots and shoots sprout from an existing plant. For example, "forests" of quaking aspen are actually stands of clones that grew from root suckers, which are shoots that sprout from the aspens' shallow, cordlike lateral roots. Suckers sprout after aboveground parts of the aspens are damaged or removed. One stand in Utah consists of about 47,000 shoots and stretches for 107 acres.

In other species, new plants sprout from modified stems. Stolons, often called runners, are stems that branch from the main stem of the plant, typically on or near the surface of the soil. Stolons may look like roots, but they have nodes, and roots do not have nodes. Adventitious roots and leafy shoots that sprout from the nodes develop into new plants (Figure 28.6**A**). A corm is a thickened underground stem that stores nutrients. It has nodes from which new plants develop. Tubers are thickened portions of underground stolons; they serve as the plant's primary storage tissue. Tubers have nodes from which new shoots and roots sprout. Potatoes are tubers; their "eyes" are nodes (Figure 28.6**B**). Cactuses and other succulents have flattened photosynthetic stems called cladodes. New plants form at nodes on the cladodes (Figure 28.6**C**).

Cuttings and Grafts For thousands of years, we humans have been taking advantage of the natural capacity of plants to reproduce asexually. Almost all houseplants, woody ornamentals, and orchard trees are clones that have been grown from stem fragments (cuttings) of a parent plant.

Propagating some plants from cuttings may be as simple as jamming a broken stem into the soil. This method uses the plant's natural ability to form roots and new shoots from stem nodes. Other plants must be grafted. Grafting means inducing a cutting to fuse with the tissues of another plant. Often, the stem of a desired plant is spliced onto the roots of a hardier one.

Propagating a plant from cuttings ensures that offspring will have the same desirable traits as the parent plant. For example, domestic apple trees (*Malus*) are typically grafted because they do not breed true for fruit color, flavor, size, or texture. Even trees grown from seeds of the same fruit produce fruits that vary, sometimes dramatically so. The genus is native to central Asia, where apple trees grow wild in forests. Apples do not breed true from seeds, so each wild tree is different from the next, and very few of the fruits are palatable (Figure 28.7).

In the early 1800s, the eccentric humanitarian John Chapman (known as Johnny Appleseed) planted millions of apple seeds in the midwestern United States. He sold the trees to homesteading settlers, who would plant orchards and make hard cider from the apples. About one of every hundred trees produced fruits that could be eaten out of hand. Its lucky owner would graft the tree and patent it. Most of the apple varieties sold in American grocery stores are clones of these trees, and they are still propagated by grafting.

Grafting is also used to increase the hardiness of a desirable plant. In 1862, the plant louse *Phylloxera* was accidentally introduced into France via imported American grapevines. European grapevines had little resistance to this tiny insect, which attacks and kills the root systems of the vines. By 1900, *Phylloxera* had destroyed two-thirds of the vineyards in Europe, and devastated the wine-making industry. Today, French vintners routinely graft their prized grapevines onto the roots of *Phylloxera*-resistant American vines.

Tissue Culture An entire plant may be cloned from a single cell. By a method called **tissue culture propagation**, a body cell is coaxed to divide and form an embryo. The technique, which can yield millions of genetically identical offspring from a single plant, is currently being used to improve food crops and to propagate rare or hybrid ornamental plants such as orchids.

FIGURE 28.7 Apples (*Malus*). **A** Commercial growers plant grafted apple trees in order to reap consistent crops. **B** Diverse fruit of 21 wild apple trees. **C** Gennaro Fazio (*left*) and Phil Forsline (*right*) are trying to maintain the genetic diversity of apple trees in the United States. They are breeding new apples with the palatability of commercial varieties and the disease resistance of wild ones.

Seedless Fruits In some plants such as figs, blackberries, and dandelions, fruits may form even in the absence of fertilization. In other species, fruit may continue to form after ovules or embryos abort. Seedless grapes and navel oranges are the result of mutations that result in arrested seed development. These plants are sterile, so they are propagated by grafting.

Seedless bananas are triploid (3*n*), so they are sterile. They are propagated by adventitious shoots that sprout from corms. Despite the ubiquity of polyploid plants in nature (Section 12.5), they rarely arise spontaneously. Plant breeders can speed up the process by treating plants with colchicine, a microtubule poison that increases the frequency of polyploidy. Tetraploid (4*n*) offspring of colchicine-treated plants are then backcrossed with diploid parent plants. The resulting triploid offspring are sterile. They make seedless fruit after pollination (but not fertilization) by a diploid plant, or on their own. Seedless watermelons are produced this way.

Take-Home Message How do plants reproduce asexually?

- Many plants propagate asexually when new shoots grow from a parent plant or pieces of it. Offspring of such vegetative reproduction are clones.
- Humans propagate plants asexually for agricultural or research purposes by grafting, tissue culture, or other methods.

28.5 Patterns of Development in Plants

In Section 28.3, we left the embryo sporophyte after its dispersal from the parent plant. What happens next? An embryonic plant complete with shoot and root apical meristems formed as part of the embryo. However, a seed typically dries out as it matures. Drying out causes the embryo's cells to stop dividing, and the embryo enters a period of temporarily suspended development (dormancy).

tissue culture propagation Laboratory method in which body cells are induced to divide and form an embryo.

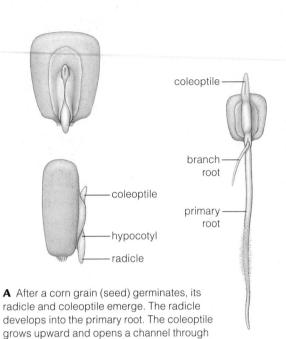

coleoptile

branch root

primary root

coleoptile

hypocotyl

radicle

A After a corn grain (seed) germinates, its radicle and coleoptile emerge. The radicle develops into the primary root. The coleoptile grows upward and opens a channel through the soil to the surface, where it stops growing.

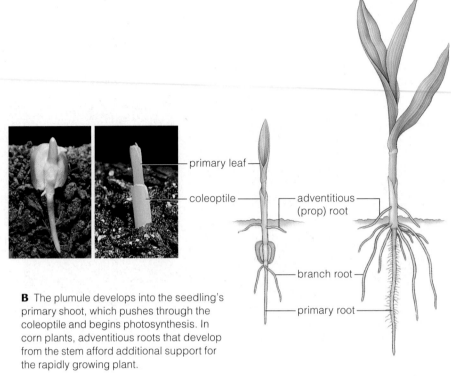

primary leaf

coleoptile

adventitious (prop) root

branch root

primary root

B The plumule develops into the seedling's primary shoot, which pushes through the coleoptile and begins photosynthesis. In corn plants, adventitious roots that develop from the stem afford additional support for the rapidly growing plant.

FIGURE 28.8 Animated! Early growth of corn, a monocot. *Below,* anatomy of a corn seed.

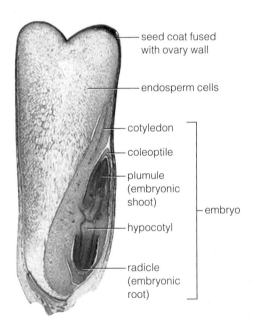

seed coat fused with ovary wall

endosperm cells

cotyledon

coleoptile

plumule (embryonic shoot)

embryo

hypocotyl

radicle (embryonic root)

An embryo may idle in its protective seed coat for years before it resumes metabolic activity. **Germination** is the process by which a mature embryo sporophyte resumes growth. The process begins with water seeping into a seed. The water activates enzymes that break down stored starches into sugars (Section 2.7). It also swells tissues inside the seed, so the coat splits open and oxygen enters. Meristem cells in the embryo use the sugars and the oxygen for aerobic respiration as they start dividing rapidly. The embryonic plant begins to grow from the meristems. Germination ends when the first part of the embryo—the embryonic root, or radicle—breaks out of the seed coat.

Seed dormancy is a climate-specific adaptation that allows germination to occur when conditions in the environment are most likely to support the growth of a seedling. For example, the weather in regions near the equator does not vary by season, so seeds of most plants native to such regions do not enter dormancy; they can germinate as soon as they are mature. By contrast, the seeds of many annual plants native to colder regions are dispersed in autumn. If they germinated immediately, the tender seedlings would not survive the coming winter. Instead, the seeds stay dormant until spring, when milder temperatures and longer daylength are more suitable for seedlings.

How does a dormant embryo sporophyte "know" when to germinate? The triggers, other than the presence of water, differ by species, and all have a genetic basis. For example, some seed coats are so dense that they must be abraded or broken (by being chewed, for example) before water can even enter the seed. Seeds of some species of lettuce must be exposed to bright light. The germination of wild California poppy seeds is inhibited by light and enhanced by smoke. The seeds of some species of pine will not germinate unless they have been previously burned. The seeds of many cool-climate plants require exposure to freezing temperatures.

Germination is just one of many patterns of development in plants. As a sporophyte grows and matures, its tissues and parts develop in other patterns characteristic of its species (Figures 28.8 and 28.9). Leaves form in predictable

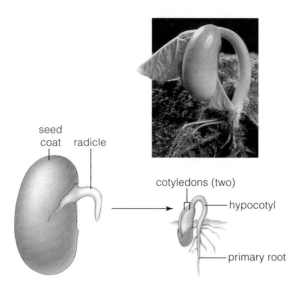

seed coat   radicle

cotyledons (two)

hypocotyl

primary root

A After a bean seed germinates, its radicle emerges and bends in the shape of a hook. Sunlight causes the hypocotyl to straighten, which pulls the cotyledons up through the soil.

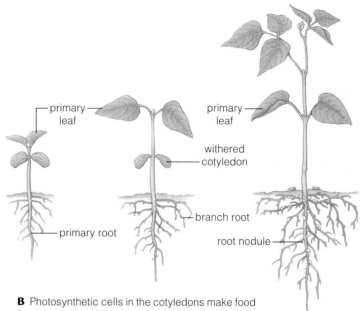

primary leaf

primary leaf

withered cotyledon

primary root

branch root

root nodule

B Photosynthetic cells in the cotyledons make food for several days. Then, the seedling's leaves take over the task and the cotyledons wither and fall off.

FIGURE 28.9 Animated! Early growth of the common bean plant, a eudicot.

shapes and sizes, stems lengthen and thicken in particular directions, flowering occurs at a certain time of year, and so on. As in germination, these patterns have a genetic basis, but they also have an environmental component.

Development includes growth, which is an increase in cell number and size. Plant cells are interconnected by shared walls, so they cannot move about within the organism. Thus, plant growth occurs primarily in the direction of cell division—and cell division occurs primarily at meristems. Behind meristems, cells differentiate and form specialized tissues. However, unlike animal cell differentiation, plant cell differentiation is often reversible, as when new shoots form on mature roots, or when new roots sprout from a mature stem.

Take-Home Message **What is plant development?**

- In plants, development includes growth and differentiation. It results in the formation of tissues and parts in predictable patterns.

- Germination and other patterns of plant development are an outcome of gene expression and environmental influences.

> Plant development depends on extensive coordination among individual cells, just as it does in animals.

28.6 Plant Hormones and Their Effects

You may be surprised to learn that plant development depends on extensive coordination among individual cells, just as it does in animals. A plant is an organism, not just a collection of cells, and as such it develops as a unit. Cells in different parts of a plant coordinate their activities by communicating with one

germination The resumption of growth after a period of dormancy.

Table 28.1 Major Plant Hormones and Some of Their Effects

Hormone	Primary Source	Effect	Site of Effect
Gibberellins			
	Stem tip, young leaves	Stimulates cell division, elongation	Stem internode
	Embryo	Stimulates germination	Seed
	Embryo (grass)	Stimulates starch hydrolysis	Endosperm
Auxins			
	Stem tip, young leaves	Stimulates cell elongation	Growing tissues
		Initiates formation of lateral roots	Roots
		Inhibits growth (apical dominance)	Lateral buds
		Stimulates differentiation of xylem	Cambium
		Inhibits abscission	Leaves, fruits
	Developing embryos	Stimulates fruit development	Ovary
Abscisic acid			
	Leaves	Closes stomata	Guard cells
		Stimulates formation of dormant buds	Stem tip
	Ovule	Inhibits germination	Seed coat
Cytokinins			
	Root tip	Stimulates cell division	Stem tip, lateral buds
		Inhibits senescence (aging)	Leaves
Ethylene			
	Damaged or aged tissue	Inhibits cell elongation	Stem
		Stimulates senescence (aging)	Leaves
		Stimulates ripening	Fruits

FIGURE 28.10 Effect of gibberellins. The three tall cabbage plants were treated with gibberellins. Two untreated cabbage plants are under the ladder.

abscisic acid Plant hormone that stimulates stomata to close in response to water stress; induces dormancy in buds and seeds.

apical dominance Growth-inhibiting effect on lateral buds, mediated by auxin produced in shoot tips.

auxin Plant hormone; stimulates cell division, elongation.

cytokinin Plant hormone that promotes cell division. Releases lateral buds from apical dominance.

ethylene Gaseous plant hormone that inhibits cell division in stems and roots; also promotes fruit ripening.

gibberellin Plant hormone; induces stem elongation, helps seeds break dormancy. Role in flowering in some species.

gravitropism Plant growth in a direction influenced by gravity.

phototropism Change in the direction of cell movement or growth in response to a light source.

tropism In plants, directional growth response to an environmental stimulus.

another. Such communication means, for example, that roots and shoots can be triggered to grow at the same time.

Plant cells use hormones to communicate with one another. Plant hormones are signaling molecules that can stimulate or inhibit plant development, including growth. Environmental cues such as the availability of water, length of night, temperature, and gravity influence plants by triggering the production and dispersal of hormones. When a plant hormone binds to a target cell, it may modify gene expression, change solute concentrations, affect enzyme activity, or activate another molecule in the cytoplasm.

Five types of plant hormones—gibberellins, auxins, abscisic acid, cytokinins, and ethylene—all interact to orchestrate plant development (Table 28.1).

Growth and other processes of development in all flowering plants, gymnosperms, mosses, ferns, and some fungi are regulated in part by **gibberellins**. These hormones induce cell division and elongation in stem tissue; thus, they cause stems to lengthen between the nodes. This effect can be demonstrated by application of gibberellin to the leaves of young plants (Figure 28.10). The short stems of Mendel's dwarf pea plants (Section 9.2) are the result of a mutation that reduces the rate of gibberellin synthesis in these plants. Gibberellins are also involved in breaking dormancy of seeds, seed germination, and the induction of flowering in biennials and some other plants.

Auxins are plant hormones that promote or inhibit cell division and elongation, depending on the target tissue. Auxins produced in apical meristems result in elongation of shoots (Figure 28.11**A,B**). They also induce cell division and differentiation in vascular cambium, fruit development in ovaries, and lateral root formation in roots. Auxins also have inhibitory effects. For example, auxin produced in a shoot tip prevents the growth of lateral buds along a lengthening stem, an effect called **apical dominance**. Gardeners routinely pinch off shoot tips to make a plant bushier. Pinching the tips ends the supply of auxin in a main stem, so lateral buds give rise to branches. Auxins also inhibit abscission, which is the dropping of leaves, flowers, and fruits from the plant.

Abscisic acid (ABA) is a hormone that was misnamed; it inhibits growth, and has little to do with abscission. ABA is part of a stress response that causes stomata to close (Section 27.6). It also diverts photosynthetic products from leaves to seeds, an effect that overrides growth-stimulating effects of other hormones as the growing season ends. ABA inhibits seed germination in some species, such as apple. Such seeds do not germinate before most of the ABA they contain has been broken down, for example by a long period of cold, wet conditions.

Plant **cytokinins** form in roots and travel via xylem to shoots, where they induce cell divisions in the apical meristems. These hormones also release lateral buds from apical dominance, and inhibit the normal aging process in leaves. Cytokinins signal to shoots that roots are healthy and active. When roots stop growing, they stop producing cytokinins, so shoot growth slows and leaves begin to deteriorate.

The only gaseous hormone, **ethylene**, is produced by damaged cells. It is also produced in autumn in deciduous plants, or near the end of the life cycle as part of a plant's normal process of aging. Ethylene inhibits cell division in stems and roots. It also induces fruit and leaves to mature and drop. Ethylene is widely used to artificially ripen fruit that has been harvested while still green.

Adjusting the Direction and Rate of Growth

Plants respond to environmental stimuli by adjusting the growth of roots and shoots. These responses are called **tropisms**, and they are typically mediated by hormones. For example, a root or shoot "bends" because of differences in auxin concentration. Auxin that accumulates in cells on one side of a shoot causes the cells to elongate more than the cells on the other side (Figure 28.11**C**). The result is that the shoot bends away from the side with more auxin. Auxin has the opposite effect in roots: It inhibits elongation of root cells. Thus, a root will bend toward the side that contains more auxin.

No matter how a seed is positioned in the soil when it germinates, the radicle always grows down, and the primary shoot always grows up. Even if a seedling is turned upside down just after germination, the primary root and shoot will curve so the root grows down and the shoot grows up. A growth response to gravity is called **gravitropism**.

How does a plant "know" which direction is up? Gravity-sensing mechanisms of many organisms are based on organelles called statoliths. Statoliths stuffed with starch grains occur in root cap cells, and also in specialized cells at the periphery of vascular tissues in the stem. Starch grains are heavier than cytoplasm, so statoliths tend to sink to the lowest region of the cell, wherever that is. The shift causes a redistribution of auxin so that this hormone is transported to the downward-facing side of roots and shoots.

Light streaming in from one direction causes a stem to curve toward its source. This response, **phototropism**, orients certain parts of the plant in the direction that will maximize the amount of light intercepted by its photosynthetic cells. Phototropism in plants occurs in response to blue light. Nonphotosynthetic pigments called phototropins absorb blue light, and translate its energy into a cascade of intracellular signals. The ultimate effect of this

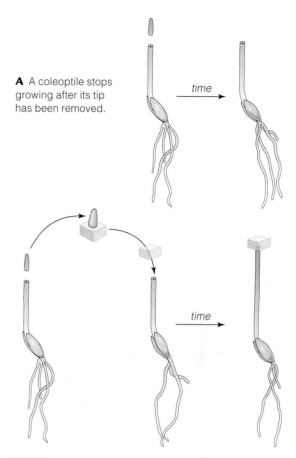

A A coleoptile stops growing after its tip has been removed.

time

time

B A block of agar will absorb auxin from the cut tip. Growth of a de-tipped coleoptile will resume when the agar block with absorbed auxin is placed on top of it.

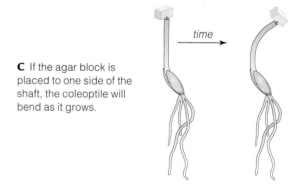

time

C If the agar block is placed to one side of the shaft, the coleoptile will bend as it grows.

FIGURE 28.11 Animated! A coleoptile lengthens in response to auxin produced in its tip.

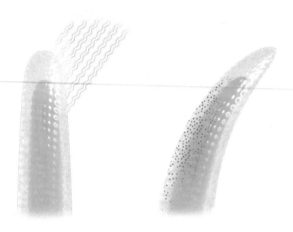

A Sunlight strikes only one side of a coleoptile.

B Auxin is transported to the shaded side, where it causes cells to lengthen.

FIGURE 28.12 Animated! Phototropism. **A,B** Auxin-mediated differences in cell elongation between two sides of a coleoptile induce bending toward light. The photo shows a shamrock plant's response to a directional light source.

cascade is that auxin is redistributed to the shaded side of a shoot or coleoptile. As a result, cells on the shaded side elongate faster than cells on the illuminated side. Differences in growth rates between cells on opposite sides of a shoot or coleoptile cause the entire structure to bend toward the light (Figure 28.12).

A plant's contact with a solid object may result in a change in the direction of its growth, a response called **thigmotropism**. The mechanism that gives rise to this response is not well understood, but it involves calcium ions and the products of at least five genes called *TOUCH*.

We see thigmotropism when a vine's tendril touches an object. The cells near the area of contact stop elongating, and the cells on the opposite side of the shoot keep elongating. The unequal growth rates of cells on opposite sides of the shoot cause it to curl around the object. A similar mechanism causes roots to grow away from contact, so they "feel" their way around rocks and other impassable objects in the soil. Mechanical stress, such as by wind exposure, inhibits stem lengthening in a touch response related to thigmotropism.

Take-Home Message **What regulates growth and development in plants?**

- Plant hormones are signaling molecules that influence plant development.
- The five main classes of plant hormones are gibberellins, auxins, abscisic acid, cytokinins, and ethylene.
- Via hormones, plants adjust the direction and rate of growth in response to environmental stimuli that include gravity, light, contact, and mechanical stress.

28.7 Responses to Recurring Change

Most organisms have a **biological clock**—an internal mechanism that governs the timing of rhythmic cycles of activity. A cycle of activity that starts anew every twenty-four hours or so is called a **circadian rhythm** (Latin *circa*, about; *dies*, day). In the circadian response called solar tracking, a leaf or flower changes position in response to the changing angle of the sun throughout the day. For example, a buttercup stem swivels so the flower on top of it always faces the sun. Unlike a phototropic response, solar tracking does not involve redistribution of auxin and differential growth. Instead, the absorption of blue light by photoreceptor proteins increases fluid pressure in cells on the sunlit side of a stem or petiole. The cells change shape, which bends the stem.

Similar mechanisms cause flowers of some plants to open only at certain times of day. For example, the flowers of many bat-pollinated plants unfurl, secrete nectar, and release fragrance only at night. Closing flowers periodically protects the delicate reproductive parts when the likelihood of pollination is low.

Like a mechanical clock, a biological one can be reset. Sunlight resets biological clocks in plants by activating and inactivating photoreceptors called **phytochromes**. These blue-green pigments are sensitive to red light (660 nanometers) and far-red light (730 nanometers). The relative amounts of these wavelengths in sunlight that reaches a given environment vary during the day and with the season. Red light causes phytochromes to change from an inactive form to an active form. Far-red light causes them to change back to their inactive form (Figure 28.13**A**). Active phytochromes bring about transcription of many genes, including some that encode components of rubisco, photosystems,

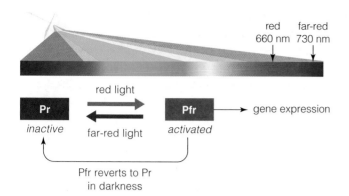

red light → Pr (inactive) ⇄ Pfr (activated) → gene expression
far-red light

Pfr reverts to Pr in darkness

A Red light changes the structure of a phytochrome from inactive (Pr) to active (Pfr) form; far-red light changes it back to the inactive form. Activated phytochromes control important processes such as germination and flowering.

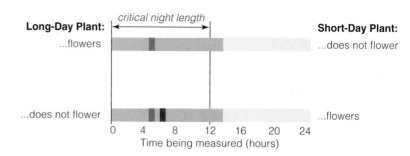

B A flash of red light interrupting a long night activates phytochrome. It causes plants to respond as if the night were short, and long-day plants flower. A pulse of far-red light, which inactivates phytochrome, cancels the effect of the red flash, and short-day plants flower. *Blue* bars indicate night length; *yellow* bars, day length.

FIGURE 28.13 Animated! Phytochrome and flowering. **Figure It Out:** What is the critical length of night that triggers a short-day plant's flowering response in the experiments shown in **B**?

Answer: 12 hours

ATP synthase, and other proteins used in photosynthesis; phototropin for phototropic responses; and molecules involved in flowering, gravitropism, and germination.

Photoperiodism is an organism's response to changes in the length of night relative to the length of day. Except at the equator, night length varies with the season. Nights are longer in winter than in summer, and the difference increases with latitude. Flowering is photoperiodic in many species of plants. *Long-day* plants such as irises flower only when the hours of darkness fall below a critical value (Figure 28.13**B**). Chrysanthemums and other *short-day* plants flower only when the hours of darkness are greater than some critical value (Figure 28.13**C**). Sunflowers and other *day-neutral* plants flower when they mature, regardless of night length.

The length of night is not the only cue for flowering. Some biennials and perennials flower only after exposure to cold winter temperatures. This process is called **vernalization** (from Latin *vernalis*, which means "to make springlike").

Senescence and Dormancy **Senescence** is the phase of a plant life cycle between full maturity and the death of plant parts or the whole plant. In many species of flowering plants, recurring cycles of growth and inactivity are responses to conditions that vary seasonally. Such plants are typically native to regions that are too dry or too cold for optimal growth during part of the year. Plants may drop leaves during such unfavorable intervals. The process by which plant parts are shed is **abscission**. It occurs in deciduous plants in response to shortening daylight hours, and year-round in evergreen plants. Abscission may also be induced by injury, water or nutrient deficiencies, or high temperatures.

Let's use deciduous plants as an example. As leaves and fruits grow in early summer, their cells produce auxin. The auxin moves into the stems, where it helps maintain growth. By midsummer, the nights are longer. Plants begin to divert nutrients away from their leaves, stems, and roots, and into flowers, fruits, and seeds. As the growing season comes to a close, nutrients are routed to twigs, stems, and roots, and auxin production declines in leaves and fruits.

abscission Process by which plant parts are shed in response to seasonal change, drought, injury, or nutrient deficiency.

biological clock Internal time-measuring mechanism by which individuals adjust their activities seasonally, daily, or both in response to environmental cues.

circadian rhythm A biological activity repeated about every 24 hours.

photoperiodism Biological response to seasonal changes in the relative lengths of day and night.

phytochrome A light-sensitive pigment that helps set plant circadian rhythms based on length of night.

senescence Phase in a life cycle from maturity until death; also applies to death of parts, such as plant leaves.

thigmotropism Redirected growth of a plant in response to contact with a solid object.

vernalization Stimulation of flowering in spring by low temperature in winter.

FIGURE 28.14 Abscission in the horse chestnut tree. *Left*, leaves change color in autumn before dropping. *Right*, the horseshoe-shaped leaf scar is all that remains of an abscission zone that formed before a leaf detached from the stem.

The auxin-deprived structures release ethylene that diffuses into nearby abscission zones—twigs, petioles, and fruit stalks. The ethylene is a signal for cells in the zone to produce enzymes that digest their own walls. The cells bulge as their walls soften, and separate from one another as the extracellular matrix that cements them together dissolves. Tissue in the zone weakens, and the structure above it drops (Figure 28.14).

For many species, growth stops in autumn as a plant enters dormancy, a period of arrested growth that is triggered by (and later ended by) environmental cues. Long nights, cold temperatures, and dry, nitrogen-poor soil are strong cues for dormancy in many plants. Dormancy-breaking cues usually operate between fall and spring. Dormant plants do not resume growth until certain conditions in the environment occur. A few species require exposure of the dormant plant to many hours of cold temperature. More typical cues include the return of milder temperatures and plentiful water and nutrients. With the return of favorable conditions, life cycles begin to turn once more as seeds germinate and buds resume growth.

Take-Home Message Do plants have biological clocks?

- Flowering plants respond to recurring cues from the environment with recurring cycles of development.

- The main environmental cue for flowering is the length of night relative to the length of day, which varies by the season in most places. Low winter temperatures stimulate the flowering of many plant species in spring.

- Abscission and dormancy are triggered by environmental cues such as seasonal changes in temperature or daylength.

28.8 Impacts/Issues Revisited:
Plight of the Honeybee

Sexual reproduction in plants involves the transfer of pollen, typically from one plant to another. Unlike animals, plants cannot move about to find a mate, so they depend on factors in the environment that can move pollen around for them. The diversity of flower form in part reflects that dependence.

A **pollination vector** is any agent that delivers pollen from an anther to a compatible stigma. Many plants are pollinated by wind, which is entirely nonspecific in where it dumps pollen. Such plants often release pollen grains by the billions, insurance in numbers that some of their pollen will reach a receptive stigma.

Other plants enlist the help of pollinators—living pollination vectors—to transfer pollen among individuals of the same species. An insect, bird, or other animal that is attracted to a particular flower often picks up pollen on a visit, then inadvertently transfers it to the flower of a different plant on a later visit. The more specific the attraction, the more efficient the transfer of pollen among plants of the same species. Given the selective advantage for flower traits that attract specific pollinators, it is not surprising that about 90 percent of flowering plants have coevolved animal pollinators.

nectar Sweet fluid exuded by some flowers; attracts pollinators.
pollination vector Any agent that moves pollen grains from one plant to another.

FIGURE 28.15 Nonreproductive traits of flowers—shape, pattern, color, and fragrance—serve to attract specific pollinators.

A flower's shape, pattern, color, and fragrance are adaptations that attract certain animal pollinators (Figure 28.15). For example, the petals of many flowers pollinated by bees usually are bright white, yellow, or blue, typically with pigments that reflect ultraviolet light. Such UV-reflecting pigments are often distributed in patterns that bees can recognize as visual guides to nectar. We can see the patterns only with special camera filters; our eyes do not have receptors that respond to UV light.

Pollinators such as bats and moths have an excellent sense of smell, and can follow concentration gradients of airborne chemicals to a flower that is emitting them. Not all flowers smell sweet; odors like dung or rotting flesh beckon beetles and flies.

An animal's reward for a visit to the flower may be **nectar** (a sweet fluid exuded by flowers), oils, nutritious pollen, or even the illusion of having sex. Nectar is the only food for most adult butterflies ❶, and it is the food of choice for hummingbirds ❷. Honeybees collect nectar ❸ and convert it to honey, which helps feed the bees through the winter. Pollen is an even richer food, with more vitamins and minerals than nectar.

Many flowers have specializations that exclude nonpollinators. For example, nectar at the bottom of a long floral tube or spur is often accessible only to a certain pollinator that has a matching feeding device. Often, stamens adapted to brush against a pollinator's body or lob pollen onto it will function only when triggered by that pollinator. Such relationships are to both species' mutual advantage: A flower that captivates the attention of an animal has a pollinator that spends its time seeking out (and pollinating) only those flowers; the animal receives an exclusive supply of the reward offered by the plant.

How Would YOU ☑ Vote?

Besides targeting pests, microencapsulated pesticides also wipe out populations of important pollinators. Should their use be restricted? See CengageNow for details, then vote online (cengagenow.com).

Summary

Section 28.1 Declines in populations of pollinators such as honeybees affect coevolved plant populations as well as other animal species that depend on the plants.

Section 28.2 Flowers consist of modified leaves (sepals, petals, stamens, and carpels) at the ends of specialized branches of angiosperm sporophytes. An ovule develops from a mass of ovary wall tissue inside carpels. Spores produced by meiosis in ovules develop into female gametophytes; those produced in anthers develop into immature male gametophytes (pollen grains).

Pollination is the arrival of pollen grains on a receptive stigma. A pollen grain germinates and forms a pollen tube that contains two sperm cells. Species-specific molecular signals guide the tube's growth down through carpel tissues to the egg. In double fertilization, one of the sperm cells in the pollen tube fertilizes the egg, forming a zygote; the other fuses with the endosperm mother cell and gives rise to endosperm.

CENGAGENOW™ Take a closer look at the life cycle of a eudicot, and investigate the plant life cycle and the structure of a flower, with the animations on CengageNow.

Section 28.3 An embryo sporophyte develops inside an ovule. A seed is a mature ovule. It consists of the embryo and its food reserves inside a protective seed coat. Fruits function in seed dispersal.

CENGAGENOW™ See how an embryo sporophyte develops inside a eudicot seed with the animation on CengageNow.

Section 28.4 Many flowering plants can reproduce asexually by sending out runners or by other types of vegetative reproduction. Many agriculturally valuable plants are produced by grafting or tissue culture propagation.

Section 28.5 Gene expression and cues from the environment coordinate plant development, which is the formation and growth of tissues and parts in predictable patterns. Germination is one pattern of development in plants.

CENGAGENOW™ Compare the growth and development of a monocot and a eudicot with the animation on CengageNow.

Section 28.6 Like animal hormones, plant hormones secreted by one cell alter the activity of a different cell. Plant hormones can promote or arrest growth of a plant by stimulating or inhibiting cell division, differentiation, elongation, and reproduction. The main plant hormones are gibberellins, auxins, cytokinins, ethylene, and abscisic acid. In addition to controlling patterns of growth and development, hormones also trigger adjustments in the direction and rate of growth in response to environmental stimuli.

CENGAGENOW™ Observe the effect of auxin on plant growth and investigate plant tropisms with the animations on CengageNow.

Section 28.7 Internal timing mechanisms such as biological clocks (including circadian rhythms) are set by daily and seasonal varia-

tions in environmental conditions. Photoperiodism is a response to changes in length of night relative to length of day. Light detection in plants involves nonphotosynthetic pigments called phytochromes (in photoperiodism) and phototropins (in phototropism).

Dormancy is a period of arrested growth that does not end until specific environmental cues occur. Dormancy is typically preceded by abscission. Senescence is the part of the plant life cycle between maturity and death of the plant or plant parts.

CENGAGENOW™ Learn how plants respond to night length with the animation on CengageNow.

Section 28.8 A flower's shape, pattern, color, and fragrance typically reflect an evolutionary relationship with a particular pollination vector, often a coevolved animal. Coevolved pollinators receive nectar, pollen, or another reward for visiting a flower.

Self-Quiz   Answers in Appendix I

1. The _____ of a flower contains one or more ovaries in which eggs develop, fertilization occurs, and seeds mature.
   a. pollen sac   c. receptacle
   b. carpel   d. sepal

2. Seeds are mature _____; fruits are mature _____.
   a. ovaries; ovules   c. ovules; ovaries
   b. ovules; stamens   d. stamens; ovaries

3. Meiosis of cells in pollen sacs forms haploid _____.
   a. megaspores   c. stamens
   b. microspores   d. sporophytes

4. Cotyledons develop as part of _____.
   a. carpels   c. embryo sporophytes
   b. accessory fruits   d. petioles

5. The seed coat forms from the _____.
   a. integuments   c. endosperm
   b. coleoptile   d. sepals

6. A new plant forms from a stem that broke off of the parent plant and fell to the ground. This is an example of _____.
   a. parthenogenesis   c. vegetative reproduction
   b. exocytosis   d. nodal growth

7. Plant hormones _____.
   a. may have multiple effects
   b. are influenced by environmental cues
   c. are active in plant embryos within seeds
   d. are active in adult plants
   e. all of the above

8. In some plants, flowering is a _____ response.
   a. phototropic   c. photoperiodic
   b. gravitropic   d. thigmotropic

9. Name one reward that a pollinator may receive in return for a visit to a flower of its coevolved plant partner.

Digging Into Data

Searching for Pollinators

Massonia depressa is a low-growing succulent plant native to the desert of South Africa. The dull-colored flowers of this monocot develop at ground level, have tiny petals, emit a yeasty aroma, and produce a thick, jelly-like nectar. These features led researchers to suspect that desert rodents such as gerbils pollinate this plant (Figure 28.16).

To test their hypothesis, the researchers trapped rodents in areas where *M. depressa* grows and checked them for pollen. They also put some plants in wire cages that excluded mammals, but not insects, to see whether fruits and seeds would form in the absence of rodents. The results are shown in Figure 28.17.

1. How many of the 13 captured rodents showed some evidence of pollen from *M. depressa*?

2. Would this evidence alone be sufficient to conclude that rodents are the main pollinators for this plant?

3. How did the average number of seeds produced by caged plants compare with that of control plants?

4. Do these data support the hypothesis that rodents are required for pollination of *M. depressa*? Why or why not?

FIGURE 28.16 The dull, petal-less, ground-level flowers of *Massonia depressa* are accessible to rodents, who push their heads through the stamens to reach the nectar at the bottom of floral cups. Note the pollen on the gerbil's snout.

Type of rodent	Number caught	# with pollen on snout	# with pollen in feces
Namaqua rock rat	4	3	2
Cape spiny mouse	3	2	2
Hairy-footed gerbil	4	2	4
Cape short-eared gerbil	1	0	1
African pygmy mouse	1	0	0

A

FIGURE 28.17 *Right*, results of experiments testing rodent pollination of *M. depressa*. **A** Evidence of visits to *M. depressa* by rodents. **B** Fruit and seed production of *M. depressa* with and without visits by mammals. Mammals were excluded from plants by wire cages with openings large enough for insects to pass through. 23 plants were tested in each group.

	Mammals allowed access to plants	Mammals excluded from plants
Percent of plants that set fruit	30.4	4.3
Average number of fruits per plant	1.39	0.47
Average number of seeds per plant	20.0	1.95

B

---

10. Which of the following statements is false?
    a. Auxins and gibberellins promote stem elongation.
    b. Cytokinins promote cell division, retard leaf aging.
    c. Abscisic acid promotes water loss and dormancy.
    d. Ethylene promotes fruit ripening and abscission.

11. Match the observation with the hormone most likely to be its cause.

    _____ ethylene   a. Your cabbage plants bolt (they
    _____ cytokinin   form elongated flowering stalks).
    _____ auxin   b. The philodendron in your room
    _____ gibberellin   is leaning toward the window.
    _____ abscisic acid   c. The last of your apples is getting
                            really mushy.
                        d. The seeds of your roommate's
                            marijuana plant do not germinate
                            no matter what he does to them.
                        e. Lateral buds on your *Ficus* plant
                            are sprouting branch shoots.

Additional questions are available on CENGAGENOW™

Critical Thinking

1. The following oat coleoptiles have been modified: either cut or placed in a light-blocking tube. Which ones will still bend toward a light source?

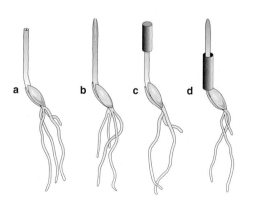

Appendix I

Chapter 1

1.	b	1.2
2.	c	1.2
3.	c	1.4
4.	energy, nutrients	1.3
5.	c	1.3
6.	b	1.4
7.	d	1.3
8.	a, d, e	1.2, 1.4
9.	a, d, f	1.2, 1.4
10.	d	1.3
11.	a	1.3
12.	b	1.5
13.	b	1.7
14.	c	1.2
	b	1.4
	d	1.6
	e	1.6
	a	1.6

Chapter 2

1.	tracer	2.2
2.	compound	2.3
3.	polar covalent	2.3
4.	hydrophobic	2.4
5.	d	2.2, 2.5
6.	acid	2.5
7.	buffer	2.5
8.	four	2.6
9.	e	2.7, 2.10
10.	double bonds	2.8
11.	False: *trans* fats are unsaturated and unhealthy	2.1, 2.8
12.	e	2.8
13.	d	2.9, 2.10
14.	d	2.9
15.	d	2.10
16.	i	2.9
	e	2.10
	b	2.8
	d	2.10
	h	2.7
	f	2.8
	a	2.4
	g	2.2
	c	2.4

Chapter 3

1.	cell	3.2
2.	c	3.2
3.	False	3.6
4.	c	3.4
5.	c	3.5
6.	c	3.4
7.	lysosomes	3.6
8.	c, b, d, a	3.6
9.	lipids; proteins	3.6
10.	False; e.g., many cells have walls on the outside of the plasma membrane	3.5, 3.7
11.	a	3.4
12.	d	3.7
13.	a	3.7
14.	c, f, a, e, d, b	3.6

Chapter 4

1.	c	4.2
2.	d	4.2
3.	d	4.3
4.	c, d	4.3, 4.4
5.	d	4.4, 4.5
6.	e	4.2, 4.4
7.	more, less	4.5
8.	water, gases, or small nonpolar molecules	4.5
9.	b	4.6
10.	a	4.5
11.	osmotic pressure or turgor	4.5
12.	e	4.6
13.	c	4.3
	d	4.4
	e	4.2
	b	4.3
	a	4.4
	f	4.5
	g	4.6
	i	4.6
	h	4.6

Chapter 5

1.	cat, bird, caterpillar are heterotrophs; weed is an autotroph	5.1
2.	carbon dioxide, sunlight	5.1
3.	a	5.3, 5.4
4.	b	5.3, 5.4
5.	c	5.4
6.	c	5.5
7.	b	5.3, 5.5
8.	oxygen	5.6
9.	False	5.6
10.	c	5.6
11.	d	5.6
12.	b	5.6
13.	d	5.7
14.	c	5.6
15.	b	5.6
	c	5.7
	a	5.6
	d	5.4, 5.6
	e	5.5

Chapter 6

1.	b	6.2
2.	d	6.3
3.	c	6.3
4.	a	6.4
5.	d	6.4
6.	d	6.4
7.	b	6.4
8.	CCAAAGAAGTTCTCT	6.4
9.	d	6.5
10.	d	6.3
	b	6.5
	c	6.4
	a	6.4

Chapter 7

1.	c	7.2
2.	b	7.3
3.	d	7.3
4.	a	7.2
5.	c	7.2
6.	15	7.4
7.	c	7.4
8.	a	7.4
9.	c	7.5
10.	a	7.3
11.	d	7.5
12.	gly-phe-leu-lys-arg	7.4
13.	val-ser-[stop]	7.4
14.	b	7.7
15.	b	7.7
16.	c	7.7
17.	c	7.4
	d	7.2
	e	7.5
	g	7.7
	f	7.4
	a	7.7
	b	7.6

Chapter 8

1.	2	8.3
2.	d	8.2
3.	a	8.2
4.	d	8.3, 8.7
5.	a	8.3
6.	c	8.3
7.	a	8.3, 8.4
8.	b	8.3
9.	c	8.2
10.	sister chromatids have just been pulled apart	8.7
11.	b	8.6
12.	alleles	8.6
13.	d	8.4
	e	8.4
	a	8.7
	c	8.7
	b	8.4
	g	8.7
	h	8.7
	f	8.7

Chapter 9

1.	b	9.2
2.	a	9.2
3.	c	9.2
4.	d	9.2
5.	c	9.2
6.	d	9.5
7.	b	9.6
8.	c	9.5
9.	d	9.6
10.	c	9.7, 9.8
11.	False	9.6
12.	True	9.7
13.	a	9.5
14.	b, d, a, c	9.2
15.	b	9.7
	a	9.5
	e	9.7
	c	9.7
	d	9.3

Chapter 10

1.	c	10.2
2.	a	10.2
3.	b	10.2
4.	b	10.2
5.	c	10.2
6.	d	10.2
7.	b	10.3
8.	d	10.3
9.	b	10.5
	d	10.2
	e	10.5
	f	10.4
10.	b	10.4
11.	b	10.5
12.	c	10.3
	g	10.4
	d	10.2
	e	10.5
	b	10.2
	a	10.1, 10.4
	f	10.1, 10.4

Chapter 11

1.	c	11.3
2.	b	11.2
3.	c	11.2
4.	d	11.3
5.	d	11.4
6.	d	11.4
7.	a	11.5
8.	65.5	11.5
9.	a, c, e, f, g, h	11.2, 11.5
10.	d	11.6
11.	d	11.6
12.	c	11.7
13.	b	11.6
14.	a	11.3
	i	11.4
	e	11.3
	f	11.4
	c	11.3
	b	11.3
	g	11.6
	d	11.4
	h	11.6

Chapter 12

1.	b	12.2
2.	c	12.2, 12.4
3.	a, d	12.3
4.	b, c	12.3
5.	e	12.3
6.	d	12.5
7.	balanced polymorphism	12.4
8.	gene flow	12.4
9.	allopatric speciation	12.5
10.	c	12.7
11.	c	12.7
12.	b	12.7
13.	c	12.4
	e	12.3
	g	12.2
	b	12.4
	f	12.6
	a	12.6
	d	12.7
	h	12.7

Chapter 13

1.	oxygen	13.2
2.	b	13.2
3.	d	13.3
4.	b	13.3
5.	c	13.4
6.	b	13.3
7.	True	13.5
8.	contractile vacuole	13.5
9.	b	13.5
10.	c	13.5
11.	c	13.5
12.	red	13.5
13.	b	13.5
14.	d	13.6
15.	g	13.5
	d	13.6
	c	13.4
	e	13.5
	i	13.4
	f	13.5
	a	13.5
	j	13.5
	b	13.5
	h	13.3

Chapter 14

1.	a	14.2
2.	c	14.5
3.	b	14.3
4.	c	14.4
5.	a	14.4
6.	d	14.5
7.	b	14.5
8.	ferns	14.4
9.	c	14.6
	d	14.2
	f	14.4
	e	14.3
	a	14.2
	b	14.7
	g	14.7
10.	c	14.8
11.	a	14.8
12.	c	14.8
13.	mycorrhiza	14.9
14.	lichen	14.9
15.	False	14.9

Chapter 15

1.	True	15.2
2.	a	15.2
3.	coelom	15.2
4.	bilateral	15.2
5.	c	15.3
6.	e	15.3

7. pharynx with gill slits, notochord, dorsal nerve cord, tail extending beyond anus — 15.4
8. pharynx with gill slits — 15.4
9. False — 15.5
10. b — 15.4
11. f — 15.6
12. False — 15.5
13. d — 15.6
14. b — 15.3
g — 15.3
l — 15.3
j — 15.3
e — 15.3
c — 15.3
d — 15.3
f — 15.3
k — 15.5
a — 15.5
h — 15.6
i — 15.6
15. 2 — 15.2
1 — 15.2
4 — 15.4
5 — 15.6
6 — 15.7
3 — 15.5

Chapter 16
1. a — 16.2
2. f — 16.3
3. a — 16.3
4. d — 16.3
5. a — 16.3
6. d — 16.4
7. True — 16.4
8. b — 16.5
9. d — 16.5
10. d — 16.3
c — 16.3
e — 16.3
a — 16.5
f — 16.3
b — 16.4

Chapter 17
1. niche — 17.2
2. False — 17.3
3. b, c, a, d — 17.3
4. b — 17.3
5. a — 17.4
6. d, a, c, b — 17.5
7. d — 17.5
8. a — 17.5
9. b, a, b, c — 17.6
10. c — 17.6
11. b — 17.6
12. a — 17.6
13. d — 17.6
14. d — 17.6
15. True — 17.6

Chapter 18
1. a — 18.2
2. d — 18.2
3. a — 18.2

4. 30 — 18.3
5. d — 18.3
6. b — 18.3
7. a — 18.3
8. a — 18.4
9. a — 18.4
10. e — 18.3
d — 18.3
b — 18.3
c — 18.3
a — 18.4
f — 18.3
g — 18.3
11. b — 18.5
12. c — 18.6
13. b — 18.5
14. a — 18.5
15. d — 18.5

Chapter 19
1. a — 19.3
2. b — 19.3
3. a — 19.3
4. a — 19.3
5. a — 19.3
6. b — 19.3
7. c — 19.3
8. b — 19.3
9. c — 19.3
10. d — 19.3
11. d — 19.3
12. neurons — 19.3
13. d — 19.4
14. d — 19.5
15. b — 19.3
h — 19.3
a — 19.4
c — 19.3
d — 19.3
e — 19.3
g — 19.3
f — 19.4

Chapter 20
1. a — 20.2
2. d — 20.2
3. d — 20.2
4. b — 20.3
5. a — 20.2
6. b — 20.4
7. d — 20.4
8. c — 20.4
9. d — 20.4
10. d — 20.5
11. b — 20.3
12. myosin — 20.4
13. a — 20.2
14. b, e, c, d, a — 20.2
15. d — 20.5
f — 20.5
g — 20.5
a — 20.5
c — 20.2
e — 20.2
j — 20.2
i — 20.2
h — 20.4
b — 20.4

Chapter 21
1. b — 21.2
2. b — 21.4
3. a — 21.5
4. c — 21.4
5. c — 21.5
6. d — 21.5
7. pulmonary — 21.3
8. a — 21.4
9. c — 21.7
10. d — 21.7
11. a — 21.7
12. a — 21.7
13. 2, 1, 3, 4, 5 — 21.7
14. b — 21.5
e — 21.7
i — 21.5, 21.7
h — 21.5
c — 21.4
g — 21.7
d — 21.7
f — 21.4
a — 21.4

Chapter 22
1. e — 22.2, 22.3
2. g — 22.4
3. c — 22.5–22.7
4. d — 22.4
5. d — 22.2
6. e.g., immediate; fixed (does not change much over a lifetime); can recognize about 1,000 pathogen-associated molecular patterns; inborn — 22.2, 22.4
7. e.g., recognizes self and nonself; targets specific antigens; changes over the course of the individual's lifetime; can recognize billions of antigens; provides lasting protection; consists of cell-mediated and antibody-mediated responses — 22.2, 22.5
8. d — 22.6
9. b — 22.5
10. e — 22.5
11. b — 22.7
12. c — 22.8
13. e, g, c, d a, f, b — 22.2
14. g — 22.8
b — 22.6
a — 22.2
e — 22.5
d — 22.8
f — 22.6
c — 22.4

Chapter 23
1. d — 23.2
2. b — 23.3
3. c — 23.3
4. b — 23.3
5. b — 23.3
6. f, b, a, d, e, c — 23.3

7. d — 23.4
8. no — 23.4
9. b — 23.4
10. d — 23.6
11. a — 23.7
12. a — 23.7
13. b — 23.7
14. c — 23.7
15. c — 23.6
a — 23.6
b — 23.6
e — 23.7
d — 23.7

Chapter 24
1. a — 24.2
2. c — 24.2
3. a — 24.2
4. b — 24.2
5. b — 24.4
6. c — 24.4
7. reflex — 24.5
8. c — 24.5
9. b — 24.5
10. c — 24.6
11. c — 24.6
12. b — 24.6
13. a — 24.6
14. c — 24.6
15. j — 24.6
c — 24.6
i — 24.5
b — 24.5
h — 24.5
f — 24.6
e — 24.3
g — 24.2
d — 24.5
a — 24.6

Chapter 25
1. a — 25.2
2. b — 25.3
3. c — 25.2
4. c, b, a, d — 25.3
5. growth hormone — 25.3
6. d — 25.4
7. b — 25.5
8. b — 25.4
9. b — 25.5
10. b — 25.4
11. False — 25.7
12. b — 25.4
13. d — 25.6
14. c — 25.7
15. e — 25.6
g — 25.4
d — 25.4
c — 25.4
h — 25.5
f — 25.7
b — 25.3
a — 25.7

Chapter 26
1. asexual — 26.2
2. a — 26.2
3. blastula, gastrula — 26.2
4. a — 26.3
5. True — 26.3
6. c — 26.3
7. c — 26.3
8. c — 26.6
9. 3, 2, 1, 4, 5 — 26.2, 26.6
10. e — 26.5
11. d — 26.6
12. e — 26.7
13. c — 26.3
h — 26.3
a — 26.6
g — 26.3
d — 26.3
e — 26.3
b — 26.3
f — 26.7

Chapter 27
1. a — 27.3
2. b — 27.2, 27.6
3. b — 27.3
4. a, b, e — 27.2
5. b — 27.4
6. c — 27.6
7. c — 27.6
8. b — 27.3
9. d — 27.5
10. humus — 27.5
11. d — 27.4
12. d — 27.6
13. b — 27.6
14. b — 27.3
d — 27.4
e — 27.2, 27.6
c — 27.2, 27.6
f — 27.3
a — 27.4

Chapter 28
1. b — 28.2
2. c — 28.3
3. b — 28.2
4. c — 28.3
5. a — 28.3
6. c — 28.4
7. e — 28.6
8. c — 28.7
9. nectar, pollen, oils, and the illusion of sex are mentioned in the text — 28.8
10. c — 28.6
11. c, e, b, a, d — 28.6

Appendix II

Answers to Genetics Problems

1. a. *AB*
   b. *AB, aB*
   c. *Ab, ab*
   d. *AB, Ab, aB, ab*

2. a. All offspring will be *AaBB*.
   b. 1/4 *AABB* (25% each genotype)
      1/4 *AABb*
      1/4 *AaBB*
      1/4 *AaBb*
   c. 1/4 *AaBb* (25% each genotype)
      1/4 *Aabb*
      1/4 *aaBb*
      1/4 *aabb*
   d. 1/16 *AABB* (6.25% of genotype)
      1/8 *AaBB* (12.5%)
      1/16 *aaBB* (6.25%)
      1/8 *AABb* (12.5%)
      1/4 *AaBb* (25%)
      1/8 *aaBb* (12.5%)
      1/16 *AAbb* (6.25%)
      1/8 *Aabb* (12.5%)
      1/16 *aabb* (6.25%)

3. A mating between a mouse from a true-breeding, white-furred strain and a mouse from a true-breeding, brown-furred strain would provide you with the most direct evidence. Because true-breeding strains of organisms typically are homozygous for a trait being studied, all F_1 offspring from this mating should be heterozygous. Record the phenotype of each F_1 mouse, then let them mate with one another. Assuming only one gene locus is involved, these are possible outcomes for the F_1 offspring:

   a. All F_1 mice are brown, and their F_2 offspring segregate: 3 brown : 1 white. *Conclusion*: Brown is dominant to white.

   b. All F_1 mice are white, and their F_2 offspring segregate: 3 white : 1 brown. *Conclusion*: White is dominant to brown.

   c. All F_1 mice are tan, and the F_2 offspring segregate: 1 brown : 2 tan : 1 white. *Conclusion*: The alleles at this locus show incomplete dominance.

4. a. Both parents are heterozygotes (*Aa*). Their children may be albino (*aa*) or unaffected (*AA* or *Aa*).

   b. All are homozygous recessive (*aa*).

   c. Homozygous recessive (*aa*) father, and heterozygous (*Aa*) mother. The albino child is *aa*, the unaffected children *Aa*.

5. A daughter could develop this muscular dystrophy only if she inherited two X-linked recessive alleles—one from each parent. Males who carry the allele are unlikely to father children because they develop the disorder and die early in life.

Appendix III   Annotations to A Journal Article

This journal article reports on the movements of a female wolf during the summer of 2002 in northwestern Canada. It also reports on a scientific process of inquiry, observation and interpretation to learn where, how and why the wolf traveled as she did. In some ways, this article reflects the story of "how to do science" told in section 1.5 of this textbook. These notes are intended to help you read and understand how scientists work and how they report on their work.

---

❶ ARCTIC

❷ VOL. 57, NO. 2 (JUNE 2004) P. 196–203

❸ Long Foraging Movement of a Denning Tundra Wolf

❹ Paul F. Frame,[1,2] David S. Hik,[1] H. Dean Cluff,[3] and Paul C. Paquet[4]

❺ (Received 3 September 2003; accepted in revised form 16 January 2004)

---

❻ ABSTRACT. Wolves (*Canis lupus*) on the Canadian barrens are intimately linked to migrating herds of barren-ground caribou (*Rangifer tarandus*). We deployed a Global Positioning System (GPS) radio collar on an adult female wolf to record her movements in response to changing caribou densities near her den during summer. This wolf and two other females were observed nursing a group of 11 pups. She traveled a minimum of 341 km during a 14-day excursion. The straight-line distance from the den to the farthest location was 103 km, and the overall minimum rate of travel was 3.1 km/h. The distance between the wolf and the radio-collared caribou decreased from 242 km one week before the excursion to 8 km four days into the excursion. We discuss several possible explanations for the long foraging bout.

❼ Key words: wolf, GPS tracking, movements, *Canis lupus*, foraging, caribou, Northwest Territories

❽ RÉSUMÉ. Les loups (*Canis lupus*) dans la toundra canadienne sont étroitement liés aux hardes de caribous des toundras (*Rangifer tarandus*). On a équipé une louve adulte d'un collier émetteur muni d'un système de positionnement mondial (GPS) afin d'enregistrer ses déplacements en réponse au changement de densité du caribou près de sa tanière durant l'été. On a observé cette louve ainsi que deux autres en train d'allaiter un groupe de 11 louveteaux. Elle a parcouru un minimum de 341 km durant une sortie de 14 jours. La distance en ligne droite de la tanière à l'endroit le plus éloigné était de 103 km, et la vitesse minimum durant tout le voyage était de 3,1 km/h. La distance entre la louve et le caribou muni du collier émetteur a diminué de 242 km une semaine avant la sortie à 8 km quatre jours après la sortie. On commente diverses explications possibles pour ce long épisode de recherche de nourriture.

Mots clés: loup, repérage GPS, déplacements, *Canis lupus*, recherche de nourriture, caribou, Territoires du Nord-Ouest

Traduit pour la revue *Arctic* par Nésida Loyer.

---

❾ Introduction

Wolves (*Canis lupus*) that den on the central barrens of mainland Canada follow the seasonal movements of their main prey, migratory barren-ground caribou (*Rangifer tarandus*) (Kuyt, 1962; Kelsall, 1968; Walton et al., 2001). However, most wolves do not den near caribou calving grounds, but select sites farther south, closer to the tree line (Heard and Williams, 1992). Most caribou migrate beyond primary wolf denning areas by mid-June and do not return until mid-to-late July (Heard et al., 1996; Gunn et al., 2001). Conse-quently, caribou density near dens is low for part of the summer.

During this period of spatial separation from the main caribou herds, wolves must either search near the homesite for scarce caribou or alternative prey (or both), travel to where prey are abundant, or use a combination of these strategies.

Walton et al. (2001) postulated that the travel of tundra wolves outside their normal summer ranges is a response to low caribou availability rather than a pre-dispersal exploration like that observed in territorial wolves (Fritts and Mech, 1981; Messier, 1985). The authors postulated this because most such travel was directed toward caribou calving grounds. We report details of such a long-distance excursion by a breeding female tundra wolf wearing a GPS radio collar. We discuss the relationship of the excursion to movements of satellite-collared caribou (Gunn et al., 2001), supporting the hypothesis that tundra wolves make directional, rapid, long-distance movements in response to seasonal prey availability.

[1] Department of Biological Sciences, University of Alberta, Edmonton, Alberta T6G 2E9, Canada
[2] Corresponding author: pframe@ualberta.ca
[3] Department of Resources, Wildlife, and Economic Development, North Slave Region, Government of the Northwest Territories, P.O. Box 2668, 3803 Bretzlaff Dr., Yellowknife, Northwest Territories X1A 2P9, Canada; Dean_Cluff@gov.nt.ca
[4] Faculty of Environmental Design, University of Calgary, Calgary, Alberta T2N 1N4, Canada; current address: P.O. Box 150, Meacham, Saskatchewan S0K 2V0, Canada

196

---

1 Title of the journal, which reports on science taking place in Arctic regions.

2 Volume number, issue number and date of the journal, and page numbers of the article.

3 Title of the article: a concise but specific description of the subject of study—one episode of long-range travel by a wolf hunting for food on the Arctic tundra.

4 Authors of the article: scientists working at the institutions listed in the footnotes below. Note #2 indicates that P. F. Frame is the *corresponding author*—the person to contact with questions or comments. His email address is provided.

5 Date on which a draft of the article was received by the journal editor, followed by date one which a revised draft was accepted for publication. Between these dates, the article was reviewed and critiqued by other scientists, a process called peer review. The authors revised the article to make it clearer, according to those reviews.

6 ABSTRACT: A brief description of the study containing all basic elements of this report. First sentence summarizes the *background* material. Second sentence encapsulates the *methods* used. The rest of the paragraph sums up the *results*. Authors introduce the main *subject* of the study—a female wolf (#388) with pups in a den—and refer to later *discussion* of possible explanations for her behavior.

7 Key words are listed to help researchers using computer databases. Searching the databases using these key words will yield a list of studies related to this one.

8 RÉSUMÉ: The French translation of the abstract and key words. Many researchers in this field are French Canadian. Some journals provide such translations in French or in other languages.

9 INTRODUCTION: Gives the background for this wolf study. This paragraph tells of known or suspected wolf behavior that is important for this study. Note that (a) major species mentioned are always accompanied by scientific names, and (b) statements of fact or *postulations* (claims or assumptions about what is likely to be true) are followed by references to studies that established those facts or supported the postulations.

10 This paragraph focuses directly on the wolf behaviors that were studied here.

11 This paragraph starts with a statement of the *hypothesis* being tested, one that originated in other studies and is supported by this one. The hypothesis is restated more succinctly in the last sentence of this paragraph. This is the *inquiry* part of the scientific process—asking questions and suggesting possible answers.

12 This map shows the study area and depicts wolf and caribou locations and movements during one summer. Some of this information is explained below.

13 STUDY AREA: This section sets the stage for the study, locating it precisely with latitude and longitude coordinates and describing the area (illustrated by the map in Figure 1).

14 Here begins the story of how prey (caribou) and predators (wolves) interact on the tundra. Authors describe movements of these nomadic animals throughout the year.

15 We focus on the denning season (summer) and learn how wolves locate their dens and travel according to the movements of caribou herds.

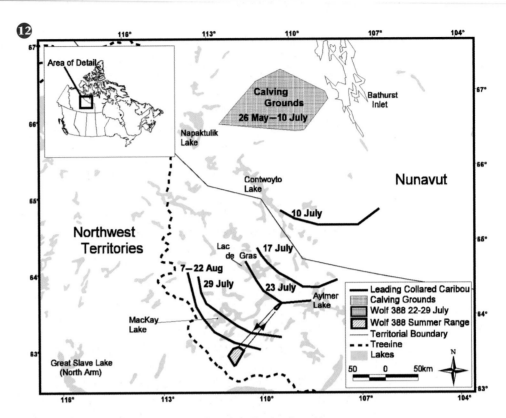

Figure 1. Map showing the movements of satellite radio-collared caribou with respect to female wolf 388's summer range and long foraging movement, in summer 2002.

🔳 Study Area

Our study took place in the northern boreal forest–low Arctic tundra transition zone (63° 30' N, 110° 00' W; Figure 1; Timoney et al., 1992). Permafrost in the area changes from discontinuous to continuous (Harris, 1986). Patches of spruce (*Picea mariana, P. glauca*) occur in the southern portion and give way to open tundra to the northeast. Eskers, kames, and other glacial deposits are scattered throughout the study area. Standing water and exposed bedrock are characteristic of the area.

🔳 *Details of the Caribou-Wolf System*

The Bathurst caribou herd uses this study area. Most caribou cows have begun migrating by late April, reaching calving grounds by June (Gunn et al., 2001;

Figure 1). Calving peaks by 15 June (Gunn et al., 2001), and calves begin to travel with the herd by one week of age (Kelsall, 1968). The movement patterns of bulls are less known, but bulls frequent areas near calving grounds by mid-June (Heard et al., 1996; Gunn et al., 2001). In summer, Bathurst caribou cows generally travel south from their calving grounds and then, parallel to the tree line, to the northwest. The rut usually takes place at the tree line in October (Gunn et al., 2001). The winter range of the Bathurst herd varies among years, ranging through the taiga and along the tree line from south of Great Bear Lake to southeast of Great Slave Lake. Some caribou spend the winter on the tundra (Gunn et al., 2001; Thorpe et al., 2001).

In winter, wolves that prey on Bathurst caribou do not behave territorially. Instead, they follow the herd throughout its winter range (Walton et al., 2001; Musiani, 2003). However, during denning (May–

Table 1. Daily distances from wolf 388 and the den to the nearest radio-collared caribou during a long excursion in summer 2002.

Date (2002)	Mean distance from caribou to wolf (km)	Daily distance from closest caribou to den
12 July	242	241
13 July	210	209
14 July	200	199
15 July	186	180
16 July	163	162
17 July	151	148
18 July	144	137
19 July[1]	126	124
20 July	103	130
21 July	73	130
22 July	40	110
23 July[2]	9	104
29 July[3]	16	43
30 July	32	43
31 July	28	44
1 August	29	46
2 August[4]	54	52
3 August	53	53
4 August	74	74
5 August	75	75
6 August	74	75
7 August	72	75
8 August	76	75
9 August	79	79

[1] Excursion starts.
[2] Wolf closest to collared caribou.
[3] Previous five days' caribou locations not available.
[4] Excursion ends.

August, parturition late May to mid-June), wolf movements are limited by the need to return food to the den. To maximize access to migrating caribou, many wolves select den sites closer to the tree line than to caribou calving grounds (Heard and Williams, 1992). Because of caribou movement patterns, tundra denning wolves are separated from the main caribou herds by several hundred kilometers at some time during summer (Williams, 1990:19; Figure 1; Table 1).

 Muskoxen do not occur in the study area (Fournier and Gunn, 1998), and there are few moose there (H.D. Cluff, pers. obs.). Therefore, alternative prey for wolves includes waterfowl, other ground-nesting birds, their eggs, rodents, and hares (Kuyt, 1972; Williams, 1990:16; H.D. Cluff and P.F. Frame, unpubl. data). During 56 hours of den observations, we saw no ground squirrels or hares, only birds. It appears that the abundance of alternative prey was relatively low in 2002.

Methods

Wolf Monitoring

We captured female wolf 388 near her den on 22 June 2002, using a helicopter net-gun (Walton et al., 2001). She was fitted with a releasable GPS radio collar (Merrill et al., 1998) programmed to acquire locations at 30-

minute intervals. The collar was electronically released (e.g., Mech and Gese, 1992) on 20 August 2002. From 27 June to 3 July 2002, we observed 388's den with a 78 mm spotting scope at a distance of 390 m.

Caribou Monitoring

In spring of 2002, ten female caribou were captured by helicopter net-gun and fitted with satellite radio collars, bringing the total number of collared Bathurst cows to 19. Eight of these spent the summer of 2002 south of Queen Maud Gulf, well east of normal Bathurst caribou range. Therefore, we used 11 caribou for this analysis. The collars provided one location per day during our study, except for five days from 24 to 28 July. Locations of satellite collars were obtained from Service Argos, Inc. (Landover, Maryland).

Data Analysis

Location data were analyzed by ArcView GIS software (Environmental Systems Research Institute Inc., Redlands, California). We calculated the average distance from the nearest collared caribou to the wolf and the den for each day of the study.

Wolf foraging bouts were calculated from the time 388 exited a buffer zone (500 m radius around the den) until she re-entered it. We considered her to be traveling when two consecutive locations were spatially separated by more than 100 m. Minimum distance traveled was the sum of distances between each location and the next during the excursion.

We compared pre- and post-excursion data using Analysis of Variance (ANOVA; Zar, 1999). We first tested for homogeneity of variances with Levene's test (Brown and Forsythe, 1974). No transformations of these data were required.

Results

Wolf Monitoring

Pre-Excursion Period: Wolf 388 was lactating when captured on 22 June. We observed her and two other females nursing a group of 11 pups between 27 June and 3 July. During our observations, the pack consisted of at least four adults (3 females and 1 male) and 11 pups. On 30 June, three pups were moved to a location 310 m from the other eight and cared for by an uncollared female. The male was not seen at the den after the evening of 30 June.

Before the excursion, telemetry indicated 18 foraging bouts. The mean distance traveled during these bouts was 25.29 km (± 4.5 SE, range 3.1–82.5 km). Mean greatest distance from the den on foraging

16 Other variables are considered—prey other than caribou and their relative abundance in 2002.

17 METHODS: There is no one scientific method. Procedures for each and every study must be explained carefully.

18 Authors explain when and how they tracked caribou and wolves, including tools used and the exact procedures followed.

19 This important subsection explains what data were calculated (average distance …) and how, including the software used and where it came from. (The calculations are listed in Table 1.) Note that the behavior measured (traveling) is carefully defined.

20 RESULTS: The heart of the report and the *observation* part of the scientific process. This section is organized parallel to the Methods section.

21 This subsection is broken down by periods of observation. Pre-excursion period covers the time between 388's capture and the start of her long-distance travel. The investigators used visual observations as well as telemetry (measurements taken using the global positioning system (GPS)) to gather data. They looked at how 388 cared for her pups, interacted with other adults, and moved about the den area.

22 The key in the lower right-hand corner of the map shows areas (shaded) within which the wolves and caribou moved, and the dotted trail of 388 during her excursion. From the results depicted on this map, the investigators tried to determine when and where 388 might have encountered caribou and how their locations affected her traveling behavior.

23 The wolf's excursion (her long trip away from the den area) is the focus of this study. These paragraphs present detailed measurements of daily movements during her two-week trip—how far she traveled, how far she was from collared caribou, her time spent traveling and resting, and her rate of speed. Authors use the phrase "minimum distance traveled" to acknowledge they couldn't track every step but were measuring samples of her movements. They knew that she went at least as far as they measured. This shows how scientists try to be exact when reporting results. Results of this study are depicted graphically in the map in Figure 2.

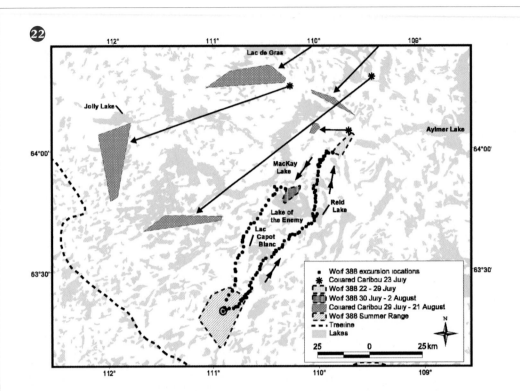

Figure 2. Details of a long foraging movement by female wolf 388 between 19 July and 2 August 2002. Also shown are locations and movements of three satellite radio-collared caribou from 23 July to 21 August 2002. On 23 July, the wolf was 8 km from a collared caribou. The farthest point from the den (103 km distant) was recorded on 27 July. Arrows indicate direction of travel.

bouts was 7.1 km (± 0.9 SE, range 1.7–17.0 km). The average duration of foraging bouts for the period was 20.9 h (± 4.5 SE, range 1–71 h).

The average daily distance between the wolf and the nearest collared caribou decreased from 242 km on 12 July, one week before the excursion period, to 126 km on 19 July, the day the excursion began (Table 1).

23 Excursion Period: On 19 July at 2203, after spending 14 h at the den, 388 began moving to the northeast and did not return for 336 h (14 d; Figure 2). Whether she traveled alone or with other wolves is unknown. During the excursion, 476 (71%) of 672 possible locations were recorded. The wolf crossed the southeast end of Lac Capot Blanc on a small land bridge, where she paused for 4.5 h after traveling for 19.5 h (37.5

km). Following this rest, she traveled for 9 h (26.3 km) onto a peninsula in Reid Lake, where she spent 2 h before backtracking and stopping for 8 h just off the peninsula. Her next period of travel lasted 16.5 h (32.7 km), terminating in a pause of 9.5 h just 3.8 km from a concentration of locations at the far end of her excursion, where we presume she encountered caribou. The mean duration of these three movement periods was 15.7 h (± 2.5 SE), and that of the pauses, 7.3 h (± 1.5). The wolf required 72.5 h (3.0 d) to travel a minimum of 95 km from her den to this area near caribou (Figure 2). She remained there (35.5 km2) for 151.5 h (6.3 d) and then moved south to Lake of the Enemy, where she stayed (31.9 km^2) for 74 h (3.1 d) before returning to her den. Her greatest distance from the den, 103 km, was recorded 174.5 h (7.3 d) after the excursion began, at 0433 on 27 July. She was 8 km from a collared caribou on 23 July, four days after the excursion began (Table 1).

The return trip began at 0403 on 2 August, 318 h (13.2 d) after leaving the den. She followed a relatively direct path for 18 h back to the den, a distance of 75 km.

The minimum distance traveled during the excursion was 339 km. The estimated overall minimum travel rate was 3.1 km/h, 2.6 km/h away from the den and 4.2 km/h on the return trip.

 Post-Excursion Period: We saw three pups when recovering the collar on 20 August, but others may have been hiding in vegetation.

Telemetry recorded 13 foraging bouts in the post-excursion period. The mean distance traveled during these bouts was 18.3 km (+ 2.7 SE, range 1.2–47.7 km), and mean greatest distance from the den was 7.1 km (+ 0.7 SE, range 1.1–11.0 km). The mean duration of these post-excursion foraging bouts was 10.9 h (+ 2.4 SE, range 1–33 h).

When 388 reached her den on 2 August, the distance to the nearest collared caribou was 54 km. On 9 August, one week after she returned, the distance was 79 km (Table 1).

Pre- and Post-Excursion Comparison

25 We found no differences in the mean distance of foraging bouts before and after the excursion period (F = 1.5, df = 1, 29, p = 0.24). Likewise, the mean greatest distance from the den was similar pre- and post-excursion (F = 0.004, df = 1, 29, p = 0.95). However, the mean duration of 388's foraging bouts decreased by 10.0 h after her long excursion (F = 3.1, df = 1, 29, p = 0.09).

26 *Caribou Monitoring*

Summer Movements: On 10 July, 5 of 11 collared caribou were dispersed over a distance of 10 km, 140 km south of their calving grounds (Figure 1). On the same day, three caribou were still on the calving grounds, two were between the calving grounds and the leaders, and one was missing. One week later (17 July), the leading radio-collared cows were 100 km farther south (Figure 1). Two were within 5 km of each other in front of the rest, who were more dispersed. All radio-collared cows had left the calving grounds by this time. On 23 July, the leading radio-collared caribou had moved 35 km farther south, and all of them were more widely dispersed. The two cows closest to the leader were 26 km and 33 km away, with 37 km between them. On the next location (29 July), the most southerly caribou were 60 km

farther south. All of the caribou were now in the areas where they remained for the duration of the study (Figure 2).

A Minimum Convex Polygon (Mohr and Stumpf, 1966) around all caribou locations acquired during the study encompassed 85 119 km².

Relative to the Wolf Den: The distance from the **27** nearest collared caribou to the den decreased from 241 km one week before the excursion to 124 km the day it began. The nearest a collared caribou came to the den was 43 km away, on 29 and 30 July. During the study, four collared caribou were located within 100 km of the den. Each of these four was closest to the wolf on at least one day during the period reported.

28 Discussion

Prey Abundance

Caribou are the single most important prey of tundra **29** wolves (Clark, 1971; Kuyt, 1972; Stephenson and James, 1982; Williams, 1990). Caribou range over vast areas, and for part of the summer, they are scarce or absent in wolf home ranges (Heard et al., 1996). Both the long distance between radio-collared caribou and the den the week before the excursion and the increased time spent foraging by wolf 388 indicate that caribou availability near the den was low. Observations of the pups' being left alone for up to 18 h, presumably while adults were searching for food, provide additional support for low caribou availability locally. Mean foraging bout duration decreased by 10.0 h after the excursion, when collared caribou were closer to the den, suggesting an increase in caribou availability nearby.

Foraging Excursion

One aspect of central place foraging theory (CPFT) **30** deals with the optimality of returning different-sized food loads from varying distances to dependents at a central place (i.e., the den) (Orians and Pearson, 1979). Carlson (1985) tested CPFT and found that the predator usually consumed prey captured far from the central place, while feeding prey captured nearby to dependants. Wolf 388 spent 7.2 days in one area near caribou before moving to a location 23 km back towards the den, where she spent an additional 3.1 days, likely hunting caribou. She began her return trip from this closer location, traveling directly to the den. While away, she may have made one or more successful kills and spent time meeting her own energetic needs before returning to the den. Alternatively, it may have taken several attempts to make a kill,

200 *P.F. Frame, et al*

24 Post-excursion measurements of 388's movements were made to compare with those of the pre-excursion period. In order to compare, scientists often use *means*, or averages, of a series of measurements—mean distances, mean duration, etc.

25 In the comparison, authors used statistical calculations (F and df) to determine that the differences between pre- and post-excursion measurements were *statistically insignificant*, or close enough to be considered essentially the same or similar.

26 As with wolf 388, the investigators measured the movements of caribou during the study period. The areas within which the caribou moved are shown in Figure 2 by shaded polygons mentioned in the second paragraph of this subsection.

27 This subsection summarizes how distances separating predators and prey varied during the study period.

28 DISCUSSION: This section is the *interpretation* part of the scientific process.

29 This subsection reviews observations from other studies and suggests that this study fits with patterns of those observations.

30 Authors discuss a prevailing *theory* (CBFT) which might explain why a wolf would travel far to meet her own energy needs while taking food caught closer to the den back to her pups. The results of this study seem to fit that pattern.

31 Here our authors note other possible explanations for wolves' excursions presented by other investigators, but this study does not seem to support those ideas.

32 Authors discuss possible reasons for why 388 traveled directly to where caribou were located. They take what they learned from earlier studies and apply it to this case, suggesting that the lay of the land played a role. Note that their description paints a clear picture of the landscape.

33 Authors suggest that 388 may have learned in traveling during previous summers where the caribou were. The last two sentences suggest ideas for future studies.

34 Or maybe 388 followed the scent of the caribou. Authors acknowledge difficulties of proving this, but they suggest another area where future studies might be done.

35 Authors suggest that results of this study support previous studies about how fast wolves travel to and from the den. In the last sentence, they speculate on how these observed patterns would fit into the theory of evolution.

36 Authors also speculate on the fate of 388's pups while she was traveling. This leads to . . .

---

which she then fed on before beginning her return trip. We do not know if she returned food to the pups, but such behavior would be supported by CPFT.

31 Other workers have reported wolves' making long round trips and referred to them as "extraterritorial" or "pre-dispersal" forays (Fritts and Mech, 1981; Messier, 1985; Ballard et al., 1997; Merrill and Mech, 2000). These movements are most often made by young wolves (1–3 years old), in areas where annual territories are maintained and prey are relatively sedentary (Fritts and Mech, 1981; Messier, 1985). The long excursion of 388 differs in that tundra wolves do not maintain annual territories (Walton et al., 2001), and the main prey migrate over vast areas (Gunn et al., 2001).

Another difference between 388's excursion and those reported earlier is that she is a mature, breeding female. No study of territorial wolves has reported reproductive adults making extraterritorial movements in summer (Fritts and Mech, 1981; Messier, 1985; Ballard et al., 1997; Merrill and Mech, 2001). However, Walton et al. (2001) also report that breeding female tundra wolves made excursions.

Direction of Movement

32 Possible explanations for the relatively direct route 388 took to the caribou include landscape influence and experience. Considering the timing of 388's trip and the locations of caribou, had the wolf moved northwest, she might have missed the caribou entirely, or the encounter might have been delayed.

A reasonable possibility is that the land directed 388's route. The barrens are crisscrossed with trails worn into the tundra over centuries by hundreds of thousands of caribou and other animals (Kelsall, 1968; Thorpe et al., 2001). At river crossings, lakes, or narrow peninsulas, trails converge and funnel towards and away from caribou calving grounds and summer range. Wolves use trails for travel (Paquet et al., 1996; Mech and Boitani, 2003; P. Frame, pers. observation). Thus, the landscape may direct an animal's movements and lead it to where cues, such as the odor of caribou on the wind or scent marks of other wolves, may lead it to caribou.

33 Another possibility is that 388 knew where to find caribou in summer. Sexually immature tundra wolves sometimes follow caribou to calving grounds (D. Heard, unpubl. data). Possibly, 388 had made such journeys in previous years and killed caribou. If this were the case, then in times of local prey scarcity she might travel to areas where she had hunted successfully before. Continued monitoring of tundra wolves may answer questions about how their food needs are met in times of low caribou abundance near dens.

34 Caribou often form large groups while moving south to the tree line (Kelsall, 1968). After a large aggregation of caribou moves through an area, its scent can linger for weeks (Thorpe et al., 2001:104). It is conceivable that 388 detected caribou scent on the wind, which was blowing from the northeast on 19–21 July (Environment Canada, 2003), at the same time her excursion began. Many factors, such as odor strength and wind direction and strength, make systematic study of scent detection in wolves difficult under field conditions (Harrington and Asa, 2003). However, humans are able to smell odors such as forest fires or oil refineries more than 100 km away. The olfactory capabilities of dogs, which are similar to wolves, are thought to be 100 to 1 million times that of humans (Harrington and Asa, 2003). Therefore, it is reasonable to think that under the right wind conditions, the scent of many caribou traveling together could be detected by wolves from great distances, thus triggering a long foraging bout.

Rate of Travel

35 Mech (1994) reported the rate of travel of Arctic wolves on barren ground was 8.7 km/h during regular travel and 10.0 km/h when returning to the den, a difference of 1.3 km/h. These rates are based on direct observation and exclude periods when wolves moved slowly or not at all. Our calculated travel rates are assumed to include periods of slow movement or no movement. However, the pattern we report is similar to that reported by Mech (1994), in that homeward travel was faster than regular travel by 1.6 km/h. The faster rate on return may be explained by the need to return food to the den. Pup survival can increase with the number of adults in a pack available to deliver food to pups (Harrington et al., 1983). Therefore, an increased rate of travel on homeward trips could improve a wolf's reproductive fitness by getting food to pups more quickly.

Fate of 388's Pups

36 Wolf 388 was caring for pups during den observations. The pups were estimated to be six weeks old, and were seen ranging as far as 800 m from the den. They received some regurgitated food from two of the females, but were unattended for long periods. The excursion started 16 days after our observations, and it is improbable that the pups could have traveled the distance that 388 moved. If the pups died, this would have removed parental responsibility, allowing the long movement.

Our observations and the locations of radio-collared caribou indicate that prey became scarce in the area of the den as summer progressed. Wolf 388 may have abandoned her pups to seek food for herself. However, she returned to the den after the excursion, where she was seen near pups. In fact, she foraged in a similar pattern before and after the excursion, suggesting that she again was providing for pups after her return to the den.

37 A more likely possibility is that one or both of the other lactating females cared for the pups during 388's absence. The three females at this den were not seen with the pups at the same time. However, two weeks earlier, at a different den, we observed three females cooperatively caring for a group of six pups. At that den, the three lactating females were observed providing food for each other and trading places while nursing pups. Such a situation at the den of 388 could have created conditions that allowed one or more of the lactating females to range far from the den for a period, returning to her parental duties afterwards. However, the pups would have been weaned by eight weeks of age (Packard et al., 1992), so nonlactating adults could also have cared for them, as often happens in wolf packs (Packard et al., 1992; Mech et al., 1999).

Cooperative rearing of multiple litters by a pack could create opportunities for long-distance foraging movements by some reproductive wolves during summer periods of local food scarcity. We have recorded multiple lactating females at one or more tundra wolf dens per year since 1997. This reproductive strategy may be an adaptation to temporally and **38** spatially unpredictable food resources. All of these possibilities require further study, but emphasize both the adaptability of wolves living on the barrens and their dependence on caribou.

Long-range wolf movement in response to caribou **39** availability has been suggested by other researchers (Kuyt, 1972; Walton et al., 2001) and traditional ecological knowledge (Thorpe et al., 2001). Our report demonstrates the rapid and extreme response of wolves to caribou distribution and movements in summer. Increased human activity on the tundra (mining, road building, pipelines, ecotourism) may influence caribou movement patterns and change the interactions between wolves and caribou in the region. Continued monitoring of both species will help us to assess whether the association is being affected adversely by anthropogenic change.

40 Acknowledgements

This research was supported by the Department of Resources, Wildlife, and Economic Development, Government of the Northwest Territories; the Department of Biological Sciences at the University of Alberta; the Natural Sciences and Engineering Research Council of Canada; the Department of Indian and Northern Affairs Canada; the Canadian Circumpolar Institute; and DeBeers Canada, Ltd. Lorna Ruechel assisted with den observations. A. Gunn provided caribou location data. We thank Dave Mech for the use of GPS collars. M. Nelson, A. Gunn, and three anonymous reviewers made helpful comments on earlier drafts of the manuscript. This work was done under Wildlife Research Permit – WL002948 issued by the Government of the Northwest Territories, Department of Resources, Wildlife, and Economic Development.

41 References

BALLARD, W.B., AYRES, L.A., KRAUSMAN, P.R., REED, D.J., and FANCY, S.G. 1997. Ecology of wolves in relation to a migratory caribou herd in northwest Alaska. Wildlife Monographs 135. 47 p.

BROWN, M.B., and FORSYTHE, A.B. 1974. Robust tests for the equality of variances. Journal of the American Statistical Association 69:364–367.

CARLSON, A. 1985. Central place foraging in the red-backed shrike (*Lanius collurio* L.): Allocation of prey between forager and sedentary consumer. Animal Behaviour 33:664–666.

CLARK, K.R.F. 1971. Food habits and behavior of the tundra wolf on central Baffin Island. Ph.D. Thesis, University of Toronto, Ontario, Canada.

ENVIRONMENT CANADA. 2003. National climate data information archive. Available online: http://www.climate.weatheroffice.ec.gc.ca/Welcome_e.html

FOURNIER, B., and GUNN, A. 1998. Musk ox numbers and distribution in the NWT, 1997. File Report No. 121. Yellowknife: Department of Resources, Wildlife, and Economic Development, Government of the Northwest Territories. 55 p.

FRITTS, S.H., and MECH, L.D. 1981. Dynamics, movements, and feeding ecology of a newly protected wolf population in northwestern Minnesota. Wildlife Monographs 80. 79 p.

GUNN, A., DRAGON, J., and BOULANGER, J. 2001. Seasonal movements of satellite-collared caribou from the Bathurst herd. Final Report to the West Kitikmeot Slave Study Society, Yellowknife, NWT. 80 p. Available online: http://www.wkss.nt.ca/HTML/08_ProjectsReports/PDF/Seasonal MovementsFinal.pdf

HARRINGTON, F.H., and ASA, C.S. 2003. Wolf communication. In: Mech, L.D., and Boitani, L., eds. Wolves: Behavior, ecology, and conservation. Chicago: University of Chicago Press. 66–103.

HARRINGTON, F.H., MECH, L.D., and FRITTS, S.H. 1983. Pack size and wolf pup survival: Their relationship under varying ecological conditions. Behavioral Ecology and Sociobiology 13:19–26.

HARRIS, S.A. 1986. Permafrost distribution, zonation and stability along the eastern ranges of the cordillera of North America. Arctic 39(1):29–38.

HEARD, D.C., and WILLIAMS, T.M. 1992. Distribution of wolf dens on migratory caribou ranges in the Northwest

37 Discussion of cooperative rearing of pups and, in turn, to speculation on how this study and what is known about cooperative rearing might fit into the animal's strategies for survival of the species. Again, the authors approach the broader theory of evolution and how it might explain some of their results.

38 And again, they suggest that this study points to several areas where further study will shed some light.

39 In conclusion, the authors suggest that their study supports the hypothesis being tested here. And they touch on the implications of increased human activity on the tundra predicted by their results.

40 ACKNOWLEDGEMENTS: Authors note the support of institutions, companies and individuals. They thank their reviewers ad list permits under which their research was carried on.

41 REFERENCES: List of all studies cited in the report. This may seem tedious, but is a vitally important part of scientific reporting. It is a record of the sources of information on which this study is based. It provides readers with a wealth of resources for further reading on this topic. Much of it will form the foundation of future scientific studies like this one.

Territories, Canada. Canadian Journal of Zoology 70:1504–1510.

HEARD, D.C., WILLIAMS, T.M., and MELTON, D.A. 1996. The relationship between food intake and predation risk in migratory caribou and implication to caribou and wolf population dynamics. Rangifer Special Issue No. 2:37–44.

KELSALL, J.P. 1968. The migratory barren-ground caribou of Canada. Canadian Wildlife Service Monograph Series 3. Ottawa: Queen's Printer. 340 p.

KUYT, E. 1962. Movements of young wolves in the Northwest Territories of Canada. Journal of Mammalogy 43:270–271.

———. 1972. Food habits and ecology of wolves on barren-ground caribou range in the Northwest Territories. Canadian Wildlife Service Report Series 21. Ottawa: Information Canada. 36 p.

MECH, L.D. 1994. Regular and homeward travel speeds of Arctic wolves. Journal of Mammalogy 75:741–742.

MECH, L.D., and BOITANI, L. 2003. Wolf social ecology. In: Mech, L.D., and Boitani, L., eds. Wolves: Behavior, ecology, and conservation. Chicago: University of Chicago Press. 1–34.

MECH, L.D., and GESE, E.M. 1992. Field testing the Wildlink capture collar on wolves. Wildlife Society Bulletin 20:249–256.

MECH, L.D., WOLFE, P., and PACKARD, J.M. 1999. Regurgitative food transfer among wild wolves. Canadian Journal of Zoology 77:1192–1195.

MERRILL, S.B., and MECH, L.D. 2000. Details of extensive movements by Minnesota wolves (Canis lupus). American Midland Naturalist 144:428–433.

MERRILL, S.B., ADAMS, L.G., NELSON, M.E., and MECH, L.D. 1998. Testing releasable GPS radiocollars on wolves and white-tailed deer. Wildlife Society Bulletin 26:830–835.

MESSIER, F. 1985. Solitary living and extraterritorial movements of wolves in relation to social status and prey abundance. Canadian Journal of Zoology 63:239–245.

MOHR, C.O., and STUMPF, W.A. 1966. Comparison of methods for calculating areas of animal activity. Journal of Wildlife Management 30:293–304.

MUSIANI, M. 2003. Conservation biology and management of wolves and wolf-human conflicts in western North America. Ph.D. Thesis, University of Calgary, Calgary, Alberta, Canada.

ORIANS, G.H., and PEARSON, N.E. 1979. On the theory of central place foraging. In: Mitchell, R.D., and Stairs, G.F., eds. Analysis of ecological systems. Columbus: Ohio State University Press. 154–177.

PACKARD, J.M., MECH, L.D., and REAM, R.R. 1992. Weaning in an arctic wolf pack: Behavioral mechanisms. Canadian Journal of Zoology 70:1269–1275.

PAQUET, P.C., WIERZCHOWSKI, J., and CALLAGHAN, C. 1996. Summary report on the effects of human activity on gray wolves in the Bow River Valley, Banff National Park, Alberta. In: Green, J., Pacas, C., Bayley, S., and Cornwell, L., eds. A cumulative effects assessment and futures outlook for the Banff Bow Valley. Prepared for the Banff Bow Valley Study. Ottawa: Department of Canadian Heritage.

STEPHENSON, R.O., and JAMES, D. 1982. Wolf movements and food habits in northwest Alaska. In: Harrington, F.H., and Paquet, P.C., eds. Wolves of the world. New Jersey: Noyes Publications. 223–237.

THORPE, N., EYEGETOK, S., HAKONGAK, N., and QITIRMIUT ELDERS. 2001. The Tuktu and Nogak Project: A caribou chronicle. Final Report to the West Kitikmeot/Slave Study Society, Ikaluktuuttiak, NWT. 160 p.

TIMONEY, K.P., LA ROI, G.H., ZOLTAI, S.C., and ROBINSON, A.L. 1992. The high subarctic forest-tundra of northwestern Canada: Position, width, and vegetation gradients in relation to climate. Arctic 45(1):1–9.

WALTON, L.R., CLUFF, H.D., PAQUET, P.C., and RAMSAY, M.A. 2001. Movement patterns of barren-ground wolves in the central Canadian Arctic. Journal of Mammalogy 82:867–876.

WILLIAMS, T.M. 1990. Summer diet and behavior of wolves denning on barren-ground caribou range in the Northwest Territories, Canada. M.Sc. Thesis, University of Alberta, Edmonton, Alberta, Canada.

ZAR, J.H. 1999. Biostatistical analysis. 4th ed. New Jersey: Prentice Hall. 663 p.

Appendix IV

A Plain English Map of the Human Chromosomes

© 2002 Susan Offner/SK45176-02

Haploid set of human chromosomes. The banding patterns characteristic of each type of chromosome appear after staining with a reagent called Giemsa. The locations of some of the 20,065 known genes (as of November, 2005) are indicated. Also shown are locations that, when mutated, cause some of the genetic diseases discussed in the text.

Appendix V

Units of Measure

Length
1 kilometer (km) = 0.62 miles (mi)
1 meter (m) = 39.37 inches (in)
1 centimeter (cm) = 0.39 inches

To convert	multiply by	to obtain
inches	2.25	centimeters
feet	30.48	centimeters
centimeters	0.39	inches
millimeters	0.039	inches

Area
1 square kilometer = 0.386 square miles
1 square meter = 1.196 square yards
1 square centimeter = 0.155 square inches

Volume
1 cubic meter = 35.31 cubic feet
1 liter = 1.06 quarts
1 milliliter = 0.034 fluid ounces = 1/5 teaspoon

To convert	multiply by	to obtain
quarts	0.95	liters
fluid ounces	28.41	milliliters
liters	1.06	quarts
milliliters	0.03	fluid ounces

Weight
1 metric ton (mt) = 2,205 pounds (lb) = 1.1 tons (t)
1 kilogram (kg) = 2.205 pounds (lb)
1 gram (g) = 0.035 ounces (oz)

To convert	multiply by	to obtain
pounds	0.454	kilograms
pounds	454	grams
ounces	28.35	grams
kilograms	2.205	pounds
grams	0.035	ounces

Temperature
Celcius (°C) to Fahrenheit (°F) :
$$°F = 1.8 \,(°C) + 32$$

Fahrenheit (°F) to Celsius:
$$°C = \frac{(°F - 32)}{1.8}$$

	°C	°F
Water boils	100	212
Human body temperature	37	98.6
Water freezes	0	32

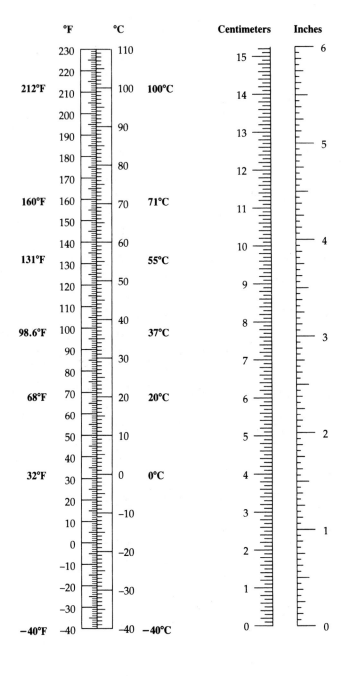

Appendix VI

Periodic Table of the Elements

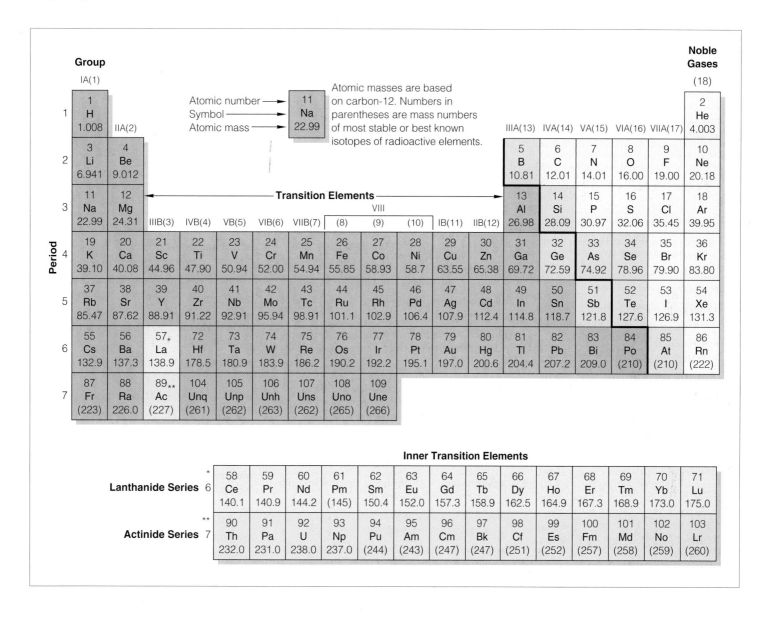

Group

IA(1)

Atomic number →	11
Symbol →	Na
Atomic mass →	22.99

Atomic masses are based on carbon-12. Numbers in parentheses are mass numbers of most stable or best known isotopes of radioactive elements.

Noble Gases (18)

Transition Elements

Inner Transition Elements

Lanthanide Series 6

58	59	60	61	62	63	64	65	66	67	68	69	70	71
Ce	Pr	Nd	Pm	Sm	Eu	Gd	Tb	Dy	Ho	Er	Tm	Yb	Lu
140.1	140.9	144.2	(145)	150.4	152.0	157.3	158.9	162.5	164.9	167.3	168.9	173.0	175.0

Actinide Series 7

90	91	92	93	94	95	96	97	98	99	100	101	102	103
Th	Pa	U	Np	Pu	Am	Cm	Bk	Cf	Es	Fm	Md	No	Lr
232.0	231.0	238.0	237.0	(244)	(243)	(247)	(247)	(251)	(252)	(257)	(258)	(259)	(260)

Glossary

abscisic acid Plant hormone; stimulates stomata to close in response to water stress, induces dormancy in buds and seeds. **577**

abscission Process by which plant parts are shed in response to seasonal change, drought, injury, or nutrient deficiency. **579**

absorption Movement of nutrient molecules from the gut into the body's internal environment. **461**

acid Substance that releases hydrogen ions in water. **28**

acid rain Rainfall with acidic pollutants. **373**

actin Protein that is the main component of thin filaments of muscle fibers. **409**

action potential Brief reversal of the charge difference across a neuron membrane. **482**

activation energy Minimum amount of energy required to start a reaction. Enzymes lower it. **65**

active site Pocket in an enzyme where substrates bind and a reaction occurs. **66**

active transport Energy-requiring mechanism by which a transport protein pumps a solute across a cell membrane against its concentration gradient. **72**

adaptation (adaptive trait) A heritable trait that enhances an individual's fitness. **200**

adaptive immunity In vertebrates, set of immune defenses tailored to specific pathogens encountered by an organism during its lifetime. Characterized by self/nonself recognition, antigen specificity, antigen receptor diversity, and immune memory. Includes antibody-mediated and cell-mediated responses. **439**

adaptive radiation A burst of genetic divergences from a lineage gives rise to many new species. **232**

adhering junction Cell junction composed of adhesion proteins; anchors cells to each other or to extracellular matrix. **57**

adhesion protein Membrane protein that helps cells stick to each other or to extracellular matrix in tissues. **48**

adipose tissue Type of connective tissue that is specialized for fat storage. **393**

adrenal cortex Outer portion of adrenal gland; secretes aldosterone and cortisol. **514**

adrenal gland Endocrine gland located above the kidney; secretes hormones with roles in urine formation and stress responses. **514**

adrenal medulla Inner portion of adrenal gland; secretes epinephrine and norepinephrine. **514**

aerobic Involving or occurring in the presence of oxygen. **88**

aerobic respiration Oxygen-requiring pathway that breaks down carbohydrates to produce ATP. **88**

age structure Of a population, distribution of individuals among various age groups. **333**

AIDS Acquired immune deficiency syndrome. A collection of diseases that develops after a virus (HIV) weakens the immune system. **453**

alcoholic fermentation Anaerobic carbohydrate breakdown pathway that produces ATP and ethanol. Begins with glycolysis; end reactions regenerate NAD^+ so glycolysis can continue. **92**

algal bloom Population explosion of single-celled aquatic organisms such as dinoflagellates. **255**

alleles Forms of a gene that encode slightly different versions of the gene's product. **144**

allele frequency Abundance of a particular allele among members of a population. **218**

allergen A normally harmless substance that provokes an immune response in some people. **452**

allergy Sensitivity to an allergen. **452**

allopatric speciation Speciation pattern in which a physical barrier that separates members of a population ends gene flow between them. **229**

allosteric Describes a region of an enzyme, other than the active site, that can bind regulatory molecules. **67**

alpine tundra High-altitude biome dominated by lichens and low plants. **367**

alveoli Tiny, thin-walled air sacs; the site of gas exchange in the lung. **428**

amino acid Small organic compound with a carboxyl group, an amine group, and a characteristic side group (R). Subunit of proteins. **33**

amnion Extraembryonic membrane that encloses an amniote embryo and the amniotic fluid. **537**

amniote Vertebrate that produces amniote eggs; a reptile (including birds), or a mammal. **310**

amniote egg Egg with four membranes that allows an embryo to develop away from water. **310**

amoeba Single-celled heterotrophic protist that moves and feeds by extending pseudopods. **258**

amphibian Tetrapod with a three-chambered heart and scaleless skin that typically develops in water, then lives on land as a carnivore with lungs. **309**

anaerobic Occurring in the absence of oxygen. **88**

analogous structures Similar body structures that evolved separately in different lineages. **209**

anaphase Stage of mitosis in which sister chromatids separate and move to opposite spindle poles. **140**

anemia Disorder in which red blood cells are few in number or impaired. **431**

aneuploidy A chromosome abnormality in which an individual's cells carry too many or too few copies of a particular chromosome. **172**

angiosperms Largest seed plant lineage. Its members make flowers and fruits. **272**

animal A eukaryotic heterotroph that is made up of unwalled cells and develops through a series of stages. Most ingest food, reproduce sexually, and can move during part or all of the life cycle. **8, 293**

animal hormone Intercellular signaling molecule secreted by an endocrine gland or cell. **505**

annelid Segmented worm with a coelom, complete digestive system, and closed circulatory system. **298**

anorexia A disorder in which a person does not eat enough to maintain a healthy weight, despite having access to food. **475**

antenna Of some arthropods, sensory structure on the head that detects touch and odors. **301**

antibody Y-shaped antigen receptor protein made only by B cells. **448**

antibody-mediated immune response Immune response in which antibodies are produced in response to an antigen. **446**

anticodon Set of three nucleotides in a tRNA; base-pairs with mRNA codon. **121**

antidiuretic hormone Hormone that is produced by the pituitary gland and increases water reabsorption by the kidney. **473**

antigen A molecule or particle that the immune system recognizes as nonself. Triggers an immune response. **439**

aorta Large artery that receives blood pumped out of the heart's left ventricle. **421**

apical dominance In plants, growth-inhibiting effect on lateral buds, mediated by auxin produced in shoot tips. **577**

apical meristem In shoot and root tips, mass of undifferentiated cells, the division of which lengthens plant parts. **553**

apicomplexan Parasitic protist that enters and lives inside the cells of its host. **255**

appendix Wormlike projection from the first part of the large intestine. **466**

aquifer Porous rock layer that holds some groundwater. **353**

arachnid Land-dwelling arthropod with four pairs of walking legs and no antennae; for example, a tick spider, or scorpion. **302**

Archaea Prokaryotic domain most closely related to eukaryotes; many members live in extreme environments. **250**

archaean A member of the prokaryotic domain Archaea. **8**

arctic tundra Northernmost biome; dominated by low plants growing over a layer of permafrost. **367**

arteriole Blood vessel that carries blood from an artery to a capillary bed. **420**

artery Large-diameter blood vessel that carries blood away from the heart. **420**

arthritis Chronic inflammation of a joint. **407**

arthropod Invertebrate with jointed legs and a hard exoskeleton that is periodically molted. **301**

asexual reproduction Reproductive mode by which offspring arise from one parent and inherit that parent's genes only. **143, 523**

atherosclerosis Narrowing of an artery's interior because of lipid deposition and inflammation. **431**

atom Particle that is a fundamental building block of all matter; consists of varying numbers of electrons, protons, and neutrons. **5, 21**

atomic number Number of protons in the atomic nucleus; determines the element. **21**

ATP Adenosine triposphate. RNA nucleotide that consists of an adenine base, five-carbon ribose sugar, and three phosphate groups. Main energy carrier between reaction sites in cells. **37, 65**

atrium Heart chamber that receives blood from veins and empties into a ventricle. **420**

australopiths Collection of now-extinct hominid lineages, some of which may be ancestral to humans. **314**

autoimmune response Immune response that mistakenly targets one's own tissues. **452**

autonomic nervous system Set of nerves that relay signals to and from internal organs and to glands. **488**

autosome Any chromosome other than a sex chromosome. **104**

autotroph Organism that makes its own food using carbon from inorganic molecules such as CO_2, and energy from the environment. **80**

auxin Plant hormone; stimulates division and elongation of plant cells. **576**

axon Cytoplasmic extension of a neuron; transmits electrical signals along its length and chemical signals at its endings. **481**

B lymphocyte (B cell) Type of white blood cell that makes antibodies. **440**

Bacteria Most diverse prokaryotic domain. **250**

bacteriophage Virus that infects prokaryotes. **259**

bacterium A member of the prokaryotic domain Bacteria. **8**

balanced polymorphism Maintenance of two or more alleles for a trait in some populations, as a result of natural selection against homozygotes. **224**

bark In woody plants, secondary phloem and periderm. **555**

base Substance that accepts hydrogen ions in water. **28**

base-pair substitution Type of mutation in which a single base-pair changes. **125**

basophil Circulating white blood cell; role in inflammation. **440**

bell curve Bell-shaped curve; typically results from graphing frequency versus distribution for a trait that varies continuously in a population. **165**

benthic province The ocean's rocks and sediments. **368**

bilateral symmetry Having right and left halves with similar parts, and a front and back that differ. **294**

bile Mix of salts, pigments, and cholesterol produced by the liver; aids in fat digestion. **464**

bioaccumulation A substance becomes more concentrated in organisms at increasingly higher levels of a food chain. **352, 374**

biodiversity Of a region, the genetic diversity within its species, variety of species, and variety of ecosystems. **379**

biofilm Community of different types of microorganisms living within a shared mass of slime. **51**

biogeochemical cycle A nutrient moves among environmental reservoirs and into and out of food webs. **353**

biogeography Study of patterns in the geographic distribution of species and communities. **197**

biological clock Internal time-measuring mechanism by which individuals adjust their activities seasonally, daily, or both in response to environmental cues. **578**

biology Systematic study of life. **4**

bioluminescent Able to use ATP to produce light. **254**

biome Type of ecosystem that can be characterized by its climate and dominant vegetation. **365**

biosphere All regions of Earth where organisms live. **5, 362**

biotic potential Maximum possible population growth rate under optimal conditions. **325**

bipedalism Standing and walking on two legs. **313**

bird An amniote with feathers. **311**

bivalve Mollusk with a hinged two-part shell. **299**

blastocyst Mammalian blastula. **536**

blastula Hollow ball of cells that forms early in animal development. **524**

blood Fluid connective tissue consisting of plasma and cells that form inside bones. **393**

blood–brain barrier Protective mechanism that prevents unwanted substances from entering cerebrospinal fluid. **490**

blood pressure Pressure exerted by blood against the walls of blood vessels. **424**

body mass index Formula used to determine a healthy weight based on height. **470**

bone tissue Connective tissue with cells surrounded by a mineral-hardened matrix of their own secretions. **393**

bony fish Jawed fish with a skeleton composed mainly of bone. **308**

bottleneck Reduction in population size so severe that it reduces genetic diversity. **225**

bronchiole In the lung, a small airway that leads from a bronchus to the alveoli. **428**

bronchus Airway connecting the trachea to a lung. **428**

brown alga Multicelled, photosynthetic protist with brown accessory pigments. **256**

bryophyte Member of an early evolving plant lineage that does not have vascular tissue; for example a moss. **271**

buffer Set of chemicals that can keep the pH of a solution stable by alternately donating and accepting ions that contribute to pH. **28**

C3 plant Type of plant that uses only the Calvin–Benson cycle to fix carbon. **87**

C4 plant Type of plant that minimizes photorespiration by fixing carbon twice, in two cell types. **87**

Calvin–Benson cycle Light-independent reactions of photosynthesis; cyclic carbon-fixing pathway that forms glucose from CO_2. **86**

CAM plant Type of C4 plant that conserves water by fixing carbon twice, at different times of day. **87**

camouflage A plant's or animal's appearance helps it blend into its surroundings. **345**

cancer Disease that occurs when a malignant neoplasm physically and metabolically disrupts body tissues.**130**

capillaries Smallest-diameter blood vessels; site of exchanges of gases and other materials between the blood and the tissues. **419**

carbohydrate Molecule that consists primarily of carbon, hydrogen, and oxygen atoms in a 1:2:1 ratio. **30**

carbon cycle Movement of carbon among rocks, water, the atmosphere, and living organisms. **355**

carbon fixation Process by which carbon from an inorganic source such as carbon dioxide becomes incorporated into an organic molecule. **86**

cardiac cycle Sequence of contraction and relaxation of heart chambers that occurs with each heartbeat. **422**

cardiac muscle tissue Striated, involuntary muscle of the heart wall. **393**

cardiac pacemaker Group of heart cells (SA node) that emits rhythmic signals calling for contraction of cardiac muscle. **423**

cardiopulmonary resuscitation (CPR) Life-saving technique that keeps oxygen flowing to tissue when the heart stops beating; involves mouth-to-mouth respiration and chest compressions. **423**

carpel Female reproductive structure of a flower; a sticky or hairlike stigma above a chamber (ovary) that contains one or more ovules. **281, 567**

carrying capacity Maximum number of individuals of a species that an environment can sustain. **326**

cartilage Connective tissue with cells surrounded by a rubbery matrix of their own secretions. **392**

cartilaginous fish Fish with a skeleton of cartilage, such as a shark. **307**

cartilaginous joint Joint where pads of cartilage hold bones together and provide cushioning, as between vertebrae. **407**

catastrophism Now-abandoned hypothesis that catastrophic geologic forces unlike those of the present day shaped Earth's surface. **199**

cDNA DNA synthesized from an RNA template by the enzyme reverse transcriptase. **182**

cell Smallest unit that has the properties of life: the capacity for metabolism, growth, homeostasis, and reproduction. **5, 44**

cell cycle A series of events from the time a cell forms until its cytoplasm divides. In eukaryotes, consists of interphase, mitosis, and cytoplasmic division. **138**

cell junction Structure that connects a cell to another cell or to extracellular matrix; e.g., gap junction, adhering junction, tight junction. **56**

cell-mediated immune response Immune response involving cytotoxic T cells and NK cells that destroy infected or cancerous body cells. **446**

cell plate After nuclear division in a plant cell, a disk-shaped structure that forms a cross-wall between the two new nuclei. **143**

cell theory Fundamental theory of biology: All organisms consist of one or more cells; the cell is the smallest unit of life; each new cell arises from another cell; and a cell passes hereditary material to its offspring. **44**

cell wall Semirigid but permeable structure that surrounds the plasma membrane of some cells. **50**

cellular slime mold Heterotrophic protist that usually lives as a single-celled predator. When conditions are unfavorable, cells aggregate into a cohesive group that can form a fruiting body. **258**

central nervous system Brain and spinal cord. **486**

centromere Constricted region in a eukaryotic chromosome where sister chromatids are attached. **103**

cephalopod Predatory mollusk with a closed circulatory system; moves by jet propulsion. **299**

cerebellum Hindbrain region that coordinates voluntary movements. **491**

cerebrospinal fluid Fluid around brain and spinal cord. **490**

cerebrum Forebrain region that controls higher functions. **491**

cervix Narrowed region of uterus; connects to vagina. **528**

chaparral Biome where mild, rainy winters and hot, dry summers support shrubs adapted to periodic fires. **365**

character Quantifiable, heritable characteristic or trait. **234**

charge Electrical property of some subatomic particles. Opposite charges attract; like charges repel. **21**

chemical bond An attractive force that arises between two atoms when their electrons interact. **24**

chemical digestion Breakdown of food molecules into smaller subunits by enzymes. **463**

chemical synapse Region where a neuron's axon endings transmit signals to another cell. **484**

chemoreceptor Sensory receptor that responds to the binding of a particular chemical. **493**

chlorophyll *a* Main photosynthetic pigment in plants. **81**

chloroplast Organelle of photosynthesis in plants and some protists. **54, 83**

chordates Animal phylum characterized by a notochord, dorsal nerve cord, pharyngeal gill slits, and a tail that extends beyond the anus. Includes some invertebrate and all vertebrate groups. **305**

chorion Outermost extraembryonic membrane of amniotes; major component of the placenta in placental mammals. **537**

choroid Pigmented middle layer of the wall of the eye. **495**

chromosome A structure that consists of DNA and associated proteins; carries part or all of a cell's genetic information. **103**

chromosome number The sum of all chromosomes in a cell of a given type. **104**

chyme Mix of food and gastric fluid. **464**

ciliate Single-celled, heterotrophic protist with many cilia. **254**

cilium Short, movable structure that projects from the plasma membrane of some eukaryotic cells. **55**

circadian rhythm A biological activity repeated about every 24 hours. **578**

clade A group of species that share a set of characters. **235**

cladistics Method of determining evolutionary relationships by grouping species into clades. **234**

cladogram Evolutionary tree diagram that shows a network of evolutionary relationships among clades. **234**

cleavage Mitotic division of an animal cell. **524**

cleavage furrow In a dividing animal cell, the indentation where cytoplasmic division will occur. **142**

climate Average weather conditions in a region over a long time period. **363**

cloaca Of most vertebrates, a body opening that functions in reproduction and elimination of urinary and digestive waste. **307**

clone A genetically identical copy; can refer to DNA, a cell, or an organism. **143, 181**

cloning vector A DNA molecule that can accept foreign DNA, be transferred to a host cell, and get replicated in it. **181**

closed circulatory system Circulatory system in which blood flows through a continuous network of vessels. **419**

cnidarian Radially symmetrical invertebrate with two tissue layers; uses tentacles with stinging cells to capture food. **296**

cochlea Coiled structure in the inner ear that holds the sound-detecting organ of Corti. **498**

codominant Refers to two alleles that are both fully expressed in heterozygous individuals. **163**

codon In mRNA, a nucleotide base triplet that codes for an amino acid or stop signal during translation. **120**

coelom A body cavity with a complete lining of tissue derived from mesoderm. **295**

coenzyme An organic cofactor. **66**

coevolution The joint evolution of two closely interacting species; each species is a selective agent that shifts the range of variation in the other. **231, 342**

cofactor A metal ion or a coenzyme that associates with an enzyme and is necessary for its function. **66**

cohesion Tendency of molecules to stick together. **27**

cohesion–tension theory Explanation of how transpiration creates a tension that pulls a cohesive column of water through xylem, from roots to shoots. **559**

cohort Group of individuals born during the same interval. **328**

collenchyma Simple plant tissue composed of living cells with unevenly thickened walls; provides flexible support. **550**

commensalism Species interaction that benefits one species and neither helps nor harms the other. **342**

community All populations of all species in a given area. **5, 340**

compact bone Dense bone with thin concentric layers of matrix. **406**

companion cell In phloem, parenchyma cell that loads sugars into sieve tubes. **550**

comparative morphology Study of body plans and structures among groups of organisms. **198**

competitive exclusion Process whereby two species compete for a limiting resource, and one drives the other to local extinction. **343**

complement A set of proteins that circulate in inactive form in blood as part of innate immunity. When activated, they destroy invaders or tag them for phagocytosis. **439**

compound Type of molecule that has atoms of more than one element. **24**

compound eye Eye that consists of many individual units, each with its own lens. **301**

concentration The number of molecules or ions per unit volume of a fluid. **69**

concentration gradient Difference in concentration between adjoining regions of fluid. **69**

condensation Process by which enzymes build large molecules from smaller subunits; water also forms. **30**

cone cell Photoreceptor that provides sharp vision and allows detection of color. **497**

conifers Cone-bearing gymnosperms. **278**

conjunctiva Mucous membrane of eyelids and around eye. **495**

connective tissue Animal tissue with an extensive extracellular matrix; provides structural and functional support. **391**

conservation biology Field of applied biology that surveys biodiversity and seeks ways to maintain and use it. **380**

consumer Organism that gets energy and carbon by feeding on tissues, wastes, or remains of other organisms; a heterotroph. **6, 350**

continuous variation In a population, a range of small differences in a shared trait. **165**

contraception Methods used to prevent pregnancy. **533**

contractile ring A thin band of actin and myosin filaments that wraps around the midsection of an animal cell; contracts and pinches the cytoplasm in two. **142**

contractile vacuole In freshwater protists, an organelle that collects and expels excess water. **253**

control group A group of objects or individuals that is identical to an experimental group except for one variable; used as a standard of comparison. **12**

coral bleaching Stress response in which a coral expels the photosynthetic protists in its tissues. **368**

coral reef In tropical sunlit seas, a formation composed of secretions of coral polyps that serves as home to many other species. **368**

cork Component of bark; waterproofs, insulates, and protects the surfaces of woody stems and roots. **555**

cork cambium In plants, a lateral meristem that gives rise to periderm. **555**

cornea Outermost layer at the front of the eye; bends light. **495**

corpus luteum Hormone-secreting structure that forms from follicle cells after ovulation. **529**

cortisol Adrenal cortex hormone that influences metabolism and immunity; secretions rise with stress. **514**

cotyledon Seed leaf; part of a flowering plant embryo. **551**

covalent bond Chemical bond in which two atoms share a pair of electrons. **24**

critical thinking Mental process of judging information before accepting it. **9**

crossing over Process in which homologous chromosomes exchange corresponding segments during prophase I of meiosis. **148**

crustaceans Mostly marine arthropods with two pairs of antennae; for example, a shrimp, crab, lobster, or barnacle. **302**

cuticle Secreted covering at a body surface. **271**

cytokinin Plant hormone that promotes cell division. Releases lateral buds from apical dominance. **577**

cytoplasm Semifluid substance enclosed by a cell's plasma membrane. **43**

cytoskeleton Dynamic framework of protein filaments that support, organize, and move eukaryotic cells and their internal structures. **54**

deciduous tree A tree that drops all its leaves annually just before a season that does not favor growth. **366**

decomposer Organism that feeds on remains and breaks organic material down into its inorganic subunits. **250, 350**

defibrillator Device that administers an electric shock to the chest wall to reset the SA node and restart the heart. **423**

deforestation Removal of all trees from a large tract of land. **372**

deletion Mutation in which one or more base pairs are lost from DNA. **124**

demographics Statistics used to describe characteristics of a population. **323**

demographic transition model Model describing the changes in human birth and death rates that occur as a region becomes industrialized. **334**

denature To unravel the shape of a protein or other large biological molecule. **35**

dendrite Neuron's signal-receiving cytoplasmic extension. **481**

dendritic cell Phagocytic white blood cell that patrols tissue fluids; presents antigen to T cells. **440**

denitrification Conversion of nitrates or nitrites to gaseous forms of nitrogen. **355**

dense connective tissue Connective tissue with many fibroblasts and fibers arranged in a random or a regular arrangement. **392**

density-dependent factor Factor that limits population growth and has a greater effect in dense populations than less dense ones. **326**

density-independent factor Factor that limits population growth and arises regardless of population density. **327**

dermal tissue system Tissue system that covers and protects exposed plant surfaces. **549**

dermis Deep layer of skin that consists of connective tissue with nerves and blood vessels running through it. **395**

desert Biome where little rain falls, humidity is low, and the main plants store water in their tissues or tap into water sources deep underground. **365**

desertification Conversion of grassland or woodlands to desertlike conditions. **372**

detritivores Consumers that feed on small bits of organic material (detritus). **350**

development Multistep process by which the first cell of a new individual becomes a multicelled adult. **7**

diaphragm Dome-shaped muscle at base of thoracic cavity that alters the size of this cavity during breathing. **429**

diastolic pressure Blood pressure when the ventricles are relaxed. **425**

diatom Single-celled photosynthetic protist with brown accessory pigments and a two-part silica shell. **256**

differentiation The process by which cells become specialized; occurs as different cell lineages begin to express different subsets of their genes. **127, 524**

diffusion Net movement of molecules or ions from a region where they are more concentrated to a region where they are less so. **69**

digestion Breakdown of food into smaller bits or molecules. **461**

dihybrid cross Experiment in which individuals with different alleles of two genes are crossed. **160**

dinoflagellate Single-celled, aquatic protist typically with cellulose plates and two flagella; may be heterotrophic or photosynthetic. **254**

diploid Having two of each type of chromosome characteristic of the species ($2n$). **104**

directional selection Mode of natural selection in which phenotypes at one end of a range of variation are favored. **219**

disease Condition that arises when a pathogen interferes with an organism's normal body functions. **262**

dislocation Bones of a joint are out of place. **407**

disruptive selection Mode of natural selection that favors two forms in a range of variation; intermediate forms are selected against. **222**

DNA Deoxyribonucleic acid. Nucleic that carries hereditary information about traits; consists of two nucleotide chains twisted in a double helix. Information encoded by its base sequence is the basis of an organism's form and function. **7, 37**

DNA cloning Set of procedures that uses living cells to make many identical copies of a DNA fragment. **181**

DNA fingerprint An individual's unique array of short tandem repeats. **184**

DNA ligase Enzyme that seals breaks or gaps in double-stranded DNA. **109**

DNA polymerase DNA replication enzyme. Assembles a new strand of DNA based on the sequence of a DNA template. **108**

DNA repair mechanism Any of several processes by which enzymes repair damaged DNA. **109**

DNA replication Process by which a cell duplicates its DNA before it divides. **108**

DNA sequence The order of nucleotide bases in a strand of DNA. **108**

dominant Refers to an allele that masks the effect of a recessive allele paired with it. **158**

dormancy Period of arrested growth. **568**

dosage compensation Theory that X chromosome inactivation equalizes gene expression between males and females. **129**

double fertilization In flowering plants, one sperm fertilizes the egg, forming the zygote, and another fertilizes a diploid cell, forming what will become endosperm. **281, 568**

eardrum Membrane of middle ear that vibrates in response to sound waves. **498**

echinoderms Invertebrates with a water–vascular system and an endoskeleton made of hardened plates and spines. **304**

ecological restoration Actively altering an area in an effort to restore or create a functional ecosystem. **381**

ecological succession A gradual change in a community in which one array of species replaces another. **347**

ecology The study of interactions among organisms, and between organisms and their environment. **322**

ecosystem A community interacting with its environment. **5, 340**

ectoderm Outermost tissue layer of an animal embryo. **524**

ectotherm Animal that gains heat from the environment; commonly called "cold-blooded." **310**

effector cell Antigen-sensitized B cell or T cell that forms in an immune response and acts immediately. **446**

egg Mature female gamete, or ovum. **148**

ejaculation Smooth muscle contractions propel semen through the penis and out of the body. **526**

electron Negatively charged subatomic particle that occupies orbitals around the atomic nucleus. **21**

electron transfer chain Array of enzymes and other molecules that accept and give up electrons in sequence, thus releasing the energy of the electrons in small, usable increments. **68**

electron transfer phosphorylation Metabolic pathway in which electron flow through electron transfer chains sets up a hydrogen ion gradient that drives ATP formation. **85**

electrophoresis Technique that separates DNA fragments by size. **184**

element A pure substance that consists only of atoms with the same number of protons. **21**

elimination Expulsion of unabsorbed material from the digestive tract. **461**

embolus Clot that forms in a blood vessel, then breaks loose. **431**

emergent property A characteristic of a system that does not appear in any of a system's component parts. **5**

emigration Movement of individuals out of a population. **324**

enamel Hard material that covers the exposed surface of teeth. **462**

endangered species A species that faces extinction in all or part of its range. **370**

endemic disease A disease that remains present at low levels in a population. **262**

endemic species A species that evolved in one place and is found nowhere else. **370**

endocrine gland Ductless gland that secretes hormones into a body fluid. **391**

endocrine system Hormone-producing glands and secretory cells of a vertebrate body. **505**

endocytosis Process by which a cell takes in a small amount of extracellular fluid by the ballooning inward of its plasma membrane. **74**

endoderm Innermost tissue layer of an animal embryo. **524**

endomembrane system Series of interacting organelles (endoplasmic reticulum, Golgi bodies, vesicles) between the nucleus and plasma membrane; produces lipids and proteins. **53**

endoplasmic reticulum (ER) Organelle that is a continuous system of sacs and tubes; extension of the nuclear envelope. Rough ER is studded with ribosomes; smooth ER is not. **53**

endorphins Natural pain-reliever molecules. **494**

endoskeleton Hard internal parts that muscles attach to and move. **304, 405**

endosperm Nutritive tissue in an angiosperm seed. **281, 568**

endosperm mother cell A cell with two nuclei ($n + n$) that is part of the female gametophyte of a flowering plant. **568**

endosymbiont A species that lives in the body of another species in a commensal, mutualistic, or parasitic relationship. **342**

endosymbiosis One species lives inside another. **245**

endotherm Animal that produces its own heat; commonly called "warm-blooded." **310**

energy The capacity to do work. **6, 63**

energy pyramid Diagram that illustrates the energy flow in an ecosystem. **352**

enzyme Protein or RNA that speeds a reaction without being changed by it. **48, 66**

eosinophil White blood cell that targets internal parasites. **440**

epidemic A disease spreads rapidly through a population. **262**

epidermis Outermost tissue layer; in animals, epithelial layer of skin. **394, 551**

epiglottis Tissue flap that folds down to prevent food from entering the airways when you swallow. **428**

epiphyte Plant that grows on the trunk or branches of another plant but does not harm it. **275**

epistasis Effect in which a trait is influenced by the products of multiple genes. **164**

epithelial tissue Sheetlike animal tissue that covers outer body surfaces and lines internal tubes and cavities. **390**

erection Penis becomes erect as a result of increased fluid pressure inside its spongy tissue. **531**

esophagus Muscular tube between the throat and stomach. **463**

essential amino acid Amino acid that the body cannot make and must obtain from food. **468**

essential fatty acid Fatty acid that the body cannot make and must obtain from the diet. **467**

estrogens Sex hormones produced by a female's ovaries. **515**

estuary A highly productive ecosystem where nutrient-rich water from a river mixes with seawater. **368**

ethylene Gaseous plant hormone that inhibits cell division in stems and roots; also promotes fruit ripening. **577**

eudicots Largest lineage of angiosperms; includes herbaceous plants, woody trees, and cacti. **280**

eugenics Idea of deliberately improving the genetic qualities of the human race. **190**

eukaryote Organism whose cells characteristically have a nucleus; a protist, plant, fungus, or animal. **8, 244**

eutrophication Nutrient enrichment of an aquatic ecosystem. **374**

evaporation Transition of a liquid to a gas. **27**

evolution Change in a line of descent. **199**

evolutionary tree Type of diagram that summarizes evolutionary relationships among a group of species. **235**

exaptation Adaptation of an existing structure for a completely different purpose; a major evolutionary novelty. **231**

exocrine gland Gland that secretes milk, sweat, saliva, or some other substance through a duct. **391**

exocytosis Process by which a cell expels a vesicle's contents to extracellular fluid. **74**

exoskeleton Hard external parts that muscles attach to and move. **301, 405**

exotic species A species that evolved in one community and later became established in a different one. **340**

experiment A test to support or falsify a prediction. **12**

experimental group A group of objects or individuals that display or are exposed to a variable under investigation; Experimental results for this group are compared with results for a control group. **12**

exponential growth A population grows by a fixed percentage in successive time intervals; the size of each increase is determined by the current population size. **324**

extinct Refers to a species that has been permanently lost. **231**

extracellular matrix (ECM) Complex mixture of secreted substances that supports cells and tissues; also has roles in cell signaling. **56**

extreme halophile Organism that lives where the salt concentration is high. **250**

extreme thermophile Organism that lives where the temperature is very high. **250**

fat Lipid with one, two, or three fatty acid tails. **32**

fatty acid Organic compound that consists of a long chain of carbon atoms with an acidic carboxyl group at one end. Carbon chain of saturated types has single bonds only; that of unsaturated types has one or more double bonds. **32**

feces Unabsorbed food material and cellular waste that is expelled from the digestive tract. **465**

feedback inhibition Mechanism by which a change that results from some activity decreases or stops the activity. **67**

fermentation An anaerobic pathway by which cells harvest energy from carbohydrates. **92**

fertilization Fusion of a sperm and an egg, the result being a single-celled zygote. **144, 524**

fetus Developing human between 9 weeks of age and birth. **539**

fever An internally induced rise in core body temperature above the normal set point; response to infection. **443**

fibroblast Main cell type in soft connective tissue; secretes collagen and other components of extracellular matrix. **391**

fibrous joint Joint where dense connective tissue holds bones firmly in place. **407**

filtration Blood pressure forces water and small solutes, but not blood cells or proteins, out across the walls of capillaries. **472**

first law of thermodynamics Energy cannot be created or destroyed. **63**

fish Aquatic vertebrate of the oldest and most diverse vertebrate group. **307**

fitness The degree of adaptation to an environment, as measured by an individual's relative genetic contribution to future generations. **200**

fixed Refers to an allele for which all members of a population are homozygous. **225**

flagellated protozoan Member of a heterotrophic lineage of protists with one or more flagella. **252**

flagellum Long, slender cellular structure used for motility. **50**

flatworm Bilaterally symmetrical invertebrate with organs but no body cavity; for example, a planarian or tapeworm. **297**

flower Specialized reproductive shoot of a flowering plant. **280**

fluid mosaic model Model of a cell membrane as a two-dimensional fluid of mixed composition. **48**

food chain Linear description of who eats whom. **350**

food web System of cross-connecting food chains. **350**

foraminiferan Heterotrophic protist that secretes a calcium carbonate shell. **253**

fossil Physical evidence of an organism that lived in the ancient past. **202**

founder effect Change in allele frequencies that occurs after a small number of individuals establish a new population. **225**

fovea Region of vertebrate eye with densest concentration of photoreceptors. **496**

fruit Mature flowering plant ovary, often with accessory parts; encloses a seed or seeds. **280, 570**

fungus Type of eukaryotic consumer with cell walls of chitin; obtains nutrients by digestion and absorption outside the body. **8, 283**

gallbladder Organ that receives bile from the liver, stores it, and expels it into the small intestine. **464**

gamete Mature, haploid reproductive cell; e.g., an egg or a sperm. **144**

gametophyte A haploid, multicelled body in which gametes form during the life cycle of plants. **148, 271**

ganglion Cluster of nerve cell bodies. **486**

gap junction Cell junction that forms a channel across the plasma membranes of adjoining animal cells. **57**

gastric fluid Fluid secreted by the stomach lining; contains enzymes, acid, and mucus. **464**

gastropod Mollusk that moves about on its enlarged foot. For example, a snail. **298**

gastrula Three-layered developmental stage formed by gastrulation in an animal. **524**

gastrulation Animal developmental process by which cell movements produce a three-layered gastrula. **524**

gene Part of a DNA base sequence; specifies an RNA or protein product. **117**

gene expression Process by which the information encoded by a gene becomes converted to an RNA or protein product. **117**

gene flow The movement of alleles into and out of a population, as by individuals that immigrate or emigrate. **226**

gene knockout A gene that has been inactivated in an organism. **128**

gene pool All of the genes in a population; a pool of genetic resources. **217**

gene therapy The transfer of a normal or modified gene into an individual with the goal of treating a genetic defect or disorder. **190**

genetic code Set of sixty-four mRNA codons, each of which specifies an amino acid or stop signal in translation. **120**

genetic drift Change in allele frequencies in a population due to chance alone. **225**

genetic engineering Process by which deliberate changes are introduced into an individual's genome. **186**

genetic equilibrium Theoretical state in which a population is not evolving. **218**

genetically modified organism (GMO) An organism whose genome has been deliberately modified. **180**

genome An organism's complete set of genetic material. **182**

genomics The study of genomes. **186**

genotype The particular alleles carried by an individual. **158**

genus Taxonomic group of species that share a unique set of traits. **8**

geologic time scale Chronology of Earth history. **204**

germination The resumption of growth after a period of dormancy. **574**

ghrelin Molecule made in the stomach and brain that increases appetite. **460**

gibberellin Plant hormone; induces stem elongation, helps seeds break dormancy. Role in flowering in some species. **576**

gills Folds or body extensions that increase the surface area for respiration. **427**

gland Cluster of secretory epithelial cells. **391**

global climate change Wide-ranging changes in rainfall patterns, average temperature, and other climate factors that are thought to result from rising concentrations of greenhouse gases. **356, 377**

glottis Opening formed when the vocal cords relax. **428**

glucagon Pancreatic hormone that raises the blood glucose level. **512**

glycolysis Set of reactions in which glucose or another sugar is broken down to two pyruvate for a net yield of two ATP. **89**

Golgi body Organelle that modifies polypeptides and lipids; also sorts and packages the finished products into vesicles. **53**

gonads Ovaries and testes; organs that produce gametes and secrete sex hormones. **515**

Gondwana Supercontinent that formed more than 500 million years ago. **207**

grasslands Biome in the interior of continents, where grasses and other low-growing plants are adapted to warm summers, cold winters, periodic fires, and grazing animals. **365**

gravitropism Plant growth in a direction influenced by gravity. **577**

gray matter Tissue in brain and spinal cord that consists of cell bodies, dendrites, and neuroglial cells. **490**

green alga Single-celled, colonial, or multicelled photosynthetic protist belonging to the group most closely related to land plants. **257**

greenhouse effect Warming of Earth's lower atmosphere and surface as a result of heat trapped by greenhouse gases. **356**

greenhouse gas Atmospheric gas such as carbon dioxide that helps keep heat from escaping into space and thus warms the Earth. **356**

ground tissue system Tissue system that makes up the bulk of the plant body; includes most photosynthetic cells. **549**

groundwater Water between soil particles and in aquifers. **353**

growth Increases in the number, size, and volume of cells in multicelled species. **7**

growth hormone Anterior pituitary hormone that promotes growth and development throughout the body. **508**

guard cell One of a pair of cells that define a stoma across the epidermis of a leaf or stem. **552**

gymnosperms Seed plant lineage that does not make flowers or fruits. **272**

habitat The type of place a species lives. **341**

half-life Characteristic time it takes for half of a quantity of a radioisotope to decay. **203**

haploid Having one of each type of chromosome characteristic of the species. **144**

heart Muscular organ that pumps fluid through a body. **419**

heart attack Heart cells die because of impaired blood flow through cardiac arteries. **431**

hermaphrodite Animal that makes both eggs and sperm. **296, 523**

heterotroph Organism that obtains energy and carbon from organic compounds assembled by other organisms. **80**

heterozygous Having two different alleles of a gene. **158**

histone Type of protein that structurally organizes eukaryotic chromosomes. **103**

homeostasis Processes by which an organism keeps internal conditions within tolerable ranges. **7, 389**

homeotic gene Type of master gene; its expression controls formation of specific body parts during development. **128**

hominid Human or extinct humanlike species. **313**

homologous Refers to the two members of a pair of chromosomes with the same length, shape, and genes. **139**

homologous structure Similar body parts that reflect shared ancestry among lineages. **208**

homozygous Having identical alleles of a gene. **157**

hot spots Threatened regions with great biodiversity that are considered a high priority for conservation efforts. **380**

humans Members of the genus *Homo*. **314**

human chorionic gonadotropin (HCG) Hormone produced by the chorion and later by the placenta; encourages maintenance of the uterus thus sustaining a pregnancy. **537**

humus Decaying organic matter in soil. **557**

hybrid The offspring of a cross between two individuals that breed true for different forms of a trait. **157**

hydrogen bond Attraction that forms between a covalently bonded hydrogen atom and another atom taking part in a separate covalent bond. **25**

hydrolysis Process by which an enzyme breaks a molecule into smaller subunits by attaching a hydroxyl group to one part and a hydrogen atom to the other. **30**

hydrophilic Describes a substance that dissolves easily in water. **26**

hydrophobic Describes a substance that resists dissolving in water. **26**

hydrostatic skeleton Of soft-bodied invertebrates, a fluid-filled chamber that muscles act on, redistributing the fluid. **405**

hydrothermal vent Place where hot, mineral-rich water streams out from an underwater opening in Earth's crust. **242, 369**

hypertension Chronically high blood pressure. **431**

hypertonic Describes a fluid with a high solute concentration relative to another fluid. **70**

hypha A single filament in a fungal mycelium. **283**

hypothalamus Forebrain region that controls processes related to homeostasis and has endocrine functions. **491, 508**

hypothesis Testable explanation of a natural phenomenon. **10**

hypotonic Describes a fluid with a low solute concentration relative to another fluid. **70**

immigration Movement of individuals into a population. **324**

immunity The body's ability to resist and fight infections. **439**

immunization A procedure designed to promote immunity to a disease. **454**

inbreeding Nonrandom mating among close relatives. **226**

incomplete dominance Condition in which one allele is not fully dominant over another, so the heterozygous phenotype is between the two homozygous phenotypes. **163**

indicator species A species that is highly sensitive to environmental changes and can be monitored to assess whether an ecosystem is threatened. **380**

inflammation A local response to tissue damage or infection; characterized by redness, warmth, swelling, and pain. **442**

ingestion Taking food into the digestive system. **461**

inheritance Transmission of DNA from parents to offspring. **7**

inhibitor Hypothalamic hormone that acts in the anterior pituitary to slow hormone secretion. **508**

innate immunity Set of inborn, fixed general defenses against infection; includes phagocytosis, inflammation, fever, complement activation. **439**

insects Land-dwelling arthropods with a pair of antennae, three pairs of legs, and—in the most diverse groups—wings. **303**

insertion Mutation in which one or more base pairs become inserted into DNA. **125**

insulin Pancreatic hormone that causes cells to take up glucose from the blood, thus lowering blood glucose level. **512**

integumentary exchange Gas exchange across the outer body surface. **427**

intercostal muscles Muscles between the ribs; help alter the size of the thoracic cavity during breathing. **429**

intermediate filament Cytoskeletal element that locks cells and tissues together. **54**

interneurons Neurons that receive signals from other neurons and integrate information. **481**

interphase In a eukaryotic cell cycle, the interval between mitotic divisions when a cell grows, roughly doubles the number of its cytoplasmic components, and replicates its DNA. **138**

interspecific competiton Two species compete for a limited resource and both are harmed by the interaction. **343**

interstitial fluid Fluid between cells of a multicelled body. **419**

intervertebral disk Cartilage disk between two vertebrae. **405**

invertebrate Animal without a backbone. **295**

***in vitro* fertilization** Infertility treatment in which gametes are combined outside the body. **534**

ion Atom that carries a charge because it has an unequal number of protons and electrons. **23**

ionic bond Type of chemical bond in which a strong mutual attraction forms between ions of opposite charge. **24**

iris Ring of smooth muscle with pupil at its center; adjusts how much light enters the eye. **495**

isotonic Describes a fluid with the same solute concentration relative to another fluid. **70**

isotopes Forms of an element that differ in the number of neutrons their atoms carry. **21**

jaws Hinged skeletal elements used in feeding. **306**

joint Region where bones meet. **407**

karyotype Image of an individual's complement of chromosomes arranged by size, length, shape, and centromere location. **105**

key innovation An evolutionary adaptation that gives its bearer the opportunity to exploit a particular environment more efficiently or in a new way. **232**

keystone species A species that has a disproportionately large effect on community structure. **348**

kidney Organ of the vertebrate urinary system that filters blood, adjusts its composition, and forms urine. **307, 471**

kidney dialysis Procedure used to cleanse blood and restore proper solute concentrations in a person with impaired kidney function. **474**

Krebs cycle Cyclic pathway that, along with acetyl–CoA formation, breaks down two pyruvate to carbon dioxide for a net yield of two ATP and many reduced coenzymes. Part of aerobic respiration pathway. **90**

K-selection Selection favoring individuals that compete well in crowded conditions; selection for offspring quality. **329**

labor Expulsion of a placental mammal from its mother's uterus by muscle contractions. **542**

lactate fermentation Anaerobic carbohydrate breakdown pathway that produces ATP and lactate. **92**

lancelets Invertebrate chordates that have a fish-like shape and retain their defining chordate traits into adulthood. **305**

large intestine Organ that receives digestive waste from the small intestine and concentrates it as feces. Also called the colon. **465**

larva Preadult stage in some animal life cycles. **296**

larynx Short airway containing the vocal cords (voice box). **428**

lateral meristem Sheetlike cylinder of meristem inside older stems and roots; vascular cambium or cork cambium. **555**

law of nature Generalization that describes a consistent natural phenomenon for which there is incomplete scientific explanation; e.g., first and second laws of thermodynamics. **11**

lens Disk-shaped structure that focuses light on an eye's photoreceptors. **495**

leptin Molecule made by adipose cells that suppresses appetite. **460**

lethal mutation Mutation that drastically alters phenotype; usually causes death. **218**

leukemia Cancer that increases white blood cell numbers. **431**

lichen Composite organism consisting of a fungus and a single-celled photosynthetic alga or bacterium. **286**

life history pattern A set of traits related to growth, survival, and reproduction such as life span, age-specific mortality, age at first reproduction, and number of breeding events. **328**

ligament Dense connective tissue that holds bones together at a joint. **407**

light-dependent reactions Metabolic pathway of photosynthesis that converts light energy to chemical energy of ATP and NADPH. **84**

light-independent reactions Metabolic pathway of photosynthesis that uses ATP and NADPH to assemble sugars from water and CO_2. **84**

lignin Material that stiffens cell walls of vascular plants. **271**

limiting factor A necessary resource, the depletion of which halts population growth. **325**

lineage Line of descent. **202**

lipid Fatty, oily, or waxy organic compound. **32**

lipid bilayer Structural foundation of cell membranes; mainly phospholipids arranged tail-to-tail in two layers. **43**

liver Organ that produces bile, stores glycogen, and detoxifies many substances. **464**

lobe-finned fish Bony fish with bony supports in its fins. **308**

logistic growth A population grows slowly, then increases rapidly until it reaches carrying capacity and levels off. **326**

loose connective tissue Connective tissue with relatively few fibroblasts and fibers scattered in its matrix. **392**

lung Of most land vertebrates and some fish, saclike organ inside which blood exchanges gases with the air. **306, 427**

lysis Breaking of a cell's plasma membrane; results in death of the cell. **260**

lysogenic pathway Bacteriophage replication pathway in which the virus becomes integrated into the host's chromosome and is passed to its descendants. **260**

lysosome Enzyme-filled vesicle that functions in intracellular digestion. **53**

lysozyme Antibacterial enzyme that occurs in body secretions such as mucus. **441**

lytic pathway Bacteriophage replication pathway in which the virus replicates in its host and quickly kills it. **260**

macroevolution Patterns of evolution that occur above the species level. **231**

macrophage Phagocytic white blood cell that patrols tissue fluids; presents antigen to T cells. **440**

mammal Vertebrate that nourishes its young with milk from mammary glands. **312**

mantle Skirtlike extension of tissue in mollusks; covers the mantle cavity and secretes the shell in species that have a shell. **298**

marsupial Mammals in which young complete development in a pouch on the mother's body. **312**

mass extinction Simultaneous loss of many lineages from Earth. **196, 231**

mass number Total number of protons and neutrons in the nucleus of an element's atoms. **21**

mast cell White blood cell that is anchored in many tissues; factor in inflammation. **440**

master gene Gene encoding a product that affects the expression of many other genes. **128**

mechanical digestion Breaking of food into smaller pieces by mechanical processes such as chewing. **462**

mechanoreceptor Sensory receptor that responds to pressure, position, or acceleration. **493**

medulla oblongata Hindbrain region that controls breathing rhythm and reflexes such as coughing and vomiting. **491**

medusa Bell-shaped, free-swimming cnidarian body form. **297**

megaspore Haploid spore that forms in ovary of seed plants; gives rise to a female gametophyte with egg cell. **277, 568**

meiosis Nuclear division process that halves the chromosome number. Basis of sexual reproduction. **137**

melatonin Hormone produced by pineal gland; affects onset of puberty, sleep–wake cycles. **516**

membrane potential Potential energy across the membrane of a resting neuron. **481**

memory cell Antigen-sensitized B or T cell that forms in a primary immune response but does not act immediately. Participates in a secondary response if the same antigen returns. **446**

meninges Membranes that surround and protect the brain and spinal cord. **490**

menopause Of a human female, the end of fertility and of menstrual cycles. **531**

menstrual cycle Approximately monthly cycle in which the uterine lining thickens in preparation for pregnancy, then is shed if pregnancy does not occur. **530**

menstruation Flow of blood and bits of shed uterine lining out through the vagina. **530**

meristem Zone of undifferentiated plant cells that can divide rapidly. **553**

mesoderm Middle tissue layer of a three-layered animal embryo. **524**

messenger RNA (mRNA) Type of RNA that carries a protein-building message. **117**

metabolic pathway Series of enzyme-mediated reactions by which cells build, remodel, or break down an organic molecule. **67**

metabolism All the enzyme-mediated chemical reactions by which cells acquire and use energy as they build and break down organic molecules. **30**

metamorphosis Dramatic remodeling of body form during the transition from larva to adult. **301**

metaphase Stage of mitosis during which the cell's chromosomes align midway between poles of the spindle. **140**

methanogen Organism that produces methane gas as a metabolic by-product. **250**

MHC markers Self-recognition protein on the surface of body cells. Triggers adaptive immune response when bound to antigen fragments. **445**

microevolution Change in allele frequencies in a population or species. **218**

microfilament Reinforcing cytoskeletal element; a fiber of actin subunits. **54**

microspore Walled haploid spore of seed plants; gives rise to a pollen grain. **277, 568**

microtubule Cytoskeletal element involved in movement; hollow filament of tubulin subunits. **54**

microvilli Thin projections that increase the surface area of brush border cells. **464**

migration Periodic movement of many individuals between one or more regions. **324**

mimicry Two or more species come to resemble one another. **345**

mineral Inorganic substance that is required in small amounts for normal metabolism. **469**

mitochondrion Double-membraned organelle that produces ATP; site of second and third stages of aerobic respiration in eukaryotes. **53**

mitosis Nuclear division mechanism that maintains the chromosome number. Basis of body growth, tissue repair and replacement in multicelled eukaryotes; also asexual reproduction in some plants, animals, fungi, and protists. **137**

model System similar to an object or event that cannot itself be tested directly. **10**

molecule An association of two or more atoms joined by chemical bonds. **5, 24**

mollusk Invertebrate with a reduced coelom and a mantle. **298**

molting Periodic shedding of an outer body layer or part. **300**

monocots Lineage of angiosperms that includes grasses, orchids, and palms. **280**

monohybrid cross Cross in which individuals with different alleles of one gene are crossed. **158**

monomers Molecules that are subunits of polymers. **30**

monophyletic group An ancestor and all of its descendants. **235**

monotreme Egg-laying mammal. **312**

morphological convergence Evolutionary pattern in which similar body parts evolve separately in different lineages. **209**

morphological divergence Evolutionary pattern in which a body part of an ancestor changes in its descendants. **208**

motor neurons Neurons that control muscles and glands. **481**

motor protein Type of energy-using protein that interacts with cytoskeletal elements to move the cell's parts or the whole cell. **54**

motor unit One motor neuron and the muscle fibers it controls. **411**

multiregional model Model that postulates *H. sapiens* populations in different regions evolved from *H. erectus* in those regions. **316**

muscle fatigue Decrease in a muscle's ability to contract despite ongoing stimulation. **412**

muscle tension Force exerted by a contracting muscle. **412**

muscle twitch Brief muscle contraction. **411**

mutation Permanent change in DNA sequence. Primary source of new alleles. **109**

mutualism Species interaction that benefits both species. **286, 342**

mycelium Mass of threadlike filaments (hyphae) that make up the body of a multicelled fungus. **283**

mycorrhiza Fungus–plant root partnership. **286, 558**

myelin Insulating material around most vertebrate axons; increases speed of signal transmission. **487**

myofibrils Threadlike, cross-banded skeletal muscle components that consist of sarcomeres arranged end to end. **409**

myosin Protein in thick filaments of muscle fibers. **409**

natural killer cell (NK cell) White blood cell that kills infected or cancerous cells. **440**

natural selection A process of evolution in which individuals of a population who vary in the details of heritable traits survive and reproduce with differing success. **200**

naturalist Person who observes life from a scientific perspective. **197**

nature Everything in the universe except what humans have manufactured. **5**

nectar Sweet fluid exuded by some flowers; attracts pollinators. **581**

nematocyst Touch-sensitive stinging organelle unique to cnidarians. **297**

negative feedback A change causes a response that reverses the change. **398**

nephron Kidney tubule and associated capillaries; filters blood and forms urine. **472**

nerve net Mesh of interacting neurons with no central control, as in the nervous system of jellies. **486**

nerves Axons bundled in connective tissues. **486**

nervous tissue Animal tissue composed of neurons and supporting cells; detects stimuli and controls responses to them. **394**

neuroglia Cells that support neurons. **481**

neuron One of the cells that make up communication lines of nervous systems; transmits electrical signals along its plasma membrane and communicates with other cells through chemical messages. **394, 481**

neurotransmitter Chemical signal released by a neuron's axon endings. **484**

neutral mutation A mutation that has no effect on survival or reproduction. **218**

neutron Uncharged subatomic particle in the atomic nucleus. **21**

neutrophil Circulating phagocytic white blood cell. **440**

niche The role of a species in its community. **341**

nitrogen cycle Movement of nitrogen among the atmosphere, soil, and water, and into and out of food webs. **354**

nitrogen fixation Formation of ammonia from nitrogen gas and hydrogen. **250, 355, 558**

nitrification Conversion of ammonium to nitrates. **355**

nondisjunction Failure of sister chromatids or homologous chromosomes to separate during meiosis or mitosis. **172**

nonpolar Having an even distribution of charge. **25**

normal flora Microorganisms that normally live in or on a healthy animal or person. **251, 440**

notochord Stiff rod of connective tissue that runs the length of the body in chordate larvae or embryos. **305**

nuclear envelope A double membrane that constitutes the outer boundary of the nucleus. **52**

nucleic acid Single- or double-stranded chain of nucleotides joined by sugar–phosphate bonds; e.g., DNA, RNA. **37**

nucleic acid hybridization Base-pairing between DNA or RNA from different sources. **182**

nucleoid Region of cytoplasm where the DNA is concentrated inside a prokaryotic cell. **43**

nucleosome A length of DNA wound around a spool of histone proteins. **103**

nucleotide Monomer of nucleic acids; has a five-carbon sugar, a nitrogen-containing base, and phosphate groups. **37**

nucleus Of an atom, core region occupied by protons and neutrons. Of a eukaryotic cell, organelle with two membranes that holds the cell's DNA. **21, 43**

nutrient Substance that an organism needs for growth and survival, but cannot make for itself. **6**

oocyte Immature egg. **528**

open circulatory system Circulatory system in which blood leaves vessels and flows among tissues. **419**

organ Structural unit composed of two or more tissues and adapted to carry out a particular task. **389**

organ of Corti Sound-detecting organ in the cochlea. **498**

organ system Organs that interact closely in some task. **389**

organelle Structure that carries out a specialized metabolic function inside a cell. **43**

organic Refers to a molecule that consists primarily of carbon and hydrogen atoms. **29**

organism Individual that consists of one or more cells. **5**

orgasm Sexual climax during which neurotransmitter release causes smooth muscle contractions accompanied by a feeling of release and pleasure. **532**

osmosis The diffusion of water across a selectively permeable membrane. **70**

osmotic pressure Amount of turgor that prevents osmosis into a hypertonic fluid. **70**

osteoporosis Disorder in which bones weaken. **406**

ovarian follicle Immature egg and surrounding cells. **528**

ovary In flowering plants, the enlarged base of a carpel, inside which one or more ovules form and eggs are fertilized. In animals, organ in which eggs form. **280, 526, 567**

oviduct Ciliated tube connecting an ovary to the uterus. **528**

ovulation Release of a secondary oocyte from an ovary. **529**

ovule Of seed plants, chamber inside which the female gametophyte forms; after fertilization, matures into a seed. **277, 567**

ovum Mature animal egg. **532**

ozone layer Atmospheric layer with a high concentration of ozone; prevents much UV radiation from reaching Earth's surface. **244, 376**

pain Perception of tissue injury. **493**

pancreas Organ that secretes digestive enzymes into the small intestine and hormones into the blood. **464**

pandemic A disease that spreads worldwide. **262**

Pangea Supercontinent that formed about 237 million years ago and broke up about 152 million years ago. **206**

parasite A species that withdraws nutrients from another species (its host), usually without killing it. **346**

parasitoid An insect that lays eggs in another insect (the host), and whose young devour their host from the inside. **346**

parasympathetic neurons Neurons of the autonomic system that encourage housekeeping tasks. **489**

parathyroid glands Four small endocrine glands whose hormone product increases the level of calcium in blood. **511**

parenchyma Simple plant tissue made up of living cells; the main component of the ground tissue system. **549**

parthenogenesis Process by which individuals develop from unfertilized eggs. **523**

passive transport Mechanism by which a concentration gradient drives the movement of a solute across a cell membrane through a transport protein. Requires no energy input. **72**

pathogen Organism that infects another organism and causes disease. **251**

PCR Polymerase chain reaction. Method that rapidly generates many copies of a specific DNA fragment. **182**

pedigree Chart showing the pattern of inheritance of a trait in a family. **167**

pelagic province The ocean's open waters. **368**

pellicle Layer of proteins that gives shape to many unwalled, single-celled protists. **252**

peptide bond A bond between the amine group of one amino acid and the carboxyl group of another. Joins amino acids in proteins. **34**

per capita growth rate For some interval, the number of individuals added divided by the initial population size. **324**

periderm Plant dermal tissue that replaces epidermis on older stems and roots. **555**

peripheral nervous system Nerves that extend through the body and carry signals to and from the central nervous system. **486**

permafrost Layer of permanently frozen soil in the Arctic. **367**

peroxisome Enzyme-filled vesicle that breaks down amino acids, fatty acids, and toxic substances. **53**

pH Measure of the number of hydrogen ions in a fluid. **27**

phagocytosis Endocytic pathway by which a cell engulfs particles such as microbes or cellular debris. **74**

pharynx Throat; opens to airways and digestive tract. **428**

phenotype An individual's observable traits. **157**

pheromones Signaling molecules that affect another member of the same species. **494**

phloem Complex vascular tissue that distributes sugars through a plant body. **271, 550**

phospholipid A lipid with a phosphate group in its hydrophilic head, and two nonpolar fatty acid tails; main constituent of cell membranes. **32**

phosphorus cycle Movement of phosphorus among rocks, water, soil, and living organisms. **354**

phosphorylation Phosphate-group transfer. **65**

photoautotroph Photosynthetic autotroph. **88**

photoperiodism Biological response to seasonal changes in the relative lengths of day and night. **579**

photoreceptor Light-sensitive sensory receptor. **493**

photorespiration Reaction in which rubisco attaches oxygen instead of carbon dioxide to ribulose bisphosphate. **87**

photosynthesis Metabolic pathway by which photoautotrophs capture light energy and use it to make sugars from CO_2 and water. **80**

photosystem Cluster of pigments and proteins that converts light energy to chemical energy in photosynthesis. **83**

phototropism Change in the direction of cell movement or growth in response to a light source. **577**

phylogeny Evolutionary history of a species or groups of species. **234**

phytochrome A light-sensitive pigment that helps set plant circadian rhythms based on length of night. **578**

pigment An organic molecule that can absorb light of certain wavelengths. **81**

pilus A protein filament that projects from the surface of some bacterial cells. **50**

pineal gland Endocrine gland in the forebrain that secretes melatonin; secretion declines when the eye is exposed to light. **516**

pioneer species Species that can colonize a new habitat. **347**

pituitary gland Pea-sized endocrine gland in the forebrain that interacts closely with the adjacent hypothalamus. **508**

placenta Of placental mammals, organ composed of maternal and embryonic tissues that allows the exchange of substances between a developing individual and its mother. **537**

placental mammal Mammal in which young are nourished within the mother's body by way of a placenta. **312**

placozoan Structurally simplest animal known, with only four types of cells and a small genome. **293**

plankton Community of tiny drifting or swimming organisms. **253**

plant A multicelled, typically photosynthetic producer, in which embryos form on a parent. **8, 271**

plaque On teeth, a thick biofilm composed of bacteria, their extracellular products, and saliva proteins. **444**

plasma Fluid portion of blood. **424**

plasma membrane A cell's outermost membrane. **43**

plasmid Of many prokaryotes, a small ring of nonchromosomal DNA replicated independently of the chromosome. **181, 249**

plasmodial slime mold Heterotrophic protist that moves and feeds as a multinucleated mass; forms a fruiting body when conditions are unfavorable. **258**

plate tectonics Theory that Earth's outer layer of rock is cracked into plates, the slow movement of which rafts continents to new locations over geologic time. **207**

platelet Cell fragment that helps blood clot. **424**

pleiotropic Refers to a gene whose product influences multiple traits. **164**

polar Having an uneven distribution of charge. **25**

polarity Any separation of charge into distinct positive and negative regions. **25**

pollen grain Male gametophyte of a seed plant. **272**

pollination Delivery of pollen to the female part of a plant. **277, 568**

pollination vector Any agent that moves pollen grains from one plant to another. **580**

pollinator Animal that moves pollen from one plant to another, thus facilitating pollination. **280, 566**

pollutant A natural or man-made substance that is released into the environment in greater than natural amounts and damages the health of organisms. **373**

polymers Molecules that consist of multiple monomers. **30**

polyp Tubular cnidarian body form that usually attaches to some surface. **297**

polypeptide Chain of amino acids linked by peptide bonds. **34**

polyploid Having three or more of each type of chromosome characteristic of the species. **172**

polysome Cluster of ribosomes that are simultaneously translating an mRNA. **123**

pons Hindbrain region between medulla oblongata and midbrain; helps control breathing. **491**

population A group of organisms of the same species who live in a specific location and breed with one another more often than they breed with members of other populations. **5, 322**

population density The number of organisms of a particular population in a given area. **323**

population distribution The way in which members of a population are spread out in their environment. **323**

population size Number of individuals in a population. **323**

predation One species feeds on another, usually killing it. **344**

predator One species that eats and kills another. **344**

prediction A statement, based on a hypothesis, about a condition that should exist if the hypothesis is not wrong. **10**

pressure flow theory Explanation of how flow of fluid through phloem is driven by differences in pressure and sugar concentration between a source and a sink. **561**

prey A species that is eaten by a predator. **344**

primary growth Plant growth from apical meristems in root and shoot tips. **553**

primary production The energy captured by an ecosystem's producers. **352**

primary succession Ecological succession occurs in an area where there was previously no soil. **347**

primate Mammalian group that includes tarsiers, monkeys, apes, and humans. **313**

primer Short, single strand of DNA designed to hybridize with a DNA fragment. **182**

prion Infectious protein. **36**

probe Short fragment of DNA labeled with a tracer; designed to hybridize with a nucleotide sequence of interest. **182**

producers Organisms that capture energy and make their own food from inorganic materials in the environment. **6, 350**

product A molecule remaining at the end of a reaction. **64**

progesterone Sex hormone produced by a female's ovaries. **515**

prokaryote Single-celled organism in which the DNA resides in the cytoplasm; a bacterium or archaean. **8, 244**

prokaryotic conjugation One prokaryotic cell transfers a plasmid to another. **249**

prokaryotic fission Method of asexual reproduction in which one prokaryotic cell divides and forms two identical descendant cells. **249**

promoter In DNA, a sequence to which RNA polymerase binds. **119**

prophase Stage of mitosis in which chromosomes condense and become attached to a newly forming spindle. **140**

prostate gland Exocrine gland that encircles a male's urethra; its secretions contribute to semen. **526**

protein Organic compound that consists of one or more chains of amino acids. **33**

protist A eukaryote that is not a plant, fungus, or animal. **8, 252**

protocells Membranous sacs that contain interacting organic molecules; hypothesized to have formed prior to the earliest life forms. **243**

proton Positively charged subatomic particle that occurs in the nucleus of all atoms. **21**

pseudocoel Body cavity not fully lined with mesoderm. **295**

pseudopod Extendable lobe of membrane-enclosed cytoplasm. **55**

puberty Period when reproductive organs mature and begin to function. **515**

pulmonary circuit Circuit through which blood flows from the heart to the lungs and back. **420**

pulse Brief stretching of artery walls that occurs when ventricular contraction forces blood into arteries. **424**

Punnett square Diagram used to predict the genetic and phenotypic outcome of a cross. **159**

pupil Opening through which light enters the eye. **495**

pyruvate Three-carbon end product of glycolysis. **89**

radial symmetry Having parts arranged around a central axis, like spokes around a wheel. **294**

radioactive decay Process by which atoms of a radioisotope spontaneously emit energy and subatomic particles when their nucleus disintegrates. **21**

radioisotope Isotope with an unstable nucleus. **21**

radiometric dating Method of estimating the age of a rock or fossil by measuring the content and proportions of a radioisotope and its daughter elements. **203**

radula Tonguelike organ of many mollusks. **298**

rain shadow Dry region on the downwind side of a coastal mountain range. **363**

ray-finned fish Bony fish with fins supported by thin rays derived from skin. **308**

reabsorption In the kidney, water and solutes leave the filtrate and enter capillaries. **472**

reactant Molecule that enters a reaction. **64**

reaction Process of chemical change. **64**

receptor Molecule or structure that can respond to a form of stimulation. **6**

receptor protein Plasma membrane protein that binds to a particular substance outside of the cell. **49**

recessive Refers to an allele with an effect that is masked by a dominant allele on the homologous chromosome. **158**

recognition protein Plasma membrane protein that tags a cell as belonging to self (one's own body). **48**

recombinant DNA A DNA molecule that contains genetic material from more than one organism. **181**

rectum Portion of the large intestine that stores feces until they are expelled. **466**

red alga Single-celled or multicelled photosynthetic protist with red accessory pigment. **256**

red blood cell Hemoglobin-filled blood cell that transports oxygen. **424**

red marrow Bone marrow that makes blood cells. **406**

reflex Automatic (involuntary) response to a stimulus. **491**

releaser Hormone that is secreted by one endocrine gland and stimulates secretion by another. **508**

replacement model Model for origin of *H. sapiens*; humans evolved in Africa, then migrated to different regions and replaced the other hominids that lived there. **317**

reproduction Process by which parents produce offspring. **7**

reproductive cloning Technology that produces genetically identical individuals. **110**

reproductive isolation Absence of gene flow between populations; part of speciation. **227**

reptile Amniote subgroup that includes lizards, snakes, turtles, crocodilians, and birds. **310**

resource partitioning Use of different portions of a limited resource; allows species with similar needs to coexist. **344**

respiration Physiological process by which gases enter and leave an animal body. **426**

respiratory cycle One inhalation and one exhalation. **429**

respiratory surface Moist surface across which gases are exchanged between animals cells and their environment. **426**

resting potential Membrane potential of a neuron that is not being stimulated. **482**

restriction enzyme Type of enzyme that cuts specific base sequences in DNA. **181**

retina Layer of eye that contains photoreceptors. **495**

reverse transcriptase A viral enzyme that uses mRNA as a template to make a strand of cDNA. **182**

rhizome Stem that grows horizontally along the ground. **274**

ribosomal RNA (rRNA) A type of RNA that becomes part of ribosomes. **121**

ribosome Organelle of protein synthesis. **50**

RNA Ribonucleic acid. Typically single-stranded nucleic acid; roles in protein synthesis. *See also* ribosomal RNA, transfer RNA, messenger RNA. **37**

RNA polymerase Enzyme that carries out transcription. **118**

RNA world Hypothetical early interval when RNA served as the material of inheritance. **242**

rod cell Photoreceptor active in dim light; provides coarse perception of image and detects motion. **497**

root hairs Hairlike, absorptive extensions of a young cell of root epidermis. **554**

root nodules Swellings of some plant roots that contain nitrogen-fixing bacteria. **558**

roundworm Unsegmented worm with a pseudocoel and a cuticle that is molted as the animal grows. **300**

r-selection Selection favoring individuals that produce many offspring fast; selection for offspring quantity. **329**

rubisco Ribulose bisphosphate carboxylase. Carbon-fixing enzyme of the Calvin–Benson cycle. **86**

salivary amylase Enzyme in saliva that begins carbohydrate digestion by breaking starch into disaccharides. **463**

salt Compound that dissolves easily in water and releases ions other than H^+ and OH^-. **26**

sampling error Difference between results derived from testing an entire group of events or individuals, and results derived from testing a subset of the group. **14**

saprobe Organism that feeds on wastes and remains. **283**

sarcomere Contractile unit of skeletal and cardiac muscle. **409**

saturated fat Fatty acid with no double bonds in its carbon tail. **32**

scales Hard, flattened elements that cover the skin of some vertebrates. **307**

science Systematic study of nature. **9**

scientific theory Hypothesis that has not been disproven after many years of rigorous testing, and is useful for making predictions about other phenomena. **10**

sclera Protective covering at the sides and back of eyeball. **495**

sclerenchyma Simple plant tissue; dead at maturity, its lignin-reinforced cell walls structurally support plant parts. **550**

scrotum Pouch that holds the testes. **526**

seamount An undersea mountain. **369**

second law of thermodynamics Energy tends to disperse. **63**

second messenger Molecule that forms inside a cell when a hormone binds at the cell surface; sets in motion reactions that alter enzyme activity inside the cell. **506**

secondary growth A thickening of older stems and roots at lateral meristems. **277, 555**

secondary succession Ecological succession occurs in an area where a community previously existed and soil remains. **347**

seed Embryo sporophyte of a seed plant packaged with nutritive tissue inside a protective coat. **272, 570**

segmentation Having a body composed of units that repeat along its length. **295**

selective permeability Membrane property that allows some substances, but not others, to cross. **69**

semen Sperm mixed with fluid secreted by exocrine glands. **526**

seminal vesicles Exocrine glands that secrete fluid into the vas deferentia; main source of semen volume. **526**

senescence Phase in a life cycle from maturity until death; also applies to death of parts, such as plant leaves. **579**

sensory neurons Neurons that detect specific stimuli such as light, touch, or a particular chemical. **481**

sensory receptor Cell or cell part that responds to a particular stimulus. **398**

sequencing Method of determining the order of nucleotides in DNA. **185**

sex chromosome Member of a pair of chromosomes that differs between males and females. **104**

sexual reproduction Reproductive mode by which offspring arise from two parents and inherit genes from both. **144, 523**

sexual selection Mode of natural selection in which some individuals outreproduce others of

a population because they are better at securing mates. **223**

sexually transmitted disease Infectious disease caused by a pathogen that can live in the reproductive tract. **534**

shell model Model of electron distribution in an atom; orbitals are shown as nested circles, electrons as dots. **23**

sister chromatid One of two attached members of a duplicated eukaryotic chromosome. **103**

sister groups The two lineages that emerge from a node on a cladogram. **235**

skeletal muscle tissue Striated, voluntary muscle that interacts with bone to move body parts. **393**

sliding-filament model Explanation of how interactions among actin and myosin filaments shorten a sarcomere and bring about muscle contraction. **410**

small intestine Longest portion of the digestive tract, and the site of most digestion and absorption. **464**

smooth muscle tissue Involuntary muscle that lines blood vessels and hollow organs; not striated. **393**

solute A dissolved substance. **26**

solvent Liquid that can dissolve other substances. **26**

somatic cell nuclear transfer (SCNT) Method of reproductive cloning in which genetic material is transferred from an adult somatic cell into an unfertilized, enucleated egg. **111**

somatic nervous system Set of nerves that control skeletal muscle and relay signals from joints and skin. **488**

somatic sensations Sensations that arise when sensory neurons in skin and near joints and muscle are activated. **493**

speciation Process by which new species arise from existing species. **227**

species A type of organism. **8**

sperm Mature male gamete. **148**

sphincter Ring of muscle that controls passage through a tubular organ or body opening. **463**

spinal cord Portion of central nervous system that connects peripheral nerves with the brain. **490**

spindle Dynamically assembled and disassembled array of microtubules that moves chromosomes during mitosis or meiosis. **140**

sponge Aquatic invertebrate that has no tissues or organs and filters food from the water. **296**

spongy bone Lightweight bone with many internal spaces. **406**

sporadic disease A disease that breaks out occasionally and affects only a small portion of a population. **262**

sporophyte Diploid, spore-producing stage of a plant life cycle. **148, 271**

sprain Ligaments of a joint are injured. **407**

stabilizing selection Mode of natural selection in which intermediate phenotypes are favored over extremes. **221**

stamen The male reproductive part of a flowering plant; consists of a pollen-producing anther on a filament. **280, 567**

stasis Macroevolutionary pattern in which a lineage persists with little or no change over evolutionary time. **231**

statistically significant Refers to a result that is statistically unlikely to have occurred by chance. **14**

stem cell A cell that can divide to form more stem cells or differentiate to become a specialized cell type. **388**

steroid A type of lipid with four carbon rings and no fatty acid tails. **33**

steroid hormone Hormone, such as testosterone, that is derived from cholesterol. **506**

stomach Muscular organ that mixes food with gastric fluid that it secretes. **464**

stomata Gaps that open between guard cells on plant surfaces; allow water vapor and gases to diffuse across the epidermis. **86, 271**

strain A subgroup within a species that has a characteristic trait or traits. **249**

stroke Event in which brain cells die because a clot or vessel rupture disrupts blood flow within the brain. **431**

stroma Semifluid matrix between the thylakoid membrane and the two outer membranes of a chloroplast; site of light-independent photosynthesis reactions. **83**

stromatolites Dome-shaped structures composed of layers of prokaryotic cells and sediments. **244**

substrate A reactant molecule that is specifically acted upon by an enzyme. **66**

surface-to-volume ratio A relationship in which the volume of an object increases with the cube of the diameter, but the surface area increases with the square. **44**

survivorship curve Graph showing the decline in numbers of a cohort over time. **328**

sustainable development Using resources to meet current human needs in a way that takes into account the needs of future generations. **382**

swim bladder Organ that adjusts buoyancy in bony fishes. **306**

symbiosis One species lives in or on another in a commensal, mutualistic, or parasitic relationship. **342**

sympathetic neurons Neurons of the autonomic system; their activity prepares the body for danger or excitement. **489**

sympatric speciation Pattern in which speciation occurs in the absence of a physical barrier. **229**

syndrome The set of symptoms that characterize a genetic disease. **167**

synovial joint Joint such as the knees that is lubricated by fluid and allows movement of bones around the joint. **407**

systemic circuit Circuit through which blood flows from the heart to the body tissues and back. **420**

systolic pressure Blood pressure when ventricles are contracting. **425**

T cell receptor (TCR) Antigen-binding receptor on the surface of T cells; also recognizes MHC markers. **445**

T lymphocyte (T cell) Circulating white blood cell central to adaptive immunity; some kinds target sick body cells. **440**

taiga Extensive northern biome that is dominated by conifers. Also called boreal forest. **366**

taxon (taxa) A grouping of organisms. **233**

taxonomy Science of naming and classifying species. **233**

telophase Stage of mitosis during which chromosomes arrive at the spindle poles and decondense, and new nuclei form. **140**

temperate deciduous forest Biome dominated by trees that drop all their leaves in fall and become dormant during a cold winter. **366**

temperature Measure of molecular motion. **27**

tendon Strap of dense connective tissue that connects a skeletal muscle to bone. **408**

testes In animals, organs that produces sperm. **526**

testosterone Sex hormone produced by a male's testes. **515**

tetrapod Vertebrate with four limbs. **306**

theory of uniformity Idea that gradual repetitive processes occurring over long time spans shaped Earth's surface. **199**

therapeutic cloning Using somatic cell nuclear transfer to produce human embryos for research. **111**

thermoreceptor Temperature-sensitive sensory receptor. **493**

thigmotropism Directional growth of a plant in response to contact with a solid object. **578**

threatened species A species likely to become endangered in the near future. **370**

threshold potential Membrane potential at which voltage-gated sodium channels in a neuron axon open, causing an action potential. **482**

thrombus Clot that forms in a vessel and remains in place. **431**

thylakoid membrane A chloroplast's highly folded inner membrane system; forms a continuous compartment in the stroma. Site of light-dependent photosynthesis reactions. **83**

thymus Endocrine gland beneath breastbone; produces hormones that help T cells mature. **511**

thyroid gland Endocrine gland at the base of the neck; produces thyroid hormone, which increases metabolism. **510**

tight junction Array of fibrous proteins that joins epithelial cells. Collectively prevent fluids from leaking between cells in tissues. **56**

tissue One or more types of cells that are organized in a specific pattern and that carry out a particular task. **294, 389**

tissue culture propagation Laboratory method in which body cells are induced to divide and form an embryo. **572**

total fertility rate Average number of children born to a woman in a particular population. **333**

tracer A molecule with a detectable label. **22**

trachea Airway to the lungs; windpipe. **428**

tracheal system Tubes that deliver air from body surface to tissues of insects and some other land arthropods. **427**

transcription Process by which an RNA is assembled from nucleotides using the base sequence of a gene as a template. **117**

transcription factor Protein that influences transcription by binding to DNA. **127**

transfer RNA (tRNA) Type of RNA that delivers amino acids to a ribosome during translation. **121**

transgenic Refers to an organism that has been genetically modified to carry a gene from a different species. **180**

translation Process by which a polypeptide chain is assembled from amino acids in the order specified by an mRNA. **117**

translocation Process that moves organic compounds through phloem. **561**

transpiration Evaporation of water from plant parts. **560**

transport protein Protein that passively or actively assists specific ions or molecules across a membrane. **49**

transposable element Small segment of DNA that can spontaneously move to a new location in a chromosome. **126**

triglyceride A lipid with three fatty acid tails attached to a glycerol backbone. **32**

trophic level Position of an organism in a food chain. **350**

tropical rain forest Species-rich tropical biome in which continual warmth and rainfall allows the dominant broadleaf trees to grow all year. **366**

tropism In plants, directional growth response to an environmental stimulus. **577**

tubular secretion Substances are moved out of capillaries and into filtrate in kidney tubules. **472**

tumor Abnormally growing and dividing mass of cells. **130**

tunicates Invertebrate chordates that lose their defining chordate traits during the transition to adulthood. **305**

turgor Pressure that a fluid exerts against a wall, membrane, or some other structure that contains it. **70**

unsaturated fat Lipid with at least one double bond in a fatty acid tail. **32**

ureter Tube that carries urine from a kidney to the bladder. **471**

urethra Tube through which urine from the bladder flows out of the body. **471**

urinary bladder Hollow, muscular organ that stores urine. **471**

urinary system Organ system that filters blood, and forms, stores, and expels urine. **471**

urine Mix of water and soluble wastes formed and excreted by the urinary system. **471**

uterus Muscular chamber where offspring develop; womb. **528**

vaccine A preparation introduced into the body in order to elicit immunity to an antigen. **454**

vacuole A fluid-filled organelle that isolates or disposes of waste, debris, or toxic materials. **53**

vagina Female organ of intercourse and birth canal. **528**

variable A characteristic or event that differs among individuals. **12**

vascular bundle Multistranded, sheathed cord of primary xylem and phloem in a stem or leaf. **551**

vascular cambium Ring of meristematic tissue that produces secondary xylem and phloem. **555**

vascular cylinder Sheathed, cylindrical array of primary xylem and phloem in a root. **554**

vascular plant A plant that has xylem and phloem. **271**

vascular tissue system Tissue system that distributes water and nutrients through a plant body. **549**

vas deferens Main duct that conveys mature sperm toward the urethra. **526**

vector Animal that transmits a pathogen between its hosts. **251**

vegetative reproduction Growth of new roots and shoots from extensions or fragments of a parent plant; form of asexual reproduction in plants. **571**

vein Large-diameter vessel that returns blood to the heart. **421**

ventricle Heart chamber that pumps blood out of the heart and into an artery. **420**

venule Small-diameter blood vessel that carries blood from capillaries to a vein. **421**

vernalization Stimulation of flowering in spring by low temperature in winter. **579**

vertebrae Bones of the backbone. **405**

vertebral column The backbone. **306, 405**

vertebrate Animal with a backbone. **295**

vesicle Small, membrane-enclosed, saclike organelle; different kinds store, transport, or degrade their contents. **53**

villi Multicelled projections from the lining of the small intestine. **464**

virus Infectious particle that consists of protein and nucleic acid and replicates only inside a living cell. **259**

visceral sensations Sensations that arise when sensory neurons in soft internal organs are stimulated. **493**

vitamin Organic substance required in small amounts for normal metabolism. **468**

warning coloration Distinctive color or pattern that makes a well-defended prey species easy to recognize. **345**

water cycle Water moves from its main reservoir—the ocean—into the air, falls as rain and snow, and flows back to the ocean. **353**

water mold Heterotrophic protist that forms a mesh of nutrient-absorbing filaments. **255**

water–vascular system Of echinoderms, a system of fluid-filled tubes and tube feet that function in locomotion. **304**

wavelength Distance between the crests of two successive waves of light. **81**

wax Water-repellent lipid with long fatty acid tails bonded to long-chain alcohols or carbon rings. **32**

white blood cell Blood cell with a role in housekeeping and defense. **424**

white matter Tissue of brain and spinal cord consisting of myelinated axons. **490**

wood Lignin-stiffened secondary xylem of some seed plants. **277, 555**

xenotransplantation Transplantation of an organ from one species into another. **189**

xylem Plant vascular tissue; distributes water and dissolved mineral ions through tubes that consist of interconnected walls of dead cells. **271, 550**

yellow marrow Bone marrow that is mostly fat; fills cavity in most long bones. **406**

yolk Nutritious material in many animal eggs. **524**

zygote Cell formed by fusion of gametes; first cell of a new individual. **144, 524**

Art Credits & Acknowledgments

CHAPTER 1: **1.1** Courtesy of Conservation International. **1.3** below, JupiterImages Corporation. **1.4** © Biosphoto/ BIOS, Bios - Auteurs (droits geres), Denis-Huot Michel & Christine/ Peter Arnold Inc. **1.5** clockwise from top left, SciMAT/ Photo Researchers, Inc.; © Dr. Harald Huber, Dr. Michael Hohn, Prof. Dr. K.O.Stetter, University of Regensburg, Germany; JupiterImages Corporation; © Biosphoto/ BIOS, Bios - Auteurs (droits geres), Denis-Huot Michel & Christine/ Peter Arnold Inc.; Courtesy of John S. Russell, Pioneer High School; Astrid Hanns-Frieder michler/ SPL/ Photo Researchers, Inc. **Page 9** JupiterImages Corporation. **1.6** © Raymond Gehman/ Corbis. **1.7** above, © Bob Jacobson/ Corbis; below, © SuperStock. **1.8** (a) © Adrian Vallin; (b), © Antje Schulte. **1.9** © Adrian Vallin. **1.10** Photo courtesy of Dr. Robert Zingg/ Zoo Zurich. **Page 16** Section 1.1, Courtesy of Conservation International; Section 1.3, © Biosphoto/ BIOS, Bios - Auteurs (droits geres), Denis-Huot Michel & Christine/ Peter Arnold Inc; Section 1.4, 1.5, JupiterImages Corporation; Section 1.6, Raymond Gehman/ Corbis; Section 1.7, © Bob Jacobson/ Corbis. **Page 17** Scientific Paper; Adrian Vallin, Sven Jakobsson, Johan Lind and Christer Wiklund, *Proc. R. Soc. B* (2005 272, 1203, 1207). Used with permission of The Royal Society and the author.

CHAPTER 2: **2.1** © ThinkStock/ SuperStock. **2.3** © Michael S. Yamashita/ Corbis. **2.7** (c) © Herbert Schnekenburger. **2.9** © JupiterImages Corporation. **2.10** Michael Grecco/ Picture Group. **Page 30** © JupiterImages Corporation. **2.13** © JupiterImages Corporation. **2.16** Tim Davis/ Photo Researchers, Inc. **Page 35** PDB ID: 1QLX; Zahn, R., Liu, A., Luhrs, T., Reik, R., Von Schroetter, C., Garcia, F.L., Billeter, M., Calzolai, L., Wider, G. Wuthrich, K.: NMR Solution Structure of the Human Prion Protein, *Proc. Nat. Acad. Sci,* USA 97 p. 145 (2000). **2.19** (b) Sherif Zaki, MD PhD; Wun-Ju Shieh, MD PhD; MPH/ CDC; (c) © Lily Echeverria/ Miami Herald. **Page 37** © JupiterImages Corporation. **Page 38** Section 2.1, 2.7, © JupiterImages Corporation; Section 2.4, © Herbert Schnekenburger; Section 2.5, Michael Grecco/ Picture Group. **2.19** This lipoprotein image was made by Amy Shih and John Stone using VMD and is owned by the Theoretical and Computational Biophysics Group, NIH Resource for Macromolecular Modeling and Bioinformatics, at the Beckman Institute, University of Illinois at Urbana-Champaign. Labels added to the original image by book author.

CHAPTER 3: **Page 42** © JupiterImages Corporation. **3.1** From left, Astrid Hanns-Frieder michler/ SPL/ Photo Researchers, Inc.; Courtesy of Allen W. H. Bé and David A. Caron; © Wim van Egmond/ Visuals Unlimited; © Dr. Dennis Kunkel/ Visuals Unlimited; © Stem Jems/

Photo Researchers, Inc. **3.3** Tony Brian, David Parker/ SPL/ Photo Researchers, Inc. **3.4** (a,b; d,e) Jeremy Pickett-Heaps, School of Botany, University of Melbourne; (c) © Prof. Franco Baldi. **3.5** Virus, CDC; Bacteria, Tony Brian, David Parker/ SPL/ Photo Researchers, Inc.; Eukaryotic cells, Jeremy Pickett-Heaps, School of Botany, University of Melbourne; Louse, Edward S. Ross; Ladybug, Image © Dole, Used under license from Shutterstock.com; Goldfish, Image © Ultrashock, Used under license from Shutterstock.com; Butterfly, From Meyer, A., Repeating Patterns of Mimicry. *PLoS Biology* Vol. 4, No. 10, e341 doi:10.1371/journal.pbio.0040341; Guinea pig, Image © Sascha Burkard, Used under license from Shutterstock.com; Dog, Image © Pavel Sazonov, Used under license from Shutterstock.com; Human, JupiterImages Corporation; Giraffe, © Ingram Publishing, SuperStock; Whale, JupiterImages Corporation; Tree, Courtesy of © Billie Chandler. **Page 46** © The Royal Society. **3.7** (a) Rocky Mountain Laboratories, NIAID, NIH; (b) © R. Calentine/ Visuals Unlimited; (c) ArchiMeDes; (d,e) © K.O. Stetter & R. Rachel, Univ. Regensburg. **3.10** © Kenneth Bart. **3.11** (a) Micrograph, Keith R. Porter; (b) Dr. Jeremy Burgess/ SPL/ Photo Researchers, Inc. **3.12** (b) © Dylan T. Burnette and Paul Forscher. **3.15** © ADVANCELL (Advanced In Vitro Cell Technologies; S.L.) www.advancell.com. **Page 57** © Stephanie Schuller/ Photo Researchers, Inc. **Page 58** Section 3.1, 3.2, Astrid Hanns-Frieder michler/ SPL/ Photo Researchers, Inc.; Section 3.3, © The Royal Society; Section 3.5, ArchiMeDes; Section 3.6, © Kenneth Bart. **Page 59** Critical Thinking #2, P.L. Walne and J. H. Arnott, *Planta*, 77:325–354, 1967.

CHAPTER 4: **4.1** above, © BananaStock/ SuperStock. **4.6** (a) right, © Scott McKiernan/ ZUMA Press. **4.8** Lisa Starr, after David S. Goodsell, RCSB, Protein Data Bank. **4.9** (a) © JupiterImages Corporation. **4.12** M. Sheetz, R. Painter, and S. Singer, *Journal of Cell Biology*, 70:193 (1976) by permission, The Rockefeller University Press. **4.13** PDB files from NYU Scientific Visualization Lab. **4.14** After David H. MacLennan, William J. Rice, and N. Michael Green, "The Mechanism of Ca^{2+} Transport by Sarco(Endo)plasmic Reticulum Ca^{2+}-ATPases." *J. Biol. Chem.* 1997, 272: 28815–28818. **4.17** Biology Media/ Photo Researchers, Inc. **4.18** (a) Courtesy Dr. Edward C. Klatt; (b) Courtesy of Downstate Medical Center, Department of Pathology, Brooklyn, NY. **Page 76** Section 4.1, © BananaStock/ SuperStock; Section 4.4, © JupiterImages Corporation; Section 4.6, M. Sheetz, R. Painter, and S. Singer, *Journal of Cell Biology*, 70:193 (1976) by permission, The Rockefeller University Press. **4.19** top, JupiterImages Corporation.

CHAPTER 5: **5.1** Photo by Peggy Greb/ USDA. **Page 82** © Larry West/ FPG/ Getty Images. **5.3**

left, JupiterImages Corporation. **5.7** (a) Courtesy of John S. Russell, Pioneer High School; (c) © Bill Boch/ FoodPix/ JupiterImages; (d) Lisa Starr. **5.12** © Ben Fink Photography. **Page 92** JupiterImages Corporation. **5.13** left, © Randy Faris/ Corbis; right, Lilly M/ http://commons.wikimedia.org; below Courtesy of © William MacDonald, M.D. **Page 94** upper, © Lois Ellen Frank/ Corbis; lower, JupiterImages Corporation. **5.15** © JupiterImages Corporation. **Page 96** Photo by Peggy Greb/ USDA. **5.15** JupiterImages Corporation. **5.16** left, © Roger W. Winstead, NC State University; right, Photo by Scott Bauer, USDA/ ARS. **Page 98** Section 5.1, Photo by Peggy Greb/ USDA; Section 5.2, © Larry West/ FPG/ Getty Images; Section 5.5, Lisa Starr; Section 5.7, © Ben Fink Photography; Section 5.8, © Lois Ellen Frank/ Corbis; Section 5.9, © JupiterImages Corporation.

CHAPTER 6: **6.1** Photos by Victor Fisher, courtesy Genetic Savings & Clone. **6.2** left, Andrew Syred/ Photo Researchers, Inc. **6.4** © University of Washington Department of Pathology. **6.6** (d) A. C. Barrington Brown, 1968 J. D. Watson. **6.7** PDB ID: 1BBB; Silva, M.M., Rogers, P.H., Arnone, A.: A third quaternary structure of human hemoglobin at 1.7-A resolution. *J. Biol. Chem* 276 pp. 17248 (1992). **6.10, 6.11** Courtesy of Cyagra, Inc., www.cyagra.com. **Page 111** left upper, © McLeod Murdo/ Corbis Sygma; lower, Photo by Victor Fisher, courtesy Genetic Savings & Clone. **Page 112** Section 6.2, Andrew Syred/ Photo Researchers, Inc.; Section 6.5, © McLeod Murdo/ Corbis Sygma. **6.12** (a) Eye of Science/ Photo Researchers, Inc.

CHAPTER 7: **7.1** Vaughan Fleming/ SPL/ Photo Researchers, Inc. **7.3** © JupiterImages Corporation. **7.4** O. L. Miller. **Page 123** © Kiseleva and Donald Fawcett/ Visuals Unlimited. **7.10** left, Dr. Gopal Murti/ SPL/ Photo Researchers, Inc.; right, © Ben Rose/ WireImage/ Getty Images. **7.11** (a) © John W. Gofman and Arthur R. Tamplin. From *Poisoned Power: The Case Against Nuclear Power Plants Before and After Three Mile Island*, Rodale Press, PA, 1979. **7.12** (a,b) David Scharf/ Photo Researchers, Inc.; (c) Eye of Science/ Photo Researchers, Inc.; (d) Courtesy of the Aniridia Foundation International, www.aniridia.net; (e) M. Bloch. **7.13** © Dr. William Strauss. **7.14** After Patten, Carlson, & others. **7.15** left, Courtesy of Robin Shoulla and Young Survival Coalition; right, From the archives of www.breastpath.com, courtesy of J.B. Askew, Jr., M.D., P.A. Reprinted with permission, copyright 2004 Breastpath.com. **Page 131** © Vaughan Fleming/ SPL/ Photo Researchers, Inc. **Page 132** Section 7.1, Vaughan Fleming/ SPL/ Photo Researchers, Inc.; Section 7.3, © JupiterImages Corporation; Section 7.6, © John W. Gofman and Arthur R. Tamplin. From *Poisoned Power: The Case Against Nuclear Power Plants Before and After Three Mile*

Island, Rodale Press, PA, 1979; Section 7.7, © Dr. William Strauss. **Page 133** © Sascha Burkard, Used under license from Shutterstock.com.

CHAPTER 8: **8.1** Micrograph, Dr. Pascal Madaule, France; inset, Courtesy of the family of Henrietta Lacks. **8.2** © Carolina Biological Supply Company/ Phototake. **8.5** left photos, Michael Clayton/ University of Wisconsin, Department of Botany; right photos, Ed Reschke. **8.6** (c) right, © D. M. Phillips/ Visuals Unlimited. **8.7** (c) right, Michael Clayton/ University of Wisconsin, Department of Botany. **8.8** Image courtesy of Carl Zeiss MicroImaging, Thornwood, NY. **8.10** top photos, With thanks to the John Innes Foundation Trustees, computer enhanced by Gary Head. **8.12** right, Courtesy of © Billie Chandler. **8.13** © Phillip B. Carpenter, Department of Biochemistry and Molecular Biology, University of Texas–Houston Medical School. **8.15** (a) © Ken Greer/ Visuals Unlimited; (b) Biophoto Associates/ Photo Researchers, Inc.; (c) James Stevenson/ Photo Researchers, Inc. **Page 151** Courtesy of the family of Henrietta Lacks. **Page 152** Sections 8.1, 8.2, © Carolina Biological Supply Company/ Phototake; Section 8.4, Michael Clayton/ University of Wisconsin, Department of Botany; Section 8.5, © D. M. Phillips/ Visuals Unlimited; Section 8.6, Image courtesy of Carl Zeiss MicroImaging, Thornwood, NY; Section 8.9, James Stevenson/ Photo Researchers, Inc. **8.17** left, SPL/ Photo Researchers, Inc.; right, Courtesy of © Dr. Thomas Ried, NIH and the American Association for Cancer Research.

CHAPTER 9: **9.1** Clockwise from top left, Courtesy of © The Cody Dieruf Benefit Foundation, www.breathinisbelievin.org; Courtesy of © Bobby Brooks and The Family of Jeff Baird; Courtesy of © Steve & Ellison Widener and Breathe Hope, http://breathehope.tamu.edu; Courtesy of © the family of Brandon Herriott; Courtesy of © The Family of Savannah Brooke Snider; Courtesy of The family of Benjamin Hill, reprinted with permission of © Chappell/Marathonfoto. **9.3** top, Jean M. Labat/ Ardea London. **Page 157** The Moravian Museum, Brno. **9.5** white pea plan flower, © George Lepp/ Corbis. **9.8** © Gary Roberts/ worldwidefeatures.com. **9.9** top, © David Scharf/ Peter Arnold, Inc. **9.10** © JupiterImages Corporation. **Page 164** Courtesy of the family of Haris Charalambous and the University of Toledo. **9.12** lower, Courtesy of Ray Carson, University of Florida News and Public Affairs. **9.13** (a) JupiterImages Corporation; (b) © age fotostock/ SuperStock; (c) © Dr. Christian Laforsch. **9.15** Courtesy of Irving Buchbinder, DPM, DABPS, Community Health Services, Hartford CT. **9.16** (b) Eddie Adams/ AP Wide World Photos. **9.17** (b) Rick Guidotti, Positive Exposure. **9.18** (a,b) Photo by Gary L. Friedman, www.Friedman Archives.com. **9.20** (b) Lauren Shear/ Photo Researchers, Inc.; (c) L. Willatt, East Anglian Regional Genetics Service/ Photo Researchers, Inc. **9.21** UNC Medical Illustration and Photography. **9.22** Saturn Stills/ SPL/ Photo

Researchers, Inc. **Page 174** Fran Heyl Associates © Jacques Cohen, computer-enhanced by © Pix Elation. **9.23** © Howard Sochurek/ The Medical File/ Peter Arnold, Inc. **Page 175** © Gary Gaugler/ The Medical File/ Peter Arnold Inc. **Page 176** Section 9.1, Courtesy of © The Cody Dieruf Benefit Foundation, www .breathinisbelievin.org; Section 9.2, © Gary Roberts/ worldwidefeatures.com; Section 9.4, JupiterImages Corporation; Section 9.7, © Lauren Shear/ Photo Researchers, Inc.; Section 9.8, © Howard Sochurek/ The Medical File/ Peter Arnold, Inc.

CHAPTER 10: **10.1** © Courtesy of Golden Rice Humanitarian Board. **10.6** Courtesy of © Genelex Corp. **10.7** Patrick Landmann/ Photo Researchers, Inc. **Page 187** Photo Courtesy of Systems Biodynamics Lab, P.I. Jeff Hasty, UCSD Department of Bioengineering, and Scott Cookson. **10.8** (4) © Lowell Georgis/ Corbis; (5) Keith V. Wood. **10.9** The Bt and Non-Bt corn photos were taken as part of field trial conducted on the main campus of Tennessee State University at the Institute of Agricultural and Environmental Research. The work was supported by a competitive grant from the CSREES, USDA titled "Southern Agricultural Biotechnology Consortium for Underserved Communities," (2000–2005). Dr. Fisseha Tegegne and Dr. Ahmad Aziz served as Principal and Co-principal Investigators respectively to conduct the portion of the study in the State of Tennessee. **10.10** (a) © Adi Nes, Dvir Gallery Ltd.; (b) Transgenic goat produced using nuclear transfer at GTC Biotherapeutics. Photo used with permission; (c) Photo courtesy of MU Extension and Agricultural Infomation. **Page 190** © Jeans for Gene Appeal. **Page 191** © Corbis/ SuperStock; © Courtesy of Golden Rice Humanitarian Board. **Page 192** Section 10.1, © Courtesy of Golden Rice Humanitarian Board; Section 10.3, Patrick Landmann/ Photo Researchers, Inc.; Section 10.4, Keith V. Wood; Section 10.5, © Jeans for Gene Appeal. **10.11** (a) Laboratory of Matthew Shapiro while at McGill University. Courtesy of Eric Hargreaves, www.pageoneuroplasticity.com.

CHAPTER 11: **11.1** © Brad Snowder. **11.2** (a,c) © Wolfgang Kaehler/ Corbis; (b) © Earl & Nazima Kowall/ Corbis; (d,e) Edward S. Ross. **11.3** (a) © Dr. John Cunningham/ Visuals Unlimited; (b) Gary Head. **Page 198** Courtesy of Daniel C. Kelley, Anthony J. Arnold, and William C. Parker, Florida State University Department of Geological Science. **11.4** (a) Courtesy George P. Darwin, Darwin Museum, Down House. **11.5** (a) © John White; (b) 2004 Arent. **11.6** Down House and The Royal College of Surgeons of England. **11.7** (a) W. B. Scott (1894); (b) above, Doug Boyer in P. D. Gingerich et al. (2001) © American Association for Advancement of Science; below, John Klausmeyer, University of Michigan Exhibit of Natural History; inset, © P. D. Gingerich, University of Michigan. Museum of Paleontology; (c) above, © P. D. Gingerich and M. D. Uhen (1996), © University of Michigan. Museum of Paleontology;

inset, Phillip Gingerich, Director, University of Michigan. Museum of Paleontology. **Pages 202, 203** Courtesy of Stan Celestian/ Glendale Comunity College Earth Science Image Archive. **11.8** (b) © PhotoDisc/ Getty Images. **11.9** © JupiterImages Corporation. **11.10** bottom, USGS. **11.11** After A.M. Ziegler, C.R. Scotese, and S.F. Barrett, *Mesozoic and Cenozoic Paleogeographic Maps* and J. Krohn and J. Sundermann (Eds.), *Tidal Frictions and the Earth's Rotation II*, Springer-Verlag, 1983. **11.13** (a) © Taro Taylor, www.flickr.com/photos/tjt195; (b) © JupiterImages Corporation; (c) © Linda Bingham. **11.14** from top, © Lennart Nilsson/ Bonnierforlagen AB; Courtesy of Anna Bigas, IDIBELL-Institut de Recerca Oncologica, Spain; From "Embryonic staging system for the short-tailed fruit bat, *Carollia perspicillata*, a model organism for the mammalian order Chiroptera, based upon timed pregnancies in captive-bred animals," C.J. Cretekos et al., *Developmental Dynamics* Volume 233, Issue 3, July 2005, Pages: 721–738. Reprinted with permission of Wiley-Liss, Inc. a subsidiary of John Wiley & Sons, Inc.; Courtesy of Prof. Dr. G. Elisabeth Pollerberg, Institut für Zoologie, Universität Heidelberg, Germany; USGS. **Page 211** © David A. Kring, NASA/ Univ. Arizona Space Imagery Center. **Page 212** Section 11.2, © Wolfgang Kaehler/ Corbis; Section 11.3, Courtesy George P. Darwin, Darwin Museum, Down House; Section 11.4, Courtesy of Stan Celestian/ Glendale Comunity College Earth Science Image Archive; Section 11.5, USGS; Section 11.6, © Lennart Nilsson/ Bonnierforlagen AB. **11.16** top, Lawrence Berkeley National Laboratory.

CHAPTER 12: **12.1** © Rollin Verlinde/ Vilda. **12.2** third from left, © Roderick Hulsbergen/ http://www.photography.euweb.nl; all others, © JupiterImages Corporation. **12.4** J. A.Bishop, L. M. Cook. **12.7** above, Peter Chadwick/ SPL/ Photo Researchers, Inc. **12.9** Thomas Bates Smith. **12.10** (a) Bruce Beehler; (b) Courtesy of Gerald Wilkinson; (c) © Ingo Arndt/ Nature Picture Library. **12.11** (a,b) After Ayala and others; (c) © Michael Freeman/ Corbis. **12.12** (a,b) Adapted from S.S. Rich, A.E. Bell, and S.P. Wilson, *Genetic drift in small populations of Tribolium* Evolution 33:579–584, Fig. 1, p. 580, 1979. Used by permission of the publisher; below, Photo by Peggy Greb/ USDA. **12.13** Dr. Victor A. McKusick. **12.14** From Meyer, A., Repeating Patterns of Mimicry. *PLoS Biology* Vol. 4, No. 10, e341 doi:10.1371/journal.pbio.0040341. **Page 227** Alvin E. Staffan/ Photo Researchers, Inc. **12.16** (a) Courtesy of Joe Decruyenaere; (b,c) © David Goodin; (d) Courtesy of Dr. James French. **12.17** G. Ziesler/ ZEFA. **12.18** above, Southeastern Regional Taxonomic Center/ South Carolina Department of Natural Resources; below, NASA. **12.20** above, © Michelle Harrington; below, Ole Seehausen and Inke van der Sluijs. **12.21** above, Courtesy of The Virtual Fossil Museum, www .fossilmuseum.net; below, Mark Erdman. **12.22** Photo by © Marcel Lecoufle. **12.23** all, © Jack Jeffrey Photography. **12.24** from left, Joaquim Gaspar; Bogdan; Opiola Jerzy;

Ravedave; Luc Viatour. **Page 235** © Rollin Verlinde/ Vilda. **Page 236** Sections 12.1, 12.2, left and center, © JupiterImages Corporation; right, © Roderick Hulsbergen/ http://www.photography.euweb.nl; Section 12.3, J. A. Bishop, L. M. Cook; Section 12.4, Courtesy of Gerald Wilkinson; Section 12.5, G. Ziesler/ ZEFA; Section 12.6, Photo by © Marcel Lecoufle.

CHAPTER 13: **13.1** © Lowell Tindell; inset, NIBSC/ SPL/ Photo Researchers, Inc. **13.3** (a) Eiichi Kurasawa/ Photo Researchers, Inc.; (b) Courtesy of the University of Washington. **13.4** (a) © Janet Iwasa; (b) Photo by Tony Hoffman, courtesy of David Deamer. **13.5** (a) Chase Studios/ Photo Researchers, Inc.; (b,c) © University of California Museum of Paleontology. **13.7** © N.J. Butterfield, University of Cambridge. **13.10** (a) CNRI/ Photo Researchers, Inc.; (b) © Dr. Dennis Kunkel/ Visuals Unlimited. **13.11** (a) Alan L. Detrick, Science Source/ Photo Researchers, Inc.; (b) © Martin Miller / Visuals Unlimited; (c) Dr. John Brackenbury/ Science Photo Library/ Photo Researchers, Inc. **13.12** (a) © R. Calentine/ Visuals Unlimited; (b) SciMAT / Photo Researchers, Inc.; (c) Stem Jems/ Photo Researchers, Inc.; (d) California Department of Health Services. **13.14** (a) © Dr. Dennis Kunkel/ Visuals Unlimited; (b) Oliver Meckes/ Photo Researchers, Inc. **13.15** Courtesy of Allen W. H. Bé and David A. Caron. **13.16** (a) Gary W. Grimes and Steven L'Hernault; (b) © Dr. David Phillips/ Visuals Unlimited. **13.17** © Frank Borges Llosa/ www.frankley.com. **13.18** Based on Fig. 1 from *Genetic linkage and association analyses for trait mapping in Plasmodium falciparum*, by Xinzhuan Su, Karen Hayton & Thomas E. Wellems, *Nature Reviews Genetics* 8, 497–506 (July 2007). **Page 255** Malaria illustration by Drew Berry, The Walter and Eliza Hall Institute of Medical Research. **13.20** (a,c) © Wim van Egmond/ Visuals Unlimited; (b) Montery Bay Aquarium; (d) © Kim Taylor/ Bruce Coleman, Inc. **13.21** (a) M I Walker/ Photo Researchers, Inc.; (b) Edward S. Ross. **13.22** (b) After Stephen L. Wolfe; (c) Dr. Richard Feldmann/ National Cancer Institute; (d) Russell Knightly/ Photo Researchers, Inc. **13.23** left, Science Photo Library/ Photo Researchers, Inc. **13.24** left, After Stephen Wolfe, *Molecular Biology of the Cell*, Wadsworth, 1993; below, Micrographs Z. Salahuddin, National Institutes of Health. **13.25** © WHO, Pierre-Michel Virot, photographer; inset, Sercomi/ Photo Researchers, Inc. **13.26** CAMR, Barry Dowsett/ Science Photo Library/ Photo Researchers, Inc. **Page 265** Kent Wood/ Photo Researchers, Inc. **Page 266** Section 13.1, NIBSC/ SPL/ Photo Researchers, Inc.; Section 13.3, © University of California Museum of Paleontology; Section 13.5, © Dr. Dennis Kunkel/ Visuals Unlimited.

CHAPTER 14: **14.1** © William Campbell/ TimePix/ Getty Images. **14.4** Courtesy of Christine Evers. **14.5** Art, Raychel Ciemma **14.6** (a) © University of Wisconsin–Madison, Department of Biology, Anthoceros CD; (b) National Park Services, Paul Stehr-Green. **14.7** Art, Raychel Ciemma; photo, A. & E. Bomford/ Ardea, London. **14.8** (a) © S. Navie; (b)

David C. Clegg/ Photo Researchers, Inc.; (c) © Klein Hubert/ Peter Arnold, Inc. **14.9** below, left, © Martin LaBar, www.flickr.com/photos/martinlabar; right, © William Ferguson. **14.11** (a) R. J. Erwin/Photo Researchers, Inc.; (b) © Stan Elems/ Visuals Unlimited; (c) © Dave Cavagnaro/ Peter Arnold, Inc. **14.12** (a) © M. Fagg, Australian National Botanic Gardens; (b) Michael P. Gadomski/ Photo Researchers, Inc.; (c) © E. Webber/ Visuals Unlimited; (d) © Gerald & Buff Corsi/ Visuals Unlimited. **14.13** (b) Courtesy of Christine Evers. **14.15** from left, Photo USDA; Photo by Scott Bauer, USDA/ ARS; Courtesy of Linn County, Oregon Sheriff's Office. **14.16** (a) © Dr. Dennis Kunkel/ Visuals Unlimited; (b) Photo by Scott Bauer/ USDA; (c) Robert C. Simpson/ Nature Stock; (d) Micrograph Garry T. Cole, University of Texas, Austin/ BPS. **14.17** After T. Rost, et al., *Botany*, Wiley, 1979. **14.18** (a) © Micrograph J. D. Cunningham/ Visuals Unlimited; inset, Micrograph Ed Reschke; (b) © age fotostock/ SuperStock; (c) © Bill Beatty/ Visuals Unlimited; (d) © Chris Worden. **14.19** (a) © Gary Braasch; (b) Gary Head; (c) © Mark E. Gibson/ Visuals Unlimited. **14.20** (a) Photo by Yue Jin/ USDA; (b) Dr. P. Marazzi/ SPL/ Photo Researchers, Inc. **Page 287** Gary A. Monroe © USDA-NRCS PLANTS Database. **Page 288** Section 14.1, © William Campbell/ TimePix/ Getty Images; Section 14.6, © Stan Elems/ Visuals Unlimited; Section 14.7, Courtesy of Christine Evers; Section 14.8, Robert C. Simpson/ Nature Stock; Section 14.9, Dr. P. Marazzi/ SPL/ Photo Researchers, Inc. **14.21** After graph from www.pfc.forestry.ca. **Page 289** Jane Burton/ Bruce Coleman, Ltd.

CHAPTER 15: **Page 292** Archaeopteryx, P. Morris/ Ardea London; others, © James Reece, Nature Focus, Australian Museum. **15.1** (a) © Ana Signorovitch; (b) Dr. Chip Clark. **15.4** (b) Image © ultimathule, Used under license from Shutterstock.com. **15.5** (b) © Wim van Egmond/ Visuals Unlimited; (c) © Brandon D. Cole/ Corbis; (d) © Jeffrey L. Rotman/ Corbis. **15.6** right, Andrew Syred/ SPL/ Photo Researchers, Inc. **15.8** Art after from Solomon, 8th edition, p. 624, figure 29-4. **15.9** (a) J. Solliday/ BPS; (b,c) J. A. L. Cooke/ Oxford Scientific Films. **15.10** (a) left, Danielle C. Zacherl with John McNulty; (b) © K.S. Matz; (c) Alex Kirstitch; (d) Frank Park/ ANT Photo Library; (e) © Dave Fleetham/Tom Stack & Associates. **15.11** (b) Courtesy of © Emily Howard Staub and The Carter Center. **15.12** (a) Jane Burton/ Bruce Coleman, Ltd; (b) After D.H. Milne, *Marine Life and the Sea*, Wadsworth, 1995. **15.13** (a) Image © Eric Isselée, Used under license from Shutterstock.com; (b) © Frans Lemmens/ The Image Bank/ Getty Images; (c) Angelo Giampiccolo / FPG / Getty Images. **15.14** (a) © David Tipling/ Photographer's Choice/ Getty Images; (b) © Peter Parks/ Imagequestmarine.com. **15.15** (a) CDC/ Piotr Naskrecki; (b) Photo by Scott Bauer/ USDA; (c) Edward S. Ross; (d) CDC/ Harvard University, Dr. Gary Alpert; (e) Courtesy of Christine Evers. **15.16** (a) Herve Chaumeton/ Agence Nature; (b) © Fred Bavendam/ Minden Pictures; (c) © George Perina, www.seapix.com.

15.17 (a) Peter Parks/ Oxford Scientific Films/ Animals Animals; (b) © California Academy of Sciences. **15.18** left, Runk & Schoenberger / Grant Heilman, Inc. **15.20** (a) Heather Angel/ Natural Visions; (b) © Jonathan Bird/ Oceanic Research Group, Inc.; (c) After E. Solomon, L. Berg, and D.W. Martin, *Biology*, Seventh Edition, Brooks/Cole. **15.21** (a) © David Nardini/ Getty Images; (b) © Norbert Wu/ Peter Arnold, Inc.; (c) Wernher Krutein/ photovault.com. **15.22** (a,c) © P. E. Ahlberg; (d) © Alfred Kamajian. **15.23** (a) Bill M. Campbell, MD; (b) Stephen Dalton/ Photo Researchers, Inc.; (c) © David M. Dennis/ Tom Stack & Associates, Inc. **15.25** (a) © S. Blair Hedges, Pennsylvania State University; (b) Z. Leszczynski/ Animals Animals; (c) © Kevin Schafer/ Corbis; (d) © Kevin Schafer/ Tom Stack & Associates. **15.26** © Gerard Lacz/ ANTPhoto.com.au. **15.27** (a) © Sandy Roessler/ FPG/ Getty Images; (b) Jean Phillipe Varin/ Jacana/ Photo Researchers, Inc.; (c) Jack Dermid. **15.28** (a) Larry Burrows/ Aspect Photolibrary; (b) Art Wolfe/ Photo Researchers, Inc.; (c) © Dallas Zoo, Robert Cabello; (d) © Kenneth Garrett/ National Geographic Image Collection. **15.29** (a,d) Pascal Goetgheluck/ Photo Researchers, Inc.; (e) Dr. Donald Johanson, Institute of Human Origins; (f) Louise M. Robbins; (g) © Kenneth Garrett/ National Geographic Image Collection. **15.31** Jean Paul Tibbles. **15.32** (a) Pascal Goetgheluck/ Photo Researchers, Inc.; (b,c) © Peter Brown. **Page 318** Section 15.2, © Ana Signorovitch; Section 15.3, Courtesy of Christine Evers; Section 15.4, Runk & Schoenberger / Grant Heilman, Inc.; Section 15.5, © Alfred Kamajian; Section 15.6, Z. Leszczynski/ Animals Animals. **15.34** right, Jane Burton/ Bruce Coleman, Ltd.

CHAPTER 16: **16.1** Courtesy of Joel Peter. **16.2** (b) © Steve Bloom/ stevebloom.com; (c) E. R. Degginger; inset, Jeff Fott Productions/ Bruce Coleman, Ltd. **16.4** (a) © G. K. Peck; (b) © Rick Leche, www.flickr.com/photos/rick_leche. **16.6** right, © Jacques Langevin/ Corbis Sygma. **16.7** (a) © Cynthia Bateman, Bateman Photography; (b) © Tom Davis. **16.8** (a) © Joe McDonald/ Corbis; (b) © Wayne Bennett/ Corbis; (c) Estuary to Abyss 2004. NOAA Office of Ocean Exploration. **16.9** (a) David Reznick/ University of California–Riverside; computer enhanced by Lisa Starr; (b,c) Hippocampus Bildarchiv; right, Helen Rodd. **16.10** © Bruce Bornstein, www.captbluefin.com. **16.11** Art by Precision Graphics, photograph, NASA. **16.13** left, © Adrian Arbib/ Corbis; right, © Don Mason/ Corbis. **Page 335** U.S. Department of the Interior, U.S. Geological Survey. **Page 336** Section 16.1, Courtesy of © Joel Peter; Section 16.2, Steve Bloom/ stevebloom.com; Section 16.3, © Jacques Langevin/ Corbis Sygma; Section 16.4, © Bruce Bornstein, www.captbluefin.com. **16.14** © Reinhard Dirscherl/ www.bciusa.com.

CHAPTER 17: **17.1** (a) Photo by Stephen Ausmus, USDA/ARS; (b) Photo by Scott Bauer/ USDA; (c) John Kabashima. **17.2** left, Donna Hutchins; right, upper, © Martin Harvey, Gallo Images/ Corbis; lower, © Len Robinson,

courses/muchinsky and www.occipita.cfa .cmu.edu. **24.22** (1) © Fabian/ Corbis Sygma; (3) Medtronic Xomed. **Page 500** Section 24.1, Manni Mason's Pictures; Section 24.5, C. Yokochi and J. Rohen, *Photographic Anatomy of the Human Body*, 2nd Ed., Igaku-Shoin, Ltd., 1979.

CHAPTER 25: **25.1** (a) © David Ryan/ SuperStock; (b) © Catherine Ledner Photography. **25.6** (a) China Daily/ Reuters; (b) Courtesy of Dr. William H. Daughaday, Washington University School of Medicine, from A. I. Mendelhoff and D. E. Smith, eds., *American Journal of Medicine*, 1956, 20:133.; (c) Courtesy of G. Baumann, MD, Northwestern University. **25.7** (a) Gary Head; (b) Scott Camazine/ Photo Researchers, Inc. **25.8** Biophoto Associates/ SPL/ Photo Researchers, Inc. **25.10** © Elizabeth Musar; inset, © Manny Hernandez/ Diabetes Hands Foundation, www.tudiabetes .com. **25.12** Photograph by Cecil Stoughton, White House, in the John F. Kennedy Presidential Library and Museum, Boston. **25.14** Dr. Carlos J. Bourdony. **Page 518** Section 25.1, © Catherine Ledner Photography; Section 25.3, China Daily/ Reuters; Section 25.4, Gary Head.

CHAPTER 26: **26.1** © Dana Fineman/ Corbis Sygma. **26.2** (a) Biophoto Associates/ Photo Researchers, Inc.; (b) © Rodger Klein/ Peter Arnold, Inc; (c) © Staebler/ JupiterImages. **26.4** (1–4) Carolina Biological Supply Company; (5) left, center, © David M. Dennis/ Tom Stack & Associates, Inc.; right, © John Shaw/ Tom Stack & Associates. **Page 531** AJPhoto/ Photo Researchers, Inc. **26.11** (a) © Heidi Specht, West Virginia University; (b) Courtesy of Elizabeth Sanders, Women's Specialty Center, Jackson, MS. **26.12** (a) Dr. E. Walker/ Photo Researchers, Inc.; (b) Western Ophthalmic Hospital/ Photo Researchers, Inc.; (c) CNRI/ Photo Researchers, Inc. **Page 537** © iStock

photo.com/ Ronnie Comeau. **26.14** © Lennart Nilsson/ Bonnierforlagen AB. **26.16** James W. Hanson, M.D. **Page 544** Section 26.1, © Dana Fineman/ Corbis Sygma; Section 26.2, © Staebler/ JupiterImages; Section 26.5, Western Ophthalmic Hospital/ Photo Researchers, Inc.

CHAPTER 27: **27.1** (a) © OPSEC Control Number #4 077-A-4; (b) Billy Wrobel, 2004. **27.3** © Donald L. Rubbelke/ Lakeland Community College. **27.4** below, © Andrew Syred/ Photo Researchers, Inc. **27.5** (a) from top, © Bruce Iverson; © Ernest Manewal/ Index Stock Imagery; Courtesy of Dr. Thomas L. Rost; © Franz Holthuysen, Making the invisible visible, Electron Microscopist, Phillips Research; (b) from top, Photo by Mike Clayton/ University of Wisconsin Department of Botany; © Darrell Gulin/ Corbis; Gary Head; Courtesy of Janet Wilmhurst, Landcare Research, New Zealand. **27.6** (a) Photo by Mike Clayton/ University of Wisconsin Botany Department; (b) Carolina Biological Supply Company. **27.9** © Dale M. Benham, Ph.D., Nebraska Wesleyan University. **27.11** (a) After Salisbury and Ross, *Plant Physiology*, Fourth Edition, Wadsworth; (b) © Brad Mogen/ Visuals Unlimited; (c) © Dr. John D. Cunningham/ Visuals Unlimited. **27.13** (b) © Peter Gasson, Royal Botanic Gardens, Kew. **Page 557** © JupiterImages Corporation. **27.14** (a) Courtesy of Mark Holland, Salisbury University; (b) Photo courtesy of Iowa State University Plant and Insect Diagnostic Clinic; (c) © Wally Eberhart/ Visuals Unlimited; (d) NifTAL Project, Univ. of Hawaii, Maui. **27.15** left, Image © Jan Martin Will, Used under license from Shutterstock.com. **27.16** (a) Micrograph by Ken Wagner/ Visuals Unlimited, computer-enhanced by Lisa Starr; (b,c) Courtesy of E. Raveh. **Page 560** Image © Nikolay Okhitin, Used under license from Shutterstock.com. **Page 561** Photo by Keith Weller, ARS, Courtesy of USDA. **Page 562**

Section 27.1, © OPSEC Control Number #4 077-A-4; Section 27.5, Courtesy of Mark Holland, Salisbury University; Section 27.6, upper, Image © Nikolay Okhitin, Used under license from Shutterstock.com; lower, Courtesy of E. Raveh. **Page 563** Edward S. Ross.

CHAPTER 28: **28.1** (a) © Alan McConnaughey, www.flickr.com/photos/engrpiman; (b) Courtesy of James H. Cane, USDA-ARS Bee Biology and Systematics Lab, Utah State University, Logan, UT. **28.2** (b) © Robert Essel NYC/ Corbis. **28.4** (a; c,d) Michael Clayton, University of Wisconsin; (b) Botanical Society of America Image Collection, used with permission. **28.5** (a) © T. M. Jones; (b) R. Carr; (c) © JupiterImages Corporation; (d) © Robert H. Mohlenbrock © USDA-NRCS PLANTS Database; (e) © Trudi Davidoff, www.WinterSown.org; (f) © iStockphoto.com/ Greggory Frieden. **28.6** (a, c) Lisa Starr; (b) Image © Daniel Gale, Used under license from Shutterstock.com. **28.7** (a) © Richard Uhlhorn Photography; (b,c) Photo by Peggy Greb, USDA, ARS. **28.8** (b) left, Barry L. Runk / Grant Heilman, Inc.; right, © James D. Mauseth, MCDB; below, © John D. Cunningham/ Visuals Unlimited. **28.9** (a) above, Herve Chaumeton/ Agence Nature. **28.10** © Sylvan H. Wittwer/ Visuals Unlimited. **28.12** below, © Cathlyn Melloan/ Stone/ Getty Images. **28.14** left, Larry D. Nooden; right, © Adrian Chalkley. **Page 580** © Alan McConnaughey, www.flickr.com/photos/engrpiman. **Page 582** Section 28.1, © Alan McConnaughey, www .flickr.com/photos/engrpiman; Section 28.2, © Robert Essel NYC/ Corbis; Section 28.3, © JupiterImages Corporation; Section 28.4, Image © Daniel Gale, Used under license from Shutterstock.com; Section 28.5, Herve Chaumeton/ Agence Nature; Section 28.6, © Sylvan H. Wittwer/ Visuals Unlimited; Section 28.7, © Adrian Chalkley. **28.16** left, © Mary Sue Ittner/ Pacific Bulb Society; right, © Steven D. Johnson.

Index

The letter *f* designates figure; *t* designates table; bold designates key terms; ■ highlights the location of applications contained in text.